Lecture Notes in Computer Science 2286
Edited by G. Goos, J. Hartmanis, and J. van Leeuwen

Springer
Berlin
Heidelberg
New York
Barcelona
Hong Kong
London
Milan
Paris
Tokyo

Sergio Rajsbaum (Ed.)

LATIN 2002: Theoretical Informatics

5th Latin American Symposium
Cancun, Mexico, April 3-6, 2002
Proceedings

Springer

Series Editors

Gerhard Goos, Karlsruhe University, Germany
Juris Hartmanis, Cornell University, NY, USA
Jan van Leeuwen, Utrecht University, The Netherlands

Volume Editor

Sergio Rajsbaum
Compaq Cambridge Research Laboratory
One Cambridge Center
Cambridge, MA 02142-1612
USA
E-mail: Sergio.Rajsbaum@compaq.com

Cataloging-in-Publication Data applied for

Die Deutsche Bibliothek - CIP-Einheitsaufnahme

Theoretical informatics : proceedings / LATIN 2002, 5th Latin American Symposium, Cancun, Mexico, April 3 - 6, 2002. Sergio Rajsbaum (ed.). - Berlin ; Heidelberg ; New York ; Barcelona ; Hong Kong ; London ; Milan ; Paris ; Tokyo : Springer, 2002
(Lecture notes in computer science ; Vol. 2286)
ISBN 3-540-43400-3

CR Subject Classification (1998): F.2, F.1, E.1, E.3, G.2, G.1, I.3.5, F.3, F.4

ISSN 0302-9743
ISBN 3-540-43400-3 Springer-Verlag Berlin Heidelberg New York

Springer-Verlag Berlin Heidelberg New York
a member of BertelsmannSpringer Science+Business Media GmbH

http://www.springer.de

Typesetting: Camera-ready by author, data conversion by Boller Mediendesign
Printed on acid-free paper SPIN: 10846327 06/3142 5 4 3 2 1 0

Preface

This volume contains the proceedings of the Latin American Theoretical INformatics (LATIN) conference to be held in Cancun, Mexico, April 3–6, 2002.

The LATIN series of symposia was launched in 1992 to foster the interaction between the Latin-American community and computer scientists around the world. This is the fifth event in the series after São Paulo, Brazil (1992), Valparaíso, Chile (1995), Campinas, Brazil (1998), and Punta del Este, Uruguay (2000). The proceedings of these conferences were also published by Springer-Verlag in the Lecture Notes in Computer Science series: volumes 583, 911, 1380, and 1776, respectively. Also as before, we expect to publish a selection of the papers in a special issue of a prestigious journal.

We received 104 submissions (one of which was withdrawn), from about 220 different authors in 26 different countries. Each paper was assigned to at least 3 Program Committee members. The Program Committee selected 44 papers from 22 countries using the help of 106 external reviewers in an electronic meeting. This is a record number of submissions and the lowest acceptance rate, compared to 86 received, 42 accepted in 2000, 53 received, 28 accepted in 1998, 68 received, 38 accepted in 1995, and 66 received, 32 accepted in 1992. Also, the conference included 6 invited talks and 2 tutorials.

The assistance of several organizations and individuals was essential for the success of this meeting. Also essential were previous LATIN meetings, because of their role in developing computer science research in Latin-America, and their valuable organizational experience. Ricardo Baeza-Yates, Daniel Panario, and Alfredo Viola provided insightful advice and shared with us their experiences as organizers of previous LATIN meetings. The conference was possible thanks to the Conference Chair, Edgar Chávez. In addition to providing advice on academic matters, he was in charge of the conference web site, the electronic submission system, local arrangements, publicity, financial administration, and fund raising. Thanks to my host institution, the Cambridge Research Laboratory of Compaq for all the support, as well to the Instituto de Matemáticas, UNAM, from which I am on leave. Finally, we thank Springer-Verlag for publishing these proceedings in the LNCS series.

LATIN has established itself as a conference for theoretical computer science research in Latin-America of the highest academic standards. It has strengthened the ties between local and international scientific communities, and fostered research in Latin-America. We believe that this volume reflects these facts.

January 2002 Sergio Rajsbaum

Invited Presentations

Tutorials

From Algorithms to Cryptography
Fabrizio Luccio and Linda Pagli — Università di Pisa, Italy

Algebraic Topology and Concurrency
Eric Goubault — Commissariat à l'énergie atomique, France
Maurice Herlihy — Brown University, USA
Martin Raussen — Aalborg University, Denmark

Invited Talks

Phase Transitions in Computer Science
Jennifer Chayes — Microsoft Research, USA

The Internet, the Web, and Algorithms
Christos H. Papadimitriou — UC Berkeley, USA

Erdős Magic
Joel Spencer — Courant Institute, USA

Open Problems in Computational Geometry
Jorge Urrutia — Instituto de Matemáticas, UNAM, México

Quantum Algorithms and Complexity
Umesh Vazirani — UC Berkeley, USA

Testing and Checking of Finite State Systems
Mihalis Yannakakis — Avaya Laboratories, USA

Organization

Conference Chair:	Edgar Chávez, Univ. Michoacana, México
Program Chair:	Sergio Rajsbaum, Compaq CRL and UNAM
Steering Committee:	Ricardo Baeza-Yates, Univ. de Chile, Chile
	Gaston Gonnet, ETH Zurich, Switzerland
	Claudio Lucchesi, Univ. de Campinas, Brazil
	Imre Simon, Univ. de Sao Paulo, Brazil

Program Committee

Amihood Amir	Bar-Ilan Univ., Israel
Mauricio Ayala Rincón	Brazilia Univ., Brazil
Ricardo Baeza-Yates	Univ. of Chile, Chile
Michael Bender	SUNY Stony Brook, USA
Leopoldo Bertossi	Carleton University, Canada
Allan Borodin	Univ. of Toronto, Canada
Bernard Chazelle	Princeton Univ., USA
Lenore Cowen	Tufts Univ., USA
Volker Diekert	Stuttgart, Germany
Javier Esparza	Univ. of Edinburgh, Scotland
Martin Farach-Colton	Rutgers Univ., USA
David Fernandez-Baca	Iowa State Univ., USA
Esteban Feuerstein	Univ. of Buenos Aires, Argentina
Juan Garay	Bell Labs, USA
Oscar H. Ibarra	UC Santa Barbara, USA
Marcos Kiwi	Univ. of Chile, Chile
Yoshiharu Kohayakawa	Univ. de Sao Paulo, Brazil
Elias Koutsoupias	UC Los Angeles, USA
Evangelos Kranakis	Carleton Univ., Canada
Daniel Leivant	Indiana Univ., USA
Alex Lopez-Ortiz	University of Waterloo, Canada
Yoram Moses	Technion, Israel
Daniel Panario	Carleton Univ., Canada
Sergio Rajsbaum (chair)	Compaq CRL and UNAM, Mexico
Alex Russell	Univ. Connecticut, USA
Maria Jos Serna	Univ. Politecnica de Catalunya, Spain
Gadiel Seroussi	Hewlet Packard, USA
Igor Shparlinski	Macquarie Univ., Australia
Imre Simon	Univ. de Sao Paulo, Brazil
Janos Simon	Univ. of Chicago, USA

Referees

Jean-Paul Allouche
Maria Alpuente
Rajeev Alur
Andris Ambainis
Sergio Antoy
Andrea Asperti
Holger Austinat
Gary Benson
Ernesto G. Birgin
Hans Bodlaender
Dan Boneh
Renato Carmo
Christian Choffrut
Michael Codish
Richard Cole
Erik Demaine
Camil Demetrescu
Vinay Deolalikar
Dan Dougherty
Guillermo Durán
Wayne Eberly
Omer Egecioglu
Andreas Enge
Kousha Etessami
Ronald Fagin
William M. Farmer
Paulo Feofiloff
Henning Fernau
Rudolf Freund
Christiane Frougny
Shuhong Gao
Zhicheng (Jason) Gao
Luca Gemignani
Blaise Genest
Mark Giesbrecht
Guillem Godoy
Eric Goubault
Bernhard Gramlich
Rachid Guerraoui
Venkat Guruswami
Michael Hanus
Tero Harju
Hugo Herbelin
Miki Hermann
Ulrich Hertrampf
Joseph D. Horton
Jarkko Kari
Sam Kim
Piotr Koszmider
Manfred Kufleitner
Hanno Lefmann
Irene Loiseau
Rita Loogen
Satya V. Lokam
Fairouz Kamareddine
David Kirkpatrick
Monika Maidl
Arnaldo Mandel
Maurice Margenstern
Isabel Mendez-Díaz
Joseph Mitchell
Larry Moss
Lucia Moura
Ian Munro
Cesar Muñoz
Robert Nieuwenhuis
Alfredo Olivero
V.Y. Pan
Gheorghe Paun
Uri Peled
David Peleg
Jordi Petit
Carl Pomerance
Haydée W. Poubel
Helmut Prodinger
Ruy de Queiroz
Femke van Raamsdonk
Klaus Reinhardt
Darío Robak
Yurii Rogojine
Peter Rossmanith
Ronny Roth
Sartaj Sahni
Kai Salomaa
Berry Schoenmakers
Nigel Smart
Jose Soares
Dina Sokol
Alin Stefanescu
Andreas Stein
A. Strejilevich de Loma
Dimitrios Thilikos
Pawel Urzyczyn
Umesh Vazirani
Helmut Veith
Alfredo Viola
Roel de Vrijer
Gabriel Wainer
Maria E. T. Walter
Marcelo Weinberger
Thomas Wilke
Jerzy Wojciechowski
Fatos Xhafa
Tao Yang
Sheng Yu
Michele Zito

Sponsors

Compaq Cambridge Research Lab
Centro Latinoam. Estudios en Inform.
European Association for TCS
Inst. de Matemáticas, UNAM
Soc. Chilena de Ciencias de la Comp.
Soc. Mexicana de Ciencias de la Comp.
Univ. Michoacana

Table of Contents

Invited Talks and Tutorials

Regular Contributions

Phase Transitions in Computer Science
Invited Talk

Jennifer Chayes
Microsoft Research
One Microsoft Way, Redmond, WA 98052
jchayes@microsoft.com

Phase transitions are familiar phenomena in physical systems. But they also occur in many probabilistic and combinatorial models, including random versions of some classic problems in theoretical computer science. In this talk, I will discuss phase transitions in several systems, including the random graph – a simple probabilistic model which undergoes a critical transition from a disordered to an ordered phase; k-satisfiability – a canonical model in theoretical computer science which undergoes a transition from solvability to insolvability; and optimum partitioning – a fundamental problem in combinatorial optimization, which undergoes a transition from existence to non-existence of a perfect optimizer.

Technically, phase transitions only occur in infinite systems. However, real systems and the systems we simulate are large, but finite. Hence the question of finite-size scaling: Namely, how does the transition behavior emerge from the behavior of large, finite systems? Results on the random graph, satisfiability and optimum partitioning locate the critical windows of these transitions and establish interesting features within these windows.

No knowledge of phase transitions will be assumed in this talk.

S. Rajsbaum (Ed.): LATIN 2002, LNCS 2286, pp. 1–1, 2002.

The Internet, the Web, and Algorithms
Invited Talk

Christos H. Papadimitriou

University of California at Berkeley, EECS
Berkeley, CA 94720
christos@cs.berkeley.edu

The Internet and the worldwide web, unlike all other computational artifacts, were not deliberately designed by a single entity, but emerged from the complex interactions of many. As a result, they must be approached very much the same way that cells, galaxies or markets are studied in other sciences: By speculative (and falsifiable) theories trying to explain how selfish algorithmic actions could have led to what we observe. I present several instances of recent work on this theme, with several collaborators.

S. Rajsbaum (Ed.): LATIN 2002, LNCS 2286, pp. 2–2, 2002.

Erdős Magic
Invited Talk

Joel Spencer

Courant Institute, NYU
251 Mercer Street, New York, NY 10012
spencer@cs.nyu.edu

The Probabilistic Method is a lasting legacy of the late Paul Erdős. We give two examples - both problems first formulated by Erdős in the 1960s with new results in the last few years and both with substantial open questions. Further in both examples we take a Computer Science vantagepoint, creating a probabilistic algorithm to create the object (coloring, packing respectively) and showing that with positive probability the created object has the desired properties.

• Given m sets each of size n (with an arbitrary intersection pattern) we want to color the underlying vertices Red and Blue so that no set is monochromatic. Erdős showed this may always be done if $m < 2^{n-1}$, we give a recent argument of Srinivasan and Radhukrishnan that extends this to $m < c2^n\sqrt{n/\ln n}$. One first colors randomly and then recolors the blemishes with a clever random sequential algorithm.

• In a universe of size N we have a family of sets, each of size k, such that each vertex is in D sets and any two vertices have only $o(D)$ common sets. Asymptotics are for fixed k with $N, D \to \infty$. We want an asymptotic packing, a subfamily of $\sim N/k$ disjoint sets. Erdős and Hanani conjectured such a packing exists (in an important special case of asymptotic designs) and this conjecture was shown by Rödl. We give a simple proof of the speaker that analyzes the random greedy algorithm.

S. Rajsbaum (Ed.): LATIN 2002, LNCS 2286, pp. 3–3, 2002.

Open Problems in Computational Geometry
Invited Talk

Jorge Urrutia

Instituto de Matemáticas Universidad Nacional Autónoma de México, México D.F. México

1 Introduction

In this paper we present a collection of problems which have defied solution for some time. We hope that this paper will stimulate renewed interest in these problems, leading to solutions to at least some of them.

1.1 Points and Circles

In 1985, in a joint paper with V. Neumann-Lara, the following result was proved: Any set P_n of n points on the plane contains two points $p, q \in P_n$ such that any circle containing them contains at least $\frac{n-2}{60}$ elements of P_n. This result was first improved to $\frac{n}{30}$ [5], then to $\lfloor \frac{n}{27} \rfloor + 2$ [10], to $\lceil \frac{5(n-3)}{84} \rceil$ [11], and to approximately $\frac{n}{4.7}$ [9]. Our first conjecture presented here is:

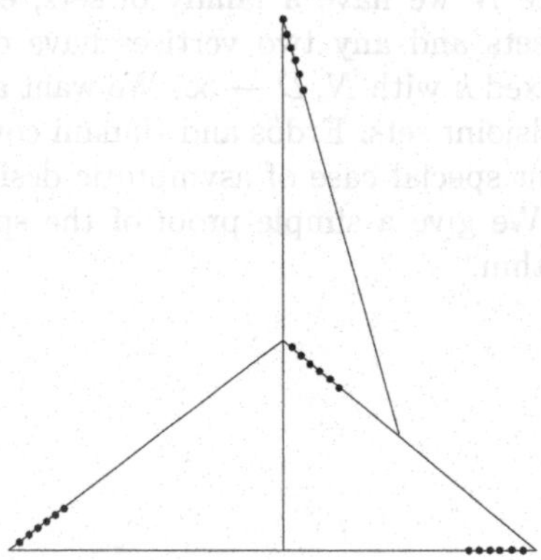

Fig. 1. A point set P_{4n} with $4n$ points such that for any two points p, q of P_{4n} there is a circle containing them that contains at most $n-1$ additional points of P_{4n}.

Conjecture 1 *Any set P_n of n points on the plane contains two elements such that any circle containing them contains at least $\frac{n}{4} \pm c$ elements of P_n.*

S. Rajsbaum (Ed.): LATIN 2002, LNCS 2286, pp. 4–11, 2002.

An example exists with $4n$ points, due to Hayward, Rappaport and Wenger [10], such that for any pair of points of P_n there is a circle containing them that contains at most $n-1$ elements of P_n, see Figure 1.

This problem has been studied for point sets in convex position (that is when the point set is the set of vertices of a convex polygon), for point sets in higher dimensions, and for families whose elements are not points, but convex sets [1,4,5]. For point sets in convex position, the problem was settled by Hayward, Rappaport and Wenger [10], who proved that the tight bound for this case is $\lceil \frac{n}{3} \rceil + 1$.

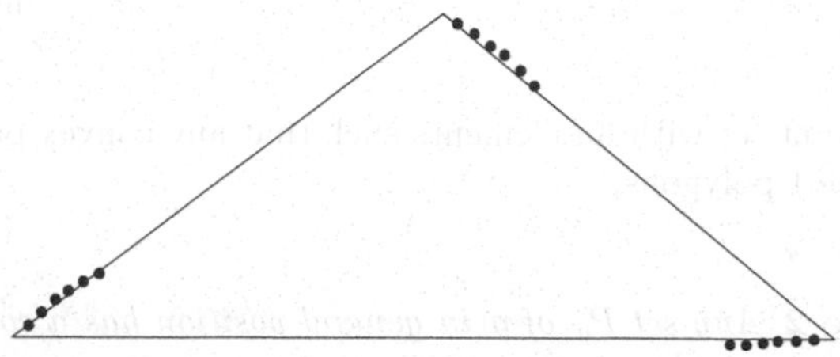

Fig. 2. A point set P_{3n} of $3n$ points in convex position such that for any two points in P_{3n} there is a circle containing them which contains at most $n-1$ extra elements of P_{3n}.

1.2 Convex Partitionings of the Convex Hull of a Point Set

The following problem arose some years ago during a series of meetings in Madrid with M. Abellanas, G. Hernandez, P. Ramos and the author. We were studying problems on quadrilaterizations of point sets. Given a set of points P_n in general position, a collection $\mathcal{F}=\{Q_1,\ldots,Q_m\}$ of convex polygons with disjoint interiors is callead a convex decomposition of the convex hull $Conv(P_n)$ of P_n, a convex decomposition of P_n for short if:

1. The union of the elements of $\mathcal{F}$ is $Conv(P_n)$
2. No element of $\mathcal{F}$ contains an element of P_n in its interior.

If all the elements of $\mathcal{F}$ are quadrilaterals (resp. triangles), $\mathcal{F}$ is called a convex quadrilaterization (triangulation) of P_n. It is well known that not all point sets admit a convex quadrilaterization (even if they contain the right number of points in their convex hull and their interior). It is easy to see that if a point set with n points, k on the boundary of its convex hull, has a convex quadrilaterization, it contains exactly $\frac{(n+k-2)}{2}$ elements. We observed that although convex quadrilaterizations of point sets do not always exist, we were always able to obtain convex partitionings of all point sets with at most $n+1$ elements that we tried. Thus we conjectured:

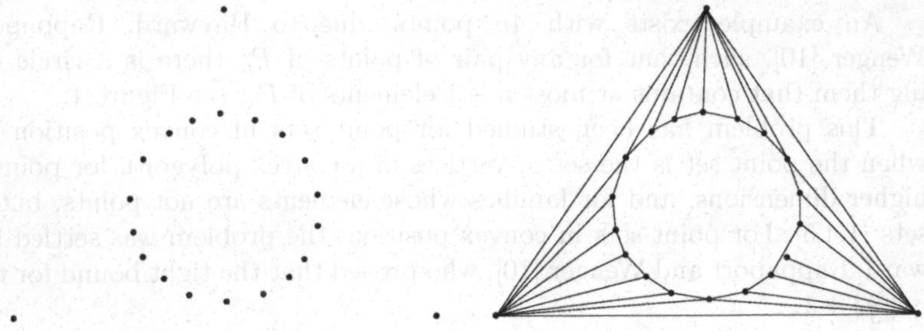

Fig. 3. A point set with $3n$ elements such that any convex partitioning of it has at least $3n + 1$ polygons.

Conjecture 2 *Any set P_n of n in general position has a convex decomposition with at most $n + 1$ elements.*

A set of points achieving this bound is shown in Figure 4. Our conjecture was proved false in 2001 by O.Aichholzer and H.Krasser [2]. They were able to construct a point set P_n such that any convex partitioning of it contains at least $n + 2$ elements.

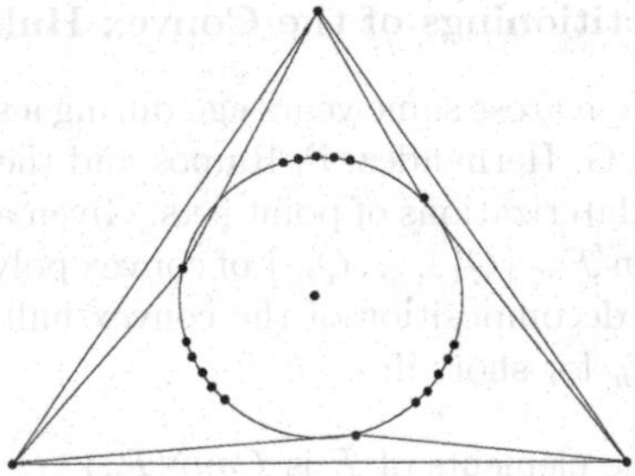

Fig. 4. Aichholzer and Krasser's point set. All the inner points are cocircular, except for the center of the circle on which the points lie, which is also in the point set.

It is known [16] that any point set P_n has a convex partitioning with at most $\lceil \frac{3n-2k}{2} \rceil$ elements, where k is, as before, the number of elements of P_n on the boundary of its convex hull. A convex partition with at most that number of elements can be obtained as follows. First calculate a triangulation of P_n. Then in a greedy way, delete as many edges as possible from our triangulation, making sure that the remaining edges induce a convex partitioning of P_n. It is proved in [16] that the remaining edges induce a convex partitioning of P_n with at most $\lceil \frac{3n-2k}{2} \rceil$ elements; see Figure 5.

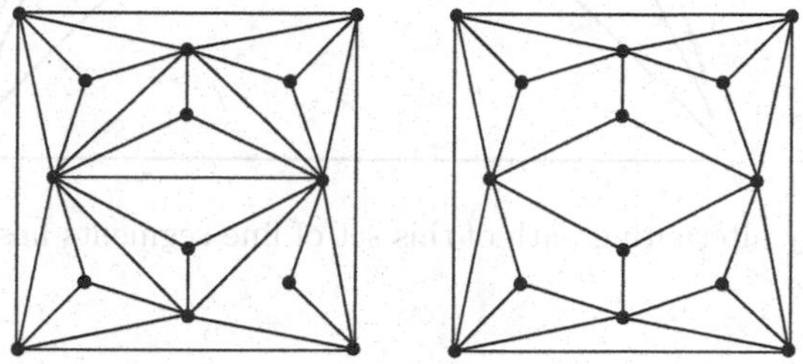

Fig. 5. Consider the triangulation on the left of the point set P_n with fourteen points. Then we can remove at most five edges of its edges to obtain a convex partitioning of P_n. The removal of exactly five edges produces a convex partitioning of this point set with $\lceil \frac{3(14)-2*4}{2} \rceil = 14$ elements.

1.3 Problems on Line Segments

The next problem is at least ten years old.

Let $\mathcal{F}= \{l_1, \ldots, l_n\}$ be a family of disjoint closed segments. A simple alternating path of $\mathcal{F}$ is a non intersecting polygonal chain with $2k+1$ line segments l_i, $i = 1, \ldots, 2k+1$ such that for all i, $k = 0, \ldots, k$, l_{2i+1} belongs to $\mathcal{F}$. Segments l_{2i} are not allowed to intersect other elements of $\mathcal{F}$, $i = 0, \ldots, k$.

Conjecture 3 *Any set $\mathcal{F}= \{l_1, \ldots, l_n\}$ of n disjoint line segments has a simple alternating path with at least $O(\ln n)$ elements.*

In Figure 6 we show a family of 2^n-1 segments such that any alternating path has $O(n)$ elements. The family consists of $2^n - 1$ segments with endpoints on a circle such that l_{2i} is visible from $\{l_j : j = i,\ 2i+1,\ 4i,\ 4i+1\}, i = 1, \ldots,\ 2^{n-1}-1$. Moreover l_{2i+1} is visible only from $\{l_j : j = i,\ 2i+2,\ 4i+2,\ 4i+3\}$, $i = 0, \ldots,\ 2^{n-1}-1$.

The following is a related problem. Given a a family $\mathcal{F}$ of n disjoint closed line segments, find a subset of it that admits a simple alternating path. That is, in our previous problem, remove the restriction that the segments l_{2i} are not allowed to intersect other elements of $\mathcal{F}$, $i = 0, \ldots,\ k$. In this version of the problem, it is not difficult to prove that any family with n line segments contains a subset with at least $n^{\frac{1}{5}}$ elements that admits a simple alternating path. This can be proved using techniques similar to those used in [14]. Observe that if $\mathcal{F}$ has $n^{\frac{1}{5}}$ elements with disjoint projections on the x-axis, this subset admits a simple alternating path, and we are done. If this is not the case, then

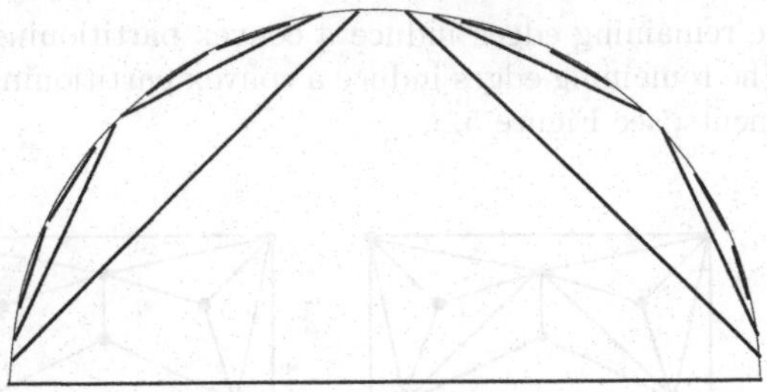

Fig. 6. Any alternating path of this set of line segments has $O(\ln n)$ elements.

there is a vertical line $\mathcal{L}$ that intersects at least $k \geq n^{\frac{4}{5}}$ segments of $\mathcal{F}$. Let $\mathcal{S}= \{m_1, \ldots, m_k\}$ be the sequence containing the slopes of the line segments in $\mathcal{F}$ intersected by $\mathcal{L}$, according to the order in which they are intersected by $\mathcal{L}$. Then by a well known result of Erdös-Szekeres, $\mathcal{S}$ contains an increasing or decreasing subsequence $\mathcal{S}'$ with at least $n^{\frac{2}{5}}$ elements. Suppose w.l.o.g. that the elements of $\mathcal{S}'$ are in increasing order. Let $\mathcal{F}'$ be the subset of $\mathcal{F}$ corresponding to the elements of $\mathcal{S}'$. Let $\mathcal{Y}$ be the sequence defined by the second coordinates of the left endpoints of the elements in $\mathcal{F}'$. Once more, there is an increasing or decreasing subsequence of $\mathcal{Y}$ with $n^{\frac{1}{5}}$ elements. It is easy to see now that the line segments of $\mathcal{F}$ corresponding to these elements admit a simple alternating path. This bound, however, seems to be far from optimal. We believe that the correct value is $O(n^{\frac{1}{2}})$. There are collections of n^2 line segments such that any subset of them that admits a simple alternating path contains at most $2n$ segments. In Figure 7 we illustrate how to obtain such family for $n = 5$. This construction is easily generalizable for any $n \geq 3$.

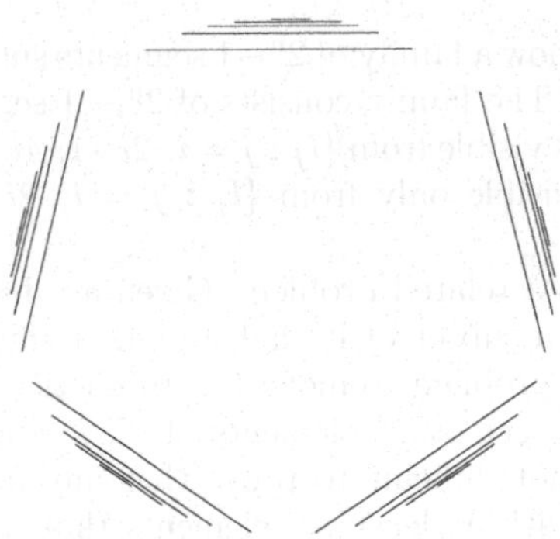

Fig. 7. A family with $5^2 = 25$ segments such that any subset of it admits a simple alternating path with at most $2 \times 5 = 10$ elements.

1.4 Separability

Given two disjoint closed convex sets, we say that a line l separates them if the convex sets are such that one is contained in each of the open half-planes defined by l. H. Tverberg [20] studied the following problem. Let $K_d(r,s) = k$ be the smallest integer k such that given n disjoint convex sets $C_1, \cdots, C_k$, there exists a closed half-plane containing at least r convex sets, and its complement contains s of them. Tverberg proved that $K_2(r,1)$ always exists. Examples found by K. Villanger show that $K_2(2,2)$ does not exist. Villanger's example consists of an arbitrarily large number of non-collinear line segments such that the convex hull of any pair of them contains the point $(1,1)$.

It is known that $K_2(r,1) \leq 12(r-1)$; see [12,7]. There are families of line segments $\mathcal{F}$ with $3m$ elements such that no element of $\mathcal{F}$ can be separated from more than $m+1$ elements [7,8].

Conjecture 4 *Any family $\mathcal{F}$ of n disjoint closed convex sets has an element that can be separated with a single line from at least $\lfloor \frac{n}{3} \rfloor \pm c$ elements of $\mathcal{F}$.*

It is known [3] that for any family $\mathcal{C}$ of congruent disks with $O(m^2 \ln m)$ elements, there always exists a direction α such that any line with direction α intersects at most m elements of $\mathcal{C}$. It then follows that there is a line that leaves at least $\frac{(m^2 \ln m)-m}{2}$ elements of $\mathcal{C}$ in each of the semiplanes which it defines. For families of n circles, not necessarily of the same size, it is known [6] that there is a line that separates a circle from at least $\frac{n-c}{2}$ other circles.

1.5 Illumination

One of my favourite areas is that of illumination. Here I will mention some open problems related to this area of research. A more extensive list of open problems and results in this area can be found in [21].

An α-floodlight is a light source that illuminates within an angular region of size α. For example, a $\frac{\pi}{3}$-floodlight illuminates an angular wedge of the plane with angular width $\frac{\pi}{3}$. The source of illumination is located at the apex of the angular region. Given a simple polygon, a floodlight is called a *vertex floodlight* if its source is located at a vertex of the polygon. The following old conjecture of mine was believed to be true up to December, 2001, when I found a counterexample:

Conjecture 5 *Any simple polygon with n vertices can be illuminated with $\lceil \frac{3n}{5} \rceil - 1$ vertex π-floodlights. We do not allow more than one floodlight on any vertex of the polygon.*

A family of polygons that requires $\lceil \frac{3n}{5} \rceil - 1$ π-vertex floodlights was obtained by F. Santos; see Figure 8.

A family of polygons with $9+8k$ vertices which require $5(k+1)$ vertex π-floodlights to illuminate them can be constructed by using the star shown in Figure 9. It is tempting to conjecture that the correct bound for the previous conjecture is $\frac{5(n-1)}{8} \pm c$. Recently Speckman and Tóth proved that any polygon

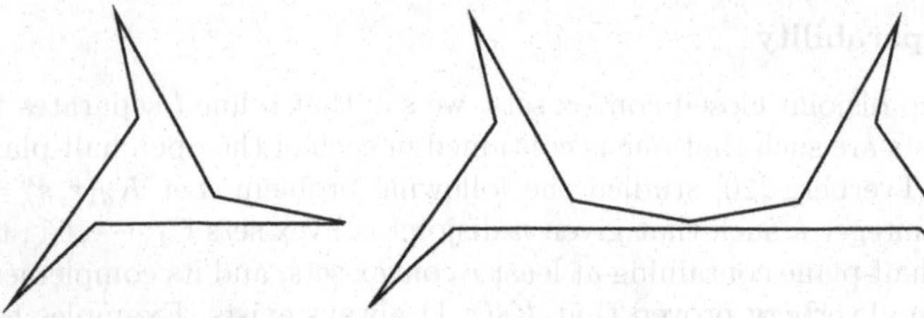

Fig. 8. A polygon with $5n + 1$ vertices which requires $3n$ vertex π-floodlights can be obtained by pasting n copies of the star polygon on the left.

with n vertices, k of which are convex, can be illuminated with $\lfloor \frac{2n-k}{3} \rfloor$ vertex π-floodlights. We close this section with two long standing conjectures on illumination, the first one due to T. Shermer:

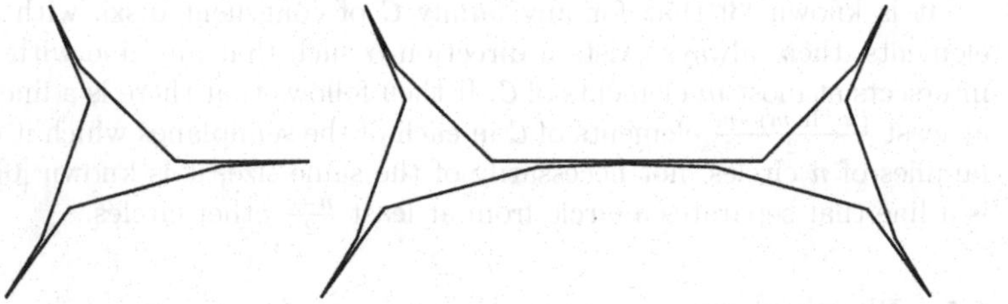

Fig. 9. Constructing a family of polygons with $8n + 1$ vertices that require $5n$ vertex π-floodlights to illuminate them.

Conjecture 6 *Any polygon with n vertices and h holes can be illuminated with $\lfloor \frac{n+h}{3} \rfloor$ vertex guards.*

In this case, the guards can illuminate all around themselves. The second conjecture is due to G. Toussaint, and was first stated in 1981:

Conjecture 7 *There is an n_0 such that any polygon $\mathcal{P}$ with $n \geq n_0$ vertices can be illuminated with $\lfloor \frac{n}{4} \rfloor$ edge guards. That is, any polygon $\mathcal{P}$ with $n \geq n_0$ vertices has a subset of $\lfloor \frac{n}{4} \rfloor$ edges such that any point in $\mathcal{P}$ is visible from one of these edges.*

Two surveys in this area [18,21] and a book by O'Rourke [17] contain most of the information concerning illumination and these problems up to 2000.

References

1. M. Abellanas, G. Hernandez, R. Klein, V. Neumann-Lara, and J. Urrutia,"A combinatorial property of convex sets". *Discrete Comput. Geom.* **17** (1997), No. 3, 307–318.
2. O. Aichholzer and H. Krasser, " The point set order type data base: A collection of applications and results". In Proc. 13th Canadian Conference on Computational Geometry CCCG 2001, pages 17-20, Waterloo, Ontario, Canada, 2001.
3. N. Alon, M. Katchalski and W.R. Pulleyblank,"Cutting disjoint disks by straight lines", *Discrete and Comp. Geom.* **4**, 239-243, (1989).
4. I. Barány and D.G. Larman, "A combinatorial property of points and ellipsoids", *Discrete Comp. Geometry* **5** (1990) 375-382.
5. I. Barány. J.H. Schmerl, S.J. Sidney and J. Urrutia, "A combinatorial result about points and balls in Euclidean space", *Discrete Comp. Geometry* **4** (1989) 259-262.
6. J. Czyzowicz, E. Rivera-Campo and J. Urrutia, "Separation of convex sets". *Discrete Appl. Math.* **51** (1994), No. 3, 325–328.
7. J. Czyzowicz, E. Rivera Campo, J. Urrutia and J. Zaks,"Separating convex sets on the plane", Proc. 2nd. Canadian Conference on Computational Geometry, University of Ottawa, (1989), pp. 50-54.
8. J. Czyzowicz, E. Rivera-Campo, J. Urrutia and J. Zaks, "Separating convex sets in the plane". *Discrete Comput. Geom.* **7** (1992), No. 2, 189–195.
9. H. Edelsbrunner, N. Hasan, R. Seidel and X.J. Shen, "Circles through two points that always enclose many points", *Geom. Dedicata,* **32** No. 1, 1-12 (1989).
10. R. Hayward, D. Rappaport y R. Wenger, "Some extremal results on circles containing points", *Disc. Comp. Geom.* **4** (1989) 253-258.
11. R. Hayward, "A note on the circle containment problem", *Disc. Comp. Geom.* **4** (1989) 263–264.
12. K. Hope and M. Katchalsk,"Separating plane convex sets", *Math. Scand.* **66** (1990), No. 1, 44–46.
13. H. Ito, H. Uehara, and M. Yokoyama,"NP-completeness of stage illumination problems", *Discrete and Computational Geometry, JCDCG'98*, pp. 158–165, *Lecture Notes in Computer Science* 1763, Springer-Verlag, (2000).
14. J. Pach, and E. Rivera-Campo, "On circumscribing polygons for line segments", *Computational Geometry, Theory and Applications* **10** (1998) 121-124.
15. V. Neumann-Lara y J. Urrutia, "A combinatorial result on points and circles in the plane", *Discrete Math.* **69** (1988) 173–178.
16. V. Neumann-Lara, E. Rivera-Campo, and J. Urrutia, "Convex partititonings of point sets", manuscript, 1999.
17. J. O'Rourke, *Art Gallery Theorems and Algorithms*, Oxford Univ. Press (1987)
18. Sherman, T., "Recent results in art galleries", Proc. IEEE (1992)1384-1399.
19. Csaba D. Tóth, "Art gallery problem with guards whose range of vision is 180°", *Computational Geometry, Theory and Applications,* **17** (2000), 121–134.
20. H. Tverberg, "A separation property of plane convex sets". *Math. Scand.* **45** (1979) No. 2, 255–260.
21. J. Urrutia, "Art gallery and illumination problems", In J.-R. Sack and J. Urrutia, (eds.), *Handbook on Computational Geometry*, North Holland (2000) 973–1127.

Quantum Algorithms
Invited Talk

Umesh Vazirani

University of California
Berkeley, CA 94720
vazirani@cs.berkeley.edu

Quantum computers are the only model of computing to credibly violate the modified Church-Turing thesis, which states that any reasonable model of computation can be simulated by a probabilistic Turing Machine with at most polynomial factor simulation overhead. This is dramatically demonstrated by Shor's polynomial time algorithms for factorization and discrete logarithms [13]. Shor's algorithm, as well as the earlier algorithm due to Simon [12] can both be cast into the general framework of the hidden subgroup problem (see for example [10]). Two recent papers [11,9] study how well this framework extends to solving the hidden subgroup problem for non-abelian groups (which includes the graph isomorphism problem).

Indeed, there are very few superpolynomial speedups by quantum algorithms that do not fit into the framework of the hidden subgroup problem. One example is the recursive fourier sampling problem [1], which provided the first formal evidence that quantum computers violate the extended Church-Turing thesis. Very recently, van Dam and Hallgren give a polynomial time quantum algorithm for solving the shifted Legendre symbol problem [3]; the algorithm does not appear to fit into the HSP framework.

There is another class of quantum algorithms that have their roots in Grover's search algorithm, which gives a quadratic speedup over brute force search. There is a matching lowerbound [2], thus showing that it will be hard to find polynomial time quantum algorithms for NP-complete problems.

Finally, a couple of years ago, an intruiging new paradigm for the design of quantum algorithms by adiabatic evolution was introduced [6]. Encouraging results from numerical simulations of this algorithm on small instances of NP-complete problems appeared in last year's March issue of Science [7]. Very recently, a few analytic results about this new paradigm have been obtained [4,5] — first, adiabatic quantum computing is truly quantum, since it gives a quadratic speedup for general search. Secondly, there is a simple class of combinatorial optimization problems on which adiabatic quantum algorithms require exponential time.

References

1. Bernstein E and Vazirani U, 1993, Quantum complexity theory, *SIAM Journal of Computation* **26** 5 pp 1411–1473 October, 1997.

S. Rajsbaum (Ed.): LATIN 2002, LNCS 2286, pp. 12–13, 2002.

2. Bennett, C. H., Bernstein, E., Brassard, G. and Vazirani, U., "Strengths and weaknesses of quantum computation," SIAM J. Computing, 26, pp. 1510-1523 (1997).
3. van Dam, W., Hallgren, H., "Efficient Quantum Algorithms for Shifted Quadratic Character Problems", quant-ph/0011067.
4. van Dam, W., Mosca, M., Vazirani, U., "How Powerful is Adiabatic Quantum Computing?" FOCS, 2001.
5. van Dam, W., Vazirani, U., "On the Power of Adiabatic Quantum Computing" manuscript, 2001.
6. E. Farhi, J. Goldstone, S. Gutmann, and M. Sipser, "Quantum Computation by Adiabatic Evolution", quant-ph report no. 0001106 (2000)
7. E. Farhi, J. Goldstone, S. Gutmann, J. Lapan, A. Lundgren, and D. Preda, "A Quantum Adiabatic Evolution Algorithm Applied to Random Instances of an NP-Complete Problem", *Science,* Vol. 292, April, pp. 472–476 (2001)
8. Grover, L., "Quantum mechanics helps in searching for a needle in a haystack," Phys. Rev. Letters, 78, pp. 325-328 (1997).
9. Grigni, M., Schulman, S., Vazirani, M., Vazirani, U., "Quantum Mechanical Algorithms for the Nonabelian Hidden Subgroup Problem", In *Proceedings of the Thirty-third Annual ACM Symposium on the Theory of Computing*, Crete, Greece, 2001.
10. L. Hales and S. Hallgren. Quantum Fourier Sampling Simplified. In *Proceedings of the Thirty-first Annual ACM Symposium on the Theory of Computing*, pages 330-338, Atlanta, Georgia, 1-4 May 1999.
11. Hallgren, S., Russell, A., Ta-Shma, A., "Normal subgroup reconstruction and quantum computation using group representations", In *Proceedings of the 32nd Annual ACM Symposium on Theory of Computing*, 627–635, 2000.
12. D. Simon. "On the power of quantum computation." In *Proc. 35th Symposium on Foundations of Computer Science (FOCS), 1994*.
13. Shor P W, Polynomial-time algorithms for prime factorization and discrete logarithms on a quantum computer, *SIAM J. Comp.*, **26**, No. 5, pp 1484–1509, October 1997.

Testing and Checking of Finite State Systems
Invited Talk

Mihalis Yannakakis

Avaya Laboratories
Basking Ridge, NJ 07920
mihalis@research.avayalabs.com

Finite state machines have been used to model a wide variety of systems, including sequential circuits, communication protocols, and other types of reactive systems, i.e., systems that interact with their environment. In testing problems we are given a system, which we may test by providing inputs and observing the outputs produced. The goal is to design test sequences so that we can deduce desired information about the given system under test, such as whether it implements correctly a given specification machine (conformance testing), or whether it satisfies given requirement properties (black-box checking).

In this talk we will review some of the theory and algorithms on basic testing problems for systems modeled by different types of finite state machines. Conformance testing of deterministic machines has been investigated for a long time; we will discuss various efficient methods. Testing of nondeterministic and probabilistic machines is related to games with incomplete information and to partially observable Markov decisions processes.

The verification of properties for finite state systems with a known structure (i.e., "white box" checking) is known as the model-checking problem, and has been thoroughly studied for many years, yielding a rich theory, algorithms and tools. Black-box checking, i.e., the verification of properties for systems with an unknown structure, combines elements from model checking, conformance testing and learning.

S. Rajsbaum (Ed.): LATIN 2002, LNCS 2286, pp. 14–14, 2002.

From Algorithms to Cryptography
Tutorial

Fabrizio Luccio and Linda Pagli

Dipartimento di Informatica, Università di Pisa
Corso Italia 40, 56125 Pisa, Italy
luccio,pagli@di.unipi.it

Style and purpose. This is a rather basic set of lectures in algorithms, with an advanced focus. Cryptography and randomization are discussed as non trivial fields of algorithm application.
Contents. Six lectures organized as follows:

1. Coding and encryption of information. Basic concepts on representing numbers, sets, and algorithms. An overview on cryptography and its development in history.
2. Algorithmic paradigms and computational complexity. Iteration and recursion. Lower and upper bounds on time and space.
3. Tractable and intractable problems. The classes P, NP, NP-hard.
4. The role of randomization. Random sources and random number generators. Hash functions. Randomized algorithms.
5. Symmetric and asymmetric cryptography. From DES to AES. Public key cryptosystems and RSA.
6. Cryptography on the Web. Identification, authentication, and digital signatures. Certification Authorities. The protocol SSL.

Prerequisites. Elementary knowledge of algorithm design, data structures, and descrete mathematics is assumed, so the bases of algorithmica will be compressed in a short and partly non conventional resume.

S. Rajsbaum (Ed.): LATIN 2002, LNCS 2286, pp. 15–15, 2002.

Dihomotopy as a Tool in State Space Analysis
Tutorial

Éric Goubault[1] and Martin Raussen[2]

[1] LIST (CEA Saclay), DTSI-SLA-LSL, 91191 Gif-sur-Yvette, France,
Eric.Goubault@cea.fr
[2] Institute of Mathematical Sciences, Aalborg University,
9220 Aalborg Øst, Denmark,
raussen@math.auc.dk

Abstract. Recent geometric methods have been used in concurrency theory for quickly finding deadlocks and unreachable states, see [14] for instance. The reason why these methods are fast is that they contain in germ ingredients for tackling the state-space explosion problem. In this paper we show how this can be made formal. We also give some hints about the underlying algorithmics. Finally, we compare with other well-known methods for coping with the state-space explosion problem.

1 Introduction

In model-checking techniques, temporal formulas, expressing important properties on traces of a concurrent system one has to verify, are checked by traversing the interleaving semantics of the program. This runs unfortunately into the "*state-space explosion problem*": the number of paths to be traversed might be exponential in the number of processes involved. It has been very tempting for a number of authors to try to use the information about the *independence* of actions to decrease this number by a possibly exponential ratio. For instance, if all actions considered are completely independent, meaning that any interleaving of actions taken in this set of actions computes the same thing, as a function from (parallel or distributed) store to store, then there is no need to consider all the interleavings to check any kind of "interesting" properties, such as safety or deadlock properties.

But this is not always as simple as we show with the transition system of Figure 1. Here we suppose that a and b are independent or "commuting" actions. The problem in Figure 1 is that we might choose to traverse only path $a.b$ since it is equivalent to $b.a$ and we will have missed the branching after b, which would have lead us into transition system C, which might contain any deadlock we want for instance. In fact, there are correct ways to infer state-space reduction methods from the independence relation. A classical one explained in Section 7.1 has been originally introduced by Valmari [46] under the name of "stubborn sets", based on a notion of independence on Petri nets. These have been ameliorated later under the name of "persistent sets" by Godefroid [21], based on the notion of independence of asynchronous transition systems. We develop in this

S. Rajsbaum (Ed.): LATIN 2002, LNCS 2286, pp. 16–37, 2002.

paper new methods for finding better state-space reduction techniques, based on *global* semantical information. This is done using geometric ideas, which have recently regained impetus after the seminal work [12] and [36]. We formalize this methodology, the "diconnected components" of the geometric semantics using a category of fractions of the fundamental category of the semantics, giving all information about all possible schedules of execution.

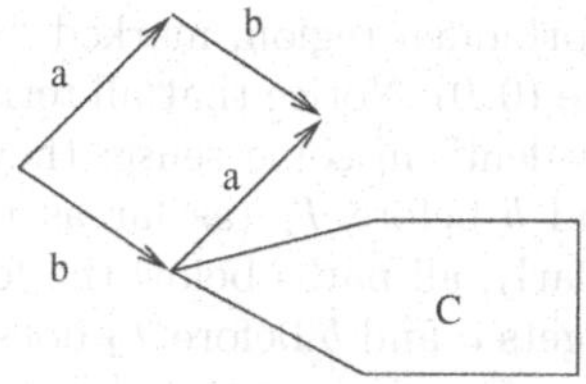

Fig. 1.

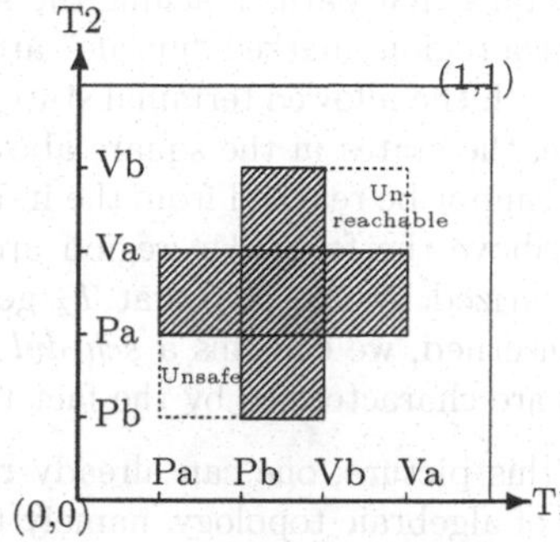

Fig. 2. Example of a progress graph

History: Towards Higher Dimensional Automata. The first "algebraic topological" model in the litterature is called *progress graph* and has appeared in operating systems theory, in particular for describing the problem of "deadly embrace" in "multiprogramming systems". *Progress graphs* are introduced in [10], but attributed to E. W. Dijkstra. The basic idea is to give a description of what can happen when several processes are modifying shared ressources. Given a shared resource a, we see it as its associated *semaphore* that rules its behaviour with respect to processes. For instance, if a is an ordinary shared variable, it is customary to use its semaphore to ensure that only one process at a time can write on it (this is mutual exclusion). Then, given n deterministic sequential processes $Q_1, \ldots, Q_n$, abstracted as a sequence of locks and unlocks on shared objects, $Q_i = R^1 a_i^1 . R^2 a_i^2 \cdots R^{n_i} a_i^{n_i}$ (R^k being P or V[1]), there is a natural way to understand the possible behaviours of their concurrent execution, by associating to each process a coordinate line in $\mathbf{R}^n$. The state of the system corresponds to a point in $\mathbf{R}^n$, whose ith coordinate describes the state (or "local time") of the ith processor.

Consider a system with finitely many processes running altogether. We assume that each process starts at (local time) 0 and finishes at (local time) 1; the P and V actions correspond to sequences of real numbers between 0 and 1, which reflect the order of the P's and V's. The initial state is $(0, \ldots, 0)$ and the final

[1] Using E. W. Dijkstra's notation P and V [12] for respectively acquiring and releasing a lock on a semaphore.

state is $(1, \ldots, 1)$. An example consisting of the two processes $T_1 = Pa.Pb.Vb.Va$ and $T_2 = Pb.Pa.Va.Vb$ gives rise to the two dimensional *progress graph* of Fig. 2. The shaded area represents states which are not allowed in any execution path, since they correspond to mutual exclusion. Such states constitute the *forbidden region.* An *execution path* is a path from the initial state $(0, \ldots, 0)$ to the final state $(1, \ldots, 1)$ avoiding the forbidden region and increasing in each coordinate - time cannot run backwards. We call these paths *directed paths* or *dipaths.* This entails that paths reaching the states in the dashed square underneath the forbidden region, marked "unsafe" are deemed to deadlock, i.e. they cannot possibly reach the allowed terminal state $(1, 1)$. Similarly, by reversing the direction of time, the states in the square above the forbidden region, marked "unreachable", cannot be reached from the initial state $(0, 0)$. Notice that all terminating paths above the forbidden region are "equivalent" in some sense: they are all characterized by the fact that T_2 gets a and b before T_1 (as far as resources are concerned, we call this a *schedule*). Similarly, all paths below the forbidden region are characterized by the fact that T_1 gets a and b before T_2 does.

In this picture, one can already recognize many ingredients that are at the center of algebraic topology, namely the classification of shapes modulo "elastic deformation". As a matter of fact, the actual coordinates that are chosen to represent the times at which Ps and Vs occur are unimportant, and these can be "stretched" in any manner, so the properties (deadlocks, schedules etc.) are invariant under some notion of deformation, or *homotopy.* This has to be a particular kind of homotopy though causing many difficulties in later work. We call it (in subsequent sections) a *directed homotopy* or *dihomotopy* in the sense that it should preserve the direction of time.

The semantics community came back to these geometric considerations with the development of "truly-concurrent" semantics, as opposed to "interleaving" semantics. The base of the argument was that interleaving semantics, i.e. the representation of parallelism by non-determinism ignores real asynchronous behaviours: $a \mid b$ where a and b are atomic is represented by the same transition system as the non-deterministic choice a then b or b then a. This fact creates problems in static analysis of (asynchronous) concurrent systems: Interleaving builds a lot of uninteresting states in the modelisation, hence induces a high cost in verification. This is called the *state-space explosion problem.* Quite a few models for true-concurrency have appeared (see in particular the account of [50]) but it is only in 1991 that geometry is proposed to solve the problem, in [36]. The diagnosis is that interleaving is only the boundary of the real picture. $a \mid b$ is really the filled-in square whose boundary is the non-deterministic choice a then b or b then a (the hollow square). The natural combinatorial notion, extension of transition systems, is that of a *cubical set,* which is a collection of points (states), edges (transitions), squares, cubes and hypercubes (higher-dimensional transitions representing the truly-concurrent execution of some number of actions). This is introduced in [36] as well as possible formalizations using n-categories, and a notion of homotopy. This is actually a combinatorial view of some kind of progress graph. Look back to Figure 2. Consider all interleavings of actions Pa,

Pb, Va and Vb: they form a subgrid of the progress graph. Take as 2-transitions (i.e. squares in the cubical set we are building) the filled-in squares. Only the forbidden region is really interleaved. Cubical sets generalize progress graphs, in that they allow any amount of non-deterministic choices as well as dynamic creation of processes. These cubical sets are called *Higher-Dimensional Automata* (HDA) following [36] because it really makes sense to consider a hypercube as some form of transition. Actually at about the same time, a bisimulation semantics was given in [47]. Notice that 2-transitions or squares are nothing but a *local commutation relation* as in Mazurkiewicz trace theory [34], *independence relation* as in asynchronous transition systems, see [2], as in trace automata, as in transition systems with independence [40], or (indirectly) as with the "confluence" relation of concurrent transition systems [45]. There are two more ingredients with HDA: the elegance and the power of the tools of geometric formalisations, and the natural generalisation to *higher* dimensions (i.e. "higher-order independence relation" or n-ary independence relations).

Example: Semaphores and progress graphs. In the rest of the paper, we will stick to one particular model which is sufficiently simple to explain, and gives sufficiently many nasty example: the shared memory model, in which asynchronous processes read and write atomically onto variables which are all in a common (shared) memory. To protect writing onto shared variables, we use mutual exclusion locks, which we put explicitely before writing a variable x, by Px, and that we release explicitely after, by Vx. It is then easy to see that writing on two distinct variables are two independant actions, as well as reading two variables (even the same one) by two processes. This model can also easily include [16] counting semaphores which are weakly synchronising objects that can be shared by n but not $n+1$ processes at the same time (for some $n > 1$). Notice that asynchronous message-passing with bounded buffers can be translated into that framework. It is therefore not only a useful example, but a quite general application indeed.

The key idea is to regard a progress graph as a topological space in which points are ordered globally through time, i.e., equipped with a (closed) partial order $\leq$. Traces of executions are continuous and increasing maps from the totally ordered unit segment to $(X, \leq)$. These are called *dipaths* for "directed paths". A dihomotopy between two dipaths f and g on X is a deformation via dipaths interpolating continuously between f and g and fixing the endpoints. The technical definitions will be given in Sect. 3. Now we can give semantics to a very simple language in which a finite number of processes can only do a deterministic sequence of lockings Px and unlockings Vx of shared resources x. So processes are just strings of P's and V's. Suppose that each semaphore x (binary or counting) is equipped with a number $s(x)$, the maximal number of processes that can share it at any time. Supposing that the length of the strings X_i $(1 \leq i \leq n)$ are integers l_i, the semantics of *Prog* is included in $[0, l_1] \times \cdots \times [0, l_n]$. A description of the progress graph $[\![Prog]\!]$ associated with *Prog* can be given by describing inductively what should be digged into this hyperrectangle. The semantics of our language can be described by the simple rule, $[k_1, r_1] \times \cdots \times [k_n, r_n] \in [\![X_1 \mid \cdots \mid X_n]\!]_2$ if there is a partition of $\{1, \cdots, n\}$ into $U \cup V$ with $card(U) = s(a) + 1$ for some object a with, $X_i(k_i) = Pa$, $X_i(r_i) = Va$ for $i \in U$ and $k_j = 0, r_j = l_j$ for $j \in V$. This language is somehow disappointing. To be able to consider looping and branching constructs, we are lead to the notion of local po-spaces in Sect. 3.1.

Goals of the present paper. After having explained the geometric semantics, the idea of deformation of paths of executions, and introduced the diconnected components approach to the state-space explosion problem, we compare (favorably) our technique with classical techniques such as persistent sets. We also review in Sect. 7.3 some orthogonal techniques which could still be used on top of our geometric technique.

2 The Fundamental Group of a Topological Space

In this section, we give a brief review of the fundamental group of a topological space, a very important concept from algebraic topology. See e.g. [1,5,27,35] for details. Hereafter, we develop a variation of this notion and apply it to state space analysis.

Topological spaces are abstractions of metric spaces. For a metric space X, nearness is expressed by a metric d measuring the distance between pairs of points. For a topological space Y, nearness is expressed with the aid of a collection of *open* subsets of Y. The usual definition for a continuous map between two metric spaces has the following generalisation for topological spaces: A map $f : Y \to Y'$ between topological spaces is *continuous* if and only if $f^{-1}(U) \subset Y$ is open for every open subset $U \subset Y'$.

In this paper, we will mainly be concerned with (different types of) *paths*, i.e., continuous maps $f : I \to X$ from an interval I into a topological space X. For the moment, we let $I = [0, 1]$ denote the unit interval with standard metric and topology. In general, one cannot compose paths in X. But if the endpoint $f_1(1)$ of f_1 coincides with the start point $f_2(0)$ of f_2, their *concatenation*

$$f_2 * f_1 : I \to X \text{ is defined by } (f_2 * f_1)(s) = \begin{cases} f_1(2s), & t \le \frac{1}{2} \\ f_2(2s-1), & t \ge \frac{1}{2}. \end{cases}$$

Both paths are thus pursued with "double speed". Concatenation defines a (non-commutative, non-associative) operation on the space $\mathcal{P}(X)$ of all paths on X.

Two points $x, y \in X$ are called *path-connected*, if there exists a path f with $f(0) = x$ and $f(1) = y$. The equivalence classes of this equivalence relation are called the *path components* of X. The image $f(X_0) \subset Y$ of a path component $X_0 \subset X$ under a continuous map $f : X \to Y$ is path-connected. As a consequence, path components are completely independent of each other, and one can investigate them "one at a time". A *loop* in a topological space X is a path $f : I \to X$ such that $f(0) = f(1)$. Loops with the same start/end-point can be concatenated. A *homotopy* of paths (loops) is a continuous map $H : I \times I \to X$ with $H(t, 0) = H(0, 0)$ and $H(t, 1) = H(0, 1)$ for all $t \in I$. It should be regarded as a one-parameter family of paths $H_t : I \to X$, $H_t(s) = H(t, s)$ (with fixed end points) connecting H_0 and H_1. Two paths $f_0, f_1 : I \to X$ with the same endpoints are called *homotopic* if there is a fixed end point homotopy $H : I \times I \to X$ with $H_0 = f_0$ and $H_1 = f_1$. Homotopy is an equivalence relation.

A continuous and strictly increasing map $\varphi : I \to I$ with $\varphi(0) = 0$ and $\varphi(1) = 1$ can be used to reparametrise a path, i.e., to pass from a path f in X to the (reparameterised) path $f \circ \varphi$ with the same image. Remark that φ is homotopic to the identity map on I; a homotopy is given by $H(t,s) = (1-t)\varphi(s) + ts$. As a consequence, the paths f and its reparametrisation $f \circ \varphi$ are homotopic via the homotopy $\bar{H}(t,s) = f(H(t,s))$.

A basic invariant of a topological space X is its *fundamental group*: Fix a base point $x_0 \in X$. The elements of the fundamental group $\pi_1(X; x_0)$ are the homotopy classes of *loops* $f : I \to X$ which start and end at $f(0) = f(1) = x_0$. Concatenation of loops at x_0 factorizes over homotopy to yield a 2-adic operation on $\pi_1(X; x_0)$. The homotopy class of the constant map $c : I \to X,\ c(s) = x_0,\ s \in I$, serves as the neutral element – since $f, f * c$ and $c * f$ are homotopic to each other. The inverse to the class of the loop f is given by the the class of the loop $f^- : I \to X, f^-(t) = f(1-t)$: $f^- * f$ and $f * f^-$ are both homotopic to c.

The size of the fundamental group has an interesting interpretation: A loop f can be regarded as a map from the unit circle $\bar{f} : S^1 \to X,\ \bar{f}(\exp(2\pi is)) = f(s)$. The loop f represents the trivial element in $\pi_1(X; x_0)$ if it is homotopic to the constant loop c. A homotopy H with $H_0 = c$ and $H_1 = f$ can be transformed into an extension $\bar{H} : D^2 \to X$ of $\bar{f}$, viz. $\bar{H}(t \exp(2\pi is)) = H(t,s)$. Conversely, an extension of $\bar{f}$ to a continuous map $\bar{H} : D^2 \to X$ gives rise to a homotopy between f and c. A homotopically trivial loop can thus be "filled in". Hence, the the fundamental group of a space "counts the numbers of holes" in it.

The fundamental group of a space does only depend on the *path component* of the base point: Let g denote an arbitrary path with $g(0) = x_0$ and $g(1) = x_1$. Then the map "conjugation with g": $\pi_1(X; x_0) \to \pi_1(X; x_1);\ [f] \mapsto [g^- * f * g]$ is a *group isomorphism* .

Examples: Proofs of the following statements can be found in almost any textbook on algebraic topology:

- The fundamental group of Euclidean space $\mathbf{R}^n$ is trivial for all n.
- The fundamental group of the unit circle S^1 is isomorphic to the integers. An explicit isomorphism $\pi_1(S^1) \to \mathbf{Z}$ associates to a loop its "winding number", i.e., it counts (with a sign) the number of times a particular value is attained. The fundamental group of an n-sphere $S^n = \{x \in \mathbf{R}^n \mid \|x\| = 1\}$ is trivial for every $n > 1$.
- The fundamental group of "the figure 8" (two circles with only a single base point in common) is the free group on two letters representing the two directed loops.
- For every group G, there is a path-connected topological space BG with $\pi_1(BG) \simeq G$.

3 The Fundamental Category of an Lpo-space

3.1 Lpo-spaces and Dipaths

There are many models for state spaces for concurrent processes and the executions on them, cf. Sect. 7. In this paper, we follow the basic idea from [16]: A po-space consists of a *topological space* X with a *partial order* $\leq \subset X \times X$. The partial order is assumed to be closed (as a subset of $X \times X$) to ensure coherence between topology and order: this makes it possible to take limits "under the $\leq$ sign". For an example of such a po-space (in fact, a progress graph) see Fig. 3; the left figure represents the state space for two processes that acquire and relinquish a lock to a single shared resource; the right figure pictures the situation where locks to two shared resources have to be acquired in reverse order by the two processes. The black areas are the "forbidden regions" of the progress graph which are *not* part of the state space.

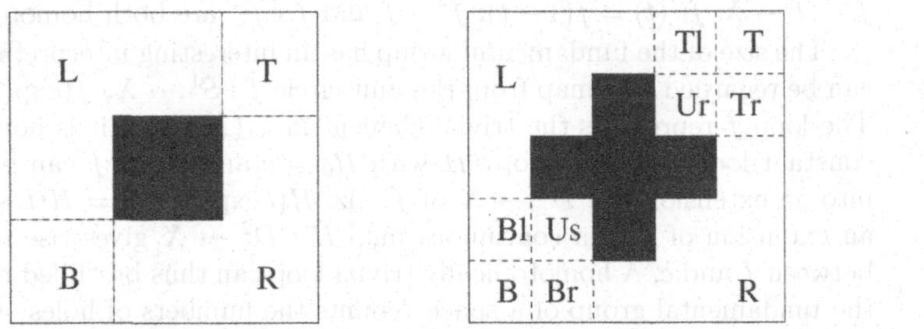

Fig. 3. Square with a hole and complement of a "Swiss flag"

If one or several of the processes contain loops, the resulting abstraction will no longer have a *global* partial order. Instead one requires for a local po-space (*lpo-space*) a relation $\leq$ on X that restricts to a partial order on sufficiently small subsets of X that form a basis for the topology. Two such relations are equivalent (and define the same local partial order) if they agree on sufficiently small open sets forming a basis for the topology. For an example, consider the relation on the unit circle $S^1 \subset \mathbf{R}^2$ given by $x \leq y \Leftrightarrow$ the angle from x to y is less than α. This relation is certainly not transitive, but it defines a local partial order if and only if $\alpha \leq \pi$; for $\alpha \leq \pi$ these are all equivalent.[2]

Traces of a concurrent system (executions) are modelled by so-called *dipaths*– di is an abbreviation for *directed.* A *short*, resp. *long* dipath in an lpo-space X is defined as an *order preserving continuous* map from the interval $\vec{I} = [0, 1]$, resp. from the non-negative reals $\mathbf{R}_{\geq 0} = \{x \in \mathbf{R} |\ x \geq 0\}$ (with the natural order) into X. A short dipath models a concurrent process from a start point to

[2] This version of the definition is due to Ulrich Fahrenberg; it is in fact equivalent to the one given in [16,17].

an end point, while a long dipath runs indefinitely (e.g., in loops) but avoiding *zeno* executions. Technically, one requires that a long dipath does not admit a limit for $t \to \infty$.

3.2 Dihomotopy

When can you be sure that two execution traces in a concurrent program provide the same result? This is the case if the corresponding dipaths $f, g : I \to X$ are *dihomotopic*. This means, that there exists a continuous order-preserving *dihomotopy* $H : I \times \vec{I} \to X$ with $H_0 = f$ and $H_1 = g$. Remark that the parameter interval is equipped with the trivial order, i.e., $(t, s) \leq (t', s') \Leftrightarrow t = t' \wedge s \leq s'$. In particular, every "intermediate" path H_t has to be a dipath. Moreover, we require a fixed start point ($H(t, 0) = H(0, 0)$) and, for short dipaths, a fixed end point ($H(t, 1) = H(0, 1)$); for long dipaths all the paths H_t have to be non-zeno.

3.3 The Fundamental Category

For an lpo-space, one should no longer watch out for a fundamental *group*. The reverse of a dipath is no longer a dipath. On a global po-space, there are no (non-trivial) directed loops at all. Instead, one has to work with the fundamental *category* of a local po-space X, or rather with two versions of it, depending on whether short or long dipaths are considered:

The *objects* of the fundamental category $\vec{\pi}_1(X)$ are the points of X. The *morphisms* between elements x and y are given as the dihomotopy classes in $\vec{\pi}_1(X; x, y)$. Composition of morphisms

$$\vec{\pi}_1(X; x, y) \times \vec{\pi}_1(X; y, z) \to \vec{\pi}_1(X; x, z)$$

is given by concatenation of dipaths – up to dihomotopy.

The category $\vec{\pi}_1^{\infty}(X)$ contains $\vec{\pi}_1(X)$. It has an additional maximal element ∞ with $Mor(x, \infty)$ consisting of the dihomotopy classes of long dipaths starting at x and $Mor(\infty, y) = \emptyset$ for $y \in X \cup \{\infty\}$. Concatenation of a (short) dipath from x to y with a (long) dipath from y yields a (long) dipath – up to dihomotopy via long dipaths.

Compared to the fundamental group, a fundamental category is an enormous gadget and it has a much less nice algebraic structure. On the other hand, in easy examples one has the impression, that the cardinality of the set of morphisms between two points is quite robust when these points are perturbed a little.

Example 1. For the square with one hole (Fig. 3), there is no morphism between the regions marked L, resp. R, and no morphism from T to any other region, neither a morphism from any other region to B. There are two morphisms from any point of B to any point of T. Moreover, from any point of B, certain points of B, L, R can be reached by (exactly one) morphism. Likewise, any point of T can be reached from (certain of the points in) L, R and T in one way.

For the complement of a "Swiss flag" (Fig. 3), the situation is a bit more complicated: There is no dipath leaving the unsafe rectangle *Us* and there is no dipath entering the unreachable rectangle *Ur* from the outside. It is possible to reach Us by a dipath from $B \cup Bl \cup Br$; from Ur, one can reach $Tl \cup Tr \cup T$. The only possibility for *two* classes of dipaths between points occurs when the first is in B and the second in T. Moreover, these classes can be represented by dipaths along the boundary, representing the two sequential executions.

The lesson to learn is that the complete "dynamics" of these state space can be described from the decomposition into the blocks studied above. It is the aim of this paper to define and describe these "dicomponents" in the general case and thus, in a realistically large model, to provide a "collapse" of the exponentially large state space into pieces that show the same behaviour with respect to execution paths between each other. It is then enough to study the "flow" between these "components" in order to capture the dynamics of the whole system.

4 Categories of Fractions and Components

4.1 Categories of Fractions

Next, we have to invest in a construction from category theory: We invert in a systematic way all those partial dipaths that never contribute to a decision along any dipath. The resulting category will then have many "zig-zag" isomorphisms giving rise to the components. Here is a general method [18,4]:

Let $\mathcal{C}$ denote a category. To keep things simple, assume $\mathcal{C}$ small, i.e., objects and morphisms are sets. Let $\Sigma \subset Mor(\mathcal{C})$ denote a *system of morphisms*, i.e., Σ includes all unit morphisms and is closed under composition. For any such system, one may construct the category of fractions $\mathcal{C}[\Sigma^{-1}]$ and the localization functor $q_\Sigma : \mathcal{C} \to \mathcal{C}[\Sigma^{-1}]$ [18,4] having the following universal property:

- For every $s \in \Sigma$, the morphism $q_\Sigma(s)$ is an *iso*morphism. - For any functor $F : \mathcal{C} \to \mathcal{D}$ such that $F(s)$ is an isomorphism for every $s \in \Sigma$, there is a unique functor $\theta : \mathcal{C}[\Sigma^{-1}] \to \mathcal{D}$ with $\theta \circ q_\Sigma = F$.

The objects of $\mathcal{C}[\Sigma^{-1}]$ are just the objects of $\mathcal{C}$. To define the morphisms of $\mathcal{C}[\Sigma^{-1}]$, one introduces a (formal) inverse s^{-1} to every morphism $s \in \Sigma(x,y)$. These inverses are collected in $\Sigma^{-1}(y,x), x, y \in Ob(\mathcal{C})$ and then in Σ^{-1}. Consider the closure of $Mor(\mathcal{C}) \cup \Sigma^{-1}$ under composition and the smallest equivalence relation containing $s^{-1} \circ s = 1_x$ and $s \circ s^{-1} = 1_y$ for $s \in \Sigma(x,y)$ that is compatible with composition. The equivalence classes correspond then to the morphisms of $\mathcal{C}[\Sigma^{-1}]$. In particular, if $t \circ \alpha = \beta \circ s$ for $s, t \in \Sigma$, then $\alpha \circ s^{-1} = t^{-1} \circ \beta$. A morphism in $\mathcal{C}[\Sigma^{-1}]$ can always be represented in the form

$$s_k^{-1} \circ f_k \circ \cdots \circ s_1^{-1} \circ f_1, \; s_j \in \Sigma, f_j \in Mor, k \in \mathbf{N}.$$

Let $Iso(\mathcal{C})$ denote the isomorphisms of the category $\mathcal{C}$, and let $\Sigma * Iso(\mathcal{C})$ denote the system of morphisms generated by Σ and by $Iso(\mathcal{C})$. The isomorphisms in $\mathcal{C}[\Sigma^{-1}]$ are the *zig-zag* morphisms, i.e.,

$$Iso(\mathcal{C}[\Sigma^{-1}]) = \{s_1^{-1} \circ s_2 \circ \cdots \circ s_{2k-1}^{-1} \circ s_{2k}, s_j \in \Sigma * Iso(\mathcal{C}), k \in \mathbf{N}\}.$$

The subcategory of $\mathcal{C}[\Sigma^{-1}]$ with all objects, the morphisms of which are given by the zig-zag morphisms in $Iso(\mathcal{C}[\Sigma^{-1}])$, forms in fact a *groupoid* [30].

4.2 Components

A "compression" of the category $\mathcal{C}[\Sigma^{-1}]$ is achieved by dividing out all isomorphisms: Two objects $x, y \in Ob(\mathcal{C})$ are Σ-isomorphic or Σ-connected – $x \simeq_\Sigma y$ – if there exists a zig-zag-morphism from x to y. This definition corresponds to usual path connectedness *with respect to paths representing isomorphisms only – but regardless orientation.* Σ-connectivity is an equivalence relation; the equivalence classes are called the *Σ-connected components* – the path components with respect to Σ-zig-zag paths, viz. the components of the groupoid above.

Next, consider the equivalence relation on the morphisms of $\mathcal{C}[\Sigma^{-1}]$ generated (under composition) by $\alpha \simeq \alpha \circ s, \quad \alpha \simeq t \circ \alpha$ with $\alpha \in Mor(x, y), s \in Inv(\mathcal{C}[\Sigma^{-1}])(x', x), t \in Inv(\mathcal{C}[\Sigma^{-1}])(y, y')$. Remark that equivalent morphisms no longer need to have the same source or target. Remark moreover, that any two zig-zag morphisms from x to y are equivalent to each other; in particular, they are equivalent to the unit-morphisms in both x and y.

Example 2. If $i_0, i_1, j_0, j_1 \in Inv(\mathcal{C}[\Sigma^{-1}])$, then $f_0, f_1, f_2 \in Mor(\mathcal{C})$ in the following diagram are equivalent to each other in $\mathcal{C}[\Sigma]^{-1}$:

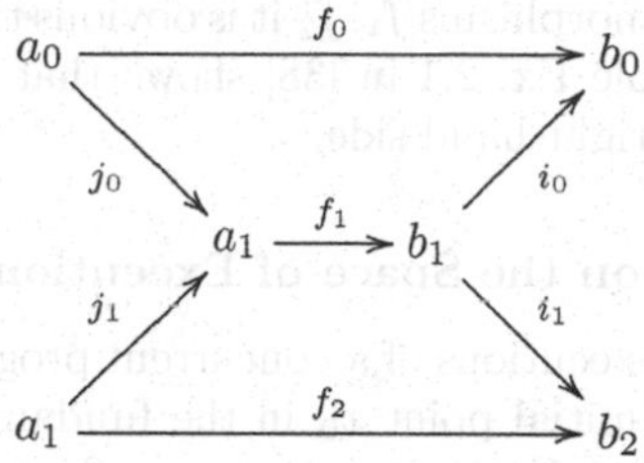

The objects of the *component category* $\pi_0(\mathcal{C}; \Sigma)$ are by definition the Σ-connected components of $\mathcal{C}$; the morphisms from $[x]$ to $[y]$, $x, y \in Ob(\mathcal{C})$, are the equivalence classes of morphisms in $\bigcup_{x' \simeq_\Sigma x, y' \simeq_\Sigma y} Mor_{\mathcal{C}[\Sigma^{-1}]}(x', y')$. The composition of $[\beta] \circ [\alpha]$ for $\alpha \in Mor_{\mathcal{C}[\Sigma^{-1}]}(x, y)$ and $\beta \in Mor_{\mathcal{C}[\Sigma^{-1}]}(y', z)$ is given by $[\beta \circ s \circ \alpha]$ with s any zig-zag morphism from y to y'. The equivalence class of that composition is independent of the choices taken.

Remark 1. These constructions decompose the study of the morphisms of $\mathcal{C}$ into two pieces: Firstly, the study of the groupoid $Inv(\mathcal{C}[\Sigma^{-1}])$ which can be performed separately on each of the Σ-connected components. For the fundamental category, all these morphisms represent executions that can be performed and/or backtracked without global effects. Secondly, certainly more important for applications, the study of the component category, which encompasses the global effects of irreversibility. In the case of the component category of the fundamental category, representatives of all non-unit dipath classes may have (different) global effects – backtracking along such a dipath class may therefore change the result of a computation.

5 Applications to State Space Analysis

In this section, we apply the preceeding constructions to our models of the state space and the space of executions (from a given initial state) of a concurrent program. The key task is to single out the relevant system of morphism Σ that is to be inverted. It should consist of morphisms that, *from a global point of view*, do not contribute with any *decision* to the outcome of the concurrent program. Here, we give the key definitions (in a general categorial framework), their motivation, and a few elementary examples. For algorithms in low dimensions, cf. Sect. 6.

5.1 Extensions of Morphisms

For a small category, let $X_0, X_1 \subset Ob(\mathcal{C})$. Let $Mor_{0,1} = \bigcup_{x_0 \in X_0, x_1 \in X_1} Mor(x_0, x_1)$ denote the set of all morphisms between X_0 and X_1. We associate to a morphism $f \in Mor(x, y)$ with $x, y \in Ob(\mathcal{C})$, the set of all its *extensions*

$$\mathcal{E}(f) = \{g \circ f \circ h |\ h \in Mor(X_0, x), g \in Mor(x, X_1)\} \subset Mor_{01}$$

from X_0 to X_1. This set consists of all morphisms from X_0 to X_1 that *factor through* f. It is empty if $Mor(X_0, x) = \emptyset$ or if $Mor(y, X_1) = \emptyset$. In the particular case $f = 1_x$, the unit at $x \in Ob(\mathcal{C})$, the set $\mathcal{E}(x) = \mathcal{E}(1_x)$ consists of all morphisms from X_0 to X_1 factoring through x.

For concatenable morphisms f_1, f_2 it is obvious that $\mathcal{E}(f_2 \circ f_1) \subseteq \mathcal{E}(f_1) \cap \mathcal{E}(f_2)$. The geometric example Ex. 2.1 in [38] shows that the left hand side may be a *proper* subset of the right hand side.

5.2 Components on the Space of Executions

The space of partial executions of a concurrent program is modelled as the set of morphisms from the initial point x_0 in the fundamental category $\vec{\pi}_1(X)$. More generally, one may associate to any category $\mathcal{C}$ and any object $x_0 \in Ob(\mathcal{C})$ the comma category $(x_0 \downarrow \mathcal{C})$ of *objects under* x_0 [33]: Its objects are the morphisms in $Mor(x_0, x)$, $x \in Ob(\mathcal{C})$, and its morphisms are the *commutative* triangles

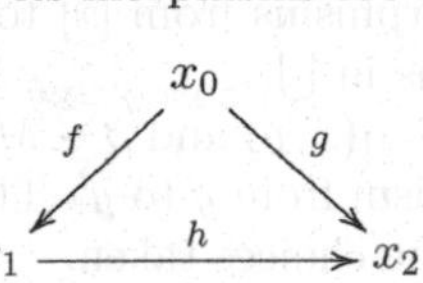

with x_0 in the top and $h \in Mor(x_1, x_2)$.

If $\mathcal{C}$ is the fundamental category $\vec{\pi}_1(X)$ and x_0 an initial element, the comma categories $(x_0 \downarrow \vec{\pi}_1(X))$ and $(x_0 \downarrow \vec{\pi}_1^\infty(X))$ have as objects the dihomotopy classes of dipaths starting at x_0 : a partial dipath $h \in \vec{\pi}_1(x_1, x_2)$ with $x_1 \in X$ and $x_2 \in X \cup \{\infty\}$ induces a map $\vec{\pi}_1(x_0, x_1) \to \vec{\pi}_1(x_0, x_2)$ by concatenation.

Assume given a (minimal) object x_0 such that $X_0 = \{x_0\}$ and a set X_1 of maximal objects in a category $\mathcal{C}$. For the fundamental category $\vec{\pi}_1(X)$, this set X_1 should be chosen as a discrete set of final accepting states[3], for the fundamental category $\vec{\pi}_1^\infty(X)$, *the* maximal object should be chosen as ∞.

[3] which could include deadlocking points

Definition 1. *A morphism s from $f \in Mor(x_0, x)$ to $g \in Mor(x_0, y)$ belongs to Σ_1 if and only if $\mathcal{E}(f) = \mathcal{E}(g) \subseteq Mor_{01}$.*

It is clear that either every or no morphism from f to g is contained in Σ_1. Obviously, Σ_1 contains the units and is closed under composition. For $\mathcal{C} = \vec{\pi}_1(X)$, a dipath s extending f to g is contained in Σ_1 if no "decision" has been made in between – all "careers" in $\vec{\pi}_1(X; x_0, X_1)$ open to f are still open to g. No (globally detectable) branching occurs between f and g.

A detection of the component category wrt. Σ_1 entails the following benefit:

Proposition 1. *Two dipaths f and g from x_0 to x_1 that proceed through the same Σ_1-components are dihomotopic.*

We illustrate the resulting component categories by two elementary examples:

Example 3. Let x_0, resp. x_1 denote the minimal, resp. the maximal element in the po-space X, the square with one hole from Fig. 3. Then Mor_{01} has two elemens represented by dipaths f_L, f_R touching R, resp. L. Any dipath within B and any dipath within $L \cup R \cup T$ is in Σ_1. No dipath starting within B and ending in L or R is in Σ_1.

The category $(x_0 \downarrow \vec{\pi}_1(X))$ consists of three Σ_1-connected components: the dipaths ending in B; those touching L and those touching R. Observe: There is *no* zig-zag path $t^{-1} \circ s$ from a dipath to R via T to a dipath to L since there are no dipaths f from x_0 to R and g from x_0 to L with $t * g$ dihomotopic to $s * f$.

The component category $\pi_0(x_0 \downarrow \vec{\pi}_1(X)), \Sigma_1)$ can – apart from the units – be represented by (end points in)

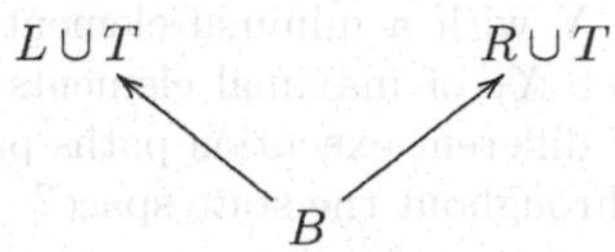

Example 4. Let Y denote the "Swiss flag" po-space from Fig. 3. Let x_0 and x_1 denote the minimal, resp. the maximal elements, and let y denote the deadlock point (maximal within the unsafe region Us). The set of accepting states is $X_1 = \{x_1, y\}$, and Mor_{01} consists of three elements – there is also a dihomotopy class with end point in y. The component category $\pi_0(x_0 \downarrow \vec{\pi}_1(Y)), \Sigma_1)$ is represented by the diagram (with obvious morphisms between the given regions/components)

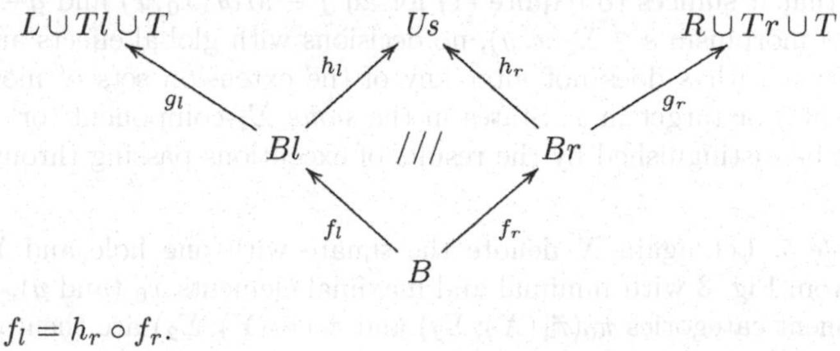

with $h_l \circ f_l = h_r \circ f_r$.

Remark 2. In [43], S. Sokołowski introduced a somehow similar approach resulting in the *fundamental poset* $\Omega_1(X)$ of a po-space X. Using our terminology, a preorder on $(x_0 \downarrow \vec{\pi}_1(X))$ is defined by:

$$f \in \vec{\pi}_1(X; x_0, x) \sqsubseteq g \in \vec{\pi}_1(X; x_0, y) \Leftrightarrow \forall h \in \vec{\pi}_1(X; y, z)$$

$$\exists\, a, j_1 \in \vec{\pi}_1(X; x, a), j_2 \in \vec{\pi}_1(X; z, a) \text{ with } j_1 * f = j_2 * h * g \in \vec{\pi}_1(X; x_0, a).$$

The equivalence classes given by "$\sqsubseteq$ and $\sqsupseteq$" are the elements of the poset $\Omega_1(X)$, equipped with the partial order induced by $\sqsubseteq$.

If one considers morphisms Mor_{01} corresponding to a set X_1 of maximal elements, it is easy to see that $f \sqsubseteq g \Leftrightarrow \mathcal{E}(f) \supseteq \mathcal{E}(g)$, and hence the Σ_1-connected components agree with the elements of $\Omega_1(X)$. The partial order between equivalence classes in $\Omega_1(X)$ corresponds to the *existence* of morphisms in $\mathcal{C}[\Sigma^{-1}]$ between elements of these classes. The component category contains more information. It allows to compare factorisations of two given morphisms and to discuss in which parts of the po-space they agree and in which they differ.

P. Gaucher [19] has a quite different categorical approach to branching and merging, not only for dipaths, but also for their higher-dimensional analoga.

5.3 Components of the State Space

Next, we shift attention to the entire state space of a concurrent program, modelled by an lpo-space X with a minimal element x_0, the only element of X_0, and a (discrete) subset X_1 of maximal elements. For an element $x \in X$, we ask: Which essentially different execution paths pass through x? How does this information develop throughout the state space?

Definition 2. *The system $\Sigma_2 \subset Mor(\mathcal{C})$ consists of all morphisms $s \in Mor(x, y)$ with $x, y \in Ob(\mathcal{C})$ satisfying*

$$\mathcal{E}(f) = \mathcal{E}(s \circ f), \mathcal{E}(g) = \mathcal{E}(g \circ s) \text{ for all } f \in Mor(-, x) \text{ and } g \in Mor(y, -). \quad (1)$$

Obviously, Σ_2 contains the units, and it is closed under composition. It is easy to see that it suffices to require (1) for all $f \in Mor(x_0, x)$ and $g \in Mor(x, X_1)$. Along a morphism $s \in \Sigma_2(x, y)$, no decisions with global effects are taken: concatenation with s does not alter any of the extension sets of morphisms with source in y or target in x. States in the *same* Σ_2-component (or *dicomponent*) cannot be distinguished by the results of executions passing through them.

Example 5. Let again X denote the square with one hole and Y the "Swiss flag" from Fig. 3 with minimal and maximal elements x_0 (and y), resp. x_1. The component categories $\pi_0(\vec{\pi}_1(X); \Sigma_2)$ and $\pi_0(\vec{\pi}_1(Y); \Sigma_2)$ are then of the form

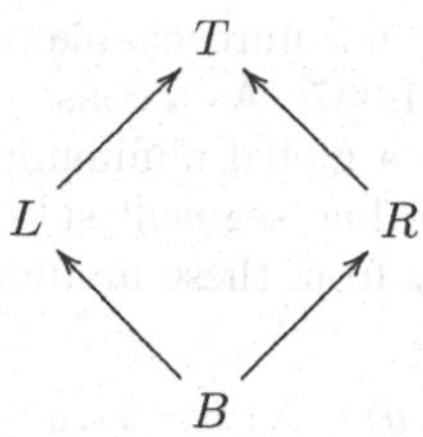

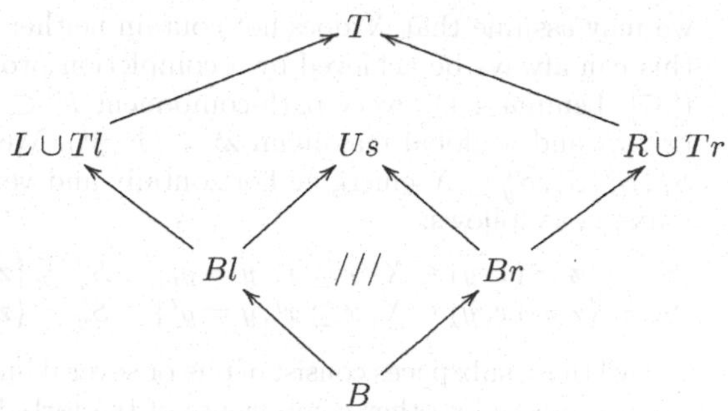

5.4 Relation to History Equivalence

The Σ_2-components refine the notion of a *dicomponent* of an lpo-space defined earlier in [16,37]. Those were only defined as sets and lacked the dynamical perspective given by the component category:

Definition 3. *The* history hf *of a morphism* $f \in Mor(X_0, X_1)$ *is defined as*

$$hf = \{x \in Ob(\mathcal{C}) | \; \exists f_0 \in Mor(X_0, x), f_1 \in Mor(x, X_1) \text{ with } f = f_1 \circ f_0\}.$$

Two objects $x, y \in Ob(\mathcal{C})$ *are* history equivalent *if and only if* $x \in hf \Leftrightarrow y \in hf$ *for all* $f \in Mor(X_0, X_1)$.

A history equivalence class $C \subset Ob(\mathcal{C})$ is thus a primitive element of the Boolean algebra generated by the histories, i.e., an intersection of histories and their complements such that for all $f \in Mor(X_0, X_1)$ either $C \subseteq hf$ or $C \cap hf = \emptyset$.

The following argument shows that a morphism $s \in \Sigma_2(x, y)$ has history equivalent source and target:

$$x \in hf \Leftrightarrow f \in \mathcal{E}(x) = \mathcal{E}(i_x) = \mathcal{E}(s) = \mathcal{E}(i_y) = \mathcal{E}(y) \Leftrightarrow y \in hf.$$

Hence, a Σ_2-component is contained in a path-component of a history equivalence class.

6 Algorithms for 2-dimensional Mutual Exclusion Models

In this section, we confine ourselves to the progress graphs described in the introduction. Classifying dipaths up to dihomotopy in these mutual exclusion models corresponds to finding out which (and how many) schedules for a given concurrent program can potentially yield different results. An algorithm arriving at such a classification in dimension two, i.e, for semaphore programs with just *two* interacting transactions, was described in [37]; the results in this section rely on the methods described there.

In this case, the state space has $X = I^2 \setminus int(F)$ as a model, i.e., a unit square from which a forbidden region F (e.g., the region in black in Fig. 3) is deleted. This region is a union of rectangles that are parallel to the axes. Since we are interested in dipaths connecting the minimal point to the maximal point, we may assume that X does not contain neither unsafe nor unreachable points; this can always be achieved by a completion process, cf. [37]. As a consequence ([37], Lemma 4.1), every path-component $F_i \subset F$ has a global minimum $\mathbf{z}_i = (x_i, y_i)$ and a global maximum $\mathbf{z}^i = (x^i, y^i)$. We define line segment subspaces $S_i^x, S_i^y, S_x^i, S_y^i \subseteq X$ emerging horizontally and vertically from these minima and maxima as follows:

$$S_i^x = \{\mathbf{z} = (x,y) \in X |\, x \le x_i, y = y_i\} \quad S_i^y = \{\mathbf{z} = (x,y) \in X |\, x = x_i, y \le y_i\}$$
$$S_x^i = \{\mathbf{z} = (x,y) \in X |\, x \ge x^i, y = y^i\} \quad S_y^i = \{\mathbf{z} = (x,y) \in X |\, x = x^i, y \ge y^i\}.$$

All these subspaces consist of one or several line segments, that may be broken up into pieces by other components of the forbidden region. Let $T_i^x \subseteq S_i^x, T_i^y \subseteq S_i^y, T_x^i \subseteq S_x^i$ and $T_y^i \subseteq S_y^i$ denote the segment touching F_i, cf. Fig. 4. The unions of these separating line segments will be called $T_{\_} = \bigcup_i (T_i^x \cup T_i^y), T^{-} = \bigcup_i (T_x^i \cup T_y^i)$ and $T = T^{-} \cup T_{\_}$. A dipath $f : I \to X$ from x to y is said *to cross* $T_{\_}$ if there exists an i such that $\emptyset \neq f^{-1}(T_{\_})$ is contained in the interior of I, i.e., if its image contains points on both sides of one of the segments. Similarly, one defines crossing wrt. T^{-} and to T. We can now detect which of the dipath classes in X are inverted in the two categories of fractions of Sect. 5:

Proposition 2. *Let $s : I \to X$ denote a (partial) dipath with $f(0) = x$ and $f(1) = y$. Its dihomotopy class $[s] \in \vec{\pi}_1(X; x, y)$ is contained in Σ_1 if and only if f does not cross $T_{\_}$; in Σ_2 if and only if f does not cross T.*

Example 6. In the example of Fig. 4 with a forbidden region consisting of four components F_i, there are six dihomotopy classes of dipaths between $\mathbf{x}_0$ and $\mathbf{x}_1$, cf. [37], Fig. 14. The upper figures contain the line segments T_*^* that, together with boundary segments of the forbidden region, cut out the components in the two cases discussed in Sect. 5. For instance, the component marked C in the rightmost figure is characterised by two ingoing non-unit morphisms that, each have two extensions. The lower figures show the associated component categories with morphisms going upward. In this example, there are no non-trivial relations.

A similar analysis in dimensions higher than two is certainly more demanding. Not only the components of F, but also their finer topological properties will certainly play a role, cf. the discussion in [37], Sect. 5.

7 Classical State-Space Reduction Techniques

7.1 Persistent Sets

Let $(S, i, E, Tran, I)$ be an asynchronous transition system [2]. This means that $(S, i, E, Tran)$ is a transition system and that $I \subseteq E \times E$ is a relation between labels E, the "independence relation" between two actions. We will not give a precise axiomatics for I here, and will keep on simple grounds. Basically, I should satisfy the following conditions (taken from [22]):

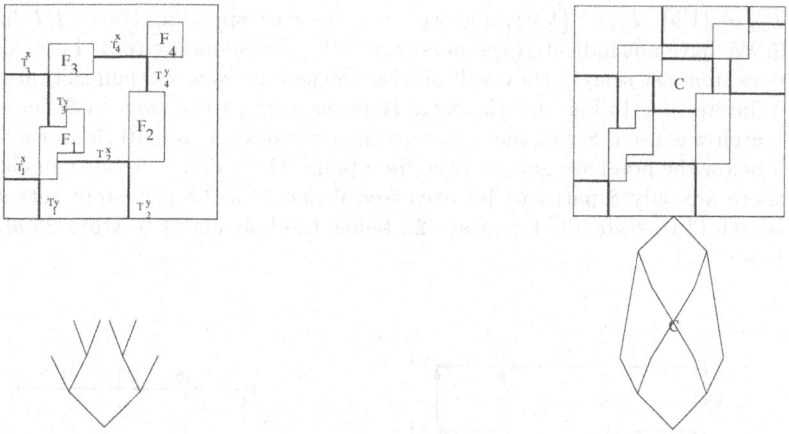

Fig. 4. Components and their categories in a 2-dimensional mutual exclusion model

- if t_1 (respectively t_2) is enabled in s and $s \rightarrow^{t_1} s'$ (respectively $s \rightarrow^{t_2} s'$) then t_2 (respectively t_1) is enabled in s if and only if t_2 (t_1) is enabled in s' (independent transitions can neither disable nor enable each other); and,

- if t_1 and t_2 are enabled in s, then there is a unique state s' such that both $s \rightarrow^{t_1} s_1 \rightarrow^{t_2} s'$ and $s \rightarrow^{t_2} s_2 \rightarrow^{t_1} s'$ (commutativity of enabled independent transitions).

In our technique, I is just a set of squares, or 2-transitions, or in the topological sense, they are elementary surfaces, enabling us to continuously deform dipaths. We extend in an intuitive manner I to sets of actions by putting AIB if for all $a \in A$, for all $b \in B$, aIb. We identify a with the singleton $\{a\}$.

Let T be a set of actions, $T \subseteq E$, and $p \in S$ be a state. We say that T is persistent in state p if, T contains only actions which are enabled at p, and, for all traces t beginning at p containing only actions q out of T, qIT. Suppose we have a set of persistent actions T_p for all states p in an asynchronous transition system. Then let us look at the following set of traces PT (identified with a series of states) in $(S, i, E, Tran, I)$ defined inductively as follows: $(i) \in PT$, and if $(p_1, \ldots, p_n) \in PT$, then $(p_1, \ldots, p_n, q) \in PT$ where $p_n \rightarrow^{t'} q \in Tran$ and $t' \notin T_{p_n}$. Deadlock detection can be performed on this subset PT of traces instead of the full set of traces of $(S, i, E, Tran, I)$. At least when $(S, i, E, Tran, I)$ is acyclic, PT is enough for checking LTL temporal formulas (and you can modify the method so that it works generally). We exemplify the method on the process $Pb.Pa.Vb.Va \mid Pa.Pb.Va.Vb$. A standard interleaving semantics would be as sketched in Figure 5, showing the presence of deadlocking state 13. One set of persistent sets is $T_1 = \{Pa\}$, $T_2 = \{Pb\}$, $T_5 = \{Pa, Pb\}$, $T_6 = \{Pb, Va\}$, $T_8 = \{Pa, Va\}$, $T_{13} = \emptyset$, $T_9 = \{Vb\}$, $T_{12} = \{Va\}$, $T_{17} = \{Pb\}$, $T_{18} = \{Va\}$, $T_{22} = \{Vb\}$, $T_{23} = \emptyset$, $T_7 = \{Pb, Vb\}$, $T_{14} = \{Vb\}$, $T_{15} = \{Pb\}$, $T_{16} = \{Pa\}$,

$T_{20} = \{Vb\}$, $T_{21} = \{Va\}$, and we show the corresponding traces PT in Figure 6. We have not indicated the persistent sets corresponding to 3, 4 etc. since in a persistent set search, they will not be reached anyway, so their actual choice is uninteresting. In Figure 5 there are 16 paths from 1 to be traversed if no selective search was used. Six of them lead to the deadlock 13, and 10 (5 above the hole, 5 below the hole) are going to the final point 23. In Figure 6, one can check that there are only 8 paths to be traversed if one uses the persistent sets selective search (3 to state 13, 1 to state 23 below the hole and 4 to state 23 above the hole).

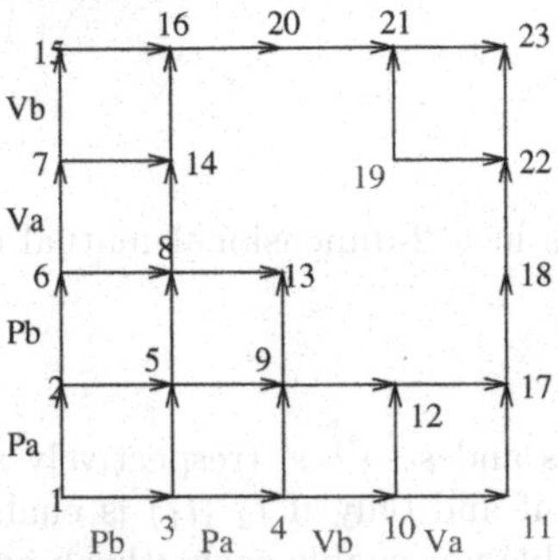

Fig. 5.

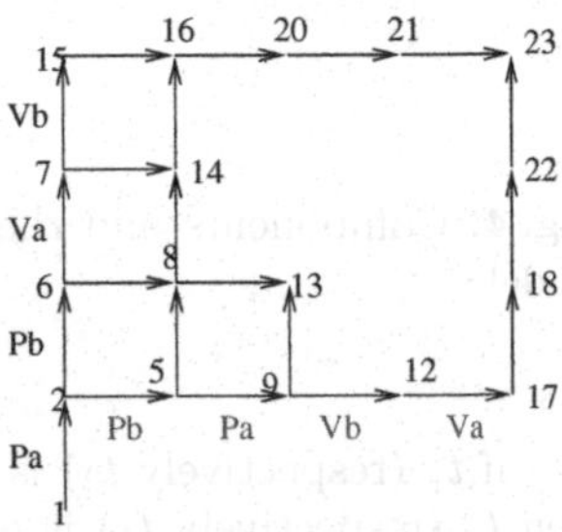

Fig. 6.

How did we find this set of persistent sets? In the PV case this can be done quite easily as follows. First the independence relation can be found out right away. Px and Py stand respectively for the query for a lock on x and y (nothing is committed yet) so they are independent actions, whatever x and y are. We should rather declare Px and Vy dependent in general: if $x = y$ this is clear, and for $x \neq y$ this can come from the fact locks on x and y are causally related (precisely as in the case of Figure 5 with $x = a$ and $x = b$). This is slightly different from the more usual case of atomic reads and writes languages in which the independance relation can be safely determined as: actions are independent if and only if they act on distinct variables. The most elaborated technique known in this framework is that of "stubborn sets" see [46], and its adaptation to the current presentation, see [22] for a precise definition. The example of persistent set we gave in Figure 6 is in fact a stubborn set. As one can see as well, the persistent set approach here reduces the 5 paths below the hole into 1, which is a representant modulo dihomotopy of these 5 dipaths. In the diconnected components approach, one finds the set of 7 diconnected components and the corresponding graph of regions pictured in Figure 7.

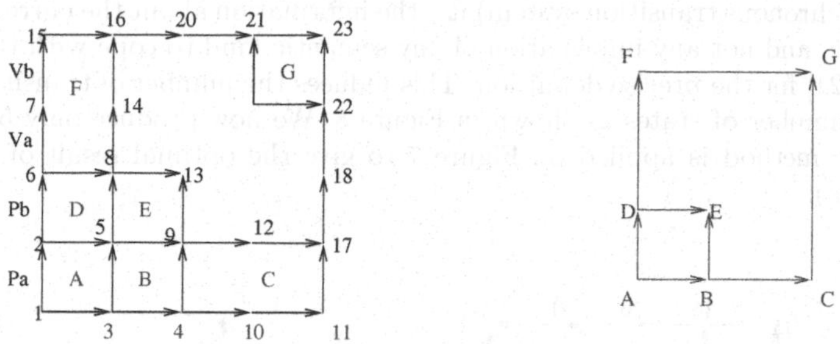

Fig. 7.

7.2 Comparison with Geometric Techniques

There are 4 dipaths to be traversed in the graph of diconnected regions to determine the behaviour of this concurrent system (two of them leading to state 13 being dihomotopic). In fact, there are two explanations why the method of diconnected components is superior to the persistent set approach. In the latter, the independence relation does not in general depend on the current state (even if this might be changed by changing the set of labels), Our notion of independence is given by a 2-transition, which depends on the current state (see for instance [25] where the link is made formal). The second and more important reason is that the diconnected graph algorithm determines regions using *global* properties, whereas the persistent sets approach uses only (in general syntactic) local criteria for reducing the state-space. Conversely, it is relatively easy to see the following: For every state p in our asynchronous transition system (or by [25], in a 2-dimensional cubical set), all traces t composed of actions outside T_p is such that all its actions are independent with T_p. So any trace from p made up of any action (those of T_p as well as those outside T_p) can be deformed (by dihomotopy, or "is equivalent to") into a trace firing first actions from T_p and then actions outside T_p. Therefore the selective search approach using only actions from T_p (for all p) is only traversing some representatives of the dihomotopy classes of paths. The persistent search approach is a particular case of dihomotopic deformation (not optimal in general).

7.3 Miscellaneous Techniques

Sleep sets. The sleep sets technique can be seen as a mere amelioration of the traversal we saw, and therefore can be combined with the method of persistent sets (as well as ours). The problem we had in Section 7.1 is that quite a few of the paths we are traversing go through the same states at some point, and have a common suffix (like paths (1,2,6,8,14,16,20,21,23) and (1,2,5,8,14,16,20,21,23) in Figure 6). It is obviously not necessary to traverse again common suffixes if we want to check future tense logical formulas. The sleep sets S_p (p a state in our asynchronous transition system) use the information about the current traversals made, and not any information of any semantic kind to cope with this problem: see [21] for the precise definition. This reduces the number of transitions but not the number of states as shown in Figure 8. We now produce only 5 paths. The same method is applied on Figure 7 to give the optimal result of Figure 9 (3 paths).

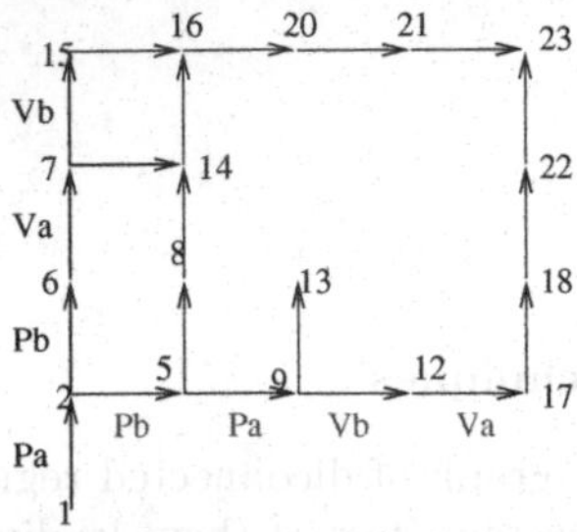

Fig. 8.

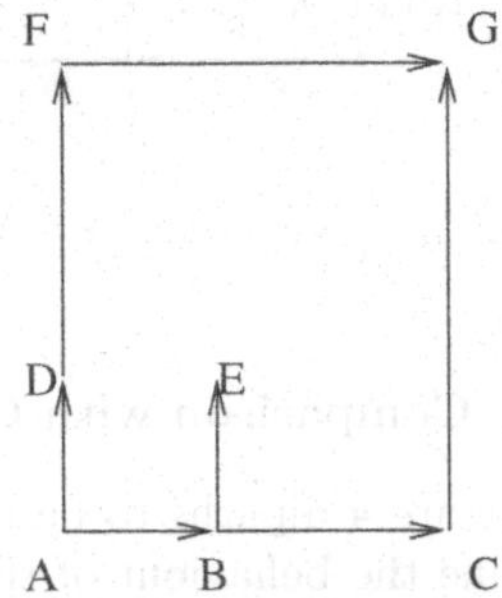

Fig. 9.

This does not entirely solve the suffix problem. A classical means to complete this method is to use "state space hashing" [20]. In this method (again orthogonal to "dihomotopy reduction"), one tries to maintain the set of states (and the set of transitions sometimes) already traversed avoiding to visit again the same sequences of states and transitions. As this database might be very big, one uses standard hashing techniques which quickly decide if two states are equal, but which might "identify" unequal states. This was used for instance in one of our simple implemented analyzers described in [14]. Again, for cyclic transition systems, this transformation does not change the deadlock(s) or LTL (future tense temporal logics) formulae that are true.

Covering steps or Virtual Coarsening. Another idea, called "covering step" in [49] or sometimes called in other situations "virtual coarsening", is to group together interleavings of independent actions by "multiset" transitions or "covering steps". For instance, when possible (basically, you should check that you are not in the situation of Figure 1), one replaces an interleaving square of two actions by the transition $s_0 \rightarrow^{\{a,b\}} s_3$. This is obviously subsumed by our notion of n-transition (formally, this can be done in the style of [39]).

Algorithmics of the representation of the state-space. A number of clever algorithmic methods have been used to reduce the representation of the state-space in memory without discarding any transition nor state, rather by compressing the representation. A very much used technique in model-checking is the representation of the transition relation with binary decision diagrams (BDDs) or QDDs as in [3] associated with symbolic representations of states [28]. Some amount of work has been devoted to "on the fly" techniques, also in model checking, see for instance [31]: Only a part of the state-space is represented during the analysis, because there is no need in general to construct first the whole state-space and then traverse it. Last but not least some techniques involving reducing the state-space using symmetry arguments have been proposed and successfully used, see [9]. All these techniques could be equally applied to our diconnected components approach, and should be exemplified in future papers. For instance, symmetry techniques are quite well studied in geometry and should apply straightforwardly to our geometric approach.

8 Concluding Remarks

Two further arguments in favor of our geometric techniques should be developped: We should be able to gain much more when the dimension of the problem (i.e. the number of processes involved) increases. The persistent sets types of methods basically use local transpositions, or in our geometric phrasing, faces of dimension 2, to equate some of the equivalent dipaths. Geometrically speaking, we can use sometime shorter deformation paths, like any hypercube, i.e. any cyclic permutation. The other argument is that geometric methods do cooperate well with abstraction mechanisms (in the sense of abstract interpretation [11]). It is in particular shown in [15] that the upper-approximation (or lower-approximation) of the forbidden regions can be carried out simply for a variety of languages, using classical abstract interpretation domains. These give lower (respectively upper) approximations of the "interesting" schedules or paths to be traversed.

References

1. M.A. Armstrong, *Basic Topology*, Springer-Verlag, 1990.
2. M. A. Bednarczyk. *Categories of asynchronous systems.* PhD thesis, University of Sussex, 1988.
3. Bernard Boigelot and Patrice Godefroid. Symbolic verification of communication protocols with infinite state spaces using QDDs. *Formal Methods in System Design: An International Journal*, 14(3):237–255, May 1999.
4. T. Borceux, *Handbook of Categorial Algebra I: Basic category theory*, Encyclopedia of Mathematics and its Applications, Cambridge University Press, 1994.
5. Glen E. Bredon, *Topology and Geometry*, GTM, vol. 139, Springer-Verlag, 1993.
6. R. Brown and P.J. Higgins, *Colimit theorems for relative homotopy groups*, J. Pure Appl. Algebra **22** (1981), 11–41.
7. ———, *On the algebra of cubes*, J. Pure Appl. Algebra **21** (1981), 233–260.
8. S.D. Carson and P.F. Reynolds, *The geometry of semaphore programs*, ACM TOPLAS **9** (1987), no. 1, 25–53.
9. E. M. Clarke, E. A. Emerson, S. Jha, and A. P. Sistla. Symmetry reductions in model checking. In *Proc. 10th International Computer Aided Verification Conference*, pages 145–458, 1998.

10. E. G. Coffman, M. J. Elphick, and A. Shoshani. System deadlocks. *Computing Surveys*, 3(2):67–78, June 1971.
11. P. Cousot and R. Cousot. Abstract interpretation: A unified lattice model for static analysis of programs by construction of approximations of fixed points. *Principles of Programming Languages 4*, pages 238–252, 1977.
12. E.W. Dijkstra, *Co-operating sequential processes*, Programming Languages (F. Genuys, ed.), Academic Press, New York, 1968, pp. 43–110.
13. L. Fajstrup, *Dicovering spaces*, Tech. Report R-01-2023, Department of Mathematical Sciences, Aalborg University, DK-9220 Aalborg Øst, November 2001.
14. L. Fajstrup, É. Goubault, and M. Raussen, *Detecting Deadlocks in Concurrent Systems*, CONCUR '98; Concurrency Theory (Nice, France) (D. Sangiorgi and R. de Simone, eds.), Lect. Notes Comp. Science, vol. 1466, Springer-Verlag, September 1998, 9th Int. Conf., Proceedings, pp. 332 – 347.
15. L. Fajstrup, E. Goubault, and M. Raussen. Detecting deadlocks in concurrent systems. Technical report, CEA, 1998.
16. ———, *Algebraic topology and concurrency*, Tech. Report R-99-2008, Department of Mathematical Sciences, Aalborg University, DK-9220 Aalborg Øst, June 1999, conditionally accepted for publication in Theoret. Comput. Sci.
17. L. Fajstrup and S. Sokolowski, *Infinitely running concurrents processes with loops from a geometric viewpoint*, Electronic Notes Theor. Comput. Sci. **39** (2000), no. 2, 19 pp., URL: http://www.elsevier.nl/locate/entcs/volume39.html.
18. P. Gabriel and M. Zisman, *Calculus of fractions and homotopy theory*, Springer-Verlag New York, Inc., New York, 1967, Ergebnisse der Mathematik und ihrer Grenzgebiete, Band 35.
19. P. Gaucher, *Homotopy invariants of higher dimensional categories and concurrency in computer science*, Math. Structures Comput. Sci. **10** (2000), no. 4, 481–524.
20. P. Godefroid, G. J. Holzmann, and D. Pirottin. State space caching revisited. In *Proc. 4th International Computer Aided Verification Conference*, pages 178–191, 1992.
21. P. Godefroid and P. Wolper, *Using partial orders for the efficient verification of deadlock freedom and safety properties*, Proc. of the Third Workshop on Computer Aided Verification, vol. 575, Springer-Verlag, Lecture Notes in Computer Science, 1991, pp. 417–428.
22. P. Godefroid and P. Wolper. Partial-order methods for temporal verification. In *Proc. of CONCUR'93*. Springer-Verlag, LNCS, 1993.
23. É. Goubault, *The Geometry of Concurrency*, Ph.D. thesis, Ecole Normale Superieure, Paris, 1995.
24. ———, *Geometry and Concurrency: A User's Guide*, Electronic Notes Theor. Comput. Sci. **39** (2000), no. 2, 16 pp., URL: http://www.elsevier.nl/locate/entcs/volume39.html.
25. ———, *Cubical sets are generalized transition systems*, avail. at `http://www.di.ens.fr/~goubault`, 2001.
26. J. Gunawardena, *Homotopy and concurrency*, Bulletin of the EATCS **54** (1994), 184–193.
27. A. Hatcher, *Algebraic Topology*, electronically avialable at `http://math.cornell.edu/~hatcher/# ATI`
28. T. A. Henzinger, O. Kupferman, and S. Qadeer. From pre-historic to post-modern symbolic model checking. In *Proc. 10th International Computer Aided Verification Conference*, pages 195–206., 1998.
29. M. Herlihy, S. Rajsbaum, and M. Tuttle, *An Overview of synchronous Message-Passing and Topology*, Electronic Notes Theor. Comput. Sci. **39** (2000), no. 2.

30. P.J. Higgins, *Categories and Groupoids*, Mathematical Studies, vol. 32, van Nortrand Reinhold, London, 1971.
31. G. Holzmann. On-the-fly model checking. *ACM Computing Surveys*, 28(4es):120–120, December 1996.
32. W. Lipski and C.H. Papadimitriou, *A fast algorithm for testing for safety and detecting deadlocks in locked transaction systems*, Journal of Algorithms **2** (1981), 211–226.
33. S. Mac Lane, *Categories for the working mathematician*, Graduate Texts in Mathematics, vol. 5, Springer-Verlag, New York, Heidelberg, Berlin, 1971.
34. A. Mazurkiewicz. Trace theory. In G. Rozenberg, editor, *Petri Nets: Applications and Relationship to Other Models of Concurrency, Advances in Petri Nets 1986, PART II, PO of an Advanced Course*, volume 255 of *LNCS*, pages 279–324, Bad Honnefs, September 1986. Springer-Verlag.
35. J.R. Munkres, *Elements of Algebraic Topology*, Addison-Wesley, 1984.
36. V. Pratt, *Modelling concurrency with geometry*, Proc. of the 18th ACM Symposium on Principles of Programming Languages. (1991).
37. M. Raussen, *On the classification of dipaths in geometric models for concurrency*, Math. Structures Comput. Sci. **10** (2000), no. 4, 427–457.
38. ______, *State spaces and dipaths up to dihomotopy*, Tech. Report R-01-2025, Department of Mathematical Sciences, Aalborg University, DK-9220 Aalborg Øst, November 2001.
39. V. Sassone and G. L. Cattani. Higher-dimensional transition systems. In *Proceedings of LICS'96*, 1996.
40. V. Sassone, M. Nielsen, and G. Winskel. Relationships between models of concurrency. In *Proceedings of the Rex'93 school and symposium*, 1994.
41. J. P. Serre, *Homologie singulière des espaces fibrés.*, Ann. of Math. **(2) 54** (1951), 425–505.
42. A. Shoshani and E. G. Coffman. Sequencing tasks in multiprocess systems to avoid deadlocks. In *Conference Record of 1970 Eleventh Annual Symposium on Switching and Automata Theory*, pages 225–235, Santa Monica, California, Oct 1970. IEEE.
43. S. Sokołowski, *Categories of dimaps and their dihomotopies in po-spaces and local po-spaces*, Preliminary Proceedings of the Workshop on Geometry and Topology in Concurrency Theory GETCO'01 (Aalborg, Denmark) (P.Cousot et al., ed.), vol. NS-01, BRICS Notes Series, no. 7, BRICS, 2001, pp. 77 – 97.
44. J. Srba, *On the Power of Labels in Transition Systems*, CONCUR 2001 (Aalborg, Denmark) (K.G. Larsen and M. Nielsen, eds.), Lect. Notes Comp. Science, vol. 2154, Springer-Verlag, 2001, pp. 277–291.
45. A. Stark. Concurrent transition systems. *Theoretical Computer Science*, 64:221–269, 1989.
46. A. Valmari, *A stubborn attack on state explosion*, Proc. of Computer Aided Verification, no. 3, AMS DIMACS series in Discrete Mathematics and Theoretical Computer Science, 1991, pp. 25–41.
47. R. van Glabbeek, *Bisimulation semantics for higher dimensional automata*, Tech. report, Stanford University, 1991.
48. R. van Glabbeek and U. Goltz. Partial order semantics for refinement of actions. *Bulletin of the EATCS*, (34), 1989.
49. F. Vernadat, P. Azema, and F. Michel. Covering step graph. *Lecture Notes in Computer Science*, 1091, 1996.
50. G. Winskel and M. Nielsen. *Models for concurrency*, volume 3 of Handbook of Logic in Computer Science. Oxford University Press, 1994.

Algorithms for Local Alignment with Length Constraints*

Abdullah N. Arslan and Ömer Eğecioğlu

Department of Computer Science
University of California, Santa Barbara
CA 93106, USA
{arslan,omer}@cs.ucsb.edu

Abstract. The local sequence alignment problem is the detection of similar subsequences in two given sequences of lengths $n \geq m$. Unfortunately the common notion of local alignment suffers from some well-known anomalies which result from not taking into account the lengths of the aligned subsequences. We introduce the *length restricted local alignment problem* which includes as a constraint an upper limit T on the length of one of the subsequences to be aligned. We propose an efficient approximation algorithm, which finds a solution satisfying the length bound, and whose score is within difference Δ of the optimum score for any given positive integer Δ. The algorithm runs in time $O(nmT/\Delta)$ using $O(mT/\Delta)$ space. We also introduce the *cyclic local alignment* problem and show how our idea can be applied to this case as well. This is a dual approach to the well-known cyclic edit distance problem.

1 Introduction

One of the central problems in computational molecular biology is the *local sequence alignment* problem (LA). LA aims to reveal similar regions in a given pair of sequences X and Y whose lengths are n and m respectively with $n \geq m$. The Smith-Waterman algorithm finds an optimal local alignment by searching for two segments with maximum similarity score by discarding poor initial and terminal fragments. However an alignment returned by the algorithm may contain a mosaic of well-conserved fragments artificially connected by poorly conserved or even unrelated fragments. As a result, a local alignment with score 1,000 and length 10,000 may be chosen over a possibly more significant shorter local alignment with score 998 and length 1,000. This deficiency causes two forms of anomalies known as the "mosaic effect" and the "shadow effect". Mosaic effect in an alignment is observed when a poor region is sandwiched between two regions with high similarity scores. Shadow effect is observed when a biologically important short alignment is not detected by the algorithm because it overlaps with a much longer yet biologically irrelevant alignment with only a slightly higher score.

* Supported in part by NSF Grant No. CCR–9821038.

S. Rajsbaum (Ed.): LATIN 2002, LNCS 2286, pp. 38–51, 2002.

In this paper we consider the *length restricted local alignment* (*LRLA*) problem in which we search for substrings I of X and J of Y that maximize the score $s(I,J)$ among all substrings satisfying $|J| \leq T$ for a given T. This is yet another attempt to eliminate problems associated with local alignment. The objective is similar to that of normalized local alignment algorithms [3], in that we aim to circumvent the undesirable mosaic and the shadow effects. The degree of fragmentation in this new problem is controlled by varying the upper bound T.

LRLA can be solved by extending the dynamic programming formulation of local alignment problem with resulting time complexity $O(Tnm)$. Since this may not be practical for large values of the parameters, we propose two approximation algorithms for *LRLA*. The first one is a simple $\frac{1}{2}$-approximation algorithm, with the same complexity as that of the local alignment problem itself. The second algorithm returns a score guaranteed to be within difference Δ of the optimum for a given $\Delta \geq 1$. The time complexity of this approximation algorithm is $O(nmT/\Delta)$, with $O(mT/\Delta)$ space. These two algorithms can also be used to approximately solve the *cyclic local alignment* problem (*CLA*) of maximizing $s(I,J)$ where I is a substring of X, and J is a substring of a cyclic shift of Y.

The outline of this paper is as follows. In section 2, we discuss related work in the literature on algorithms for restricted versions of local alignment and cyclic edit distance. In section 3 we present the requisite notions and give the notation we use. Section 4 contains the description of the approximation algorithms *HALF* and *APX-LRLA*. Concluding remarks are presented in section 5.

2 Previous Work

The anomalies of mosaic and shadow effects that exist in the ordinary formulation of local alignment and the Smith-Waterman algorithm lead to problems in comparison of long genomic sequences and comparative gene prediction. This was recently pointed out by Zhang et al., 1999 [15] who proposed to decompose a discovered local alignment into sub-alignments that avoid the mosaic effect. However, the post-processing approach may miss the alignments with the best degree of similarity if the Smith-Waterman algorithm missed them in the first place. As a result, highly similar fragments may be ignored if they are not parts of longer alignments dominating other local similarities. Another approach to fixing the problems with the Smith-Waterman algorithm is based on the notion of *X-drop*, a region within an alignment that scores below X. The alignments that contain no X-drops are called *X-alignments*. Although X-alignments are expensive to compute in practice, Altschul et al., 1997 [2] and Zhang et al., 1998 [14] used some heuristics for searching databases with this approach. Alexandrov and Solovyev, 1998 [1] proposed to normalize the alignment score by its length and demonstrated that this new approach leads to better protein classification. Arslan et al., 2001 [3] defined the *normalized local alignment problem* where the goal is to find substrings I and J that maximize $s(I,J)/(|I|+|J|)$ among all substrings I and J with $|I|+|J| \geq t$, where $s(I,J)$ is the score, and t is a threshold for the minimal overall length of I and J. Because of the cubic time complexity of the exact algorithm as an approximation to the original problem they proposed a solution to the maximization of $s(I,J)/(|I|+|J|+L)$ for a given parameter L. This can be done in time $O(nm \log n)$ and using $O(m)$ space [3].

The length restricted local alignment problem considered in this paper tries to eliminate problems associated with local alignment, and it has an efficient approximation algorithm which allows for a control over the length of the optimal local alignment sought. The limit is placed on only the substring J of Y. We believe that the underlying scoring scheme should limit the length of the other substring involved in an optimal alignment automatically, and therefore having two limits, one for $|I|$ and another for $|J|$ is redundant.

An application of length restricted local alignment is the formulation of the *cyclic local alignment* problem as a special case of *LRLA*. The cyclic local alignment is the problem of maximizing $s(I,J)$ over all I and J, where I is a substring of X, and J is a substring of a *cyclic shift* of Y. The cyclic local alignment problem is the length restricted local alignment problem with strings X and YY, and limit $T = |Y|$. The approximation algorithms we develop can readily be used for this problem. Moreover it defines a dual approach to well-known *cyclic edit distance problem* which aims to find the minimum edit distance between string X and a cyclic shift of Y over all possible cyclic shifts of Y.

Cyclic edit distance problem appears in many applications, and there is extensive literature on the subject. Bunke and Bühler, 1993 [4] presented cyclic edit distance as a method for two-dimensional shape recognition. Uliel et al., 1999 [12] suggested using it for detecting circular permutations in proteins. There are many algorithms for the problem. Fu and Lu [6] presented an $O(nm^2)$-time algorithm, Maes [10] proposed an algorithm with $O(nm \log m)$ time and $O(nm)$ space complexity, and described how to reduce the space complexity to $O((n+m)\log m)$. There are also $O(nm)$-time suboptimal algorithms developed by Gorman et al., 1988 [7], Bunke and Bühler, 1993 [4], and Uliel et al., 1999 [12]. Gregor and Thomason, 1996 [8] presented an output-size sensitive algorithm whose time complexity ranges from $O(nm)$ to $O(nm^2)$. There are faster algorithms for the case of unit-cost edit operations in which each edit operation has weight 1 except for a match (substitution of the same symbol, or no operation) whose weight is 0. For this case, Chung, 1998 [5] proposed an algorithm for a generalized version of the problem called *banded cyclic string-to-string correction problem* whose time complexity ranges between $O(nm)$ and $O(nm \log m)$ for cyclic edit distance computation. Landaue et al., 1998 [9] described an algorithm for incremental string comparison which can be used to solve cyclic edit distance problem with unit costs in time $O(m^2)$.

Since cyclic local alignment problem is a special case of *LRLA*, the approximation algorithms we describe in this paper can also be used to solve the cyclic local alignment problem with the same complexity. This makes cyclic local alignment an alternative to cyclic edit distance in applications using cyclic string comparison.

3 Preliminaries and Definitions

Given two strings $X = x_1x_2 \ldots x_n$ and $Y = y_1y_2 \ldots y_m$ with $n \geq m$, the *alignment graph* $G_{X,Y}$ (*edit graph* in the context of string editing) is used to represent all possible *alignments* between X and Y. It is a directed acyclic graph having $(n+1)(m+1)$ lattice points (u, v) as vertices for $0 \leq u \leq n$, and $0 \leq v \leq m$ (Figure 1). An *alignment path* for substrings $x_i \cdots x_k$, and $y_j \cdots y_l$ is a directed path from the vertex $(i-1, j-1)$ to (k, l) in $G_{X,Y}$ where $i \leq k$ and $j \leq l$. Horizontal and vertical arcs correspond to insert and delete operations respectively. The diagonal arcs correspond to substitutions which are either matching (if the corresponding symbols are the same), or mismatching (otherwise). If we trace the arcs of an alignment path for substrings I and J and perform the indicated edit operations in the given order on I, we obtain J.

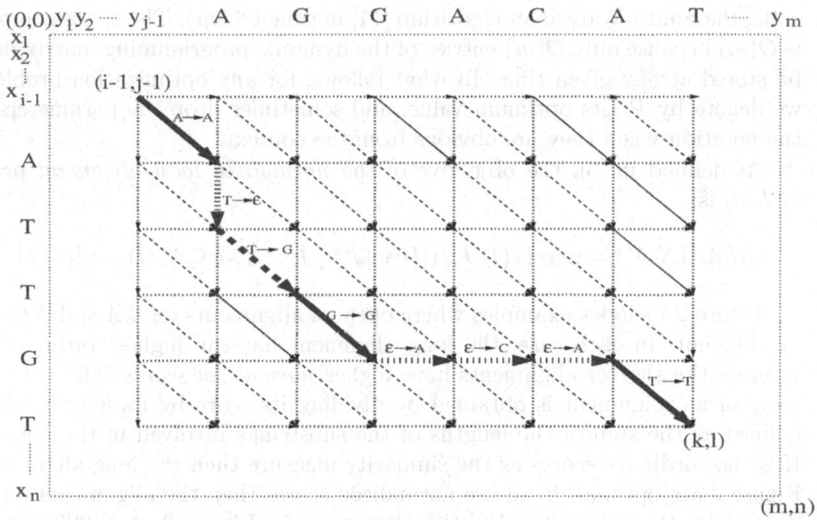

Fig. 1. Alignment graph $G_{X,Y}$ where $x_i \cdots x_k = ATTGT$ and $y_j \cdots y_l = AGGACAT$. Matching diagonal arcs are drawn as solid lines while mismatching diagonal arcs are shown by dashed lines. Dotted lines are used for horizontal and vertical arcs. An example alignment path is shown. Labels of the arcs on this path are the corresponding edit operations where ϵ denotes the null string.

The objective of sequence alignment is to quantify the similarity between two strings. There are various scoring schemes for this purpose. In the *basic scoring scheme*, the arcs of $G_{X,Y}$ are assigned weights determined by non-negative reals δ (*mismatch penalty*) and μ (*indel* or *gap penalty*). We assume that $s(x_i, y_j)$ is the similarity score between the symbols x_i, and y_j which is normally 1 for a match $(x_i = y_j)$ and $-\delta$ for a mismatch $(x_i \neq y_j)$. We will use the terms alignment and alignment path interchangeably.

The following is the classical dynamic programming formulation [13] to compute the maximum local alignment score $\mathcal{S}_{i,j}$ ending at each vertex (i,j):

$$\mathcal{S}_{i,j} = \max\{0,\ \mathcal{S}_{i-1,j} - \mu,\ \mathcal{S}_{i-1,j-1} + s(x_i, y_j),\ \mathcal{S}_{i,j-1} - \mu\} \tag{1}$$

for $1 \le i \le n$, $1 \le j \le m$, with the boundary conditions $\mathcal{S}_{i,j} = 0$ whenever $i = 0$ or $j = 0$.

Let $\subseteq$ indicate the substring relation. The *local alignment* (*LA*) problem seeks for substrings $I \subseteq X$ and $J \subseteq Y$ with the highest similarity score. The optimal value $LA^*(X,Y)$ for this problem is given by

$$LA^*(X,Y) = \max\{s(I,J) \mid I \subseteq X, J \subseteq Y\} = \max_{i,j} \mathcal{S}_{i,j} \tag{2}$$

where $s(I,J)$ is the best alignment score between I and J. LA^* can be computed using the Smith-Waterman algorithm [11] in time $O(nm)$. The space complexity is $O(m)$ because only $O(m)$ entries of the dynamic programming matrix need to be stored at any given time. In what follows, for any optimization problem $\mathcal{P}$, we denote by $\mathcal{P}^*$ its optimum value, and sometimes drop the parameters from the notation when they are obvious from the context.

As defined in [3], the objective of the *normalized local alignment* problem (*NLAt*) is:

$$NLAt^*(X,Y) = \max\{s(I,J)/(|I| + |J|) \mid I \subseteq X, J \subseteq Y, |I| + |J| \ge t\} \tag{3}$$

Figure 2 includes examples where optimal alignments for *LA* and *NLAt* may be different. In each case, the long alignment has the highest ordinary score whereas the shorter alignments have higher *normalized scores*. The normalized score of an alignment is obtained by dividing its score by its length, which is defined as the sum of the lengths of the substrings involved in the alignment. If we use ordinary scores as the similarity measure then the long alignments in Figure 2 are optimal. If we use normalized scores then the alignments returned depend on the value of t. For the alignments in Figure 2 $t = 200$ is a separating value in determining the optimality of short and long alignments. The need to have control over the alignment lengths becomes apparent when we use normalized scores. Without controlling the desired alignment lengths, with normalized scores short alignments overshadow the true long alignments causing yet another anomaly. Arslan et al., 2001 [3] changed the objective function to $s(I,J)/(|I|+|J|+L)$ by introducing parameter L which gives a degree of control over the total length of the optimal subsequences. In this way, the length constraint can be dropped [3]. This gave rise to an efficient algorithm which runs in time $O(nm \log n)$ and using $O(m)$ space. However an adequate control over the length through parameter L is difficult to describe. Given a limit T, we define the *length restricted local alignment* (*LRLA*) score between X and Y as

$$LRLA^*(X,Y,T) = \max\{s(I,J) \mid I \subseteq X, J \subseteq Y, \text{ and } |J| \le T\} \tag{4}$$

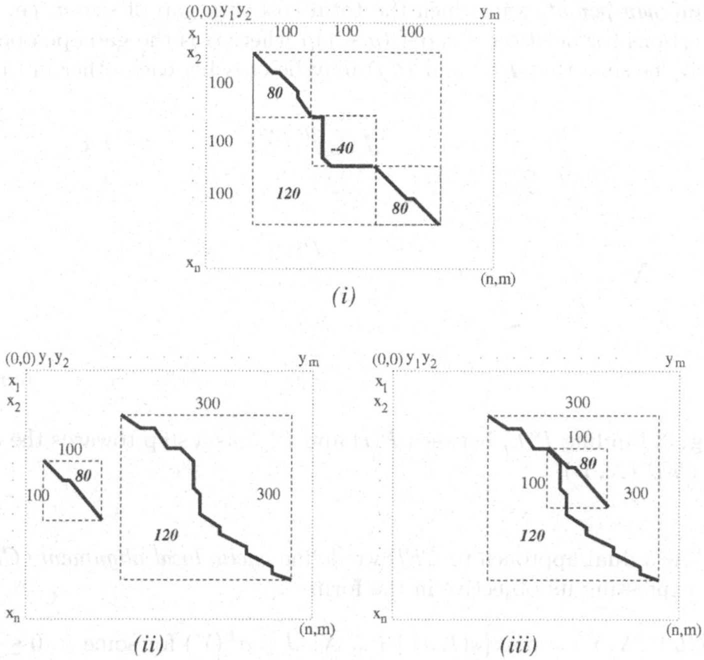

Fig. 2. Some example alignments. The numbers written in italic are the ordinary scores of the alignments. The normalized score of the shorter alignment(s) is $80/200 = 0.4$ while that of the longer alignment is $120/600 = 0.2$.

In *LRLA* problem, the horizontal lengths of the resulting alignments are controlled by upper limit T on one of the substrings, which in practice will be determined by biological considerations. For the alignments in Figure 2, setting $T = 100$ or $T = 300$ changes the optimality of the short and long alignments when the ordinary scores are used.

The *cyclic edit distance* (*CED*) between X and Y is the minimum edit distance between X and any cyclic shift of Y:

$$CED^*(X,Y) = \min\{ed(X,\sigma^k(Y)) \mid 0 \le k < m\} \tag{5}$$

where *ed* denotes the edit distance, and $\sigma^k(Y)$ is the cyclic shift of Y by k which is defined as follows: $\sigma^0(Y)= Y$, and for $0 < k < m$, $\sigma^k(Y) = y_{k+1}\cdots y_m y_1 \cdots y_k$.

Maes' algorithm [10] computes $CED^*(X,Y)$ in $O(nm \log m)$ steps. For any k, $ed(X,\sigma^k(Y))$ is the cost of the shortest (least-cost) path $P(k)$ between the vertices $(0,k)$ and $(n, m+k)$ in $G_{X,YY}$ as shown in Figure 3. Maes' idea is to make use of the "non-crossing" property of shortest paths, which restricts the candidate $P(k)$ to be squeezed between $P(i)$ and $P(j)$ where $i < k < j$ as illustrated in the figure. However this idea cannot be generalized to the case of *affine gap penalties* in which the total cost of a gap of size a, i.e. a block of a insertions (or deletions), is $\alpha + (a-1)\mu$ where α is the gap open penalty. It can easily be seen that $P(i)$ and $P(j)$ may be crossing each other in this case.

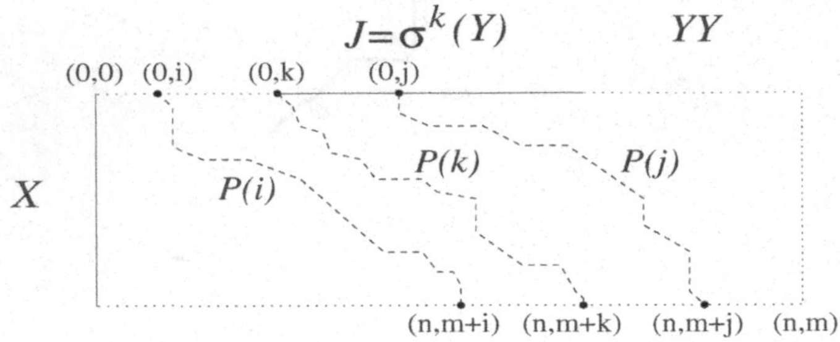

Fig. 3. Finding $P(k)$ between $P(i)$ and $P(j)$ as a step towards the computation of $CED^*(X,Y)$.

As a dual approach to *CED* we define *cyclic local alignment* (*CLA*) problem by expressing its objective in the form

$$CLA^*(X,Y) = \max\{s(I,J) \mid I \subseteq X,\ J \subseteq \sigma^k(Y) \text{ for some } k,\ 0 \le k < m\} \quad (6)$$

Note that $CLA^*(X,Y) = LRLA^*(X,YY,|Y|)$, and therefore *CLA* is a special case of *LRLA*.

4 Approximation Algorithms for *LRLA*

We first give a simple $\frac{1}{2}$-approximation algorithm *HALF* for the *LRLA* problem. Clearly, we can assume $T < m$, for otherwise we can run the local alignment algorithm without alteration on $G_{X,Y}$ and obtain the exact solution to *LRLA* in time $O(nm)$. Let $u = \lceil m/T \rceil$ and put $Y_j = y_{(j-1)T+1} \cdots y_{jT}$ for $1 \le j < u$ with $Y_u = y_{(u-1)T+1} \cdots y_m$. Thus $Y = Y_1 Y_2 \cdots Y_u$, and the Y_j partition Y into blocks of length T each (except possibly for Y_u, which may be shorter). Let

$$HALF^* = \max_{1 \le j < u} \{s(I,J) \mid I \subseteq X,\ J \subseteq Y_j Y_{j+1}\} \quad (7)$$

Finding an optimal alignment a for *HALF* requires solving $u-1$ ordinary local alignment problems among strings X and $Y_j Y_{j+1}$ for $1 \le j < u$ as schematically described in Figure 4 . Total time taken for this is $O(nm)$, as each Y_j needs to be considered at most twice during the computations. The space complexity of *HALF* is $O(m)$.

Let a be an optimal alignment found by *HALF*. We obtain an approximation to $LRLA^*$ from a as follows. Suppose a involves a substring of $Y_j Y_{j+1}$ for some j. Let a_1, a_2 denote the two halves of a that lie in Y_j and Y_{j+1}, respectively as

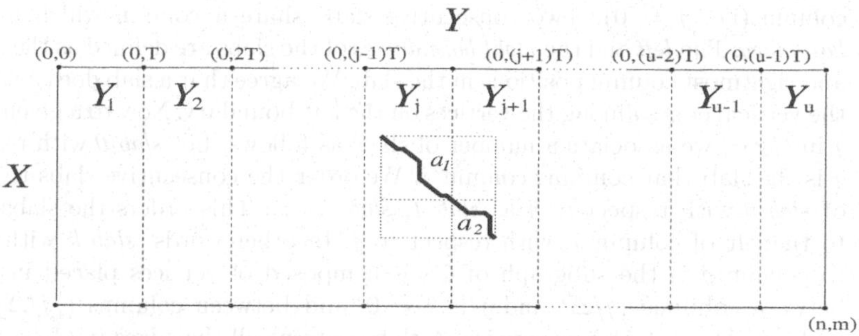

Fig. 4. Regions of alignment graph explored in computing *HALF**, and parts a_1 and a_2 of an optimal alignment a.

shown in Figure 4. Without loss of generality, assume that a_1 is the one with the higher score, say $\widehat{S}$. Then $2\widehat{S} \geq LRLA^*$, and therefore $\frac{1}{2}LRLA^* \leq \widehat{S} \leq LRLA^*$. Clearly the horizontal length of a_1 is $\leq T$. The same approximation and complexity results hold for the cases of arbitrary scoring matrices and affine gap penalties. To solve the individual local alignment problems, Algorithm *HALF* uses the existing algorithms in the underlying scoring scheme.

We give next an approximation algorithm *APX-LRLA*, which computes a local alignment score $\widehat{S} \leq LRLA^*$ within a prescribed difference Δ from $LRLA^*$, i.e.

$$LRLA^* - \Delta \leq \widehat{S} \leq LRLA^*.$$

If desired, the position of an alignment achieving $\widehat{S}$, and consequently the substrings $I \subseteq X$ and $J \subseteq Y$ with $|J| \leq T$ achieving this score can also be identified. The complexity of the algorithm is $O(nmT/\Delta)$ time and $O(mT/\Delta)$ space.

For simplicity, we assume a basic scoring scheme: i.e. score between the symbols x_i, and y_j is 1 for a match ($x_i = y_j$) and $-\delta$ for a mismatch ($x_i \neq y_j$), and the indel score is $-\mu$. We argue later that the algorithm can be easily modified within the same complexity results for the case of arbitrary scoring matrices which allows a different score for each individual operation.

Our approximation idea is that instead of a single score, we maintain at each node (i, j) of $G_{X,Y}$, a list of scores with the property that for any given optimum score achieved by an alignment ending at (i, j) and starting within a past horizontal window of size T of (i, j) at least one element of the list lies within Δ of this score. We show that the dynamic programming formulation can be extended to preserve this property through the nodes. In particular, an alignment with maximum score $\widehat{S}$ in the list of scores computed in one of the nodes (i, j) will be between $LRLA^* - \Delta$ and $LRLA^*$. We assume that Δ is integral, otherwise we use the largest integer smaller than the given value for Δ.

Similar to the case of *HALF*, we imagine the columns of the graph $G_{X,Y}$ as grouped into vertical slabs of $\Delta + 1$ columns each, starting with the leftmost column (i.e. $j = 0$). Two consecutive slabs share a column which we call a *boundary.* The *left* and the *right boundaries* of the slabs are defined as the leftmost and rightmost column positions in the slab. We agree that a slab does not contain the vertical edges among the vertices on the left boundary. Now to a given column j in $G_{X,Y}$, we associate a number of slabs as follows. Let *slab 0* with respect to j is the slab that contains column j. We order the consecutive slabs to the left of *slab 0* with respect to j as *slab 1, slab 2, ...*. This orders the slabs weakly to the left of column j, with respect to j. In other words, *slab k* with respect to column j is the subgraph of $G_{X,Y}$ composed of vertices placed inclusively between columns $\lfloor j/\Delta \rfloor$ and j if $k = 0$, and between columns $(\lfloor j/\Delta \rfloor - k)\Delta$ and $(\lfloor j/\Delta \rfloor - k + 1)\Delta$, otherwise. A slab contains all the edges in $G_{X,Y}$ incident to the vertices it contains except for the vertical edges on the left boundary which belong to the preceding slab. Figure 5 includes sample slabs with respect to column j, and alignments ending at some node (i, j) . From the dynamic programming formulation (1), we note the following observation for optimal alignments with the basic scoring scheme assumed: any optimal alignment (with positive score) ending at a given node (i, j) has to start with a match since only the matches have positive scores. Let $S_{i,j,k}$ for $0 \leq k \leq \lfloor T/\Delta \rfloor - 1$ represent the

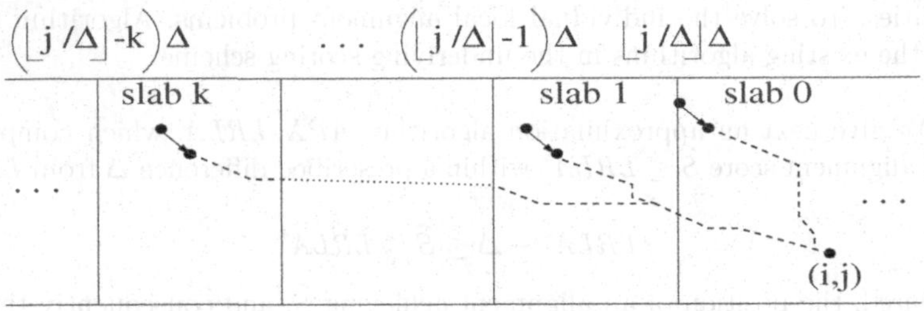

Fig. 5. Slabs with respect to column j, and alignments ending at node (i, j) starting at different slabs.

optimum score achievable by any alignment ending at node (i, j) and starting at slab k with respect to column j. For *LRLA*, we are only interested in alignments with horizontal length not exceeding T. A single slab can contribute at most Δ to the score of any alignment. Consequently, we need to store at each node at most $\lfloor T/\Delta \rfloor$ scores $S_{i,j,k}$, for $0 \leq k \leq \lfloor T/\Delta \rfloor - 1$ corresponding to the $\lfloor T/\Delta \rfloor$ slabs that include (i, j) and span a past horizontal window of length at most T. Figure 6 shows the steps of our approximation algorithm *APX-LRLA*. The processing is done row-by-row starting with the top row ($i = 0$) of $G_{X,Y}$.

We can modify the Smith-Waterman algorithm such that it breaks the ties in scores by selecting alignments with smaller horizontal lengths. This modified algorithm can be used in Step 1 to check if the maximum score over all the alignments is achieved by an alignment whose horizontal length does not exceed T. Step 2 of the algorithm performs the initialization of the lists of the nodes in the top row ($i = 0$). Step 3 implements computation of scores as dictated by the dynamic programming formulation in (1):

- If the current node (i, j) is not on the first column after a boundary then nodes $(i-1, j)$, $(i-1, j-1)$ and $(i, j-1)$ share the same slabs with node (i, j). In this case $\mathcal{S}_{i,j,k}$ is calculated by using $\mathcal{S}_{i-1,j,k}$, $\mathcal{S}_{i-1,j-1,k}$, and $\mathcal{S}_{i,j-1,k}$ as

$$\mathcal{S}_{i,j,k} = \max\{0, \mathcal{S}_{i-1,j,k} - \mu, \mathcal{S}_{i-1,j-1,k} + s(x_i, y_j), \mathcal{S}_{i,j-1,k} - \mu\}.$$

- If the current node is on the first column following a boundary (i.e. $j \bmod \Delta = 1$) then the slabs for the nodes involved in the computations for node (i, j) differ as shown in Figure 7. In this case slab k for node (i, j) is slab $k-1$ for the nodes at column $j-1$. Moreover any alignment ending at (i, j) starting at slab 0 for (i, j) can either only include one of the edges $((i-1, j), (i, j))$, $((i-1, j-1), (i, j))$, or $((i, j-1), (i, j))$, or extend an alignment from node $(i-1, j)$. The edges $((i-1, j), (i, j))$ and $((i, j-1), (i, j))$ both have negative weight $-\mu$. Therefore, $\mathcal{S}_{i,j,0}$ is set to $\max\{0, s(x_i, y_j), \mathcal{S}_{i-1,j,0} - \mu\}$. For slab $k > 0$, $\mathcal{S}_{i,j,k}$ is calculated by

$$\mathcal{S}_{i,j,k} = \max\{0, \mathcal{S}_{i-1,j,k} - \mu, \mathcal{S}_{i-1,j-1,k-1} + s(x_i, y_j), \mathcal{S}_{i,j-1,k-1} - \mu\}$$

During these computations, the running maximum score is also updated whenever a newly computed score $\mathcal{S}_{i,j,k}$ is larger than the current maximum, and the final value is returned in Step 3. The alignment position achieving this score may also be desired. This can be done by maintaining for each optimal alignment its start and end positions besides its score. In this case in addition to the running maximum score, the start and end positions of a maximal alignment should be stored and updated.

If there is an alignment with the maximum score and with horizontal length not exceeding T then the algorithm returns this score in Step 1. Otherwise, we first show that for any node (i, j) and slab k, $\mathcal{S}_{i,j,k}$ calculated by the algorithm is the optimum score achievable over the set of all the alignments ending at node (i, j) and starting at slab k with respect to column j. This claim is proved by induction. If we assume that the claim is true for nodes $(i-1, j)$, $(i-1, j-1)$ and $(i, j-1)$, and for their slabs, then we can easily see by following Step 3 of the algorithm that the claim holds for node (i, j) and its slabs. Consider the alignments with horizontal length at most T. If there is an optimal alignment with score $LRLA^*$ and with length at most $T - \Delta$, then this alignment is captured during the calculations at its right end point. In this case, by the previous claim, the algorithm returns the optimum score $LRLA^*$. If all optimal alignments with score $LRLA^*$ have horizontal length $> T - \Delta$, we show that during the computations the algorithm observes a score which does not differ more than Δ from the optimum score $LRLA^*$. To see this, let an optimal alignment start at node (i', j') and end at node (i, j). We know that its horizontal length is larger than $T - \Delta$. Let (i'', j'') be the node the alignment crosses at the boundary $(\lfloor T/\Delta \rfloor - 1)\Delta$. This is the right boundary of the slab (i', j') lies in, or equivalently, the left boundary of slab $k = \lfloor T/\Delta \rfloor - 1$ relative to column j as shown in Figure 8. Note that the score of the part of the alignment between nodes (i', j') and

Algorithm $APX\text{-}LRLA(\delta,\mu)$

1. **Run a modified Smith-Waterman algorithm. If the maximum score is achieved within horizontal length** $\leq T$ **then return this score and exit**
2. **Initialization:**
    set $LRLA^* = 0$
    set $\mathcal{S}_{0,j,k} = 0$ for all $j,k,\quad 0 \leq j \leq m,$ and $0 \leq k \leq \lfloor T/\Delta \rfloor - 1$
3. **Main computations :**
    for $i = 1$ to n do {
      set $\mathcal{S}_{i,0,k} = 0$ for all $k,\ 0 \leq k \leq \lfloor T/\Delta \rfloor - 1$
      for $j = 1$ to m do {
        if (j mod $\Delta = 1$) then
          {
            set $\mathcal{S}_{i,j,0} = \max\{0, s(x_i,y_j), \mathcal{S}_{i-1,j,0} - \mu\}$
            set $LRLA^* = \max\{LRLA^*, \mathcal{S}_{i,j,0}\}$
            for $k = 1$ to $\lfloor T/\Delta \rfloor - 1$ do {
              set $\mathcal{S}_{i,j,k} = \max\{0, \mathcal{S}_{i-1,j,k} - \mu, \mathcal{S}_{i-1,j-1,k-1} + s(x_i,y_j), \mathcal{S}_{i,j-1,k-1} - \mu\}$
              set $LRLA^* = \max\{LRLA^*, \mathcal{S}_{i,j,k}\}$
            }
          } else
          {
            for $k = 0$ to $\lfloor T/\Delta \rfloor - 1$ do {
              set $\mathcal{S}_{i,j,k} = \max\{0, \mathcal{S}_{i-1,j,k} - \mu, \mathcal{S}_{i-1,j-1,k} + s(x_i,y_j), \mathcal{S}_{i,j-1,k} - \mu\}$
              set $LRLA^* = \max\{LRLA^*, \mathcal{S}_{i,j,k}\}$
            }
          }
      }
    }
3. **Return** $LRLA^*$

Fig. 6. Algorithm $APX\text{-}LRLA$ which approximates $LRLA^*$ within difference Δ.

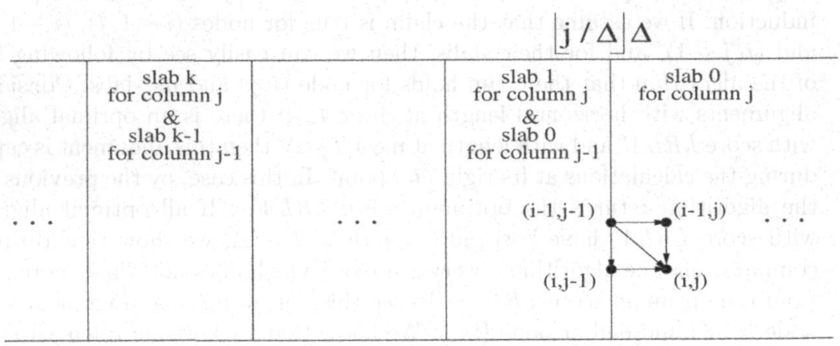

Fig. 7. Left boundary of slab 0 with respect to column j, and numbering of slabs with respect to columns j and $j-1$.

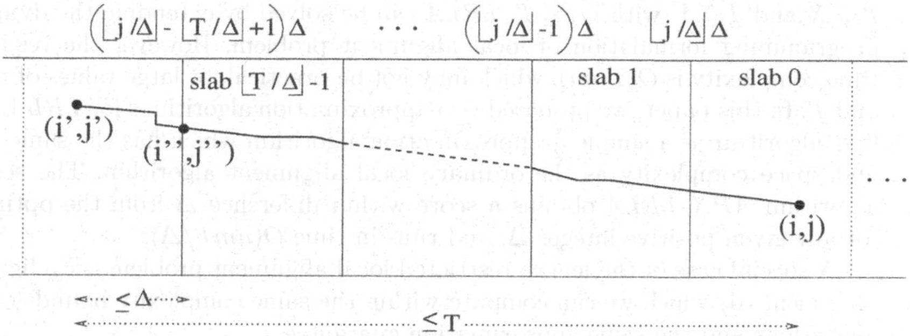

Fig. 8. Orientation of an optimal alignment ending at node (i, j) when its horizontal length is larger than $T - \Delta$.

(i'', j'') is at most Δ according to our basic scoring scheme. Since (i'', j'') is one of the nodes in slab k relative to j $\mathcal{S}_{i,j,k}$ is larger than or equal to the score of the part of the given optimal alignment between (i'', j'') and (i, j) by our previous claim on the optimality of $\mathcal{S}_{i,j,k}$. Thus $\mathcal{S}_{i,j,k}$ does not differ from $LRLA^*$ by more than Δ, i.e.

$$LRLA^* - \Delta \leq \mathcal{S}_{i,j,k}$$

Note that it is likely that a score higher than the guaranteed lower bound $LRLA^* - \Delta$ is returned frequently in practice as the algorithm explores all the nodes as possible end points. Algorithm *APX-LRLA* essentially implements the dynamic programming formulation (1). It is similar to the Smith-Waterman algorithm except at each node instead of a single score, $\lfloor T/\Delta \rfloor$ scores are stored and manipulated. Therefore the resulting complexity exceeds that of the Smith-Waterman algorithm by a factor of $\lfloor T/\Delta \rfloor$. That is, the time complexity of *APX-LRLA* is $O(nmT/\Delta)$. The algorithm requires $O(mT/\Delta)$ space since we need the entries in the previous and the current row to calculate the entries in the current row.

Algorithm *APX-LRLA* can easily be generalized to other common scoring schemes with simple modifications. For example, varying penalties (or scores) can easily be incorporated for the computation of optimum scores at each node for arbitrary scoring matrices. For the approximation result to hold we assume that the maximum positive score for any individual operation is at most 1. In the scoring schemes we study in this paper this can be satisfied by normalizing all the scores by dividing them by the maximum individual positive score which does not affect the optimality of the alignments.

5 Concluding Remarks

We considered the length restricted local alignment problem *LRLA* in which we search for substrings I and J that maximize the score $s(I, J)$ among all

$I \subseteq X$ and $J \subseteq Y$ with $|J| \leq T$. *LRLA* can be solved by extending the dynamic programming formulation of local alignment problem. However the resulting time complexity is $O(Tnm)$, which may not be practical for large values of n, m, and T. In this paper, we proposed two approximation algorithms for *LRLA*. The first algorithm is a simple $\frac{1}{2}$-approximation algorithm which has the same time and space complexity as the ordinary local alignment algorithm. The second algorithm *APX-LRLA* obtains a score within difference Δ from the optimum for any given positive integer Δ, and runs in time $O(nmT/\Delta)$.

A special case of the length restricted local alignment problem is cyclic local alignment (6) which we can compute within the same complexity bounds of our algorithms and the same approximation guarantee.

The algorithms can be generalized to the more common scoring scheme of affine gap penalties which is widely used in practice. The same approximation and performance guarantee hold for the affine gaps case as well. These will be reported elsewhere.

The algorithms presented are simple to implement and provably efficient. The degree of approximation can be controlled with a reasonable trade-off of optimality versus time and space. There is also reason to believe that the approximate score returned is on the average much closer to the actual optimum than the worst case error bound of Δ.

References

1. N. N. Alexandrov and V. V. Solovyev. Statistical significance of ungapped alignments. *Pacific Symposium on Biocomputing (PSB-98), (eds. R. Altman, A. Dunker, L. Hunter, T. Klein)*, pages 463–472, 1998.
2. S. F. Altschul, T. L. Madden, A. A. Schaffer, J. Zhang, Z. Zhang, W. Miller, and D. J. Lipman. Gapped Blast and Psi-Blast: a new generation of protein database search programs. *Nucleic Acids Research*, 25:3389–3402, 1997.
3. A. N. Arslan, Ö. Eğecioğlu, and P. A. Pevzner. A new approach to sequence comparison: Normalized local alignment. *Bioinformatics*, 17(4):327–337, 2001.
4. H. Bunke and U. Bühler. Applications of approximate string matching to 2d shape recognition. *Pattern Recognition*, 26(12):1797–1812, 1993.
5. K-L. Chung. An improved algorithm for solving the banded cyclic string-to-string correction problem. *Theoretical Computer Science*, 201:275–279, 1998.
6. K. S. Fu and S. Y. Lu. Size normalization and pattern orientation problems in syntactics clustering. *IEEE Trans. Systems Man Cybernet*, 9:55–58, 1979.
7. J. W. Gorman, O. R. Mitchell, and F. P. Kuhl. Partial shape recognition using dynamic programming. *IEEE Trans. Pattern Anal. Mech. Intell. PAMI-10*, pages 257–266, 1988.
8. J. Gregor and M. G. Thomason. Efficient dynamic programming alignment of cyclic shifts by shift elimination. *Pattern Recognition*, 29(7):1179–1185, 1996.
9. G. M. Landau, E. W. Myers, and J. P. Schmidt. Incremental string comparison. *Siam J. Comput.*, 27(2):557–582, 1998.
10. M. Maes. On a cyclic string-to-string correction problem. *Information Processing Letters*, 35:73–78, 1990.
11. T. F. Smith and M. S. Waterman. The identification of common molecular subsequences, *J. of Molecular Biology*, 147:195–197, 1981.

12. S. Uliel, A. Fliess, A. Amir, and R. Unger. A simple algorithm for detecting circular permutations in proteins. *Bioinformatics*, 15(11):930–936, 1999.
13. M. S. Waterman. *Introduction to computational biology.* Chapman & Hall, 1995.
14. Z. Zhang, P. Berman, and W. Miller. Alignments without low-scoring regions. *J. Comput. Biol.*, 5:197–200, 1998.
15. Z. Zhang, P. Berman, T. Wiehe, and W. Miller. Post-processing long pairwise alignments. *Bioinformatics*, 15:1012–1019, 1999.

An Algorithm That Builds a Set of Strings Given Its Overlap Graph*

Marília D.V. Braga and João Meidanis

Institute of Computing, University of Campinas, P.O.Box 6176, 13084-971, Campinas, São Paulo, Brazil.
{mvbraga, meidanis}@ic.unicamp.br

Abstract. The k-overlap graph for a set of strings is a graph such that each vertex corresponds to a string and there is a directed edge between two vertices if there is an overlap of at least k characters between their corresponding strings. Given a directed graph G, an integer $k \geq 1$, and a finite alphabet Σ of at least two symbols, we propose an algorithm to obtain a set of strings C, written over Σ, such that G is its k-overlap graph. The algorithm runs in exponential time on the maximum degree of G, due to the size of the returned strings, but in polynomial time on k, $|\Sigma|$, and the size of the graph. A practical application of this algorithm is its use to prove the NP-hardness of *Minimum Contig Problems* family (*MCP*) and its variation *MCPr*, which are based on the DNA Fragment Assembly problem.

1 Introduction

Given an integer $k \geq 1$ and a set C of strings, written over an alphabet Σ, the k-overlap graph of C, denoted by $G_k(C)$, is the directed graph G such that $V(G) = C$ and $E(G) = \{uv \mid \text{there is an integer } \ell \geq k \text{ such that } last(u,\ell) = first(v,\ell)\}$, where we denote by $last(s,\ell)$ the last ℓ characters of s and by $first(s,\ell)$ the first ℓ characters of s [3].

We say that there is an overlap of ℓ characters between strings s and t when $last(s,\ell) = first(t,\ell)$. An example of overlap graph can be seen in Figure 1.

In 1992, Gusfield, Landau and Schieber [4] presented an efficient algorithm for the following problem. Given a set $C = \{s_1, s_2, s_3, \ldots, s_n\}$ of strings, find the greatest integer ℓ_{ij} such that $last(s_i, \ell_{ij}) = first(s_j, \ell_{ij})$, for all pair of strings $s_i, s_j \in C$, with $i \neq j$. Notice that this algorithm can be used to build $G_k(C)$ for any $k \geq 0$, because $s_i s_j \in E(G_k(C))$ if and only if $\ell_{ij} \geq k$.

Here we propose an algorithm that solves the inverse problem: given a directed graph G we build a set C' of strings over an alphabet Σ_G of size $|V(G)| + |E(G)|$ with the property that $G = G_1(C')$. For any $k \geq 1$ and for any finite alphabet Σ with at least two symbols, we also show how it is possible to modify C' and obtain a set of strings C, written over Σ, such that $G = G_k(C)$.

* Research supported by Brazilian agencies FAPESP (Grant 97/11629-2), and CNPq (Grant Pronex 664107/1997-4).

S. Rajsbaum (Ed.): LATIN 2002, LNCS 2286, pp. 52–63, 2002.

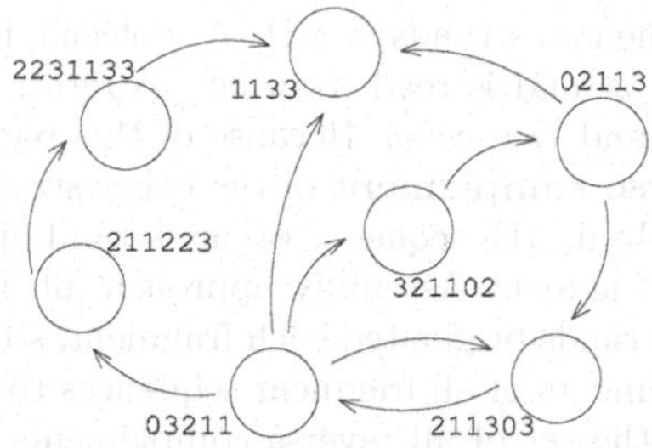

Fig. 1. 2-overlap graph for a set of strings written over the alphabet $\{0, 1, 2, 3\}$

A practical application of this algorithm is its use to prove the NP-hardness of *Minimum Contig Problems* family (*MCP*) and its variation *MCPr* [3, 1], which are based on the DNA Fragment Assembly problem [6].

An ordered set of strings $S = (s_1, s_2, \ldots, s_{|S|})$ is called a k-path, for an integer $k \geq 0$, if each pair of consecutive strings s_{i-1}, s_i (where i is an integer between 2 and $|S|$) is such that there is an integer $\ell \geq k$ with $last(s_{i-1}, \ell) = first(s_i, \ell)$. We say that S is a k-path in a set of strings C if S is a subset of C.

Observe that if a graph G is a k-overlap graph for a set of strings C, a path in G corresponds to a k-path in C.

Given an integer k and a set of strings C written over an alphabet Σ, with $|\Sigma| = n$, the $MCP_n(k)$ is the problem of finding the smallest set of k-paths in C that *covers* all strings in C, that is, each string in C must be in exactly one k-path if it is not a proper substring of another string in C, or at most in one k-path if it is a proper substring of another string in C. Figure 2 shows a solution of an instance of *MCP*.

```
s TTACGCTA........
t TTACGC..........
u ....GCTAGAGATC...
v ......TAGAGAT....
w ..........ATCGTT
  ----------------
  TTACGCTAGAGATCGTT
```

All input strings can be covered by one 2-path, (s, u, w), whose consensus is TTACGCTAGAGATCGTT.

Fig. 2. A solution of an instance of $MCP_4(2)$

We know that a DNA molecule is a double strand of nucleotides, identified by their bases A, C, G, T, organized in a way where an A in one strand is always linked to a T in the other, while a C in one strand is always linked to a G in the other. Moreover, the two strands in a DNA molecule have opposite orientations, that is, if the first strand is read from left to right, the second strand is read from right to left, and *vice-versa*. Because of this regular structure, one strand is called the **reverse complement** of the other strand in a molecule, meaning that it is easy to obtain the sequence of one strand given the other [6].

On the DNA Fragment Assembly approach [6], it is not possible to know which of the two strands originated each fragment, so it is necessary to consider the reverse complements of all fragment sequences to assemble the molecule.

Defining as $\overline{C}$ the set of all reverse complements of the strings in a set C, written over the alphabet {A, C, G, T}, the *MCPr* family of problems is a variation of *MCP* that finds the smallest set of k-paths in $C \cup \overline{C}$, where each string in C or its reverse complement must be in exactly one k-path if it is not a proper substring of another string in $C \cup \overline{C}$, or at most in one k-path if it is a proper substring of another string in $C \cup \overline{C}$. The solutions of an instance of *MCPr* can be seen in figure 3.

```
s  TTACGCTA.........   s̄  .........TAGCGTAA
u  ....GCTAGAGATC...   ū  ...GATCTCTAGC....
w  ...........ATCGTT   w̄  AACGAT...........
   -----------------      -----------------
   TTACGCTAGAGATCGTT      AACGATCTCTAGCGTAA
```

All input strings can be covered by one 2-path, that can be the same (s, u, w) of figure 2 or $(\overline{w}, \overline{u}, \overline{s})$, whose consensus is AACGATCTCTAGCGTAA.

Fig. 3. Solutions of an instance of *MCPr* (2)

2 Directed Edge Colorings

We will now introduce a notion of coloring for directed graphs. We assume the basic definitions of standard graph theory [2].

Given a directed graph G, we define a *directed matching* as a set $M \subseteq E(G)$, such that for any two distinct edges $uv, xw \in M$, we have $u \neq x$ and $v \neq w$. Notice that the edges of M form a collection of disjoint directed paths and cycles in G. If a set $\mathcal{L} = \{M_1, M_2, \ldots, M_p\}$ of disjoint directed matchings covers all the edges of $E(G)$ we say that $\mathcal{L}$ is a *directed edge coloring* of G.

We denote by $\Delta(H)$ the maximum degree in an undirected graph H and by $\Delta^+(G)$ and $\Delta^-(G)$, respectively, the maximum indegree and maximum outdegree of a directed graph G. Observe that $p \geq \max\{\Delta^+(G), \Delta^-(G)\}$, where p is the cardinality of any directed edge coloring of G.

Theorem 1 *For every directed graph G there is a directed edge coloring $\mathcal{L} = \{M_1, M_2, \ldots, M_p\}$, such that $p = \max\{\Delta^+(G), \Delta^-(G)\}$.*

Proof:

Given a directed graph G we build $H = bip(G)$ as follows. For each vertex $v \in V(G)$ we include two vertices v', v'' in $V(H)$, so that $|V(H)| = 2|V(G)|$. For each edge $uv \in E(G)$, we include an edge $u'v''$ in $E(H)$, so that $|E(H)| = |E(G)|$. Notice that H is a bipartite undirected graph.

Given an undirected edge coloring $\mathcal{C} = \{M'_1, M'_2, \ldots, M'_p\}$ for $bip(G)$, consider $M_i = \{uv \mid u'v'' \in M'_i\}$. Since M'_i is a matching in $bip(G)$, we have that M_i is a directed matching in G. The matchings M'_i are disjoint and cover all the edges of $bip(G)$, and therefore the same is true for the directed matchings M_i with respect to G.

König's Theorem [2, p.103] states that the minimum number of colors (matchings) in an undirected coloring of a bipartite graph H is its maximum degree $\Delta(H)$. Since $\Delta(bip(G)) = \max\{\Delta^+(G), \Delta^-(G)\}$, there exists a directed edge coloring for G with $\max\{\Delta^+(G), \Delta^-(G)\}$ colors.

□

Figure 4 shows an example of $bip(G)$ and a directed edge coloring for a directed graph G.

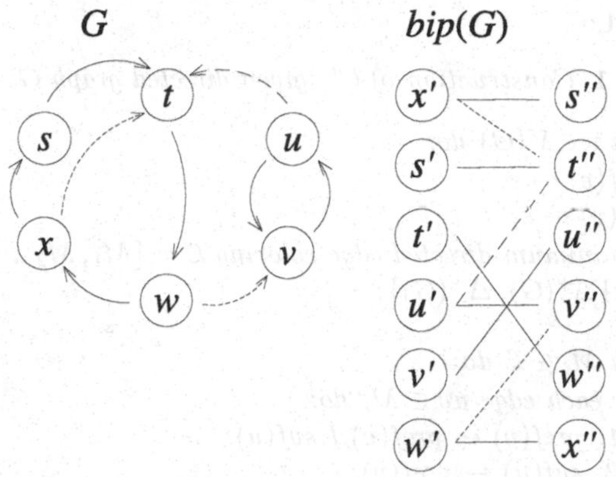

Fig. 4. A directed graph G, its bipartite version $bip(G)$, and a directed edge coloring

3 The Algorithm

Given a directed graph G with $|V(G)| = n$ and $|E(G)| = m$, and an integer $k \geq 1$, we propose an algorithm that runs in time polynomial in n and m and exponential in $p = \max\{\Delta^+(G), \Delta^-(G)\}$ and constructs a set C of strings, written over any alphabet with at least two symbols, such that $G = G_k(C)$ and $|s|$ is $O(k.\log_2(m+n).2^p)$ for every $s \in C$.

The algorithm is composed of two parts:

1. First we build a set C' of strings, written over alphabet $\Sigma_G = \{0, 1, 2, ..., m + n - 1\}$, such that $G = G_1(C')$ and $|s'| \leq 2^{p+1} - 1$ for every $s' \in C'$. For each $v \in V(G)$, the string generated is of the form *pref*$(v).B(v).$*suf*(v), where $B(v)$ is a unique character, not occuring in *pref*(v) or in *suf*(v) or in any other string of C'. This implies that any overlap between strings corresponding to two vertices u and v is in fact an overlap between *suf*(u) and *pref*(v). This also guarantees that no string in C' is a substring of another string in C'.
2. Then we use a function f to code the strings in C', forming a set C of strings written over an arbitrary alphabet with at least two symbols, satisfying $G = G_k(C)$, because the properties of C' with respect to G are maintained in C, that is, for each $v \in V(G)$, the new string is of the form $f(pref(v)).f(B(v)).f(suf(v))$, where $f(B(v))$ is a string, not occuring in $f(pref(v))$ or in $f(suf(v))$ or in any other string of C. This implies that any overlap between strings corresponding to two vertices u and v is in fact an overlap between $f(suf(u))$ and $f(pref(v))$. This also guarantees that no string in C is a substring of another string in C.

3.1 Part 1

Algorithm 1 *Construction of C', given directed graph G:*

1. *For each $v \in V(G)$ do:*
    1.1. *pref*$(v) \leftarrow \varepsilon$;
    1.2. *suf*$(v) \leftarrow \varepsilon$;
2. *Find a minimum directed edge coloring $\mathcal{L} = \{M_1, M_2, \ldots, M_p\}$ for G, where $p = \max\{\Delta^+(G), \Delta^-(G)\}$;*
3. $I \leftarrow 0$;
4. *For each $M_i \in \mathcal{L}$ do:*
    4.1. *For each edge $uv \in M_i$ do:*
        4.1.1. *pref*$(v) \leftarrow$ *pref*$(v).I.$*suf*(u);
        4.1.2. *suf*$(u) \leftarrow$ *pref*(v);
        4.1.3. $I \leftarrow I + 1$;
5. $C' \leftarrow \emptyset$;
6. *For each $v \in V(G)$ do:*
    6.1. $C' \leftarrow C' \cup \{pref(v).I.suf(v)\}$;
    6.2. $I \leftarrow I + 1$;
7. *Return C';*

Figure 5 shows an example of execution of the algorithm.

Denote by $A(u, v)$ the value of I when edge uv is being processed in step 4.1.1. For every $uv \in E(G)$, we have $0 \leq A(u, v) \leq m - 1$ and $A(u, v) \neq A(w, x)$ if $uv \neq wx$.

We will prove the correctness of the algorithm and then analyze the sizes of the strings generated. To prove the correctness, we need to show that:

(i) If $uv \in E(G)$, then there is an integer $\ell \geq 1$ such that $last(suf(u), \ell) = first(pref(v), \ell)$.

Given edge $uv \in E(G)$, at the moment it was processed in step 4.1.2 we created an overlap between $suf(u)$ and $pref(v)$ consisting of at least one character: $A(u, v)$. After that point in the execution of the algorithm, $suf(u)$ grows only to the left, while $pref(v)$ grows only to the right, maintaining the overlap throughout the end of the algorithm. Therefore, there is an integer $\ell \geq 1$ with $last(suf(u), \ell) = last(pref(v), \ell)$.

(ii) If $last(suf(u), \ell) = first(pref(v), \ell)$ for some integer $\ell \geq 1$ and two vertices $u, v \in V(G)$, then $uv \in E(G)$.

Let i be the outdegree of u, and let uv_1, uv_2, ..., uv_i be the edges going out of u, in the order they are processed by the algorithm, so that $A(u, v_i) > \ldots > A(u, v_2) > A(u, v_1)$. Character $A(u, v_i)$ appears only once in $suf(u)$ and is the largest (viewed as an integer) among all characters in $suf(u)$. In addition, for each i' such that $1 \leq i' < i$, character $A(u, v_{i'})$ appears only once to the right of $A(u, v_{i'+1})$ and is the largest integer in that portion of $suf(u)$. It follows that, for any suffix s of $suf(u)$, the largest integer in s is a character of the form $A(u, v_{i'})$ with $1 \leq i' \leq i$.

Analogously, if j is the indegree of v and u_1v, u_2v, ..., u_jv are the edges going into v in the order they are processed by the algorithm, we have that the largest integer in any prefix of $pref(v)$ is a character of the form $A(u_{j'}, v)$ with $1 \leq j' \leq j$.

If $last(suf(u), \ell) = first(pref(v), \ell)$ then the largest character in this string is $A(u, v_{i'})$, for some i' between 1 and i, and also $A(u_{j'}, v)$, for some j' between 1 and j. It follows that $A(u, v_{i'}) = A(u_{j'}, v)$, so $u = u_{j'}$ and $v = v_{i'}$, that is, edge uv exists in G.

This completes the correctness proof.

The next step is to prove that $|s| \leq 2^{p+1} - 1$ for every $s \in C'$, where $p = \max\{\Delta^+(G), \Delta^-(G)\}$. Recall that $\mathcal{L} = \{M_1, M_2, \ldots, M_p\}$ is a directed coloring of G.

Denote by $size(i)$ the length of the longest string obtained after processing all the edges of M_i. We have $size(0) = 0$ because $pref(.)$ and $suf(.)$ are empty in the beginning.

For every vertex $v \in V(G)$, it is clear that each of the strings $pref(v)$ and $suf(v)$ is modified at most once during the processing of each M_i, because of the way directed matchings were defined. The modification consists in concatenating a suffix with a single character and with a prefix, hence we have $size(i) \leq 2.size(i-1) + 1$. It is easy to prove by induction that $size(i) \leq 2^i - 1$, implying that after all edges are processed we have $size(p) \leq 2^p - 1$.

The representative of a vertex v has length $|pref(v)| + |suf(v)| + 1 \leq 2.size(p) + 1 \leq 2^{p+1} - 1$, completing this analysis.

Computational Complexity Let us compute the complexity of the algorithm for an instance G with $|V(G)| = n$, $|E(G)| = m$ and $p = \max\{\Delta^+(G), \Delta^- G\}$.

In step 2 we find a directed edge coloring of G. As we have seen in the proof of Theorem 1, this can be done by solving an undirected bipartite edge coloring problem, which has complexity $O(pm)$ [5].

The other time-stressing steps are step 4, which requires $\sum_{i=1}^{|\mathcal{L}|} |M_i|.(2^{p+1} - 1) = O(2^p m)$ and step 6, which takes $O(2^p n)$ time. Therefore, the total running time is $O(2^p(m+n))$.

Notice that the exponential component of the running time is due solely to the processing of exponentially large strings. If the strings could be implicitly represented as a concatenation of pointers to other strings, the output and the algorithm would be polynomial.

3.2 Part 2

In the second part of the algorithm, we use a function f, with special properties, to code the strings of C', forming a set C of strings written over an arbitrary, given alphabet with at least two symbols, such that $G = G_k(C)$ and $|s|$ is $O(k.\log_2(m+n).2^p)$ for each $s \in C$, where $m = |E(G)|$ and $n = |V(G)|$.

Before constructing function f, we introduce the concepts of homomorphism and occurrence.

Given alphabets Σ and Υ, a **homomorphism** $g : \Sigma^* \longrightarrow \Upsilon^*$ is a function such that, for every $u, v \in \Sigma^*$, we have $g(uv) = g(u)g(v)$. As a consequence of this property, we have $g(\varepsilon) = \varepsilon$.

Given two strings s and t, we say that there is an occurrence of t in s when either t is a substring of s or there is an integer $\ell \geq 1$ such that $last(s, \ell) = first(t, \ell)$.

Therefore, we say that a homomorphism $g : \Sigma^* \longrightarrow \Upsilon^*$ **preserves occurrences** when g is a one-to-one function such that, for all $c \in \Sigma$, $g(c)$ has a fixed size, denoted by $|g|$ and, given $s, t \in \Sigma^*$, we have:

- Given any integer $\ell \geq 0$, we have $last(s, \ell) = first(t, \ell)$ if, and only if, $last(g(s), \ell|g|) = first(g(t), \ell|g|)$;
- t is a proper substring of s if, and only if, $g(t)$ is a proper substring of $g(s)$.

Algorithm 2 *Codification of the strings in C' into C, written over a given alphabet Σ, with $|\Sigma_G| \geq |\Sigma| \geq 2$, based on the given integer $k \geq 1$.*

1. *Take a homomorphism $g : \Sigma_G^* \longrightarrow \Sigma^*$, such that g preserves occurrences and $|g| \leq 2\lceil \log_{|\Sigma|}(m+n) \rceil + 1$;*
2. *Calculate the integer $k' = \lceil k/|g| \rceil$;*
3. *Take the homomorphism $h_{k'} : \Sigma_G^* \longrightarrow \Sigma_G^*$, such that $h_{k'}(c) = c^{k'}$ for each $c \in \Sigma_G$;*
4. *Construct the function $f : \Sigma_G^* \longrightarrow \Sigma^*$, defined as $f(s) = g(h_{k'}(s))$ for each $s \in \Sigma_G^*$;*
5. *Return the set $C = \{f(s) \mid s \in C'\}$;*

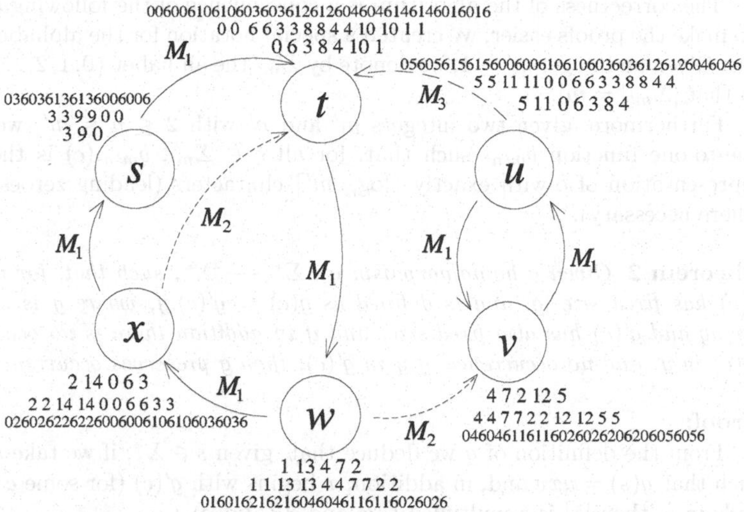

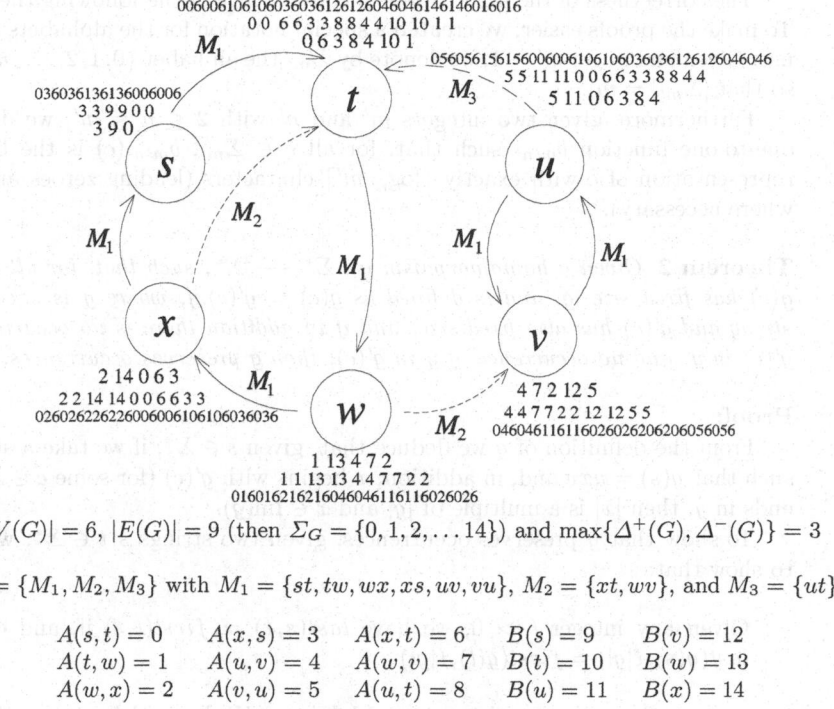

$|V(G)| = 6$, $|E(G)| = 9$ (then $\Sigma_G = \{0, 1, 2, \ldots 14\}$) and $\max\{\Delta^+(G), \Delta^-(G)\} = 3$

$\mathcal{L} = \{M_1, M_2, M_3\}$ with $M_1 = \{st, tw, wx, xs, uv, vu\}$, $M_2 = \{xt, wv\}$, and $M_3 = \{ut\}$

$A(s,t) = 0$	$A(x,s) = 3$	$A(x,t) = 6$	$B(s) = 9$	$B(v) = 12$
$A(t,w) = 1$	$A(u,v) = 4$	$A(w,v) = 7$	$B(t) = 10$	$B(w) = 13$
$A(w,x) = 2$	$A(v,u) = 5$	$A(u,t) = 8$	$B(u) = 11$	$B(x) = 14$

Iterations on step 4 of algorithm 1:

Processing M_1:	Processing M_2:
(i) $suf(s) = pref(t) = 0$	(vii) $suf(x) = pref(t) = 0\,6\,3$
(ii) $suf(t) = pref(w) = 1$	(viii) $suf(w) = pref(v) = 4\,7\,2$
(iii) $suf(w) = pref(x) = 2$	
(iv) $suf(x) = pref(s) = 3$	
(v) $suf(u) = pref(v) = 4$	Processing M_3:
(vi) $suf(v) = pref(u) = 5$	(ix) $suf(u) = pref(t) = 0\,6\,3\,8\,4$

Defining g on step 1 of algorithm 2, for $|\Sigma| = 7$ and $k = 4$:

$g(0) = 006$	$g(3) = 036$	$g(6) = 106$	$g(9) = 136$	$g(12) = 206$
$g(1) = 016$	$g(4) = 046$	$g(7) = 116$	$g(10) = 146$	$g(13) = 216$
$g(2) = 026$	$g(5) = 056$	$g(8) = 126$	$g(11) = 156$	$g(14) = 226$

Calculating the integer k' on step 2 of algorithm 2 as $k' = \lceil k/|g| \rceil = \lceil 4/3 \rceil = 2$, the homomorphism $h_{k'}$ on step 3 of algorithm 2 is $h_2(c) = c^2$ for each $c \in \Sigma_G$.

Fig. 5. Executing the algorithm for $|\Sigma| = 7$ and $k = 4$

The correctness of the algorithm is a consequence of the following theorems. To make the proofs easier, we created a special notation for the alphabets we will use. Given an integer $m' \geq 1$, we denote by $\Sigma_{m'}$ the alphabet $\{0, 1, 2, \ldots, m'-1\}$, so that $|\Sigma_{m'}| = m'$.

Furthermore, given two integers m' and n' with $2 \leq n \leq m'$, we define a one-to-one function $b_{m'n'}$ such that, for all $c \in \Sigma_{m'}$, $b_{m'n'}(c)$ is the base n' representation of c with exactly $\lceil \log_{n'} m' \rceil$ characters (leading zeroes are used where necessary).

Theorem 2 *Given a homomorphism $g : \Sigma^* \longrightarrow \Sigma^*$, such that, for all $c \in \Sigma$, $g(c)$ has fixed size $|g|$ and is defined as $g(c) = g'(c).y$, where y is a constant string and $g'(c)$ has also fixed size, and if in addition there is no occurrence of $g'(c)$ in y, and no occurrence of y in $g'(c)$, then g preserves occurrences.*

Proof:

From the definition of g we deduce that, given $s \in \Sigma^*$, if we take a string x such that $g(s) = uxv$ and, in addition, x begins with $g'(c)$ (for some $c \in \Sigma$) and ends in y, then $|x|$ is a multiple of $|g|$ and $x \in \text{Im}(g)$.

To show that g preserves occurrences, given two strings $s, t \in \Sigma^*$, we need to show that:

– Given any integer $\ell > 0$, we have $last(s, \ell) = first(t, \ell)$ if, and only if, $last(g(s), \ell|g|) = first(g(t), \ell|g|)$.

Suppose that $s = uv$ and $t = vw$, with $|v| = \ell$. We know that $g(s) = g(u)g(v)$ and $g(t) = g(v)g(w)$. Therefore, there exists the overlap $g(v)$ between $g(s)$ and $g(t)$. Thus, we have $|g(v)| = |v||g| = \ell|g|$.

On the other hand, suppose that $g(s) = uv$ and $g(t) = vw$. We can be certain that v begins with $g'(t_1)$, because $g(t)$ begins with $g'(t_1)$, and ends in y, because $g(s)$ ends in y. We conclude that $|v|$ is a multiple of $|g|$ and $v \in \text{Im}(g)$. Furhtermore, we have $u, w \in \text{Im}(g)$, with $s = g^{-1}(u)g^{-1}(v)$ and $t = g^{-1}(v)g^{-1}(w)$. Therefore, if $|v|/|g| = \ell$, we have $last(g(s), \ell|g|) = first(g(t), \ell|g|)$ and, in consequence, $last(s, \ell) = first(t, \ell)$.

– t is a proper subtring of s if, and only if, $g(t)$ is a proper substring of $g(s)$.

Now suppose that $s = utv$. We know that $g(s) = g(u)g(t)g(v)$ and, therefore, we can say that $g(t)$ is a substring of $g(s)$.

Analogously, suppose that $g(s) = u.g(t).v$. Certainly v ends in y, because $g(s)$ ends in y, and v begins with $g'(s_j)$, for some integer j, because $g(t)$ ends in y. In addition, u also ends in y, because $g(t)$ begins with $g'(t_1)$ and u begins with $g'(s_1)$, because $g(s)$ begins with $g'(s_1)$. With this, we show that $|u|$ e $|v|$ are multiples of $|g|$ and we have $u, v \in \text{Im}(g)$. Therefore, we conclude that $s = g^{-1}(u)g^{-1}(g(t))g^{-1}(v) = g^{-1}(u).t.g^{-1}(v)$, which implies that t is a substring of s.

□

Theorem 3 *For every pair of integers m', n' such that $m' \geq n' \geq 2$, there is a homomorphism $g : \Sigma_{m'}^* \longrightarrow \Sigma_{n'}^*$ preserving occurrences.*

Proof:

Take integers m', n', with $m' \geq n'$, and consider two cases: $n' \geq 3$, and $n' = 2$.

1. In the first case, for any pair of integers m', n', with $m' \geq n' \geq 3$, we take $y = n' - 1$. For all $c \in \Sigma_{m'}$, define $g(c) = b_{m'(n'-1)}(c).y$, so that $|g| = \lceil \log_{n'-1} m' \rceil + 1$. Since $b_{m'(n'-1)}$ is one-to-one, g is also one-to-one, because y is a constant string.
   Note that, for all $c \in \Sigma_{m'}$, the string $b_{m'(n'-1)}(c)$ has fixed size and y is a constant string, with no occurrence of $b_{m'(n'-1)}(c)$ in y, and no occurence of y in $b_{m'(n'-1)}(c)$.
   Therefore, by Theorem 2, the homomorphism g preserves occurrences.
2. In the second case, for any $m' \geq 2$, define $g(c) = 1.b_{m'2}(c).1.0^{i+1}$, for all $c \in \Sigma_{m'}$, where $i = \lceil \log_2 m' \rceil$, resulting in $|g| = 2i + 3$. Since $b_{m'2}$ is one-to-one, so is g, because the strings 1 and 1.0^{i+1} are constant.
   Note that, for all $c \in \Sigma_{m'}$, the string $1.b_{m'2}(c).1$ has fixed size and 0^{i+1} is a constant string, with no occurrence of $1.b_{m'2}(c).1$ in 0^{i+1}, and no occurrence of 0^{i+1} in $1.b_{m'2}(c).1$.
   Therefore, also by Theorem 2, the homomorphism g preserves occurrences.

In both cases we have $|g| \leq 2\lceil \log_{n'} m' \rceil + 1$.

□

Theorem 4 *Given any alphabet Σ, the homomorphism $h_k : \Sigma^* \longrightarrow \Sigma^*$ such that $h_k(c) = c^k$ for all $c \in \Sigma$, preserves occurrences, for all $k \geq 1$.*

Proof:

Taking $k \geq 1$, we know that h_k is one-to-one, with $|h_k| = k$.

Consider strings $s, t \in \Sigma^*$ with $s = c_1 c_2 \ldots c_{|s|}$ and $t = d_1 d_2 \ldots d_{|t|}$. Then, we have $h_k(s) = h_k(c_1)h_k(c_2)\ldots h_k(c_{|s|})$ and $h_k(t) = h_k(d_1)h_k(d_2)\ldots h_k(d_{|t|})$.

To show that h_k preserves occurrences, we need to show that:

- Given an integer $\ell > 0$, we have $last(s, \ell) = first(t, \ell)$ if, and only if, $last(h_k(s), \ell k) = first(h_k(t), \ell k)$.

Suppose that $s = uv$ and $t = vw$, with $|v| = \ell$. Then $h_k(s) = h_k(u)h_k(v)$ and $h_k(t) = h_k(v)h_k(w)$. It follows that there is overlap $h_k(v)$ between $h_k(s)$ e $h_k(t)$. Therefore, we have $last(h_k(s), \ell k) = first(h_k(t), \ell k)$.

On the other hand, suppose $h_k(s) = uv$ and $h_k(t) = vw$. If $u, v, w \notin \text{Im}(h_k)$, we would have $u = c_1^k c_2^k \ldots c_{i-1}^k c_i^j$, $v = c_i^{k-j} c_{i+1}^k c_{i+2}^k \cdots c_{|s|}^k = d_1^k d_2^k \ldots d_{\ell-1}^k d_\ell^m$ and $w = d_\ell^{k-m} d_{\ell+1}^k d_{\ell+2}^k \ldots d_{|t|}^k$, where $1 \leq j < k$ and $1 \leq i < |s|$, and, analogously, $1 \leq m < k$ and $1 \leq \ell < |t|$.

Then, we would have necessarily $\ell - 1 = |s| - (i+1) + 1 = |s| - i$ and $m = k - (j+1) + 1 = k - j$, implying that $v = c_i^m c_{i+1}^k c_{i+2}^k \ldots c_{|s|}^k = d_1^k d_2^k \ldots d_{|s|-i}^k d_{|s|-i+1}^m$

and, therefore, we would have $c_i = d_1 = c_{i+1} = d_2 = c_{i+2} = \ldots = d_{|s|-i} = c_{|s|} = d_{|s|-i+1}$. But then we could take strings $u' = c_1^k c_2^k \ldots c_{i-1}^k$, $v' = c_i^k c_{i+1}^k c_{i+2}^k \ldots c_{|s|}^k = d_1^k d_2^k \ldots d_{\ell-1}^k d_\ell^k$ and $w' = d_{\ell+1}^k d_{\ell+2}^k \ldots d_{|t|}^k$, which satisfy $u', v', w' \in \mathrm{Im}(h_k)$. In conclusion, if $|v'|/k = \ell$, we have $last(h_k(s), \ell k) = first(h_k(t), \ell k)$ and hence $last(s, \ell) = first(t, \ell)$.

– t is a proper substring of s if, and only if, $h_k(t)$ is a proper substring of $h_k(s)$.

Now suppose that $s = utv$. We know that $h_k(s) = h_k(u)h_k(t)h_k(v)$ and, therefore, $h_k(t)$ is a proper substring of $h_k(s)$.

Conversely, suppose that $h_k(s) = u.h_k(t).v$. If $u, v \notin \mathrm{Im}(h_k)$, we would have $u = c_1^k c_2^k \ldots c_i^k c_{i+1}^j$ and $v = c_{\ell-1}^{k-j} c_\ell^k c_{\ell+1}^k \ldots c_{|s|}^k$, where $1 \leq j < k$ and $i < \ell$, with $i + |t| + \ell + 1 = |s|$.

Then, $h_k(s) = c_1^k c_2^k \ldots c_i^k c_{i+1}^j .d_1^k d_2^k \ldots d_{|t|}^k c_{\ell-1}^{k-j} c_\ell^k c_{\ell+1}^k \ldots c_{|s|}^k$ and, therefore, we would have $c_{i+1} = d_1 = d_2 = \ldots = d_{|t|} = c_{\ell-1} = c$. But then we could take strings $u' = c_1^k c_2^k \ldots c_i^k$, $x = c^k$ and $v' = c_\ell^k c_{\ell+1}^k \ldots c_{|s|}^k$, satisfying $h_k(s) = u' h_k(t) x v' = u' x h_k(t) v'$. It follows that is is always possible to choose u, v in $\mathrm{Im}(h_k)$ with $s = h_k^{-1}(u).t.h_k^{-1}(v)$. In other words, t is a substring of s.

□

Given a set of strings C, we denote by $T(C)$ the set formed by every string of C that is not a proper substrings of other string in C. The following result is important in the proof of NP-hardness of the family MCP [3, 1].

Theorem 5 *Given integers $m' \geq n' \geq 2$, a homomorphism $g : \Sigma_{m'}^* \longrightarrow \Sigma_{n'}^*$ that preserves occurrences, and a set of strings C', written over $\Sigma_{m'}$, the set $C = g(C')$, satisfies:*

1. *$T(C) = g(T(C'))$ and*
2. *for all pairs of integers k and k' with $(k' - 1)|g| < k \leq k'|g|$, we have $G_k(C) = G_{k'}(C')$.*

Proof:

Since $C = g(C')$, we can restrict the homomorphism g to the bijection $g : C' \longrightarrow C$.

(1) Since g preserves occurrences, g preserves in C the same substring relationships that existed in C'. Therefore, $T(g(C')) = g(T(C'))$, that is, g commutes with T. It follows that $T(C) = g(T(C'))$.

(2) Take $u, v \in C'$ and consider $g(u), g(v) \in C$. By definition we know that $(u, v) \in E(G_{k'}(C'))$ implies $last(u, \ell) = first(v, \ell)$ and $\ell \geq k'$. Since g preserves occurrences, we also have $last(g(u), \ell|g|) = first(g(v), \ell|g|)$ and $\ell|g| \geq k'|g|$, that is, for all integers $k \leq k'|g|$, we have $(g(u), g(v)) \in E(G_k(C))$. On the other hand, if $(u, v) \notin E(G_{k'}(C'))$, and there exists ℓ such that $last(u, \ell) = first(v, \ell)$, then $\ell \leq k' - 1$. Consequently, if there is ℓ' such that $last(g(u), \ell'|g|) = first(g(v), \ell'|g|)$, then $\ell' \leq k' - 1$, because otherwise we would have $last(u, \ell') = first(v, \ell')$, and the edge (u, v) would be present in $G_{k'}(C')$, which is a contradiction. Therefore, for all integers $k > (k' - 1)|g|$, we have $(g(u), g(v)) \notin E(G_k(C))$.

We conclude that, for all pairs of integers k e k' such that $(k'-1)|g| < k \leq k'|g|$, we have $G_k(C) = G_{k'}(C')$.

□

This guarantees the correctness of the algorithm 2, once we proved that functions g (on step 1) and $h_{k'}$ (on step 3) preserve occurrences and, by Theorem 5, we can say that $G = G_1(C') = G_{k'}(h_{k'}(C')) = G_k(g(h_{k'}(C'))) = G_k(C)$.

Computational Complexity Notice that each string had its size multiplied by $|g||h_{k'}|$, where $|g| \leq 2\lceil \log_{|\Sigma|}(n+m)\rceil + 1$ and $|h_{k'}| = \lceil h/|g|\rceil \leq k/|g|+1$. Then $|h_{k'}||g| \leq k+|g| \leq k+2\lceil \log_{|\Sigma|}(n+m)\rceil + 1 \leq k+2\lceil(\log_2(n+m))/(\log_2|\Sigma|)\rceil + 1$.

The total running time is given by the size of the string set multiplied by the size of each string, that is $O(n2^p(k+(\log_2(n+m))/(\log_2|\Sigma|)))$.

4 Final Remarks

Although the algorithm runs in exponential time on the maximum degree of the input graph, it can be useful on studying problems that involve strings and graphs.

Considering specifically the *MCP* family of problems [3], we used this algorithm to reduce the Hamiltonian path problem, restricted to graphs with maximum degree smaller than or equal to 3, to the decision version of each problem in *MCP*, proving that *MCP* is NP-hard [1].

In addition, we applied occurrence-preserving homomorphisms to reduce the problems in *MCP* to the problems in *MCPr* [3], showing that this family is also NP-hard [1].

References

[1] M. D. V. Braga. Grafos de Seqüências de DNA. Master's thesis, Instituto de Computação, UNICAMP, Campinas, SP, November 2000. In Portuguese at http://www.ic.unicamp.br/~meidanis/research/fa.html.

[2] R. Diestel. *Graph Theory*. Springer-Verlag New York, Inc, 1997.

[3] C. E. Ferreira, C. C. de Souza, and Y. Wakabayashi. Rearrangement of DNA fragments: a branch-and-cut algorithm. *Discrete Applied Mathematics*, 2001. To appear.

[4] D. Gusfield, G. M. Landau, and B. Schieber. An efficient algorithm for the all pairs suffix-prefix problem. *Information Processing Letters*, 41(4):181–185, March 1992.

[5] A. Schrijver. Bipartite edge coloring in $O(\Delta m)$ time. *SIAM Journal on Computing*, 28(3):841–846, 1998.

[6] J. C. Setubal and J. Meidanis. *Introduction to Computational Molecular Biology*. PWS Publishing Company, 1997.

Conversion between Two Multiplicatively Dependent Linear Numeration Systems

Christiane Frougny[1,2]

[1] L.I.A.F.A.
Case 7014, 2 place Jussieu, 75251 Paris Cedex 05, France
Christiane.Frougny@liafa.jussieu.fr
[2] Université Paris 8

Abstract. We consider two linear numeration systems, with characteristic polynomial equal to the minimal polynomial of two Pisot numbers β and γ respectively, such that β and γ are multiplicatively dependent. It is shown that the conversion between one system and the other one is computable by a finite automaton.

1 Introduction

This work is about the conversion of integers represented in two different numeration systems, linked in a certain sense. Recall that the conversion between base 4 and base 2 is computable by a finite automaton, but that conversion between base 3 and base 2 is not. More generally, two numbers $p > 1$ and $q > 1$ are said to be *multiplicatively dependent* if there exist integers k and ℓ such that $p^k = q^\ell$. A set of natural integers is said to be *p-recognizable* if the set of representations in base p of its elements is recognizable by a finite automaton. Bűchi has shown that the set $\{q^n \mid n \geq 0\}$ is p-recognizable only if p and q are multiplicatively dependent integers [5]. In contrast, the famous theorem of Cobham [7] states that the only sets of natural integers that are both p- and q-recognizable, when p and q are two multiplicatively independent integers > 1, are unions of arithmetic progressions, and thus are k-recognizable for any integer $k > 1$. Several generalizations of Cobham's Theorem have been given, see [18,10,6,8]. In particular this result has been extended by Bès [4] to non-standard numeration systems.

The most popular non-standard numeration system is probably the Fibonacci numeration system. Recall that every non-negative integer can be represented as a sum of Fibonacci numbers, which can be chosen non-consecutive (see Section 2.2). It is also possible to represent an integer as a sum of Lucas numbers. Since Fibonacci and Lucas numbers satisfy the same recurrence relation, the question of the conversion between Lucas representations and Fibonacci representations is very natural.

A *linear numeration system* is defined by an increasing sequence of integers satisfying a linear recurrence relation. The generalization of the Cobham's Theorem by Bès [4] is the following: let two linear numeration systems such that their characteristic polynomials are the minimal polynomials of two multiplicatively

S. Rajsbaum (Ed.): LATIN 2002, LNCS 2286, pp. 64–75, 2002.

independent Pisot numbers[1]; the only sets of natural integers such that their representations in these two systems are recognizable by a finite automaton are unions of arithmetic progressions.

From the result of Bès follows that the conversion between two linear numeration systems U and Y linked to two multiplicatively independent Pisot numbers cannot be realized by a finite automaton. In this paper, we prove that the conversion between two linear numeration systems U and Y such that their characteristic polynomials are the minimal polynomials of two multiplicatively *dependent* Pisot numbers is computable by a finite automaton. This implies that a set of integers which is U-recognizable is then also Y-recognizable. Note that in [6] it is shown that if U and V are two linear numeration systems with the same characteristic polynomial which is the minimal polynomial of a Pisot number, then a U-recognizable set is also V-recognizable.

The paper is organized as follows. First we recall several results which will be of use in this paper. In particular, the normalization in a linear numeration system consists in converting a representation on a "big" alphabet onto the so-called *normal* representation, obtained by a greedy algorithm. Here the system U is fixed. It is shown in [15] that, basically, when the sequence U is linked to a Pisot number, like the Fibonacci numbers are linked to the golden mean, then normalization is computable by a finite automaton on any alphabet of digits. In the present work we first construct a finite automaton realizing the conversion from Lucas representations to Fibonacci representations. Then we consider two sequences of integers U and V. If the elements of V can be linearly expressed (with rational coefficients) in those of U, and if the normalization in the system U is computable by a finite automaton, then so it is for the conversion from V-representations to U-representations. From this result we deduce that if U and V have for characteristic polynomial the same minimal polynomial of a Pisot number, with different initial conditions, then the conversion from V-representations to U-representations is computable by a finite automaton.

Next we introduce two different linear numeration systems associated with a Pisot number β of degree m. The first one, U_β, is defined from the point of view of the symbolic dynamical system defined by β. We call it *Fibonacci-like*, because when β is equal to the golden mean, it is the Fibonacci numeration system. The second one, V_β, is defined from the algebraic properties of β. More precisely, for $n \geq 1$, the n-th term of V_β is $v_n = \beta^n + \beta_2^n + \cdots + \beta_m^n$, where $\beta_2, \ldots, \beta_m$ are the algebraic conjugates of β. We call it *Lucas-like*, because when β is equal to the golden mean, it is the Lucas numeration system. The conversion from V_β to U_β (or any sequence with characteristic polynomial equal to the minimal polynomial of β) is shown to be computable by a finite automaton.

Then we consider two linear numeration systems, U and Y, such that their characteristic polynomial is equal to the minimal polynomial of a Pisot number

[1] A Pisot number is an algebraic integer such that its algebraic conjugates are strictly less than 1 in modulus. The golden mean and the natural integers are Pisot numbers.

β, or γ respectively, where β and γ are multiplicatively dependent. Then the conversion from Y to U is shown to be computable by a finite automaton.

In the appendix we give some connections between the Lucas-like numeration system V_β and the base β numeration system for the case where β is a quadratic Pisot unit. Note that in [14] we have proved that the conversion from U_β-representations to folded β-representations is computable by a finite automaton, and in [16], that this is possible only if β is a quadratic Pisot unit.

2 Preliminaries

Words. An *alphabet* A is a finite set. A finite sequence of elements of A is called a *word*, and the set of words on A is the free monoid A^*. The *empty word* is denoted by ε. The set of infinite sequences or infinite words on A is denoted by $A^{\mathbb{N}}$. Let v be a non-empty word of A^*, denote by v^n the concatenation of v to itself n times, and by v^ω the infinite concatenation $vvv\cdots$. An infinite word of the form uv^ω is said to be *eventually periodic*. A *factor* of a (finite or infinite) word w is a finite word f such that $w = ufv$.

***U*-representations.** The definitions recalled below and related results can be found in the survey [13]. We consider a generalization of the usual notion of numeration system, which yields a representation of the natural numbers. The base is replaced by an infinite sequence of integers. The basic example is the well-known Fibonacci numeration system.

Let $U = (u_n)_{n\geq 0}$ be a strictly increasing sequence of integers with $u_0 = 1$. A *U-representation* of a non-negative integer N is a finite sequence of integers $(d_i)_{k\geq i\geq 0}$ such that $N = \sum_{i=0}^{k} d_i u_i$. Such a representation will be written $(N)_U = d_k \cdots d_0$, most significant digit first.

Among all possible U-representations of a given non-negative integer N one is distinguished and called the *normal U-representation* of N; it is also called the *greedy* representation, since it can be obtained by the following greedy algorithm [11]: given integers m and p let us denote by $q(m,p)$ and $r(m,p)$ the quotient and the remainder of the Euclidean division of m by p. Let $k \geq 0$ such that $u_k \leq N < u_{k+1}$ and let $d_k = q(N, u_k)$ and $r_k = r(N, u_k)$, and, for $i = k-1$, $\ldots$, 0, $d_i = q(r_{i+1}, u_i)$ and $r_i = r(r_{i+1}, u_i)$. Then $N = d_k u_k + \cdots + d_0 u_0$. The normal U-representation of N is denoted by $\langle N\rangle_U$. The normal U-representation of 0 is the empty word ε. The set of greedy or normal U-representations of all the non-negative integers is denoted by $G(U)$. In this work, we consider only the case where the sequence U is linearly recurrent. Then the numeration system associated with U is said to be a *linear numeration system*. The digits of a normal U-representation are contained in a *canonical* finite alphabet A_U associated with U.

Let D be a finite alphabet of integers. The *normalization* in the system U on D^* is the partial function $\nu_{U,D^*} : D^* \to A_U^*$ that maps a word $d_k \cdots d_0$ of D^* such that $N = \sum_{i=0}^{k} d_i u_i$ is non-negative onto the normal U-representation of N.

Let U and V be two sequences of integers, and let D be a finite alphabet of integers. The *conversion* from the numeration system V to the numeration system U on D^* is the partial function $\chi : D^* \to A_U^*$ that maps a V-representation $d_k \cdots d_0$ in D^* of a non-negative integer $N = \sum_{i=0}^{k} d_i v_i$ onto the normal U-representation of N. In fact the alphabet D plays no peculiar role, and we will simply speak of the conversion from V to U.

Beta-Expansions. We now consider numeration systems where the base is a real number $\beta > 1$. Representations of real numbers in such systems were introduced by Rényi [17] under the name of *beta-expansions*. Let the base $\beta > 1$ be a real number. First let x be a real number in the interval $[0, 1]$. A *representation in base* β of x is an infinite sequence of integers $(x_i)_{i \geq 1}$ such that $x = \sum_{i \geq 1} x_i \beta^{-i}$. A particular beta-representation, called the *beta-expansion*, can be computed by the "greedy algorithm" : denote by $\lfloor y \rfloor$ and $\{y\}$ the integer part and the fractional part of a number y. Set $r_0 = x$ and let for $i \geq 1$, $x_i = \lfloor \beta r_{i-1} \rfloor$, $r_i = \{\beta r_{i-1}\}$. Then $x = \sum_{i \geq 1} x_i \beta^{-i}$, where the x_i's are elements of the canonical alphabet $A_\beta = \{0, \ldots, \lfloor \beta \rfloor\}$ if β is not an integer, or $A_\beta = \{0, \ldots, \beta - 1\}$ if β is an integer. The beta-expansion of x is denoted by $d_\beta(x)$. Let D_β be the set of β-expansions of numbers of $[0, 1[$. The closure of D_β is called the *β-shift* S_β. It is a symbolic dynamical system, that is to say a closed shift invariant subset of $A_\beta^{\mathbb{N}}$.

Let D be a finite alphabet of integers. The *normalization* in base β on $D^{\mathbb{N}}$ is the partial function $\nu_{\beta, D^{\mathbb{N}}} : D^{\mathbb{N}} \to A_\beta^{\mathbb{N}}$ that maps a word $(x_i)_{i \geq 1}$ of $D^{\mathbb{N}}$ such that $x = \sum_{i \geq 1} x_i \beta^{-i} \in [0, 1[$ onto the β-expansion of x.

Secondly, we consider a real number x greater than 1. There exists $k \in \mathbb{N}$ such that $\beta^k \leq x < \beta^{k+1}$. Hence $0 \leq x/\beta^{k+1} < 1$, thus it is enough to represent numbers from the interval $[0, 1]$, since by shifting we will get the representation of any positive real number. A β-representation of an $x = \sum_{k \leq i \leq -\infty} x_i \beta^i$ will be denoted by $(x)_\beta = x_k \cdots x_0 . x_{-1} x_{-2} \cdots$.

If a representation ends in infinitely many zeros, like $v0^\omega$, the ending zeros are omitted and the representation is said to be *finite*.

A *Pisot number* is an algebraic integer such that its algebraic conjugates are strictly less than 1 in modulus. It is known that if β is a Pisot number then $d_\beta(1)$ is finite or infinite eventually periodic [2].

Automata. We refer the reader to [9]. An *automaton over* A, $\mathcal{A} = (Q, A, E, I, T)$, is a directed graph labelled by elements of A. The set of vertices, traditionally called *states*, is denoted by Q, $I \subset Q$ is the set of *initial* states, $T \subset Q$ is the set of *terminal* states and $E \subset Q \times A \times Q$ is the set of labelled *edges*. If $(p, a, q) \in E$, we denote $p \xrightarrow{a} q$. The automaton is *finite* if Q is finite. A subset H of A^* is said to be *recognizable by a finite automaton* if there exists a finite automaton $\mathcal{A}$ such that H is equal to the set of labels of paths starting in an initial state and ending in a terminal state. A *2-tape automaton* with input alphabet A and output alphabet B is an automaton over the non-free monoid $A^* \times B^*$:

$\mathcal{A} = (Q, A^* \times B^*, E, I, T)$ is a directed graph the edges of which are labelled by elements of $A^* \times B^*$. The automaton is finite if Q and E are finite. The finite 2-tape automata are also known as *transducers*. A relation R of $A^* \times B^*$ is said to be *computable by a finite automaton* if there exists a finite 2-tape automaton $\mathcal{A}$ such that R is equal to the set of labels of paths starting in an initial state and ending in a terminal state. A function is computable by a finite automaton if its graph is computable by a finite 2-tape automaton. These definitions extend to relations (and functions) of infinite words as follows: a relation R of infinite words is computable by a finite automaton if there exists a finite 2-tape automaton such that R is equal to the set of labels of infinite paths starting in an initial state and going infinitely often through a terminal state. Recall that the set of relations computable by a finite automaton is closed under composition and inverse.

Previous Results. In this work we will make use of the following results. Let U be a linearly recurrent sequence of integers such that its characteristic polynomial is exactly the minimal polynomial of a Pisot number. Then the set $G(U)$ of normal U-representations of non-negative integers is recognizable by a finite automaton, and, for every alphabet of positive or negative integers D, normalization ν_{U,D^*} is computable by a finite automaton [15]. Normalization in base β, when β is a Pisot number, is computable by a finite automaton on any alphabet D [12]. Addition and multiplication by a fixed positive integer constant are particular cases of normalization, and thus are computable by a finite automaton, in the system U and in base β.

3 Fibonacci and Lucas

Let us recall that the *Fibonacci numeration system* is defined by the sequence F of Fibonacci numbers

$$F = \{1, 2, 3, 5, 8, 13, \dots\}.$$

The canonical alphabet is $A_F = \{0, 1\}$ and the set of normal representations is equal to $G(F) = 1(\{0,1\}^* \setminus \{0,1\}^* 11 \{0,1\}^*) \cup \varepsilon$. Words containing a factor 11 are forbidden.

The *Lucas numeration system* is defined by the sequence L of Lucas numbers

$$L = \{1, 3, 4, 7, 11, 18, \dots\}.$$

The canonical alphabet is $A_L = \{0, 1, 2\}$ and the set of normal representations is equal to $G(L) = G(F) \cup (G(F) \setminus \varepsilon)\{02\} \cup \{2\}$. We give in Table 1 below the normal Fibonacci and Lucas representations of the first natural integers.

The Fibonacci and the Lucas sequences both have for characteristic polynomial

$$P(X) = X^2 - X - 1.$$

The root > 1 of P is denoted by φ, the golden mean, and its algebraic conjugate by φ'. Since $\varphi + \varphi' = 1$, for coherence of notations with the general case, we

N	Fibonacci	Lucas
1	1	1
2	10	2
3	100	10
4	101	100
5	1000	101
6	1001	102
7	1010	1000
8	10000	1001
9	10001	1002
10	10010	1010
11	10100	10000

Table 1. Normal Fibonacci and Lucas representations of the 11 first integers

denote $F = (F_n)_{n \geq 0}$ and $L = (L_n)_{n \geq 1}$. Recall that for every $n \geq 1$, $L_n = \varphi^n + \varphi'^n$. The associated dynamical system is the golden mean shift, which is the set of bi-infinite sequences on $\{0,1\}$ having no factor 11.

Although the following result is a consequence of the more general one below (Theorem 1), we give here a direct construction.

Proposition 1. *The conversion from a Lucas representation of an integer to the normal Fibonacci representation of that integer is computable by a finite automaton.*

Proof. First, for every $n \geq 3$, we get $L_n = F_{n-1} + F_{n-3}$. Take N a positive integer and a L-representation $(N)_L = d_k \cdots d_1$, where the d_i's are in an alphabet $B \supseteq \{0,1,2\}$, and $k \geq 4$. Then $N = d_k L_k + \cdots + d_1 L_1$, thus $N = d_k F_{k-1} + d_{k-1} F_{k-2} + (d_{k-2} + d_k) F_{k-3} + \cdots + (d_3 + d_5) F_2 + (d_2 + d_4) F_1 + (d_1 + d_2 + d_3) F_0$, hence the word $d_k d_{k-1} (d_{k-2} + d_k) \cdots (d_3 + d_5)(d_2 + d_4)(d_1 + d_2 + d_3)$ is a Fibonacci representation of N on a certain finite alphabet of digits D.

The conversion from a word of the form $d_k \cdots d_1$ in B^*, where $k \geq 4$, onto a word of the form $d_k d_{k-1} (d_{k-2} + d_k) \cdots (d_3 + d_5)(d_2 + d_4)(d_1 + d_2 + d_3)$ on D^* is computable by a finite automaton $\mathcal{A} = (Q, B \times C, E, \{\varepsilon\}, \{t\})$: the set of states is $Q = \{\varepsilon\} \cup B \cup (B \times B) \cup \{t\}$ where $\{t\}$ is the unique terminal state. The initial state is ε. For each d in B, there is an edge $\varepsilon \xrightarrow{d/d} d$. For each d and c in B, there is an edge $d \xrightarrow{c/c} (d,c)$. For each $(d,c) \in B \times B$ and a in B, there is an edge $(d,c) \xrightarrow{a/a+d} (c,a)$. For each $(d,c) \in B \times B$ and a in B, there is a terminal edge $(d,c) \xrightarrow{a/a+c+d} t$. Words of length less than 4 are handled directly.

Then it is enough to normalize in the Fibonacci system on D^*, and it is known that this is realizable by a finite automaton, see Section 2.6. □

On Figure 1 we give an automaton realizing the conversion from normal Lucas representations to Fibonacci representations on $\{0,1,2\}^*(\{\varepsilon\} \cup \{3\})$. States of the form (d,c) are denoted by dc. Note that this automaton is not deterministic on inputs. Since we are dealing with *normal* Lucas representations, the automaton has less states than the one constructed in the proof of Proposition 1 above. To decrease the complexity of the drawing, we introduce more than one terminal state. Terminal states are indicated by an outgoing arrow. The result must be normalized afterwards.

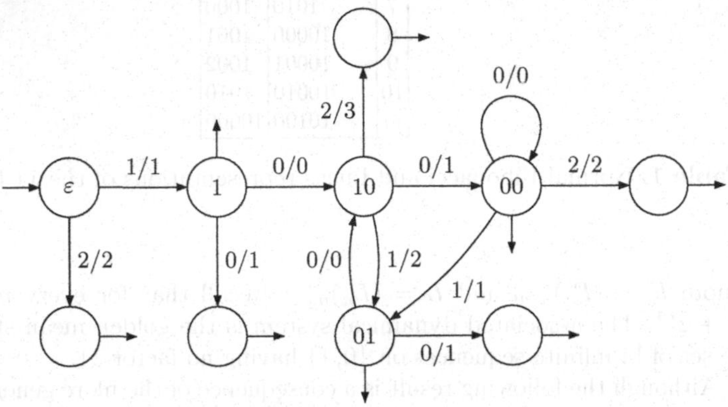

Fig. 1. Conversion from normal Lucas representations to Fibonacci representations

4 A Technical Result

We now consider two linearly recurrent sequences $U = (u_n)_{n \geq 0}$ and $V = (v_n)_{n \geq 0}$ of positive integers. The result below is the generalization of Proposition 1.

Proposition 2. *If there exist r rational constants λ_i's for $1 \leq i \leq r$ and $K \geq 0$ such that for every $n \geq K$, $v_n = \lambda_1 u_{n+r-1} + \cdots + \lambda_r u_n$, and if the normalization in the system U is computable by a finite automaton on any alphabet, then the conversion from a V-representation of an integer to the normal U-representation of that integer is computable by a finite automaton.*

Proof. One can assume that the λ_i's are all of the form p_i/q where the p_i's belong to $\mathbb{Z}$ and q belongs to $\mathbb{N}$, $q \neq 0$. Let N be a positive integer and consider a V-representation $(N)_V = b_j \cdots b_0$, where the b_i's are in an alphabet of digits $B \supseteq A_V$. Then $qN = b_j q v_j + \cdots + b_0 q v_0$. Since for $n \geq K$, $q v_n = p_1 u_{n+r-1} + \cdots + p_r u_n$, and $v_0, v_1, \ldots, v_{K-1}$ can be expressed in the system U, we get that qN is of the form $qN = d_{j+r-1} u_{j+r-1} + \cdots + d_0 u_0$ where the d_i's are elements of a finite alphabet of digits $D \supset A_U$. By assumption, ν_{U,D^*} is computable by a finite automaton. It remains to show that the function which maps $\nu_{U,D^*}(d_{j+r-1}\cdots d_0) = < qN >_U$ onto $< N >_U$ is computable by a finite automaton, and this is due to the fact that it is the inverse of the multiplication by the natural q, which is computable by a finite automaton in the system U, see Section 2.6. □

5 Common Characteristic Polynomial

The Fibonacci and the Lucas numeration systems are examples of different numeration systems having the same characteristic polynomial, but different initial conditions.

Theorem 1. *Let P be the minimal polynomial of a Pisot number of degre m. Let U and V be two sequences with common characteristic polynomial P and different initial conditions. The conversion from a V-representation of a positive integer to the normal U-representation of that integer is computable by a finite automaton.*

Proof. Since the polynomial P is the minimal polynomial of a Pisot number, normalization in the system U is computable by a finite automaton on any alphabet (see Section 2.6). On the other hand, the family $\{u_n, u_{n+1}, \cdots, u_{n+m-1} \mid n \geq 0\}$ is free, because the annihilator polynomial is the minimal polynomial. Since U and V have the same characteristic polynomial, it is known from standard results of linear algebra that there exist rational constants λ_i such that, for each $n \geq 0$, $v_n = \lambda_1 u_{n+m-1} + \cdots + \lambda_m u_n$. The result follows then from Proposition 2. □

6 Two Numeration Systems Associated with a Pisot Number

Let β be a Pisot number. We define two numeration systems associated with β.

Fibonacci-like Numeration System. Recall that the β-expansion of 1 is finite or eventually periodic [2].

First suppose that the β-expansion of 1 is finite, $d_\beta(1) = t_1 \cdots t_N$. A linear recurrent sequence $U_\beta = (u_n)_{n\geq 0}$ is canonically associated with β as follows

$$u_n = t_1 u_{n-1} + \cdots + t_N u_{n-N} \quad \text{for } n \geq N$$

$$u_0 = 1, \text{ and for } 1 \leq i \leq N-1, \; u_i = t_1 u_{i-1} + \cdots + t_i u_0 + 1.$$

The characteristic polynomial of U_β is thus

$$K(X) = X^N - t_1 X^{N-1} - \cdots - t_N.$$

Suppose now that the β-expansion of 1 is infinite eventually periodic,

$$d_\beta(1) = t_1 \cdots t_N (t_{N+1} \cdots t_{N+p})^\omega$$

with N and p minimal. The sequence $U_\beta = (u_n)_{n\geq 0}$ is the following one

$$u_n = t_1 u_{n-1} + \cdots + t_{N+p} u_{n-N-p} + u_{n-p} - t_1 u_{n-p-1} - \cdots - t_N u_{n-N-p}$$

for $n \geq N+p$,

$$u_0 = 1, \text{ and for } 1 \leq i \leq N+p-1, u_i = t_1 u_{i-1} + \cdots + t_i u_0 + 1.$$

The characteristic polynomial of U_β is now

$$K(X) = X^{N+p} - \sum_{i=1}^{N+p} t_i X^{N+p-i} - X^N + \sum_{i=1}^{N} t_i X^{N-i}.$$

Note that in general $K(X)$ may be reducible. Since $K(X)$ is defined from the β-expansion of 1, we will say that it is the *beta-polynomial* of β.

The system U_β is said to be the *canonical* numeration system associated with β. In [3] it is shown that the set of normal representations of the integers $G(U_\beta)$ is exactly the set of finite factors of the β-shift S_β. The numeration system U_β is the natural one from the point of view of symbolic dynamical systems.

Lucas-like Numeration System. Now we introduce another linear recurrent sequence $V_\beta = (v_n)_{n\geq 0}$ associated with β a Pisot number of degree m as follows. Denote by $\beta_1 = \beta$, β_2, ... , β_m the roots of the minimal polynomial $P(X) = X^m - a_1 X^{m-1} - \cdots a_m$ of β. Set

$$v_0 = 1, \text{ and for } n \geq 1, \; v_n = \beta_1^n + \cdots + \beta_m^n.$$

Then the characteristic polynomial of V_β is equal to $P(X)$.

As an example let us take $\beta = \varphi$ the golden mean. Then U_φ is the set of Fibonacci numbers, and V_φ is the set of Lucas numbers (for $n \geq 1$). If β is an integer, then the two systems U_β and V_β are the same, the standard β-ary numeration system.

Proposition 3. *Let β be a Pisot number such that its beta-polynomial $K(X)$ is equal to its minimal polynomial. Let U be any linear sequence with characteristic polynomial equal to $K(X)$ (in particular U_β). The conversion from the linear numeration system V_β to the linear numeration system U (and conversely) is computable by a finite automaton.*

Proof. It comes from the fact that U and V_β have the same characteristic polynomial, which is the minimal polynomial of a Pisot number. Thus normalization in both systems is computable by a finite automaton on any alphabet, and the result follows by Theorem 1. □

7 Multiplicatively Dependent Numeration Systems

First recall that if β is a Pisot number of degree m then, for any positive integer k, β^k is a Pisot number of degree m (see [1]). Two Pisot numbers β and γ will be said to be *multiplicatively dependent* if there exist two positive integers k and ℓ such that $\beta^k = \gamma^\ell$. Then β and γ have the same degree m.

Theorem 2. *Let β and γ be two multiplicatively dependent Pisot numbers. Let U and Y be two linear sequences with characteristic polynomial equal to the minimal polynomial of β and γ respectively. Then the conversion from the Y-numeration system to the U-numeration system is computable by a finite automaton.*

Proof. Set $\delta = \beta^k = \gamma^\ell$. As above, let $V_\beta = (v_n)_{n \geq 0}$ with $v_0 = 1$ and $v_n = \beta_1^n + \cdots + \beta_m^n$ for $n \geq 1$. Let us denote the conjugates of δ by $\delta = \delta_1, \delta_2, \ldots, \delta_m$. Note that for $1 \leq i \leq m$, we have that $\delta_i = \beta_i^k$. Set $W = (w_n)_{n \geq 0}$ with $w_n = \delta_1^n + \cdots + \delta_m^n$ for $n \geq 1$. Then W is the Lucas-like numeration system associated with δ. Now, for $n \geq 1$, $w_n = v_{kn}$. Thus any W-representation of an integer N of the form $(N)_W = d_k \cdots d_0$ gives a V_β-representation $(N)_{V_\beta} = d_k 0^{k-1} d_{k-1} 0^{k-1} \cdots d_1 0^{k-1} d_0$, and thus the conversion from W-representations to Lucas-like V_β-representations is computable by a finite automaton. The same is true for the conversion from W-representations to V_γ-representations. By Proposition 3 the conversion from Y to V_γ, and that from V_β to U are computable by a finite automaton, and the result follows. □

A set S of natural integers is said to be *U-recognizable* if the set $\{< n >_U \mid n \in S\}$ is recognizable by a finite automaton.

Corollary 1. *Let β and γ be two multiplicatively dependent Pisot numbers. Let U and Y be two linear sequences with characteristic polynomial equal to the minimal polynomial of β and γ respectively. Then a set which is U-recognizable is Y-recognizable as well.*

References

1. M.-J. Bertin, A. Decomps-Guilloux, M. Grandet-Hugot, M. Pathiaux-Delefosse, J.- P. Schreiber, *Pisot and Salem numbers*, Birkhäuser, 1992.
2. A. Bertrand, Développements en base de Pisot et répartition modulo 1. *C.R.Acad. Sc.*, Paris **285** (1977), 419–421.
3. A. Bertrand-Mathis, Comment écrire les nombres entiers dans une base qui n'est pas entière. *Acta Math. Acad. Sci. Hungar.* **54** (1989), 237–241.
4. A. Bès, An extension of the Cobham-Semënov Theorem. *Journal of Symbolic Logic* **65** (2000), 201–211.
5. J. R. Bűchi, Weak second-order arithmetic and finite automata. *Z. Math. Logik Grundlagen Math.* **6** (1960), 66–92.
6. V. Bruyère and G. Hansel, Bertrand numeration systems and recognizability. *Theoret. Comp. Sci.* **181** (1997), 17–43.

7. A. Cobham, On the base-dependence of sets of numbers recognizable by finite automata. *Math. Systems Theory* **3** (1969), 186–192.
8. F. Durand, A generalization of Cobham's Theorem. *Theory of Computing Systems* **31** (1998), 169–185.
9. S. Eilenberg, *Automata, Languages and Machines*, vol. A, Academic Press, 1974.
10. S. Fabre, Une généralisation du théorème de Cobham. *Acta Arithm.* **67** (1994), 197–208.
11. A.S. Fraenkel, Systems of numeration. *Amer. Math. Monthly* **92(2)** (1985), 105–114.
12. Ch. Frougny, Representation of numbers and finite automata. *Math. Systems Theory* **25** (1992), 37–60.
13. Ch. Frougny, Numeration Systems, Chapter 7 in M. Lothaire, *Algebraic Combinatorics on Words*, Cambridge University Press, to appear, available at `http://www-igm.univ-mlv.fr/~berstel/Lothaire/index.html`.
14. Ch. Frougny and J. Sakarovitch, Automatic conversion from Fibonacci representation to representation in base φ, and a generalization. *Internat. J. Algebra Comput.* **9** (1999), 351–384.
15. Ch. Frougny and B. Solomyak, On Representation of Integers in Linear Numeration Systems, In *Ergodic theory of* $\mathbf{Z}^d$*-Actions*, edited by M. Pollicott and K. Schmidt, London Mathematical Society Lecture Note Series **228** (1996), Cambridge University Press, 345–368.
16. Ch. Frougny and B. Solomyak, On the context-freeness of the θ-expansions of the integers. *Internat. J. Algebra Comput.* **9** (1999), 347–350.
17. A. Rényi, Representations for real numbers and their ergodic properties. *Acta Math. Acad. Sci. Hungar.* **8** (1957), 477–493.
18. A. L. Semënov, The Presburger nature of predicates that are regular in two number systems. *Siberian Math. J.* **18** (1977), 289–299.

Appendix: Quadratic Pisot Units

Here we are interested only in the case where β is a quadratic Pisot unit, that is to say the root > 1 of the polynomial $P(X) = X^2 - aX - 1$, with $a \geq 1$, or of the polynomial $P(X) = X^2 - aX + 1$, with $a \geq 3$. We denote the conjugate of β by β', $|\beta'| < 1$. In that case there are nice properties relating the numeration in the systems U_β and V_β and in base β. It is known that, when β is a quadratic Pisot unit, every positive integer has a finite β-expansion [15], the conversion from U_β-representations to β-representations folded around the radix point is computable by a finite automaton [14], and this property is characteristic of quadratic Pisot units [16].

As an example, we give in Table 2 the φ-expansions of the first integers.

We now make the link with the Lucas-like numeration V_β.

Case $\beta^2 = a\beta + 1$. Then $d_\beta(1) = a1$.

First suppose that $a \geq 2$. The following result is a simple consequence of the fact that for $n \geq 1$, $v_n = \beta^n + {\beta'}^n$ and that $\beta' = -\beta^{-1}$. The signed digit $-d$ is denoted by $\bar{d}$.

N	φ-expansions
1	1.
2	10.01
3	100.01
4	101.01
5	1000.1001
6	1010.0001
7	10000.0001
8	10001.0001
9	10010.0101
10	10100.0101
11	10101.0101

Table 2. φ-expansions of the 11 first integers

Lemma 1. *Let B be a finite alphabet of digits containing A_{V_β}. If $(N)_{V_\beta} = d_k \cdots d_0$, with $d_i \in B$, then $(N)_\beta = d_k \cdots d_0 . \bar{d}_1 d_2 \bar{d}_3 \cdots (-1)^k d_k$.*

Note that the digits in $(N)_\beta$ are elements of the alphabet $\tilde{B} = \{d, \bar{d} \mid d \in B\}$. Then the β-expansion of N is obtained by using the normalization $\nu_{\beta,\tilde{B}^{\mathbb{N}}}$ (which is computable by a finite automaton).

Now we treat the case $a = 1$. The connection between Lucas representations and representations in base the golden mean φ is the following one.

Lemma 2. *Let B be a finite alphabet of digits containing A_L. If $(N)_L = d_k \cdots d_1$, with $d_i \in B$, then $(N)_\varphi = d_k \cdots d_1 0 . \bar{d}_1 d_2 \cdots (-1)^k d_k$.*

As above, the φ-expansion of N is obtained by using the normalization $\nu_{\varphi,\tilde{B}^{\mathbb{N}}}$.

Case $\beta^2 = a\beta - 1$. Then $d_\beta(1) = (a-1)(a-2)^\omega$.

The following lemma is just a consequence of the fact that for $n \geq 1$, $v_n = \beta^n + \beta'^n$ and that $\beta' = \beta^{-1}$.

Lemma 3. *Let B be a finite alphabet of digits containing A_{V_β}. If $(N)_{V_\beta} = d_k \cdots d_0$, with $d_i \in B$, then $(N)_\beta = d_k \cdots d_0 . d_1 \cdots d_k$.*

Proposition 4. *If $d_k \cdots d_0$ is the normal V_β-representation of N then $d_k \cdots d_0 . d_1 \cdots d_k$ is the β-expansion of N.*

Proof. Note that $G(V_\beta) = \{w \in G(U_\beta) \mid w \neq w'(a-1)(a-2)^n,\ n \geq 1\}$. Now, it is enough to show that if $w = d_k \cdots d_0$ is in $G(V_\beta)$, then $d_k \cdots d_1 d_0 d_1 \cdots d_k$ contains no factor in $I = \{(a-1)(a-2)^n(a-1) \mid n \geq 0\}$. First, w has no factor in I since $G(V_\beta) \subset G(U_\beta)$. Second, $d_0 d_1 \cdots d_k$ has no factor in I either, because I is symmetrical. Third, suppose that $g = d_k \cdots d_1 d_0 d_1 \cdots d_k$ is of the form $g = g'(a-1)(a-2)^j(a-2)^{n-j}(a-1)g''$, with $w = g'(a-1)(a-2)^j$. Then $w \notin G(V_\beta)$, a contradiction. □

Star Height of Reversible Languages and Universal Automata

Sylvain Lombardy and Jacques Sakarovitch

Laboratoire Traitement et Communication de l'Information (CNRS / ENST)
Ecole Nationale Supérieure des Télécommunications
46, rue Barrault, 75634 Paris Cedex 13, France
{lombardy,sakarovitch}@enst.fr

Abstract. The *star height* of a regular language is an invariant that has been shown to be effectively computable in 1988 by Hashiguchi. But the algorithm that corresponds to his proof leads to impossible computations even for very small instances. Here we solve the problem (of computing star height) for a special class of regular languages, called *reversible languages*, that have attracted much attention in various areas of formal language and automata theory in the past few years. These reversible languages also strictly extend the classes of languages considered by McNaughton, Cohen, and Hashiguchi for the same purpose, and with different methods.

Our method is based upon the definition (inspired by the reading of Conway's book) of an automaton that is effectively associated to every language — which we call the *universal automaton* of the language — and that contains the image of any automaton that accepts the language. We show that the universal automaton of a reversible language contains a subautomaton where the star height can be computed.

Key words: Finite automata, regular expressions, star height, reversible automata, reversible languages, universal automata.

Introduction

Among all invariants attached to regular languages, the *star height* introduced by Eggan in 1963 proved to be the most puzzling one. The formal definition of star height will be recalled below but, in one word, one can say that the star height of a (regular) language K is the minimum of *nested* star operations that has to be used in order to describe K by a regular expression. As this star operation is the only[1] regular operation that goes from finite to infinite, star height is a very sensible complexity measure.

In a manner of a parallel to Kleene's theorem, Eggan ([7]) showed that the star height of a rational expression is related to another quantity that is defined

[1] Regular languages are *closed* under complement but complement is not considered as a regular operation.

S. Rajsbaum (Ed.): LATIN 2002, LNCS 2286, pp. 76–90, 2002.

on a finite automaton which produces the expression, a quantity which he called *rank* and which was later called *loop complexity*. And he stated the following two problems:

i) Does there exist, on a fixed finite alphabet, regular languages of arbitrary large star height?

ii) Is the star height of a regular language computable?

The first problem was solved, positively, in 1966 by Dejean and Schützenberger ([6]). Soon afterwards, in 1967, McNaughton ([14]) gave a conceptual proof of what Dejean and Schützenberger had established by means of combinatorial virtuosity (one of the "jewels" of formal language theory, as stated in [17]). He proved that the loop complexity, and thus the star height, of a *pure-group language*, *i.e.* a language whose syntactic monoid is a finite group, is computable. And that this family contains languages of arbitrary large loop complexity (the languages considered by Dejean and Schützenberger were — of course (?) — pure-group languages).

In addition, McNaughton's gave simple evidence of the fact that the star height of a regular language *is not a syntactic invariant*, or, which is roughly equivalent, the minimal automaton of a language is not of minimal loop complexity. As we explain below, the work of McNaughton was pushed further on, and, to some extend, our present result can be seen as its ultimate stage, which is reached by new methods.

Before that, we have to mention first that the second problem (the computability of star height) remained open for long and was considered as one of the most difficult in the theory of automata. It was eventually solved (positively) by Hashiguchi in 1988 ([10]). The method of Hashiguchi is completely different from the previous work and can be (briefly) described as follow. Given an expression E that denotes a language L on A^*, a bound M (super-exponential in the size of E) is computed. If L is of star height 1, it is denoted by an expression of height 1 which is shorter than M. If no such expression denotes L, then *all languages* denoted by those expressions are taken as *letters of a new alphabet*; and if L is of star height 2, it is denoted by an expression of star height 1 on this new alphabet and which is shorter than M. And so on. The process stops at the latest when the star height of E is reached.

Although very impressive and recognized as a tour de force, this solution presents unsatisfactory aspects and this opinion can be explained in two ways. First, the algorithm that this solution embodies is not only of high complexity (it is known to be PSPACE-hard) but also leads to computations that are by far impossible, even for very small examples. For instance, if L is accepted by a 4 state automaton of loop complexity 3 (and with a small 10 element transition monoid), then a *very low minorant* of the number of languages to be compared with L for equality is:

$$\left(10^{10^{10}}\right)^{\left(10^{10^{10}}\right)^{\left(10^{10^{10}}\right)}} .$$

Second, and this is not independent from the previous observation, the algorithm does not deal with *any structural aspect* of the language, *i.e.* the data are just *numbers*, here the number of elements of the syntactic monoid, but not the monoid itself, nor the automaton, whose form or properties could be taken into account. This is the reason why we think that the problem deserves to be considered again.

We address here the problem of computing the star height of the family of reversible languages, that are a very natural generalization of pure-group languages. A regular language is *reversible* if it is recognized by a *reversible automaton*. An automaton is reversible if every letter, and thus every word, induces a 1-1 partial mapping on the set of states of the automaton, *i.e.* if its transition function is both deterministic and co-deterministic. These reversible languages have already attracted much attention in automata theory, and more widely in theoretical computer science. They have been studied in connexion with inverse monoids (Silva [18]) and for their topological properties (Pin [15], Héam [13]).

It is noteworthy that the minimal automaton of a reversible language is not necessarily a reversible automaton — a part of the originality and the intricacy of our result lays in that discrepancy. Pin has shown in [15] that it is decidable whether a regular language is reversible or not. This makes the study of reversible languages meaningful.

The starting point of our work is the definition of the *universal automaton* of a language. The universal automaton $\mathcal{U}_L$ of a language L is finite if and only if L is regular (if the minimal automaton of L has n states, $\mathcal{U}_L$ is effectively computable and has between n and 2^n states) and has the property that any automaton $\mathcal{A}$ which recognizes L has a morphic image in $\mathcal{U}_L$. Somehow, $\mathcal{U}_L$ plays the same role with respect to *any* automaton which recognizes L as the role played by the *minimal automaton* of L with respect to *any deterministic* automaton which recognizes L. In particular, $\mathcal{U}_L$ contains as a subautomaton any minimal automaton (even non-deterministic ones) that recognizes L. The definition of the universal automaton is directly derived from the one of the *factor matrix* of a language, given by Conway in [5, Chap. 6]. The definition of the universal automaton has also been mentionned in [1].

The aim of this paper is the presentation and the proof of the following result.

Theorem 1. *The universal automaton of a reversible language K contains a subautomaton of minimal loop complexity that recognizes K.*

As the universal automaton $\mathcal{U}_K$ is effectively computable, this result directly yields, by means of a simple inspection of all its subautomata and their loop complexity, an algorithm computing the star height of K which is far more faster than the one derived from Hashiguchi's method. The mere statement of that algorithm yields a procedure which has a doubly exponential complexity in the number of the states of the minimal automaton recognizing K.

Theorem 1 clearly extends the new version we have given to McNaughton's theorem by means of the universal automaton ([12]). Along the same line as

McNaughton, star height was shown to be computable by Cohen [3] for "(special) reset-free events" and by Hashiguchi [9] for "reset-free events" in general. As reset-free events are reversible languages whose minimal automaton is reversible (the special ones are those for which this minimal automaton has only one final state), Theorem 1 stricly encompasses these developments as well. It should be noted also that the way Theorem1 is proved is reverse to the one used in McNaughton's or Hashiguchi's proofs. They start from an automaton which has the desired properties (it is the minimal automaton of the language) and they show they are able to build from it an automaton of minimal loop complexity for the language. We do not (need to) know explicitely the reversible automaton that accepts the language but we show that an automaton of minimal loop complexity can be found in an automaton that we can compute from the language (it is the universal automaton).

The paper is organized as follow. We first recall the definitions of star height and loop complexity, and give in Section 2 the one of universal automaton of a language. In Section 3, we present McNaughton result on pure-group languages within our framework, as it is an introduction to the main theorem. In the next section, we define the reversible languages and state the main property of their universal automaton by means of the subset expansion. The proof of Theorem1 is given in the last section. Due to space limitation, proofs are somewhat sketchy.

1 From Star Height of Expressions to Loop Complexity of Automata

We follow [8] for the standard definitions and notation for automata.

Regular expressions (over A^*) are the well-formed formulae built from the atomic formulae that are $\mathsf{0}$, $\mathsf{1}$ and the letters of A and using the binary operators "+" and "·" and the unary operator "∗" . The *star height* of an expression E, denoted by $\mathsf{h}(\mathsf{E})$, is defined recursively by:

$$\text{if } \mathsf{E}=\mathsf{0},\ \mathsf{E}=\mathsf{1} \text{ or } \mathsf{E}=a\in A, \qquad \mathsf{h}(\mathsf{E})=0\,,$$

$$\text{if } \mathsf{E}=\mathsf{E}'+\mathsf{E}'' \text{ or } \mathsf{E}=\mathsf{E}'\cdot\mathsf{E}'', \qquad \mathsf{h}(\mathsf{E})=\max(\mathsf{h}(\mathsf{E}'),\mathsf{h}(\mathsf{E}''))\,,$$

$$\text{if } \mathsf{E}=\mathsf{F}^*, \qquad \mathsf{h}(\mathsf{E})=1+\mathsf{h}(\mathsf{F})\,.$$

Example 1. The expressions $(a+1)(a^2+b)^*a+1$ and $(b^*a+1)\,(a\,b^*a)^*$ have star height 1 and 2 respectively. As they both denote the language K_1 accepted by the automaton $\mathcal{A}_1$, this shows that two equivalent expressions may have different star heights.

Definition 1. *The star height of a regular language K of A^*, which we note as $\mathsf{h}(\mathrm{K})$, is the minimum of the star height of the expressions that denote*[2] *the language K:*

[2] We write $|\mathsf{E}|$ for the language denoted by the expression E. Similarly, we write $|\mathcal{A}|$ for the language accepted by the automaton $\mathcal{A}$.

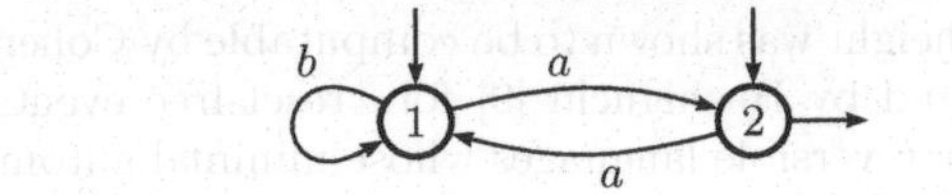

Fig. 1. The automaton $\mathcal{A}_1$

$$\mathsf{h}(K) = \min\{\mathsf{h}(\mathsf{E}) \mid \mathsf{E} \in \mathsf{Rat}\, A^* \quad |\mathsf{E}| = K\}\,.$$

The star height of an expression reflects also a structural property of an automaton (more precisely, of the underlying graph of an automaton) which corresponds to that expression. In order to state it, we define the notion of *a ball* of a graph: a ball in a graph is a strongly connected component that contains at least one arc.

Definition 2. *The* loop complexity[3] *of a graph $\mathcal{G}$ is the integer $\mathsf{lc}(\mathcal{G})$ recursively defined by:*

$$\mathsf{lc}(\mathcal{G}) = 0 \qquad \textit{if } \mathcal{G} \textit{ contains no ball (in particular, if } \mathcal{G} \textit{ is empty);}$$
$$\mathsf{lc}(\mathcal{G}) = \max\{\mathsf{lc}(\mathcal{P}) \mid \mathcal{P} \textit{ ball of } \mathcal{G}\} \qquad \textit{if } \mathcal{G} \textit{ is not a ball itself;}$$
$$\mathsf{lc}(\mathcal{G}) = 1 + \min\{\mathsf{lc}(\mathcal{G} \setminus \{\mathsf{s}\}) \mid s \textit{ vertex of } \mathcal{G}\} \qquad \textit{if } \mathcal{G} \textit{ is a ball.}$$

As Eggan showed, star height and loop complexity are the two faces of the same notion:

Theorem 2. *[7] The star height of a language K is equal to the minimal loop complexity of an automaton that recognizes K.*

In a previous paper [12], we showed an even stronger connection between star height of an expression and loop complexity of an automaton.

Proposition 1. *The loop complexity of a trim automaton $\mathcal{A}$ is equal to the infimum of the star height of the expressions that are obtained by the different possible runs of the McNaughton-Yamada algorithm on $\mathcal{A}$.*

Theorem 2 allows to deal with automata instead of expressions, and to look for automata of minimal loop complexity instead of expressions of minimal star height. This is what we do in the sequel.

---

[3] Eggan [7] as well as Cohen [3] and Hashiguchi [9] call it "cycle rank", Bchi calls it "feedback complexity". McNaughton [14] calls loop complexity of a language the minimum cycle rank of an automaton that accepts the language. We have taken this terminology and made it parallel to star height, for "rank" is a word of already many different meanings.

2 The Universal Automaton of a Language

Let $\mathcal{A} = \langle Q, M, E, I, T\rangle$ be an automaton[4] over a monoid M. For any state q of $\mathcal{A}$ let us call "*past of q (in $\mathcal{A}$)* " the set of labels of computations that go from an initial state of $\mathcal{A}$ to q, let us denote it by $\mathsf{Past}_{\mathcal{A}}(q)$; *i.e.*

$$\mathsf{Past}_{\mathcal{A}}(q) = \{m \in M \mid \exists i \in I \quad i \xrightarrow[\mathcal{A}]{m} q\}.$$

In a dual way, we call "*future of q (in $\mathcal{A}$)*" the set of labels of computations that go from q to a final state of $\mathcal{A}$, and we denote it by $\mathsf{Fut}_{\mathcal{A}}(q)$; *i.e.*

$$\mathsf{Fut}_{\mathcal{A}}(q) = \{m \in M \mid \exists t \in T \quad q \xrightarrow[\mathcal{A}]{m} t\}.$$

For every q in Q it then obviously holds:

$$[\mathsf{Past}_{\mathcal{A}}(q)]\ [\mathsf{Fut}_{\mathcal{A}}(q)] \subseteq |\mathcal{A}| \,. \tag{1}$$

Moreover, if one denotes by $\mathsf{Trans}_{\mathcal{A}}(p, q)$ the set of labels of computations that go from p to q, it then holds:

$$[\mathsf{Past}_{\mathcal{A}}(p)]\ [\mathsf{Trans}_{\mathcal{A}}(p, q)]\ [\mathsf{Fut}_{\mathcal{A}}(q)] \subseteq |\mathcal{A}| \,. \tag{2}$$

It can also be observed that a state p of $\mathcal{A}$ is initial (resp. final) if and only if 1_{A^*} belongs to $\mathsf{Past}_{\mathcal{A}}(p)$ (resp. to $\mathsf{Fut}_{\mathcal{A}}(p)$).

Hence, if K is the subset of M recognized by $\mathcal{A}$, every state of $\mathcal{A}$ induces a *subfactorization* of K: this is how equation (1) will be called. It is an idea due to J. Conway [5, chap. 6] to take the converse point of view, that is to build an automaton from the factorizations of a subset (in any monoid).

More specifically, let K be any subset of a monoid M and let us call *factorization* of K a pair (L, R) of subsets of M such that $L\,R \subseteq K$ and (L, R) is *maximal*[5] for that property in $M \times M$. We denote by Q_K the set of factorizations of K and for every p, q in Q_K *the factor* $F_{p,q}$ of K is the subset of M such that

$$L_p\, F_{p,q}\, R_q \subseteq K$$

and $F_{p,q}$ is *maximal* for that property in M. If $\alpha\colon M \longrightarrow N$ is a morphism that recognizes K, *i.e.* $K\alpha\alpha^{-1} = K$, and if (L, R) is a factorization and F is a factor of K then:

i) $L = L\alpha\alpha^{-1}$, $R = R\alpha\alpha^{-1}$, and $F = F\alpha\alpha^{-1}$;

ii) $(L\alpha, R\alpha)$ is factorization and $F\alpha$ is factor of $K\alpha$;

or, in other words, factorizations and factors are *syntactic objects* with respect to K. As a consequence, Q_K is finite if and only if K is recognizable.

In [5], the $F_{p,q}$ are organized as a $Q_K \times Q_K$-matrix of subsets of A^*, called the *factor matrix* of the language K. A further step consists in building an

[4] An automaton over M is a labelled graph where the label of edges are taken in M.
[5] In the partial order induced by the inclusion in M.

automaton, which we call *the universal automaton* of K, denoted by $\mathcal{U}_K$, and based on the factorizations and the factors of K:

$$\mathcal{U}_K = \langle Q_K, A, E_K, I_K, T_K \rangle ,$$

$$\text{where} \qquad I_K = \{p \in Q_K \mid 1_{A^*} \in L_p\} , \quad T_K = \{q \in Q_K \mid 1_{A^*} \in R_q\}$$

$$\text{and} \qquad E_K = \{(p, a, q) \in Q_K \times A \times Q_K \mid p, q \in Q_K , \ a \in F_{p,q}\} ,$$

and, obviously, $|\mathcal{U}_K| = K$. What makes $\mathcal{U}_K$ *universal* is expressed in the following result.

Theorem 3. *[16] If $\mathcal{A} = \langle Q, A, E, I, T \rangle$ is a trim automaton that accepts K, then there exists an* automaton morphism *from $\mathcal{A}$ into $\mathcal{U}_K$, and $\mathcal{U}_K$ is minimal for this property.*

In particular, $\mathcal{U}_K$ contains as a subautomaton every *minimal* automaton (deterministic, or non deterministic) that recognizes K.

Example 2. The factorizations of $K_1 = |\mathcal{A}_1|$ are (they have to be computed in the syntactic monoid):

$$\begin{gathered}(a^*, (1 + b^*a)(ab^*a)^*), \\ (a^*b^*(a^2 + b)^*, (a^2 + b)^*a), \\ (1 + a^*b^*(a^2 + b)^*a, (ab^*a)^*).\end{gathered}$$

The universal automaton of K_1 is shown opposite.

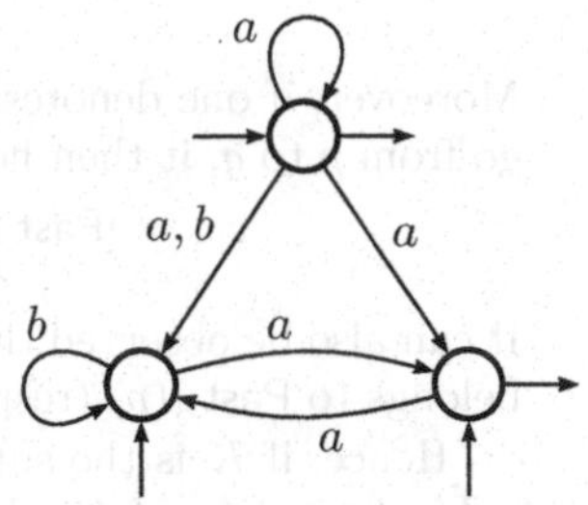

Fig. 2. The universal automaton of K_1

3 The Star Height of Pure-Group Languages

Before dealing with reversible languages in whole generality, we present the statement and the proof of McNaughton's theorem on pure-group languages by means of the universal automaton (*cf.* [12] for a complete exposition).

Theorem 4. *The universal automaton of a pure-group language K contains a subautomaton of minimal loop complexity that recognizes K.*

For any automaton $\mathcal{B}$ that recognizes a language K — and in particular for one of minimal loop complexity — there exists a morphism from $\mathcal{B}$ into $\mathcal{U}_K$. If an (automaton) morphism were preserving loop complexity or, at least, were not increasing it, the theorem would follow immediately, and not only for group languages but for any language. But this is not the case, by far. With that idea in mind, one has to consider morphisms of special kind.

Definition 3. *A morphism* $\varphi\colon \mathcal{B} \to \mathcal{A}$ *is said to be* conformal[6] *if any computation in* $\mathcal{A}$ *is the image of (at least) one computation in* $\mathcal{B}$.

Theorem 5. *[14, Theorem 3] If* $\varphi\colon \mathcal{B} \to \mathcal{A}$ *is a conformal morphism, then the loop complexity of* $\mathcal{B}$ *is greater than or equal to the one of* $\mathcal{A}$: $\mathsf{lc}(\mathcal{B}) \geqslant \mathsf{lc}(\mathcal{A})$. □

Even in the case of a pure-group language K, the morphism φ from an automaton $\mathcal{B}$ (that recognizes K) into $\mathcal{U}_K$ is not necessarily conformal. The proof of Theorem 5 boils down to show that, nevertheless, φ is conformal on those balls of $\mathcal{B}$ that are crucial for the loop complexity. This is proved *via* the following two results, and this is where our method differs from McNaughton's proof.

Proposition 2. *The balls of the universal automaton of a pure-group language are deterministic and complete*[7].

The result follows from the fact that every state of $\mathcal{U}_K$ can be identified with a subset of the syntactic group G of K. The balls of $\mathcal{U}_K$ are exactly the orbits of these subsets under the action of the group G.[8]

Lemma 1. *Let* K *be a language of* A^* *whose syntactic monoid is a group* G. *For every* g *in* G, *let* H_g *be the set of words whose image in* G *is* g *and let* $W = H_{1_G}$. *Let* $\mathcal{A}_K = \langle Q, A, E, \{i\}, T\rangle$ *be the minimal automaton of* K *and* $\mathcal{B}$ *any equivalent automaton.*

For every g *in the image of* K *in* G, *there exists a state* r *in* $\mathcal{B}$ *such that:*

i) $W \cap \mathsf{Past}_{\mathcal{B}}(r) \neq \emptyset$ *and* $H_g \cap \mathsf{Fut}_{\mathcal{B}}(r) \neq \emptyset$.

ii) For every loop[9], *labelled by a word* y, *around the initial state* i *in* $\mathcal{A}_K$, *there exists a loop, labelled by a word* $x\,y\,z$, *around* r *in* $\mathcal{B}$, *where* x *is in* W.

Proof. Let k be the order of G and n the number of states of $\mathcal{B}$. Let l be an integer and C_l the set of words of length smaller than l and labelling a loop around i in $\mathcal{A}_K$. Let w_l be the product of all k-th power of words in C_l:

$$w_l = \prod_{v \in C_l} v^k .$$

Every v^k is in W and so is w_l. The image of w_l in G is 1_G. Therefore, for every u in H_g, ${w_l}^n u$ is in the language K. Hence, there is a successful computation,

---

6 McNaughton call them "*pathwise*" (*in* [14]) but his definition of morphism is slightly different from ours.

7 *i.e.* for every state p in every ball, there exists for every letter of the alphabet exactly one transition whose origin is p, that is labelled by a, and whose end belongs to the same ball as p.

8 Let us note that we give here a statement which is sligtly different from the corresponding lemma (Lemma 6) in [12]. The forthcomming Lemma 3 appears thus as a natural generalization of Lemma 1 for reversible languages.

9 We call loop around s any path whose origin and end are s.

labelled by ${w_l}^n u$ in $\mathcal{B}$. As $\mathcal{B}$ has only n states, there is a loop labelled by a power of w_l around a state r_l of $\mathcal{B}$. For r_l, i) holds and ii) holds for y shorter than l.

If we consider the infinite sequence $r_0, r_1, r_2, \dots$, we can find a state r that occurs infinitely often. Thus i) and ii) are met by this state r. □

Proof. (of Theorem 4) Let φ be a morphism from $\mathcal{B}$ into $\mathcal{U}_K$, $\mathcal{C}_g$ the ball of $\mathcal{B}$ containing r and $\mathcal{D}_g$ the ball of $\mathcal{U}_K$ containing $r\varphi$. Proposition 2 and Lemma 1 imply together that the restriction of φ to $\mathcal{C}_g$ is *conformal* and *surjective* onto $\mathcal{D}_g$ and that $\mathcal{D}_g$ accepts all words in H_g. The union of the $\mathcal{D}_g$, for g in the image of K in G recognizes K and its loop complexity is smaller than or equal to the one of $\mathcal{B}$ (which could have been chosen of minimal loop complexity). □

4 Reversible Languages and Their Universal Automata

Definition 4. *An automaton (on a free monoid) is* reversible *if its transition function is deterministic and codeterministic. A regular language is* reversible *if it is accepted by a finite reversible automaton.*

The minimal automaton of a reversible language is not necessarily reversible, *cf.* on Fig. 3, the minimal automaton of the reversible automaton $\mathcal{A}_1$. However, it is decidable whether a given regular language is reversible by performing some computations on its syntactic monoid (*cf.* [15]).

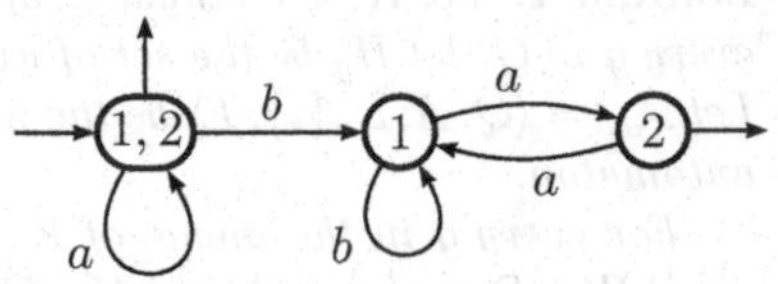

Fig. 3. The minimal automaton of $\mathcal{A}_1$

By comparison, the "reset-free events" of Hashiguchi ([9]) are those reversible languages whose minimal automaton is reversible and the "special[10] reset-free events" of Cohen ([3]) are those for which this minimal automaton has only one final state.

As for group languages, the proof of Theorem 1 will be based on a property of universal automata.

Proposition 3. *The balls of the universal automaton of a reversible language are reversible.*

The proof of Proposition 3 is far more elaborate than the corresponding property for group language. It is first based on the construction of a (good) approximation of $\mathcal{U}_K$ from an automaton $\mathcal{A}$ which accepts K without computing its transition monoid. The states of this new automaton will be *sets of subsets* of the state set of $\mathcal{A}$.

[10] They are not called special in [3] for they are the only ones considered.

Let $\mathcal{A} = \langle Q, A, E, I, T\rangle$ be an automaton that accepts K and (λ, μ, ν) the corresponding Boolean representation:

$$\lambda \in \mathbb{B}^Q \text{ is a row vector: } \lambda_p = 1 \Leftrightarrow p \in I\,,$$
$$\nu \in \mathbb{B}^Q \text{ is a column vector: } \nu_p = 1 \Leftrightarrow p \in T\,,$$
$$\text{and} \qquad \mu\colon A^* \to \mathbb{B}^{Q\times Q} \text{ is a morphism: } (a\mu)_{p,q} = 1 \Leftrightarrow (p, a, q) \in E\,.$$

For every u in A^*, $\lambda{\cdot}u\mu$ is a (row) vector of $\mathbb{B}^Q$ and thus represents a subset of Q. A set of vectors, *i.e.* a set of subsets of Q, is an *antichain* if its elements are incomparable (for the inclusion order). For instance, the set of antichains in $\mathfrak{P}(\mathbb{B}^{\{1,2\}})$ is $\Big\{\{(0,0)\}, \{(1,0)\}, \{(0,1)\}, \{(1,1)\}, \{(1,0),(0,1)\}\Big\}$.

One can associate to any factorization (L, R) of K, indeed to L, an antichain of $\mathfrak{P}(\mathbb{B}^Q)$: if u and v are in L and R respectively, then: $\lambda{\cdot}u\mu{\cdot}v\mu{\cdot}\nu = 1$; if u' is such that $\lambda{\cdot}u'\mu = \lambda{\cdot}u\mu$, then $u' \in L$ since (L, R) is maximal. Moreover, if $\lambda{\cdot}u\mu \subseteq \lambda{\cdot}u'\mu$, then u' is in L as well and therefore L is characterized by an antichain.

Definition 5. *Let (λ, μ, ν) be the Boolean representation of an automaton $\mathcal{A}$ on A^*. Let S be the set of subsets of $\lambda{\cdot}A^*\mu$ that are antichains. We call* subset expansion *of $\mathcal{A}$ the automaton $\mathcal{V}_\mathcal{A} = \langle S, A, F, J, U\rangle$ defined by*

$$J = \{Y \in S \mid \exists \xi \in Y \quad \xi \subseteq \lambda\}\,, \qquad U = \{X \in S \mid \forall \theta \in X \quad \theta{\cdot}\nu = 1\}$$
$$F = \{(X, a, Y) \in S \times A \times S \mid \forall \theta \in X\,,\ \exists \xi \in Y \quad \xi \subseteq \theta{\cdot}a\mu\}\,.$$

It is immediate that $\mathcal{V}_\mathcal{A}$ is equivalent to $\mathcal{A}$, and the following lemma is not difficult to prove.[11]

Lemma 2. *If an automaton $\mathcal{A}$ accepts K, $\mathcal{U}_K$ is a subautomaton of $\mathcal{V}_\mathcal{A}$.* □

Proposition 4. *The balls of the subset expansion of a reversible automaton are deterministic.*

Proof. Let $\mathcal{V}_\mathcal{A}$ be the subset expansion of $\mathcal{A}$ represented by (λ, μ, ν) of dimension Q. Let X and Y be two states in a given ball of $\mathcal{V}_\mathcal{A}$. There exist then u in $\mathsf{Trans}_{\mathcal{V}_\mathcal{A}}(X, Y)$ and v in $\mathsf{Trans}_{\mathcal{V}_\mathcal{A}}(Y, X)$. Recall that X and Y are sets of subsets of Q; we denote by X_i (resp. by Y_i) the subset of elements of X (resp. of Y) which contain i states of Q.

Claim: for every integer i, there is a bijection from X_i into Y_i (resp. from Y_i into X_i) induced by u (resp. by v). Suppose the claim does not hold and let i be the smallest integer such that u does not induce a bijection from X_i into Y_i.

[11] It can be noted that the definition of the subset expansion does not come completely out of the blue: in the case where $\mathcal{A}$ is the minimal automaton of a language K, $\mathcal{V}_\mathcal{A}$ is "the subset automaton of order 0 of K" described by Cohen and Brzozowski in [4].

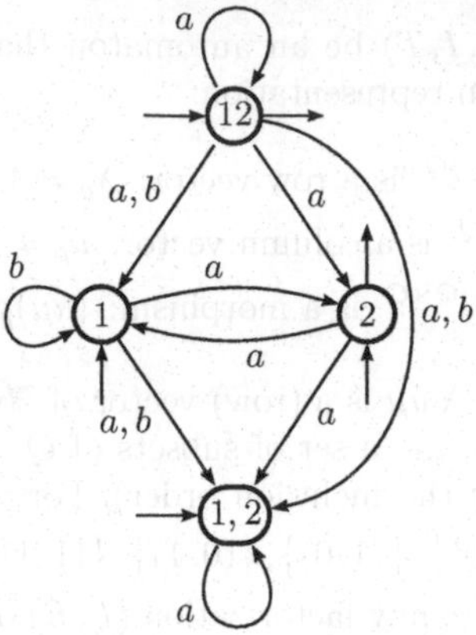

Fig. 4. The subset expansion of $\mathcal{A}_1$

i) If there exists θ in X_i such that $\theta{\cdot}u\mu$ is not in Y_i, there exist $j < i$, ξ in Y_j and θ' in X_j such that $\theta'{\cdot}u\mu = \xi \subseteq \theta{\cdot}u\mu$. As $u\mu$ is a 1-to-1 mapping, $\theta' \subseteq \theta$, which is a contradiction because θ and θ' are incomparable. Thus u induces a function from X_i into Y_i.

ii) If there exist θ and θ' in X_i such that $\theta{\cdot}u\mu = \theta'{\cdot}u\mu$, then, again since $u\mu$ is a 1-to-1 mapping, $\theta = \theta'$. Thus u induces an injective function from X_i into Y_i.

iii) For the same reason, v induces an injective function from Y_i into X_i. Thus both u and v induce a bijection between X_i and Y_i.

Therefore, the state that can be reach in a ball by a path labelled by u from a state X is completely defined by $\{\theta{\cdot}u\mu \mid \theta \in X\}$. The balls of $\mathcal{V}_\mathcal{A}$ are deterministic.

Proposition 3 is then the direct consequence of Lemma 2 and of Proposition 4. Indeed, if the balls of $\mathcal{V}_\mathcal{A}$ are deterministic, so are those of $\mathcal{U}_K$. If K is reversible, $\widetilde{K}$, the mirror image of K is reversible as well, and the balls of $\mathcal{U}_{\widetilde{K}}$, which are *the transposed* of those of $\mathcal{U}_K$, are deterministic. Therefore, the balls of $\mathcal{U}_K$ are codeterministic, thus reversible. □

5 Star Height of Reversible Languages

We can now proceed to the proof of Theorem 1. We begin with a property which is the equivalent of Lemma 1 for reversible languages. In order to state that property, we define a decomposition of a language according to an automaton that accepts it; this is an adaptation of a method devised by Hashiguchi in [9].

Let $\mathcal{A} = \langle Q, A, E, I, T\rangle$ be an automaton that accepts K. Every computation in $\mathcal{A}$ can be decomposed, in a unique way, into a sequence:

$$i \xrightarrow{v_0} p_1 \xrightarrow{u_1} q_1 \xrightarrow{v_1} p_2 \xrightarrow{u_2} q_2 \longrightarrow q_{m-1} \xrightarrow{v_{m-1}} p_m \xrightarrow{u_m} q_m \xrightarrow{v_m} t$$

where i (resp. t) is an initial (resp. final) state, p_i (resp. q_i) is the first (resp. the last) state of the i-th ball crossed by the path, the words v_i label paths between balls and the words u_i, possibly empty, label paths in balls.

Let W_{p_i} be the set of words whose image in the transition monoid of $\mathcal{A}$ is an idempotent *and* that label a loop around p_i; let G_i be the set of words whose image in this monoid is the same as u_i, and let $H_i = W_{p_i} G_i$. The language $v_0 H_1 v_1 H_2 \ldots v_{m-1} H_m v_m$ is a subset of K that consists of words which label paths with the same "behaviour" in the automaton. We call such a set an $\mathcal{A}$-*constituent* of K. The states p_1, q_1, p_2, ... ,q_m are *the markers* of this $\mathcal{A}$-constituent. There are finitely many distinct $\mathcal{A}$-constituents of K and their union is equal to K.

We are now in a position to state the generalization of Lemma 1.

Lemma 3. *Let $\mathcal{A}$ be a reversible automaton and $\mathcal{B}$ any equivalent automaton. Let $v_0 H_1 v_1 H_2 \ldots v_{m-1} H_m v_m$ be any $\mathcal{A}$-constituent of $|\mathcal{A}| = K$ and p_1, q_1, p_2, ... , q_m its markers. Then there exist m states r_1, r_2, ... ,r_m in $\mathcal{B}$ such that:*

i) $v_0 W_{p_1} \cap \mathsf{Past}_{\mathcal{B}}(r_1) \neq \emptyset$, $\quad H_m v_m \cap \mathsf{Fut}_{\mathcal{B}}(r_m) \neq \emptyset$,
and, $\forall i \in [1; m-1] \quad (H_i v_i W_{p_i}) \cap \mathsf{Trans}_{\mathcal{B}}(r_i, r_i + 1) \neq \emptyset$.

ii) For every i in $[1; m]$, for every loop around p_i labelled by a word v, there exists a loop around r_i, labelled by a word $u\,v\,w$, where u is in W_{p_i} and w in A^*.

Proof. There exists an integer k such that, for every word $v \in A^*$, the image of v^k in the transition monoid of $\mathcal{A}$ is an idempotent. Let n be the number of states of $\mathcal{B}$. Let l be an integer that will silently index the sets we define now. For every $i \in [1; m]$, let C_i be the set of words of length smaller than l that label a loop around p_i in $\mathcal{A}$. Let w_i be the product of all k-th power of words in C_i:

$$w_i = \prod_{v \in C_i} v^k .$$

For every $v_0 u_1 v_1 \cdots v_m$ in the $\mathcal{A}$-constituent,

$$w = v_0 (w_1)^n u_1 v_1 (w_2)^n u_2 ... (w_m)^n u_m v_m$$

is in the $\mathcal{A}$-constituent as well. Hence, there is a successful computation labelled by w in $\mathcal{B}$. As $\mathcal{B}$ has only n states, this path contains, for every i, a loop labelled by a power of w_i around a state r_i of $\mathcal{B}$. The m-tuple $r^{(l)} = (r_1, r_2, \ldots, r_m)$ verifies i) and ii) for y shorter than l. If we consider the infinite sequence $r^{(1)}, r^{(2)}, \ldots$, we can find an m-tuple that occurs infinitly often and that verifies the lemma. □

Proof. (of Theorem 1, Sketch) Let $\mathcal{A}$ be a reversible automaton that accepts K, $\mathcal{B}$ an equivalent automaton with minimal loop complexity, and $\mathcal{U}_K$ the universal automaton of K.

For every $\mathcal{A}$-constituent of K — with markers p_1, q_1, p_2, ... , q_m —, there exist m states r_1, r_2, ... , r_m in $\mathcal{B}$ verifying the Lemma 3 and a morphism φ from $\mathcal{B}$ into $\mathcal{U}_K$ that maps r_1, r_2, ... , r_m onto s_1, s_2, ... , s_m.

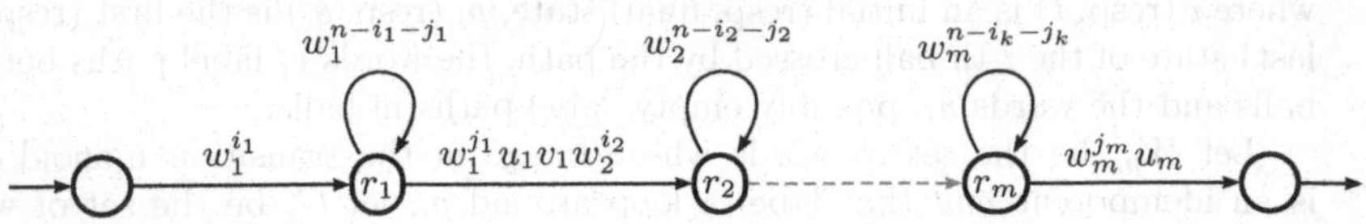

Fig. 5. A witness word for a $\mathcal{A}$-constituent.

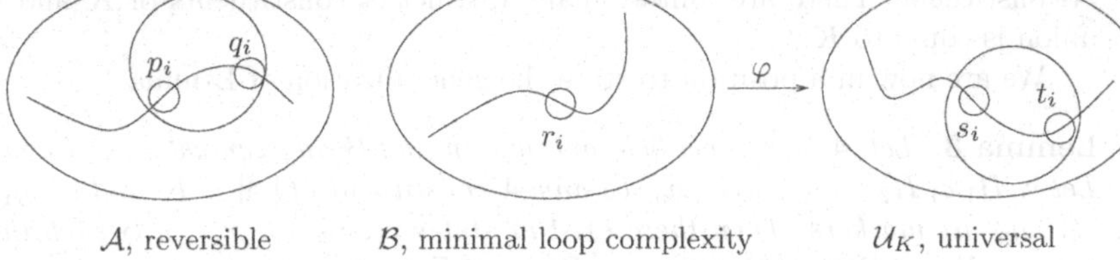

Fig. 6. Proof of Theorem 1

For every i, let $\mathcal{S}_i$ be the ball of $\mathcal{U}_K$ that contains s_i. For every path of $\mathcal{S}_i$, there exists a loop around s_i that contains this path and that is labelled by an idempotent[12] y.

As $\mathcal{U}_K$ is a subautomaton of $\mathcal{V}_\mathcal{A}$, s_i can be seen as a set of subsets of Q. One of these subsets contains p_i. Since there is a bijection from these sets into themselves induced by y, and as y is an idempotent, this bijection is an identity. Therefore, there is a loop labelled by y around p_i.

By Lemma 3, there is a loop in $\mathcal{B}$ labelled by $x\,y\,z$ around r_i, with x idempotent. The image by φ of this loop is a loop around s_i in $\mathcal{U}_K$. Since x is an idempotent, it labels a loop around s_i (in $\mathcal{S}_i$) by itself; thus, there is a path in $\mathcal{S}_i$, beginning in s_i and labelled by y which is the image of a path in the ball of r_i. As the balls of $\mathcal{U}_K$ are deterministic, this path is the one which contains the path chosen in $\mathcal{S}_i$ which is therefore the image by φ of a path of the ball of r_i.

Hence, the morphism φ is conformal from the ball of r_i onto $\mathcal{S}_i$, whose loop complexity is smaller than or equal to the loop complexity of the ball of r_i.

Then it can easily be verified that there exists a state t_i in $\mathcal{S}_i$ such that: i) H_i is a subset of $\mathsf{Trans}_{\mathcal{U}_K}(s_i, t_i)$; ii) if $i < m$, there is a path in $\mathcal{U}_K$ without any loop between t_i and s_{i+1} labelled by v_i, and if $i = m$, t_i is terminal.

This establishes that every $\mathcal{A}$-constituent is accepted by a subautomaton of $\mathcal{U}_K$ with a loop complexity smaller than or equal to the minimal loop complexity for K and this subautomaton contains entire balls of $\mathcal{U}_K$. Thus, the union of these subautomata in $\mathcal{U}_K$ accepts K with minimal loop complexity. □

[12] of the transition monoid of $\mathcal{A}$.

6 A Word about Complexity

Theorem 1 yields naturally an algorithm which computes the star height of a reversible language.

As the size of the universal automaton of a language K can be exponential in the size of the minimal automaton of K, this algorithm can not be of a lower complexity than exponential. For reasons we shall briefly explain below, the algorithm is indeed doubly exponential. However, the properties that have been established in order to prove Theorem 1 may be taken into account in order to avoid expensives operations (such as, for instance, the enumeration of all subautomata of the universal automaton).

Theorem 1 can be restated as follows: the universal automaton $\mathcal{U}_K$ of a reversible language K contains a subautomaton

i) which recognizes K,
ii) of minimal loop complexity,
ii) whose balls are balls of $\mathcal{U}_K$.

And this statement can be translated into the following algorithm.

First start from $\mathcal{A}_K$, the minimal automaton of K with n states and test whether K is reversible or not; this operation has polynomial complexity [15]. Next, build the universal automaton $\mathcal{U}_K$, identify every ball of $\mathcal{U}_K$ and compute the loop complexity of each of them.

Let k be the loop complexity of $\mathcal{A}_K$. For every integer i from 1 to k, build the largest subautomaton $\mathcal{D}_i$ of $\mathcal{U}_K$ which contains no state of any ball of $\mathcal{U}_K$ of loop complexity larger than i, and test whether $\mathcal{D}_i$ accepts K. The first i for which this condition holds is the star height of K.

There are two critical sections in that algorithm. The first one is the computation of the loop complexity of the balls of $\mathcal{U}_K$. The size of the balls can be exponential in n and the computation of the loop complexity of an automaton is exponential in the size of the automaton. The second critical section is the test for equivalence of $\mathcal{D}_i$. This requires *a priori* the determinization of $\mathcal{D}_i$, which is already of exponential size.

References

1. ARNOLD A., DICKY A., AND NIVAT M., A Note about Minimal Non-deterministic Automata. *Bull. of E.A.T.C.S.* **47** (1992), 166–169.
2. BÜCHI J. R., *Finite Automata, their Algebras and Grammars: Toward a Theory of formal Expressions.* D. Siefkes (Ed.). Springer-Verlag, 1989.
3. COHEN R., Star height of certain families of regular events. *J. Computer System Sci.* **4** (1970), 281–297.
4. COHEN R. AND BRZOZOWSKI R., General properties of star height of regular events. *J. Computer System Sci.* **4** (1970), 260–280.
5. CONWAY J. H., *Regular algebra and finite machines.* Chapman and Hall, 1971.
6. DEJEAN F. AND SCHÜTZENBERGER M. P., On a question of Eggan. *Inform. and Control* **9** (1966), 23–25.
7. EGGAN L. C., Transition graphs and the star-height of regular events. *Michigan Mathematical J.* **10** (1963), 385–397.

8. EILENBERG S., *Automata, Languages and Machines* vol. A, Academic Press, 1974.
9. HASHIGUCHI K., The star height of reset-free events and strictly locally testable events. *Inform. and Control* **40** (1979), 267–284.
10. K. HASHIGUCHI, Algorithms for determining relative star height and star height. *Inform. and Computation* **78** (1988), 124–169.
11. LOMBARDY S., On the construction of reversible automata for reversible languages, submitted.
12. LOMBARDY S. AND SAKAROVITCH J., On the star height of rational languages: a new version for two old results, *Proc. 3rd Int. Col. on Words, Languages and Combinatorics*, (M. Ito, Ed.) World Scientific, to appear. Available at the URL: `www.enst.fr/~jsaka`.
13. HÉAM P.-C., Some topological properties of rational sets. *J. of Automata, Lang. and Comb.*, to appear.
14. MCNAUGHTON R., The loop complexity of pure-group events. *Inform. and Control* **11** (1967), 167–176.
15. PIN J.-E., On reversible automata. In *Proc. 1st LATIN Conf., (I. Simon, Ed.), Lecture Notes in Comput. Sci.* **583** (1992), 401–416.
16. SAKAROVITCH J., *Eléments de théorie des automates.* Vuibert, to appear.
17. SALOMAA A., *Jewels of formal language theory.* Computer Science Press, 1981.
18. SILVA P., On free inverse monoid languages *Theoret. Informatics and Appl.* **30** (1996), 349–378.

Weakly Iterated Block Products of Finite Monoids

Howard Straubing[1] and Denis Thérien[2*]

[1] Boston College
straubin@cs.bc.edu
[2] McGill University
denis@cs.mcgill.ca

Abstract. The block product of monoids is a bilateral version of the better known wreath product. Unlike the wreath product, block product is not associative. All decomposition theorems based on iterated block products that have appeared until now have assumed right-to-left bracketing of the operands. We here study what happens when the bracketing is made left-to-right. This parenthesization is in general weaker than the traditional one. We show that weakly iterated block products of semilattices correspond exactly to the well-known variety **DA** of finite monoids: if groups are allowed as factors, the variety **DA*G** is obtained. These decomposition theorems allow new, simpler, proofs of recent results concerning the defining power of generalized first-order logic using two variables only.

1 Introduction

Rhodes and Tilson [3] introduced the bilateral semidirect product of monoids, and the related block product, and used them to develop the notion of the kernel of a homomorphism of monoids. The underlying idea behind such products is quite old; its precursors can be found in the "triple products" of Eilenberg [1], in the work of Schützenberger on the Schützenberger product [4] and on sequential bimachines [5]. In [3] we find the following bilateral version of the Krohn-Rhodes theorem: Every finite monoid M divides an iterated bilateral semidirect product

$$(M_n ** \cdots (M_3 **(M_2 ** M_1)) \cdots),$$

where each M_i is either a semilattice or a simple group that divides M. In particular, if M is aperiodic, then M divides an iterated product of semilattices. While bilateral products are somewhat unwieldy to work with, they result in decompositions with simpler factors than are possible with unilateral products, making them especially suitable for some applications. For example, Straubing [9] applies

* contact author: School of Computer Science, McGill University, 3480 University, Montréal, QC, Canada, H3A 2A7. Research supported by NSERC, FCAR and the Von Humboldt Foundation.

S. Rajsbaum (Ed.): LATIN 2002, LNCS 2286, pp. 91–104, 2002.

the bilateral Krohn-Rhodes theorem to find logical characterizations of classes of regular languages.

In the present paper we consider what happens when we bracket the iterated bilateral semidirect product in the opposite direction. We find that the monoids that divide an iterated bilateral semidirect product

$$(\cdots((M_1 ** M_2) ** M_3) ** \cdots ** M_n)$$

of semilattices are precisely the members of the pseudovariety **DA**, and those that divide an iterated product of groups and semilattices are precisely the members of $\mathbf{DA} * \mathbf{G}$. This constitutes a kind of inside-out Krohn-Rhodes theorem. We also give an application of our decomposition result: A new, transparent proof of theorems of Thérien, Wilke and Straubing [11,8] on the definability of regular languages by generalized first-order sentences with two variables.

2 Algebraic Preliminaries

We suppose that the reader is familiar with the fundamental notions concerning the connections between semigroups and automata: division of monoids, recognition of regular languages by finite monoids, and the definition and basic properties of the syntactic monoid. We refer the reader to Chapter 1 of Pin [2] for an introduction to this material.

2.1 One-Sided and Bilateral Products of Monoids

Let M and N be finite monoids. Following a convention introduced by Eilenberg, we will write the product in M additively. Thus we write the identity of M as 0, and the kth power of $m \in M$ as $k \cdot m$. This is done to make the notation more readable, and not to suggest that M is commutative. A *left action* of N on M associates to each pair $(n, m) \in N \times M$ an element nm of M, subject to the following laws:

$$n(m + m') = nm + nm'$$

$$(nn')m = n(n'm)$$

$$n0 = 0$$

$$1m = m$$

for all $m, m' \in M$, $n, n' \in N$. Given such a left action we define the *semidirect product* $M * N$ with respect to this left action as the monoid whose underlying set is $M \times N$ with multiplication given by

$$(m, n)(m', n') = (m + nm', nn'),$$

for all $m, m' \in M$, $n, n' \in N$. It is straightforward to verify that this multiplication is associative, and that $(0, 1)$ is the identity for this multiplication; thus

$M * N$ is indeed a monoid. There may be many different left actions of M on N, giving rise to nonisomorphic semidirect products $M * N$.

We define a *right action* of N on M analogously, and define the *reverse semidirect product* $N *_r M$ with respect to this action as the monoid structure on $N \times M$ with multiplication given by

$$(n, m)(n', m') = (nn', mn' + m').$$

Suppose we have both a left and a right action of N on M, and that these two actions satisfy

$$(nm)n' = n(mn'),$$

for all $m \in M$, $n, n' \in N$. We can then define another monoid structure on $M \times N$ with multiplication given by

$$(m, n)(m', n') = (mn' + nm', nn').$$

Once again, it is straightforward to verify that this is an associative multiplication with identity $(0, 1)$. We call the resulting monoid a *bilateral semidirect product* and denote it $M * *N$. Observe that every ordinary and reverse semidirect product is a special instance of the bilateral semidirect product, since we can define one of the two actions to be the identity map on M for all $n \in N$.

We now describe two related products. Once again, let M and N be finite monoids. We return to using the standard multiplicative notation for the product in M. The *wreath product* $M \circ N$ is a monoid structure on $M^N \times N$, with multiplication given by

$$(F, n)(F', n') = (G, nn'),$$

where for all $n'' \in N$,

$$G(n'') = F(n'')F'(n''n).$$

The *block product* $M \square N$ is a monoid structure on $M^{N \times N} \times N$, with multiplication given by

$$(F, n)(F', n') = (G, nn'),$$

where for all $(n_1, n_2) \in N \times N$,

$$G(n_1, n_2) = F(n_1, n'n_2)F'(n_1 n, n_2).$$

The following proposition summarizes the essential facts about these products. The proofs (which are all quite simple) can be found in Eilenberg [1] or in Rhodes and Tilson [3]. In what follows, if M is a finite monoid, then M^r denotes the reversed monoid.

Proposition 1. *Let M, N be finite monoids.*
(a) If M and N are groups, then every semidirect product $M * N$ is a group.
(b) If $M * *N$ is a bilateral semidirect product with N a group, then $M * *N$ is isomorphic to a semidirect product $M * N$.

(c) Given a bilateral semidirect product $M**N$ there exist a left action of N on M and a right action of N on the resulting semidirect product $M*N$ such that $M**N$ is isomorphic to a submonoid of $N *_r (M*N)$.
(d) For every bilateral semidirect product $M**N$ there is a bilateral semidirect product M^r**N^r such that $(M**N)^r$ is isomorphic to M^r**N^r.
(e) $M \circ N$ is isomorphic to a semidirect product $M' * N$, where M' is the direct product of $|N|$ copies of M.
(f) Every semidirect product $M*N$ divides $M \circ N$.
(g) $M\Box N$ is isomorphic to a bilateral semidirect product $M''**N$, where M'' denotes the direct product of $|N|^2$ copies of M.
(h) Every bilateral semidirect product $M**N$ divides $M\Box N$.
(i)If $M_1 \prec M_2$, $N_1 \prec N_2$, then $M_1 \circ N_1 \prec M_2 \circ N_2$, and $M_1 \Box N_1 \prec M_2 \Box N_2$.
(j) Let 1 denote the trivial monoid. Then $M \circ 1$, $1 \circ M$, $M\Box 1$, and $1\Box M$ are all isomorphic to M.
(k) $M \circ N \prec M\Box N$.

2.2 Product Pseudovarieties

A *pseudovariety* of finite monoids is a collection of finite monoids that contains all divisors of its members, and the direct products of any two of its members. We use standard names for certain important pseudovarieties: $\mathbf{J}_1$ denotes the pseudovariety of finite semilattices (*i.e.,* idempotent and commutative monoids), $\mathbf{R}$ the pseudovariety of $\mathcal{R}$-trivial monoids, $\mathbf{G}$ the pseudovariety of finite groups, and $\mathbf{1}$ the pseudovariety whose only member is the trivial monoid. Another important pseudovariety, denoted $\mathbf{DA}$, will be discussed at length below.

If $\mathbf{V}$ and $\mathbf{W}$ are pseudovarieties, then $\mathbf{V} * \mathbf{W}$ denotes the family of finite monoids that divide a semidirect product $M * N$, where $M \in \mathbf{V}$ and $N \in \mathbf{W}$. We define $\mathbf{V} *_r \mathbf{W}$ analogously. We denote by $\mathbf{V}\Box\mathbf{W}$ the family of finite monoids that divide a bilateral semidirect product $M**N$, with $M \in \mathbf{V}$ and $N \in \mathbf{W}$. By Proposition 1, $\mathbf{V} * \mathbf{W}$ is also the family of divisors of wreath products $M \circ N$, with $M \in \mathbf{V}$ and $N \in \mathbf{W}$, and similarly $\mathbf{V}\Box\mathbf{W}$ is the family of divisors of block products $M\Box N$. $\mathbf{V} * \mathbf{W}$, $\mathbf{V} *_r \mathbf{W}$, and $\mathbf{V}\Box\mathbf{W}$ are themselves all pseudovarieties.

If $\mathbf{V}$ is a pseudovariety, we set $\mathbf{V}^r = \{M^r : M \in \mathbf{V}\}$. This, too, is a pseudovariety.

The following proposition summarizes properties of these product varieties. Most of these are direct consequences of Proposition 1. See [1] or [3] for the proofs.

Proposition 2. *Let* $\mathbf{U}$,$\mathbf{V}$,$\mathbf{W}$ *be pseudovarieties of finite monoids.*
(a) If $\mathbf{V}, \mathbf{W} \subseteq \mathbf{G}$, then $\mathbf{V} * \mathbf{W} \subseteq \mathbf{G}$.
(b) If $\mathbf{W} \subseteq \mathbf{G}$, then $\mathbf{V}\Box\mathbf{W} = \mathbf{V} * \mathbf{W}$.
(c) $\mathbf{V}\Box\mathbf{W} \subseteq \mathbf{W} *_r (\mathbf{V} * \mathbf{W})$.
(d) $\mathbf{V}^r\Box\mathbf{W}^r = (\mathbf{V}\Box\mathbf{W})^r$.
(e)$(\mathbf{U} * \mathbf{V}) * \mathbf{W} = \mathbf{U} * (\mathbf{V} * \mathbf{W})$.

Let $\mathcal{C}$ be a collection of finite monoids, and let $\mathbf{V}$ be the smallest pseudovariety containing $\mathcal{C}$. We denote by $pc(\mathcal{C})$ the union of the pseudovarieties $\mathbf{V}$, $\mathbf{V} * \mathbf{V}$, $\mathbf{V} * \mathbf{V} * \mathbf{V}$, etc. Note that we are implicitly using part (e) of the above proposition in this definition. $pc(\mathcal{C})$ is the smallest pseudovariety containing $\mathcal{C}$ that is closed under $*$; *pc* stands for "product closure".

For pseudovarieties $\mathbf{U}, \mathbf{V}, \mathbf{W}$, we have

$$(\mathbf{U}\Box\mathbf{V})\Box\mathbf{W} \subseteq \mathbf{U}\Box(\mathbf{V}\Box\mathbf{W}),$$

however the converse inclusion is in general false. Because of this non-associativity, we must be careful about how we define iterated product varieties for the block product. We denote by $wb^k(\mathbf{V})$ the pseudovariety

$$(\cdots((\mathbf{V}\Box\mathbf{V})\Box\mathbf{V})\Box\cdots\Box\mathbf{V}),$$

with k occurrences of $\mathbf{V}$. We denote by $wbpc(\mathbf{V})$ the union of the $wb^k(\mathbf{V})$ over all $k \geq 0$. We also denote (not without some abuse of notation) by $wbpc(\mathbf{V}\cup\mathbf{W})$ the union of all the pseudovarieties

$$[\mathbf{V}_1,\ldots,\mathbf{V}_k] = (\cdots((\mathbf{V_1}\Box\mathbf{V_2})\Box\mathbf{V_3})\Box\cdots\Box\mathbf{V_k}),$$

where for each i, $\mathbf{V}_i = \mathbf{V}$ or $\mathbf{V}_i = \mathbf{W}$. It may not be obvious that this is a pseudovariety. To see that it is, observe that parts (i) and (j) of Proposition 1 can be used to show that the pseudovarieties $[\mathbf{V}_1,\ldots,\mathbf{V}_k]$ and $[\mathbf{W}_1\ldots,\mathbf{W}_m]$ are both contained in the pseudovariety $[\mathbf{V}_1,\ldots,\mathbf{V}_k,\mathbf{W}_1,\ldots,\mathbf{W}_m]$, and thus $wbpc(\mathbf{V}\cup\mathbf{W})$ contains the direct product of any two of its members. *wbpc* stands for "weak block product closure".

2.3 The Pseudovariety DA

If M is a finite monoid and $m \in M$, then we denote by m^ω the unique power of m that is idempotent. **DA** consists of all finite monoids M such that for all $x, y, z \in M$,

$$(xyz)^\omega y(xyz)^\omega = (xyz)^\omega.$$

It follows directly from this characterization that **DA** is a pseudovariety. If we take $x = z = 1$, we find that every monoid in **DA** satisfies the identity

$$y^\omega y = y^\omega,$$

and thus is aperiodic (i.e., contains no nontrivial groups). **DA** was introduced by Schützenberger [6].

We give an equivalent characterization of **DA** in terms of congruences on finitely generated free monoids. This is due to Thérien and Wilke [11]. Let Σ be a finite alphabet, with $|\Sigma| = n$. We define a family of equivalence relations $\sim_{n,k}$, $k \geq 0$, on Σ^* as follows. If either $n = 0$ or $k = 0$, $\sim_{n,k}$ is the trivial congruence that identifies all words of Σ^*. Now suppose n and k are both positive. We will define $\sim_{n,k}$ inductively: If $w \in \Sigma^*$ then we denote by $c(w)$ the set of letters appearing in w. If $\sigma \in c(w)$, then w has a unique factorization

$$w = w_0 \sigma w_1,$$

where $\sigma \notin c(w_0)$. We define

$$l_\sigma^{n,k}(w) = ([w_0]_{n-1,k}, [w_1]_{n,k-1}),$$

where $[v]_{m,r}$ denotes the $\sim_{m,r}$-equivalence class of v. Observe that the definition above makes sense, because w_0 is a word over the $(n-1)$-letter alphabet $\Sigma - \{\sigma\}$. Similarly, there is a unique factorization

$$w = w_0' \sigma w_1',$$

such that $\sigma \notin c(w_1')$. We set

$$r_\sigma^{n,k}(w) = ([w_0']_{n-1,k}, [w_1']_{n,k-1}).$$

We now set, for $v, w \in \Sigma^*$, $v \sim_{n,k} w$ if and only if $c(v) = c(w)$, and for all $\sigma \in c(v)$,

$$l_\sigma^{n,k}(v) = l_\sigma^{n,k}(w), r_\sigma^{n,k}(v) = r_\sigma^{n,k}(w).$$

It is easy to verify that every $\sim_{n,k}$ is a congruence of finite index on Σ^*. Thérien and Wilke show that the quotient monoids $\Sigma^*/\sim_{n,k}$ are all in **DA**, and that every monoid in **DA** is a homomorphic image of some $\Sigma^*/\sim_{n,k}$.

3 The Decomposition Theorem

Our main result is:

Theorem 1. *Let* **H** *be any pseudovariety of finite groups. Then*

$$wbpc(\mathbf{J}_1 \cup \mathbf{H}) = \mathbf{DA} * pc(\mathbf{H}).$$

In particular, taking $\mathbf{H} = \mathbf{1}$, we have

$$wbpc(\mathbf{J}_1) = \mathbf{DA}.$$

We give the proof in the next two subsections.

3.1 Proof that $wbpc(\mathbf{J}_1 \cup \mathbf{H}) \subseteq \mathbf{DA} * pc(\mathbf{H})$

We first show:

Lemma 1. $\mathbf{DA} \square \mathbf{J}_1 \subseteq \mathbf{DA}$.

Proof. By Proposition 2,

$$\mathbf{DA} \square \mathbf{J}_1 \subseteq \mathbf{J}_1 *_r (\mathbf{DA} * \mathbf{J}_1).$$

Since **DA** is closed under reversal, it suffices, again by applying Proposition 2 to prove the lemma with $\square$ replaced by $*$. We show this using the defining identity for **DA**:

$$(xyz)^\omega y(xyz)^\omega = (xyz)^\omega.$$

That is, we will show that if M is a monoid satisfying this identity, and $N \in \mathbf{J}_1$, then any semidirect product $M*N$ satisfies the same identity. Let $x, y, z \in M*N$, with

$$x = (x_1, x_2), y = (y_1, y_2), z = (z_1, z_2).$$

Since M and N are both aperiodic, there exists $k > 0$ such that

$$u^\omega = u^k = u^{k+1}$$

for all $u \in M * *N$, and (using additive notation in M)

$$k \cdot v = (k+1) \cdot v$$

for all $v \in M$. Since $N \in \mathbf{J}_1$ we have

$$(x_2y_2z_2)^2 = x_2y_2z_2x_2 = x_2y_2z_2y_2 = x_2y_2z_2z_2 = x_2y_2z_2,$$

so that

$$\begin{aligned}(xyz)^\omega &= (xyz)^{k+1} \\ &= (x_1 + x_2y_1 + x_2y_2z_1 + x_2y_2z_2 \cdot k \cdot (x_1 + y_1 + z_1), x_2y_2z_2)\end{aligned}$$

Thus

$$\begin{aligned}(xyz)^\omega y(xyz)^\omega = (&x_1 + x_2y_1 + x_2y_2z_1 + x_2y_2z_2 \cdot (k \cdot (x_1 + y_1 + z_1) + \\ &y_1 + (k+1) \cdot (x_1 + y_1 + z_1)), x_2y_2z_2)\end{aligned}$$

Since M satisfies the identity for **DA**,

$$k \cdot (x_1 + y_1 + z_1) + y_1 + (k+1) \cdot (x_1 + y_1 + z_1) = k \cdot (x_1 + y_1 + z_1),$$

from which it follows that

$$(xyz)^\omega y(xyz)^\omega = (xyz)^\omega.$$

To complete the proof that $wbpc(\mathbf{J}_1 \cup \mathbf{H}) \subseteq \mathbf{DA} * pc(\mathbf{H})$, it is enough to show

$$(\mathbf{DA} * pc(\mathbf{H}))\square\mathbf{H} \subseteq \mathbf{DA} * pc(\mathbf{H}),$$

and

$$(\mathbf{DA} * pc(\mathbf{H}))\square\mathbf{J}_1 \subseteq \mathbf{DA} * pc(\mathbf{H}).$$

The first inclusion follows from Proposition 2.

$$\begin{aligned}(\mathbf{DA} * pc(\mathbf{H}))\square\mathbf{H} &= (\mathbf{DA} * pc(\mathbf{H})) * \mathbf{H} \\ &= \mathbf{DA} * (pc(\mathbf{H}) * \mathbf{H}) \\ &= \mathbf{DA} * pc(\mathbf{H}).\end{aligned}$$

For the second, we argue as in the proof of Lemma 1: It is enough to show this inclusion when $\Box$ is replaced by $*$. We now use the following two facts, first proved in Stiffler [7]:

$$\mathbf{R} = pc(\mathbf{J}_1),$$

and

$$\mathbf{H} * \mathbf{J}_1 \subseteq \mathbf{R} * \mathbf{H},$$

for any pseudovariety of groups $\mathbf{H}$. This gives

$$\begin{aligned}(\mathbf{DA} * pc(\mathbf{H})) * \mathbf{J}_1 &= \mathbf{DA} * (pc(\mathbf{H}) * \mathbf{J}_1)\\ &\subseteq \mathbf{DA} * (\mathbf{R} * pc(\mathbf{H}))\\ &= (\mathbf{DA} * pc(\mathbf{J}_1)) * pc(\mathbf{H})\\ &\subseteq \mathbf{DA} * pc(\mathbf{H}),\end{aligned}$$

by Proposition 2 and Lemma 1.

3.2 Proof that $\mathbf{DA} * pc(\mathbf{H}) \subseteq wbpc(\mathbf{J}_1 \cup \mathbf{H})$.

We first note that it is sufficient to show $\mathbf{DA} \subseteq wbpc(\mathbf{J}_1)$. For suppose that this is true. If $M \in \mathbf{DA} * pc(\mathbf{H})$ we have

$$M \prec N \circ G$$

for some $N \in \mathbf{DA}$, $G \in \mathbf{H}$. Our assumption and Proposition 1 then give:

$$\begin{aligned}M &\prec N \circ H_k \circ \cdots \circ H_1\\ &\prec (\cdots((N \Box H_k)\Box H_{k-1})\Box \cdots \Box H_1)\\ &\prec (\cdots((V_r \Box V_{r-1})\Box V_{r-2})\Box \cdots \Box V_1)\Box H_k \Box \cdots \Box H_1),\end{aligned}$$

where $H_1, \ldots, H_k \in \mathbf{H}$, and $V_1, \ldots, V_r \in \mathbf{J}_1$. Thus $M \in wbpc(\mathbf{J}_1 \cup \mathbf{H})$.

To prove $\mathbf{DA} \subseteq wbpc(\mathbf{J}_1)$, we use the generating family of congruences for $\mathbf{DA}$, defined in 2.3. We thus need to show that for every finite alphabet Σ, with $|\Sigma| = n$, and every $k \geq 0$, $\Sigma^*/\sim_{n,k} \in wbpc(\mathbf{J}_1)$. This is trivially true if $n = 0$ or $k = 0$. If n and k are both positive, then we claim

$$\Sigma^*/\sim_{n,k} \in wb^{n+k-1}(\mathbf{J}_1).$$

We prove this by induction on $n + k - 1$. It is enough to show that for each $w \in \Sigma^*$ and each $\sigma \in c(w)$, the languages

$$\begin{aligned}L_1 &= \{v \in \Sigma^* : c(v) = c(w)\},\\ L_2 &= \{v \in \Sigma^* : l_\sigma^{n,k}(v) = l_\sigma^{n,k}(w)\},\\ L_2 &= \{v \in \Sigma^* : r_\sigma^{n,k}(v) = r_\sigma^{n,k}(w)\},\end{aligned}$$

are all recognized by monoids in $wb^{n+k-1}(\mathbf{J}_1)$, since each class of $\sim_{n,k}$ is a boolean combination of these languages.

L_1 is obviously recognized by the monoid whose elements are the subsets of Σ, with union as multiplication—this is in $\mathbf{J}_1$. We now need only show that L_2 is recognized by a monoid in $wb^{n+k-1}(\mathbf{J}_1)$, since the result for L_3 follows from the reversal closure of the block product of pseudovarieties (Proposition 2). We will now show that L_2 is recognized by the monoid

$$M = ((\Sigma - \{\sigma\})^*/\sim_{n-1,k} \times \Sigma^*/\sim_{n,k-1}) \circ U_1,$$

which, coupled with the inductive hypothesis, gives the desired result. To show recognition, we define a map $\phi : \Sigma \to M$ as follows: We set $\phi(\sigma) = (G_\sigma, 0)$, and $\phi(\tau) = (F_\tau, 1)$ for $\tau \neq \sigma$, where

$$F_\tau(1) = ([\tau]_{n-1,k}, 1),$$

$$G_\sigma(1) = (1, 1),$$

$$F_\tau(0) = (1, [\tau]_{n,k-1}),$$

$$G_\sigma(0) = (1, [\sigma]_{n,k-1}).$$

The map ϕ extends to a unique homomophism from Σ^* into M. Let $w \in \Sigma^*$, and let $\phi_1(w)$ and $\phi_2(w)$ denote, respectively, the left and right co-ordinates of $\phi(w)$. It follows readily that if $w \in \Sigma^*$ and $\sigma \notin c(w)$, then $\phi_2(w) = 1$, and $(\phi_1(w))(1) = ([w]_{n-1,k}, 1)$. Otherwise, $\phi_2(w) = 0$, and

$$(\phi_1(w))(1) = ([w_0]_{n-1,k}, [w_1]_{n,k-1}),$$

where $w = w_0\sigma w_1$ is the unique factorization with $\sigma \notin c(w_0)$.Thus $L_2 = \phi^{-1}(X)$, where

$$X = \{(G, 0) : G(1) = ([w_0]_{n-1,k}, [w_1]_{n,k-1})\}.$$

Thus M recognizes L_2, as required.

4 An Application to Logic

4.1 Regular Languages and Generalized First-Order Logic

Regular languages can be defined by sentences of first-order logic, using the following scheme: Variables in a sentence denote positions in a word over the underlying input alphabet Σ. There are two kinds of atomic formulas: $x < y$, which is interpreted to mean that position x is to the left of position y, and $Q_\sigma x$, which means that the letter in position x is σ. A sentence such as

$$\exists x(\forall y(\neg y < x) \wedge Q_\sigma x)$$

is satisfied by all words having at least one letter, and whose first letter is σ. Thus the sentence *defines* the regular language $\sigma\Sigma^*$, consisting of all the words that satisfy the sentence. We will allow our sentences to contain, in addition to the usual existential and universal quantifiers, *modular* quantifiers $\exists^{r \bmod n}$, where $0 \leq r < n$. A formula of the form

$$\exists^{r \bmod n} x\phi(x)$$

is interpreted to mean that the number of positions x for which $\phi(x)$ holds is congruent to r modulo n.

See Straubing [9] for an extensive treatment of this method of defining formal languages with formulas of logic. In practice, it has been found that classes of languages defined by "natural"-looking boolean-closed classes of sentences can usually be characterized in terms of the syntactic morphisms and syntactic monoids of the languages.

In this section we apply Theorem 1 to give a new proof of the following result, due to Thérien and Wilke [11] for the case $n = 1$ (no modular quantifiers) and to Straubing and Thérien [8] for the general case. Let $\mathbf{G}_{com}^{(n)}$ denote the pseudovariety of finite abelian groups whose exponents divide n.

Theorem 2. *Let Σ be a finite alphabet, and let $n \geq 1$. A language $L \subseteq \Sigma^*$ is definable by a sentence that uses only two variables, ordinary quantifiers and modular quantifiers of modulus n if and only if $M(L) \in \mathbf{DA} * pc(\mathbf{G}_{com}^{(n)})$.*

In the case $n = 1$, the modular quantifier is superfluous, so the theorem says that L is definable by a two-variable sentence if and only if $M(L) \in \mathbf{DA}$.

4.2 Languages Recognized by Bilateral Semidirect Products

Let Σ be a finite alphabet, M a finite monoid, and $\alpha : \Sigma^* \to M$ a homomorphism. We set $\Gamma = M \times \Sigma \times M$, and define a length-preserving map $\tau_\alpha : \Sigma^* \to \Gamma^*$ by

$$\tau_\alpha(\sigma_1 \cdots \sigma_n) = \gamma_1 \cdots \gamma_n,$$

where $\sigma_i \in \Sigma$ and

$$\gamma_i = (\alpha(\sigma_1 \cdots \sigma_{i-1}), \sigma_i, \alpha(\sigma_{i+1} \cdots \sigma_n)) \in \Gamma,$$

for $i = 1, \ldots, n$. (In the above equation, we take the left component of γ_1 and the right component of γ_n to be the identity of M.) The following facts about homomorphisms from finitely-generated free monoids into bilateral semidirect products are from Thérien [10], where they are stated in terms of congruences on finitely generated free monoids.

Lemma 2. *(a) Let Σ, M, Γ be as above. Let N be a finite monoid, and let $\beta : \Sigma^* \to N ** M$ be a homomorphism into a bilateral semidirect product. Let $(n, m) \in N ** M$. Then there is a homomorphism $\alpha : \Sigma^* \to M$ and a language $L \subseteq \Gamma^*$ recognized by N such that*

$$\beta^{-1}(n, m) = \alpha^{-1}(m) \cap \tau_\alpha^{-1}(L).$$

(b) Let $\alpha, \Sigma, M, \Gamma$ be as above. Let $L \subseteq \Gamma^*$ be a regular language recognized by a finite monoid N. Then $\tau_\alpha^{-1}(L)$ is recognized by $N \Box M$.

4.3 Construction of Two-Variable Sentences

We now prove Theorem 2 in detail. We begin by showing that every language whose syntactic monoid is in $\mathbf{DA} * pc(\mathbf{G}_{com}^{(n)})$ is definable by a two-variable sentence of the required kind. Let $L \subseteq \Sigma^*$, with $M(L) \in \mathbf{DA} * pc(\mathbf{G}_{com}^{(n)})$. By Theorem 1, there exists a homomorphism

$$\phi : \Sigma^* \to N = (\cdots((M_1 \Box M_2) \Box M_3) \Box \cdots \Box M_k,$$

where each M_i is either a semilattice or an abelian group with exponent dividing n, such that $L = \phi^{-1}(X)$ for some $X \subseteq N$. We construct a defining sentence for L by induction on k. First, suppose that M_k is a semilattice. By Lemma 2, L is a boolean combination of languages of the form $\alpha^{-1}(m)$ and $\tau_\alpha^{-1}(K)$, where $\alpha : \Sigma^* \to M_k$ is a homomorphism, $m \in M_k$, and $K \subseteq (M_k \times \Sigma \times M_k)^*$ is recognized by $(\cdots((M_1 \Box M_2) \Box M_3) \Box \cdots \Box M_{k-1}$. We need to show that both $\alpha^{-1}(m)$ and $\tau_\alpha^{-1}(K)$ are two-variable definable.

Observe that the value of $\alpha(w)$ depends only on $c(w)$. Thus if $\Sigma' \subseteq \Sigma$ we can write $\alpha(\Sigma')$ to denote the image under α of any w such that $c(w) = \Sigma'$. Consequently $\alpha^{-1}(m)$ is defined by a boolean combination of *one*-variable sentences of the form $\exists x Q_\sigma x$.

By the inductive hypothesis, K is defined by a two-variable sentence over $\Gamma = M_k \times \Sigma \times M_k$. We obtain a sentence for $\tau_\alpha^{-1}(K)$ by replacing each subformula $Q_{(m_1,\sigma,m_2)} x$ of the sentence by a disjunction of formulas that say 'the letter in position x is σ, the set of letters in positions to the left of x is Σ_1, and the set of letters in positions to the right of x is Σ_2', where the disjunction is over all subsets Σ_1, Σ_2 of Σ such that $\alpha(\Sigma_1) = m_1$ and $\alpha(\Sigma_2) = m_2$. Such a formula is

$$\begin{aligned} Q_\sigma x \wedge \forall y (y < x \to &\bigvee_{\tau \in \Sigma_1} Q_\tau y) \\ \wedge &\bigwedge_{\tau \in \Sigma_1} \exists y (y < x \wedge Q_\tau y) \\ \wedge\ \forall y (y > x \to &\bigvee_{\tau \in \Sigma_2} Q_\tau y) \\ \wedge &\bigwedge_{\tau \in \Sigma_2} \exists y (y > x \wedge Q_\tau y). \end{aligned}$$

The resulting defining sentence for $\tau_\alpha^{-1}(K)$ still uses only two variables, since the new variable y that we introduced is used only within the scopes of the new quantifiers; thus we are able to re-use a variable of the same name already occurring in the defining sentence for K.

We proceed very similarly in the case where $M_k \in \mathbf{G}_{com}^{(n)}$. In this case, the value of a homomorphism $\alpha : \Sigma^* \to M_k$ on a word w is determined by the number of times, modulo n, that each letter $\tau \in \Sigma$ appears in w; that is, by a $|\Sigma|$-tuple $(n_\sigma)_{\sigma \in \Sigma}$ of elements of $\mathbf{Z}_n$. Thus $\alpha^{-1}(m)$ is defined by a boolean combination of one-variable sentences of the form $\exists^{r \bmod n} x Q_\sigma x$. We obtain a two-variable sentence for $\tau_\alpha^{-1}(K)$ from a defining sentence for K upon replacing each $Q_{(m_1,\sigma,m_2)}$ by a disjunction of sentences of the form

$$Q_\sigma x \wedge \bigwedge_{\tau\in\Sigma} \exists^{n_\tau \bmod n}(y < x \wedge Q_\tau y)$$
$$\wedge \bigwedge_{\tau\in\Sigma} \exists^{n'_\tau \bmod n}(y < x \wedge Q_\tau y)$$

for some $|\Sigma|$-tuples $(n_\sigma)_{\sigma\in\Sigma}$ and $(n'_\sigma)_{\sigma\in\Sigma}$.

4.4 The Syntactic Monoid of Two-Variable Definable Languages

We complete the proof of Theorem 2 by showing that if $L \subseteq \Sigma^*$ is definable by a two-variable sentence θ all of whose modular quantifiers are of modulus n, then $M(L) \in \mathbf{DA} * pc(\mathbf{G}_{com}^{(n)})$. We first describe a kind of normal form for such sentences. Let $\mathcal{Q}$ be an innermost quantifier symbol in θ; that is, a quantifier symbol such that no other quantifier appears within its scope. We may assume that $\mathcal{Q}$ is either an ordinary existential quantifer or a modular quantifier, since the universal quantifier can be defined in terms of the existential quantifier. Let us suppose that the quantifier $\mathcal{Q}$ quantifies the variable x. We can write the subformula appearing after the quantifier as the disjunction of mutually exclusive formulas of the form

$$x\mathcal{R}y \wedge Q_\sigma x \wedge Q_\tau y,$$

where $\mathcal{R}$ is one of $<$, $>$, or $=$. We can then rewrite the entire quantified subformula as a boolean combination of formulas of the form

$$\mathcal{Q}x(x\mathcal{R}y \wedge Q_\sigma x \wedge Q_\tau y).$$

This is clear if $\mathcal{Q}$ is existential, since the existential quantifier commutes with disjunction. If $\mathcal{Q}$ is modular, then we note that

$$\exists^{r \bmod n} x(\theta_1 \vee \cdots \vee \theta_s),$$

where the θ_i are mutually exclusive, is equivalent to

$$\bigvee \bigwedge_{j=1}^{s} \exists^{r_i \bmod n} x\theta_i,$$

where the disjunction is over all s-tuples $(r_1, \ldots, r_s)$ of elements of $\mathbf{Z}_n$ whose sum is r. Now note that we can move the atomic formula $Q_\tau y$ from within the scope of the quantifer; that is,

$$\mathcal{Q}x(x\mathcal{R}y \wedge Q_\sigma x \wedge Q_\tau y)$$

is equivalent to

$$Q_\tau y \wedge \mathcal{Q}x(x\mathcal{R}y \wedge Q_\sigma x),$$

unless $\mathcal{Q}$ is the modular quantifier $\exists^{0 \bmod n}$, in which case the formula is equivalent to

$$\neg Q_\tau y \vee \neg\exists x(x\mathcal{R}y \wedge Q_\sigma x)$$
$$\vee (Q_\tau y \wedge \exists^{0 \bmod n} x(x\mathcal{R}y \wedge Q_\sigma x)).$$

Finally, we note that $\mathcal{Q}x(x = y \wedge Q_\sigma x)$ is equivalent to $Q_\sigma y$ if $\mathcal{Q}$ is either an existential quantifier or the modular quantifier $\exists^{1 \bmod n}$, and is never satisfied otherwise. We may thus assume that in θ, every innermost quantified subformula has the form

$$\mathcal{Q}x(x < y \wedge Q_\sigma x)$$

or

$$\mathcal{Q}x(x > y \wedge Q_\sigma x).$$

This is our normal form.

We prove $M(L) \in \mathbf{DA} * pc(\mathbf{G}_{com}^{(n)})$ by induction on the depth of nesting of the quantifiers and by the number of innermost quantifiers at this depth. At each step we will either decrease the number of quantifiers at the maximal depth, or decrease the depth. The base case is depth 0, in which the sentence simply says TRUE or FALSE, and $M(L)$ is the trivial monoid. Let us accordingly pick a quantifier at maximal depth. If the quantifier is an existential quantifier, then we set M to be the quotient of Σ^* by the congruence that identifies words v and w if and only if $c(v) = c(w)$. If the quantifier is modular, then M is the quotient of Σ^* by the congruence that identifies v and w if and only if, for all $\sigma \in \Sigma$,

$$|v|_\sigma \equiv |w|_\sigma \pmod{n},$$

where $|v|_\sigma$ denotes the number of occurrences of σ in v. In either case we let $\alpha : \Sigma^* \to M$ denote the projection onto the quotient by the congruence. In the former case, elements of M can be identified with subsets of Σ, and $M \in \mathbf{J}_1$. In the latter case, elements of M are $|\Sigma|$-tuples of elements of $\mathbf{Z}_n$, and $M \in \mathbf{G}_{com}^{(n)}$.

We now transform θ into a sentence over $\Gamma = M \times \Sigma \times M$. If the selected innermost quantified formula is

$$\exists x(x < y \wedge Q_\sigma x),$$

we replace it by

$$\bigvee Q_{(\Sigma_1,\tau,\Sigma_2)}x,$$

where the disjunction is over all $\Sigma_1 \subset \Sigma$ such that $\sigma \in \Sigma_1$, and all $\tau \in \Sigma$, $\Sigma_2 \subseteq \Sigma$. In case the selected quantified subformula is

$$\exists^{r \bmod n} x(x < y \wedge Q_\sigma x),$$

we replace it by the disjunction of all

$$Q_{((n_\tau)_{\tau\in\Sigma},\tau,(n'_\tau)_{\tau\in\Sigma})}x,$$

such that $n_\sigma = r$. We replace all other subformulas of θ of the form $Q_\sigma x$ by the disjunction of all $Q_{(\Sigma_1,\sigma,\Sigma_2)}x$, where $\Sigma_1, \Sigma_2 \subseteq \Sigma$. We proceed analogously if the quantified subformula contains $x > y$ instead of $x < y$.

The result is a sentence η over Γ that has either smaller depth than θ, or fewer quantifiers at maximal depth. By the inductive hypothesis, the language $K \subseteq \Gamma^*$ defined by η has its syntactic monoid in $\mathbf{DA}*pc(\mathbf{G}_{com}^{(n)})$. We have $L = \tau_\alpha^{-1}(K)$, so $M(L) \prec M(K)\Box M$, by Lemma 2. Thus by Theorem 1, $M(L) \in \mathbf{DA}*pc(\mathbf{G}_{com}^{(n)})$.

4.5 Concluding Remarks

If we take the union over all moduli $n > 0$, we obtain Theorem 2 in the form originally stated in [8]: A language is definable by a two-variable sentence if and only if it belongs to $\mathbf{DA} * \mathbf{G}_{sol}$, where $\mathbf{G}_{sol}$ is the pseudovariety of finite solvable groups. In [8] we discuss the open decision problem for $\mathbf{DA} * \mathbf{G}_{sol}$.

It is interesting to compare our proof of Theorem 2 with the logical applications of the bilateral Krohn-Rhodes theorem in Straubing [9]. In the latter, we build up our logical formulas by quantifying over existing formulas. In the present paper, however, we construct formulas by replacing atomic formulas with quantified formulas of depth 1, and that's why we need the inside-out Krohn-Rhodes theorem!

References

1. S. Eilenberg, *Automata, Languages and Machines,* vol. B, Academic Press, New York, 1976.
2. J. E. Pin, *Varieties of Formal Languages,* Plenum, London, 1986.
3. J. Rhodes and B. Tilson, "The Kernel of Monoid Morphisms", *J. Pure and Applied Algebra* **62** (1989) 227–268.
4. M. P. Schützenberger, "On finite monoids having only trivial subgroups", *Information and Control* **8** (1965) 190-194.
5. M. P. Schützenberger, "A remark on finite transducers", *Information and Control* **4** (1961), 185-196.
6. M. P. Schützenberger, "Sur le Produit de Concatenation Non-ambigu", *Semigroup Forum* **13** (1976), 47-76.
7. P. Stiffler, "Extensions of the Fundamental Theorem of Finite Semigroups", *Advances in Mathematics,* **11** 159-209 (1973).
8. H. Straubing and D. Thérien, "Regular languages defined by generalized first-order formulas with a bounded number of bound variables", *Proc. 18th Symposium on Theoretical Aspects of Computer Science* 551-562 (2001).
9. H. Straubing, *Finite Automata, Formal Logic and Circuit Complexity,* Birkhäuser, Boston, 1994.
10. D. Thérien, "Two-sided wreath products of categories", *J. Pure and Applied Algebra* **74** (1991) 307-315.
11. D. Thérien and T. Wilke, "Over Words, Two Variables are as Powerful as One Quantifier Alternation," *Proc. 30th ACM Symposium on the Theory of Computing* 256-263 (1998).

The Hidden Number Problem in Extension Fields and Its Applications

María Isabel González Vasco[1], Mats Näslund[2], and Igor E. Shparlinski[3]

[1] Dept. of Mathematics, University of Oviedo
Oviedo 33007, Spain
mvasco@orion.ciencias.uniovi.es
[2] Ericsson Research
SE-16480 Stockholm, Sweden
mats.naslund@era-t.ericsson.se
[3] Dept. of Computing, Macquarie University
Sydney, NSW 2109, Australia
igor@ics.mq.edu.au

Abstract. We present polynomial time algorithms for certain generalizations of the *hidden number problem* which has played an important role in gaining understanding of the security of commonly suggested one way functions.

Namely, we consider an analogue of this problem for a certain class of polynomials over an extension of a finite field; recovering a hidden polynomial given the values of its trace at randomly selected points. Also, we give an algorithm for a variant of the problem in free finite dimensional modules. This result can be helpful for studying security of analogues of the RSA and Diffie–Hellman cryptosystems over such modules.

The *hidden number problem* is also related to the so called *black-box field* model of computation. We show that simplified versions of the above recovery problems can be used to derive positive results on the computational power of this model.

1 Introduction

Modern cryptography is based on the (presumed) difficulty of solving certain problems, mainly algebraic ones. The computational hardness of the problem should hopefully imply the security of a cryptographic primitive such as a public-key cryptosystem, a pseudo random number generator, etc. However, the relation between "security" and "hardness" is sometimes difficult to sort out.

In particular, the fact that a problem is hard, often means that it is, on average, hard to find an *exact* solution. If, for example, finding the solution corresponds to finding a secret key, it is desirable to be able to measure the hardness of computing *partial* or *approximate* solutions. For instance, let p be a prime number and $\mathbb{F}_p$ denote a field of p elements which we assume to consist of the elements $\{0, 1, \ldots, p-1\}$. Then, although the discrete logarithm problem in $\mathbb{F}_p$ is considered hard, it is well-known that at least one bit of the logarithm

S. Rajsbaum (Ed.): LATIN 2002, LNCS 2286, pp. 105–117, 2002.

is completely trivial to compute. Therefore, one would like to understand how finding approximate or partial solutions is related to the hardness of the "exact" version of the problem.

Here we study some variants of the *hidden number problem* introduced in 1996 by Boneh and Venkatesan [4,5]. This problem can be stated as follows: *recover a number* $\alpha \in \mathbb{F}_p$ *such that for polynomially many (in terms of* $\log p$*) known random* $\tau \in \mathbb{F}_p$ *certain approximations to the values of* $\alpha\tau$ *are known.*

In [4,5] Boneh and Venkatesan proposed a polynomial time algorithm for this problem when the absolute approximating error is at most $p \exp\left(-c \log^{1/2} p\right)$ with some absolute constant $c > 0$. This result has found a large number of applications to studying the bit security of Diffie-Hellman, Shamir and several other cryptosystems [18,19], as well as rigorous results on attacking (following the heuristic arguments of [21,31]) the DSA and DSA-like signature schemes [7,32,33]. In fact it has turned out that for the above applications the condition that τ is selected uniformly at random from $\mathbb{F}_p$ is too restrictive. Accordingly, various modifications of the original *hidden number problem* have been considered in [7,18,19,32,33,36,37,38].

In particular, in [36] an algorithm has been given for a similar problem: *let* $f(X) \in \mathbb{F}_p[X]$ *be a polynomial over the finite field* $\mathbb{F}_p$ *of* p *elements. Assume* f *belongs to a certain large class of polynomials, including the polynomials of small degree. Knowing approximations to* $f(t)$ *at polynomially many (in terms of* $\log p$*) random points* $t \in \mathbb{F}_p$*, recover* f.

For another variant of polynomial interpolation problem and several relevant references see [1,14].

The mentioned algorithm is efficient only for sufficiently large p. In Section 2.1 we consider a different situation when $f(X)$ is a polynomial over a field $\mathbb{F}_{q^m}$ of q^m elements and the exact values of the trace $\mathrm{Tr}_{\mathbb{F}_{q^m}/\mathbb{F}_q}(f(t))$ are given at polynomially many random $t \in \mathbb{F}_{q^m}$. Our algorithm works for any underlying field $\mathbb{F}_q$ (for instance, it applies to the case $q = 2$), but requires that exact values of the trace are known.

Using traces of polynomials of bounded degree over finite fields for constructing *hash functions* has recently been studied in [20,30]. In our contribution we consider a more general class of *sparse polynomials*. Moreover, the ground field need not be a prime field; namely, we consider polynomials over extensions $\mathbb{K} = \mathbb{F}_{q^m}$ of degree $m \geq 1$ of the finite field $\mathbb{F} = \mathbb{F}_q$ of $q = p^s$ elements.

Section 2.2 is devoted to the study of a certain variant of the *hidden number problem* in some very general algebraic settings. Given a free finite dimensional module $\mathcal{M}$ over a commutative ring, we give a probabilistic algorithm that recovers a hidden element α making use of an oracle $\mathcal{O}_\alpha$, which chooses at random several linear mappings τ and outputs the elements $\tau(\alpha)$. An interesting example of such a module is the residue ring $\mathcal{M} = \mathbb{F}_q[X]/(g)$, with $g \in \mathbb{F}_q[X]$.

Having established that solving a partial problem is essentially as hard to solve as the "full" problem itself, it of course remains an issue to find out what the hardness of the full problem really is. For the algebraic problems usually considered, complexity theory still has a long way to go here, but different restricted computational models have been proposed, hoping to prove lower (and upper) bounds under such restrictions.

The so called *black-box fields*, see [3], constitute such a model of computation, related to the *hidden number problem*. By disallowing "representation specific" operations, one hopes to be able to deduce complexity bounds for the problem under study. A lower bound could for instance be taken as evidence for the security of some cryptographic primitive, *if* the model is not too computationally restrictive.

In Section 3 we use a variant of the above recovery problems to derive a positive result for the power of the black-box model.

We conclude this note by pointing out several generalizations and special cases which could be of interest for further study.

2 Variants of the Hidden Number Problem

We give polynomial time algorithms for different modifications of the *hidden number problem*. They make use of oracles which, by assumption, always provide correct outputs.

2.1 Trace Interpolation Problem over Finite Fields

Let $\mathbb{F}$ be the finite field $\mathbb{F}_q$ of $q = p^s$ elements and $\mathbb{K} = \mathbb{F}_{q^m}$, for $m \in \mathbb{N}$. As usual, we denote by

$$\mathrm{Tr}_{\mathbb{K}/\mathbb{F}}(z) = z + z^q + \ldots + z^{q^{m-1}} \tag{1}$$

the trace of $z \in \mathbb{K}$ in $\mathbb{F}$.

Our results are based on a certain lower bound for the number of non-zeros of sparse polynomials. Recall that $f(X) \in \mathbb{K}[X]$ is called an *n-sparse polynomial* if it contains at most n monomials, that is

$$f(X) = \sum_{j=1}^{n} A_j X^{\nu_j}, \qquad 0 \le \nu_1 < \ldots < \nu_n. \tag{2}$$

Denote by $R(f)$ the number of non-zeros of $f(X) \in \mathbb{K}[X]$ over $\mathbb{K}^*$, that is, the number of $a \in \mathbb{K}^*$ with $f(a) \neq 0$. The following lower bound is well known.

Lemma 1. *Assume that $f(X) \in \mathbb{K}[X]$ is a non-zero n-sparse polynomial of the form (2) with $n \ge 1$, and of degree $\deg f \le q^m - 2$. Then the bound*

$$R(f) \ge (q^m - 1)/n$$

holds.

Proof. Let g be a primitive root of $\mathbb{K}$. We denote by $R(f)$ the number of integers $x = 0, \ldots, q^m - 2$ with $f(g^x) \neq 0$. Noticing that $g^x = g^{x+k(q^m-1)}$ for all integers x and k, we see that $R(f) = T/n$, where T is the number of $z = 0, \ldots, n(q^m-1)-1$ with $f(g^z) \neq 0$.

Now we show that for every integer $z \in \{0, \ldots, n(q^m-1)-1\}$, amongst n consecutive values $f(g^{z+i})$, $i = 0, \ldots, n-1$, there exists at least one non-zero element. Indeed, assume that

$$f(g^{z+i}) = 0, \qquad i = 0, \ldots, n-1.$$

Then from (2) we see that the system of linear equations

$$\sum_{j=1}^{n} Z_j g^{i\nu_j} = 0, \qquad i = 0, \ldots, n-1$$

has a non-zero solution $Z_j = A_j g^{z\nu_j}$, $j = 1, \ldots, n$. However, the matrix of this system is a Vandermonde matrix corresponding to pairwise distinct elements $g^{z\nu_j}$, $j = 1, \ldots, n$ (because $\deg f \leq q^m - 2$). Therefore $T \geq q^m - 1$ and the result follows. □

A similar bound is also known for multivariate polynomials.

For a vector $\nu = (\nu_1, \ldots, \nu_n)$ such that the following nm numbers

$$\nu_j q^i, \qquad j = 1, \ldots, n, \ i = 0, \ldots, m-1, \tag{3}$$

are pairwise distinct modulo $q^m - 1$, we denote by $\mathcal{P}_n(\nu)$ the class of n-sparse polynomials $f(X) \in \mathbb{K}[X]$ of the form (2) of degree $\deg f \leq q^m - 2$.

For a polynomial $f(X) \in \mathbb{K}[X]$, let $\mathcal{O}_f$ denote an oracle which for every call selects (uniformly at random) $t \in \mathbb{K}^*$ and outputs t and $\mathrm{Tr}_{\mathbb{K}/\mathbb{F}}(f(t))$.

We show how any polynomial $f \in \mathcal{P}_n(\nu)$ can be reconstructed in probabilistic polynomial time given an oracle $\mathcal{O}_f$. Note that the condition $f \in \mathcal{P}_n(\nu)$ is necessary, as it ensures the oracle calls can determine uniquely a polynomial, whereas if, for instance, for some $j > k$ we have $\nu_1 q^j \equiv \nu_2 q^k \pmod{q^m - 1}$ then the two distinct polynomials

$$f_1(X) = \sum_{j=1}^{n} A_j X^{\nu_j} \qquad \text{and} \qquad f_2(X) = \left(A_1 + A_2^{q^{m+k-j}}\right) X^{\nu_1} + \sum_{j=3}^{n} A_j X^{\nu_j}$$

satisfy $\mathrm{Tr}_{\mathbb{K}/\mathbb{F}}(f_1(t)) = \mathrm{Tr}_{\mathbb{K}/\mathbb{F}}(f_2(t))$ for all $t \in \mathbb{K}^*$ and therefore $\mathcal{O}_{f_1} = \mathcal{O}_{f_2}$.

Assume that given a system of N linear equations in N variables over $\mathbb{K}$, we can solve it (either find the unique solution, or identify the number of solutions) in $O(N^\vartheta)$ arithmetic operations in $\mathbb{K}$. For example, we can take $\vartheta = 2.38$, see [10].

Theorem 1. *There exists a probabilistic algorithm which, for any exponent vector $\nu = (\nu_1, \ldots, \nu_n)$ satisfying (3) and any polynomial $f \in \mathcal{P}_n(\nu)$, having access to the oracle $\mathcal{O}_f$, makes the expected number of $O(m^2n^2)$ oracle calls and finds f in in expected number of $O((mn)^{2+\vartheta})$ arithmetic operations in $\mathbb{K}$.*

Proof. Let $N = mn$ and let e_k, $k = 1, \ldots, N$, be the non-negative residues modulo $q^m - 1$ of the integers (3) (ordered in an arbitrary order). For $s = 1, \ldots, N$ we consider the polynomial matrices

$$M_s(X_1, \ldots, X_s) = (X_r^{e_k})_{r,k=1}^{s} .$$

If for some $t_1, \ldots, t_N \in \mathbb{K}^*$, the values $\mathrm{Tr}_{\mathbb{K}/\mathbb{F}}(f(t_1)), \ldots, \mathrm{Tr}_{\mathbb{K}/\mathbb{F}}(f(t_N))$ are known and also

$$\det M_N(t_1, \ldots, t_N) \neq 0, \tag{4}$$

then we obtain a non-singular system of linear equations for the coefficients of f (and their conjugates) which can be solved in polynomial time.

Now we show how to design a sequence of calls of the oracle $\mathcal{O}_f$ so that the resulting sequence $t_1, \ldots, t_N$ satisfies the condition (4).

As a first step we make an oracle call which returns some $t \in \mathbb{K}^*$ and $\mathrm{Tr}_{\mathbb{K}/\mathbb{F}}(f(t))$. We put $t_1 = t$. In particular

$$M_1(t_1) = t_1^{e_1} \neq 0.$$

Let $s \geq 2$. Assume $t_1, \ldots, t_{s-1}$, selected from the oracle calls at the first $s-1$ steps, satisfy

$$\det M_{s-1}(t_1, \ldots, t_{s-1}) \neq 0.$$

As $f \in \mathcal{P}_n(\nu)$, the integers (3) are pairwise distinct, which implies that the polynomial $\det M_s(t_1, \ldots, t_{s-1}, X)$ is nonzero, and thus from Lemma 1 we see that there are at least $(q^m - 1)/s$ values of $t \in \mathbb{K}^*$ with

$$\det M_s(t_1, \ldots, t_{s-1}, t) \neq 0.$$

Thus such a value of t will be returned by the oracle $\mathcal{O}_f$, together with the value of $\mathrm{Tr}_{\mathbb{K}/\mathbb{F}}(f(t))$ after the expected number of s calls.

Note that for each call, the condition (4) can be checked using $O(N^{\vartheta})$ arithmetic operations in $\mathbb{K}$. Therefore, after the expected number of

$$\sum_{s=1}^{N} s = \frac{mn(mn-1)}{2}$$

oracle calls we obtain $t_1, \ldots, t_N \in \mathbb{K}$ satisfying the condition (4) with known values of $\mathrm{Tr}_{\mathbb{K}/\mathbb{F}}(f(t_1)), \ldots, \mathrm{Tr}_{\mathbb{K}/\mathbb{F}}(f(t_N))$. The corresponding system of equations can be solved using $O(N^{\vartheta})$ arithmetic operations in $\mathbb{K}$. The result now follows. □

We remark that, information-theoretically, any algorithm needs at least mn oracle calls.

In the spirit of [18,19], it may be worth investigating the case when the values of t are not uniformly distributed and the oracle outputs are only correct with a certain probability. Then, if a public key scheme allows a representation so that the secret key corresponding to a public key ρ can be expressed as a polynomial $f_\rho(t)$, very interesting statements about the security can be proved (see proof of Theorem 4.1 of [18]).

2.2 Hidden Element Problem in Modules

In the case of linear polynomials $f(X) = \alpha X$ the same problem can be considered in certain general algebraic settings. This is in fact a variant of the *hidden number problem* of [4,5].

Let $\mathcal{R}$ be a commutative ring with identity and let $\mathcal{M}$ be a free module of dimension $n \geq 1$ over $\mathcal{R}$. Let $\mathrm{Hom}_{\mathcal{R}}(\mathcal{M}, \mathcal{R})$ denote the module of $\mathcal{R}$-linear homomorphisms of $\mathcal{M}$, see Section 2 of Chapter 3 of [24].

We consider the following *hidden element problem: let ψ be a probability distribution on $\mathrm{Hom}_{\mathcal{R}}(\mathcal{M}, \mathcal{R})$ and α any element in $\mathcal{M}$. Given an oracle $\mathcal{O}_\alpha$ which for each call selects a random element $\tau \in \mathrm{Hom}_{\mathcal{R}}(\mathcal{M}, \mathcal{R})$ according to the probability distribution ψ and returns $\tau(\alpha)$ and τ, find the element α.*

We show that very simple linear algebra arguments lead to a probabilistic algorithm for this problem. This result could be helpful for designing new cryptographic constructions in modules over rings. It provides valuable information about the security of analogues of the RSA and Diffie–Hellman scheme over such modules.

It can also be useful for constructing new hard core functions, for instance through generalizations of the results in [15,16,17,28,29,30].

Let $\eta_1, \ldots, \eta_n$ be a basis of $\mathcal{M}$ over $\mathcal{R}$, then $\mathrm{Hom}_{\mathcal{R}}(\mathcal{M}, \mathcal{R})$ is a free module of dimension $n \geq 1$ over $\mathcal{R}$. Moreover, it is known that one can select a *dual basis* $\omega_1, \ldots, \omega_n$ of $\mathrm{Hom}_{\mathcal{R}}(\mathcal{M}, \mathcal{R})$ over $\mathcal{R}$, that is, a basis such that $\omega_j(\eta_i) = \delta_{ij}$, where as usual

$$\delta_{ij} = \begin{cases} 1, \text{ if } i = j; \\ 0, \text{ if } i \neq j; \end{cases} \qquad 0 \leq i, j \leq n-1,$$

see Section 6 of Chapter 13 of [24]. We refer to such two bases as a pair of dual bases.

We now denote by $P_\psi(n)$ the probability that for n^2 elements $t_{i,k} \in \mathcal{R}$, $i, k = 1, \ldots, n$, chosen independently at random accordingly to the probability distribution ψ, the system of equations

$$\sum_{i=1}^{n} z_i t_{i,k} = b_k, \qquad 1 \leq k \leq n,$$

has at most one solution for any right hand side $b_1, \ldots, b_n$.

Also, we denote by $M(n)$ the number of arithmetic operations in $\mathcal{R}$ which are required to solve a system of n linear equations in n variables over $\mathcal{R}$ (that is, either to find the unique solution or to identify that it has multiple solution, or it has no solution).

Theorem 2. *There exists an algorithm which, for any $\alpha \in \mathcal{M}$, having access to the oracle $\mathcal{O}_\alpha$ and representing the elements of $\mathcal{M}$ and $\mathrm{Hom}_{\mathcal{R}}(\mathcal{M}, \mathcal{R})$ through a pair of dual bases $\eta_1, \ldots, \eta_n$ and $\omega_1, \ldots, \omega_n$, makes the expected number of $nP_\psi(n)^{-1}$ oracle calls and finds α in expected number of at most $P_\psi(n)^{-1}M(n)$ arithmetic operations in $\mathcal{R}$.*

Proof. The algorithm can be described as follows:

Step 1: Make n calls to the oracle $\mathcal{O}_\alpha$ which returns $\tau_k(\alpha)$, $k = 1, \dots, n$, for n elements $\tau_1, \dots, \tau_n \in \mathrm{Hom}_\mathcal{R}(\mathcal{M}, \mathcal{R})$ selected independently at random with the probability distribution ψ.

Step 2: Try to solve the system of linear equations

$$\sum_{i=1}^{n} a_i t_{i,k} = \tau_k(\alpha), \tag{5}$$

where

$$\tau_k = \sum_{i=1}^{n} t_{i,k}\omega_i, \qquad 1 \le k \le n. \tag{6}$$

Step 3: If Step 2 is unsuccessful then repeat Steps 1 and 2, otherwise output $(a_1, \dots, a_n)$ as the coordinates of α in the basis $\eta_1, \dots, \eta_n$.

Note that (5) is derived from the definition of the dual basis, the equation (6) and the equality

$$\alpha = \sum_{i=1}^{n} a_i \eta_i.$$

The expected number of calls to the oracle $\mathcal{O}_\alpha$ the algorithm makes is $nP_\psi(n)^{-1}$. Obviously Steps 1 and 2 are repeated the expected number of $P_\psi(n)^{-1}$ times and each time Step 2 costs $M(n)$ arithmetic operations in $\mathcal{R}$. □

One of the most interesting special cases is that of polynomial rings over a finite field $\mathbb{F}_q$ of $q = p^s$ elements, p prime. Namely, let $g \in \mathbb{F}_q[X]$ be a polynomial of degree n, and take $\mathcal{R} = \mathbb{F}_q$, $\mathcal{M} = \mathbb{F}_q[X]/(g)$. If g is irreducible then $\mathbb{F}_q[X]/(g) \cong \mathbb{F}_{q^n}$ and the problem can be formulated as a problem of recovering $A \in \mathbb{F}_{q^n}$ from an oracle which gives a prescribed coefficient of $AT \in \mathbb{F}_{q^n}$ for a random element $T \in \mathbb{F}_{q^n}$, which can be reduced to the trace interpolation problem of Section 2.1 with linear polynomials.

For arbitrary g, not necessarily irreducible, we know (see Section 6 of Chapter 3 of [24]) that $\mathrm{Hom}_\mathcal{R}(\mathcal{M}, \mathcal{R})$ is isomorphic to $\mathcal{M}$. Also, given an $\mathbb{F}_q$-basis $1, X, \dots, X^{n-1}$ of $\mathcal{M}$, there is a polynomial time algorithm for finding the corresponding dual basis of $\mathrm{Hom}_\mathcal{R}(\mathcal{M}, \mathcal{R}) \cong \mathcal{M}$. In fact one easily sees that this is a linear algebra problem. In addition, $\mathrm{Hom}_\mathcal{R}(\mathcal{M}, \mathcal{R})$ is isomorphic to the group of functions $\tau_T(A)$, where $T \in \mathbb{F}_q[X]/(g)$, which give the constant coefficient of AT. Indeed, there are exactly q^n such functions and all of them are linear, therefore they give the whole $\mathrm{Hom}_\mathcal{R}(\mathcal{M}, \mathcal{R})$. Certainly any other coefficient can be nominated as well. Thus we obtain a problem of recovering a polynomial $A \in \mathbb{F}_q[X]/(g)$ from an oracle which gives a prescribed coefficient of $AT \in \mathbb{F}_q[X]/(g)$ for a random polynomial $T \in \mathbb{F}_q[X]/(g)$. Note that in this case the algorithm runs in probabilistic polynomial time.

It is also useful to note that when ψ is the uniform distribution,

$$P_\psi(n) = \prod_{i=1}^{n}(1 - q^{-i}) \geq \prod_{i=1}^{\infty}(1 - 2^{-i}) = 0.2888\ldots.$$

is the proportion of non-singular matrices over $\mathbb{F}_q$, see Section 1 of Chapter 8 of [26]. For a more general statement, which applies to a much more general class of distributions, see [22].

In another interesting case, where $\mathcal{R} = \mathbb{Z}/(M)$ is a residue ring modulo an integer M, $P_\psi(n)$ is evaluated in [6].

3 On the Power of the Black-Box Model

A *generic*, or *black-box* algorithm, is a model of computation where algebraic operations on "objects" belonging to an algebraic structure can be done through oracles, but where "representation specific" things, for example, relating "the ith bit" of the binary representation of an object to direct algebraic properties, are not possible. The representation need not even be unique. More precisely we say that a finite field $\mathbb{F}_q$, $q = p^n$, is given by a *black-box model* (c.f. [3]) if

- elements $x \in \mathbb{F}_q$ are represented through a surjective representation
$$\rho : \{0,1\}^m \to \mathbb{F}_q \ (m \geq \log q);$$
- there are addition and multiplication oracles
$$A, M : \{0,1\}^m \times \{0,1\}^m \to \{0,1\}^m,$$
such that
$$\rho(A(x,y)) = \rho(x) + \rho(y), \qquad \rho(M(x,y)) = \rho(x)\rho(y);$$
- there is an identity oracle $E : \{0,1\}^m \to \{0,1\}$, such that $E(x,y) = 1$ if and only if $\rho(x) = \rho(y)$;
- there is a sampling oracle S that on request returns representations of independently and randomly chosen elements of the field.

We assume that for $p = O(1)$, we can on request obtain representations of elements of the ground field $\mathbb{F}_p$ (this can be simulated by E, M, and S, as for instance, "0" can be obtained by adding any element to itself p times). We focus on the case of binary fields, $p = 2$, though the results are easily extended.

Hitherto, the main applications of black-box algorithms have been to either show lower bounds for number theoretic problems (within the model), for example, see [35], or to show that upper bounds (efficient algorithms) for black-box problems would give efficient means to break number theoretic cryptosystems, for example, see [3] (similar ideas in fact appear already in [25]).

It has been observed that many natural things are computable in the black-box model, but to our knowledge, so far no results have been obtained showing that basic cryptographic primitives are possible. Surely, if cryptography itself is not possible within a model, then lower bounds obtained in the same model have little bearing on "reality".

One of the most basic cryptographic primitive is the *pseudo random generator* (PRG). In the normal Turing-machine model, a sufficient condition for the existence of a PRG is the existence of a one way permutation. Can an analogue be proven in the black-box model? In order to answer this question positively, we assume the existence of a one way permutation, denoted f, computable in the black-box model.

By the results of Blum and Micali [2] it would suffice to show the existence of a family of *hard core predicates* for f, computable in the black-box model. Such can be thought of as a set of boolean functions, B, having the property that if given $f(x), b$ (for random $b \in B$ and $x \in \mathbb{F}_{2^n}$), it is possible to compute $b(x)$ with success $1/2 + \varepsilon$, then it is also possible to recover x in time $(n/\varepsilon)^{O(1)}$.

Goldreich and Levin [13] have shown that (in the Turing model) every one way function has such a family, defined by

$$\{b_r^n(x) = \langle r, x \rangle \bmod 2 \mid r \in \mathbb{F}_{2^n}\}_{n \geq 1}.$$

This is however not valid in the black-box model, as computing the inner product requires fixing a basis for the field, which would make it representation dependent.

As observed by several authors (see, for example [30]) it follows from the proof of the Goldreich-Levin theorem that

$$\{B_r^n(x) = \mathrm{Tr}(rx) \mid r \in \mathbb{F}_{2^n}\}_{n \geq 1}$$

(where we write Tr for $\mathrm{Tr}_{\mathbb{F}_{2^n}/\mathbb{F}_2}$) is also a family of hard cores for any one way function. This is more promising, as by definition (1), the trace *is* (efficiently) computable by black-box operations. Still, a straightforward translation of their result does not solve the problem. The proof of Goldreich-Levin (and all subsequent proofs) consists of a reduction from inverting f to predicting $B_r^n(x)$, and that *reduction* is not performed in the black-box model as it uses a fixed representation/basis of $\mathbb{F}_{2^n}$ and the ability to manipulate x "coordinatewise" with respect to this basis. We show how to avoid this fixed representation. Or rather, how to *enable* such manipulations without assuming a certain representation.

We recall that a function ν is called *negligible* if for any constant $c > 0$, $\nu(n) = o(n^{-c})$.

The following result can be proved by using some arguments from the proof of Theorem 8 of [30].

Theorem 3. *The family of functions*

$$\{B_r^n(x) = \mathrm{Tr}(\rho(r)\rho(x)) \mid \rho(r) \in \mathbb{F}_{2^n}\}_{n \geq 1}$$

is a family of hard core functions in the black-box model for any one way function f. That is, if there is a probabilistic algorithm P such that for a certain non-negligible function ε,

$$\Pr_{x,r,\$}\left[\rho(P(\rho(r),\rho(f(x)))) = \rho(B_r^n(x))\right] \geq \frac{1}{2} + \varepsilon(n), \qquad \text{for all } n \geq 1,$$

then there exists a probabilistic algorithm I^P which recovers randomly selected x with non negligible probability from input $f(x)$. Such algorithm makes polynomially many operations and calls to P. Probability is obtained choosing x and r uniformly at random, and taking into account the internal coinflips of P, denoted by \$.

Proof. For simplicity, we identify the representation α, by the element itself, $\rho(\alpha) \in \mathbb{F}_{2^n}$, for fixed n. The main concern is that all operations are confined to the black-box model; thus we do not refer to any basis representation of x (and we replace $\rho(x)$ with merely x for the sake of brevity of notation). The same arguments work word-by-word for any ρ.

By our assumption, P predicts $\mathrm{Tr}(rx)$ with probability $1/2+\varepsilon(n)$ over choices of r and x. By Markov's inequality, there is a set of x of density $\varepsilon(n)/2$, for which P is correct with probability $1/2+\varepsilon(n)/2$ over the choice of r. It suffices to work with such "good" x.

We use standard arguments, but adapted to the black-box model. Pick a random $\alpha \in \mathbb{F}_{2^n}$ and assume that α generates a normal basis. Then, any element $x \in \mathbb{F}_{2^n}$ can be expressed as

$$x = \sum_{i=0}^{n-1} x_i \alpha^{2^i}, \tag{7}$$

for some $x_i \in \mathbb{F}_2$. Our goal is, given $f(x)$ and an algorithm predicting $\mathrm{Tr}(rx)$, to find these x_i (or "objects" representing them), which then give us a representation of x.

We now construct a dual basis $\{\beta_j\}_{j=0}^{n-1}$. As one can easily see, that construction is a linear algebra problem over $\mathbb{F}_2$ that can be solved in (black-box) polynomial time. Note that since the trace is $\mathbb{F}_2$-linear, when expressing x as in (7), we obtain

$$\mathrm{Tr}(\beta_j x) = \mathrm{Tr}\left(\beta_j\left(\sum_{i=0}^{n-1} x_i\alpha^{2^i}\right)\right) = \sum_{i=0}^{n-1}\mathrm{Tr}\left(\beta_j x_i \alpha^{2^i}\right) = \sum_{i=0}^{n-1} x_i \mathrm{Tr}\left(\beta_j \alpha^{2^i}\right) = x_j.$$

This means that if we could obtain $\{\mathrm{Tr}(\beta_j x)\}_{j=0}^{n-1}$ then we are done. However, the predictor P need not be correct on these "special" β_j, and we need to do some more work. We follow an old idea of Rackoff [12].

Pick $\gamma_0, \gamma_1, \ldots, \gamma_k$ independently and uniformly at random in $\mathbb{F}_{2^n}$. Let

$$r_l = \gamma_0 + \sum_{i=1}^{k} t_i^{(l)} \gamma_i, \qquad l = 1, \ldots, 2^k,$$

where $(t_1^{(l)}, \ldots, t_k^{(l)})$ is the lexicographically lth k-bit string. It is easy to see that for each l, r_l is uniformly distributed. Also, they are pairwise independent as for

$l_1 \neq l_2$, the difference $r_{l_1} - r_{l_2}$ contains at least one term of form γ_i and γ_i is uniformly distributed, hence the difference is uniformly distributed as well.

Next, guess $\mathrm{Tr}(\gamma_i x)$ for all $i = 0, 1, \ldots, k$. There are 2^{k+1} possibilities. Note that for the correct guess, by linearity, each $\mathrm{Tr}(r_l x)$ is also known. As long as $k = O(\log \varepsilon(n)^{-1})$, we can try all of the possibilities.

Now observe that

$$\mathrm{Tr}((r_l + \beta_j)x) - \mathrm{Tr}(r_l x) = \mathrm{Tr}(\beta_j x) = x_j.$$

Thus, the knowledge of $\mathrm{Tr}(r_l x)$, and an oracle prediction for $\mathrm{Tr}((r_l + \beta_j)x)$ give one "vote" for (a representation of) x_j.

As the r_l, $l = 1, \ldots, 2^k$, are uniformly distributed and pairwise independent, so are $\{r_l + \beta_j\}$. By applying Chebyshev's inequality, we see that $k \sim \log(n\varepsilon(n)^{-2})$ is sufficient to bring the error probability of a majority vote down to $1/(2n)$, and the probability that any of the n x_js are wrong is bounded by $1/2$. This produces a list of 2^{k+1} candidate representations for x, each of which can be verified by applying f.

It follows from more general results of [8] (see also [9,11]) that there is an absolute constant $c > 0$ such that α generates a normal basis of $\mathbb{F}_{2^n}$ over $\mathbb{F}_2$ with probability at least $c \log^{-1/2} n$. Therefore, repeating the above procedure for polynomially many random α we obtain the desired result. □

4 Remarks

Our results may be helpful for constructing new hard core functions, for both the usual Turing and the black-box model; in particular, to generalizations of the results of [28,29], see also [15,16,17]. They can also be used to study the security of *polynomial generators of pseudorandom numbers* [27,34] and *polynomial hash functions* [20,30].

A very challenging problem is to extend the results of Sections 2.1, 2.2 to so-called *noisy* oracles which for each call return the correct value only with certain probability (c.f. the special case of Section 3). For example, it would be very important to obtain analogues of Theorem 1 for oracles $\widetilde{\mathcal{O}}_{f,\varepsilon}$ which return the correct value of $\mathrm{Tr}_{\mathbb{K}/\mathbb{F}}(f(t))$ with probability at least $1/q + \varepsilon$, where $0 \le \varepsilon \le (q-1)/q$. This is related to studying *hard core bits* of the trace function of polynomials. Surveys of similar results for linear functions and polynomials of bounded degree can be found in [15,17,28,29,30].

It is also very interesting to extend our results to the case where the exponent vector $\nu = (\nu_1, \ldots, \nu_n)$ is not known.

There is also a much simpler modification of this problem, when the value of t can be given to the oracle (instead of being chosen at random by it). This is indeed the natural setting for constructing hard-core functions based on the trace, for then such functions are deterministic (for further details on such constructions, see [16]). In addition, in that case the standard amplification probability technique suffices to solve the problem.

Multivariate polynomials can be considered as well. For multivariate polynomials an analogue of our main tool, the lower bound on the number of non-zeros of Lemma 1, can be found in [23].

Acknowledgement. The authors would like to thank Consuelo Martínez and Alex Russell for their interest as well as for several helpful discussions.

References

1. D. Bleichenbacher and P. Q. Nguyen, 'Noisy polynomial interpolation and noisy Chinese remaindering', *Lect. Notes in Comp. Sci., Springer-Verlag*, Berlin, **1807** (2000), 53–69.
2. M. Blum and S. Micali, 'How to generate cryptographically strong sequences of pseudo-random bits', SIAM J. Comp., **13** (1984), 850–864.
3. D. Boneh and R. J. Lipton, 'Algorithms for Black-Box Fields and their Application to Cryptography', in *Lect. Notes in Comp. Sci.*, Springer-Verlag, Berlin, **1109** (1996), 283–297.
4. D. Boneh and R. Venkatesan, 'Hardness of computing the most significant bits of secret keys in Diffie–Hellman and related schemes', *Lect. Notes in Comp. Sci.*, Springer-Verlag, Berlin, **1109** (1996), 129–142.
5. D. Boneh and R. Venkatesan, 'Rounding in lattices and its cryptographic applications', *Proc. 8th Annual ACM-SIAM Symp. on Discr. Algorithms*, ACM, NY, 1997, 675–681.
6. R. P. Brent and B. D. McKay 'Determinants and ranks of random matrices over $\mathbb{Z}_m$', *Discr. Math.*, **66** (1987), 123–137.
7. E. El Mahassni, P. Q. Nguyen and I. E. Shparlinski, 'The insecurity of some DSA-like signature schemes with partially known nonces', *Proc. Workshop on Lattices and Cryptography, Boston, MA, 2001*, Springer-Verlag, Berlin, (to appear).
8. G. S. Frandsen, 'On the density of normal bases in finite fields', *Finite Fields and Their Appl.*, **6** (2000), 23–38.
9. S. Gao and D. Panario, 'Density of normal elements', *Finite Fields and Their Appl.*, **3** (1997), 141–150.
10. J. von zur Gathen and J. Gerhard, *Modern computer algebra*, Cambridge University Press, Cambridge, 1999.
11. J. von zur Gathen and M, Giesbrecht, 'Constructing normal bases in finite fields', *J. Symbol. Comp.*, **10** (1990), 547–570.
12. O. Goldreich, *Modern Cryptography, Probabilistic Proofs and Pseudo-randomness*, Springer-Verlag, Berlin, 1999.
13. O. Goldreich and L. A. Levin, 'A Hard Core Predicate for any One Way Function', in *Proc., 21st ACM STOC*, 1989, 25–32.
14. O. Goldreich and R. Rubinfeld and M. Sudan, 'Learning polynomials with queries: the highly noisy case', *Proc. of the 36th Annual Symposium on Foundations of Computer Science*, IEEE Computer Society Press, Los Alamitos, CA, 1995, 294–303.
15. M. Goldmann and M. Näslund, 'The complexity of computing hard core predicates', *Lect. Notes in Comp. Sci.*, Springer-Verlag, Berlin, **1294** (1997), 1–15.
16. M. Goldman, M. Näslund and A. Russell 'Complexity bounds on general hard-core predicates', *J. Cryptology*, **14** (2001), 177–195.

17. M. I. González Vasco and M. Näslund, 'A survey of hard core functions', *Proc. Workshop on Cryptography and Computational Number Theory*, Singapore 1999, Birkhäuser, 2001, 227–256.
18. M. I. González Vasco and I. E. Shparlinski, 'On the security of Diffie–Hellman bits', *Proc. Workshop on Cryptography and Computational Number Theory*, Singapore 1999, Birkhäuser, 2001, 257–268.
19. M. I. González Vasco and I. E. Shparlinski, 'Security of the most significant bits of the Shamir message passing scheme', *Math. Comp.*, **71** (2002), 333–342.
20. T. Helleseth and T. Johansson, 'Universal hash functions from exponential sums over finite fields and Galois rings', *Lect. Notes in Comp. Sci.*, Springer-Verlag, Berlin, **921** (1996), 31–44.
21. N. A. Howgrave-Graham and N. P. Smart, 'Lattice attacks on digital signature schemes', *Designs, Codes and Cryptography*, **23** (2001), 283–290.
22. J. Kahn and J. Komlós, 'Singularity probabilities for random matrices over finite fields', *Combinatorics, Probability and Computing*, **10** (2001), 137–157.
23. M. Karpinski and I. E. Shparlinski, 'On some approximation problems concerning sparse polynomials over finite fields', *Theor. Comp. Sci.*, **157** (1996), 259–266.
24. S. Lang, *Algebra*, Addison-Wesley, MA, 1965.
25. A. Lempel, G. Seroussi and J. Ziv, 'On the power of straight-line algorithms over finite fields', *IEEE Trans. on Information Theory*, **IT-28** (1982), 875–880.
26. R. Lidl and H. Niederreiter, *Finite fields*, Cambridge University Press, Cambridge, 1997.
27. S. Micali and C. P. Schnorr, 'Efficient, perfect polynomial random number generators', *J. Cryptology*, **3** (1991), 157–172.
28. M. Näslund, 'Universal hash functions & hard core bits', *Lect. Notes in Comp. Sci.*, Springer-Verlag, Berlin, **921** (1995), 356–366.
29. M. Näslund, 'All bits in $ax+b$ are hard', *Lect. Notes in Comp. Sci.*, Springer-Verlag, Berlin, **1109** (1996), 114–128.
30. M. Näslund and A. Russell, 'Hard core functions: Survey and new results', *Proc. of NordSec'99*, 1999, 305–322.
31. P. Nguyen, 'The dark side of the Hidden Number Problem: Lattice attacks on DSA', *Proc. Workshop on Cryptography and Computational Number Theory*, Singapore 1999, Birkhäuser, 2001, 321–330.
32. P. Nguyen and I. E. Shparlinski, 'The insecurity of the Digital Signature Algorithm with partially known nonces', *J. Cryptology*, (to appear).
33. P. Nguyen and I. E. Shparlinski, 'The insecurity of the elliptic curve Digital Signature Algorithm with partially known nonces', *Preprint*, 2000, 1–24.
34. H. Niederreiter and C. P. Schnorr, 'Local randomness in polynomial random number and random function generators', *SIAM J. Comp.*, **13** (1993), 684–694.
35. V. Shoup, 'Lower bounds for discrete logarithms and related problems', in *Lect. Notes in Comp. Sci.*, Springer-Verlag, Berlin, **1233** (1997), 256–266.
36. I. E. Shparlinski, 'Sparse polynomial approximation in finite fields', *Proc. 33rd ACM Symp. on Theory of Comput.*, Crete, Greece, July 6-8, 2001, 209–215.
37. I. E. Shparlinski, 'On the generalised hidden number problem and bit security of XTR', *Lect. Notes in Comp. Sci.*, Springer-Verlag, Berlin, **2227** (2001), 268–277.
38. I. E. Shparlinski, 'Security of polynomial transformations of the Diffie–Hellman key', *Preprint*, 2000, 1–8.

The Generalized Weil Pairing and the Discrete Logarithm Problem on Elliptic Curves

Theodoulos Garefalakis

Department of Mathematics
Royal Holloway
University of London,
Egham, Surrey, TW20 0EX, UK
theo.garefalakis@rhul.ac.uk

Abstract. We review the construction of a generalization of the Weil pairing, which is non-degenerate and bilinear, and use it to construct a reduction from the discrete logarithm problem on elliptic curves to the discrete logarithm problem in finite fields, which is efficient for curves with trace of Frobenius congruent to 2 modulo the order of the base point. The reduction is as simple to construct as that of Menezes, Okamoto, and Vanstone [16], and is provably equivalent to that of Frey and Rück [10].

1 Introduction

Since the seminal paper of Diffie and Hellman [8], the discrete logarithm problem (DLP) has become a central problem in algorithmic number theory, with direct implications in cryptography. For arbitrary finite groups the problem is defined as follows: Given a finite group G, a base point $g \in G$ and a point $y \in \langle g \rangle$ find the smallest non-negative integer ℓ such that $y = g^{\ell}$.

In their paper, Diffie and Hellman proposed a method for key agreement, whose security required that DLP be hard for the group $(\mathbb{Z}/p)^*$ of integers modulo a prime p. This is the multiplicative group of the finite field $\mathbb{F}_p$. Considering an arbitrary finite field $\mathbb{F}_q$ instead, the method can almost trivially be extended to work in the multiplicative group of $\mathbb{F}_q$, where q is a prime power. The security of the protocol now requires DLP to be hard in this group.

The result of the efforts of a number of researchers was the development of the index calculus method [1,3,6,11,15,18] and later the number field sieve and the function field sieve [2,4,7,12]. The methods are designed to compute discrete logarithms in any finite field, and are particularly efficient for finite fields of the form $\mathbb{F}_q$ with $q = p$ a prime, or $q = p^n$ with p a small prime and n large. In both these cases, the above methods run in subexponential time: the index calculus method in time $\exp((c_1 + o(1))(\log q)^{1/2}(\log\log q)^{1/2})$, and the number field and function field sieves in time $\exp((c_2 + o(1))(\log q)^{1/3}(\log\log q)^{2/3})$, where c_1 and c_2 are small constants.

S. Rajsbaum (Ed.): LATIN 2002, LNCS 2286, pp. 118–130, 2002.

The above developments, led Miller [17] and Koblitz [14] to consider alternative groups, where the group operation can be efficiently computed, but the DLP is hard. Their proposal was the group of points of an elliptic curve E over a finite field $\mathbb{F}_q$, denoted $E(\mathbb{F}_q)$. Traditionally, the group operation here is denoted additively. Thus the elliptic curve discrete logarithm problem (ECDLP) is defined as follows: Given an elliptic curve $E/\mathbb{F}_q$, a base point $P \in E(\mathbb{F}_q)$ and a point $Q \in \langle P \rangle$ find the smallest non-negative integer ℓ such that $Q = \ell \cdot P$.

ECDLP in general remains of exponential time complexity to this day. However, it was the work of Menezes, Okamoto and Vanstone [16], that showed that not all elliptic curves offer the same level of security. The authors used the well known Weil pairing, e_m, to translate the ECDLP from $E(\mathbb{F}_q)$ to the DLP in an extension field $\mathbb{F}_{q^k}^*$, which can subsequently be solved using one of the subexponential methods discussed earlier (MOV reduction). A necessary condition for it to be efficient is the existence a small integer k such that

1. $E[m] \subseteq E(\mathbb{F}_{q^k})$, where $m = \#\langle P \rangle$,
2. $m | q^k - 1$.

The authors were able to prove that for *supersingular* curves both conditions hold for $k \leq 6$. Subsequently, Frey and Rück [10] proposed another reduction, based on the Tate pairing ϕ_m. The advantage of this method is that $\phi_m(P, S)$ is an m-th root of unity for an easily computable point S (in most interesting cases $S = P$). Then the only requirement for the reduction to go through is that $m | q^k - 1$ for a small k. Clearly, this is a less restrictive condition. In fact, one cannot avoid this condition, as *any* isomorphism from $\langle P \rangle$ to a subgroup of $\mathbb{F}_{q^k}$ implies that $\#\langle P \rangle = m | q^k - 1$.

Later, Harasawa, Shikata, Suzuki, and Imai [13] attempted to generalize the method of Menezes, Okamoto, and Vanstone to apply to a larger class of elliptic curves. Their generalization appeared to be very limited. The main reason is that no efficient method is known to find a point $S \in E[m]$ such that $e_m(P, S)$ is a primitive m-th root of unity, if E is non-supersingular.

The purpose of this paper is to bridge the gap between the MOV reduction and the Frey-Rück reduction. We start from a well known generalization of the Weil pairing, e_ψ (see [5, p.45] [19, p.107]). The construction of the pairing is as simple as that of the Weil pairing, but has the nice property of the (more involved) Tate pairing, namely $e_\psi(P, P)$ is a suitable primitive root of unity. We show how to construct a group isomorphism between $\langle P \rangle$ and μ_r, where $r = \#\langle P \rangle$ is a prime, and μ_r is the group of r-th roots of unity. Our construction applies to elliptic curves $E/\mathbb{F}_q$ such that $r | q - 1$, i.e., $a_q \equiv 2 \pmod r$. For the cases of interest in cryptography, the order r of P is very close to the order of $E(\mathbb{F}_q)$ (and certainly greater than $2\sqrt{q}$). Then, the condition $r | q - 1$ is equivalent to $a_q = 2$. We note that our construction can be generalized to work for $r | q^k - 1$ for any $k \geq 1$. If the degree of the extension k is reasonably small, the resulting reduction is efficient. We want to stress that the reduction presented in this work is not a new attack to elliptic curve cryptosystems. It is an alternative, elementary construction of the reduction of Frey and Rück.

The paper is structured as follows. In Section 2, we review the construction of the generalized Weil pairing e_ψ parameterized by an isogeny ψ, and state the properties that will be used later. In Section 3, we specialize the isogeny ψ to $1-\phi$, where ϕ is the Frobenius endomorphism. In Section 4, we consider curves with trace of Frobenius $a_q = 2$, and show how to find a point P', such that $e_\psi(P,P')$ is a primitive r-th root of unity. In Section 5, we give an algorithm to compute the pairing in the case of interest. It turns out that for $Q \in \langle P \rangle$, the value $e_\psi(Q,P)$ is the multiplicative inverse of the value $\phi_r(Q,P)$ of the Tate pairing used by Frey and Rück. Finally, in Section 6 we show how to obtain a reduction in the more general case $a_q \equiv 2 \pmod r$.

2 The Pairing

In this section, we review a generalization of the Weil pairing. As for the rest of the paper, p is prime, and $q = p^k$.

Let E be an elliptic curve over $\mathbb{F}_q$. Also let $\psi : E \to E$ be a non-zero endomorphism of E, and denote its dual by $\widehat{\psi}$. Let $T \in \ker(\widehat{\psi})$ — such a point exists, since $\widehat{\psi}$ is onto. We denote by m the degree of ψ. Then, the divisor $D = m(T) - m(O)$ is principal. Let $f_T \in \overline{\mathbb{F}}_q(E)$ be a function such that

$$\mathrm{div}(f_T) = m(T) - m(O).$$

We consider now the divisor of $f_T \circ \psi$.

$$\begin{aligned}\mathrm{div}(f_T \circ \psi) &= \mathrm{div}(\psi^* f_T) = \psi^* \mathrm{div}(f_T)\\ &= m\left(\psi^*(T) - \psi^*(O)\right),\end{aligned}$$

the last equality being true by the definition of ψ^* ($\mathbb{Z}$-linearity). We note that

$$\begin{aligned}\psi^*(T) - \psi^*(O) &= \sum_{\psi P = T} e_\psi(P)(P) - \sum_{\psi R = O} e_\psi(R)(R)\\ &= \deg_i \psi \left(\sum_{\psi R = O} (T' + R) - (R) \right),\end{aligned}$$

where $\psi T' = T$. Here we used the fact that ψ is an isogeny, and therefore $e_\psi(P)$ does not depend on P, and equals to $\deg_i(\psi)$. The last line of the derivation shows that the divisor is principal, since it has degree zero, and it sums to

$$[\deg_i \psi] \sum_{\psi R = O} T' = [\deg \psi] T' = \widehat{\psi} \circ \psi(T') = \widehat{\psi} T = O.$$

So it must be the divisor of some function $g_T \in \overline{\mathbb{F}}_q(E)$. Thus,

$$(f_T \circ \psi) = m\mathrm{div}(g_T) = \mathrm{div}(g_T^m),$$

which implies that

$$g_T^m = f_T \circ \psi. \tag{1}$$

g_T is defined up to a multiplicative constant of course. Let now $S \in \ker(\psi)$, and X any point of $E(\overline{\mathbb{F}}_q)$.

$$g_T(X+S)^m = f_T(\psi X + \psi S) = f_T(\psi X) = f_T \circ \psi X = g_T(X)^m.$$

We define the pairing

$$e_\psi : \ker(\psi) \times \ker(\widehat{\psi}) \to \mu_m$$

as

$$e_\psi(S,T) = \frac{g_T(X+S)}{g_T(X)}. \tag{2}$$

The above definition does not depend on the choice of X. Indeed, if τ_S denotes the translation by S map

$$\begin{aligned} \tau_S : E &\to E \\ X &\mapsto X+S \end{aligned}$$

then we can write $e_\psi(S,T)$ as

$$e_\psi(S,T) = \frac{g_T \circ \tau_S}{g_T}(X),$$

and the function $g_T \circ \tau_S / g_T$ is constant. To see that, we need to note that $\psi = \psi \circ \tau_S$ because $S \in \ker(\psi)$. Then,

$$\begin{aligned} \operatorname{div}(g_T \circ \tau_S) &= \tau_S^* \operatorname{div}(g_T) \\ &= \tau_S^* \circ \psi^*((T)-(O)) \\ &= (\psi \circ \tau_S)^*((T)-(O)) \\ &= \psi^*((T)-(O)) \\ &= \operatorname{div}(g_T), \end{aligned}$$

therefore e_ψ is well-defined. Furthermore, it is an easy exercise to show that the generalized Weil pairing is bilinear and non-degenerate. The proofs are essentially the same as in the case of the Weil pairing.

Theorem 1. *Let p be a prime, and $q = p^k$. Let $E/\mathbb{F}_q$ be an elliptic curve, $\psi : E \to E$ be an endomorphism of E of degree m prime to p, and $\widehat{\psi}$ its dual. Then there exist a pairing*

$$e_\psi : \ker(\psi) \times \ker(\widehat{\psi}) \to \mu_m,$$

with the following properties

1. *Bilinear:*

$$\begin{aligned} e_\psi(S_1+S_2,T) &= e_\psi(S_1,T)e_\psi(S_2,T), \\ e_\psi(S,T_1+T_2) &= e_\psi(S,T_1)e_\psi(S,T_2). \end{aligned}$$

2. *Non-degenerate:*

$$\textit{If} \quad e_\psi(S,T) = O \quad \textit{for all} \quad T \in \ker(\psi), \quad \textit{then} \quad T = O.$$

Remark 1. The pairing in Theorem 1 is defined for any endomorphism ψ with the property $p \not| \deg(\psi)$. If we specialize ψ to be the multiplication by n map, and $p \not| n$, then we recover the Weil paring. This justifies the name "generalized Weil pairing".

3 A Special Pairing

In this section, we use the generalized Weil pairing to construct an isomorphism between a subgroup of $E(\mathbb{F}_q)$ and a suitable group of roots of unity in $\overline{\mathbb{F}}_q$. Our goal is to reduce the DLP on certain elliptic curves to the DLP in the multiplicative group of finite fields. The notation throughout the paper is as follows: A point $P \in E(\mathbb{F}_q)$ is given, of prime order r. We wish to solve the DLP in $\langle P \rangle$ by constructing an efficiently computable isomorphism

$$\langle P \rangle \longrightarrow \mu_r.$$

Most of the ingredients for the proposed isomorphism are present. In particular, e_ψ maps pairs of points to roots of unity, which form a group. We need to specialize the isogeny ψ, so that $\ker(\psi)$ is related to the group $E(\mathbb{F}_q)$. Let $\psi = 1 - \phi$, where ϕ is the q-th power Frobenius automorphism. Then we have $\ker(\psi) = E(\mathbb{F}_q)$, and $\widehat{\psi} = 1 - \widehat{\phi}$. Also

$$\# \ker(\widehat{\psi}) \mid \deg(\widehat{\psi}) = \deg(\psi) = \# E(\mathbb{F}_q) = N,$$

where the divisibility comes from the fact that

$$\ker(\widehat{\psi}) = \deg_s(\widehat{\psi}) \quad \text{and} \quad \deg(\widehat{\psi}) = \deg_s(\widehat{\psi}) \cdot \deg_i(\widehat{\psi}).$$

Assuming that p does not divide N, we have a bilinear, non-degenerate pairing

$$e_\psi : E(\mathbb{F}_q) \times \ker(\widehat{\psi}) \to \mu_N.$$

We stress that this pairing exists and is bilinear and non-degenerate for *any* elliptic curve E and *any* finite field $\mathbb{F}_q$.

The group of r-th roots of unity, μ_r, is contained in the smallest extension of $\mathbb{F}_q$, say in $\mathbb{F}_{q^k}$ such that $r | q^k - 1$. We will mainly be concerned with the case $r | q - 1$, i.e., when all the r-th roots of unity are contained in $\mathbb{F}_q$. Then, the condition reads $r | q - 1$, or equivalently

$$a_q \equiv 2 \pmod{r}. \tag{3}$$

In cryptography, the point P is chosen to have very large order r, close to the order of the whole group $E(\mathbb{F}_q)$. Thus r is of order q, which implies that Equation (3) is equivalent (for such a choice of P) to $a_q = 2$. This will be the main case in our investigation.

4 Curves with Trace Equal to 2

In this section, we consider elliptic curves with trace of Frobenius $a_q = 2$. Let ϕ be the q-th power Frobenius map. Let $Q \in \#E(\mathbb{F}_q)$. We wish to find the point $\widehat{\phi}(Q)$. For that we consider the following

$$(1-\phi)\circ(1-\widehat{\phi}) = 1 - \phi - \widehat{\phi} + [q].$$

From the above observation, we have that

$$(1-\phi)\circ(1-\widehat{\phi})Q = O.$$

Therefore,

$$Q - \phi(Q) - \widehat{\phi}(Q) + [q]Q = O,$$

which implies

$$\widehat{\phi}(Q) = [q]Q.$$

We know that $q+1-a_q = \#E(\mathbb{F}_q)$, therefore $[q]Q = [a_q - 1]Q$. Thus,

$$\widehat{\phi}(Q) = [a_q - 1]Q.$$

Suppose now that the curve has $a_q = 2$. Then, for every point $Q \in E(\mathbb{F}_q)$ we have $(1-\widehat{\phi})Q = O$, i.e.,

$$E(\mathbb{F}_q) \subseteq \ker(1-\widehat{\phi}).$$

Furthermore,

$$\#\ker(1-\widehat{\phi}) = \deg_s(1-\widehat{\phi}) \leq \deg(1-\widehat{\phi}) = \deg(1-\phi) = \#E(\mathbb{F}_q).$$

This implies that

$$\ker(1-\widehat{\phi}) = E(\mathbb{F}_q).$$

To summarize, for a curve E with trace of Frobenius $a_q = 2$, we have a pairing

$$e_\psi : E(\mathbb{F}_q) \times E(\mathbb{F}_q) \to \mu_N,$$

where $N = \#E(\mathbb{F}_q)$. Note that $N = q-1$, and p (the characteristic) does not divide N. Therefore from Theorem 1 it must be bilinear and non-degenerate.

4.1 A Structure Theorem

We need to introduce some more notation for this section. The group $E(\mathbb{F}_q)$ is isomorphic to $\mathbb{Z}/n_1\mathbb{Z} \oplus \mathbb{Z}/n_2\mathbb{Z}$, with $n_2|n_1$, and $n_2|q-1$. This means that $\#E(\mathbb{F}_q) = N = n_1 n_2$. We denote by (T_1, T_2) a pair of generators of $E(\mathbb{F}_q)$. We recall that P is a point in $E(\mathbb{F}_q)$ of prime order r. For the remainder of this paper, we assume that $n_1 = lr^k$, $r \not| l$ and that $n_2|l$, i.e., r does not divide n_2. This is usually the case in cryptography, as the point P is chosen to have very large order. Then $\langle P\rangle$ is contained in $\langle T_1\rangle$. Our goal is to show that $e_\psi(P,P)$ is a primitive r-th root of unity.

Lemma 1. *There exist points $T, S \in E(\mathbb{F}_q)$ such that $e_\psi(T,S)$ is a primitive n_1-st root of unity.*

Proof. The image of $e_\psi(T,S)$ as T and S range over $E(\mathbb{F}_q)$ is a subgroup of μ_N, say equal to μ_d. Then it follows that for all $(T,S) \in E(\mathbb{F}_q) \times E(\mathbb{F}_q)$,

$$1 = e_\psi(T,S)^d = e_\psi([d]T, S).$$

The non-degeneracy of the e_ψ pairing implies that $[d]T = O$ for all $T \in E(\mathbb{F}_q)$. In particular, if $T = T_1$ then it must be $d = n_1$.

Lemma 2. *The order of $e_\psi(T_1,T_1)$ is divisible by r^k.*

Proof. Let $T = [x_1]T_1 + [x_2]T_2$ and $S = [y_1]T_1 + [y_2]T_2$ be one pair of points such that $e_\psi(T,S)$ is a primitive n_1-st root of unity, which exists by Lemma 1. Suppose now to the contrary, that r^k does not divide $e_\psi(T_1,T_1)$. Then

$$e_\psi(T,S) = e_\psi(T_1,T_1)^{x_1y_1} e_\psi(T_1,T_2)^{x_1y_2} e_\psi(T_2,T_1)^{x_2y_1} e_\psi(T_2,T_2)^{x_2y_2}$$

Note now that the order of $e_\psi(T_1,T_1)$ divides $n_1 = lr^k$, but by assumption r^k does not divide it. Therefore, the order of $e_\psi(T_1,T_1)$ divides lr^{k-1}. Obviously, the orders of $e_\psi(T_1,T_2)$, $e_\psi(T_2,T_1)$, and $e_\psi(T_2,T_2)$ divide l. Thus we have,

$$e_\psi(T_1,T_1)^{lr^{k-1}} = e_\psi(T_1,T_2)^{lr^{k-1}} = e_\psi(T_2,T_1)^{lr^{k-1}} = e_\psi(T_2,T_2)^{lr^{k-1}} = 1.$$

Therefore,

$$e_\psi(T,S)^{lr^{k-1}} = 1,$$

which is a contradiction, since $lr^{k-1} < n_1$.

Theorem 2. *Let $P' \in E(\mathbb{F}_q)$, be a point of order r^d. Then, $e_\psi(P,P')$ is a primitive r-th root of unity if and only if $k < d+1$.*

Proof. It is clear that $e_\psi(P,P')$ is either a primitive r-th root of unity or 1. This is because

$$e_\psi(P,P')^r = e_\psi([r]P,P') = e_\psi(O,P') = 1.$$

We recall that $\langle P\rangle$, and $\langle P'\rangle$ are subgroups of $\langle T_1\rangle$. It follows that $P = [lr^{k-1}]T_1$ and $P' = [lr^{k-d}]T_1$. Then we have

$$\begin{aligned} e_\psi(P,P') &= e_\psi([lr^{k-1}]T_1, [lr^{k-d}]T_1) \\ &= e_\psi(T_1,T_1)^{l^2r^{2k-d-1}}. \end{aligned}$$

Then Lemma 2 implies,

$$\begin{aligned} e_\psi(P,P') = 1 &\iff \\ 2k-d-1 \geq k &\iff \\ k \geq d+1. \end{aligned}$$

We note, that if r^d is the exact power of r dividing N, then the point P' of the previous theorem can be computed efficiently using the probabilistic method described by Frey, Múller, and Rück in [9]. More importantly, in cryptography the point P is chosen to have very large order r (practically on the same order as q). For that reason, we state the following corollary.

Corollary 1. *Let $P \in E(\mathbb{F}_q)$ be a point of order r, such that r^2 does not divide $\#E(\mathbb{F}_q)$. Then $e_\psi(P,P)$ is a primitive r-th root of unity.*

We want to emphasize that Corollary 1 is in sharp contrast with the properties of the Weil pairing. For the Weil pairing, $e_r(P,P)$ for every $P \in E[N]$. In our case, when $k=1$ the value $e_\psi(P,P)$ is not trivial, and in fact is a primitive r-th root of unity. This eliminates a major obstacle of the Weil pairing approach: The point that makes $e_\psi(P,\cdot)$ a primitive r-th root of unity is defined over $\mathbb{F}_q$ (in the case of the Weil pairing it exists in a *very* large extension, unless the curve is supersingular). Furthermore, it is known in advance. We have the following theorem.

Theorem 3. *Let P be a point in $E(\mathbb{F}_q)$ of prime order r, such that r^d does not divide N. Then there is an efficiently computable point P' such that the map*

$$V : \langle P \rangle \to \mu_r, \quad Q \mapsto e_\psi(Q, P')$$

is a group isomorphism. In particular, if $d=2$, then $P'=P$.

5 Computing the Pairing

We turn now to the computation of the generalized Weil pairing. A computation using the definition directly would result an exponential time algorithm. Thus, we need some other formula suitable for the computation. Such a formula can be found using Galois cohomology. This formula, not surprisingly, also provides the connection between our construction and the Frey-Rück construction that uses the Tate pairing. Although part of the material of this section is well known, we choose to include it, in order to keep the paper as self contained as possible.

Let $E/\mathbb{F}_q$ be an elliptic curve, and let $\psi : E \to E$ be an isogeny. We start from the following exact sequence.

$$0 \longrightarrow \ker(\psi) \longrightarrow E(\overline{\mathbb{F}}_q) \xrightarrow{\psi} E(\overline{\mathbb{F}}_q) \longrightarrow 0. \tag{4}$$

Taking $\mathrm{Gal}(\overline{\mathbb{F}}_q/\mathbb{F}_q)$ cohomology, we obtain the following long sequence

$$\begin{aligned} 0 \longrightarrow E(\mathbb{F}_q) \cap \ker(\psi) \longrightarrow E(\mathbb{F}_q) &\xrightarrow{\psi} E(\mathbb{F}_q) \\ \xrightarrow{\delta} H^1(G, \ker(\psi)) \longrightarrow H^1(G, E(\overline{\mathbb{F}}_q)) &\xrightarrow{\psi} H^1(G, E(\overline{\mathbb{F}}_q)), \end{aligned}$$

where $G = \mathrm{Gal}(\overline{\mathbb{F}}_q/\mathbb{F}_q)$. We can extract now the short exact sequence, sometimes called the *Kummer sequence* for $E/\mathbb{F}_q$,

$$0 \longrightarrow \frac{E(\mathbb{F}_q)}{\psi E(\mathbb{F}_q)} \xrightarrow{\delta} H^1(G, \ker(\psi)) \longrightarrow H^1(G, E(\overline{\mathbb{F}}_q))[\psi] \longrightarrow 0, \tag{5}$$

where $H^1(G, E(\overline{\mathbb{F}}_q))[\psi]$ denotes the subgroup of $H^1(G, E(\overline{\mathbb{F}}_q))$ that is sent to the zero cocycle class by ψ. The connecting homomorphism δ is defined as follows. Let $P \in E(\mathbb{F}_q)$, and let $Q \in E(\overline{\mathbb{F}}_q)$ such that $\psi(Q) = P$. Then a 1-cocycle representing $\delta(P)$ is given by

$$\begin{aligned} G &\to \ker(\psi) \\ \sigma &\mapsto Q^\sigma - Q; \end{aligned}$$

that is

$$\delta(P)(\sigma) = Q^\sigma - Q.$$

From this point on, we specialize $\psi = 1 - \phi$, the case of interest here. Then we know that $\ker(\psi) = E(\mathbb{F}_q)$, so the action of G on $\ker(\psi)$ becomes trivial, and therefore

$$H^1(G, \ker(\psi)) = Hom(G, \ker(\psi)).$$

Furthermore, Hilbert's Theorem 90 provides the isomorphism

$$\frac{\mathbb{F}_q^*}{(\mathbb{F}_q^*)^r} \cong H^1(G, \mu_r).$$

Assume further, that $a_q \equiv 2 \pmod r$, for a prime r. Then we know that $q - 1 \equiv 0 \pmod r$, and therefore, $\mathbb{F}_q$ contains all the r-th roots of unity. Denote by μ_r the group of r-th roots of unity in $\mathbb{F}_q$. Then G acts trivially on μ_r, so

$$H^1(G, \mu_r) = Hom(G, \mu_r),$$

and we have the isomorphism

$$\begin{aligned} \delta_K : \mathbb{F}_q^*/(\mathbb{F}_q^*)^r &\to Hom(G, \mu_r) \\ b \cdot (\mathbb{F}_q^*)^r &\mapsto (\sigma \mapsto \beta^\sigma/\beta), \end{aligned}$$

where $b \in \mathbb{F}_q^*$, $\beta \in \overline{\mathbb{F}}_q^*$, and $\beta^r = b$. In other words, for some $b \in \mathbb{F}_q^*$, $\delta_K(b)$ is a homomorphism from G to μ_r, and

$$\delta_K(b)(\sigma) = \frac{\beta^\sigma}{\beta}. \tag{6}$$

Then it can be shown (see [19, Section X.1] or [5, Section V.2]), that there exists a pairing

$$B : \frac{E(\mathbb{F}_q)}{\psi E(\mathbb{F}_q)} \times \ker(\psi) \to \frac{\mathbb{F}_q^*}{(\mathbb{F}_q^*)^r},$$

such that

$$e_\psi(\delta(S), T) = \delta_K(B(S, T)).$$

We note, that $\delta(S)$ is not a point in $\ker(\psi)$, and $\delta_K(B(S,T))$ is not an r-th root of unity. The above relation is to be interpreted as:

$$\text{For any} \quad \sigma \in G, \quad e_\psi(\delta(S)(\sigma), T) = \delta_K(B(S,T))(\sigma). \tag{7}$$

The crucial thing is that the bilinear pairing B can be computed efficiently, at least in the case of interest. In fact, if $T \in \ker(\psi)$ is a point of order r, and $S \neq T$, then

$$B(S,T) \equiv f_T(S) \pmod{(\mathbb{F}_q^*)^r},$$

where f_T is a function with divisor

$$\operatorname{div}(f_T) = r(T) - r(O).$$

If $T = S$, then we can use bilinearity to obtain

$$B(T,T) = f_T(-T)^{-1}.$$

More generally, for any point $X \neq T$ we have

$$\begin{aligned} B(S,T) &= B(S+X-X,T) = B(S+X,T)\,B(-X,T) \\ &= B(S+X,T)\,B(X,T)^{-1} = \frac{f_T(S+X)}{f_T(X)}. \end{aligned}$$

We recall that now that our problem is the following: Given points $P, Q \in E(\mathbb{F}_q)$, with $\#\langle P\rangle = r$, we want to compute $e_\psi(Q,P)$. We deal with elliptic curves with $a_q \equiv 2 \pmod r$. From Equation (7) we have

$$e_\psi(\delta(S)(\sigma),P) = \delta_K(B(S,P))(\sigma), \tag{8}$$

where $\delta(S)(\sigma) = R^\sigma - R$ for some point R such that $\psi(R) = S$.

If we choose σ to be the q-power Frobenius automorphism in $G = \operatorname{Gal}(\overline{\mathbb{F}}_q/\mathbb{F}_q)$, and $S = -Q$, then we have

$$\phi(R) = R^\sigma \quad \text{for any } R \in E. \tag{9}$$

Also,

$$\begin{aligned} \psi(R) = S &\implies \\ R - \phi(R) = S &\implies \\ \phi(R) - R = -S &\implies \\ R^\sigma - R = -S &\implies \\ \delta(S)(\sigma) = Q&. \end{aligned}$$

Thus, Equation (8) has become

$$e_\psi(Q,P) = \delta_K(B(-Q,P))(\sigma), \tag{10}$$

where σ is now fixed (and equal to the Frobenius automorphism).

It remains to compute $\delta_K(B(-Q,P))(\sigma)$. We recall from Equation (6) that

$$\delta_K(B(-Q,P))(\sigma) = \frac{\beta^\sigma}{\beta},$$

where
$$\beta^r = B(-Q,P) = B(Q,P)^{-1} \equiv f_P(Q)^{-1} \pmod{(\mathbb{F}_q^*)^r}.$$
We have,
$$\frac{\beta^\sigma}{\beta} = \frac{\beta^q}{\beta} = \beta^{q-1}.$$
Therefore, $\delta_K(B(-Q,P))(\sigma)$ can be computed as
$$\delta_K(B(-Q,P))(\sigma) = \left(\frac{f_P(X)}{f_P(X+Q)}\right)^{(q-1)/r},$$
for any point $X \in E(\overline{\mathbb{F}}_q)$, $X \neq P$. Putting everything together, we have
$$e_\psi(Q,P) = \left(\frac{f_P(X)}{f_P(X+Q)}\right)^{(q-1)/r}. \tag{11}$$
Equation (11) can now be used to compute the value $e_\psi(Q,P)$. One first computes $f_P(X+Q)$ and $f_P(X)$ using repeated doubling. The point X has to be chosen suitably, so that the points X and $X+Q$ do not appear in the support of the divisors of the functions that appear in the computation. Those functions have divisors with support contained in $\langle P\rangle$, so one wants to avoid $X \in \langle P\rangle$. Thus one may choose $X \in E(\mathbb{F}_q)$, which in the case $q-1 > r$ yields a useful point with probability at least $1/2$, or one may even choose $X \in E(\mathbb{F}_{q^2})$, which yields a useful point with probability at least $1 - 1/q$. The algorithm for computing the classical Weil pairing was first given by Miller. An elegant presentation of the same algorithm is contained in [9]. The value $\left(\frac{f_P(X)}{f_P(X+Q)}\right)^{(q-1)/r}$ is computed using repeated squaring in $\mathbb{F}_q$.

Finally, it is interesting to note that for elliptic curves with $a_q \equiv 2 \pmod r$, and if P is a point of order r, then
$$e_\psi(Q,P) = \phi_r(Q,P)^{-1},$$
where ϕ_r is the Tate pairing, used by Frey, Müller, and Rück in [9].

6 Curves with Trace Congruent to 2

We can relax the requirement $a_q = 2$ a little, and assume only that $a_q \equiv 2 \pmod r$. This is equivalent to say $r|q-1$. Then, it is not in general the case that $\ker(1-\widehat{\phi}) = E(\mathbb{F}_q)$. However, if in the above derivation we take $Q \in \langle P\rangle$. Then we conclude that
$$\widehat{\phi}(Q) = [a_q - 1]Q = Q,$$
because $a_q \equiv 2 \pmod n$ and $[r]Q = O$. Thus, we have
$$\langle P\rangle \subseteq \ker(1-\widehat{\phi}).$$

For simplicity, we will only consider the case that no higher power of r divides $N = \#E(\mathbb{F}_q)$ — which is the only interesting case in cryptography. Then we claim that $e_\psi(P,P)$ is again a primitive r-th root of unity.

Lemma 3. *There exist a point $S \in \ker(1-\widehat{\phi})$, such that $e_\psi(P,S)$ is a primitive r-th root of unity.*

Proof. It is clear that $e_\psi(P,S)$ is an r-th root of unity. Furthermore, as the point S ranges over $\ker(1-\widehat{\phi})$, the values $e_\psi(P,S)$ are in a subgroup of μ_N, say μ_d. It follows that for all $S \in \ker(1-\widehat{\phi})$, we have

$$1 = e_\psi(P,S)^d = e_\psi([d]P,S).$$

The non-degeneracy of e_ψ then implies that $[d]P = O$, i.e., r divides d. It follows that the order of $e_\psi(P,S)$ is exactly r for some point S.

As we pointed out in Section 3, we have $\widehat{N} = \#\ker(1-\widehat{\phi})\,|N$. We also showed that $r|\#\ker(1-\widehat{\phi})$. We adopt the following notation: $N = lr$, and $\widehat{N} = \widehat{l}r$, with $\widehat{l}|l$. Also, $\ker(1-\widehat{\phi})$ is the product of at most two cyclic groups, one of which contains $\langle P\rangle$. If (S_1,S_2) is a pair of generators for $\ker(1-\widehat{\phi})$, it follows that the order of $e_\psi(P,S_1)$ divides r. If the order was 1, then it would violate the non-degeneracy of e_ψ (the argument is virtually the same as in Lemma 2 followed by Theorem 2 for $k=1$). Then, since $P \in \langle S_1\rangle$, it will be $P = [l']S_1$. Therefore,

$$1 = e_\psi(P,P)^d = e_\psi(P,S_1)^{l'd},$$

which implies that d has to be r (since $r^2 \not| \widehat{N}$). Therefore, we have the theorem.

Theorem 4. *Let $E/\mathbb{F}_q$ be an elliptic curve, $P \in E(\mathbb{F}_q)$ a point of prime order r such that $r^2 \not| N$, and assume that $a_q \equiv 2 \pmod r$. Then $e_\psi(P,P)$ is a primitive r-th root of unity.*

We also note that the proof given in Section 5 goes through in this more general case word by word. Therefore, the algorithm of the previous section works in the case $a_q \equiv 2 \pmod r$ as it is.

Acknowledgments. I am grateful to Allan Borodin for encouraging my research, Kumar Murty, Daniel Panario, and Ian Blake for numerous enlightening discussions on the subject, and to Nigel Boston and Igor Shparlinski for reading the manuscript and providing many helpful comments.

References

1. L.M. Adleman. A subexponential algorithm for the discrete logarithm problem with applications to cryptography. In *Proc. 20th IEEE Found. Comp. Sci. Symp.*, pages 55–60, 1979.
2. L.M. Adleman. The function field sieve. In *ANTS: 1st International Algorithmic Number Theory Symposium (ANTS)*, volume 877 of *LNCS*, pages 108–121, Berlin, Germany, 1994. Springer.
3. L.M. Adleman and J. DeMarrais. A subexponential algorithm for discrete logarithms over all finite fields. In Douglas R. Stinson, editor, *Proc. CRYPTO 93*, pages 147–158. Springer, 1994. Lecture Notes in Computer Science No. 773.

4. L.M. Adleman and M.D. Huang. Function field sieve method for discrete logarithms over finite fields. *INFCTRL: Information and Computation (formerly Information and Control)*, 151:5-16, 1999.
5. I. Blake, G. Seroussi, and N. Smart. *Elliptic curves in Cryptography*, volume 265 of *London Mathematical Society, Lecture Note Series.* Cambridge University Press, 1999.
6. I. F. Blake, R. Fuji-Hara, R. C. Mullin, and S. A. Vanstone. Computing logarithms in finite fields of caracteristic two. *SIAM J. Alg. Disc. Methods*, 5:276–285, 1985.
7. D. Coppersmith. Fast evaluation of logarithms in fields of characteristic two. *IEEE Trans. Inform. Theory*, IT-30:587–594, 1984.
8. W. Diffie and M. Hellman. New directions in cryptography. *IEEE Trans. Inform. Theory*, 22:472–492, 1976.
9. G. Frey, M. Müller, and H.G. Rück. The tate pairing and the discrete logarithm applied to elliptic curve cryptosystems. *IEEE Trans. Inform. Theory*, 45(5):1717–1719, 1999.
10. G. Frey and H.G. Rück. A remark concerning m-divisibility and the discrete logarithm in the divisor class group of curves. *Mathematics of Computation*, 62(206):865–874, 1994.
11. T. Garefalakis and D. Panario. The index calculus method using non-smooth polynomials. *Mathematics of Computation*, 70(235):1253–1264, 2001.
12. D.M. Gordon. Discrete logarithms in GF(p) using the number field sieve. *SIAM J. Disc. Math.*, 6(1):124–138, February 1993.
13. R. Harasawa, J. Shikata, J. Suzuki, and H. Imai. Comparing the MOV and FR reductions in elliptic curve cryptography. In *Advances in Cryptology: EUROCRYPT '99*, volume 1592 of *Lecture Notes in Computer Science*, pages 190–205. Springer, 1999.
14. N. Koblitz. Elliptic curve cryptosystems. *Mathematics of Computation*, 48(177):203–209, 1987.
15. K. S. McCurley. The discrete logarithm problem. *Proc. of Symp. in Applied Math.*, 42:49–74, 1990.
16. A. Menezes, E. Okamoto, and S. Vanstone. Reducing elliptic curve logarithms to logarithms in a finite field. *IEEE Transactions on Information Theory*, 39, 1993.
17. V. S. Miller. Uses of elliptic curves in cryptography. In Hugh C. Williams, editor, *Advances in cryptology — CRYPTO '85: proceedings*, volume 218 of *Lecture Notes in Computer Science*, pages 417–426. Springer-Verlag, 1986.
18. C. Pomerance. Fast, rigorous factorization and discrete logarithm algorithms. In *Discrete Algorithms And Complexity, Proc. of the Japan-US Joint Seminar, Academic Press*, pages 119–143, 1986.
19. J. H. Silverman. *The Arithmetic of Elliptic Curves*, volume 106 of *Graduate Texts in Mathematics.* Springer-Verlag, 1986.

Random Partitions with Non Negative r^{th} Differences

Rod Canfield[1], Sylvie Corteel[2], and Pawel Hitczenko[3]

[1] Department of Computer Science,
University of Georgia Athens, GA 30602, USA
erc@cs.uga.edu

[2] PRiSM, Université de Versailles,
45 Av. des Etats Unis, 78035 VERSAILLES, France
syl@prism.uvsq.fr

[3] Department of Maths and Computer Science,
Drexel University, Philadelphia, PA 19104, USA
phitczen@mcs.drexel.edu

Abstract. Let $P_r(n)$ be the set of partitions of n with non negative r^{th} differences. Let λ be a partition of an integer n chosen uniformly at random among the set $P_r(n)$ Let $d(\lambda)$ be a positive r^{th} difference chosen uniformly at random in λ. The aim of this work is to show that for every $m \geq 1$, the probability that $d(\lambda) \geq m$ approaches the constant $m^{-1/r}$ as $n \to \infty$ This work is a generalization of a result on integer partitions [7] and was motivated by a recent identity by Andrews, Paule and Riese's Omega package [3]. To prove this result we use bijective, asymptotic/analytic and probabilistic combinatorics.

1 Introduction

A partition λ of n is a sequence of integers

$$\lambda = (\lambda_1, \lambda_2, \ldots, \lambda_k) \text{ with } \lambda_1 \geq \lambda_2 \ldots \geq \lambda_k \geq 1 \text{ and } |\lambda| = \sum_{j=1}^{k} \lambda_j = n.$$

The same partition λ can also be written in its frequential notation, that is :

$$\lambda = n^{m_n}(n-1)^{m_{n-1}} \ldots 1^{m_1} \text{ with } m_j = |\{i \mid \lambda_i = j\}|, 1 \leq j \leq n.$$

The number m_j is called the *multiplicity* of the part j in λ. We will use both representations. The *part sizes* of a partition λ are the indices j such that m_j is positive. The number of part sizes can therefore be defined as : $|\{j \mid m_j > 0\}|$. The *conjugate* of a partition λ is the partition $\lambda' = (\lambda'_1, \ldots)$ where $\lambda'_i = |\{j \mid \lambda_j \geq i\}|$, $1 \leq i \leq \lambda_1$. We now define the rth differences. Let $\lambda = (\lambda_1, \lambda_2, \ldots, \lambda_k)$ be a partition of n and $\Delta^r(\lambda) = (\Delta^r_1(\lambda), \ldots, \Delta^r_k(\lambda))$ be its rth differences. The rth *differences* can be computed by the following recurrence :

$$\Delta^r_i(\lambda) = \begin{cases} \lambda_i & \text{if } i = k \text{ or } r = 0 \\ \Delta^{r-1}_i(\lambda) - \Delta^{r-1}_{i+1}(\lambda) & \text{otherwise} \end{cases}$$

S. Rajsbaum (Ed.): LATIN 2002, LNCS 2286, pp. 131–140, 2002.

In what follows we will write $\Delta_i^r(\lambda) : \Delta_i^r$ for short. Let P_r be the set of partitions with non negative r^{th} differences, that is to say, $(\lambda_1, \lambda_2, \ldots, \lambda_k) \in P_r$ if and only if $\Delta_i^r \geq 0$ for $1 < i < k$. Let $P_r(n)$ be the set of partitions $\lambda \in P_r$ with $|\lambda| = n$ and let $p_r(n) = |P_r(n)|$.

This work was motivated by two previous results. The first result is an identity on partitions with non negative r^{th} differences. It was discovered by Andrews, Paule and Riese's Omega package. Let $F_r(n)$ be the set of partitions of n into parts in the set $S_r = \{\binom{i+r}{r}, i \geq 0\}$.

Theorem 1. [2,3] *There is a one-to-one correspondence between the partitions in $P_r(n)$ and the partitions in $F_r(n)$.*

The second result is on ordinary integer partitions (partitions with non negative 1st differences) :

Theorem 2. [7] *Let m be a fixed positive integer. Let λ be a partition of an integer n chosen uniformly at random among the set of all partitions of n. Let $s(\lambda)$ be a part size chosen uniformly at random from the set of all part sizes that occur in λ. The probability that the multiplicity of $s(\lambda)$ is equal to m approaches the constant $1/(m(m+1))$ as $n \to \infty$.*

This result can easily be restated in the following way :

Corollary 1. *Let m be a fixed positive integer. Let λ be a partition of an integer n chosen uniformly at random among the set of all partitions of n. Let $s(\lambda)$ be a part size chosen uniformly at random from the set of all part sizes that occur in λ. The probability that the multiplicity of $s(\lambda)$ is greater than or equal to m approaches the constant $1/m$ as $n \to \infty$.*

Note that if $r = 1$ there exists a one-to-one correspondence between the number of positive 1st differences in any given partition and the number of part sizes of its conjugate λ'. For example if $\lambda = (7,7,6,3,2,2,2)$ then the 1st differences are $\Delta^1(\lambda) = (0,1,3,1,0,0,2)$. Hence λ has four positive 1st differences. The conjugate λ' is $(7,7,4,3,3,3,2)$ and has four part sizes: 7,4,3,2. Moreover this correspondance can be refined. It is easy to see that the ith 1st difference (i.e. $\lambda_i - \lambda_{i+1}$) is equal to the multiplicity of the part i in λ'. If we use again our example we have $\Delta_7^1(\lambda) = 2$ which is the multiplicity of the part 7 in λ'. Therefore Corollary 1 is equivalent to :

Corollary 2. *Let m be a fixed positive integer. Let λ be a partition of an integer n chosen uniformly at random among the set of all partitions of n. Let $d(\lambda)$ be a 1st difference chosen uniformly at random from the set of all positive 1st differences that occur in λ. The probability that $d(\lambda)$ is greater than or equal to m approaches the constant $1/m$ as $n \to \infty$.*

Our aim is therefore to generalize the result of Corollary 2 by using the identity of Theorem 1. Let us now state our generalization.

Theorem 3. *Let m and r be fixed positive integers. Let λ be a partition of an integer n chosen uniformly at random among the set $P_r(n)$ of all partitions of n with non negative r^{th} differences. Let $d(\lambda)$ be a positive r^{th} difference chosen uniformly at random from the set of all positive r^{th} differences that occur in λ. The probability that $d(\lambda) \geq m$ approaches the constant $m^{-1/r}$ as $n \to \infty$.*

The purpose of this paper is to prove Theorem 3. We now present the organization of the paper. We will first give in Section 2 a simple bijection of the identity of Theorem 1 that gives several refinements of the identity. This bijection was advertised/announced by Zeilberger in his very own journal [6]. We then present in Section 3 some asymptotic results on the number of partitions in $P_r(n)$ and some asymptotics results on the average number of rth differences greater or equal to m in the partitions in $P_r(n)$. Finally in Section 4 we use some probabilistic arguments which generalize the works on ordinary partitions [8,7]. The association of the three parts gives us the proof of our result. We conclude the paper by presenting some future work in Section 5.

2 Bijective Combinatorics

In this section we are going to present a bijection f between the partitions in $P_r(n)$ and the partitions of n into parts $\binom{i+r}{r}$, $i \geq 0$. Let $\lambda = (\lambda_1, \lambda_2, \ldots, \lambda_k)$ be a partition in $P_r(n)$ then its image by the bijection f in its frequential notation is :

$$f(\lambda) = \binom{k-1+r}{r}^{\Delta_k^r} \binom{k-2+r}{r}^{\Delta_{k-1}^r} \cdots \binom{r}{r}^{\Delta_1^r}$$

where the Δ_i^r $(1 \leq i \leq k)$ are the rth differences of λ. It is clear that $f(\lambda)$ is a partition into parts $\binom{r+i}{r}$ with $i \geq 0$. Now let us prove that $f(\lambda)$ is a partition of n. From the definition of $\Delta^r(\lambda)$, it is easy to see that

$$\lambda_i = \sum_{j=i}^{k} \binom{r+j-i-1}{r-1} \Delta_j^r.$$

As $\binom{r+i-1}{r} = \sum_{j=0}^{i-1} \binom{r+j-1}{r-1}$, we have $|f(\lambda)| = |\lambda|$. We can reconstruct λ from $\Delta^r(\lambda)$. The reverse mapping f^{-1} is then easy to define. Let μ be a partition of n into parts $\binom{r+i}{r}$ with $\mu_1 = \binom{k+r-1}{r}$. Let $\mu^{(i)}$ be the multiplicity of the part $\binom{i-1+r}{r}$ in μ, $1 \leq i \leq k$. Then

$$f^{-1}(\mu) = \left(\sum_{j=1}^{k} \binom{r+k-j-1}{r-1} \mu^{(j)}, \sum_{j=2}^{k} \binom{r+k-j-1}{r-1} \mu^{(j)}, \ldots, \binom{r-1}{r-1} \mu^{(k)} \right).$$

It is then easy to see that f is a bijection. Note that if $r = 1$ then $f(\lambda) = \lambda'$, the conjugate of λ.

As we said in the introduction, this bijection gives some refinements of the identity. Let us now present these refinements.

Theorem 4. *There is a one-to-one correspondence between the partitions in $P_r(n)$ with k parts and j positive r^{th} differences and the partitions of n into parts $\binom{i+r}{r}$, $i \geq 0$, whose largest part is $\binom{k-1+r}{r}$ and with j parts sizes.*

Proof. Straightforward with the bijection.

Let us now illustrate the refinements thanks to generating functions and recurrences. Let $p_r(n,k,m)$ be the number of partitions in $P_r(n)$ with k parts and $\sum_{i=1}^{k} \Delta_i^r = m$. Then

$$\sum_{n,k\geq 0} p_r(n,k,m) q^n y^k x^m = 1 + \sum_{k\geq 1} \frac{y^k x q^{\binom{k-1+r}{r}}}{(1-xq^{\binom{r}{r}})(1-xq^{\binom{r+1}{r}})\dots(1-xq^{\binom{r+k-1}{r}})}$$

Let $p_{r,\leq}(n,k,m)$ be the number of partitions in $P_r(n)$ with at most k parts and $\sum_{i=1}^{k} \Delta_i^r = m$, then

$$\sum_{n\geq 0} p_{r,\leq}(n,k,m) q^n x^m = \prod_{i=0}^{k} (1-xq^{\binom{i+r}{r}})^{-1}.$$

Let $p_{r,\leq}(n,k)$ be the number of partitions in $P_r(n)$ with at most k parts. We can get an easy recurrence to compute this number :

$$p_{r,\leq}(n,k) = \begin{cases} 0 & \text{if } n < 0 \text{ or } k = 0 \text{ and } n > 0 \\ 1 & \text{if } n = 0 \text{ and } k = 0 \\ p_{r,\leq}(n - \binom{r+k-1}{r}, k) + p_{r,\leq}(n, k-1) & \text{otherwise.} \end{cases}$$

3 Asymptotic Combinatorics

Now we are going to derive an asymptotic expansion of the number $p_r(n)$ of partitions in $P_r(n)$:

Theorem 5. *As $n \to \infty$, we have:*

$$p_r(n) \sim c n^{-(r+1)/2} \exp[C(1+r) n^{\frac{1}{1+r}}] \tag{1}$$

where the constants C and c are given by

$$C = \left[r!^{1/r} \frac{1}{r} \zeta(1+r^{-1}) \Gamma(1+r^{-1})\right]^{\frac{r}{r+1}} \tag{2}$$

and

$$c = \prod_{j=0}^{r-1} j! \left(\frac{C}{2\pi}\right)^{\frac{r+1}{2}} (r!)^{-r/2} (1+r^{-1})^{-1/2} \tag{3}$$

Proof. The proof of the previous is based on a more general recent result :

Theorem 6. [4] *Let* $P(x) = Ax^d + Bx^{d-1} + \cdots$ *be a polynomial of degree* d *which is positive for* $x \geq 1$, *and which assumes integral values for integral* x. *Let* S *be the set* $\{P(1), P(2), \ldots\}$, *and assume that the set* S *has gcd* 1. *Let* $p_S(n)$ *be the number of partitions of* n *whose parts lie in the set* S. *Let* ρ_j *be the negatives of the roots of* $P(x)$, *so that*

$$P(x) = A\prod_{j=1}^{d}(x+\rho_j)$$

(in particular, $B/A = \sum \rho_j$*). Then,*

$$p_S(n) \quad = \quad c\,n^{-\kappa-\frac{1}{2}}\exp\{(1+d)Cn^{\frac{1}{d+1}} + o(1)\},$$

where

$$\kappa = \frac{Ad+B}{A(d+1)}$$
$$\alpha = 1/d$$
$$C = \left[A^{-\alpha}\alpha\zeta(1+\alpha)\Gamma(1+\alpha)\right]^{\frac{d}{d+1}}$$

and

$$c = \prod_{j=1}^{d}\Gamma(1+\rho_j)\,C^{1+B/Ad}\,A^{\frac{1}{2}+B/Ad}\,(1+\alpha)^{-1/2}\,(2\pi)^{-(d+1)/2} \tag{4}$$

This theorem follows with a little effort from Ingham's Tauberian theorem for partitions [11]. To prove Theorem 5 we just have to set $P(x) = \binom{x+r-1}{r}$. Hence we have $d = r$, $A = 1/r!$, $B = 1/(2(r-2)!)$, and $\rho_j = j-1$.

Let us now compute the asymptotic behavior of the average number of positive r^{th} difference of the partitions in $P_r(n)$. This number $\delta_r(n)$ can be defined as follows for any n : $\delta_r(n) = \frac{1}{p_r(n)}\sum_{\lambda\in P_r(n)} |\{i \mid \Delta_i^r(\lambda) > 0\}|$. Thanks to the bijection it is straightforward to compute this value for any n and indeed :

$$\delta_r(n) = \frac{1}{p_r(n)}\sum_{i\geq 0} p_r\left(n - \binom{r+i}{r}\right)$$

Proposition 1. *For suitable constant* A

$$\delta_r(n) \sim An^{\frac{1}{1+r}} \tag{5}$$

namely, A *is given by*

$$A = \Gamma(1+r^{-1})\,r!^{1/r}\,C^{-1/r}$$

where the constant C *is given in (2).*

Proof. Given our asymptotic formula for $p_r(n)$, it is not hard to find that $p_r(n-K)/p_r(n)$ is the K-th power of a certain number τ_n which is less than one, but converges to one as n grows. The sum over K of these K-th powers, only for K in the set of allowed parts, can be computed by use of Mellin's formula. This yields the stated result.

Let us now compute the asymptotic behavior of the average number of rth differences greater or equal to m in the partitions in $P_r(n)$. This number $\delta_{r,m}(n)$ can be defined as follows for any n : $\frac{1}{p_r(n)}\sum_{\lambda\in P_r(n)}|\{i \mid \Delta_i^r(\lambda)\geq m\}|$. Thanks to the bijection it is straightforward to compute this value for any n and $m\geq 1$:

$$\delta_{r,m}(n) = \frac{1}{p_r(n)}\sum_{i>0} p_r\left(n - m\binom{r+i}{r}\right).$$

Proposition 2.

$$\delta_{r,m}(n) \sim m^{-1/r} A n^{\frac{1}{1+r}}$$

Proof. Similar to the proof of Proposition 1.

A straightforward consequence of Propositions 1 and 2 is then :

Proposition 3.

$$\lim_{n\to\infty}\frac{\delta_{r,m}(n)}{\delta_r(n)} = \frac{1}{m^{1/r}}$$

4 Probabilistic Combinatorics

In this section we complete the proof of Theorem 3. The probability in question is the average value of the ratio

$$\mathbf{E}\frac{D_{n,r,m}}{D_{n,r}},$$

where $D_{n,r,m} = \sum_k I(\Delta_k^r \geq m)$ and $D_{n,r} = D_{n,r,1}$. To compute this average ratio we find it convenient to consider $F_r(n)$, the image of $P_r(n)$ under the bijection f. The explicit form of that bijection tells us that the Δ_k^r's are multiplicities of parts in the partitions of n whose part sizes are in the set $S_r = \{\binom{i+r}{r},\ i = 0, 1, \ldots\}$. To prove our Theorem 3, we will use probabilistic arguments which are the generalization of the work on ordinary partitions [7,8].

Let $\mathbf{Q}$ be the uniform probability measure on the set $P(n)$ of all partitions of n and $\mathbf{Pr}$ the uniform probability measure on $P_r(n)$. Since $P_r(n) \subset P(n)$ the measure **Pr** is a restriction of **Q** to $P_r(n)$ or, in other words it's the conditional measure on *all* partitions given that a partition is in $P_r(n)$. That is, for any subset A of $P_r(n)$:

$$\mathbf{Pr}(\lambda_r \in A) = \mathbf{P}((\Gamma_{\binom{\ell+r}{r}}) \in A | \sum_\ell \binom{\ell+r}{r} \Gamma_{\binom{\ell+r}{r}} = n),$$

with Γ_k independent geometric random variables with the parameters $1 - q^k$.

Our argument will follow that of [7, Theorem 3]. First we establish a lower bound on the probability

$$\mathbf{P}(\sum_\ell \binom{\ell+r}{r} \Gamma_{\binom{\ell+r}{r}} = n).$$

In order to do so, we will closely follow an argument of Fristedt [8]. We will choose the value of $q = q_n$ which makes the expected value of the sum $\sum_\ell \binom{\ell+r}{r} \Gamma_{\binom{\ell+r}{r}}$ asymptotic to n and then we establish a (local) limit theorem for the (normalized) random variables

$$X = \sum_{\ell \geq 0} \binom{\ell+r}{r} \Gamma_{\binom{\ell+r}{r}},$$

to show that the probability in question is of order 1 over the standard deviation of the sum. Since Γ's are geometric, the expected value of the sum is

$$\mu = \sum_{\ell=0}^{\infty} \binom{\ell+r}{r} \frac{q^{\binom{\ell+r}{r}}}{1 - q^{\binom{\ell+r}{r}}},$$

and since they, in addition, are independent the variance is

$$\sigma^2 = \sum_{\ell=0}^{\infty} \binom{\ell+r}{r}^2 \frac{q^{\binom{\ell+r}{r}}}{\left(1 - q^{\binom{\ell+r}{r}}\right)^2}.$$

We can show that :

$$\mu \sim \frac{(r!)^{1/r}}{r} \frac{1}{\ln^{(r+1)/r}(1/q)} \int_0^\infty u^{1/r} \frac{e^{-u}}{1-e^{-u}} du$$
$$= \frac{(r!)^{1/r}}{r} \frac{C_r}{\ln^{(r+1)/r}(1/q)},$$

where C_r is taken from [9, formula 3.411-7]. Thus, the choice

$$q = q_n = \exp(-\frac{1}{Cn^{r/(r+1)}}),$$

where the C comes from the Theorem 5, makes $\mu \sim n$ and we also get that there exists a constant H_r such that :

$$\sigma^2 \sim H_r n^{\frac{2r+1}{r+1}}.$$

Now for the local limit theorem. Let Y be $(X-\mu)/\sigma$ and $\phi(t) = \mathbf{E}e^{itY}$, be the characteristic function of Y. We can show that $\forall t \in \mathbf{R}$

$$\phi(t) \to \exp\left(-\frac{t^2}{2}\right)$$

which establishes the local limit theorem. It remains to strenghten it to the local limit theorem. To this end we appeal to [5, Theorem 2.9] with $h = 1/\sigma$. We can find an integrable function $f^* = \exp(-c_r t^2)$ and a sequence $(\beta_n) = \gamma\sigma^{1/(2r+1)}$, $\beta_n \to \infty$, such that

$$\forall t \qquad |\phi(t)| I(|t| \leq \beta_n) \leq f^*(t),$$

and

$$\sup_{\beta_n \leq |t| \leq \pi\sigma} |\phi(t)| = o(1/\sigma).$$

Therefore we have established our local limit theorem.

We now appeal to Theorem 2.9 of [5]. Since $n - \mathbf{E}X = O(n^{r/(r+1)})$ then if we set $y = (X - \mathbf{E}X)/\sigma$ we have $y \to 0$ and thus

$$\lim_{n\to\infty} n^{\frac{2r+1}{2(r+1)}} \mathbf{P}(X = n) = \lim_{n\to\infty} \sigma \mathbf{P}(Y = y) = \frac{1}{\sqrt{2\pi}},$$

i.e. there exists a constant c_r with

$$\mathbf{P}(X = n) \geq n^{-\frac{2r+1}{2(r+1)}}/c_r,$$

for large n.

To complete the proof, let $I_j = \{\Gamma_{\binom{j+r}{r}} \geq 1\}$. Then, denoting for simplicity a set and its indicator by the same symbol, we have

$$\begin{aligned}\mathbf{Pr}(|D_{n,r} - \mathbf{E}D_{n,r}| \geq t) &= \frac{\mathbf{P}(|\sum_j (I_j - \mathbf{E}I_j)| \geq t \cap \sum_l \binom{l+r}{r}\Gamma_{\binom{l+r}{r}} = n)}{\mathbf{P}(\sum_l \binom{l+r}{r}\Gamma_{\binom{l+r}{r}} = n)} \\ &\leq c_r n^{\frac{2r+1}{2(r+1)}} \mathbf{P}(|\sum_j (I_j - \mathbf{E}I_j)| \geq t).\end{aligned}$$

Since the random variables $(I_j - \mathbf{E}I_j)$ are independent, mean – zero, and uniformly bounded by 1 we get by using the "arcsinh" inequality due to Prokhorov ([12, Theorem 5.2.2(ii)]) :

$$\mathbf{P}\left(\sum_j |I_j - \mathbf{E}I_j| \geq t\right) \leq 2\exp\left\{-\frac{t}{2}\text{arcsinh}\left(\frac{t}{2\text{var}(\sum_j I_j)}\right)\right\}$$

As there exists a constant d_r such that :

$$\text{var}(\sum_j I_j) \sim d_r n^{1/(r+1)},$$

we finally have, using the fact that $\text{arcsinh}x \geq x/2$ for x close to 0 :

$$\mathbf{Pr}(|D_{n,r} - \mathbf{E}D_{n,r}| \geq t_n) \leq 2c_r n^{(2r+1)/(2(r+1))} \exp\left(-\frac{t_n^2}{2d_r n^{1/(r+1)}}\right)$$

Thus, selecting $t = t_n$ so that $t_n = o(n^{1/(r+1)})$ and $t_n^2/n^{1/(r+1)} \to \infty$ not too slow, say, $t_n = \Theta(n^{3/(2(r+1))})$, we see that

$$\mathbf{Pr}(|D_{n,r} - \mathbf{E}D_{n,r}| \geq t_n) \to 0, \quad n \to \infty.$$

Hence, since $t_n = o(\mathbf{E}D_{n,r})$, integrating $D_{n,r,m}/D_{n,r}$ over the set $\{|D_{n,r} - \mathbf{E}D_{n,r}| \leq t_n\}$ and its complement, we obtain

$$\mathbf{E}\frac{D_{n,r,m}}{D_{n,r}} \sim \frac{\mathbf{E}D_{n,r,m}}{\mathbf{E}D_{n,r}}, \qquad n \to \infty.$$

By Proposition 3 we know that $\mathbf{E}D_{n,r,m}/\mathbf{E}D_{n,r} \sim m^{-1/r}$, $n \to \infty$. Hence we get our result :

$$\mathbf{E}\frac{D_{n,r,m}}{D_{n,r}} \sim \frac{1}{m^{1/r}}, \qquad n \to \infty.$$

5 Future Work

One interesting question would be to show how fast the probability converges. It appears that our proof does come with specific rates. The question is therefore to show if they are optimal or not. Let us note we also showed that the probability that a random part size of a random partition into parts $\binom{i+r}{r}$ of n has multiplicity at least m is $m^{-1/r}$ when $n \to \infty$. Our aim is now to identify the sets S such that the probability that a random part size of a random partition into parts in S of n has multiplicity at least m is a "nice" constant when $n \to \infty$.

References

1. G.E. Andrews, The Theory of partitions, Encycl. Math. Appl. vol. 2, Addison-Weley, 1976.
2. G.E. Andrews, MacMahon's Partition analysis: II, Fundamental Theorems, *Annals of Combinatorics*, to appear.
3. G.E. Andrews, P. Paule and A. Riese, MacMahon's Partition Analysis III: The Omega Package, *Preprint.*
4. E.R. Canfield, S. Corteel, P. Hitczenko, *preprint.*
5. N. R. Chaganty, J. Sethuraman, Strong large deviation and local limit theorems, *Ann. Probab.* **21** (1993), 1671-1690.
6. Sylvie Corteel's One-Line Proof of a Partition Theorem Generated by Andrews-Paule-Riese's Computer *Shalosh B. Ekhad's and Doron Zeilberger's Very Own Journal.*
7. S. Corteel, B. Pittel, C.D. Savage and H.S. Wilf, On the multiplicity of parts in a random partition, *Random Structures and Algorithms*, 14, No.2, 185-197, 1999.

8. B. Fristedt, The structure of random partitions of large integers, *Trans. Amer. Math. Soc.* **337** (1993), 703-735.
9. I. S. Gradshteyn, I. M. Ryzhik, *Table of Integrals, Series, and Products*, 4th ed., Academic Press, New York, 1965.
10. W. K. Hayman, On a generalisation of Stirling's formula *J. für die reine und angewandte Mathematik* , 196, 1956, 67–95.
11. A.E. Ingham, A Tauberian theorem for partitions. *Ann. of Math.***42** (1941), 1075-1090.
12. W. Stout, *Almost Sure Convergence*, Academic Press, New York, 1974.

Beta-Expansions for Cubic Pisot Numbers

Frédérique Bassino

I.G.M., Université de Marne La Vallée
77454 Marne-la-Vallée Cedex 2. France
bassino@univ-mlv.fr

Abstract. Real numbers can be represented in an arbitrary base $\beta > 1$ using the transformation $T_\beta : x \to \beta x \pmod 1$ of the unit interval; any real number $x \in [0, 1]$ is then expanded into $d_\beta(x) = (x_i)_{i\geq 1}$ where $x_i = \lfloor \beta T_\beta^{i-1}(x) \rfloor$.
The closure of the set of the expansions of real numbers of $[0, 1[$ is a subshift of $\{a \in \mathbb{N} \mid a < \beta\}^{\mathbb{N}}$, called the beta-shift. This dynamical system is characterized by the beta-expansion of 1; in particular, it is of finite type if and only if $d_\beta(1)$ is finite; β is then called a simple beta-number.
We first compute the beta-expansion of 1 for any cubic Pisot number. Next we show that cubic simple beta-numbers are Pisot numbers.

Introduction

Representations of real numbers in an arbitrary base $\beta > 1$, called beta-expansions, have been introduced by Rényi ([14]). They arise from the orbits of the piecewise-monotone transformation of the unit interval : $T_\beta : x \to \beta x \pmod 1$. Such transformations were extensively studied in ergodic theory ([13]).

More precisely, any real number $x \in [0, 1]$ is expanded into $d_\beta(x) = (x_i)_{i\geq 1}$ where $x_i = \lfloor \beta T_\beta^{i-1}(x) \rfloor$. The nonnegative integers d_i are elements of the digit alphabet $A = \{a \in \mathbb{N} \mid a < \beta\}$. These representations generalize standard representations in an integral base to a real base; indeed the beta-expansion of any real number of $[0, 1[$ can equivalently be obtained by the greedy algorithm. Only the beta-expansion of 1 differs.

Properties of beta-expansions are strongly related to symbolic dynamics ([4]). The closure of the set of infinite sequences, appearing as beta-expansions of numbers of the interval $[0, 1[$, is a dynamical system, that is, a closed shift-invariant subset of $A^{\mathbb{N}}$, called the *beta-shift*.

An important property of the beta-shift is that its nature is entirely determined, in a combinatorial manner, by the beta-expansion of 1: the beta-shift is sofic, that is to say the set of its finite factors is recognized by a finite automaton, if and only the beta-expansion of 1 is eventually periodic ([3]); it is of finite type, that is to say the set of its finite factors is defined by forbidding a finite set of words, if and only if the beta-expansion of 1 is finite ([12]).

When the beta-expansion of 1 is eventually periodic, β is called a *beta-number* and when the beta-expansion of 1 is finite, β is said to be a *simple beta-number*.

S. Rajsbaum (Ed.): LATIN 2002, LNCS 2286, pp. 141–152, 2002.

The eventually periodic beta-expansions were extensively studied by Bertrand ([3]) and by Schmidt ([15]). In particular, it is known that Pisot numbers are beta-numbers. Concerning Salem numbers, we only know that if β is a Salem number of degree 4, then the beta-expansion of 1 is eventually periodic ([5]). It is conjectured that Salem numbers of degree 6 are still beta-numbers, but not all Salem numbers of degree 8 ([7]).

The domain of the Galois conjugates of all beta-numbers was also investigated independently by Solomyak ([16]) and by Flatto, Lagarias and Poonen ([8]).

For a general presentation of the beta-shift one can refer to [9].

In the following, we summarize properties of beta-numbers. We compute the beta-expansion of 1 for any cubic Pisot number and we establish a characterization of cubic simple beta-numbers, showing that they are Pisot numbers.

A very close problem, seen from the point of view of numeration systems, was studied by Akiyama ([1]). He showed that in the cubic case, the real numbers of the set $\mathbb{N}[\beta^{-1}]$ have a finite beta-expansion if and only β is a Pisot unit and 1 has a finite beta-expansion. This finiteness problem is equivalent to a problem of fractal tiling generated by Pisot numbers.

1 Beta-Numbers

Real numbers can be represented in an arbitrary base $\beta > 1$ using the transformation $T_\beta : x \to \beta x \pmod 1$ of the unit interval; any real number $x \in [0, 1]$ is then expanded into $d_\beta(x) = (x_i)_{i\geq 1}$ where $x_i = \lfloor \beta T_\beta^{i-1}(x) \rfloor$. When a beta-expansion is of the form uv^ω, the expansion is said to be *eventually periodic.* If a representation ends with infinitely many zeros, like $u0^\omega$, it is said to be *finite* and the ending zeros are omitted.

Let us denote by S_β the closure of all beta-expansions of real numbers of $[0, 1[$ and by σ the (one-sided) shift defined by $\sigma((x_i)_{i\geq 1}) = (x_{i+1})_{i\geq 1}$. The set S_β endowed with the shift is called the beta-shift, it is a subshift of $A^{\mathbb{N}}$, A being the digit set, *i.e.*, $A = \{a \in \mathbb{N} \mid a < \beta\}$.

An important property ([13]) of the beta-shift S_β is that its nature is entirely determined by $d_\beta(1)$ the beta-expansion of 1. Indeed, setting $d^*(1) = d_\beta(1)$ if $d_\beta(1)$ is infinite and $d^*(1) = (d_1 d_2 \ldots d_{n-1}(d_n - 1))^\omega$ if $d_\beta(1) = d_1 d_2 \ldots d_{n-1} d_n$, a sequence x of nonnegative integers belongs to S_β if and only if it satisfies the following lexicographical order conditions: $\forall p \geq 0, \quad \sigma^p(x) \leq d^*(1)$.

Recall that the beta-expansion of 1 also can be characterized ([13]) by lexicographical order conditions: let $d = (d_i)_{i\geq 1}$ be a sequence of nonnegative integers different from 10^ω, such that $\sum_{i\geq 1} d_i \beta^{-i} = 1$, with $d_1 \geq 1$ and for $i \geq 2$, $d_i \leq d_1$, then d is the beta-expansion of 1 if and only if for all $p \geq 1$, $\sigma^p(d) < d$.

We recall that an algebraic integer β strictly greater than 1 is called a *Perron number* if all its Galois conjugates have modulus strictly less than β, a *Pisot number* if all its Galois conjugates have modulus strictly less than 1, and a *Salem number* if all its conjugates are less than 1 in modulus and at least one conjugate has modulus 1.

Let β be a beta-number. Denote by $d_\beta(1) = d_1 \dots d_n(d_{n+1} \dots d_{n+p})^\omega$, where n and p are chosen minimal, the beta-expansion of 1. Then the adjacency matrix $\mathcal{M}_\beta$ of the finite automaton recognizing the set of its finite factors (Fig.1) is a *primitive* (*i.e.*, its associated graph is strongly connected and the lengths of its cycles are relatively prime) nonnegative integral matrix whose spectral radius is β; so, from the Perron-Frobenius theorem, β is a Perron number.

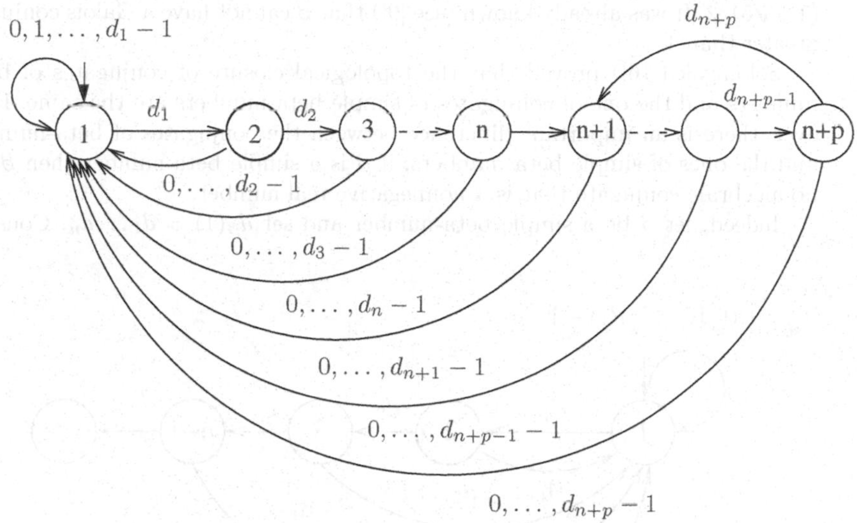

Fig. 1. Automaton recognizing the set of the finite factors of S_β

The characteristic polynomial of $\mathcal{M}_\beta$

$$P(X) = X^{n+p} - \sum_{i=1}^{n+p} d_i X^{n+p-i} - X^n + \sum_{i=1}^{n} d_i X^{n-i}$$

is called, following the terminology introduced by Hollander ([11]), the associated *beta-polynomial.*

As P is a multiple of the minimal polynomial M_β of β, $P(0) = d_{n+p} - d_n$ is a multiple of $|M_\beta(0)| = |\prod \beta_i|$, where β_i runs over the set of algebraic conjugates of β. So, we get that $|\prod \beta_i|$ has to be smaller than $\lfloor \beta \rfloor$.

As a consequence, in the quadratic case, the only beta-numbers are the Pisot numbers. Conversely, it is known that if β is a Pisot number then β is a beta-number ([2]). An important gap remains between Pisot and Perron numbers.

Example 1. The quadratic number $\beta = (1 + \sqrt{13})/2$ is not a beta-number since $M_\beta(X) = X^2 - X - 3$ and $M_\beta(0) > \lfloor \beta \rfloor$.

Let β be the Pisot number $(3+\sqrt{5})/2$, then β is a beta-number and $d_\beta = 21^\omega$.

Let β be the golden ratio $(1+\sqrt{5})/2$, then β is a simple beta-number and $d_\beta(1) = 11$.

On the other hand, the domain of the Galois conjugates of beta-numbers was studied by Solomyak ([16]) and independently by Flatto, Lagarias and Poonen ([8]). They showed in particular that if the beta-expansion of 1 is eventually periodic then the Galois conjugates of β have modulus less than the golden ratio $(1+\sqrt{5})/2$. It was already known (see [9]) that β cannot have a Galois conjugate greater than 1.

Solomyak ([16]) proved that the topological closure of conjugates of beta-numbers and the one of conjugates of simple beta-numbers are the same. However, there is an important difference between the conjugates of beta-numbers and the ones of simple beta numbers: if β is a simple beta-number then β has no algebraic conjugate that is a nonnegative real number.

Indeed, let β be a simple beta-number and set $d_\beta(1) = d_1 \dots d_n$. Consider

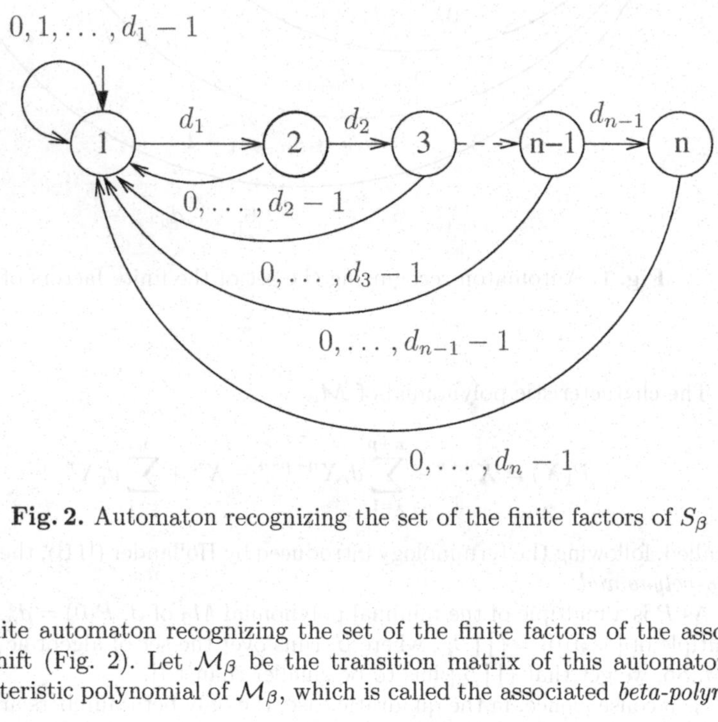

Fig. 2. Automaton recognizing the set of the finite factors of S_β

the finite automaton recognizing the set of the finite factors of the associated beta-shift (Fig. 2). Let $\mathcal{M}_\beta$ be the transition matrix of this automaton. The characteristic polynomial of $\mathcal{M}_\beta$, which is called the associated *beta-polynomial*,

$$P(X) = X^n - \sum_{i=1}^{n} d_i X^{n-i}$$

has only one positive real root.

Example 2. Salem numbers are roots of reciprocal polynomials. Thus if β is a Salem number, $1/\beta > 0$ is a Galois conjugate of β, and so β is not a simple beta-number.

The previous conditions are sufficient for a quadratic algebraic integer to be a simple beta-number.

Proposition 1. *[10] The simple beta-numbers of degree 2 are exactly the quadratic Pisot numbers without a positive real Galois conjugate. They are the positive roots of the polynomials*

$$X^2 - aX - b \quad \text{with} \quad a \geq b \geq 1,$$

The beta-expansion of 1 *is then* $d_\beta(1) = ab$.

Example 3. The minimal polynomial of $(1+\sqrt{5})/2$ is $X^2 - X - 1$, $(1+\sqrt{5})/2$ is a simple beta-number and $d_\beta(1) = 11$.

The minimal polynomial of $(3+\sqrt{5})/2$ is $X^2 - 3X + 1$, therefore $(3+\sqrt{5})/2$ is not a simple beta-number.

2 Beta-Expansions of 1 for Cubic Pisot Numbers

Let us recall the characterization of cubic Pisot numbers due to Akiyama ([1])

Theorem 1 (Akiyama [1]). *Let* $\beta > 1$ *be a cubic number and let*

$$M_\beta(x) = X^3 - aX^2 - bX - c$$

be its minimal polynomial.

Then β *is a Pisot number if and only if it both inequalities*

$$|b-1| < a + c \quad \text{and} \quad (c^2 - b) < sgn(c)(1 + ac)$$

hold.

Remark 1. Note that a must be a nonnegative integer.

The following theorem gives the β-expansion of 1 for any cubic Pisot number.

Theorem 2. *Let* β *be a cubic Pisot number and let*

$$M_\beta(x) = X^3 - aX^2 - bX - c$$

be its minimal polynomial. Then the beta-expansion of 1 *is*

- *Case 1 : When* $b \geq a$, *then* $d_\beta(1) = (a+1)(b-1-a)(a+c-b)(b-c)c$.
- *Case 2: When* $0 \leq b \leq a$, *if* $c > 0$, $d_\beta(1) = abc$, *otherwise,*

$$d_\beta(1) = a[(b-1)(c+a)]^\omega.$$

- *Case 3: When* $-a < b < 0$, *if* $b + c \geq 0$, *then* $d_\beta(1) = (a-1)(a+b)(b+c)c$, *otherwise* $d_\beta(1) = (a-1)(a+b-1)(a+b+c-1)^\omega$
- *Case 4: When* $b \leq -a$, *let* k *be the integer of* $\{2, 3, \ldots, a-2\}$ *such that, denoting* $e_k = 1 - a + (a-2)/k$, $e_k \leq b + c < e_{k-1}$.
  - *If* $b(k-1) + c(k-2) \leq (k-2) - (k-1)a$, $d_\beta(1) = d_1 \ldots d_{2k+2}$ *with*

$$
\begin{aligned}
d_1 &= a - 2, \\
d_{k+2-i} &= -(k+3-i) + a(k+2-i) + b(k+1-i) + c(k-i), 3 \leq i \leq k \\
d_k &= -k + ak + b(k-1) + c(k-2) \\
d_{k+1} &= -(k-1) + ak + bk + c(k-1) \\
d_{k+2} &= -(k-2) + a(k-1) + bk + ck \\
d_{2k+2-i} &= -(i-2) + a(i-1) + bi + c(i+1) \quad k \geq 3, 2 \leq i \leq (k-1) \\
d_{2k+1} &= b + 2c \quad \text{and} \quad d_{2k+2} = c.
\end{aligned}
$$

  - *If* $b(k-1) + c(k-2) > (k-2) - (k-1)a$, *let* m *be the integer defined by* $m = \min\{i \in \mathbb{N}$ *such that* $(i+1)b + ic > i - (i+1)a\}$.

When $m = 1$, $d_\beta(1) = (a-2)(2a+b-2)(2a+2b+c-2)(2a+2b+2c-2)^\omega$.

When $m > 1$, $d_\beta(1) = d_1 d_2 \ldots d_{m+2} d_{m+3}^\omega$, *with*

$$
\begin{aligned}
d_1 &= a - 2, \quad d_2 = 2a + b - 3, \\
d_{m+3-i} &= 2a + b - 3 + (m+1-i)(a+b+c-1) \quad m \geq 3, 3 \leq i \leq m, \\
d_{m+1} &= 2a + b - 2 + (m-1)(a+b+c-1), \\
d_{m+2} &= a + b - 1 + m(a+b+c-1), \\
d_{m+3} &= (m+1)(a+b+c-1).
\end{aligned}
$$

Example 4. When $a \geq b \geq 0$ and $c > 0$, we obtain the only beta-expansion of 1 of length 3.

The smallest Pisot number has $M_\beta = X^3 - X - 1$ as minimal polynomial, it is a simple beta-number and $d_\beta(1) = 10001$.

The positive root β of $M_\beta = X^3 - 3X^2 + 2X - 2$ is a simple beta-number and $d_\beta(1) = 2102$.

The case where $b \leq -a$ shows that from a cubic simple beta-number, we can obtain an arbitrary long beta-expansion of 1. For any integer k greater than or equal to 2, the real root β of the irreducible polynomial $X^3 - (k+2)X^2 + 2kX - k$, is a simple beta number whose integer part is equal to k, and the beta-expansion of 1 has length $2k+2$. For $k = 2$, we get $d_\beta(1) = 221002$; for $k = 3$, we get $d_\beta(1) = 31310203$.

Example 5. The greatest positive root β of $M_\beta = X^3 - 2X^2 - X + 1$ is a beta-number and $d_\beta(1) = 2(01)^\omega$.

If β is the positive root of $X^3 - 5X^2 + 3X - 2$, then $d_\beta(1) = 413^\omega$. When β is the greatest positive root of $X^3 - 5X^2 + X + 2$, then $d_\beta(1) = 431^\omega$.

For any integer k greater than or equal to 3, the real root β of the irreducible polynomial $X^3 - (k+2)X^2 + (2k-1)X - (k-1)$, is a beta number whose integer part is equal to k, and the beta-expansion of 1 is eventually periodic of period 1, the length of its preperiod k. For $k = 3$, we get $d_\beta(1) = 3302^\omega$; for $k = 4$, we get $d_\beta(1) = 42403^\omega$.

Proof. It is known that Pisot numbers are beta-numbers, thus, for any cubic Pisot number β, the beta-expansion of 1 is finite or eventually periodic. In any case, we first compute the associated beta-polynomial P. Next we prove that the sequence $d = (d_i)_{i\geq 1}$ of nonnegative integers obtained from the beta-polynomial satisfy lexicographical order conditions: for all $p \geq 1$, $\sigma^p(d) < d$.

First of all, we recall that, from Theorem 1, a cubic number β, greater than 1 and having

$$M_\beta(X) = X^3 - aX^2 - bX - c$$

as minimal polynomial, is a cubic Pisot number if and only if it both

$$|b-1| < a+c \quad \text{and} \quad (c^2 - b) < sgn(c)(1+ac)$$

hold.

Denote by Q the *complementary factor* of the beta-polynomial P defined by $P(X) = M_\beta(X)Q(X)$. As we shall see in what follows, the value of Q depends upon the value of the coefficients of M_β.

Case 1: When $b > a$, as β is a Pisot number, from Theorem 1, c is a positive integer. In this case, the complementary factor is $Q(X) = X^2 - X + 1$ and $d_\beta(1) = (a+1)(b-1-a)(a+c-b)(b-c)c$.

Indeed, as $(c^2 - b) < sgn(c)(1+ac)$ and $c > 0$, we get $c \leq a+1$. As $|b-1| < a+c$, we get $b-1-a \leq a$ and $0 \leq a-b+c$. From $b > a$, we get that $0 \leq b-a-1$ and, as $c \leq a+1$, that $a-b+c \leq a$. Finally as $0 \leq a-b+c \leq a$, we obtain $0 \leq b-c \leq a$.

Case 2: When $0 \leq b \leq a$, the complementary factor is then $Q(X) = 1$ and the associated beta-polynomial is equal to the minimal polynomial.

If $c > 0$, then $d_\beta(1) = abc$. Indeed, as $(c^2 - b) < sgn(c)(1+ac)$, we get $c \leq a$.

If $c < 0$, then $d_\beta(1) = a[(b-1)(a+c)]^\omega$. As $|b-1| < a+c$, we get $b-1 \leq a-2$. As $(c^2 - b) < sgn(c)(1+ac)$, we get that $c \geq -a$ and, consequently, $0 \leq c+a \leq a-1$.

Case 3: When $-a < b < 0$, if $b+c \geq 0$ then the complementary factor is $Q(X) = X+1$ and $d_\beta(1) = (a-1)(a+b)(b+c)c$. Indeed, as $-a < b < 0$, we obtain $1 \leq a+b \leq a-1$. Since $b+c \geq 0$, c is a positive integer. From $(c^2 - b) < sgn(c)(1+ac)$, we get that $c \leq a-1$ and $b+c \leq a-2$.

If $b+c < 0$, then $Q(X) = 1$ and $d_\beta(1) = (a-1)(a+b-1)(a+b+c-1)^\omega$. As $-a < b < 0$, we get $0 \leq a+b-1 \leq a-2$. From $|b-1| < a+c$, we get that $1 \leq a+b+c-1$ and as $b+c < 0$, we obtain $a+b+c-1 \leq a-2$.

Case 4: First of all, since $|b-1| < a+c$, we get $-a+2 \leq b+c$. Moreover as $b \leq -a$, we get $c \geq 2$ and as $(c^2 - b) < sgn(c)(1+ac)$, we obtain $c \leq a-2$, thus $b+c \leq -2$. So, there exists an integer k in $\{2, 3, \ldots, a-2\}$, such that, denoting $e_k = 1 - a + (a-2)/k$, $e_k \leq b+c < e_{k-1}$.

When $b(k-1)+c(k-2) \leq (k-2)-(k-1)a$, the complementary factor is

$$Q(X) = \frac{(X^k-1)(X^{k+1}-1)}{(X-1)^2}$$

and $d_\beta(1) = d_1 \ldots d_{2k+2}$ with

$$\begin{aligned}
d_1 &= a-2,\\
d_{k+2-i} &= -(k+3-i)+a(k+2-i)+b(k+1-i)+c(k-i), k \geq 3, 3 \leq i \leq k\\
d_k &= -k+ak+b(k-1)+c(k-2)\\
d_{k+1} &= -(k-1)+ak+bk+c(k-1)\\
d_{k+2} &= -(k-2)+a(k-1)+bk+ck\\
d_{2k+2-i} &= -(i-2)+a(i-1)+bi+c(i+1) \quad k \geq 3, 2 \leq i \leq (k-1)\\
d_{2k+1} &= b+2c \quad \text{and} \quad d_{2k+2} = c.
\end{aligned}$$

We now verify that the lexicographical order conditions on $d_\beta(1)$ are satisfied.

As $2 \leq c \leq a-2$ and $b+c \leq -2$, we get $d_{2k+1} \leq a-4$. From $e_k \leq b+c$ and $b(k-1)+c(k-2) \leq (k-2)-(k-1)a$, we get $d_{2k+1} \geq 0$.

For $k \leq 3$ and $2 \leq i \leq k-1$, $d_{2k+2-i} = -(i-2)+a(i-1)+bi+c(i+1)$. As $b+c < e_i$, we get $d_{2k+2-i} < c$. As $-a+2 \leq b+c$ and $b+2c \geq 0$, we get $d_{2k+2-i} \geq i$.

As $e_k \leq b+c$, we obtain $d_{k+2} \geq 0$. Since $c \leq a-2$, $d_{k+1} > d_{k+2}$ and since $b+c \leq -2$, $d_k > d_{k+1}$. Moreover from $b(k-1)+c(k-2) \leq (k-2)-(k-1)a$, we get $d_k \leq a-2$.

For $k \leq 3$, as $|b-1| < a+c$, we obtain $d_2 < \cdots < d_{k-1}$. As $b+c < e_{k-1}$ and $b+2c \leq 0$, we get $d_{k-1} < a-2$. Moreover from $c \leq a-2$ and $a+b+c-1 > 0$, we get that $d_2 = 2a+b-3$ is nonnegative.

All d_i's are smaller than d_1, only d_{2k+2} and d_k can be equal to d_1. Therefore we have to verify that $d_2 \geq d_{k+1}$ when $k \geq 3$ (otherwise $d_2 = d_k$ and $d_k > d_{k+1}$). If $d_k = a-2$, then $b+c = e_k$, and $d_{k+1} = a-c-1$. As $a+b+c-1 > 0$, we obtain $d_{k+1} \leq d_2$. In case of equality, if $k=3$, then $d_3 = d_k$ and $d_k > d_{k+2}$, otherwise $d_3 > d_2$ and $d_{k+1} > d_{k+2}$, therefore $d_3 > d_{k+2}$.

So lexicographical order conditions are satisfied and $d_1 \ldots d_{2k+2}$ is the beta-expansion of 1.

When $b(k-1)+c(k-2) > (k-2)-(k-1)a$, as $b \leq -a$, we get $k \geq 3$. Let m be the integer defined by $m = \min\{i \in \mathbb{N} \text{ such that } (i+1)b+ic > i-(i+1)a\}$. Note that by definition of m, $m \leq k-2$ and since $b \leq -a$, $m \geq 1$. In this case, the complementary factor is

$$Q(X) = \sum_{i=0}^{m} X^i.$$

The beta-expansion of 1 is then eventually periodic with period 1, the length of the preperiod is $m+2$.

When $m=1$, $P(X) = X^4-(a-1)X^3-(a+b)X^2-(b+c)X-c$ and

$$d_\beta(1) = (a-2)(2a+b-2)(2a+2b+c-2)(2a+2b+2c-2)^\omega.$$

Here $d_3 = d_{m+2} = a+b-1+m(a+b+c-1)$ and $d_4 = d_{m+3} = (m+1)(a+b+c-1)$.
When $m > 1$,

$$P(X) = X^{m+3} - (a-1)X^{m+2} - (a+b-1)X^{m+1} - \sum_{i=3}^{m}(a+b+c-1)X^i$$

$$-(a+b+c)X^2 - (b+c)X - c$$

and $d_\beta(1) = d_1 d_2 \ldots d_{m+2} d_{m+3}^\omega$, with

$$\begin{aligned}
&d_1 = a-2, \quad d_2 = 2a+b-3,\\
&d_{m+3-i} = 2a+b-3+(m+1-i)(a+b+c-1) \quad m \geq 3, 3 \leq i \leq m,\\
&d_{m+1} = 2a+b-2+(m-1)(a+b+c-1),\\
&d_{m+2} = a+b-1+m(a+b+c-1),\\
&d_{m+3} = (m+1)(a+b+c-1).
\end{aligned}$$

In both cases, $d_1 = a-2$. Since $b(k-1)+c(k-2) > (k-2)-(k-1)a$ and $c \leq a-2$, we get $-2a+3 \leq b$. Moreover as $b \leq -a$, $1 \leq d_2 \leq a-2$ when $m=1$, and $0 \leq d_2 \leq a-3$ otherwise. By definition of m, $(m+1)b+mc > m-(m+1)a$, thus $d_{m+2} \geq 0$ and $d_{m+3} \geq c$. Since $e_k \leq b+c < e_{k-1}$ and $m \leq k-2$, we obtain $d_{m+3} \leq a-3$ and $d_{m+2} \leq a-c-3$.

When $m > 1$, since $mb+(m-1)c \leq (m-1)-ma$, we get $d_{m+1} \leq a-2$. As $0 \leq 2a+b-2$ and $a+b+c-1 > 0$, $d_{m+1} > 0$. Moreover as $a+b+c-1 > 0$, one has $d_2 < d_3 < \ldots < d_{m+1}$. Note that, when $m \geq 3$, $d_2 \neq a-2$.

We now study the cases where d_i is not strictly smaller than d_1. When $m=1$, only d_2 may be equal to $a-2$, then $b=-a$ and $d_3 = c-2$, thus $d_3 < d_2$. When $m > 1$, only d_{m+1} may be equal to $a-2$, then $mb = -ma-(m-1)c+(m-1)$, and thus $d_2 - d_{m+2} = a-1-c$ is a positive integer.

We have proved that the lexicographical order conditions on $d_\beta(1)$:

$$d_1 d_2 \ldots d_{m+3}^\omega >_{lex} d_i d_{i+1} \ldots d_{m+3}^\omega \quad \text{for } 2 \leq i \leq m+3,$$

are satisfied, showing in this way that the announced beta-expansions of 1 are right.

Remark 2. The polynomials Q that appear in the cubic case are cyclotomic. In the general case, Q can be noncyclotomic and even nonreciprocal ([6]).

3 Cubic Simple Beta-Numbers

In the following, we establish that cubic simple beta-numbers are Pisot numbers. Next we give necessary and sufficient conditions on the coefficients of the minimal polynomial of β for β to be a simple beta-number.

Theorem 3. *If β is a cubic simple beta-number then β is a Pisot number.*

Remark 3. This is no longer true for simple beta-numbers of degree 4. For example, the positive root of $X^4 - 3X^3 - 2X^2 - 3$ is a simple beta-number, but is not a Pisot number.

Proof. Let β be a cubic simple beta-number and let

$$M_\beta(X) = X^3 - aX^2 - bX - c$$

be its minimal polynomial. Then β has no positive real algebraic conjugate and c is a positive integer smaller than $\lfloor \beta \rfloor$.

The condition on the product c of the roots of the polynomial M_β, *i.e.*, $|c| \leq \lfloor \beta \rfloor$, directly implies, when the Galois conjugates of β are not real numbers, that β is a Pisot number.

The only other case is the case where both Galois conjugates γ_1 and γ_2 of β are negative real numbers. We then assume that β is a cubic simple beta-number that is not a Pisot number, and show that these hypotheses are contradictory. Let γ_1 and γ_2 be the Galois conjugates of β. As $0 < c \leq \lfloor \beta \rfloor$, if one of the γ_i's is smaller than -1 the other one is greater than -1. Moreover, as the modulus of a Galois conjugate of a beta-number is smaller than the golden ratio, one can suppose, for example, that

$$\frac{1+\sqrt{5}}{2} < \gamma_2 < -1 < \gamma_1 < 0 < \beta$$

Consequently, $M_\beta(-1) > 0$, in other words, $b > a + c + 1$. Note that here $a \in \{\lfloor \beta \rfloor - 2, \lfloor \beta \rfloor - 1\}$.

As β is a simple beta-number, $d_\beta(1) = d_1 d_2 \ldots d_n$. Denote by P the associated β-polynomial:

$$P(X) = X^n - \sum_{i=1}^{n} d_i X^{n-i}$$

and denote by $Q = \sum_{i \geq 0} q_i X^i$ the quotient of the division upon the increasing powers of P by M_β. In other words,

$$P(X) = M_\beta(X) Q(X)$$

We shall show, by induction, that $q_0 \geq 1$, and that for all $i \geq 0$, $|q_{i+1}| > |q_i|$ with $sgn(q_{i+1}) = -sgn(q_i)$. We shall conclude from the growth of the moduli of its coefficients that Q is an infinite series, and thus that $d_\beta(1)$ is not finite.

In what follows, we mainly use the fact that the d_i's are nonnegative integers smaller than $\lfloor \beta \rfloor$ and the inequality $b \geq a + c + 2$.

First of all, as $d_n = q_0 c$ and d_n and c are positive integers, $q_0 \geq 1$. Since $d_{n-1} = q_0 b + q_1 c$ and $q_0 \geq 1$, $d_{n-1} \geq q_0 a + 2q_0 + (q_0 + q_1)c$. When $a = \lfloor \beta \rfloor - 1$, we directly get from $d_{n-1} \leq \lfloor \beta \rfloor$, that $q_1 < -q_0$. When $a = \lfloor \beta \rfloor - 2$, the lexicographical order conditions on $d_\beta(1)$ imply that

$$d_{n-1} d_n < d_1 d_2 \ldots d_n.$$

By definition of beta-expansions, $d_1 = \lfloor\beta\rfloor$ and here $d_2 < d_n$. Indeed as

$$\gamma_2 = \frac{1}{2}\left(a - \beta + \sqrt{(a-\beta)^2 - \frac{4c}{\beta}}\right),$$

and $\gamma_2 > -(1+\sqrt{5})/2$, we get that

$$c > \frac{\sqrt{5}-1}{2}\beta + \frac{1+\sqrt{5}}{2}\beta\{\beta\},$$

and in particular, that $c > \beta/2$, consequently $d_n = c$ and that $\beta\{\beta\} < c$. Thus $d_2 = \lfloor\beta\{\beta\}\rfloor$ is strictly smaller than d_n. Therefore the previous lexicographical order condition implies that $d_{n-1} < \lfloor\beta\rfloor$. So, as $d_{n-1} \geq \lfloor\beta\rfloor + (q_0+q_1)c$, $q_1 < -q_0$.

As $d_{n-2} = q_0a + q_1b + q_2c$ and $q_1 < -q_0 < 0$, $d_{n-2} \leq (q_1+q_0)a + 2q_1 + (q_1+q_2)c$, that is $d_{n-2} < -\lfloor\beta\rfloor + (q_1+q_2)c$, so $q_2 > -q_1$.

For all positive integers i, $d_{n-(2i+1)} = -q_{2i-2} + q_{2i-1}a + q_{2i}b + q_{2i+1}c$. From $q_{2i} > 0$, we get $d_{n-(2i+1)} \geq (q_{2i-1} + q_{2i})a + q_{2i} + (q_{2i} - q_{2i-2}) + (q_{2i} + q_{2i+1})c$. From $(q_{2i-1} + q_{2i}) \geq 1$, $q_{2i} > 2i$ and $(q_{2i} - q_{2i-2}) > 1$, we obtain $d_{n-(2i+1)} > \lfloor\beta\rfloor + (q_{2i} + q_{2i+1})c$, and thus $q_{2i+1} < -q_{2i}$.

For all positive integers i, $d_{n-(2i+2)} = -q_{2i-1} + q_{2i}a + q_{2i+1}b + q_{2i+2}c$. From $q_{2i+1} < 0$, we get $d_{n-(2i+1)} \leq (q_{2i} + q_{2i+1})a + q_{2i+1} + (q_{2i+1} - q_{2i-1}) + (q_{2i+1} + q_{2i+2})c$. As $(q_{2i} + q_{2i+1}) \leq -1$, $q_{2i+1} < -(2i+1)$ and $(q_{2i+1} - q_{2i-1}) < -1$, we get $d_{n-(2i+2)} < -\lfloor\beta\rfloor + (q_{2i+1} + q_{2i+2})c$, thus $q_{2i+2} > -q_{2i+1}$.

So Q is an infinite series; consequently if β is not a Pisot number, $d_\beta(1)$ is not finite.

As a consequence of Theorems 2 and 3, we obtain the above characterization of cubic simple beta-numbers.

Proposition 2. *Let β be a cubic Pisot number and let*

$$M_\beta(x) = X^3 - aX^2 - bX - c$$

be its minimal polynomial.

Then β is a simple beta-number if and only it satisfies one of the following conditions:

- *Case 1: $b \geq 0$ and $c > 0$*
- *Case 2: $-a < b < 0$ and $b + c \geq 0$*
- *Case 3: $b \leq -a$ and $b(k-1) + c(k-2) \leq (k-2) - (k-1)a$, where k is the integer in $\{2, 3, \ldots, a-2\}$ such that, denoting $e_k = 1 - a + (a-2)/k$, $e_k \leq b + c < e_{k-1}$.*

The problem of finding such a characterization remains open for simple beta-numbers of higher degree.

References

1. S. Akiyama. Cubic Pisot units with finite beta expansions. In F.Halter-Koch and R.F. Tichy, editors, *Algebraic Number Theory and Diophantine Analysis*, 11–26. de Gruyter, 2000.
2. A. Bertrand. Développements en base de Pisot et répartition modulo 1. *C. R. Acad. Sci. Paris*, 285:419–421, 1977.
3. A. Bertrand-Mathis. Développement en base θ, répartition modulo 1 de la suite $(x\theta^n)_{n\geq 0}$, langages codés et θ-shift. *Bull. Soc. Math. France*, 114:271–323, 1986.
4. F. Blanchard. β-expansions and symbolic dynamics. *Theor. Comput. Sci.*, 65:131–141, 1989.
5. D. W. Boyd. Salem numbers of degree four have periodic expansions. In *Number theory*, pages 57–64. de Gruyter, 1989.
6. D. W. Boyd. On beta expansions for Pisot numbers. *Mathematics of Computation*, 65(214):841–860, 1996.
7. D. W. Boyd. On the beta expansion for Salem numbers of degree 6. *Mathematics of Computation*, 65(214):861–875, 1996.
8. L. Flatto, J. Lagarias, and B. Poonen. The zeta function of the beta transformation. *Ergodic Theory Dynamical Systems*, 14:237–266, 1994.
9. C. Frougny. Numeration Systems, chapter 7, in M. Lothaire, *Algebraic Combinatorics on Words*. Cambridge University Press, to appear, available at `http://www-igm.univ-mlv.fr/~berstel/Lothaire/`.
10. C. Frougny and B. Solomyak. Finite β-expansions. *Ergodic Theory Dynamical Systems*, 12:713–723, 1992.
11. M. Hollander. Greedy numeration systems and regularity. *Theory of Computing Systems*, 31:111–133, 1998.
12. S. Ito and Y. Takahashi. Markov subshifts and realization of β-expansions. *J. Math. Soc. Japan*, 26:33–55, 1974.
13. W. Parry. On the beta expansions of real numbers. *Acta Math. Acad. Sci. Hung.*, 11:401–416, 1960.
14. A. Rényi. Representations for real numbers and their ergodic properties. *Acta Math. Acad. Sci. Hung.*, 8:477–493, 1957.
15. K. Schmidt. On periodic expansions of Pisot numbers and Salem numbers. *Bull. London Math. Soc.*, 12:269–278, 1980.
16. B. Solomyak. Conjugates of beta-numbers and the zero-free domain for a class of analytic functions. *Proc. London Math. Soc.*, 68(3):477–498, 1994.

Facility Location Constrained to a Polygonal Domain

Prosenjit Bose[1]* and Qingda Wang[2]

[1] School of Computer Science, Carleton University, Ottawa, Ontario, Canada.
jit@scs.carleton.ca

[2] Platform Computing, Toronto, Ontario, Canada.
qiwang@platform.com

Abstract. We develop efficient algorithms for locating an obnoxious facility in a simple polygonal region and for locating a desirable facility in a simple polygonal region. Many realistic facility location problems require the facilities to be constrained to lie in a simple polygonal region. Given a set S of m demand points and a simple polygon R of n vertices, we first show how to compute the location of an obnoxious facility constrained to lie in R, in $O((m+n)\log m + m\log n)$ time. We then show how to compute the location of a desirable facility constrained to lie in R, also in $O((m+n)\log m + m\log n)$ time. Both running times are an improvement over the known algorithms in the literature. Finally, our results generalize to the setting where the facility is constrained to lie within a set of simple polygons as opposed to a single polygon at a slight increase in complexity.

1 Introduction

Facility location is a classical problem of operations research that has also been examined in the computational geometry community. In classical facility location problems, we try to locate *desirable* facilities, such as hospitals, supermarkets, post-offices and warehouses, such that the distance function between the facilities and the demand sites (e.g., customers) is minimized. In *obnoxious* facility location problems, we try to locate obnoxious facilities, such as nuclear power plants, garbage dump sites, mega-airports, and chemical plants, such that the distance function between the facilities and the demand points is maximized. Notice that for the obnoxious facility location problem to be meaningful and practical, some constraints on the location of the facility should be specified, e.g., forcing it to lie in some bounded region.

Most facility location problems dealt with in the literature assumed convex feasible regions (including rectangular regions). However, this assumption is usually not very realistic since the feasible regions are often not convex due to geographical reasons and zoning laws. Therefore in this paper we study facility location problems under the assumption of simple polygonal feasible region. We

* Research supported by NSERC and MITACS.

S. Rajsbaum (Ed.): LATIN 2002, LNCS 2286, pp. 153–164, 2002.

have included all of the details that are deemed essential, however, due to space constraints, some details may be omitted in this extended abstract. Full details can be found in [Wan01].

In Section 2 and 3 we present improved algorithms for locating an obnoxious facility and a desirable facility in a simple polygonal region, respectively, under Euclidean distance. Both our algorithms run in subquadratic time. Section 4 concludes the paper.

2 Locating an Obnoxious Facility in a Simple Polygon

2.1 Related Work

The problem of locating a single obnoxious facility for a set of demand points is also known as the largest empty circle problem. Toussaint[Tou83] was the first to have studied the obnoxious facility location problem constrained to a simple polygonal region. Given n demand points, he proposed an $O(n \log n)$ algorithm if the facility is constrained to lie in a convex n-gon, and an $O(n \log n + k \log n)$ algorithm if the facility is constrained to lie in a simple n-gon P, where k denotes the number of intersections occurring between edges of P and edges of the Voronoi diagram of the demand points. Note k could be $\Omega(n^2)$.

In this section we improve Toussaint's algorithm. Our algorithm runs in $O(n \log n)$ time even if the constraining polygon is not convex, eliminating the dependence of the running time on k.

2.2 Preliminaries

The problem is stated as the following:
Problem 1 *Given a set $S = s_1, s_2, ..., s_m$ of m demand points and an n vertex simple polygon R, locate a single obnoxious facility within R such that the smallest Euclidean distance between each demand point and the facility is maximized.*

It is assumed that the demand points are in general position, i.e. no three points are collinear and no four points are co-circular.

Toussaint [Tou83] proved the following lemma:

Lemma 1. *The location of the single obnoxious facility under the constraints stipulated in Problem 1 lies on one of the following points that yields the largest empty circle:*

1. *A vertex of the Voronoi diagram of S (denoted as $VD(S)$) contained in R.*
2. *A vertex of the polygon R.*
3. *A proper intersection point of $VD(S)$ and the boundary of R.*

The quadratic running time in Toussaint's algorithm occurs in computing the third part of the above. We refine the characterization of the location of the optimal solution allowing us to circumvent this problem.

For convenience, i.e. to avoid problems caused by unbounded edges of $VD(S)$, let B be a large rectangle that contains S, R and all vertices of $VD(S)$. Denote by $BVD(S)$ the graph induced by $B \cap VD(S)$. Given $VD(S)$, $BVD(S)$ can be computed in linear time.

Lemma 2. *If the location of the single obnoxious facility under the constraints stipulated in problem 1 lies on a proper intersection point c between an edge e of $BVD(S)$ and the boundary of R, then among all intersection points between e and the boundary of R, c is the closest to one of the two endpoints of e.*

Proof. Let p and q be the two endpoints of e. For the sake of contradiction, suppose both $[cp]$ and $[cq]$ contains at least one intersection point between $BVD(S)$ and the boundary of R, denoted by x and y, respectively. Let $V(b)$ be one of the two Voronoi cells incident on e. We have either $\angle bcp$ or $\angle bcq$ is $\geq \pi/2$. Without loss of generality suppose $\angle bcp \geq \pi/2$. Now the circle centered at x with radius $d(x, b)$ is empty. However, $d(x, b) > d(c, b)$. Hence, we have a contradiction.

■

Lemma 2 indicates that we need not consider all the proper intersection points between $VD(S)$ and R. Rather, for all intersection points occurring between a given edge $e = [pq]$ of $BVD(S)$ and the boundary of R, we need consider at most two points, i.e. the points closest to p and q. The intersection points with this property can be found efficiently by ray shooting queries over R [HS95] as follows. For each vertex v of $BVD(S)$ and each edge e incident on v, we shoot a ray emanating from v and containing e. Let p be the hit point of the boundary of R by the ray. If p is on e, it is the intersection point of e and R that is the closest to v, a candidate with the property mentioned in Lemma 2.

We can preprocess R in $O(n)$ time such that a ray shooting query over R can be answered in $O(\log n)$ time [HS95]. Using this technique allows us to avoid the expensive computation of all intersection points between $VD(S)$ and R. To summarize, the following lemma characterizes the desired location of the obnoxious facility.

Lemma 3. *The location of the single obnoxious facility under the constraints stipulated in Problem 1 lies on one of the following points that yields the largest empty circle:*

1. *A vertex of $VD(S)$ contained in R;*
2. *A vertex of the polygon R;*
3. *A point of the boundary of R hit by a ray emanating from a vertex v of $BVD(S)$ and containing an edge of $BVD(S)$ incident on v.*

2.3 The Algorithm

Based on Lemma 3, we outline the following algorithm to compute the location of the obnoxious facility constrained to a simply polygon.

Algorithm 1 *Single obnoxious facility location (S, R)*

Input: A set of demand points $S = s_1, s_2, ..., s_m$, and a simple polygon R of n vertices.

Output: Point(s) for the location of a single obnoxious facility within R that maximizes the minimum distance between the facility and the demand points in S.

1. Compute $VD(S)$ and $BVD(S)$. Preprocess R for point inclusion test using the algorithms of Kirkpatrick [Kir83] or Edelsbrunner, Guibas and Stolfi [EGS86]. Compute the set of vertices of $VD(S)$ contained in R using the point location structure of R, and compute their largest empty circles.
2. Preprocess $VD(S)$ for point inclusion test using the algorithms of [EGS86] or [Kir83]. For each vertex v of R, determine which Voronoi cell it is in using the point location structure of $VD(S)$, and compute its largest empty circle.
3. Preprocess R in $O(n)$ time for ray shooting query over R [HS95]. For each vertex v of $BVD(S)$ and each edge e of $BVD(S)$ incident on v, perform a ray shooting query over R, with a ray emanating from v and containing e [HS95]. If the hit point of the boundary of R by the ray is on e, compute its largest empty circle.
4. Select and output the largest empty circles found in steps 1, 2 and 3.

Theorem 1. *Given a set of demand points $S = s_1, s_2, ..., s_m$, and an n-vertex simple polygon R, we can compute the location of a single obnoxious facility constrained to lie in R, under Euclidean distance, in time $O((m+n)\log m + m\log n)$.*

Proof. The correctness of the algorithm follows from Lemma 3.

Let us analyze the complexity of the algorithm:

1. In step 1, $VD(S)$ can be computed in $O(m\log m)$ time [PS85] and $BVD(S)$ can then be computed in $O(m)$ time. Preprocess of R for point inclusion test can be done in $O(n)$ time ([Kir83] or [EGS86]). There are $O(m)$ vertices in $VD(S)$, each of which costs $O(\log n)$ time to decide if it is contained in R. Thus step 1 can be done in total $O(m\log m + m\log n)$ time.
2. In step 2, preprocessing of $VD(S)$ for point inclusion test can be done in $O(m)$ time ([Kir83] or [EGS86]). Each of the n vertices of R costs $O(\log m)$ time to locate the Voronoi cell it is in. Therefore step 2 can be done in total $O(m + n\log m)$ time.
3. In step 3, preprocessing of R for ray shooting query can be done in $O(n)$ time and each ray shooting query takes $O(\log n)$ time [HS95]. Since the number of edges in $BVD(S)$ is $O(m)$, and we perform at most two ray shooting queries for each edge, the total time spent in step 3 is $O(n + m\log n)$.
4. Step 4 can be done in $O(m+n)$.

Therefore, the total complexity of the algorithm is $O((m+n)\log m + m\log n)$.

■

By reducing to the MAX-GAP problem, the largest empty circle problem has a known lower bound of $\Omega(m\log m)$ [PS85]. Thus Algorithm 1 runs in optimal

$\Theta(m \log m)$ time when m is $\Omega(n)$, and $\Theta(n)$ time when m is $O(1)$. It remains an open problem whether the algorithm is optimal when m is $o(n)$ and $\omega(1)$.

Algorithm 1 can be extended to the case where the constraining region R is a set of r disjoint simple polygons $\{R_1, R_2, \ldots, R_r\}$ with a total of n vertices. Hershberger and Suri [HS95] show how to compute a data structure in $O(n\sqrt{r} + n \log n + r^{3/2} \log r)$ time and $O(n)$ space to answer a ray shooting query in $O(\sqrt{r} \log n)$ time. This leads to the following corollary:

Corollary 1. *Given a set of demand points $S = s_1, s_2, ..., s_m$, and a region R consisting of r disjoint simple polygons with a total of n vertices, we can compute the location of a single obnoxious facility constrained to lie in R, under Euclidean distance, in time $O((m + n) \log m + (m\sqrt{r} + n) \log n + n\sqrt{r} + r^{3/2} \log r)$.*

3 Euclidean Center Constrained to a Simple Polygon

3.1 Related Work

The problem of locating a single desirable facility for a set of demand points is also known as the Euclidean center problem. The problem of computing the unconstrained Euclidean center has a rich history, which is summarized in [PS85].

Bose and Toussaint[BT00] solved the problem of computing the Euclidean center constrained to a simple polygon. Their algorithm finds the Euclidean center of a set of m points constrained to lie in a simple polygon P of n vertices in $O((m+n) \log(m+n))$ time when P is convex, and $O((m+n) \log(m+n)+k)$ time when P is not convex, where k denotes the number of intersections occurring between edges of P and edges of the farthest point Voronoi diagram of the demand points. Notice k could be $\Omega(mn)$.

Extending the technique of Megiddo [Meg83], Hurtado, Sacristan and Toussaint [HST00] obtained an $O(n + m)$ time algorithm for the problem when the Euclidean center is constrained to lie in a convex polygon, reducing by a logarithmic factor the running time of the algorithm presented in [BT00].

It can be shown that the Euclidean center problem constrained to a simple polygon is equivalent to the problem in the presence of polygonal obstacles. D. Halperin, M. Sharir and K. Goldberg studied the Euclidean 2-center problem in the presence of polygonal obstacles and obtained efficient algorithms [HSG00]. In the paper they mentioned that D. Halperin and C. Linhart have solved the 1-center problem with obstacles in $O((m+n) \log(mn))$ time [HL99]. However, their algorithm with the above mentioned time bound solved the decision problem only, and an approximation approach based on binary search is used. Thus the algorithm is not exact.

In this section we present an efficient algorithm that is based on Bose and Toussaint's approach [BT00] to solve the Euclidean center problem constrained to a simple polygon. Our algorithm runs in $O((m + n) \log m + m \log n)$ time even if the constraining polygon is not convex, eliminating the dependency of the running time on k. Although the algorithms in [HL99] has the advantage that it can be generalized to solve Euclidean 2-center problem, the merits of our algorithm compared to [HL99] are the following: (i) it is exact; (ii) it is more efficient, considering the approximation scheme in [HL99] adds a multiplicative logarithmic factor to the running time. (iii) it is simpler to implement, since it involves well-known structures only.

3.2 Preliminaries

The problem is stated as the following:

Problem 2 *Given a set $S = s_1, s_2, ..., s_m$ of m demand points and an n vertex simple polygon region R, locate a single desirable facility within R such that the largest Euclidean distance between each demand point and the facility is minimized.*

It is assumed that the demand points are in general position, i.e. no three points are collinear and no four points are co-circular.

Bose and Toussaint's algorithm [BT00] makes use of the farthest point Voronoi diagram of the set S, denoted as $FPVD(S)$. Given a point $x \in E^2$, we let $\phi(x)$ denote the farthest neighbors of x in S, that is, the set of points in S such that $d(x, \phi(x)) = max_{y \in S} d(x, y)$ where d is the Euclidean distance. The $FPVD(S)$ partitions the plane into unbounded convex cells, $V(s_i)$, such that for any point $p \in V(s_i), s_i \in \phi(p)$.

The following lemma is adopted from [BT00] characterizing the location of the Euclidean center:

Lemma 4. *The Euclidean center under the constraints stipulated in Problem 2 is either the midpoint of the diameter of S provided that the diametrical circle contains the set S and the midpoint is contained in R, or lies on one of the following points that yields the smallest enclosing circle:*

1. *A vertex of $FPVD(S)$ contained in R;*
2. *A vertex of the polygon R;*
3. *A proper intersection point of $FPVD(S)$ and the boundary of R;*
4. *A point x on an edge e of R with the property that $\forall y \in e$, if $\phi(y) = \phi(x)$ then $d(y, \phi(x)) \geq d(y, \phi(x))$.*

The quadratic runtime occurs in computing part 3 and part 4 of the above. In the following we are going to prove Lemmas 5 to 8, which will allow us to improve Bose and Toussaint's algorithm, reducing the quadratic runtime, by improving the characterization of the Euclidean center.

Improving Part 3 of Lemma 4 A similar approach as in Section 2.2 is used to improve part 3 of Lemma 4.

Lemma 5. *If the Euclidean center c under the constraints stipulated in Problem 2 lies on a proper intersection point c between an edge e of $FPVD(S)$ and the boundary of R, and if c is not the midpoint of $DIAM(S)$, then among all intersection points between e and the boundary of R, c is closest to an endpoint of e.*

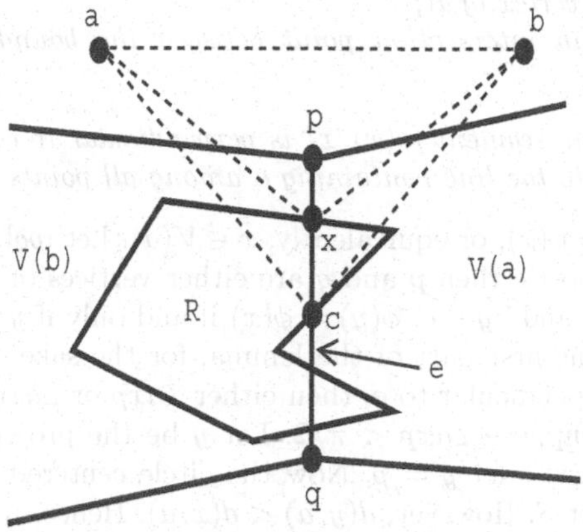

Fig. 1. Illustration of proof for Lemma 5

Proof. See Figure 1. Let $V(a)$ and $V(b)$ be the two cells incident on e. The edge e must lie on the bisector of line segment $[ab]$, for the points on e are equidistant from a and b. The points a, b, and c must form a triangle because otherwise $[ab]$ would be the diameter. Thus neither $[ca]$ nor $[cb]$ is perpendicular to e. If e is unbounded, let p be the unique endpoint of e. Otherwise, let p be the endpoint of e that is closer to a (and b). Then we have $\angle acp < \pi/2$. For the sake of contradiction, suppose c is not the closest to p among all intersection points between e and the boundary of R. Let x be the closest to p. Now the circle centered at x with radius $d(x, a)$ encloses the set S. However, $d(x, a) < d(c, a)$. Hence, we have a contradiction. ■

Thus, for all intersection points occurring between a given edge e of $FPVD(S)$ and the boundary of R, we need only consider the points closest to an endpoint of e, of which there are at most two.

Similar as in Section 2.2, the intersection points with this property can be found efficiently by a ray shooting query over R. Next we develop another technique to efficiently compute part 4 of Lemma 4.

Improving Part 4 of Lemma 4

Lemma 6. *If the Euclidean center x under the constraints stipulated in Problem 2 lies on an edge e of R and has the following properties:*

1. $\forall y \in e$, *if* $\phi(y) = \phi(x)$ *then* $d(y, \phi(x)) \geq d(y, \phi(x))$;
2. x *is not the midpoint of* $DIAM(S)$;

3. x is not a vertex of R;

4. x is not an intersection point between the boundary of R and edges of $FPVD(S)$;

then the line segment $[\phi(x), x]$ is perpendicular to e, and $\phi(x)$ is the unique farthest point to the line containing e among all points in S.

Proof. Let $a = \phi(x)$, or equivalently, $x \in V(a)$. Let $[pq]$ be the portion of e that lies in $V(a)$. Notice then p and q are either vertices of R or intersection points of e and $V(a)$, and $\forall y \in e$, $\phi(y) = \phi(x)$ if and only if $y \in [pq]$.

To prove the first part of the lemma, for the sake of contradiction suppose $[ax]$ is not perpendicular to e, then either $\angle axp$ or $\angle axq$ is $< \pi/2$. Without loss of generality suppose $\angle axp < \pi/2$. Let y be the projection point of a on e, if it exists, otherwise let $y = p$. Now the circle centered at y with radius $d(y, a)$ encloses the set S. However, $d(y, a) < d(x, a)$. Hence, we have a contradiction.

Let l denote the line containing e. To prove the second part (see Figure 2), for the sake of contradiction suppose there exists a point b of S, such that the distance between b and l is no less than that between a and l. Let y be the projection point of b on l. Given $d(y, b) \geq d(x, a)$, in the following we will prove $d(x, b) > d(x, a)$, which contradicts that $x \in V(a)$.

First, if $d(y, b) > d(x, a)$, or $d(y, b) = d(x, a)$ and $x \neq y$, then obviously $d(x, b) > d(x, a)$. Secondly, we show it is impossible that $d(y, b) = d(x, a)$ and $x = y$. In this case b is the mirror image of a with respect to l and x is the midpoint of $[ab]$. Since both a and b are the farthest points of x, x lies on the common edge of $V(a)$ and $V(b)$. In other words, $[ab]$ crosses and is perpendicular to an edge of $FPVD(S)$. Thus $[ab]$ is $DIAM(S)$ and x is its midpoint. However, we know x is not a midpoint of $DIAM(S)$. Hence we have a contradiction.

■

Lemma 6 indicates that for each edge e of R, there is at most one point that can be qualified as a candidate for the constrained Euclidean center with the property mentioned in the lemma, despite the fact that e may span many cells of $FPVD(S)$. When there are two points in S that are both the farthest to l, there is no point on e that can be qualified.

To find the point in S that is the farthest to a line, we first observe the following: the point in a given set of points S that is the farthest to a given line l is a vertex of the convex hull of S, denoted as $CH(S)$ (Proof omitted). Let h be the number of edges of $CH(S)$. In the following we show that by preprocessing $CH(S)$ in $O(h)$ time, we can find the farthest point of S to a given query line in $O(\log h)$ time.

Let $ang(e)$ be the angle of an edge e of $CH(S)$, measured as the angle between a right oriented horizontal ray x and e. The edges are directed in such a way that a walk along the directed edges constitutes a counterclockwise traversal of the boundary of $CH(S)$ (Figure 3). We store $ang(e)$ for each edge e of $CH(S)$ in sorted order in a set A. This can be done in $O(h)$ time given that edges in $CH(S)$ are in sorted order.

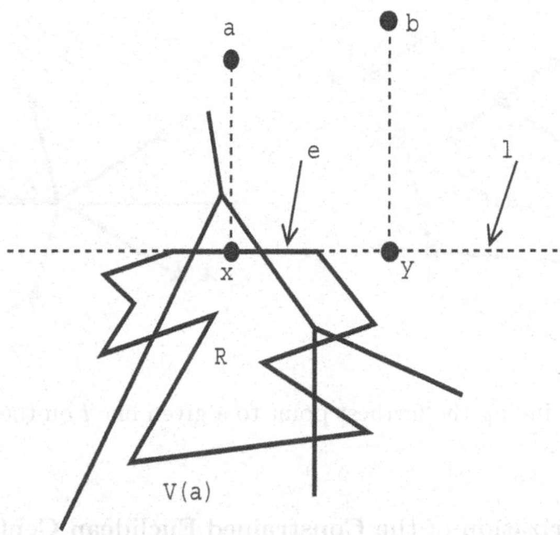

Fig. 2. Illustration of proof for Lemma 6

Given a line l, to answer the query of the farthest point in $CH(S)$, we first compute the angle of l for both directions of l, denoted as $ang(l_\alpha)$ and $ang(l_\beta)$. For each of $ang(l_k)$ ($k = \alpha$ or β), we do a binary search in A to find out the two angels in A in between which $ang(l_k)$ falls. The two angles correspond to two neighboring edges in $CH(S)$, whose common vertex is then obtained. Denote by u and v the vertices obtained corresponding to $ang(l_\alpha)$ and $ang(l_\beta)$, respectively (Figure 3). Compute their respective distance to l. The vertex corresponding to the larger distance is then returned. If the two distances happen to be equal, by Lemma 6 we need not return a vertex. Moreover, during the binary search, if $ang(l_k)$ ($k = \alpha$ or β) happens to be equal to one of the angles in A, which means there are two vertices in $CH(S)$ on the same side of l that are both the farthest to l, then again we need not return a vertex corresponding to $ang(l_k)$ according to Lemma 6.

The query time is bounded by the binary search time, which is $O(\log h)$. Thus to compute Part 4 of Lemma 4, after preprocessing $CH(S)$ in $O(h)$ time, for each edge e of R, in $O(\log h)$ time we can find the unique point $p \in S$ that is the farthest from the line containing e. We then compute the projection point of p on e. Given $h = O(m)$, we have

Lemma 7. *Given $CH(S)$, part 4 of Lemma 4 can be computed in $O(m+n\log m)$ time.*

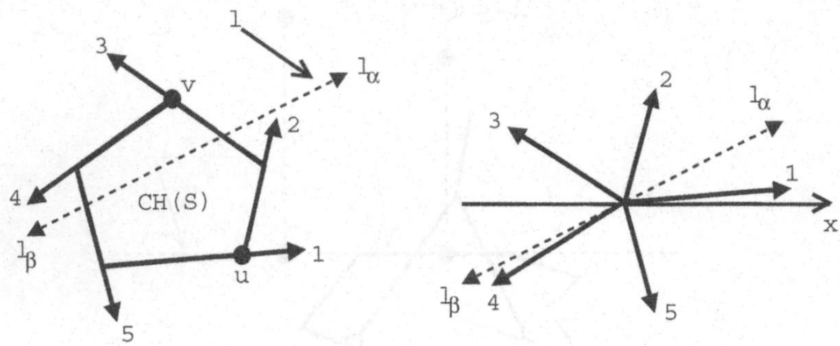

Fig. 3. Finding the farthest point to a given line l on the convex hull of S.

Characterization of the Constrained Euclidean Center Lemma 4, 5 and 6 lead to the following lemma characterizing the location of the Euclidean center constrained to a simple polygon.

Lemma 8. *The Euclidean center under the constraints stipulated in Problem 2 is either the midpoint of $DIAM(S)$ provided that the diametrical circle contains the set S and the midpoint is contained in R, or lies on one of the following points that yields the smallest enclosing circle:*

1. *A vertex of $FPVD(S)$ contained in R;*
2. *A vertex of R;*
3. *A point of the boundary of R hit by a ray emanating from a vertex v of $FPVD(S)$ and containing an edge of $FPVD(S)$ incident on v;*
4. *A projection point on an edge e of R of a point in S that is the farthest to e.*

3.3 The Algorithm

Based on Lemma 8, we outline the following algorithm to compute the Euclidean center constrained to a simply polygon.

Algorithm 2 *Euclidean Center (S,R)*
Input: A set of demand points $S = s_1, s_2, ..., s_m$, and a simple polygon R of n vertices.

Output: Euclidean center of S constrained to lie in R.

1. Compute $CH(S)$ and $DIAM(S)$. Compute the circle C having $DIAM(S)$ as diameter. Preprocess R for point inclusion test using the algorithms of Kirkpatrick [Kir83] or Edelsbrunner, Guibas and Stolfi [EGS86]. If the midpoint of $DIAM(S)$ is contained in R and all points of S are contained in C then exit with the midpoint of $DIAM(S)$ and radius of C.

2. Compute $FPVD(S)$. Compute the set of vertices of $FPVD(S)$ contained in R using the point location structure of R, and compute their smallest enclosing circles.
3. Preprocess $FPVD(S)$ for point inclusion test using the algorithms of [EGS86] or [Kir83]. For each vertex v of R, determine which Voronoi cell it is in using the point location structure of $FPVD(S)$, and compute its smallest enclosing circle.
4. Preprocess R in $O(n)$ time for ray shooting query over R [HS95]. For each vertex v of $FPVD(S)$ and each edge e of $FPVD(S)$ incident on v, perform a ray shooting query over R, with a ray emanating from v and containing e [HS95]. If the point of the boundary of R hit by the ray is on e, compute its smallest enclosing circle.
5. Store in A in sorted order the set of angles of each edge of $CH(S)$. For each edge e of R, find the unique farthest point p of $CH(S)$ to e, if it exists, and compute the projection point of p on e, using the techniques of Lemma 7. Compute the smallest enclosing circle of the projection point.
6. Select and output the smallest enclosing circles found in steps 2-5.

Theorem 2. *Given a set of demand points $S = s_1, s_2, ..., s_m$, and an n-vertex simple polygon R, we can compute the Euclidean center of S constrained to lie in R in time $O((m+n)\log m + m\log n)$.*

Proof. The correctness of the algorithm follows from Lemma 8. The analysis is similar to that given in Theorem 1 and will not be duplicated here. The full proof can be found in [Wan01]

■

When m is $O(1)$, Algorithm 2 runs in optimal $\Theta(n)$ time. It remain open problems whether Algorithm 2 is optimal when m is $\omega(1)$ or whether there exists linear time solution to the problem.

As with the case of obnoxious facility location, the placement of single desirable facility can be generalized.

Corollary 2. *Given a set of demand points $S = s_1, s_2, ..., s_m$, and a region R consisting of r disjoint simple polygons with a total of n vertices, we can compute the location of a single desirable facility constrained to lie in R, under Euclidean distance, in time $O((m+n)\log m + (m\sqrt{r}+n)\log n + n\sqrt{r} + r^{3/2}\log r)$.*

4 Conclusion

We developed an efficient algorithm for locating a single obnoxious facility in a simple polygonal region. We also presented an efficient algorithm for locating a single desirable facility in a simply polygonal region. The algorithms make use of the Voronoi diagram or the farthest-point Voronoi diagram. They are based on and are improvements of the algorithms in [HST00] and [BT00]. Given a set S of m demand points and a simple polygon R of n vertices, we have shown how to compute the location of either an obnoxious or a desirable facility to lie in R, in $O((m+n)\log m + m\log n)$ time. The previous best known algorithms solving the above problems both had worst case quadratic running times.

References

[BT00] P. Bose and G. T. Toussaint. Computing the constrained euclidean, geodesic and link centers of a simple polygon with applications. *Studies of Location Analysis, Special Issue on Computational Geometry*, 15:37–66, 2000.

[EGS86] H. Edelsbrunner, L. J. Guibas, and J. Stolfi. Optimal point location in a monotone subdivision. *SIAM J. Comput.*, 15:317–340, 1986.

[HL99] D. Halperin and C. Linhart. The smallest enclosing disks with obstacles. Manuscript, Dep. of Comp. Science, Tel-Aviv University, Israel, 1999.

[HS95] J. Hershberger and S. Suri. A pedestrian approach to ray shooting: Shoot a ray, take a walk. *Journal of Algorithms*, 18(3):403–431, 1995.

[HSG00] D. Halperin, M. Sharir, and K. Goldberg. The 2-center problem with obstacles. In *Proc. of the 16th ACM Symp. on Comp. Geom.*, pages 80–90, 2000.

[HST00] F. Hurtado, V. Sacristn, and G. Toussaint. Constrained facility location. *Studies of Location Analysis, Special Issue on Computational Geometry*, 15:17–35, 2000.

[Kir83] D. Kirkpatrick. Optimal search in planar subdivisions. *SIAM J. Comput.*, 12(1):28–35, 1983.

[Meg83] N. Megiddo. Linear-time algorithms for linear programming in r3 and related problems. *SIAM J. Comput.*, 12:759–776, 1983.

[PS85] F. P. Preparata and M. I. Shamos. *Computational Geometry: an Introduction.* Springer-Verlag, New York, 1985.

[Tou83] G. T. Toussaint. Computing largest empty circles with location constraints. *Int. J. of Comp. and Information Science*, 12(5):347–358, 1983.

[Wan01] Q. Wang. *Facility Location Constrained to a Simple Polygon.* Master's thesis, Carleton University, 2001.

A Deterministic Polynomial Time Algorithm for Heilbronn's Problem in Dimension Three
(Extended Abstract)

Hanno Lefmann[1] and Niels Schmitt[2]

[1] Fakultät für Informatik, Technische Universität Chemnitz, Straße der Nationen 62, D-09107 Chemnitz, Germany.
lefmann@informatik.tu-chemnitz.de

[2] Lehrstuhl Mathematik und Informatik, Fakultät für Mathematik, Ruhr-Universität Bochum, D-44780 Bochum, Germany.
nschmitt@lmi.ruhr-uni-bochum.de

Abstract. Heilbronn conjectured that among any n points in the 2-dimensional unit square $[0,1]^2$, there must be three points which form a triangle of area at most $O(1/n^2)$. This conjecture was disproved by Komlós, Pintz and Szemerédi [15] who showed that for every n there exists a configuration of n points in the unit square $[0,1]^2$ where all triangles have area at least $\Omega(\log n/n^2)$. Here we will consider a 3-dimensional analogue of this problem and we will give a deterministic polynomial time algorithm which finds n points in the unit cube $[0,1]^3$ such that the volume of every tetrahedron among these n points is at least $\Omega(\log n/n^3)$.

1 Introduction

An old conjecture of Heilbronn stated that among any n points in the 2-dimensional unit square $[0,1]^2$ (or unit disc) there are three distinct points which form a triangle of area at most c/n^2 for some constant $c > 0$. Erdös observed that this conjecture, if true, would be best possible, as the points $1/n \cdot (i \bmod n, i^2 \bmod n)_{i=0,\ldots,n-1}$ on the moment-curve in the $n \times n$ grid would show, cf. [2]. However, Komlós, Pintz and Szemerédi [15] disproved Heilbronn's conjecture by showing for every n the existence of a configuration of n points in the unit square $[0,1]$ with every three points forming a triangle of area at least $\Omega(\log n/n^2)$. Using derandomization techniques, this existence argument was made constructive in [6], where a polynomial time algorithm was given, which finds n points in $[0,1]^2$ achieving the lower bound $\Omega(\log n/n^2)$ on the minimum triangle area. Upper bounds on Heilbronn's triangle problem were given in a series of papers by Roth [17,18,19,20,21] and Schmidt [23], cf. Rothschild and Straus [22] for related results, and the currently best upper bound is due to Komlós, Pintz and Szemerédi [14] and is of the order $n^{-8/7+\varepsilon}$ for every fixed $\varepsilon > 0$.

If n points are dropped uniformly at random and independently of each other in $[0,1]^2$, then the expected value of the smallest area of a triangle among these n points is $\Theta(1/n^3)$, as was recently proved by Jiang, Li and Vitány [12].

S. Rajsbaum (Ed.): LATIN 2002, LNCS 2286, pp. 165–180, 2002.

A k-dimensional version of Heilbronn's problem was considered recently by Barequet [3]. For given dimension $k \geq 3$ he showed, that for every $n \in \mathbb{N}$ there exist n points in the k-dimensional unit cube $[0,1]^k$ such that the minimum volume of every simplex spanned by any $(k+1)$ of these n points is at least $\Omega(1/n^k)$. Barequet gave two different approaches for proving this lower bound. One approach uses a random argument. The other approach is similar to Erdös' construction, namely taking the points $P_l = (l^j \bmod n/n)_{j=1,\ldots,k}$ for $l = 0, 1, \ldots, n-1$ on the moment-curve. According to Bollobás [5], this construction for the k-dimensional case was already mentioned by Erdös some time ago.
In [16], Barequet's lower bound was improved by a factor $\Theta(\log n)$ for dimensions $k \geq 3$ by using a probabilistic existence argument, based on a variant of Theorem 2.2. For the proof continuity arguments were crucial.

Theorem 1.1. [16] *For every fixed integer $k \geq 3$ and for every $n \in \mathbb{N}$ there exists a configuration of n points in the k-dimensional unit cube $[0,1]^k$ such that the volume of any simplex spanned by any $(k+1)$ points is at least $\Omega(\log n/n^k)$.*

Here we will give for dimension $k = 3$ a deterministic polynomial time algorithm for the result in Theorem 1.1:

Theorem 1.2. *For every positive integer n one can find in polynomial time a configuration of n points in the unit cube $[0,1]^3$ such that the volume of each tetrahedron spanned by any four of these points is at least $\Omega(\log n/n^3)$.*

The proof of Theorem 1.2 is based on techniques from combinatorics and number theory. Some of our arguments are given for the case of arbitrary dimension $k \geq 3$, where appropriate. Our techniques possibly might be extended to the case of arbitrary dimension $k \geq 3$. However, so far we are only able to provide a polynomial time algorithm for the case $k = 3$.

2 Hypergraphs

In our arguments we use hypergraphs. The *independence number of a hypergraph* and the 2-cycles is important in our considerations:

Definition 2.1. *Let $\mathcal{G} = (V, \mathcal{E})$ be a* hypergraph *where each edge $E \in \mathcal{E}$ satisfies $E \subseteq V$. The hypergraph $\mathcal{G}$ is k-*uniform *if every edge $E \in \mathcal{E}$ contains exactly k vertices.*
A subset $I \subseteq V$ is called independent *if I contains no edge $E \in \mathcal{E}$. The largest size of an independent set in $\mathcal{G}$ is called the* independence number $\alpha(\mathcal{G})$.
In a k-uniform hypergraph $\mathcal{G} = (V, \mathcal{E})$, $k \geq 3$, a 2-cycle is a pair $\{E_1, E_2\}$ of distinct edges $E_1, E_2 \in \mathcal{E}$ with $|E_1 \cap E_2| \geq 2$. A 2-cycle *$\{E_1, E_2\}$ in $\mathcal{G}$ is called* $(2, j)$-cycle *if $|E_1 \cap E_2| = j$, where $j = 2, \ldots, k-1$.*

Let $B_k(T) = \{x \in \mathbb{R}^k \mid \|x\| \leq T\} \subset \mathbb{R}^k$ be the k-dimensional ball around the origin with radius T. We reformulate our geometrical problem as a problem of finding a large independent set in a suitably defined hypergraph. To do so, we discretize the 3-dimensional search space, namely we consider only grid points from the set $B_3(T) \cap \mathbb{Z}^3$, where T is of suitable size, i.e. polynomial in n. With this discretization, we must take care of degenerate tetrahedra with volume 0. For some parameter $B > 0$ and for the points in $B_3(T) \cap \mathbb{Z}^3$ we form a hypergraph $\mathcal{G} = \mathcal{G}(B) = (V, \mathcal{E})$ with the vertex set $V = B_3(T) \cap \mathbb{Z}^3$ of $\Theta(T^3)$ grid points. The hypergraph $\mathcal{G}$ contains 4-element edges and, for technical reasons, also 3-element edges. The 4-element edges $E \in \mathcal{E}$ are given by all subsets of four points from the set $B_3(T) \cap \mathbb{Z}^3$, no three of them on a line, which form a tetrahedron of volume at most B. The 3-element edges are given by all collinear triples of points from $B_3(T) \cap \mathbb{Z}^3$. An independent set in this hypergraph $\mathcal{G}(B)$ yields a subset of grid points from $B_3(T) \cap \mathbb{Z}^3$, where all tetrahedra have volume bigger than B. To find a large independent set, we use a result due to Ajtai, Komlós, Pintz, Spencer and Szemerédi [1], see also [10], stated here in an algorithmic variant obtained by derandomization techniques proven in [4], see [11]:

Theorem 2.2. [1],[10],[4],[11],[7] *Let $k \geq 3$ be a fixed integer. Let $\mathcal{G} = (V, \mathcal{E})$ be a k-uniform hypergraph on $|V| = N$ vertices and with average degree $t^{k-1} := k \cdot |\mathcal{E}|/|V|$. If for some constant $\gamma > 0$ the hypergraph $\mathcal{G}$ contains at most $N \cdot t^{2k-j-1-\gamma}$ many (2,j)-cycles for $j = 2, \ldots, k-1$, then one can find in $\mathcal{G}$ in polynomial time an independent set of size at least $\Omega(N/t \cdot (\log t)^{1/(k-1)})$.*
The conclusion also holds, if (i) the hypergraph $\mathcal{G} = (V, \mathcal{E})$ contains besides the k-element edges at most $N \cdot t^{i-1-\delta}$ many i-element edges for $i = 2, \ldots, k-1$ for some $\delta > 0$, or (ii) if t^{k-1} is only an upper bound on the average degree for the k-element edges of $\mathcal{G}$.

In recent years, several applications of Theorem 2.2 have been found, see [4]. Here we give another application of this deep result. To prove Theorem 1.2, we will count the degenerate and non-degenerate tetrahedra in $B_3(T) \cap \mathbb{Z}^3$.

3 Grids in $\mathbb{Z}^k$

In this section we describe some of the algebraic, geometric tools which we will use frequently in our arguments.

Definition 3.1. *A* grid *L of $\mathbb{Z}^k$ is a subset of $\mathbb{Z}^k$, which is generated by all linear combinations of some linearly independent vectors $q_1, \ldots, q_m \in \mathbb{R}^k$, where all coefficients are integers, i.e. $L = \mathbb{Z}q_1 + \ldots + \mathbb{Z}q_m$. The parameter m is called the* rank *of the grid L and the set $Q = \{q_1, \ldots, q_m\}$ is called a* basis *of L.*

Definition 3.2. *Let $Q = \{q_1, \ldots, q_m\} \subset \mathbb{Z}^k$ be a set of linear independent vectors. (i) The $k \times m$* generator matrix *of Q (up to the ordering of the vectors) is defined by $G(Q) := (q_1, \ldots, q_m)_{k \times m}$.*
(ii) The fundamental parallelepiped *F_Q of Q is the following set*

$$F_Q := \Big\{ \sum_{i=1}^{m} \alpha_i \cdot q_i \mid 0 \leq \alpha_1, \ldots, \alpha_m \leq 1 \Big\} \subseteq \mathbb{R}^k ,$$

and its volume *$vol(F_Q)$ is given by $vol(F_Q) := (\det(G(Q)^\top \cdot G(Q)))^{1/2}$.*

If Q and Q' are two bases of a grid $L \subseteq \mathbb{Z}^k$, then the volumes of the corresponding fundamental parallelepipeds are equal, i.e. $\text{vol}(F_Q) = \text{vol}(F_{Q'})$. The parameter $d(L) := \text{vol}(F_Q)$ is called the *determinant* of the grid L, see [8].
For integers $a_1, \ldots, a_k \in \mathbb{Z}$, which are not all equal to 0, let $\gcd(a_1, \ldots, a_k)$ be their *greatest common divisor*.

Theorem 3.3. *Let $a = (a_1, \ldots, a_k)^\top \in (\mathbb{Z} \setminus \{0\})^k$ with $\gcd(a_1, \ldots, a_k) = 1$. Then the set L of all solutions in $\mathbb{Z}^k$ of the equation $a_1 \cdot X_1 + \ldots + a_k \cdot X_k = 0$ is a grid in $\mathbb{Z}^k$ with $rank(L) = k - 1$.*

A recursive procedure to find a basis of a grid L in $\mathbb{Z}^k$ can be found in [9].
We use the standard scalar product $< a, b > := \sum_{i=1}^k a_i \cdot b_i$ for vectors $a = (a_1, \ldots, a_k)^\top \in \mathbb{R}^k$ and $b = (b_1, \ldots, b_k)^\top \in \mathbb{R}^k$. The Euclidean distance $\text{dist}(a, b)$ of the corresponding points is defined by $\text{dist}(a, b) = (\sum_{i=1}^k (a_i - b_i)^2)^{1/2}$. The length of a vector $a \in \mathbb{R}^k$ is defined by $\|a\| := \sqrt{< a, a >}$. For a point $p \in \mathbb{R}^k$ and a real subspace $V \subseteq R^k$ let $\text{dist}(p, V) := \min\{\text{dist}(p, v) \mid v \in V\}$. For a set $\{q_1, \ldots, q_m\} \subseteq \mathbb{R}^k$ of vectors let $\text{span}(\{q_1, \ldots, q_m\})$ be the linear space over $\mathbb{R}$, generated by $q_1, \ldots, q_m$.
The following two results can be found in [13]:

Lemma 3.4. *Let $V \subseteq \mathbb{R}^k$ be a $(k-1)$-dimensional linear subspace and let $a \in \mathbb{R}^k \setminus \{0^k\}$ be a nonzero vector which is orthogonal to V. Then every point $p \in \mathbb{R}^k$ satisfies $dist(p, V) = | < p^\top, a > | / \|a\|$.*

Lemma 3.5. *Let $q_1, \ldots, q_m \in \mathbb{R}^k$ be linearly independent vectors over $\mathbb{R}$. Let $U := span(\{q_1, \ldots, q_{m-1}\})$. Then the fundamental parallelepiped $F_{\{q_1, \ldots, q_m\}}$ fulfills $vol(F_{\{q_1, \ldots, q_m\}}) = dist(q_m, U) \cdot vol(F_{\{q_1, \ldots, q_{m-1}\}})$.*

Theorem 3.6. *Let $k \in \mathbb{N}$ be fixed. Let L be a grid in $\mathbb{Z}^k$ with $rank(L) = m$ and let $a_1, \ldots, a_m \in L$ be linearly independent. Then there exists a basis $b_1, \ldots, b_m$ of L with $\|b_i\| = O(\max_j \|a_j\|)$ for $i = 1, \ldots, m$.*

Proof. The proof is similar to that in [8], Lemma 8, p. 135–136, which is based on Theorem I, p. 11–13 in [8], which can be adapted to our situation. □

3.1 Maximal Grids in $\mathbb{Z}^k$

Definition 3.7. *A grid L in $\mathbb{Z}^k$ is called m-maximal, if $rank(L) = m$ and for every grid L' in $\mathbb{Z}^k$ with $rank(L') = m$ and $L \subseteq L'$ it is $L = L'$.*

A vector $a = (a_1, \ldots, a_k)^\top \in \mathbb{Z}^k \setminus \{0^k\}$ is called *primitive*, if $\gcd(a_1, \ldots, a_k) = 1$ and $a_j > 0$ for $j = \min\{i \mid a_i \neq 0\}$.

Lemma 3.8. *There is a bijection between the set of all $(k-1)$-maximal grids L in $\mathbb{Z}^k$ and the set of all primitive vectors $a_L \in \mathbb{Z}^k \setminus \{0^k\}$, i.e. a_L is the unique primitive normal vector of the grid L.*

Proof. Omitted. □

Lemma 3.9. *Let L be a $(k-1)$-maximal grid in $\mathbb{Z}^k$ with primitive normal vector $a_L = (a_1, \ldots, a_k)^\top \in \mathbb{Z}^k \setminus \{0^k\}$. Then there exists a vector $v \in \mathbb{Z}^k \setminus L$, such that $\mathbb{Z}^k$ can be partitioned into residue classes $s \cdot v + L$, $s \in \mathbb{Z}$, of L, i.e., $\mathbb{Z}^k = \biguplus_{s \in \mathbb{Z}} (s \cdot v + L)$, where each $x \in L$ satisfies $dist(s \cdot v + x, span(L)) = |s| / \|a_L\|$.*

Here, all *residue classes* L' of a $(k-1)$-maximal grid L in $\mathbb{Z}^k$ are of the form $L' = x + L$ for some $x \in \mathbb{Z}^k$.

Proof. By the lemma of Bezout, choose $v \in \mathbb{Z}^k$ with $< v, a_L >= 1$ and apply Lemma 3.4. □

Lemma 3.10. *Let L be a $(k-1)$-maximal grid in $\mathbb{Z}^k$ with primitive normal vector $a_L \in \mathbb{Z}^k \setminus \{0^k\}$ and with basis Q. Then the determinant $d(L)$ of L fulfills $d(L) = vol(F_Q) = \|a_L\|$.*

Proof. Let $Q = \{q_1, \ldots, q_{k-1}\}$ be a basis of L. By Lemma 3.9, there exists a vector $v = q_k \in \mathbb{Z}^k$ such that $\mathbb{Z}^k = \biguplus_{s \in \mathbb{Z}} (s \cdot q_k + L)$, hence $q_1, \ldots, q_k$ is a basis of $\mathbb{Z}^k$. Using $\text{dist}(q_k, \text{span}(L)) = 1/\|a_L\|$ we obtain with Lemma 3.5 that

$$1 = d(\mathbb{Z}^k) = \text{vol}(F_{\{q_1, \ldots, q_k\}}) = d(L)/\|a_L\| .$$

□

3.2 Simplices and Maximal Grids in $\mathbb{Z}^k$

Definition 3.11. *For a subset $S = \{p_0, \ldots, p_k\} \subset \mathbb{R}^k$ of $(k+1)$ points, the set $S^* = \{p_0 + \sum_{i=1}^k \lambda_i \cdot (p_i - p_0) \mid \sum_{i=1}^k \lambda_i \le 1; \lambda_1, \ldots, \lambda_k \in [0,1]\}$ is called a* simplex. *We call each set S and S^* a* simplex *and specifically we call the points $p_0, \ldots, p_k$* extreme points *of the simplex.*
(i) The rank *of the simplex S is defined by $rank(S) = \dim(span(\{p_1 - p_0, \ldots, p_k - p_0\}))$. (ii) The simplex S is* non-degenerate, *if $rank(S) = k$. If $rank(S) < k$, we call S a* degenerate simplex.

Lemma 3.12. *Let $k \in \mathbb{N}$ be fixed. Let $S \subseteq B_k(T) \cap \mathbb{Z}^k$ be a set of points with $rank(S) \le k-1$. Then there exists a $(k-1)$-maximal grid L of $\mathbb{Z}^k$ such that S is contained in some residue class $v + L$ of L for some $v \in \mathbb{Z}^k$, and L has a basis $q_1, \ldots, q_{k-1} \subset \mathbb{Z}^k$ with $\max_i \|q_i\| = O(T)$.*

Proof. For $S = \{p_0, \ldots, p_m\} \subseteq B_k(T) \cap \mathbb{Z}^k$ the vectors $p_1 - p_0, \ldots, p_m - p_0$ span a grid L' in $\mathbb{Z}^k$ with $\text{rank}(L') = r \le k-1$, and have length $\|p_i - p_0\| \le 2 \cdot T$ for $i = 1, \ldots, m$. Take $(k-1-r)$ unit vectors from $\mathbb{Z}^k \setminus L'$, add them to L', and obtain a grid L'' of $\mathbb{Z}^k$ with $\text{rank}(L'') = k-1$ and $L' \subseteq L''$. The grid L'' uniquely determines a $(k-1)$-maximal grid L of $\mathbb{Z}^k$ with $L' \subseteq L'' \subseteq L$ and with $S \subseteq p_0 + L$. By Theorem 3.6, there exists a basis $q_1, \ldots, q_{k-1}$ of L with $\max_i \|q_i\| = O(T)$. □

Theorem 3.13. *Let $k \in \mathbb{N}$ be fixed. Let L be a $(k-1)$-maximal grid of $\mathbb{Z}^k$ with primitive normal vector $a_L \in \mathbb{Z}^k$ and let $Q = \{q_1, \ldots, q_{k-1}\}$ be a basis of L with $\max_i \|q_i\| = O(T)$. Then the following holds:*

i) The primitive normal vector a_L satisfies $\|a_L\| = O(T^{k-1})$.
ii) There are at most $O(T \cdot \|a_L\|)$ distinct residue classes $s \cdot v + L$ with $s \in \mathbb{Z}$ and $v \in \mathbb{Z}^k \setminus L$ such that $(s \cdot v + L) \cap B_k(T) \neq \emptyset$.
iii) For every residue class $v + L$ with $v \in \mathbb{Z}^k$ it holds $|(v + L) \cap B_k(T)| = O(T^{k-1}/\|a_L\|)$.

Proof. (*i*): Let F_Q be the fundamental parallelepiped of Q. By Lemma 3.10, we have $\|a_L\| = d(L) = \text{vol}(F_Q)$ and $F_Q \subseteq B_k(c \cdot T)$ for some constant $c > 0$. Since $\dim(F_Q) = k-1$, the volume of the parallelepiped F_Q is at most $O(T^{k-1})$, hence $\|a_L\| = \text{vol}(F_Q) = O(T^{k-1})$.
(*ii*): By Lemma 3.9, distances between distinct residue classes of L are multiples of $1/\|a_L\|$. The distance between two points in $B_k(T)$ is at most $2 \cdot T$, hence at most $O(T \cdot \|a_L\|)$ residue classes of L have a nonempty intersection with $B_k(T)$.
(*iii*): The volume of a $(k-1)$-dimensional space S intersected with $B_k(T)$ is at most $O(T^{k-1})$. Since $F_Q \subseteq B_k(c \cdot T)$ and $\text{vol}(F_Q) = \|a_L\|$, we can cover the set $S \cap B_k(c \cdot T)$ by at most $O(T^{k-1}/\|a_L\|)$ distinct translates of F_Q. As L is maximal, the interior of F_Q contains no points from L. □

3.3 Representations by Sums of Squares

We mention without proofs some estimates concerning the number of representations of an integer as a sum of squares. Let $r_k(d)$ be the number of vectors $(x_1, \ldots, x_k)^\top \in \mathbb{Z}^k$ with $x_1^2 + \ldots + x_k^2 = d$.

Lemma 3.14. *For fixed $k, r \in \mathbb{N}$ and for every $n \in \mathbb{N}$ it is $\sum_{d=1}^n r_k(d) = \Theta(n^{k/2})$, and*

$$\sum_{d=1}^{n} \frac{r_k(d)}{d^r} = \begin{cases} O(n^{k/2-r}) & \text{if } k/2 - r > 0 \\ O(\log n) & \text{if } k/2 - r = 0 \\ O(1) & \text{if } k/2 - r < 0. \end{cases}$$

For a $(k-1)$-maximal grid L in $\mathbb{Z}^k$ with primitive normal vector $a_L \in \mathbb{Z}^k$ and $d \in \mathbb{N}$, we denote by $r_k(d; a_L)$ the number of points $P \in L$ with $(\text{dist}(O, P))^2 = d$, where O is the origin.

Lemma 3.15. *Let $k, r \in \mathbb{N}$ be fixed. Let L be a $(k-1)$-maximal grid in $\mathbb{Z}^k$ with primitive normal vector $a_L \in \mathbb{Z}^k$. For every $n \in \mathbb{N}$ it is $\sum_{d=1}^n r_k(d; a_L) = O(n^{(k-1)/2}/\|a_L\|)$, and*

$$\sum_{d=1}^{n} \frac{r_k(d; a_L)}{d^r} = \begin{cases} O\left(\frac{n^{\frac{k-1}{2}-r}}{\|a_L\|}\right) & \text{if } \frac{k-1}{2} - r > 0 \\ O\left(\frac{\log n}{\|a_L\|}\right) & \text{if } \frac{k-1}{2} - r = 0 \\ O\left(\frac{1}{\|a_L\|}\right) & \text{if } \frac{k-1}{2} - r < 0. \end{cases}$$

4 Counting Tetrahedra in $B_3(T) \cap \mathbb{Z}^3$

First we estimate the number of degenerate simplices in $B_k(T) \cap \mathbb{Z}^k$ for arbitrary but fixed dimension $k \geq 3$.

Theorem 4.1. *For fixed $k \in \mathbb{N}$ the number $D_k(T)$ of degenerate simplices in $B_k(T) \cap \mathbb{Z}^k$ satisfies $D_k(T) = O(T^{k^2} \cdot \log T)$.*

Proof. By Lemma 3.12, each degenerate $(k+1)$-element subset of points in $B_k(T) \cap \mathbb{Z}^k$ is contained in a residue class L' of some $(k-1)$-maximal grid L in $\mathbb{Z}^k$, where L has a basis $q_1, \ldots, q_{k-1} \in \mathbb{Z}^k$ with $\|q_i\| = O(T)$ for $i = 1, \ldots, k-1$. By Theorem 3.13 (i), it suffices to consider all primitive normal vectors $a_L \in \mathbb{Z}^k$ of length $\|a_L\| = O(T^{k-1})$. By Theorem 3.13 (ii), there are at most $O(T \cdot \|a_L\|)$ residue classes L' of L with $L' \cap B_k(T) \neq \emptyset$.
Fix a residue class L' of a $(k-1)$-maximal grid L in $\mathbb{Z}^k$, determined by its primitive normal vector $a_L \in \mathbb{Z}^k$ with $\|a_L\| = O(T^{k-1})$. By Theorem 3.13 (iii), the set $L' \cap B_k(T)$ contains at most $O(T^{k-1}/\|a_L\|)$ points, hence from $L' \cap B_k(T)$ we can select $(k+1)$ points in at most $O(\binom{T^{k-1}/\|a_L\|}{k+1})$ ways, to obtain a degenerate simplex. This implies

$$D_k(T) = O\left(\sum_{a \in \mathbb{Z}^k; \|a\| = O(T^{k-1})} T \cdot \|a\| \cdot \binom{\frac{T^{k-1}}{\|a\|}}{k+1} \right) = O\left(T^{k^2} \cdot \sum_{d=1}^{O(T^{2k-2})} \frac{r_k(d)}{d^{k/2}} \right)$$

$$= O(T^{k^2} \cdot \log T)\,,$$

as $\sum_{d=1}^{n} r_k(d)/d^{k/2} = O(\log n)$ by Lemma 3.14. □

From now on we consider only the case of dimension $k = 3$, i.e. we count the non-degenerate tetrahedra $S = \{p_0, p_1, p_2, p_3\} \subseteq B_3(T) \cap \mathbb{Z}^3$. The *volume* of a tetrahedron $S = \{p_0, \ldots, p_3\}$ is given by $\text{vol}(S) = 1/3 \cdot h \cdot \text{area}(p_0, p_1, p_2)$, where h is the distance of p_3 to the affine real space generated by p_0, p_1, p_2 and $\text{area}(p_0, p_1, p_2)$ is the area of the triangle.

Theorem 4.2. *The number $N_3(T; B)$ of non-degenerate tetrahedra $S \subseteq B_3(T) \cap \mathbb{Z}^3$ with vol$(S) \leq B$ satisfies $N_3(T; B) = O(B \cdot T^9)$.*

To prove this, we first count the non-degenerate triangles S in $L \cap B_3(T)$ for a 2-maximal grid L in $\mathbb{Z}^3$ with $\text{area}(S) \geq v$ resp. $\text{area}(S) \leq v$.

Lemma 4.3. *Let L' be a residue class of a 2-maximal grid L in $\mathbb{Z}^3$ with primitive normal vector $a_L \in \mathbb{Z}^3$. For any positive reals $v, B > 0$, the number of non-degenerate tetrahedra $S = \{p_0, p_1, p_2, p_3\}$ in $B_3(T) \cap \mathbb{Z}^3$ with $S \setminus \{p_3\} \subseteq L'$ and area$(S \setminus \{p_3\}) \geq v$ and vol$(S) \leq B$ is at most $O(B \cdot T^8/(v \cdot \|a_L\|^3))$.*

Proof. By Lemma 3.12 we can assume that the 2-maximal grid L in $\mathbb{Z}^3$ with primitive normal vector $a_L \in \mathbb{Z}^3$ has a basis $q_1, q_2 \in \mathbb{Z}^3$ with $\|q_1\|, \|q_2\| = O(T)$. For every residue class L' of L the set $L' \cap B_3(T)$ contains by Theorem 3.13 (iii) at most $O(T^2/\|a_L\|)$ points. From the set $L' \cap B_3(T)$ we can choose three points (with triangle-area at least v) in at most $O(\binom{T^2/\|a_L\|}{3})$ ways. Since the tetrahedra have volume at most B, the distance between a corresponding fourth point and the span of L' is at most $O(B/v)$. By Lemma 3.9, the distance between distinct residue classes of L is a multiple of $1/\|a_L\|$, and since $|L'' \cap B_3(T)| = O(T^2/\|a_L\|)$ for every residue class L'' of L, a fourth point can be chosen in at most

$$O(B/v \cdot \|a_L\| \cdot T^2/\|a_L\|) = O(B \cdot T^2/v)$$

ways, and we obtain for the number of tetrahedra

$$O\left(\binom{T^2/\|a_L\|}{3} \cdot \frac{B \cdot T^2}{v}\right) = O\left(\frac{B \cdot T^8}{v \cdot \|a_L\|^3}\right). \qquad \square$$

Lemma 4.4. *Let L' be a residue class of a 2-maximal grid in $\mathbb{Z}^3$ with primitive normal vector $a_L \in \mathbb{Z}^3$. Let P and Q be distinct points in L'. The number of non-degenerate triangles S in the set $L' \cap B_3(T)$ with extreme points P and Q and with area$(S) \leq v$ is at most $O(v \cdot T/(dist(P,Q) \cdot \|a_L\|))$.*

Proof. By an affine mapping $f: L \longrightarrow \mathbb{Z}^2$ with $P' := f(P)$ for $P \in L$, we transform the 2-maximal grid L in $\mathbb{Z}^3$ with primitive normal vector $a_L \in \mathbb{Z}^3$ (or any residue class L' of L) into the standard 2-dimensional rectangular grid $\mathbb{Z}^2$ with basis $(0,1)^\top$ and $(1,0)^\top$. Points $P, Q, R \in L' \cap B_3(T)$ become the grid points $P', Q', R' \in E \cap \mathbb{Z}^2$ within an ellipsoid E. If area$(P,Q,R) = v$, then area$(P',Q',R') = v/\|a_L\|$, as can be easily seen.
Let $a = (a_1,a_2,a_3)^\top \in \mathbb{Z}^3$, $b = (b_1,b_2,b_3)^\top \in \mathbb{Z}^3$ be a basis of the grid L. We can assume that $L' = L$ and that $P = (0,0,0)$ and $Q = \lambda \cdot a + \mu \cdot b$ are the two given points where $\lambda, \mu \in \mathbb{Z}$. We obtain the points $f(P) = P' = (0,0)$ and $f(Q) = Q' = (\lambda, \mu)$ which are contained in the ellipsoid E. Let $g = \gcd(\lambda, \mu)$ and set $\lambda' := \lambda/g$ and $\mu' := \mu/g$, where $\mu' > 0$. The line L_1 through the points P' and Q' in $\mathbb{Z}^2$ has the primitive normal vector $a_N := (\mu', -\lambda')^T$.
To estimate the number of points $R \in L' \cap B_3(T)$ such that area$(P,Q,R) \leq v$, we compute the number of points $R' \in E \cap \mathbb{Z}^2$ such that area$(P',Q',R') \leq v/\|a_L\|$. The distance of R' to the line L_1 is at most $2 \cdot v/(\|a_L\| \cdot \text{dist}(P',Q'))$. By Lemma 3.9, lines L_1' (residue classes of L_1) parallel to the line L_1 have distance a multiple of $1/\|a_N\|$, thus we consider at most $O(v \cdot \|a_N\|/(\|a_L\| \cdot \text{dist}(P',Q'))$ such lines L_1'. As two points in $B_3(T)$ have distance at most $2 \cdot T$, the line L_1 intersects the ellipsoid E in two points with distance D, where

$$D = O(T \cdot \text{dist}(P',Q')/\text{dist}(P,Q)) \ ,$$

and this also holds for each line L_1' parallel to L_1. The distance between two points in $L_1' \cap \mathbb{Z}^2$ is a multiple of $\|a_N\|$, hence $|L_1' \cap E \cap \mathbb{Z}^2| = O(D/\|a_N\|)$. Thus, the number of non-degenerate triangles in $L' \cap B_3(T)$ with area at most v and with extreme points P and Q is at most

$$O\left(\frac{v \cdot \|a_N\|}{\|a_L\| \cdot \text{dist}(P',Q')} \cdot \frac{D}{\|a_N\|}\right) = O\left(\frac{v \cdot T}{\text{dist}(P,Q) \cdot \|a_L\|}\right). \qquad \square$$

Corollary 4.5. *Let L be a 2-maximal grid in $\mathbb{Z}^3$ with primitive normal vector $a_L \in \mathbb{Z}^3$. For every residue class L' of L the number of non-degenerate triangles S in the set $L' \cap B_3(T)$ with area$(S) \leq v$ is at most $O(v \cdot T^4/||a_L||^3)$.*

Proof. Without loss of generality let $L' = L$. There are at most $O(T^2/||a_L||)$ choices to fix a point $P \in L \cap B_3(T)$. For every $d \in \mathbb{N}$ there are $r_3(d; a_L)$ points $Q \in L$ with $(\text{dist}(P,Q))^2 = d$. By Lemmas 4.4 and 3.15, the number of non-degenerate triangles $S \subset L \cap B_3(T)$ with area$(S) \leq v$ is at most

$$O\left(\frac{T^2}{||a_L||} \cdot \sum_{d=1}^{O(T^2)} \frac{v \cdot T}{d^{1/2} \cdot ||a_L||} \cdot r_3(d; a_L)\right)$$
$$= O\left(\frac{v \cdot T^3}{||a_L||^2} \cdot \sum_{d=1}^{O(T^2)} \frac{r_3(d; a_L)}{d^{1/2}}\right) = O\left(\frac{v \cdot T^4}{||a_L||^3}\right).$$

□

Lemma 4.6. *Let $L \subseteq \mathbb{Z}^3$ be a 2-maximal grid with primitive normal vector $a_L \in \mathbb{Z}^3$. Then every non-degenerate triangle $S \subset L$ fulfills area$(S) \geq ||a_L||/2$.*

Proof. The minimum area of a non-degenerate triangle in L is half of the volume of a fundamental parallelepiped F_Q, where vol$(F_Q) = ||a_L||$. □

Lemma 4.7. *Let L' be a residue class of a 2-maximal grid L in $\mathbb{Z}^3$ with primitive normal vector $a_L \in \mathbb{Z}^3$. For any positive reals $B, v > 0$, the number of non-degenerate tetrahedra $S = \{p_0, \ldots, p_3\}$ in $B_3(T) \cap \mathbb{Z}^3$ with $S \setminus \{p_3\} \subseteq L'$, with area$(S \setminus \{p_3\}) \leq v$ and with vol$(S) \leq B$ is at most $O(B \cdot v \cdot T^6/||a_L||^4)$.*

Proof. By Corollary 4.5 there are at most $O(v \cdot T^4/||a_L||^3)$ non-degenerate triangles $S' = \{p_0, p_1, p_2\}$ in $L' \cap B_3(T)$ with area$(S') \leq v$. Since vol$(S' \cup \{p_3\}) \leq B$ and, by Lemma 4.6, area$(S') \geq ||a_L||/2$, the distance between a fourth point p_3 of a tetrahedron with volume at most B and the real space generated by S' is at most $O(B/||a_L||)$. By Lemma 3.9, distances between residue classes L' of L are a multiple of $1/||a_L||$, and also $|L' \cap B_3(T)| = O(T^2/||a_L||)$. Thus the number of choices for the point p_3 is at most

$$O(B/||a_L|| \cdot T^2/||a_L|| \cdot ||a_L||) = O(B \cdot T^2/||a_L||) .$$

With Corollary 4.5, the number of tetrahedra $S = \{p_0, \ldots, p_3\}$ in $B_3(T) \cap \mathbb{Z}^3$ with vol$(S) \leq B$ and $S \setminus \{p_3\} \subseteq L'$ and area$(S \setminus \{p_3\}) \leq v$ is at most

$$O\left(\frac{v \cdot T^4}{||a_L||^3} \cdot \frac{B \cdot T^2}{||a_L||}\right) = O\left(\frac{B \cdot v \cdot T^6}{||a_L||^4}\right) .$$

□

The upper bounds in Lemmas 4.3 and 4.7 do have for $v := (T^2 \cdot ||a_L||)^{1/2}$ the same growth rate, namely $O(B \cdot T^7/||a_L||^{7/2})$, hence we summarize as follows:

Lemma 4.8. *Let L' be a residue class of a 2-maximal grid L in $\mathbb{Z}^3$ with primitive normal vector $a_L \in \mathbb{Z}^3$. The number of non-degenerate tetrahedra $S = \{p_0, p_1, p_2, p_3\} \subseteq B_3(T) \cap \mathbb{Z}^3$ with $S \setminus \{p_3\} \subseteq L'$ and $vol(S) \leq B$ is at most $O(B \cdot T^7/||a_L||^{7/2})$.*

Proof (of Theorem 4.2). For a fixed primitive normal vector $a_L \in \mathbb{Z}^3$ there are at most $O(T \cdot ||a_L||)$ residue classes L' of the grid L with $L' \cap B_3(T) \neq \emptyset$. By Lemmas 4.8 and 3.14, we infer

$$N_3(T; B) = O\left(\sum_{\substack{a \in \mathbb{Z}^3 \\ ||a|| = O(T^2)}} T \cdot ||a|| \cdot \frac{B \cdot T^7}{||a||^{7/2}} \right) = O\left(B \cdot T^8 \cdot \sum_{d=1}^{O(T^4)} \frac{r_3(d)}{d^{5/4}} \right) = O\left(B \cdot T^9\right).$$

□

5 2-Cycles

As mentioned before, for integers $n \in \mathbb{N}$ and for some value $B > 0$ and for the given set $B_3(T) \cap \mathbb{Z}^3$ of points we construct a hypergraph $\mathcal{G} = \mathcal{G}(B) = (V, \mathcal{E})$ with the vertex set being the set $B_3(T) \cap \mathbb{Z}^3$ of $\Theta(T^3)$ grid-points. The 4-element edges are given by all subsets of four points, no three of them on a line, which form a tetrahedron of volume at most B, including degenerate tetrahedra. The 3-element edges are given by all triples of points from $B_3(T) \cap \mathbb{Z}^3$ on a line. This hypergraph $\mathcal{G}(B)$ can easily be constructed in time $O(T^{12})$ by computing in time $O(1)$ the volume of every quadruple of points in $B_3(T) \cap \mathbb{Z}^3$. An independent set in $\mathcal{G}(B)$ corresponds to a set of points in $B_3(T) \cap \mathbb{Z}^3$, where all tetrahedra have volume bigger than B.

To apply Theorem 2.2, we have to show that its assumptions are satisfied. Set

$$B := T^3 \cdot \log n / n^3 \quad \text{and} \quad T := n^{1+\varepsilon} \tag{1}$$

for some $\varepsilon > 0$, thus $B = n^{3\varepsilon} \cdot \log n$. By Theorems 4.1 and 4.2, we bound the number of 4-element edges of the hypergraph $\mathcal{G} = \mathcal{G}(B) = (V, \mathcal{E})$ with $|\mathcal{E}| = O(T^9 \cdot \log T + B \cdot T^9) = O(B \cdot T^9)$. With $|V| = \Theta(T^3)$ the average degree t^3 of $\mathcal{G}(B)$ for the 4-element edges satisfies

$$t^3 = O(B \cdot T^9/T^3) = O(B \cdot T^6) .$$

We will use in our further computations an upper bound for t and for some constant $c > 0$ we set

$$t^3 := c \cdot B \cdot T^6 . \tag{2}$$

First we count the number of triples of collinear points in $B_3(T) \cap \mathbb{Z}^3$, that is, the number $tr(\mathcal{G})$ of 3-element edges in $\mathcal{G}$. Two distinct points P and Q in $B_3(T) \cap \mathbb{Z}^3$ can be chosen in at most $O(T^6)$ ways. A third point on the line through P and Q intersected with $B_3(T) \cap \mathbb{Z}^3$ can be chosen in at most $O(T)$ ways, thus $tr(\mathcal{G}) = O(T^7)$. The assumptions of Theorem 2.2 are fulfilled for $0 < \delta < 2\varepsilon/(2+3\varepsilon)$ as the following shows:

$$\begin{aligned} tr(\mathcal{G}) &= O(T^7) = O(T^7 \cdot (\log n)^{2/3-\delta/3} \cdot n^{2\varepsilon-3\varepsilon\delta-2\delta}) \\ &= O(B^{2/3-\delta/3} \cdot T^{7-2\delta}) = O(T^3 \cdot t^{2-\delta}) . \end{aligned} \tag{3}$$

Next we count the 2-cycles in our hypergraph $\mathcal{G}$, where we distinguish $(2,2)$-cycles and $(2,3)$-cycles. For both estimates we distinguish three cases: (a) both tetrahedra are degenerate, or (b) one tetrahedron is degenerate and the other one is non-degenerate or (c) both tetrahedra are non-degenerate. The corresponding numbers of $(2,i)$-cycles, $i = 2,3$, are denoted by $s_{2,i}(\mathcal{G}; dd)$, $s_{2,i}(\mathcal{G}; dn)$, $s_{2,i}(\mathcal{G}; nn)$, respectively.

5.1 (2, 2)-Cycles

Here we estimate the number $s_{2,2}(\mathcal{G})$ of $(2,2)$-cycles in our hypergraph $\mathcal{G} = \mathcal{G}(B)$, that is the number of pairs of tetrahedra in $B_3(T) \cap \mathbb{Z}^3$, which have exactly two extreme points in common, and both tetrahedra have volume at most B.
Case (a): Both tetrahedra are degenerate. By Theorem 4.1, there are at most $O\left(T^9 \cdot \log T\right)$ degenerate tetrahedra in the set $B_3(T) \cap \mathbb{Z}^3$. Fix one of these tetrahedra. The second degenerate tetrahedron is contained in a 2-maximal grid in $\mathbb{Z}^3$ and has two extreme points, say $P = (p_1, p_2, p_3)$ and $Q = (q_1, q_2, q_3)$, in common with the first one. Fix a primitive normal vector $b_M := (b_1, b_2, b_3) \in \mathbb{Z}^3$ with $\|b_M\| = O(T^2)$ by Theorem 3.13 (i), which belongs to a 2-maximal grid M in $\mathbb{Z}^3$. There is at most one residue class M' of M such that $P, Q \in M'$, and with $y_i := p_i - q_i$ for $i = 1,2,3$ it must hold that

$$b_1 \cdot y_1 + b_2 \cdot y_2 + b_3 \cdot y_3 = 0 . \tag{4}$$

By Theorem 3.13 (iii), the set $M' \cap B_3(T)$ contains at most $O\left(T^2/\|b_M\|\right)$ points. Two further points can be chosen from $M' \cap B_3(T)$ in at most $O(\binom{T^2/\|b_M\|}{2})$ ways. Hence we infer for the number $s_{2,2}(\mathcal{G}; dd)$ of pairs of degenerate tetrahedra in $B_3(T) \cap \mathbb{Z}^3$ which have two extreme points in common, where $y = (y_1, y_2, y_3)^\top \neq 0^3$, say $y_3 \neq 0$, refers to the chosen points P and Q of the first tetrahedron:

$$\begin{aligned} s_{2,2}(\mathcal{G}; dd) &= O\left(T^9 \cdot \log T \cdot \sum_{\substack{b \in \mathbb{Z}^3; \|b\| = O(T^2) \\ <b,y>=0}} \binom{\frac{T^2}{\|b\|}}{2} \right) \\ &= O\left(T^{13} \cdot \log T \cdot \sum_{\substack{b \in \mathbb{Z}^3; \|b\| = O(T^2) \\ <b,y>=0}} \frac{1}{\|b\|^2} \right) = O(T^{13} \cdot (\log T)^2) , \end{aligned} \tag{5}$$

since with Lemma 3.14 we have

$$\sum_{\substack{b\in\mathbb{Z}^3;\|b\|=O(T^2)\\ <b,y>=0}} \frac{1}{\|b\|^2} = O\left(\sum_{\substack{b_1,b_2\in\mathbb{Z}\\ |b_1|,|b_2|=O(T^2)}} \frac{1}{b_1^2+b_2^2+\left(\frac{b_1\cdot y_1}{y_3}+\frac{b_2\cdot y_2}{y_3}\right)^2}\right)$$

$$= O\left(\sum_{\substack{b_1,b_2\in\mathbb{Z}\\ |b_1|,|b_2|=O(T^2)}} \frac{1}{b_1^2+b_2^2}\right) = O\left(\sum_{d=1}^{O(T^4)} \frac{r_2(d)}{d}\right) = O(\log T)\,.$$

Case (b): One tetrahedron is degenerate and the other one is non-degenerate with volume at most B. By Theorem 4.2, the number of non-degenerate tetrahedra with volume at most B in the set $B_3(T)\cap\mathbb{Z}^3$ is at most $O(B\cdot T^9)$. Fix such a tetrahedron and fix two of its extreme points $P=(p_1,p_2,p_3)$ and $Q=(q_1,q_2,q_3)$ where $y_i := p_i - q_i$ for $i=1,2,3$. As in case (a), see (5), we obtain

$$s_{2,2}(\mathcal{G};dn) = O\left(B\cdot T^9\cdot \sum_{\substack{b\in\mathbb{Z}^3;\|b\|=O(T^2)\\ <b,y>=0}} \binom{\frac{T^2}{\|b\|}}{2}\right) = O(B\cdot T^{13}\cdot\log T)\,. \qquad (6)$$

Case (c): Both simplices have volume at most B, are non-degenerate and have two extreme points in common. To estimate $s_{2,2}(\mathcal{G};nn)$, we fix a 2-maximal grid L in $\mathbb{Z}^3$ with primitive normal vector $a_L\in\mathbb{Z}^3$ where $\|a_L\|=O(T^2)$. Fix a point $P\in B_3(T)\cap\mathbb{Z}^3$. There is exactly one residue class L' of L with $P\in L'$. For $d\in\mathbb{N}$ there are at most $r_3(d;a_L)$ many points $Q\in L'\cap B_3(T)$ with $(\mathrm{dist}(P,Q))^2=d$. Let P and Q be the two common extreme points of both tetrahedra.
By Lemma 4.4, there are at most $O(v\cdot T/(\|a_L\|\cdot d^{1/2}))$ points $R\in L'\cap B_3(T)$ with $\mathrm{area}(P,Q,R)\le v$. A fourth point in $B_3(T)\cap\mathbb{Z}^3$ of a tetrahedron with volume at most B can be chosen in at most $O(B\cdot T^2/\|a_L\|)$ ways. Moreover, there are at most $O(T^2/\|a_L\|)$ third points $R\in L'\cap B_3(T)$ such that $\mathrm{area}(P,Q,R)>v$, and a fourth point in $B_3(T)\cap\mathbb{Z}^3$ of a tetrahedron with volume at most B can be chosen in at most $O(B\cdot T^2/v)$ ways. With $v:=T^{1/2}\cdot\|a_L\|^{1/2}\cdot d^{1/4}$ the number of choices for a third and fourth point of the first tetrahedron is at most

$$O\left(\frac{B\cdot v\cdot T^3}{\|a_L\|^2\cdot d^{1/2}}+\frac{B\cdot T^4}{v\cdot\|a_L\|}\right) = O\left(\frac{B\cdot T^{7/2}}{\|a_L\|^{3/2}\cdot d^{1/4}}\right)\,. \qquad (7)$$

Concerning the second tetrahedron with the extreme points $P=(p_1,p_2,p_3)$ and $Q=(q_1,q_2,q_3)$, let $y_i := p_i - q_i$ for $i=1,2,3$. For a primitive normal vector $b_M=(b_1,b_2,b_3)\in\mathbb{Z}^3$, if $P,Q\in M'$ for some residue class M' of a 2-maximal grid M in $\mathbb{Z}^3$ then (4) holds. Since a third point $R\in M'$ of the tetrahedron satisfies $R\in B_3(T)\cap\mathbb{Z}^3$, there are at most $O(T^3)$ choices for $b_M\in\mathbb{Z}^3$. Let $C\subseteq\mathbb{Z}^3$ be the set of all these primitive normal vectors b_M. Using $\sum_{e=1}^{N} r_2(e)=\Theta(N)$ by Lemma 3.14, the number of choices to extend these two points P and Q to a second tetrahedron in $\mathbb{Z}^3\cap B_3(T)$ is by (7) at most

$$O\left(\frac{B\cdot T^{7/2}}{d^{1/4}}\cdot\sum_{b\in C}\frac{1}{\|b\|^{3/2}}\right)=O\left(\frac{B\cdot T^{7/2}}{d^{1/4}}\cdot\sum_{(b_1,b_2,b_3)\in C}\frac{1}{(b_1^2+b_2^2)^{3/4}}\right)$$

$$=O\left(\frac{B\cdot T^{7/2}}{d^{1/4}}\cdot\sum_{e=1}^{O(T^3)}\frac{r_2(e)}{e^{3/4}}\right)=O\left(\frac{B\cdot T^{17/4}}{d^{1/4}}\right). \tag{8}$$

Combining (7) and (8) and summing over the choices of the points P and Q the number of choices for the two tetrahedra is at most

$$s_{2,2}(\mathcal{G};nn)=O\left(T^3\cdot\sum_{a\in\mathbb{Z}^3;\|a\|=O(T^2)}\sum_{d=1}^{O(T^2)}r_3(d;a)\cdot\frac{B^2\cdot T^{31/4}}{\|a\|^{3/2}\cdot d^{1/2}}\right)$$

$$=O\left(B^2\cdot T^{43/4}\cdot\sum_{a\in\mathbb{Z}^3;\|a\|=O(T^2)}\frac{1}{\|a\|^{3/2}}\cdot\sum_{d=1}^{O(T^2)}\frac{r_3(d;a)}{d^{1/2}}\right)$$

$$=O\left(B^2\cdot T^{47/4}\cdot\sum_{d=1}^{O(T^4)}\frac{r_3(d)}{d^{5/4}}\right)=O(B^2\cdot T^{51/4}). \tag{9}$$

To apply Theorem 2.2, we must have for some suitable constant $\gamma>0$ that

$$\begin{aligned}s_{2,2}(\mathcal{G})&=s_{2,2}(\mathcal{G};dd)+s_{2,2}(\mathcal{G};dn)+s_{2,2}(\mathcal{G};nn)\\&=O(T^3\cdot t^{5-\gamma})=O(T^{13-2\gamma}\cdot B^{5/3-\gamma/3}),\end{aligned} \tag{10}$$

since $t=c'\cdot B^{1/3}\cdot T^2$ by (2). Recall that $B:=T^3\cdot\log n/n^3$ by (1) with $T:=n^{1+\varepsilon}$ for fixed $\varepsilon>0$. For $0<\varepsilon<1/11$ and $0<\gamma<2\varepsilon/(3+2\varepsilon)$, using (5), (6) and (9) we infer

$$T^{13}\cdot(\log T)^2+B^2\cdot T^{51/4}=O(B\cdot T^{13}\cdot\log T)\,,$$

hence it suffices to consider only case (b). Here, by (6) and (1), we have

$$\frac{B\cdot T^{13}\cdot\log T}{T^{13-2\gamma}\cdot B^{5/3-\gamma/3}}=O\left(\frac{(\log n)^{1/3+\gamma/3}}{n^{2\varepsilon-3\varepsilon\gamma-2\gamma}}\right)=o(1)\,,$$

for $0<\gamma<2\varepsilon/(3+2\varepsilon)$ and (10) holds.

5.2 (2, 3)-Cycles

Next we estimate the number $s_{2,3}(\mathcal{G})$ of (2, 3)-cycles in our hypergraph $\mathcal{G}=\mathcal{G}(B)$, that is, the number of pairs of tetrahedra in $B_3(T)\cap\mathbb{Z}^3$, each with volume at most B, having exactly three extreme points in common. We can assume that these three points are not collinear, since we consider such triples separately. As in the case of (2, 2)-cycles we distinguish three cases.

Case (a): Both tetrahedra are degenerate and have three extreme points in common. The three common points are not on a line, hence they uniquely determine a residue class L' of a 2-maximal grid L in $\mathbb{Z}^3$ with $L' \cap B_3(T) \neq \emptyset$ and with primitive normal vector $a_L \in \mathbb{Z}^3$ where $\|a_L\| = O(T^2)$ and all extreme points are contained in $L' \cap B_3(T)$. Since $|L' \cap B_3(T)| = O(T^2/\|a_L\|)$, we can choose five extreme points of the two tetrahedra in at most $O(\binom{T^2/\|a_L\|}{5})$ ways. By Theorem 3.13 (ii), there are at most $O(T \cdot \|a_L\|)$ residue classes L' of L with $L' \cap B_3(T) \neq \emptyset$ and with Lemma 3.14 we infer

$$s_{2,3}(\mathcal{G}; dd) = O\left(\sum_{a \in \mathbb{Z}^3; \|a\| = O(T^2)} T \cdot \|a\| \cdot \binom{\frac{T^2}{\|a\|}}{5} \right) = O\left(T^{11} \cdot \sum_{d=1}^{O(T^4)} \frac{r_3(d)}{d^2} \right)$$
$$= O(T^{11}) . \tag{11}$$

Case (b): One of the two tetrahedra is non-degenerate with volume at most B, the other is degenerate. By Theorem 4.2, there are at most $O(B \cdot T^9)$ non-degenerate tetrahedra with volume at most B in $B_3(T) \cap \mathbb{Z}^3$. Fix one of these tetrahedra and choose three of its extreme points as the common points of both tetrahedra. These three points determine uniquely a residue class L' of a 2-maximal grid in $\mathbb{Z}^3$. A fourth point of the second degenerate tetrahedron is also contained in $L' \cap \mathbb{Z}^3$. With $|L' \cap B_3(T)| = O(T^2)$ we infer

$$s_{2,3}(\mathcal{G}; dn) = O(B \cdot T^9 \cdot T^2) = O(B \cdot T^{11}) . \tag{12}$$

Case (c): Both tetrahedra are non-degenerate, each with volume at most B. The three common extreme points are contained in a residue class L' of a 2-maximal grid L in $\mathbb{Z}^3$ with primitive normal vector $a_L \in \mathbb{Z}^3$.
By Corollary 4.5, the number of triangles in $L' \cap B_3(T)$ with area at most v is at most $O(v \cdot T^4/\|a_L\|^3)$. The two fourth points of the tetrahedra can be chosen in at most $O((B \cdot T^2/\|a_L\|)^2)$ ways.
There are at most $O(\binom{T^2/\|a_L\|}{3})$ triangles in $L' \cap B_3(T)$ with area at least v. Two fourth points of the tetrahedra can be chosen in at most $O((B \cdot T^2/v)^2)$ ways. With $v := T^{2/3} \cdot \|a_L\|^{2/3}$, the number of such pairs of tetrahedra is at most

$$O\left(\frac{v \cdot T^4}{\|a_L\|^3} \cdot \left(\frac{B \cdot T^2}{\|a_L\|} \right)^2 + \frac{T^6}{\|a_L\|^3} \cdot \left(\frac{B \cdot T^2}{v} \right)^2 \right) = O\left(\frac{B^2 \cdot T^{26/3}}{\|a_L\|^{13/3}} \right) .$$

Summing over all 2-maximal grids L in $\mathbb{Z}^3$ and all its $O(T \cdot \|a_L\|)$ residue classes L' with $L' \cap B_3(T) \neq \emptyset$, we obtain with Lemma 3.14:

$$s_{2,3}(\mathcal{G}; nn) = O\left(\sum_{a \in \mathbb{Z}^3; \|a\| = O(T^2)} T \cdot \|a\| \cdot \frac{B^2 \cdot T^{26/3}}{\|a\|^{13/3}} \right) = O(B^2 \cdot T^{29/3}) . \tag{13}$$

As in the case of $(2,2)$-cycles considering the estimates (11), (12) and (13), with $t = (c \cdot B \cdot T^6)^{1/3}$ and $B := T^3 \cdot \log n / n^3$ and $T := n^{1+\varepsilon}$ for $0 < \varepsilon < 4/5$ and $0 < \gamma < \varepsilon/(2 + 3\varepsilon)$ we obtain

$$\begin{aligned} s_{2,3}(\mathcal{G}) &= s_{2,3}(\mathcal{G}; dd) + s_{2,3}(\mathcal{G}; dn) + s_{2,3}(\mathcal{G}; nn) \\ &= O(T^3 \cdot t^{4-\gamma}) = O(T^{11-2\gamma} \cdot B^{4/3-\gamma/3}) . \end{aligned}$$

With (10) and (3) the assumptions of Theorem 2.2 are fulfilled and we apply it to our hypergraph $\mathcal{G} = \mathcal{G}(B) = (V, \mathcal{E})$ with the upper bound $t^3 = c \cdot B \cdot T^6$ on the average degree for the 4-element edges, and we find in $\mathcal{G}(B)$ in time polynomial in $T = n^{1+\varepsilon}$, hence in n, an independent set of size at least

$$\Omega\left(\frac{T^3}{B^{1/3} \cdot T^2} \cdot (\log t)^{1/3}\right) = \Omega\left(\frac{T}{(\log n)^{1/3} \cdot n^{\varepsilon}} \cdot (\log n)^{1/3}\right) = \Omega(n) \ .$$

Thus, we have found at least $\Omega(n)$ points in time polynomial in n such that the volume of every tetrahedron is at least $\Omega(\log n/n^3)$.

References

1. M. Ajtai, J. Komlós, J. Pintz, J. Spencer, and E. Szemerédi, Extremal uncrowded hypergraphs, *J. Comb. Th. A*, 32, 1982, 321–335.
2. N. Alon and J. Spencer, *The Probabilistic Method,* Wiley & Sons, 1992.
3. G. Barequet, A lower bound for Heilbronn's triangle problem in d dimensions, *SIAM Journal on Discrete Mathematics,* 14, 2001, 230–236.
4. C. Bertram–Kretzberg and H. Lefmann, The algorithmic aspects of uncrowded hypergraphs, *Proc. '8th ACM-SIAM Symp. on Discrete Algorithms SODA'97',* 1997, 296–304.
5. B. Bollobás, personal communication, 2001.
6. C. Bertram-Kretzberg, T. Hofmeister, and H. Lefmann, An algorithm for Heilbronn's problem, *SIAM Journal on Computing,* 30, 2000, 383–390.
7. C. Bertram-Kretzberg, H. Lefmann, V. Rödl, and B. Wysocka, Proper bounded edge-colorings, *Combinatorics, Complexity & Logic, Proc. '1st Int. Conf. on Discrete Mathematics and Theoretical Computer Science DMTCS'96',* eds. D. S. Bridges et al., Springer, 1996, 121–130.
8. J. W. S. Cassels, *An Introduction to the Geometry of Numbers,* Springer, 1971.
9. H. Cohen, *A Course in Computational Algebraic Number Theory,* Springer, 1993.
10. R. A. Duke, H. Lefmann, and V. Rödl, On uncrowded hypergraphs, *Rand. Struct. & Alg.,* 6, 1995, 209–212.
11. A. Fundia, Derandomizing Chebychev's inequality to find independent sets in uncrowded hypergraphs, *Rand. Struct. & Alg.,* 8, 1996, 131–147.
12. T. Jiang, M. Li, and P. Vitány, Kolmogorov complexity and a triangle problem of the Heilbronn type, *preprint,* 2000.
13. M. Koecher, *Lineare Algebra und analytische Geometrie,* 4th ed., Springer, 1997.
14. J. Komlós, J. Pintz, and E. Szemerédi, On Heilbronn's triangle problem, *J. of the London Math. Soc.,* 24, 1981, 385–396.
15. J. Komlós, J. Pintz, and E. Szemerédi, A lower bound for Heilbronn's problem, *J. of the London Math. Soc.,* 25, 1982, 13–24.
16. H. Lefmann, On Heilbronn's problem in higher dimension, *Proc. '11th ACM-SIAM Symp. on Discrete Algorithms SODA'00',* 2000, 60–64.
17. K. F. Roth, On a problem of Heilbronn, *J. of the London Math. Soc.,* 26, 1951, 198–204.
18. K. F. Roth, On a problem of Heilbronn, II, *Proc. of the London Math. Soc. (3),* 25, 1972, 193–212.
19. K. F. Roth, On a problem of Heilbronn, III, *Proc. of the London Math. Soc. (3),* 25, 1972, 543–549.

20. K. F. Roth, Estimation of the area of the smallest triangle obtained by selecting three out of n points in a disc of unit area, *Proc. of Symp. in Pure Math.*, 24, 1973, AMS, Providence, 251–262.
21. K. F. Roth, Developments in Heilbronn's triangle problem, *Adv. in Math.*, 22, 1976, 364–385.
22. B. L. Rothschild and E. G. Straus, On triangulations of the convex hull of n points, *Combinatorica,* 5, 1985, 167–179.
23. W. M. Schmidt, On a problem of Heilbronn, *J. of the London Math. Soc. (2),* 4, 1972, 545–550.

A Metric Index for Approximate String Matching*

Edgar Chávez[1]** and Gonzalo Navarro[2]

[1] Escuela de Ciencias Físico-Matemáticas, Universidad Michoacana. Edificio "B", Ciudad Universitaria, Morelia, Mich. México 58000.
elchavez@fismat.umich.mx

[2] Depto. de Ciencias de la Computación, Universidad de Chile. Blanco Encalada 2120, Santiago, Chile.
gnavarro@dcc.uchile.cl

Abstract. We present a radically new indexing approach for approximate string matching. The scheme uses the metric properties of the edit distance and can be applied to any other metric between strings. We build a metric space where the sites are the nodes of the suffix tree of the text, and the approximate query is seen as a proximity query on that metric space. This permits us finding the R occurrences of a pattern of length m in a text of length n in average time $O(m \log^2 n + m^2 + R)$, using $O(n \log n)$ space and $O(n \log^2 n)$ index construction time. This complexity improves by far over all other previous methods. We also show a simpler scheme needing $O(n)$ space.

1 Introduction and Related Work

Indexing text to permit efficient approximate searching on it is one of the main open problems in combinatorial pattern matching. The approximate string matching problem is: Given a long text T of length n, a (comparatively short) pattern P of length m, and a threshold r, retrieve all the pattern occurrences, that is, text substrings whose *edit distance* to the pattern is at most r. The *edit distance* between two strings is defined as the minimum number of character insertions, deletions and substitutions needed to make them equal. This distance is used in many applications, but several other distances are of interest.

In the on-line version of the problem, the pattern can be preprocessed but the text cannot. There are numerous solutions to this problem [25], but none is acceptable when the text is too long since the search time is proportional to the text length. Indexing text for approximate string matching has received attention only recently. Despite some progress in the last decade, the indexing schemes for this problem are still rather immature.

There exist some indexing schemes specialized to word-wise searching on natural language text [21,3]. These indexes perform quite well in that case but

* Supported by CYTED VII.19 RIBIDI project (both authors), CONACyT grant 36911 (first author), and Fondecyt grant 1-000929 (second author).

** On leave of absence at Universidad de Chile.

S. Rajsbaum (Ed.): LATIN 2002, LNCS 2286, pp. 181–195, 2002.

they cannot be extended to handle the general case. Extremely important applications such as DNA, proteins, music or oriental languages fall outside this case.

The indexes that solve the general problem can be divided into three classes. *Backtracking* [17,34,11,15] uses the suffix tree [2], suffix array [20] or DAWG [12] of the text in order to factor out its repetitions. A sequential algorithm on the text is simulated by backtracking on the data structure. These algorithms take time exponential on m or r but in many cases independent of n, the text size. This makes them attractive when searching for very short patterns.

Partitioning [31,30,5] partitions the pattern into pieces to ensure that some of the pieces must appear without alterations inside every occurrence. An index able of exact searching is used to detect the pieces and the text areas that have enough evidence of containing an occurrence are checked with a sequential algorithm. These algorithms work well only when r/m is small.

The third class [24,6] is a hybrid between the other two. The pattern is divided into large pieces that can still contain (less) errors, they are searched for using backtracking, and the potential text occurrences are checked as in the partitioning methods. The hybrid algorithms are more effective because they can find the right point between length of the pieces to search for and error level permitted. Using the appropriate partition of the pattern, these methods achieve on average $O(n^\lambda)$ search time, for some $0 < \lambda < 1$ that depends on r. They tolerate moderate error ratios r/m.

We propose in this paper a brand new approach to the problem. We take into account that the edit distance satisfies the triangle inequality and hence it defines a metric space on the set of text substrings. We can re-express the approximate search problem as a range search problem on this metric space. This approach has been attempted before [8,4], but in those cases the particularities of the problem made it possible to index $O(n)$ elements. In the general case we have $O(n^2)$ text substrings.

The main contribution of this paper is to devise a method (based on the suffix tree of the text) to meaningfully collapse the $O(n^2)$ text substring into $O(n)$ sets, and to find a way to build a metric space out of those sets. The result is an indexing method that, at the cost of requiring on average $O(n \log n)$ space and $O(n \log^2 n)$ construction time, permits finding the R approximate occurrences of the pattern in $O(m \log^2 n + m^2 + R)$ average time. This is a complexity breakthrough over previous work, and it is easier than in other approaches to extend the idea to other distance functions such as reversals. Moreover, it represents an original approach to the problem that opens a vast number of possibilities for improvements. We consider also a simpler version of the index needing $O(n)$ space and that, despite not involving a complexity breakthrough, promises to be better in practice.

We use the following notation in the paper. Given a string $s \in \Sigma^*$ we denote its length as $|s|$. We also denote s_i the i-th character of s, for an integer $i \in \{1..|s|\}$. We denote $s_{i...j} = s_i s_{i+1} \ldots s_j$ (which is the empty string if $i > j$) and

$s_{i\ldots} = s_{i\ldots|s|}$. The empty string is denoted as ε. A string x is said to be a *prefix* of xy, a *suffix* of yx and a *substring* of yxz.

2 Metric Spaces

We describe in this section some concepts related to searching metric spaces. We have concentrated only in the part that is relevant for this paper. There exist recent surveys if more complete information is desired [10].

A metric space is, informally, a set of black-box objects and a distance function defined among them, which satisfies the triangle inequality. The problem of *proximity searching* in metric spaces consists of indexing the set such that later, given a query, all the elements of the set that are close enough to the query can be quickly found. This has applications in a vast number of fields, such as non-traditional databases (where the concept of exact search is of no use and we search for similar objects, e.g. databases storing images, fingerprints or audio clips); machine learning and classification (where a new element must be classified according to its closest existing element); image quantization and compression (where only some vectors can be represented and those that cannot must be coded as their closest representable point); text retrieval (where we look for documents that are similar to a given query or document); computational biology (where we want to find a DNA or protein sequence in a database allowing some errors due to typical variations); function prediction (where we want to search for the most similar behavior of a function in the past so as to predict its probable future behavior); etc.

Formally, a *metric space* is a pair $(\mathbb{X}, d)$, where $\mathbb{X}$ is a "universe" of objects and $d : \mathbb{X} \times \mathbb{X} \longrightarrow \mathbb{R}^+$ is a distance function defined on it that returns non-negative values. This distance satisfies the properties of reflexivity ($d(x,x) = 0$), strict positiveness ($x \neq y \Rightarrow d(x,y) > 0$), symmetry ($d(x,y) = d(y,x)$) and triangle inequality ($d(x,y) \leq d(x,z) + d(z,y)$).

A finite subset $\mathbb{U}$ of $\mathbb{X}$, of size $n = |\mathbb{U}|$, is the set of objects we search. Among the many queries of interest on a metric space, we are interested in the so-called *range queries*: Given a query $q \in \mathbb{X}$ and a tolerance radius r, find the set of all elements in $\mathbb{U}$ that are at distance at most r to q. Formally, the outcome of the query is $(q,r)_d = \{u \in \mathbb{U},\ d(q,u) \leq r\}$. The goal is to preprocess the set so as to minimize the computational cost of producing the answer $(q,r)_d$.

From the plethora of existing algorithms to index metric spaces, we focus on the so-called *pivot-based* ones, which are built on a single general idea: Select k elements $\{p_1, \ldots, p_k\}$ from $\mathbb{U}$ (called *pivots*), and identify each element $u \in \mathbb{U}$ with a k-dimensional point $(d(u,p_1), \ldots, d(u,p_k))$ (i.e. its distances to the pivots). The index is basically the set of kn coordinates. At query time, map q to the k-dimensional point $(d(q,p_1), \ldots, d(q,p_k))$. With this information at hand, we can filter out using the triangle inequality any element u such that $|d(q,p_i) - d(u,p_i)| > r$ for some pivot p_i, since in that case we know that $d(q,u) > r$ without need to evaluate $d(u,q)$. Those elements that cannot be filtered out using this rule are directly compared against q.

An interesting feature of pivot-based algorithms is that they can reduce the number of final distance evaluations by increasing the number of pivots. Define $D_k(x,y) = \max_{1\leq j\leq k} |d(x,p_j) - d(y,p_j)|$. Using the pivots $p_1,...,p_k$ is equivalent to discarding elements u such that $D_k(q,u) > r$. As more pivots are added we need to perform more distance evaluations (exactly k) to compute $D_k(q,*)$ (these are called *internal* evaluations), but on the other hand $D_k(q,*)$ increases its value and hence it has a higher chance of filtering out more elements (those comparisons against elements that cannot be filtered out are called *external*). It follows that there exists an optimum k.

If one is not only interested in the number of distance evaluations performed but also in the total CPU time required, then scanning all the n elements to filter out some of them may be unacceptable. In that case, one needs *multidimensional range search* methods, which include data structures such as the *kd*-tree, *R*-tree, *X*-tree, etc. [36,14]. Those structures permit indexing a set of objects in k-dimensional space in order to process range queries.

In this paper we are interested in a metric space where the universe is the set of strings over some alphabet, i.e. $\mathbb{X} = \Sigma^*$, and the distance function is the so-called *edit distance* or *Levenshtein distance*. This is defined as the minimum number of character insertions, deletions and substitutions necessary to make two strings equal [19,25]. The edit distance, and in fact any other distance defined as the best way to convert one element into the other, is reflexive, strictly positive (as long as there are no zero-cost operations), symmetric (as long as the operations allowed are symmetric), and satisfies the triangle inequality.

The algorithm to compute the edit distance $ed()$ is based on dynamic programming. Imagine that we need to compute $ed(x,y)$. A matrix $C_{0..|x|,0..|y|}$ is filled, where $C_{i,j} = ed(x_{1..i}, y_{1..j})$, so $C_{|x|,|y|} = ed(x,y)$. This is computed as

$$\begin{aligned} C_{i,0} &= i, \quad C_{0,j} = j, \\ C_{i,j} &= \text{if } (x_i = y_j) \text{ then } C_{i-1,j-1} \text{ else } 1 + \min(C_{i-1,j}, C_{i,j-1}, C_{i-1,j-1}) \end{aligned}$$

The algorithm takes $O(|x||y|)$ time. The matrix can be filled column-wise or row-wise (there are more sophisticated ways as well). For reasons that will be made clear later, we prefer the row-wise filling. The space required is only $O(|y|)$, since only the previous row must be stored in order to compute the new one, and therefore we just keep one row and update it.

3 Text Indexing

Suffix trees are widely used data structures for text processing [2,1]. Any position i in a text T defines a *suffix* of T, namely $T_{i...}$. A *suffix trie* is a trie data structure built over all the suffixes of T. At the leaf nodes the pointers to the suffixes are stored. Every substring of T can be found by traversing a path from the root. Roughly speaking, each suffix trie leaf represents a suffix and each internal node represents a different substring of T.

To improve space utilization, this trie is compacted into a Patricia tree [23] by compressing unary paths. The edges that replace a compressed path store the whole string that they represent (via two pointers to their initial and final text position). Once unary paths are not present the trie, now called *suffix tree*, has $O(n)$ nodes instead of the worst-case $O(n^2)$ of the trie. The suffix tree can be directly built in $O(n)$ time [22,35]. Any algorithm on a suffix trie can be simulated at the same cost in the suffix tree.

We call *explicit* those suffix trie nodes that survive in the suffix tree, and *implicit* those that are collapsed. Figure 1 shows the suffix trie and tree of the text "abracadabra". Note that a special endmarker "$", smaller than any other character, is appended to the text so that all the suffixes are external nodes.

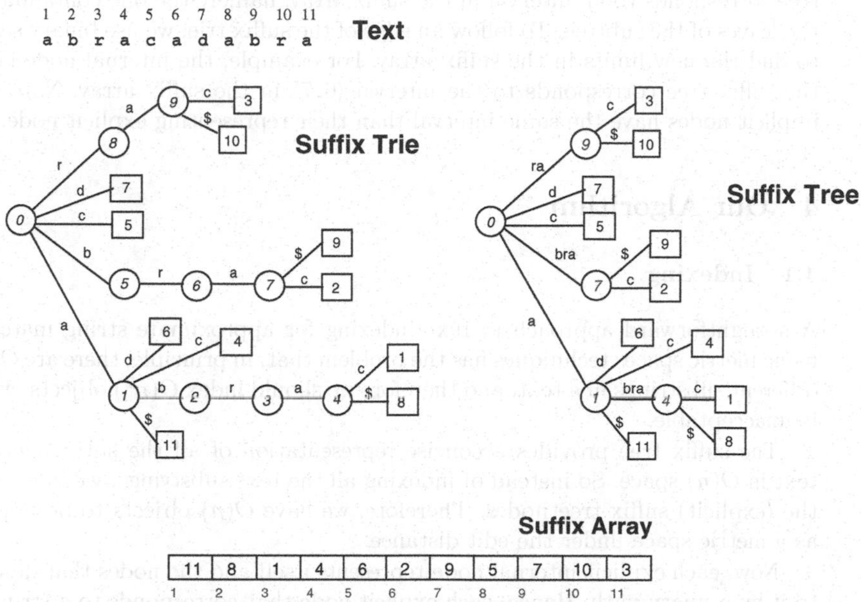

Fig. 1. The suffix trie, suffix tree and suffix array of the text "abracadabra".

The figure shows the *internal* nodes of the trie (numbered 0 to 9 in italics inside circles), which represent text substrings that appear more than once, and the *external* nodes (numbered 1 to 11 inside squares), which represent text substrings that appear just once. Those leaves do not only represent the unique substrings but all their extensions until the full suffix. In the suffix tree, only some internal nodes are left, and they represent the same substring as before plus the prefixes that may have been collapsed. For example the internal node (7) of the suffix tree represents now the compressed nodes (5) and (6), and hence the strings "b", "br" and "bra". The external node (1) represents "abrac", but also "abraca", "abracad", etc. until the full suffix "abracadabra".

Finally, the suffix array [20] is a more compact version of the suffix tree, which requires much less space and poses a small penalty over the search time. If the leaves of the suffix tree are traversed in left-to-right order, all the suffixes of the text are retrieved in lexicographical order. A suffix array is simply an array containing all the pointers to the text suffixes listed in lexicographical order, as shown in Figure 1. The suffix array stores one pointer per text position.

The suffix array can be directly built (without building the suffix tree) in $O(n \log n)$ worst case time and $O(n \log \log n)$ average time [20]. While suffix trees are searched as tries, suffix arrays are binary searched. However, almost every algorithm on suffix trees can be adapted to work on suffix arrays at an $O(\log n)$ penalty factor in the time cost. This is because each subtree of the suffix tree corresponds to an interval in the suffix array, namely the one containing all the leaves of the subtree. To follow an edge of the suffix trie, we use binary search to find the new limits in the suffix array. For example, the internal node (7) in the suffix tree corresponds to the interval $\langle 6, 7 \rangle$ in the suffix array. Note that implicit nodes have the same interval than their representing explicit node.

4 Our Algorithm

4.1 Indexing

A straightforward approach to text indexing for approximate string matching using metric spaces techniques has the problem that, in principle, there are $O(n^2)$ different substrings in a text, and therefore we should index $O(n^2)$ objects, which is unacceptable.

The suffix tree provides a concise representation of all the substrings of a text in $O(n)$ space. So instead of indexing all the text substrings, we index only the (explicit) suffix tree nodes. Therefore, we have $O(n)$ objects to be indexed as a metric space under the edit distance.

Now, each explicit internal node represents itself and the nodes that descend to it by a unary path. Hence, each explicit node that corresponds to a string xy and its parent corresponds to the string x represents the following set of strings

$$x[y] \quad = \quad \{xy_1,\ xy_1y_2,\ \ldots,\ xy\}$$

where $x[y]$ is a notation we have just introduced. For example, the internal node (4) in Figure 1 represents the strings "a[bra]" = {"ab", "abr", "abra"}.

The leaves of the suffix tree represent a unique text substring and all its extensions until the full text suffix is obtained. Hence, if $T = zxy$ and x is a unique text substring (whose prefixes are not unique), then the corresponding suffix tree node is an explicit leaf, which for us represents the set $\{x\} \cup x[y]$. Table 1 shows the substrings represented by each node in our running example. Note that the external nodes that descend by the terminator character "$", i.e. $e(8$–$11)$, represent a substring that is also represented at its parent and hence it can be disregarded.

Node	Suffix trie	Suffix tree	Node	Suffix trie/tree
$i(0)$	ε	ε	$e(1)$	abra[cadabra]
$i(1)$	a	[a]	$e(2)$	bra[cadabra]
$i(2)$	ab		$e(3)$	ra[cadabra]
$i(3)$	abr		$e(4)$	a[cadabra]
$i(4)$	abra	a[bra]	$e(5)$	[cadabra]
$i(5)$	b		$e(6)$	a[dabra]
$i(6)$	br		$e(7)$	[dabra]
$i(7)$	bra	[bra]	$e(8)$	abra
$i(8)$	r		$e(9)$	bra
$i(9)$	ra	[ra]	$e(10)$	ra
			$e(11)$	a

Table 1. The text substrings represented by each node of the suffix trie and tree of Figure 1. Internal nodes are represented as $i(x)$ and externals as $e(x)$.

Hence, instead of indexing all the $O(n^2)$ text substrings, we index $O(n)$ sets of strings, which are the sets represented by the explicit internal and the external nodes of the suffix tree. In our example, this set is $\mathbb{U} = \{\varepsilon$, [a], a[bra], [bra], [ra], abra[cadabra], bra[cadabra], ra[cadabra], a[cadabra], [cadabra], a[dabra], [dabra]$\}$.

We have now to decide how to index this metric space formed by $O(n)$ sets of strings. Many options are possible, but we have concentrated on a pivot based approach. We select at random k different text substrings that will be our *pivots*. For reasons that are made clear later, we choose to select pivots of lengths 0, 1, 2, $\cdots$, $k-1$. For each explicit suffix tree node $x[y]$ and each pivot p_i, we compute the distance between p_i and all the strings represented by $x[y]$. From the set of distances from a node $x[y]$ to p_i, we store the minimum and maximum ones. Since all these strings are of the form $\{xy_1...y_j,\ 1 \le j \le |y|\}$, all the edit distances can be computed in $O(|p_i||xy|)$ time.

Following our example, let us assume that we have selected $k = 5$ pivots $p_0 =$ "", $p_1 =$ "a", $p_2 =$ "br", $p_3 =$ "cad" and $p_4 =$ "raca". Figure 2 (left) shows the computation of the edit distances between $i(4) =$ "a[bra]" and $p_3 =$ "cad". The result shows that the minimum and maximum values of this node with respect to this pivot are 2 and 4, respectively.

In the case of external suffix tree nodes, the string y tends to be quite long ($O(n)$ length on average), which yields a very high computation time for all the edit distances and anyway a very large value for the maximum edit distance (note that $ed(p_i, xy) \ge |xy| - |p_i|$). We solve this by pessimistically assuming that the maximum distance is n when the suffix tree node is external. The minimum edit distance can be found in $O(|p_i| \max(|p_i|, |x|))$ time, because it is not necessary to consider arbitrarily long strings $xy_1...y_j$: If we compute the matrix row by row, then after having processed x we have a minimum value seen up to now, v. Then there is no point in considering rows j such that $|x| + j - |p_i| > v$. Hence we work until row $j = v + |p_i| - |x| \le |p_i|$.

		c	a	d
	0	1	2	3
a	1	1	1	2
b	2	2	2	**2**
r	3	3	3	**3**
a	4	4	3	**4**

		c	a	d
abra	4	4	3	4
c	5	4	4	**4**
a	6	5	4	**5**
d	7	6	5	**4**
a	8	7	6	**5**
b	9	8	7	**6**
r	10	9	8	**7**
a	11	10	9	**8**

Fig. 2. The dynamic programming matrix to compute the edit distance between `"cad"` and `"a[bra]"` (left) or `"abra[cadabra]"` (right). The emphasized area is where the minima and maxima are taken from.

Figure 2 (right) illustrates this case with $e(1)$ = `"abra[cadabra]"` and the same p_4 = `"cad"`. Note that to compute the new set of edit distances we have started from $i(4)$, which is the parent node of $e(1)$ in the suffix tree. This can always be done in a depth first traversal of the suffix tree and saves construction time. Note also that it is not necessary to compute the last 4 rows, since they measure the edit distance between strings of length 8 or more against one of length 3. The distance cannot be smaller than 5 and we have found at that point a minimum equal to 4. In fact we just assume that the maximum is 11, so the minimum and maximum value for this external node and this pivot are 4 and 11. In particular, since when indexing external nodes $x[y]$ we always have $ed(p_i, x)$ already computed, they can be indexed in $O(|p_i|^2)$ time.

Once this is done for all the suffix tree nodes and all the pivots we have a set of k minimum and maximum values for each explicit suffix tree node. This can be regarded as a hyperrectangle in k dimensions:

$$x[y] \;\rightarrow\; \langle \quad \begin{array}{l} (\min(ed(x[y],p_0)),\ldots,\min(ed(x[y],p_{k-1}))), \\ (\max(ed(x[y],p_0)),\ldots,\max(ed(x[y],p_{k-1}))) \quad \rangle \end{array}$$

where we are sure that all the strings in $x[y]$ lie inside the rectangle. In our example, the minima and maxima for $i(4)$ with respect to p_0 to p_4 are $\langle 2,4\rangle$, $\langle 1,3\rangle$, $\langle 1,2\rangle$, $\langle 2,4\rangle$ and $\langle 3,3\rangle$. Therefore $i(4)$ is represented by the hyperrectangle $\langle(2,1,1,2,3),(4,3,2,4,3)\rangle$. On the other hand, the ranges for $e(1)$ are $\langle 5,11\rangle$, $\langle 4,11\rangle$, $\langle 3,11\rangle$, $\langle 4,11\rangle$ and $\langle 2,11\rangle$ and its hyperrectangle is therefore $\langle(5,4,3,4,2),(11,11,11,11,11)\rangle$.

4.2 Searching

Let us now consider a given query P searched for with at most r errors. This is a range query with radius r in the metric space of the suffix tree nodes. As for pivot based algorithms, we compare the pattern P against the k pivots and obtain a k-dimensional coordinate $(ed(P,p_1),\ldots,ed(P,p_k))$.

Let p_i be a given pivot and $x[y]$ a given node. If it holds that

$$ed(P, p_i) + r < \min(ed(x[y], p_i)) \quad \vee \quad ed(P, p_i) - r > \max(ed(x[y], p_i))$$

then, by the triangle inequality, we know that $ed(P, xy') > r$ for any $xy' \in x[y]$. The elimination can be done using any pivot p_i. In fact, the nodes that are not eliminated are those whose rectangle has nonempty intersection with the rectangle $\langle(ed(P,p_1) - r, \ldots, ed(P,p_k) - r), (ed(P,p_1) + r, \ldots, ed(P,p_k) + r)\rangle$.

Figure 3 illustrates. The node contains a set of points and we store its minimum and maximum distance to two pivots. These define a (2-dimensional) rectangle where all the distances from any substring of the node to the pivots lie. The query is a pattern P and a tolerance r, which defines a circle around P. After taking the distances from P to the pivots we create a hypercube (a square in this case) of width $2r + 1$. If the square does not intersect the rectangle, then no substring in the node can be close enough to P.

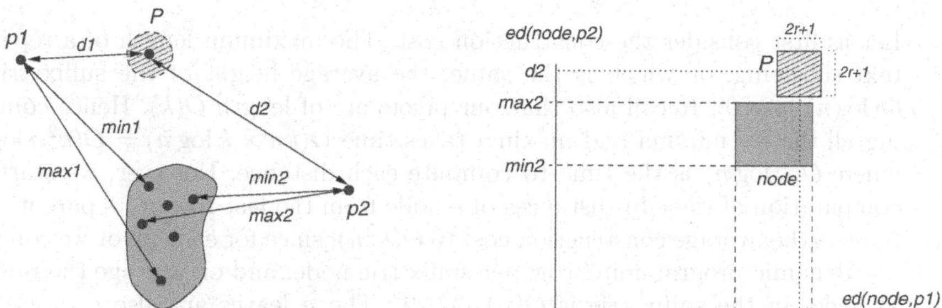

Fig. 3. The elimination rule using two pivots.

We have to solve the problem of finding all the k-dimensional rectangles that intersect a given query rectangle. This is a classical multidimensional range search problem [36,14]. We could for example use some variant of R-trees [16,7], which would also yield a good data structure to work on secondary memory.

Those nodes $x[y]$ that cannot be eliminated using any pivot must be directly compared against P. For those whose minimum distance to P is at most r, we report all their occurrences, whose starting points are written in the leaves of the subtree rooted by the node that has matched. In our running example, if we are searching for "`abr`" with tolerance $r = 1$, then node $i(4)$ qualifies, so we report the text positions in the corresponding tree leaves: 1 and 8.

Observe that in order to compare P against a given suffix tree node $x[y]$, the edit distance algorithm forces us to compare it against every prefix of x as well. Those prefixes correspond to suffix tree nodes in the path from the root to $x[y]$. In order not to repeat work, we mark in the suffix tree the nodes that we have to compare explicitly against P, and also mark every node in their path to the root. Then, we backtrack on the suffix tree entering every marked node and keeping track of the edit distance between P and the node. The new row is computed using the row of the parent, just as done with the pivots. This avoids recomputing the same prefixes for different suffix tree nodes, and incidentally is similar to the simplest backtracking approach [15], except that in this case we only follow marked paths. In this respect, our algorithm can be thought of as a preprocessing to a backtracking algorithm, which filters out some paths.

As a practical matter, note that this is the only step where the suffix tree is required. We can even print the text substrings that match the pattern without the help of the suffix tree, but we need it in order to report all their text positions. For this sake, a suffix array is much cheaper and does a better job (because all the text positions are listed in a contiguous interval). In fact, the suffix array can also replace the suffix tree at indexing time.

5 Analysis

Let us first consider the construction cost. The maximum length of a repeated text substring, or which is the same, the average height of the suffix trie, is $O(\log n)$ [32,29]. Recall also that our pivots are of length $O(k)$. Hence computing all the kn minima and maxima takes time $O(kn \times k \log n) = O(k^2 n \log n)$, where $O(k \log n)$ is the time to compute each distance. However, we start the computation of the edit distances of a node from the last row of its parent. This reduces the average construction cost to $O(k^2 n)$, since for each pivot we compute one dynamic programming row per suffix trie node, and on average the number of nodes in the suffix trie is $O(n)$ [32,29]. The n leaves are also computed in $O(|p_i|^2 n) = O(k^2 n)$ time.

The total space required by the data structure is $O(kn)$, since we need to store for each explicit node a pointer to the suffix tree and its k coordinates. The suffix tree itself takes $O(n)$ space.

It remains to determine the average search time. A key element of the analysis is a constant α, which is the probability that, for a random hyperrectangle of the set, along some fixed coordinate, the corresponding segment of the query hypercube intersects with the corresponding segment of the hyperrectangle. Another way to put it is that, along that coordinate, the query point falls inside the hyperrectangle projection onto that coordinate after it is enlarged in r units along each dimension. In operational terms, α is the probability that some given pivot (that corresponding to the selected coordinate) does *not* permit discarding a given element. Note that α does not depend on k, only on r.

The first part of the search is the computation of the edit distances between the k pivots and the pattern P of length m. This takes $O(k^2 m)$ time.

The second part is the search for the rectangles that intersect the query rectangle. Many analyses of the performance of R-trees exist in the literature [33,18,26,27,13]. Despite that most of them deal with the exact number of disk accesses, their abstract result is that the expected amount of work on the R-tree (and variants such as the KDB-tree [28]) is $O(n\alpha^k \log n)$.

The third part, finally, is the direct check of the pattern against the suffix tree nodes whose rectangles intersect the query rectangle. Since we discard using any of k random pivots, the probability of not discarding a node is α^k. As there are $O(n)$ suffix tree nodes, we check on average $\alpha^k n$ nodes, with a total cost of $O(\alpha^k n \times m^2)$. The m^2 is the cost to compute the edit distance between a pattern of length m and a candidate whose length must be between $m-r$ and $m+r$. This is because the pivot ε removes all shorter or longer candidates.

At the end, we report the R results in $O(R)$ time using a suffix tree traversal. Hence our total average cost is bounded by $k^2m + n\alpha^k \log n + n\alpha^k m^2 + R$ for $0 \le \alpha \le 1$. This is optimized for $k^* = \log_{1/\alpha} n + O(\log\log n) \ge \log_{1/\alpha} n = \Theta(\log n)$.

If we use $\log_{1/\alpha} n$ pivots the search cost becomes $O(m \log^2 n + m^2 + R)$ on average. Note that the influence of the search radius r is embedded in α. This is much better complexity than all previous work, which obtains $O(mn^\lambda)$ time for some $0 < \lambda < 1$. Moreover, much of previous work requires $m = \Omega(\log n)$ to obtain sublinearity, while our approach does not.

The price is in the construction time and space, which become $O(n \log^2 n)$ and $O(n \log n)$, respectively. Especially the latter can be prohibitive and we may have to content ourselves with a smaller k. There seems to be no good tradeoff between space and time, e.g., to obtain $O(n^\lambda)$ time we also need $\Theta(\log n)$ pivots. Most other indexes require $O(n)$ space and construction time.

Finally, it is worth mentioning that, since we automatically discard any internal node not in the length $[m-r, m+r]$ thanks to the pivot $p_0 = \varepsilon$, there is a worst-case limit σ^{m+r} on the number of suffix tree nodes to consider for the last phase. Although this limit is exponential on m and r, it is independent of n. Other indexing schemes based on the suffix tree share the same property.

6 Towards a Practical Implementation

Despite that we have obtained an important reduction in time complexity with respect to n and m, our result is hiding a multiplying factor that depends on the search radius. It is possible that this constant is too large (that is, α too close to 1) and makes the whole approach useless. Also, the extra space requirement (which also increases as α tends to 1) can be unmanageable. In this section we consider an alternative approach that is simpler and likely to obtain better results in practice, despite not involving a complexity breakthrough.

6.1 Indexing Only Suffixes

A simpler index that derives from the same ideas of the paper considers only the n text suffixes and no internal nodes. Each suffix $[T_{j...}]$ represents all the text substrings starting at i, and it is indexed according to the minimum distance between those substrings and each pivot.

The good point of the approach is reduced space. Not only the set $\mathbb{U}$ can have up to half the elements of the original approach, but also only k values (not $2k$)

are stored for each element, since all the maximum values are the same. This permits using up to four times the number of pivots of the previous approach at the same memory requirement. Note that we do not even need to build or store the suffix array: We just read the suffixes from the text and index them. Our only storage need is that of the metric index.

The bad point is that the selectivity of the pivots is reduced and some redundant work is done. The first is a consequence of storing only minimum values, while the second is a consequence of not factoring out repeated text substrings. That is, if some substring P' of T is close enough to P and it appears many times in T, we will have to check all its occurrences one by one.

Without using a suffix tree structure, the construction of the index can be done in time $O(k|p_i|n)$ as follows. The algorithm depicted in Section 2 to compute edit distance can be modified so as to make $C_{0,j} = 0$, in which case $C_{i,j}$ becomes the minimum edit distance between $x_{1...i}$ and a suffix of $y_{1...j}$. If x is the reverse of $|p_i|$ and y the reverse of T, then $C_{|p_i|,j}$ will be the minimum edit distance between $|p_i|$ and a prefix of $T_{n-j+1...}$, which is precisely $\min(ed(p_i, T_{n-j+1...}))$. So we need $O(|p_i|n)$ time per pivot. The space to compute this is just $O(|p_i|)$ by doing the computation column-wise.

6.2 Using an Index for High Dimensions

The space of strings has a distance distribution that is rather concentrated around its mean μ. The same happens to the distances between a pivot p_i and suffixes $[T_{j...}]$ or the pattern P. Since we can only discard suffixes $[T_{j...}]$ such that $ed(p_i, P) + r < \min(ed(p_i, [T_{j...}]))$, only the suffixes with a large $\min(ed(p_i, [T_{j...}]))$ value are likely to be discarded using p_i. Storing all the other $O(n)$ distances to p_i is likely to be a waste of space. Moreover, we can use that memory to introduce more pivots. Figure 4 illustrates.

The idea is to fix a number s and, for each pivot p_i, store only the s largest $\min(ed(p_i, [T_{j...}]))$ values. Only those suffixes can be discarded using pivot p_i. The space of this index is $O(ks)$ and its construction time is unchanged. We can still use an R-tree for the search, although the rectangles will cover all the space except on s coordinates. The selectivity is likely to be similar since we have discarded uninteresting coordinates, and we can tune number k versus selectivity s of the pivots for the same space usage $O(ks)$.

One can go further to obtain $O(n)$ space as follows. Choose the first pivot and determine its s farthest suffixes. Store a list (in increasing distance order) of those suffixes and their distance to the first pivot and remove them from further consideration. Then choose a second pivot and find its s farthest suffixes from the remaining set. Continue until every suffix has been included in the list of some pivot. Note that every suffix appears exactly in one list. At search time, compare P against each pivot p_i, and if $ed(P, p_i) + r$ is smaller than the smallest (first) distance in the list of p_i, skip the whole list. Otherwise traverse the list until its end or until $ed(P, p_i) + r$ is smaller than the next element. Each traversed suffix must be directly compared against P. A variant of this idea has proven extremely useful to deal with concentrated histograms [9]. It also permits

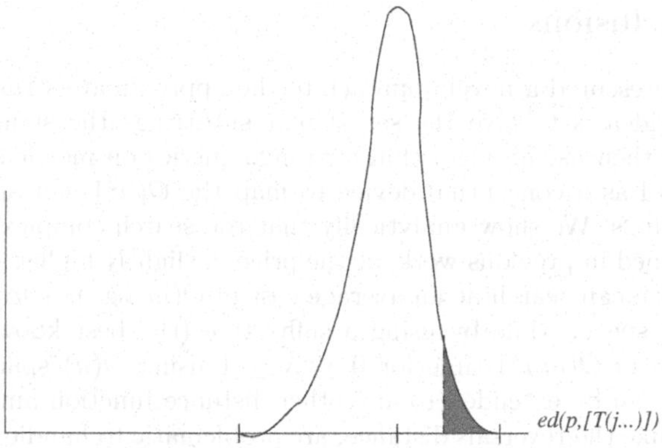

Fig. 4. The distance distribution to a pivot p, including that of pattern P. The grayed area represents the suffixes that can be discarded using p.

efficient secondary storage implementation by packing the pivots in disk pages and storing the lists consecutively in the same order of the pivots.

Since we choose $k = n/s$ pivots, the construction time is high, namely $O(n^2|p_i|/s)$. However, the space is $O(n)$, with a low constant (close to 5 in practice) that makes it competitive against the most economical structures for the problem. The search time is $O(|p_i|mn/s)$ to compare P against the pivots, while the time to traverse the lists is difficult to analyze.

The pivots chosen must not be very short, because their minimum distance to any $[T_{j...}]$ is at most $|p_i|$. In fact, any pivot not longer than $m + r$ is useless.

6.3 Using Specific Strings Properties

We can complement the information given by the metric index with knowledge of the string properties we are indexing. For example, if suffix $[T_{j...}]$ is proven to be at distance $r + t$ from P, then we can also discard suffixes starting in the range $j - t + 1 \ldots j + t - 1$.

Another idea is to compute the edit distance between the reverse pivot and the reverse pattern. Although the result is the same, we learn also the distances between the pivot and suffixes of the pattern. This can also be useful to discard suffixes at verification time: If $d' = ed(P_{1...\ell}, T_{i...i'})$ and we know from the index that $ed(P_{\ell+1...}, T_{i'+1...}) > r - d'$, then a match is not possible.

Other ideas, such as hybrid algorithms that partition the pattern and search for the pieces permitting less errors [6], can be implemented over our metric index instead of over a suffix tree or array. Indeed, our data structure should compete in the area of backtracking algorithms, as the others are orthogonal.

7 Conclusions

We have presented a novel approach to the approximate string matching problem. The idea is to give the set of text substrings the structure of a metric space and then use an algorithm for range queries on metric spaces. The suffix tree is used as a conceptual device to map the $O(n^2)$ text substrings to $O(n)$ sets of strings. We show analytically that the search complexity is better than that obtained in previous work, at the price of slightly higher space usage. More precisely we can search at an average cost of $O(m \log^2 n + m^2 + R)$ time using $O(n \log n)$ space, while by using a suffix tree (the best known technique) one can search in $O(mn^\lambda)$ time for $0 \leq \lambda \leq 1$ using $O(n)$ space. Moreover, our technique can be extended to any other distance function among strings, some of which, like the reversals distance, are problematic to handle with the previous approaches.

The proposal opens a number of possibilities for future work. We plan to explore other methods to reduce the number of substrings (we have used the suffix tree nodes and the suffixes), other metric space indexing methods (we have used pivots), other multidimensional range search techniques (we have used *R*-trees), other pivot selection techniques (we took them at random), etc. A more practical setup needing $O(n)$ space has been described in Section 6.

Finally, the method promises an efficient implementation on secondary memory (e.g., with *R*-trees), which is a weak point in most current approaches.

References

1. A. Apostolico. The myriad virtues of subword trees. In *Combinatorial Algorithms on Words*, NATO ISI Series, pages 85–96. Springer-Verlag, 1985.
2. A. Apostolico and Z. Galil. *Combinatorial Algorithms on Words*. Springer-Verlag, New York, 1985.
3. R. Baeza-Yates and G. Navarro. Block-addressing indices for approximate text retrieval. In *Proc. ACM CIKM'97*, pages 1–8, 1997.
4. R. Baeza-Yates and G. Navarro. Fast approximate string matching in a dictionary. In *Proc. SPIRE'98*, pages 14–22. IEEE Computer Press, 1998.
5. R. Baeza-Yates and G. Navarro. A practical q-gram index for text retrieval allowing errors. *CLEI Electronic Journal*, 1(2), 1998. `http://www.clei.cl`.
6. R. Baeza-Yates and G. Navarro. A hybrid indexing method for approximate string matching. *J. of Discrete Algorithms (JDA)*, 1(1):205–239, 2000. Special issue on Matching Patterns.
7. N. Beckmann, H. Kriegel, R. Schneider, and B. Seeger. The R*-tree: an efficient and robust access method for points and rectangles. In *Proc. ACM SIGMOD'90*, pages 322–331, 1990.
8. E. Bugnion, T. Roos, F. Shi, P. Widmayer, and F. Widmer. Approximate multiple string matching using spatial indexes. In *Proc. 1st South American Workshop on String Processing (WSP'93)*, pages 43–54, 1993.
9. E. Chávez and G. Navarro. An effective clustering algorithm to index high dimensional metric spaces. In *Proc. SPIRE'2000*, pages 75–86. IEEE CS Press, 2000.
10. E. Chávez, G. Navarro, R. Baeza-Yates, and J. Marroquín. Searching in metric spaces. *ACM Comp. Surv.*, 2001. To appear.

11. A. Cobbs. Fast approximate matching using suffix trees. In *Proc. CPM'95*, pages 41–54, 1995. LNCS 937.
12. M. Crochemore. Transducers and repetitions. *Theor. Comp. Sci.*, 45:63–86, 1986.
13. C. Faloutsos and I. Kamel. Beyond uniformity and independence: analysis of R-trees using the concept of fractal dimension. In *Proc. ACM PODS'94*, pages 4–13, 1994.
14. V. Gaede and O. Günther. Multidimensional access methods. *ACM Comp. Surv.*, 30(2):170–231, 1998.
15. G. Gonnet. A tutorial introduction to Computational Biochemistry using Darwin. Technical report, Informatik E.T.H., Zuerich, Switzerland, 1992.
16. A. Guttman. R-trees: a dynamic index structure for spatial searching. In *Proc. ACM SIGMOD'84*, pages 47–57, 1984.
17. P. Jokinen and E. Ukkonen. Two algorithms for approximate string matching in static texts. In *Proc. MFCS'91*, volume 16, pages 240–248, 1991.
18. I. Kamel and C. Faloutsos. On packing R-trees. In *Proc. ACM CIKM'93*, pages 490–499, 1993.
19. V. Levenshtein. Binary codes capable of correcting spurious insertions and deletions of ones. *Problems of Information Transmission*, 1:8–17, 1965.
20. U. Manber and E. Myers. Suffix arrays: a new method for on-line string searches. *SIAM J. on Computing*, pages 935–948, 1993.
21. U. Manber and S. Wu. GLIMPSE: A tool to search through entire file systems. In *Proc. USENIX Technical Conference*, pages 23–32, Winter 1994.
22. E. McCreight. A space-economical suffix tree construction algorithm. *J. of the ACM*, 23(2):262–272, 1976.
23. D. Morrison. PATRICIA — practical algorithm to retrieve information coded in alphanumeric. *J. of the ACM*, 15(4):514–534, 1968.
24. E. Myers. A sublinear algorithm for approximate keyword searching. *Algorithmica*, 12(4/5):345–374, Oct/Nov 1994.
25. G. Navarro. A guided tour to approximate string matching. *ACM Comp. Surv.*, 33(1):31–88, 2001.
26. B. Pagel, H. Six, H. Toben, and P. Widmayer. Towards an analysis of range queries. In *Proc. ACM PODS'93*, pages 241–221, 1993.
27. D. Papadias, Y. Theodoridis, and E. Stefanakis. Multidimensional range queries with spatial relations. *Geographical Systems*, 4(4):343–365, 1997.
28. J. Robinson. The K-D-B-tree: a search structure for large multidimensional dynamic indexes. In *Proc. ACM PODS'81*, pages 10–18, 1981.
29. R. Sedgewick and P. Flajolet. *Analysis of Algorithms*. Addison-Wesley, 1996.
30. F. Shi. Fast approximate string matching with q-blocks sequences. In *Proc. WSP'96*, pages 257–271. Carleton University Press, 1996.
31. E. Sutinen and J. Tarhio. Filtration with q-samples in approximate string matching. In *Proc. CPM'96*, LNCS 1075, pages 50–61, 1996.
32. W. Szpankowski. Probabilistic analysis of generalized suffix trees. In *Proc. CPM'92*, LNCS 644, pages 1–14, 1992.
33. Y. Thedoridis and T. Sellis. A model for the prediction of R-tree performance. In *Proc. ACM PODS'96*, pages 161–171, 1996.
34. E. Ukkonen. Approximate string matching over suffix trees. In *Proc. CPM'93*, pages 228–242, 1993.
35. E. Ukkonen. Constructing suffix trees on-line in linear time. *Algorithmica*, 14(3):249–260, 1995.
36. D. White and R. Jain. Algorithms and strategies for similarity retrieval. Technical Report VCL-96-101, Visual Comp. Lab., Univ. of California, July 1996.

On Maximal Suffices and Constant-Space Linear-Time Versions of KMP Algorithm

Wojciech Rytter[1,2]

[1] Department of Computer Science, Liverpool University, Chadwick Building, Peach Street, Liverpool L69 7ZF, U.K.

[2] Instytut Informatyki, Uniwersytet Warszawski, Poland

rytter@mimuw.edu.pl

Abstract. We investigate several **simple** ways of transforming Knuth-Morris-Pratt algorithm (KMP) into a constant-space and still linear time string-matching algorithm. We also identify a class of very special patterns for which the transformation is particularly simple and show usefulness of the class. Constant-space linear-time string-matching algorithms are usually very sophisticated. Most of them consist of two phases: (very technical) *preprocessing phase* and *searching phase*. An exception is *one-phase* Crochemore's algorithm [2]. It is an on-line version of KMP algorithm with "*on-the-fly*" computation of pattern shifts (as *approximate* periods). We explore further Crochemore's approach, and construct alternative algorithms which are differently structured. In Crochemore's algorithm the approximate-period function is restarted *from inside*, which means that several internal variables of this function are changing globally, also Crochemore's algorithm strongly depends on the concrete implementation of approximate-periods computation. We present a simple modification of KMP algorithm which works in $O(1)$-space, $O(n)$-time for **any** function which computes periods or approximate periods in $O(1)$-space, and linear time. The approximate-period function can be treated as a *black box*. We show also that lexicographically self-maximal patterns are especially well suited for Crochemore-style string matching. A new $O(1)$ space string-matching algorithm, *MaxSuffix-Matching*, is proposed in the paper, which gives yet another example of applicability of maximal suffices.

1 Introduction

The two very classical algorithms in "*stringology*" are Knuth-Morris-Pratt (KMP) algorithms and algorithm computing lexicographically maximal suffix *MaxSuf*(w) of the word w. In [2] Crochemore has shown how to *combine* KMP with *MaxSuf* to achieve *single-phase* constant-space linear-time string-matching on-line algorithm. "*On-line*" means here that only the symbols of the longest prefix of P occurring in T are read. We pursue Crochemore's approach. Assume that the pattern and the text are given as *read-only* tables P and T, where $|P| \leq |T| = n$. We count space as the number of additional integer registers (each in the range $[0 \ldots n]$) used in the algorithm. The *string-matching problem* consists in finding

S. Rajsbaum (Ed.): LATIN 2002, LNCS 2286, pp. 196–208, 2002.

all occurrences of P in T. The algorithms solving this problem with linear cost and (simultaneously) constant space are the most interesting and usually the most sophisticated. The first constant-space linear time algorithm was given by Galil and Seiferas in [6]. Later Crochemore and Perrin in [3] have shown how to achieve $2n$ comparisons algorithm preserving small amount of memory. Alternative algorithms were presented in [7] and [8]. The KMP algorithm is $O(1)$-space algorithm except the table for shifts (or, equivalently, periods). The natural approach to constant-space algorithms is to get rid of these tables. We show how it can be done in different ways by employing Crochemore's approach [2]:

$O(n)$ space table $\Rightarrow$ function computable in $O(1)$-space and $O(n)$ time.

The above approach lacks details and is not a full receipt, some unsophisticated *algorithmic engineering* related to technicalities of KMP algorithm is needed. In this paper we do not pursue the issue of the exact number of symbol comparisons, which is usually not a dominating part in the whole complexity. We will be satisfied with linear time, as long as the algorithm is reasonably simple. From a teaching point of view it is desirable to have a linear time $O(1)$ space algorithm which is easily understandable.

2 KMP Algorithm with Exact and Approximate Shifts

KMP algorithm aligns P with T starting at some position i and finds the longest partial match: $P[1 \dots j] = T[i+1 \dots i+j]$. Then KMP makes a *shift*, determined by the size j of the partial match.

```
algorithm KMP1;
  i := 0; j := 0;
  while i ≤ n − m do
    begin
     while j < m and P[j + 1] = T[i + j + 1] do  j = j + 1;
     {MATCH:} if j = m then report match at i;
       i := i + max{1, Period(j)};  j := j − Period(j);
    end ;
```

The shifts are precomputed and stored in an array, and they are exactly the periods of $P[1 \dots j]$ for $j \geq 1$. Consequently we use periods in place of shifts. If $j = 0$ then the shift is forced to be 1. Let *period*(x) be the size of the shortest period of a word x.

Denote *Period*(j) = *period*($P[1 \dots j]$) for $j \geq 1$ and *Period*(0) = 0. It is much easier to compute in $O(1)$ space an *approximate period.* Define:

$$AprPeriod(j) \;=\; \begin{cases} Period(j), & \text{if } Period(j) \leq \frac{j}{2} \\ nil, & \text{if } Period(j) > \frac{j}{2} \end{cases}$$

We say that a text x is *periodic* iff $period(x) \leq |x|/2$. Hence the function *ApprPeriod* gives the value of periods for periodic texts only. The function *ApprPeriod* is computed in very simple *MaxSuf-and-Period* algorithm presented in the last section for completeness. We substitute the function of exact period by *ApprPeriod* and modify the KMP1 algorithm as follows.

```
ALGORITHM KMP2;
  i:=0; j:=0;
  while i ≤ n − m do
    begin
      while j < m and P[j + 1] = T[i + j + 1] do   j = j + 1;
        MATCH: if j = m then report  match at i;
        period := ApprPeriod(j);
        if period = nil then  begin i := i + ⌈(j+1)/2⌉; j := 0 end
        else   {Periodic-Case:}
            begin i := i + max{1, period}; j = j − period end
  end;
```

The following fact is well known and used in many constant space string-matching algorithms.

Theorem 1. *Algorithms KMP1 and KMP2 work in linear time if the values of ApprPeriod and Period are available in constant time.*

3 Constant Space Version of KMP without Preprocessing for Very Special Patterns: *SpecialCase-KMP*

There is one *very special* class of patterns for which exact period $O(1)$-space computation is trivial: **self-maximal patterns**. These are the patterns P such that $MaxSuf(P) = P$. *Maximal suffices* play important role in the computation of periods for two reasons:

(1) if P is periodic then $period(MaxSuf(P)) = period(P)$,

(2) if P is *self-maximal* then each of its prefices is.

(3) if P is *self-maximal* then $period(P)$ can be trivially computed by the following function:

```
function Naive-Period(j);
  period := 1;
  for i := 2 to j do
    if P[i] ≠ P[i − period]  then period := i;
  return (period);
```

Example. The function *Naive-Period* usually gives incorrect output for non-self-maximal words, for example consider the string:

$$P = (aba)^4ab = abaabaabaabaabaabaa.$$

The consecutive values of *period* computed by the function for consecutive positions are:

a	b	a	a	b	a	a	b	a	a	b	a	a	b	a	a	b	a	a
1	2	2	4	5	5	7	8	8	10	11	11	13	14	14	16	17	17	19

Hence *Naive-Period*(19)) = 19, for $P = (aba)^6a$, while *Period*(19) = 3.

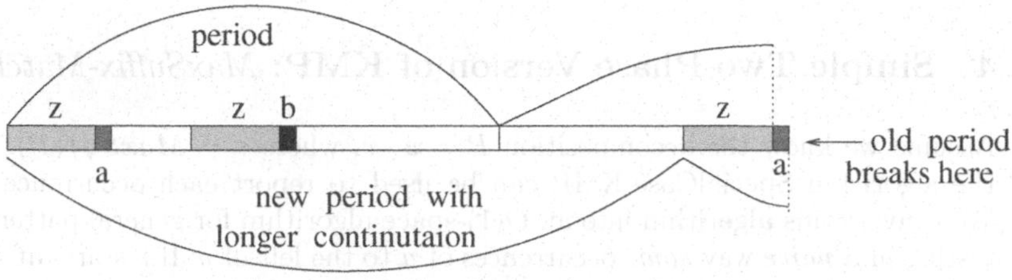

Fig. 1. Assume in the algorithm *Naive-Period* $P[j - Period(j-1)] \neq P[j]$. Then $a = P[j]$, $b = P[j - period]$, and $a < b$, where $period = Period(j-1)$. If $Period(j) < j$ then, due to two periods, zb is a subword of $P[1 \ldots j]$ and za is a prefix of P. Then zb is a proper subword of $P[1 \ldots j-1]$ which is lexicographically greater than P. This contradicts self-optimality of P. Hence $Period(j) = j$.

A very informal justification is given in Figure 1 for the proof of the following lemma.

Lemma 1. *Assume P is a self-maximal string. If $P[j - P(j-1)] \neq P[j]$ then $Period(j) = j$. The function Naive-Period computes correctly the exact period of P.*

We can embed the computation of *Naive-Period(j)* directly into KMP algorithm using Crochemore's approach. $Period(j) = period$ is computed here "*on-the-fly*" in especially simple way.

```
ALGORITHM SpecialCase-KMP;
 i := 0 ; j := 0; period := 1;
 while i ≤ n − m do
   begin
    while j < m and P[j + 1] = T[i + j + 1] do
        begin
        j = j + 1;  if j > 1 and P[j] ≠ P[j − period]
        then period := j end;
    {MATCH:} if j = m then report match at i;
    i := i + period;
     if j ≥ 2 · period then j := j − period;
    else  begin j := 0; period := 1 end;
  end ;
```

4 Simple Two-Phase Version of KMP: *MaxSuffix-Matching*

Assume we know the decomposition $P = u \cdot v$, where $v = MaxSuf(P)$. Then the algorithm SpecialCase-KMP can be used to report each occurrence of v. We convert this algorithm into an $O(1)$-space algorithm for general patterns by testing in a *naive* way *some* occurrences of u to the left of v. If v starts at i then uv can start at $i - |u|$. We do not need to do it for each occurrence of v due to the following fact. Assume that 0 is a *special* default occurrence of v.

Lemma 2. [Key-Lemma] *Assume i is an actual occurrence of v in T and the previous occurrence is prev. If $i - prev < |u|$ then there is no occurrence of uv at position $i - |u|$.*

Proof. The maximal suffix v of P can start only in one place in P.

```
ALGORITHM Informal-MaxSuffix-Matching ;
 Let P = uv, where v = MaxSuf(P);
 Search for v with algorithm SpecialCase-KMP;
 for each occurrence i of v in T do
     Let prev be the previous occurrence of v
     if  i − prev > |u| then
       if u occurs to the left of v then report match;
       { occurrence of u is tested in a naive way}
```

We embed the tests for u into the algorithm *SpecialCase-KMP* and obtain less informal description of the algorithm. The tests $u = T[i-|u|+1\ldots i]$ are being done in a *naive* way. If $v = MaxSuf(P)$ is known then it is a particularly simple $O(1)$-space matching algorithm.

```
ALGORITHM MaxSuffix-Matching;
  i := |u| ; j := 0; period := 1; prev := 0;
  while i ≤ n − |v| do
    begin
     while j < |v| and v[j + 1] = T[i + j + 1] do
       begin
       j = j + 1;  if j > period and v[j] ≠ v[j − period]
       then period := j end;
    {MATCH OF v:} if j = |v| then  begin
         if i − prev > |u| and u  =  T[i − |u| + 1 . . . i]
         then report match at i − |u|;  prev := i;
         end
       i := i + period;
        if j ≥ 2 · period then j := j − period
        else  begin j := 0; period := 1 end;
     end ;
```

Theorem 2. *The algorithm MaxSuffix-Matching works in linear time and constant space. It makes at most n extra symbol comparisons with respect to the total costs of searching for v.*

Proof. Additional *naive* tests for u are related to disjoint segments of T, due to Lemma 2, and altogether take at most n extra symbol comparisons.

Example. Consider the decomposition of the 7-th Fibonacci word and history of the algorithm *MaxSuffix-matching* on an example text, see Figure 2.

5 Simple $O(1)$-space Versions of KMP without Preprocessing

Let us treat *Period* and *ApprPeriod* as functions computable in linear time with $O(1)$ space. This converts algorithms KMP1 and KMP2 into constant-space algorithms.

Are the algorithms KMP1 and KMP2 still linear time algorithms ? The answer is "NO". The counterexamples are given, for example, by highly periodic texts:

$$P = a^n, \; T = a^{n-1}ba^n.$$

$$P \;=\; Fib_7 \;=\; abaababaabaababaababa \;=\; \overbrace{abaababaabaa}^{u} \mid \overbrace{babaababa}^{v}$$

$$T = abaabaa\,\overbrace{babaababa}^{v}\,baa\,\overbrace{\underline{babaababa}}^{v}\,\underline{abaa}\,\overbrace{babaababa}^{v}\,ba$$

Fig. 2. The decomposition of the 7-th Fibonacci word and history of *MaxSuffix-Matching* on an example text T. Only for the third occurrence of v we check if there is an occurrence of u (underlined segment) to the left of this occurrence of v, because gaps between previous occurrences of v are too small. The first occurrence is too close to the beginning of T.

To deal with periodic cases we add *memory* to remember the last value of *period* and the range R of its applicability. We use the following fact:

Same-Period-Property:

$$Period(R) = period \;\Rightarrow\; \forall\ (i \in [2 \cdot period \ldots R])\ Period(i) = period$$

The value of j is *remembered* in the additional register R. Whenever we need later $Period(j)$ for a new value of j which is still in the range $[2 \cdot period \ldots R]$ we do not need to recompute it due to *Same-Period-Property.* In KMP1 we simply insert:

if $j \notin [2 \cdot period \ldots R]$ **then**
  begin $period := Period(j)$; $R := j$ **end**;

In KMP2 we have to change *Period* to *ApprPeriod* and take care additionally of situation $period = nil$. Observe that our modifications of KMP1 and KMP2 look rather cosmetic. The only new variables are *period* and R and the tables of periods are treated as functions. This is rather naive approach, though effective, and to reduce constant coefficients the algorithms should be tuned up further, resulting possibly in great loss of simplicity. Surprisingly such a *rough* approach guarantees linear time, assuming that computation of periods each time takes linear time. The modified algorithms KMP1 and KMP2 are named *ConstantSpace-KMP1* and *Constant Space-KMP2*. For technical reasons assume $Period(0) = ApprPeriod(0) = 0$.

We obtain now two very simple algorithms whose description and correctness are trivial but analysis is not.

```
ALGORITHM Constant-Space-KMP1;
 i:=0; j:=0; period := n; R := 0;
 while i ≤ n − m do
  begin
   while j < m and P[j + 1] = T[i + j + 1] do   j = j + 1;
    {MATCH:} if j = m then report  match at i;
    if j ∉ [2 · period ... R] then
          begin period := Period(j); R := j end;
    i := i + max{1, period}; j = j − period
  end end;
```

```
ALGORITHM Constant-Space-KMP2;
 i:=0; j:=0; period := n; R := 0;
 while i ≤ n − m do
  begin
   while j < m and P[j + 1] = T[i + j + 1] do  j = j + 1;
    MATCH: if j = m then  report  match at i;
    if period = nil or j ∉ [2 · period ... R] then
          begin period := ApprPeriod(j); R := j end;
    if period > nil then  begin i := i + ⌈(j+1)/2⌉; j := 0 end
    else  begin i := i + max{1, period}; j = j − period end
 end;
```

We analyse both algorithms together. The complexity of the algorithms differs from KMP1 and KMP2 by the cost of computing the periods, and this costs can be *charged* to the lengths j of *partial matches* for which the period functions are called. A *partial match* of P is an alignment of P with T over a segment $[i{+}1 \ldots i{+}j]$ in T such that: $P[1 \ldots j] = T[i{+}1 \ldots i{+}j]$, and ($P[j{+}1] \neq T[i{+}j{+}1]$ or $j = m$). We identify in this proof partial match with the segment of T equal to $[i+1 \ldots i+j]$ as well as with the text $T[i+1 \ldots i+j]$.
Define partial match as an **expensive match** if for this partial match we compute one of the functions *Period* or *ApprPeriod*. We say that position $j{+}1$ breaks periodicity if $\mathit{Period}(j) < \mathit{Period}(j+1)$, in other words $P[j+1-\mathit{Period}(j)] \neq P[j+1]$.

Lemma 3. *If $j+1$ breaks periodicity then for any $r > j$:*
$$\mathit{Period}(r) \geq j - \mathit{Period}(j).$$

Proof. (By contradiction) Assume $Period(r) \leq j - Period(j)$, and $period = Period(j)$, $period' = Period(r)$. Then $j+1-period-period' \geq j+1-period-(j-period) \geq 1$, so $P[j+1-period'] = P[j+1]$ and $P[j+1-period-period'] = P[j+1-period]$.

Since $j+1-period' < j$ we obtain

$$\begin{aligned} P[j+1] &= P[j+1-period'] = P[j+1-period'-period] \\ &= P[j+1-period] \end{aligned}$$

which is a contradiction.

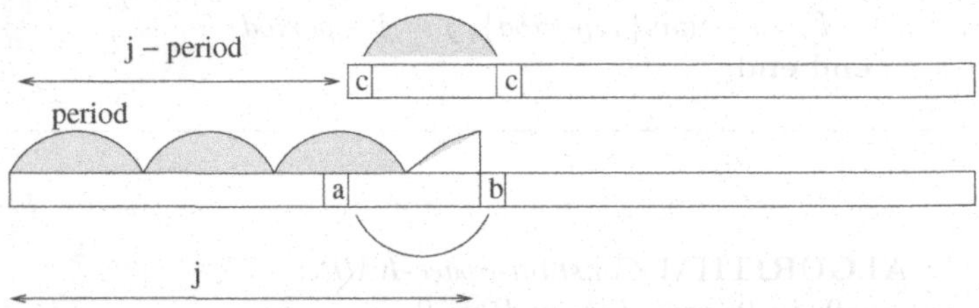

Fig. 3. If $j+1$ breaks periodicity ($a \neq b$) then next period should be large.

If an *expensive match* x starts at i and next one starts at j then we say that $j-i$ is the shift of x and write $EShift(x) = j - i$. If y is the expensive match following x then write $NextEShift(x) = EShift(y)$. If x is the last match then $EShift(x) = NextEShift(x) = |x|$.

Lemma 4.
If x is an expensive match then

$$EShift(x) < |x|/3 \Rightarrow NextEShift(x) \geq |x|/2.$$

Proof. We can assume that $Period(j) = period \leq |x|/3$, otherwise the first shift would be at least $|x|/3$. Consider first three cases when $|x| = j < m$. x corresponds in T to interval $[i, i+j+1]$, see Figure 4. We have: $P[1,j] = T[i, i+j]$, $P[j+1] \neq T[i+j+1]$.

Case 1 (Figure 4) $|x| < m$, $T[i+j+1] \neq P[j+1-period]$. The pattern is shifted several times by distance *period* until the partial match shrinks to the size less than $2 \cdot period$. All partial matches until this moment are not expensive since the same period occured at least twice. Since $period \leq |x|/3$ we have $EShift(x) \geq |x| - 2 \cdot period \geq |x|/3$.

Case 2 (Figure 4) $|x| < m$, $T[i+j+1] = P[j+1-period]$, and $T[i+j+r] \neq P[i+j+r]$ for $r \leq period$. Next partial match (which is not necessarily an expensive one) is at most $|x|$.

After the first shift the situation is the same as in Case 1. All partial matches are non-expensive, till the moment when the match shrinks and the shift becomes large enough.

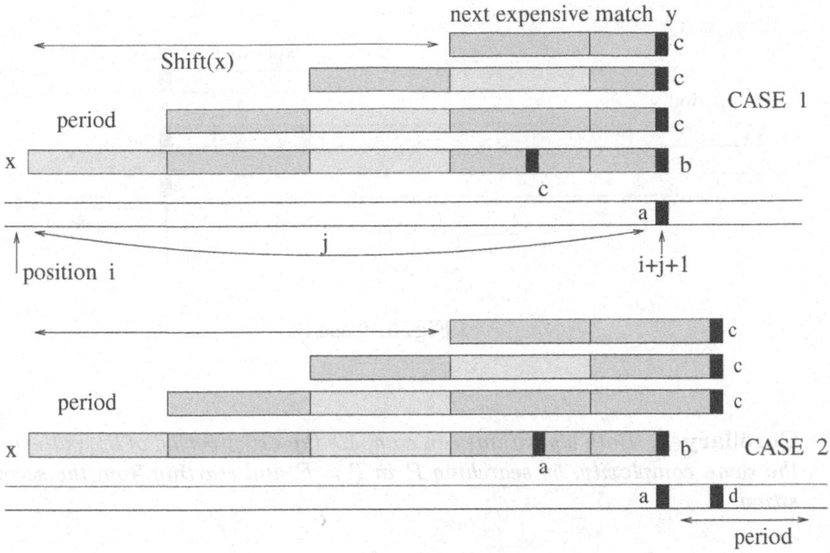

Fig. 4. Cases 1 and 2.

Case 3 (Figure 5) $|x| < m$, $T[i+j+1] = P[j+1-\mathit{period}]$, $|y| \geq j+1$, where y is next partial match. $|y| > |x|$ and y is necessarily expensive since its length exceeds R.

Now position $j+1$ breaks periodicity and Lemma 3 Applies. The new period is at least $j - \mathit{period}$, and possibly approximate period does not give correct value. In this case $NextEShift(x) \geq (j - \mathit{period})/2 \geq |x|/3$.

Case 4 $|x| = m$. The algorithm makes several shifts of the whole pattern over itself (without necessity to recompute the period) until a mismatch is hit. Several shifts without recomputing the period are done until the current partial match shrinks to the word no longer than $2 \cdot \mathit{Period}(m)$. The same argument as in Case 1 applies.

Theorem 3. *Assume Period(j) and ApprPeriod(j) can be computed in O(j) time with O(1) space. Then the algorithms* Constant-Space-KMP1 *and* Constant-Space-KMP2 *report each occurrence of the pattern in linear time with constant space. Both algorithms are on-line: they only read the symbols of the longest prefix of P occurring in the text.*

Proof. We only need to prove that the total cost of all calls to *Period* or *ApprPeriod* is linear, due to Theorem 1. Due to Lemma 4 expensive matches are amortized by their shifts or the next shits. Hence the cost is $O(n)$ since total size of all shifts is linear.

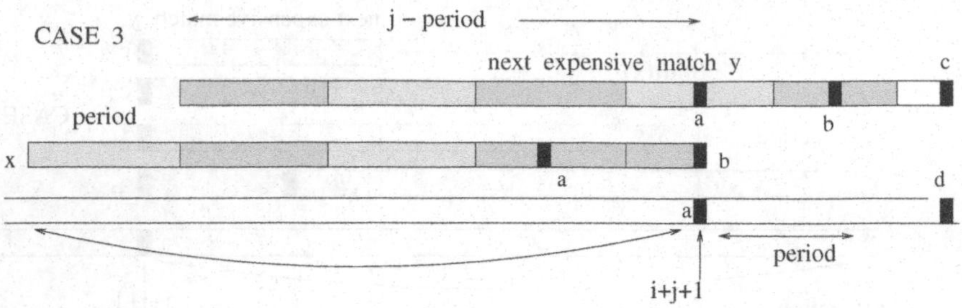

Fig. 5. Case 3.

Corollary 1. *Both algorithms can compute the exact period of the pattern within the same complexity, by searching P in $T = P$ and starting from the second position.*

Corollary 2. *The algorithm Constant-Space-KMP1 with the function Period replaced by Naive-Period solves string-matching problem for self-maximal patterns in $O(1)$-space and linear time.*

6 Maximal Suffices and Approximate Periods

For completeness we include small space computation of approximate periods and maximal suffices. Our presentation differs from previously known ones, see [9] and [5] , its main component is our (very simple) function *Naive-Period*. We convert first the function *Naive-Period* into a computation of the length of the longest self-maximal prefix of a given text x together with its shortest period.

```
function Longest-SelfMax-Prefix(x);
  period :=1;
  for i := 2 to |x| do
      if x[i] < x[i – period] then period := i
      else if x[i] > x[i – period] then
            return (i – 1, period)
  return (|x|, period) {there was no return earlier};
```

We use the computation of *Longest-SelfMax-Prefix* as a key component in maximal suffix computation for the whole text.

```
function MaxSuf-and-Period(P); {|P| = m}
  j := 1;
  repeat
      (k, period) := Longest-SelfMax-Prefix( P[j ...m] );
      if k = m − j + 1 then return (P[j ...m], period)
      else  j := j + k − (k mod period)
```

Technically the last function returns only the starting position of the maximal suffix, to economize space. We use the function *MaxSuf-and-Period* to compute approximate periods.

```
function ApprPeriod(x);
  (v, period) := MaxSuf-and-Period(x);
  if |v| ≥ |x|/2 and period is a period of x
  { test naively if positions in [1 ... |x| − |v|] obey periodicity}
  then return period else return nil;
```

The algorithms in this section are different presentations of classical algorithms, the proof of the following result can be found for example in [5].

Lemma 5. *MaxSuffix-and-Period algorithm runs in* $O(|x|)$ *time with* $O(1)$ *space. It makes less than* $2.|x|$ *letter comparisons. If* x *is periodic then the exact period of* x *is returned.*

7 Conclusion

We contributed to the text algorithmics by providing a new costant space algorithm and showing a very simple way of transforming the KMP algorithm into a constant-spce linear-time algorithm, assuming we have a function computing periods, or approximate periods, wich works within similar complexity. Our algorithm *MaxSuffix-Matching* is probably the easiest-to-understand constant-space linear time pattern matching algorithm. Its correctness is trivial as well as analysis of its complexity. The algorithm is well suited to be used in a course of algorithms and data structures and it shows the power of the concept of a maximal suffix. The two-way pattern-matching algorithm of Crochemore and Perrin has a similar stucture and uses smaller number of comparisons, but its correctness if very complicated, it is one of the most misterious algorithms. The shift in the algorithm, when finding a mismatch of v at position $j+1$ is simply $j+1$. However v is a much more complicated object than the maximal suffix, it is the shorter of two maximal suffixes with repect to the reversed orderings of the alphabet. Hence the maximal suffix computation has to be performed twice in the preprocessing phase. Our approach can also give a simplified version of Crochemore-Perrin algorithm. In the original description they considered cases depending if the whole pattern is highly periodic. We can avoid it, constructing a similar algorithm as *MaxSuffix-Matching* with the shift function $j + 1$ we can obtain a simlified version of Crochemore-Perrion algorithm due to the trick based on our Key-Lemma (Lemma 2). We call this algorithm *Modified-CP* algorithm.

References

1. D. Breslauer, Saving Comparisons in the Crochemore–Perrin String Matching Algorithm. In Proc. of *1st European Symp. on Algorithms*, p. 61–72, 1993.
2. M. Crochemore, String-matching on ordered alphabets. *Theoret. Comput. Sci.*, 92, p. 33–47, 1992.
3. M. Crochemore and D. Perrin, Two-way string-matching. *J. Assoc. Comput. Mach.*, 38(3), p. 651–675, 1991.
4. M. Crochemore and W. Rytter, Cubes, squares and time space efficient string matching, Algorithmica 13,5 (1995) 405-425
5. M. Crochemore and W. Rytter, Text algorithms, *Oxford University Press*, New York, 1994
6. Z. Galil and J. Seiferas, Time-space-optimal string matching. *J. Comput. System Sci.*, 26, p. 280–294, 1983.
7. L. Gąsieniec, W. Plandowski and W. Rytter, The zooming method: a recursive approach to time-space efficient string-matching. *Theoretical Computer Science* 1995
8. L. Gąsieniec, W. Plandowski and W. Rytter, String matching in small time and space, Combinatorial Pattern Matching 1995, Lecture Notes in Computer Science 1995
9. J-P. Duval, Factorizing words over an ordered alphabet, J. Algorithms 4 (1983): 363-381.
10. D.E. Knuth, J.H. Morris and V.R. Pratt, Fast pattern matching in strings. *SIAM J. Comput.*, 6, p. 322–350, 1977.
11. M. Lothaire, Combinatorics on Words. Addison-Wesley, Reading, MA., U.S.A., 1983.

On the Power of BFS to Determine a Graphs Diameter
(Extended Abstract)

Derek G. Corneil[1], Feodor F. Dragan[2], and Ekkehard Köhler[3]

[1] Dept. of Computer Science, University of Toronto, Toronto, Ontario, Canada, dgc@cs.toronto.edu

[2] Dept. of Computer Science, Kent State University, Kent, Ohio, U.S.A., dragan@cs.kent.edu

[3] Fachbereich Mathematik, Technische Universität Berlin, Berlin, Germany, ekoehler@math.TU-Berlin.DE

Abstract. Recently considerable effort has been spent on showing that Lexicographic Breadth First Search (LBFS) can be used to determine a tight bound on the diameter of graphs from various restricted classes. In this paper, we show that in some cases, the full power of LBFS is not required and that other variations of Breadth First Search (BFS) suffice. The restricted graph classes that are amenable to this approach all have a small constant upper bound on the maximum sized cycle that may appear as an induced subgraph. We show that on graphs that have no induced cycle of size greater than k, BFS finds an estimate of the diameter that is no worse than $\operatorname{diam}(G) - \lfloor k/2 \rfloor - 2$.

1 Introduction

Recently considerable attention has been given to the problem of developing fast and simple algorithms for various classical graph problems. The motivation for such algorithms stems from our need to solve these problems on very large input graphs, thus the algorithms must be not only fast, but also easily implementable.

Determining the diameter of a graph is a fundamental and seemingly quite time consuming operation. For arbitrary graphs (with n vertices and m edges), the current fastest algorithm runs in $O(nm)$ time which is too slow to be practical for very large graphs. This naive algorithm examines each vertex in turn and performs a Breadth First Search (BFS) starting at the particular vertex. Such a sweep starting at vertex x immediately determines $\operatorname{ecc}(x)$, the eccentricity of vertex x. Recall that the *eccentricity* of vertex x, $\operatorname{ecc}(x) = \max_{y \in V} \operatorname{d}(x, y)$, where $\operatorname{d}(x, y)$ denotes the distance between x and y; the *diameter* of G equals the maximum eccentricity of any vertex in V. It is clear that this algorithm actually computes the entire distance matrix; clearly knowing the distance matrix immediately yields the diameter of the graph.

For dense graphs, the best result known is by Seidel [12], who showed that the all pairs shortest path problem (and hence the diameter problem) can be solved

S. Rajsbaum (Ed.): LATIN 2002, LNCS 2286, pp. 209–223, 2002.

in $O(M(n) \log n)$ time where $M(n)$ denotes the time complexity for fast matrix multiplication involving small integers only. The current best matrix multiplication algorithm is due to Coppersmith and Winograd [2] and has $O(n^{2.376})$ time bound. Unfortunately, fast matrix multiplication algorithms are far from being practical and suffer from large hidden constants in the running time bound.

Note that no efficient algorithm for the diameter problem in general graphs, avoiding the computation of the whole distance matrix, has been designed. Thus, the question whether for a graph the diameter can be computed easier than the whole distance matrix remains still open.

Clearly, performing a BFS starting at a vertex of maximum eccentricity easily produces the graph's diameter. Thus one way to approximate the diameter of a graph is to find a vertex of high eccentricity; this is the approach taken in this paper. This is not, however, the only approach. For example, Aingworth et al [1] obtain a ratio 2/3 approximation to the diameter in time $O(m\sqrt{nlogn}+n^2logn)$. Note that a ratio of 1/2 can easily be achieved by choosing an arbitrary vertex (the eccentricity of any vertex is at least one half the diameter of the graph) and performing a BFS starting at this vertex. It follows also from results in [1,6] (see also paper [13] which surveys recent results related to the computation of exact and approximate distances in graphs) that the diameter problem in unweighted, undirected graphs can be solved in $\tilde{O}(\min\{n^{3/2}m^{1/2}, n^{7/3}\})$ time with an additive error of at most 2 without matrix multiplication. Here, $\tilde{O}(f)$ means $O(f \text{ polylog}(n))$. The motivation behind the work of Aingworth et al is to find a fast, easily implementable algorithm (they avoid using matrix multiplication), a motivation that we share.

Our approach is to examine the naive algorithm of choosing a vertex, performing some version of BFS from this vertex and then showing a nontrivial bound on the eccentricity of the last vertex visited in this search. In fact, this algorithm is one of the "classical" algorithms in graph theory; if one restricts one's attention to trees, then this algorithm produces a vertex of maximum eccentricity (see e.g. [9]). This approach has already received considerable attention. (In the following we let v denote the vertex that appears last in a particular search; the definition of the various searches and families of graphs will be presented in the next section.) For example, Dragan et al [8] have shown that if LBFS is used for chordal graphs, then $\text{ecc}(v) \geq \text{diam}(G) - 1$ whereas for interval graphs $\text{ecc}(v) = \text{diam}(G)$. It is clear from the work of Corneil et al [4], that by using LBFS on AT-free graphs, one has $\text{ecc}(v) \geq \text{diam}(G) - 1$. Dragan [7], again using LBFS, has shown that $\text{ecc}(v) \geq \text{diam}(G) - 2$ for HH-free graphs, $\text{ecc}(v) \geq \text{diam}(G) - 1$ for HHD-free graphs and $\text{ecc}(v) = \text{diam}(G)$ for graphs that are both HHD-free and AT-free.

It is interesting to note that Corneil et al [4] have looked at double sweep LBFSs (i.e. start an LBFS from a vertex that is last in a previous arbitrarily chosen LBFS) on chordal and AT-free graphs. They have provided a forbidden subgraph structure on graphs where $\text{ecc}(v) = \text{diam}(G) - 1$. They also presented both chordal and AT-free graphs where for no c, is the c-sweep LBFS algorithm guaranteed to find a vertex of maximum eccentricity. Furthermore, they showed that for any c, there is a graph G where $\mathrm{ecc}(v) \leq \mathrm{diam}(G) - c$, where v is the vertex visited last in a 2-sweep LBFS. This graph G, however, has a large induced cycle whose size depends on c.

These results motivate a number of interesting questions:

- Is it an inherent property of LBFS to end in a vertex of high eccentricity for the various restricted graph families mentioned above? What happens if we use other variants of BFS?
- Why do AT-free and chordal graphs, two families with very disparate structure, exhibit such similar behaviour with respect to the efficacy of LBFS to find vertices of high eccentricity?
- Although LBFS "fails" to find vertices of high eccentricity for graphs in general, all known examples that exhibit such failure have large induced cycles. If we bound the size of the largest induced cycle, can we get a bound on the eccentricity of the vertex that appears last in an LBFS?
- If the previous question is answered in the affirmative, is the full power of LBFS needed? What happens if we just use BFS?

This paper addresses these questions. In the next section, we present the various forms of BFS and define the graph theoretic terminology used throughout the paper. In Section 3 we examine the behaviour of the different versions of BFS on various restricted graph families. We establish some new bounds and show, by example, that all stated bounds on $\mathrm{ecc}(v)$ are tight. In Section 4, we examine families of graphs where the size of the largest induced cycle is bounded and show that BFS does succeed in getting vertices of high, with respect to k, eccentricity.

2 Notation and Definitions

First we formalize the notion of BFS and then discuss various variations of it. We caution the reader that there is some confusion in the literature between BFS and what we call LL, defined below. In defining the various versions of BFS, we are only concerned with identifying the last vertex visited by the search; straightforward modifications produce the list of vertices in the order that they are visited by the search. It should be noted that none of the orderings are unique; instead, each search identifies one of the possible end-vertices.

Algorithm BFS: Breadth First Search
Input: graph $G(V, E)$ and vertex u
Output: vertex v, the last vertex visited by a BFS starting at u

Initialize queue Q to be $\{u\}$ and mark u as "visited".
while $Q \neq \emptyset$ **do**
  Let v be the first vertex of Q and remove it from Q.
  Each unvisited neighbour of v is added to the end of Q and marked as "visited".

Note that the above algorithm can easily be modified to obtain the "layers" of V with respect to u. In particular, for each $0 \le i \le \mathrm{ecc}(u)$, the *i-th layer of V with respect to u*, denoted $L_i(u)$, $= \{v : \mathrm{d}(u, v) = i\}$. This motivates the next algorithm, LL.

Algorithm LL: Last Layer
Input: graph $G(V, E)$ and vertex u
Output: vertex v, a vertex in the last layer of u

Run BFS to get the layering of V with respect to u.
Choose v to be an arbitrary vertex in the last layer.

Clearly any vertex returned by BFS can also be returned by LL; the converse is not true as shown by the graph in Figure 1.

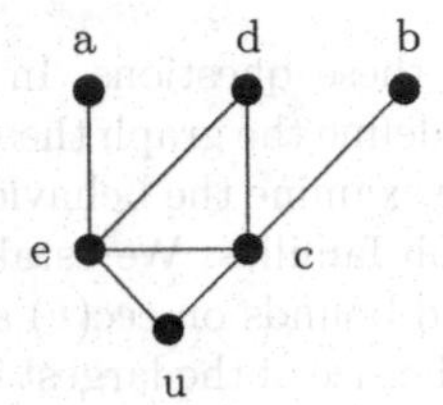

Fig. 1. No BFS starting at u can return vertex d.

Now we modify this algorithm to obtain a vertex in the last layer that has minimum degree with respect to the vertices in the previous layer.

Algorithm LL+: Last Layer, Minimum Degree
Input: graph $G(V, E)$ and vertex u
Output: vertex v, a vertex in the last layer of u, that has minimum degree with respect to the vertices in the previous layer

Run BFS to get the layering of V with respect to u.
Choose v to be an arbitrary vertex in the last layer that has minimum degree with respect to the vertices in the previous layer.

Finally we introduce Lexicographic Breadth First Search (LBFS). This search paradigm was discovered by Rose et al [11] and was shown to yield a simple linear time algorithm for the recognition of chordal graphs. In light of the great deal of work currently being done on LBFS, it is somewhat surprising that interest in LBFS lay dormant for quite a while after [11] appeared.

Algorithm LBFS: Lexicographic Breadth First Search
Input: graph $G(V, E)$ and vertex u
Output: vertex v, the last vertex visited by an LBFS starting at u

Assign label $\emptyset$ to each vertex in V.
for $i = n$ **downto** 1 **do**
  Pick an unmarked vertex v with the largest (with respect to lexicographic order) label.
  Mark v "visited".
  For each unmarked neighbour y of v, add i to the label of y.

If vertex e is removed from the graph in Figure 1, we have a graph where a vertex, namely d, can be visited last by a BFS from u but by no LBFS from u.

We now turn to the definitions of the various graph families introduced in the previous section. A graph is *chordal* if it has no induced cycle of size greater than 3. An *interval* graph is the intersection graph of intervals of a line. Lekkerkerker and Boland [10] defined an *asteroidal triple* to be a triple of vertices such that between any two there is a path that avoids the neighbourhood of the third and showed that a graph is an interval graph iff it is both chordal and asteroidal triple-free (AT-free). A *claw* is the complete bipartite graph $K_{1,3}$, a *hole* is an induced cycle of length greater than 4, a *house* is a 4-cycle with a triangle added to one of the edges of the C_4 and a *domino* is a pair of C_4s sharing an edge. A graph is *HH-free* if it contains no induced houses or holes and is *HHD-free* if it contains no induced houses, holes or dominos.

Finally, in order to capture the notion of "small" induced cycles, we define a graph to be *k-chordal* if it has no induced cycles of size greater than k. Note that chordal graphs are precisely the 3-chordal graphs and AT-free graphs are 5-chordal. We define the *disk* of radius r centered at u to be the set of vertices of distance at most r to u, i.e., $D_r(u) = \{v \in V : \mathrm{d}(u, v) \leq r\} = \bigcup_{i=0}^{r} L_i(u)$.

3 Restricted Families of Graphs

We now see how the four search algorithms mentioned in the previous section behave on the following families of graphs: chordal, AT-free, {AT,claw}-free, interval, and hole-free. The results are summarized in the following table. In this table the references refer to the paper where the lower bound was established; a [*] indicates that the result is new. A figure reference refers to the appropriate figure where it is shown that the lower bound is tight. In each of the figures the vertex pair a,b forms a diametral pair, i.e. $\mathrm{d}(a, b) = \mathrm{diam}(G)$. Below each figure a BFS, LBFS, LL, or LL+ ordering is given that achieves the corresponding bounds; vertex u is always the start-vertex and v the end-vertex of the appropriate search; different BFS-layers are separated by a $|$.

GRAPH CLASS	LL	LL+	BFS	LBFS
chordal graphs	$\geq D-2$ [3] Fig. 4	$\geq D-2$ [3] Fig. 5	$\geq D-1$ [*] Fig. 2	$\geq D-1$ [8] Fig.6
AT-free graphs	$\geq D-2$ [*] Fig. 3	$\geq D-1$ [*] Fig. 7	$\geq D-2$ [*] Fig. 3	$\geq D-1$ [4] Fig. 7
{AT,claw}-free graphs	$\geq D-1$ [*] Fig. 2	$= D$ [*]	$\geq D-1$ [*] Fig. 2	$= D$ [*]
interval graphs	$\geq D-1$ [*] Fig. 2	$= D$ [*]	$\geq D-1$ [*] Fig. 2	$= D$ [8]
hole-free graphs	$\geq D-2$ [*] Fig. 8	$\geq D-2$ [*] Fig. 8	$\geq D-2$ [*] Fig. 8	$\geq D-2$ [*] Fig. 8

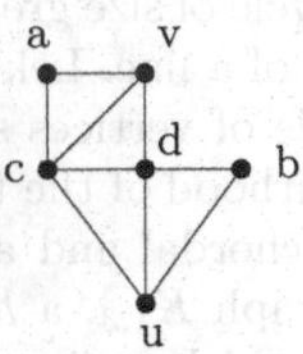

Fig. 2. BFS: u|bcd|av

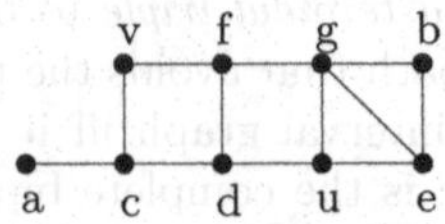

Fig. 3. BFS: u|dge|cfb|av

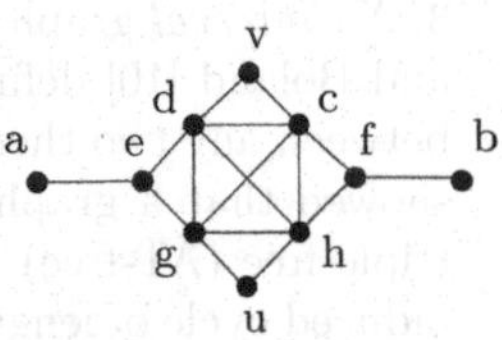

Fig. 4. LL: u|gh|edcf|abv

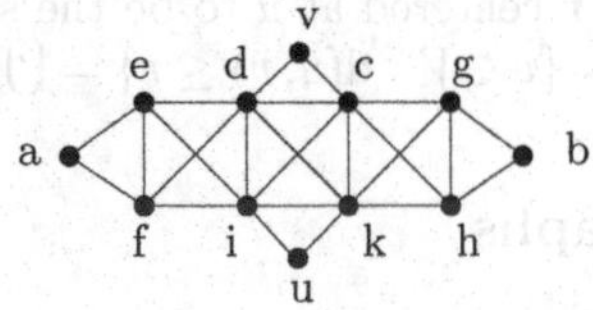

Fig. 5. LL+: u|ik|fedcgh|abv

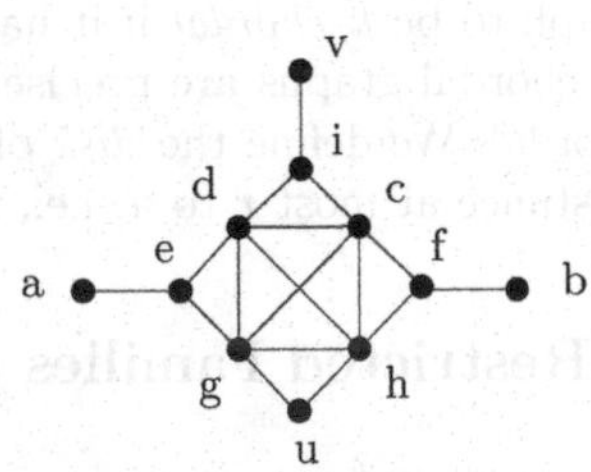

Fig. 6. LBFS: u|gh|edcf|aib|v

To illustrate the types of techniques that are used to establish lower bounds in the table, we now show that $\text{ecc}(v) \geq \text{diam}(G) - 1$ for chordal graphs when BFS is used. This result subsumes the result shown in [8] that this lower bound holds when LBFS is used. The journal version of the paper will contain proofs for all new results mentioned in the table.

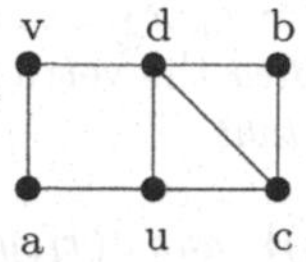

Fig. 7. LBFS: u|cda|bv

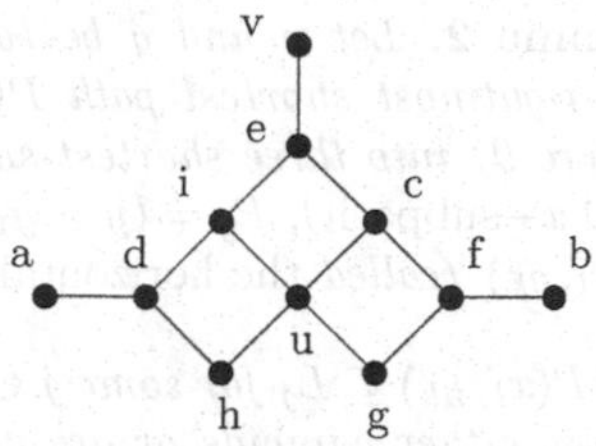

Fig. 8. LBFS: u|ghic|dfe|abv

First we comment on the BFS algorithm. In particular, we may regard BFS as having produced a numbering from n to 1 in decreasing order of the vertices in V where vertex u is numbered n. As a vertex is placed on the queue, it is given the next available number. The last vertex visited, v, is given the number 1. Thus BFS may be seen to generate a rooted tree T with vertex u as the root. A vertex y is the *father* in T of exactly those neighbours in G which are inserted into the queue when y is removed.

An ordering $\sigma = [v_1, v_2, \ldots, v_n]$ of the vertex set of a graph G generated by a BFS will be called a *BFS–ordering* of G. Let $\sigma(y)$ be the number assigned to a vertex y in this BFS–ordering σ. Denote also by $f(y)$ the father of a vertex y with respect to σ. The following properties of a BFS–ordering will be used in what follows. Since all layers of V considered here are with respect to u, we will frequently use notation L_i instead of $L_i(u)$.

(P1) If $y \in L_q$ $(q > 0)$ then $f(y) \in L_{q-1}$ and $f(y)$ is the vertex from $N(y) \bigcap L_{q-1}$ with the largest number in σ.
(P2) If $x \in L_i$, $y \in L_j$ and $i < j$, then $\sigma(x) > \sigma(y)$.
(P3) If $x, y \in L_j$ and $\sigma(x) > \sigma(y)$, then either $\sigma(f(x)) > \sigma(f(y))$ or $f(x) = f(y)$.
(P4) If $x, y, z \in L_j$, $\sigma(x) > \sigma(y) > \sigma(z)$ and $f(x)z \in E$, then $f(x) = f(y) = f(z)$ (in particular, $f(x)y \in E$).

In what follows, by $P(x, y)$ we will denote a path connecting vertices x and y. Proof of the following lemma is omitted.

Lemma 1. *If vertices a and b of a disk $D_r(u)$ of a chordal graph are connected by a path $P(a, b)$ outside of $D_r(u)$ (i.e., $P(a, b) \bigcap D_r(u) = \{a, b\}$), then a and b must be adjacent.*

Let σ be a BFS–ordering of a chordal graph G started at a vertex u. Let also $P(a, b) = (a = x_1, x_2, \ldots, x_{k-1}, x_k = b)$ be a shortest path of G connecting vertices a and b. We say that $P(a, b)$ is a *rightmost shortest path* if the sum $\sigma(x_1) + \sigma(x_2) + \ldots + \sigma(x_{k-1}) + \sigma(x_k)$ is the largest among all shortest paths connecting a and b.

Lemma 2. *Let x and y be two arbitrary vertices of a chordal graph G. Every rightmost shortest path $P(x,y)$ between x and y can be decomposed (see Figure 9) into three shortest subpaths $P_x = (x = x_1, x_2, \ldots, x_l)$ (called* the vertical x–subpath*), $P_y = (y = y_1, y_2, \ldots, y_k)$ (called* the vertical y–subpath*) and $P(x_l, y_k)$ (called* the horizontal subpath*) such that*

1. *$P(x_l, y_k) \subseteq L_j$ for some $j \in \{0, 1, \ldots, \mathrm{ecc}(u)\}$, and $\mathrm{d}(x_l, y_k) \le 2$ (i.e., x_l and y_k either coincide or are adjacent or have a common neighbour in $G[L_j] \cap P(x_l, y_k)$);*
2. *$x_{l-i} \in L_{j+i}$ for $0 \le i \le l-1$;*
3. *$y_{k-i} \in L_{j+i}$ for $0 \le i \le k-1$.*

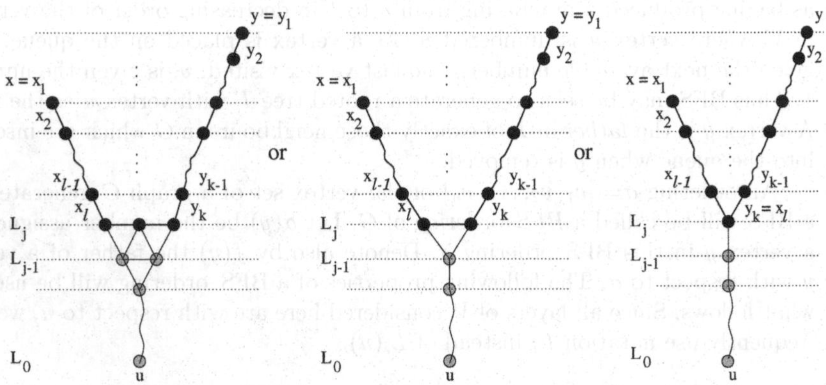

Fig. 9. The structure of rightmost shortest path in chordal graphs.

Proof. First we prove that $|P(x,y) \bigcap L_i| \le 3$ for any $i = 1, 2, \ldots, \mathrm{ecc}(s)$. Assume that the intersection of $P(x,y)$ and a layer L_q, for some index q, contains at least four vertices. Let a, b, c, d be the first four vertices of $P(x,y) \bigcap L_q$ on the way from x to y. We claim that $ab, bc, cd \in E$.

If $ab \notin E$ then, by Lemma 1, subpath $P(a,b)$ of the path $P(x,y)$ is completely contained in disk $D_q(s)$. In particular, the neighbor b' of b on $P(a,b)$ belongs to $D_{q-1}(s)$. Using the same arguments, we conclude that $bc \in E$ or subpath $P(b,c)$ of the path $P(x,y)$ is contained in $D_q(s)$. If $bc \in E$ then a neighbor v of c in L_{q-1} must be adjacent with b' (by Lemma 1). Since $\sigma(v) > \sigma(b)$, we get a contradiction to $P(x,y)$ is a rightmost path (we can replace vertex b of $P(x,y)$ with v and get a shortest path between x and y with larger sum). If $bc \notin E$ then the neighbor b'' of b on $P(b,c)$ is also contained in $D_{q-1}(s)$. By Lemma 1, vertices b' and b'' must be adjacent, but this is impossible since $P(x,y)$ is a shortest path.

Thus, vertices a and b have to be adjacent. If vertices b and c are not adjacent, then the neighbor b'' of b on $P(b,c)$ belongs to $D_{q-1}(s)$ and, by Lemma 1, it must be adjacent to any neighbor u of a in L_{q-1}. Since $\sigma(u) > \sigma(b)$ holds, again a contradiction to $P(x,y)$ is a rightmost path arises. Consequently, vertices b and c are adjacent, too. Completely in the same way one can show that c and d have to be adjacent. Note also that the adjacency of a with b and b with c is proved without using the existence of the vertex d.

We have now $ab, bc, cd \in E$ and the induced path (a,b,c,d) is a rightmost shortest path (as a subpath of the rightmost shortest path $P(x,y)$). Consider neighbors a' and d' in L_{q-1} of a and d, respectively. Since $\sigma(a')+\sigma(d') > \sigma(b)+\sigma(c)$ and the path (a,b,c,d) is rightmost and shortest, vertices a' and d' can neither coincide nor be adjacent. But then we get a path (a',a,b,c,d,d') connecting two non-adjacent vertices of $D_{q-1}(s)$ outside of the disk. A contradiction to Lemma 1 obtained shows that $|P(x,y) \bigcap L_i| \leq 3$ holds for any $i = 1,2,\ldots,\mathrm{ecc}(s)$.

Now let $|P(x,y) \bigcap L_q| = 3$ and let a,b,c be the vertices of $P(x,y) \bigcap L_q$ on the way from x to y. It follows from the discussion above that $ab, bc \in E$ and both the neighbor a' of a on subpath $P(x,a)$ and the neighbor c' of c on subpath $P(c,y)$ (if they exist) belong to the layer L_{q+1}. If for example $a' \in L_{q-1}$, then, by Lemma 1, a' must be adjacent to any neighbor v of b in L_{q-1}. Again, since $\sigma(v) > \sigma(a)$, a contradiction to $P(x,y)$ is a rightmost path arises.

Furthermore, if for some index q, $P(x,y) \bigcap L_q = \{a,b\}$, and $P(x,y) = (x,\ldots,a',a,b,b',\ldots,y)$, then both a' and b' belong to the layer L_{q+1}.

Summarizing all these we conclude that, while moving from x to y along the path $P(x,y)$, we can have only one horizontal edge or only one pair of consecutive horizontal edges. Here by horizontal edge we mean an edge with both end-vertices from the same layer. All other vertices of the path $P(x,y)$ belong to higher layers.

Having the structure of a rightmost shortest path established, we can now prove the main result for chordal graphs. In presenting a rightmost shortest path, we use "/"s to differentiate the appropriate subpaths.

Theorem 1. *Let v be the vertex of a chordal graph G last visited by a BFS. Then* $\mathrm{ecc}(v) \geq \mathrm{diam}(G) - 1$.

Proof. Let x,y be a pair of vertices such that $\mathrm{d}(x,y) = \mathrm{diam}(G)$, and consider two rightmost shortest paths

$$P(x,v) = (x = x_1, x_2, \ldots, x_{l-1}/x_l, \ldots, v'_h/v'_{h-1}, \ldots, v'_2, v'_1 = v)$$

and

$$P(y,v) = (y = y_1, y_2, \ldots, y_{k-1}/y_k, \ldots, v''_g/v''_{g-1}, \ldots, v''_2, v''_1 = v)$$

connecting vertex v with x and y, respectively (see Figure 10). By Lemma 2, each of these paths consists of two (perhaps of length 0) vertical subpaths and one horizontal path of length not greater than 2. Assume, without loss of generality, that $h \leq g$ and let $x_l, \ldots, v'_h \in L_q$. Since $v \in L_{\mathrm{ecc}(u)}$ we also have $l \leq h$ and $k \leq g$.

By Lemma 1, vertices v'_h, v''_h in L_q either coincide or are adjacent. Note that, if $d(x_l, v''_h) \leq 1$, then $d(x,y) \leq d(x,x_l) + 1 + d(v''_h, y) \leq d(v, v''_h) + 1 + d(v''_h, y) = d(v,y) + 1 \leq ecc(v) + 1$. That is, $ecc(v) \geq d(x,y) - 1 = diam(G) - 1$, and we are done. Hence, we may assume that $d(x_l, v''_h) \geq 2$ and, therefore, $x_l \neq v'_h$.

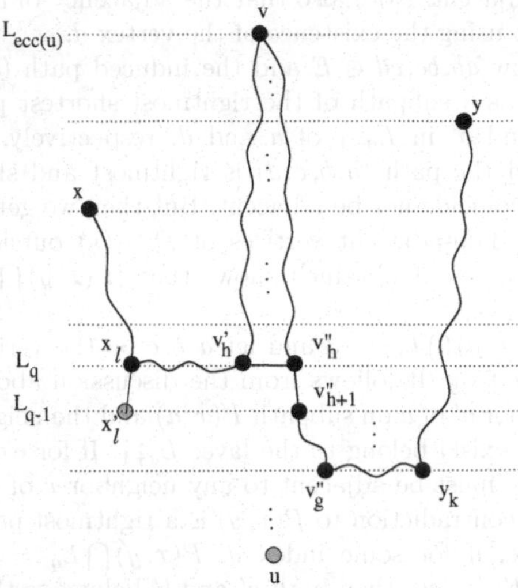

Fig. 10. Rightmost shortest paths $P(x,v)$ and $P(y,v)$.

We distinguish between two cases. The first one is simple. Only for the second case we will need to use the special properties of a BFS–ordering.

Case $g > h$.
In this case there exists a vertex v''_{h+1} in the intersection $P(y,v) \bigcap L_{q-1}$. Consider also a neighbour x'_l of x_l in L_{q-1}. Since vertices x'_l and v''_{h+1} are connected by path $(x'_l, x_l, \ldots, v'_h, v''_h, v''_{h+1})$ outside of the disk $D_{q-1}(u)$, by Lemma 1, they are adjacent if they do not coincide. Hence, $d(x_l, v''_{h+1}) \leq 2$ and, therefore, $d(x,y) \leq d(x,x_l) + 2 + d(v''_{h+1}, y) \leq d(v, v''_h) + 2 + d(v''_{h+1}, y) = d(v,y) + 1 \leq ecc(v) + 1$, i.e., again $ecc(v) \geq d(x,y) - 1 = diam(G) - 1$.

Case $g = h$.
From the discussion above (now, since $g = h$ we have a symmetry), we may assume that $d(y_k, v'_h) \geq 2$ and $y_k \neq v''_h$. Consider neighbours x'_l and y'_k in L_{q-1} of vertices x_l and y_k, respectively (see Figure 11(a)). By Lemma 1, they are adjacent if do not coincide, i.e., $d(x_l, y_k) \leq 3$. Now, if at least one of the equalities $d(x_l, v'_h) = 2$, $d(y_k, v''_h) = 2$ holds, then we are done. Indeed, if for example $d(x_l, v'_h) = 2$, then $d(x,v) = d(x,x_l) + 2 + d(v'_h, v)$ and, therefore, $d(x,y) \leq d(x,x_l) + d(x_l, y_k) + d(y_k, y) \leq d(x,x_l) + 3 + d(v'_h, v) = d(x,v) + 1 \leq ecc(v) + 1$.

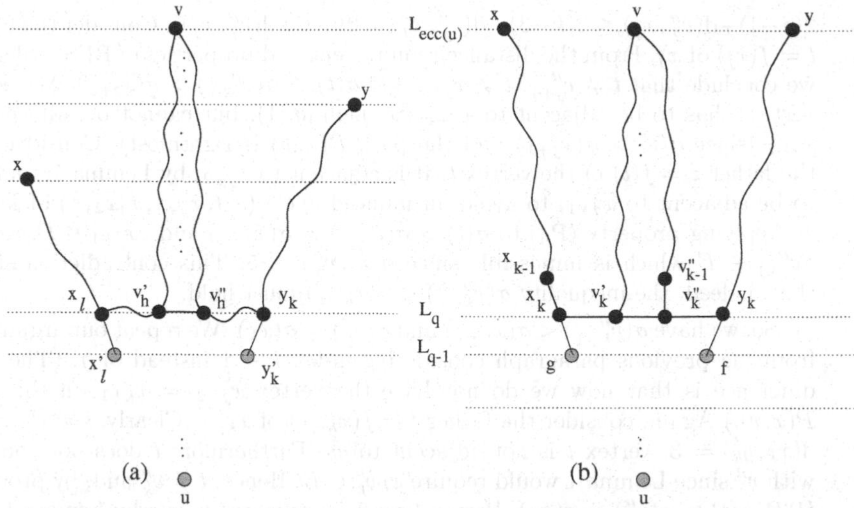

Fig. 11. (a) Horizontal subpaths of $P(x,v)$ and $P(y,v)$ are in the same layer, (b) Paths $P(x,v)$ and $P(y,v)$ have similar shape and the same length $2k-1$.

So, we may assume that $x_l v'_h, y_k v''_h \in E$. Moreover, since $d(y_k, v'_h) \geq 2$, vertices v'_h and v''_h cannot coincide, i.e., they are adjacent. If $l < h$ or $k < h$ or $d(x_l, y_k) < 3$, again we will get $d(x,y) \leq ecc(v)+1$ by comparing distances $d(v,y) = h+k-1$, $d(v,x) = h+l-1$ with $d(x,y) \leq l-1+d(x_l,y_k)+k-1 \leq l+1+k$. Thus, we arrive at a situation when $l = k = h$, $x_l v'_h, y_k v''_h, v'_h v''_h \in E$, and $d(x_l, y_k) = 3$, $d(v,x) = d(v,y) = 2k-1$ (see Figure 11(b)). We may also assume that $d(x,y) = 2k+1$, since otherwise, $d(x,y) \leq 2k = d(v,y)+1 \leq ecc(v)+1$ and we are done. We show that this final configuration (with $d(x,y) = 2k+1$) is impossible because of the properties of BFS–orderings.

Assume, without loss of generality, that $\sigma(y_k) > \sigma(x_k)$ and consider the fathers $f = f(y_k)$, $g = f(x_k)$ of y_k and x_k, respectively. Since $d(x_k, y_k) = 3$, we have $f \neq g$ and $f x_k, g y_k \notin E$. By Lemma 1 and property (P3) of BFS–ordering, vertices f and g are adjacent and $\sigma(f) > \sigma(g)$. Chordal graphs cannot contain an induced cycle of length greater than 3. Therefore, in the cycle formed by $g, x_k, v'_k, v''_k, y_k, f$, at least chords $g v'_k$ and $f v''_k$ must be present. Since for the father $f(v''_k)$ of v''_k we have $\sigma(f(v''_k)) \geq \sigma(f) > \sigma(g)$, inequality $\sigma(v''_k) > \sigma(x_k)$ must hold (here we used properties (P1) and (P3) of BFS–orderings). We will need the inequality $\sigma(v''_k) > \sigma(x_k)$ later to get our final contradiction.

Now consider vertices x_{k-1} and v''_{k-1}. We claim that $\sigma(v''_{k-1}) < \sigma(x_{k-1})$. Assume that this is not the case, and let j ($j \in \{1, 2, \ldots, k-2\}$) be the largest index such that $\sigma(v''_j) < \sigma(x_j)$ (recall that $\sigma(v''_1) = \sigma(v) = 1 < \sigma(x) = \sigma(x_1)$). Then $\sigma(v''_{j+1}) > \sigma(x_{j+1})$ holds, and since $j \leq k-2$ and $d(v,x) = 2k-1$, we obtain $d(x_j, v''_j) \geq 5$ (because of $2k-1 = d(v,x) \leq d(v,v''_j) + d(v''_j, x_j) + d(x_j, x) =$

$2(j-1)+d(v_j'',x_j) \le 2(k-3)+d(v_j'',x_j) = 2k-6+d(v_j'',x_j)$). Consider the father $t = f(x_j)$ of x_j. From the distance requirement and properties of BFS–orderings we conclude that $t \ne v_{j+1}''$, $t \ne x_{j+1}$ and $\sigma(t) > \sigma(v_{j+1}'') > \sigma(x_{j+1})$. Moreover, vertex t has to be adjacent to x_{j+1} (by Lemma 1), but cannot be adjacent to x_{j+2} (since $\sigma(t) > \sigma(x_{j+1})$ and the path $P(x,v)$ is rightmost). Consider now the father $z = f(t)$ of the vertex t. It is adjacent to x_{j+2}, by Lemma 1, and has to be adjacent to x_{j+1}, to avoid an induced cycle (z,t,x_{j+1},x_{j+2},z) of length 4. Applying property (P4) to $\sigma(t) > \sigma(v_{j+1}'') > \sigma(x_{j+1})$ and $zx_{j+1} \in E$, we get $zv_{j+1}'' \in E$ which is impossible since $d(x_j, v_j'') \ge 5$. This contradiction shows that, indeed, the inequality $\sigma(v_{k-1}'') < \sigma(x_{k-1})$ must hold.

So, we have $\sigma(v_{k-1}'') < \sigma(x_{k-1})$ and $\sigma(v_k'') > \sigma(x_k)$. We repeat our arguments from the previous paragraph considering index $k-1$ instead of j. (The only difference is that now we do not have the vertex $x_{j+2} = x_{k+1}$ on the path $P(x,v)$.) Again, consider the father $t = f(x_{k-1})$ of x_{k-1}. Clearly, $t \ne v_k'$. Since $d(x_k, y_k) = 3$, vertex t is not adjacent to y_k. Furthermore t does not coincide with v_k'' since Lemma 1 would require $x_k v_k'' \in E$. Hence, $t \ne v_k''$ and, by property (P3), $\sigma(t) > \sigma(v_k'') > \sigma(x_k)$. Vertex t must be adjacent to x_k, by Lemma 1, but cannot be adjacent to v_k', since the path $P(v,x)$ is rightmost and $\sigma(t) > \sigma(x_k)$. To avoid an induced cycle of length 4, vertex t is not adjacent to v_k'' as well.

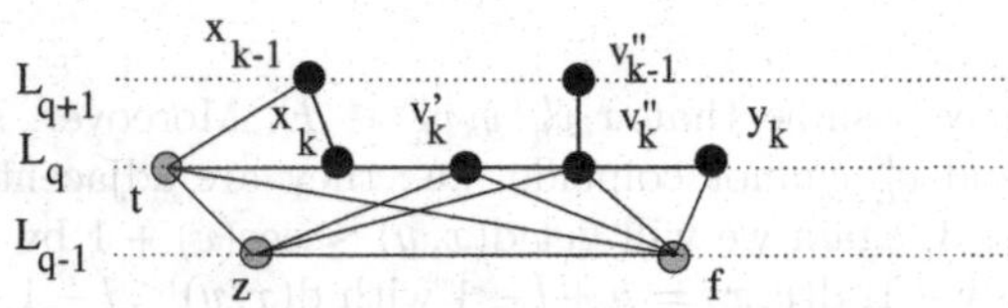

Fig. 12. Illustration to the proof of Theorem 1.

Consider also the father $z = f(t)$ of the vertex t (see Figure 12). If $zy_k \in E$ then $d(x,y) \le d(y,y_k)+1+d(z,x_{k-1})+d(x_{k-1},x) \le k-1+1+2+k-2 = 2k$, and a contradiction to the assumption $d(x,y) = 2k+1$ arises. Therefore, z and y_k are not adjacent and hence $z \ne f$ (recall that f is the father of y_k and $fv_k'' \in E$, $fx_k \notin E$). By Lemma 1, $zf \in E$. Since $\sigma(t) > \sigma(v_k'')$ and $fv_k'' \in E$, by properties (P3) and (P1), we get $\sigma(z) \ge \sigma(f(v_k'')) \ge \sigma(f)$, i.e. $\sigma(z) > \sigma(f)$. Consequently, $\sigma(t) > \sigma(y_k)$. Now, vertex z cannot be adjacent to x_k, since this would apply the adjacency of z with y_k, too (by $\sigma(t) > \sigma(y_k) > \sigma(x_k)$ and property (P4)). But then, in the cycle (z,t,x_k,v_k',v_k'',f,z) only chords zv_k', zv_k'', fv_k', ft are possible, which are not enough to avoid an induced cycle of length greater than 3 in G. A contradiction with the chordality of G completes the proof of the theorem.

4 *k*-chordal Graphs

As mentioned in the introduction, the examples that show that LBFS fails to find vertices of high eccentricity all have large induced cycles. Furthermore, both chordal and AT-free graphs have constant bounds on the maximum size of induced cycles. Thus one would hope that for k-chordal graphs where k is a constant, some form of BFS would succeed in finding a vertex whose eccentricity is within some function of k of the diameter. In fact, we show that LL is sufficiently strong to ensure this. First, a lemma that is used in the proof.

Lemma 3. *If vertices a and b of a disk $D_r(u)$ of a k-chordal graph are connected by a path $P(a,b)$ outside of $D_r(u)$ (i.e., $P(a,b) \bigcap D_r(u) = \{a,b\}$), then* $\mathrm{d}(a,b) \leq \lfloor k/2 \rfloor$.

Proof. Assume $\mathrm{d}(a,b) > \lfloor k/2 \rfloor$, and let P be an induced subpath of $P(a,b)$ connecting vertices a and b. Consider shortest paths $P(a,u)$ and $P(b,u)$ (connecting a with u and b with u, respectively). Using vertices of these paths we can construct an induced path $Q(a,b)$ with the property that all its vertices except a and b are contained in $D_{r-1}(u)$. By our construction, the cycle C obtained by the concatenation of P and $Q(a,b)$ is induced. Since $\mathrm{d}(a,b) > \lfloor k/2 \rfloor$, both paths P and $Q(a,b)$ must be of length greater than $\lfloor k/2 \rfloor$. Therefore, the cycle C has the length at least $\lfloor k/2 \rfloor + 1 + \lfloor k/2 \rfloor + 1 > k$, that is impossible.

Theorem 2. *Let G be a k-chordal graph. Consider a LL, starting in some vertex u of G and let v be a vertex of the last BFS layer. Then* $\mathrm{ecc}(v) \geq \mathrm{diam}(G) - \lfloor k/2 \rfloor - 2$.

Proof. Let x, y be a pair of vertices such that $\mathrm{d}(x,y) = \mathrm{diam}(G)$, and consider two shortest paths $P(x,v)$ and $P(y,v)$ connecting vertex v with x and y, respectively. Let also q be the minimum index such that

$$L_q \bigcap (P(x,v) \bigcup P(y,v)) \neq \emptyset.$$

Consider a vertex $z \in L_q \bigcap (P(x,v) \bigcup P(y,v))$, and assume, without loss of generality, that z belongs to $P(y,v)$, i.e., we have $\mathrm{d}(v,z) + \mathrm{d}(z,y) = \mathrm{d}(v,y)$. Assume, for now, that $z \neq u$. Consider also a shortest path $Q(x,u)$ between x and u, vertices $x' \in L_q \bigcap Q(x,u)$, $x'' \in L_{q-1} \bigcap Q(x,u)$, and a neighbour w of z in L_{q-1}. Since vertices x'' and w belong to $D_{q-1}(u)$ and can be connected outside of $D_{q-1}(u)$, by Lemma 3, $\mathrm{d}(x'', w) \leq \lfloor k/2 \rfloor$ must hold. We have also $\mathrm{d}(x,x') \leq \mathrm{d}(v,z)$ because $v \in L_{\mathrm{ecc}(u)}$. Therefore, $\mathrm{d}(x,y) \leq \mathrm{d}(x,x') + 1 + \mathrm{d}(x'',w) + 1 + \mathrm{d}(z,y) \leq \mathrm{d}(v,z) + \lfloor k/2 \rfloor + 2 + \mathrm{d}(z,y) = \mathrm{d}(v,y) + \lfloor k/2 \rfloor + 2 \leq \mathrm{ecc}(v) + \lfloor k/2 \rfloor + 2$, i.e., $\mathrm{ecc}(v) \geq \mathrm{diam}(G) - \lfloor k/2 \rfloor - 2$. Finally, if $z = u$, then a similar argument as above establishes a tighter bound on $\mathrm{ecc}(v)$.

This result can be strengthened further for 4-chordal graphs and 5-chordal graphs. We can prove the following.

Theorem 3. *Let G be a 5-chordal graph. Consider a LL, starting in some vertex u of G and let v be a vertex of the last BFS layer. Then* $\mathrm{ecc}(v) \geq \mathrm{diam}(G) - 2$.

Again this bound on $\mathrm{ecc}(v)$ is tight. Figure 8 represents a 4-chordal graph G for which a LBFS exists such that the vertex v, last visited by this LBFS, has eccentricity equal to $\mathrm{diam}(G) - 2$.

Examples of k-chordal graphs with larger difference between diameter and the eccentricity of the vertex last visited in some variant of BFS are given in the following two figures. Figure 13 shows a 6-chordal graph. For this graph G, there is an LL, starting in vertex u, that ends in vertex v, where $\mathrm{ecc}(v) = \mathrm{diam}(G) - 3$. This shows that the bound of $\mathrm{diam}(G) - 2$ which holds for 4- and 5-chordal graphs (for LL and LL+) can not hold for 6-chordal graphs as well.

In Figure 14 each of the dashed edges stands for a path of length k. Thus this graph G is $4k$-chordal. The diameter of G is also $4k$, the eccentricity of v, the vertex last visited by some LBFS started in u, is $\mathrm{ecc}(v) = 2k + 1$. Hence the difference between the diameter and the eccentricity $\mathrm{ecc}(v)$ is $2k - 1$. This shows that, at least for the $4k$-chordal graphs, the bound on $ecc(v)$ (for LL) given in Theorem 2 is close to the best possible. It is within 3 of the bound that could be achieved by LBFS.

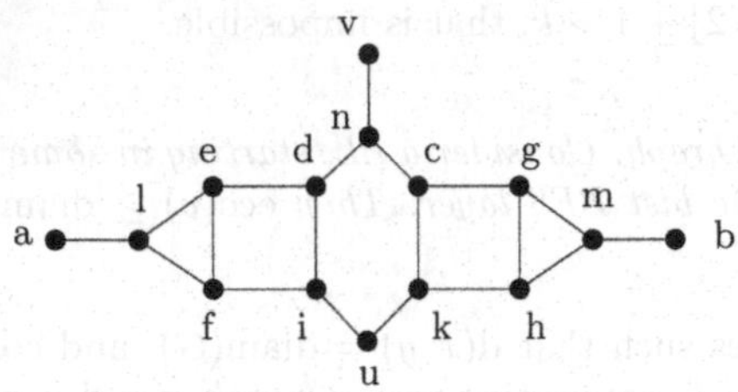

Fig. 13. LL: u|ik|fdch|lengm|abv

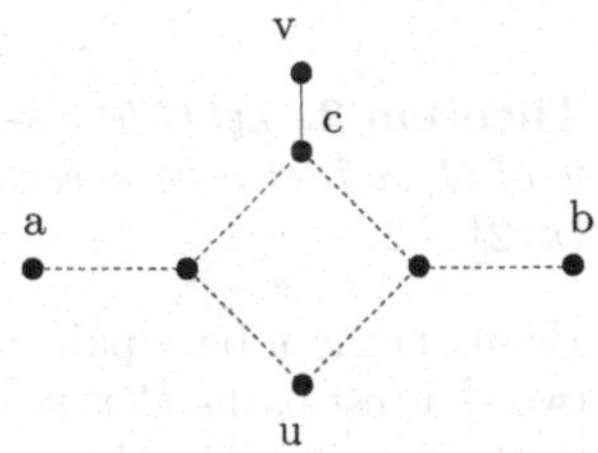

Fig. 14. LBFS: u|...|acb|v

Acknowledgements:

D.G.C. and E.K. wish to thank the Natural Science and Engineering Research Council of Canada for their financial support.

References

1. D. Aingworth, C. Chekuri, P. Indyk and R. Motwani, Fast estimation of diameter and shortest paths (without matrix multiplication), SIAM J. on Computing, 28 (1999), 1167–1181.

2. D. Coppersmith and S. Winograd, Matrix multiplication via arithmetic progression, Proceedings of the 19th ACM Symposium on Theory of Computing, 1987, 1–6.
3. V.D. Chepoi and F.F. Dragan, Linear-time algorithm for finding a central vertex of a chordal graph, "Algorithms - ESA'94 " Second Annual European Symposium, Utrecht, The Netherlands, September 1994, Springer, LNCS 855 (Jan van Leeuwen, ed.), (1994), 159–170.
4. D.G. Corneil, F.F. Dragan, M. Habib and C. Paul, Diameter determination on restricted graph families, Discrete Appl. Math., 113 (2001), 143–166.
5. D.G. Corneil, S. Olariu and L. Stewart, Linear time algorithms for dominating pairs in asteroidal triple–free graphs, SIAM J. on Computing, 28 (1999), 1284–1297.
6. D. Dor, S. Halperin, and U. Zwick, All pairs almost shortest paths, Proceedings of the 37th Annual IEEE Symposium on Foundations of Computer Science, 1996, 452–461.
7. F.F. Dragan, Almost diameter of a house-hole-free graph in linear time via LexBFS, Discrete Appl. Math., 95 (1999), 223–239.
8. F.F. Dragan, F. Nicolai and A. Brandstädt, LexBFS–orderings and powers of graphs, Proc. of the WG'96, LNCS 1197, (1997), 166–180.
9. G. Handler, Minimax location of a facility in an undirected tree graph, Transportation Sciences, 7 (1973), 287–293.
10. C.G. Lekkerkerker and J.C. Boland, Representation of a finite graph by a set of intervals on the real line, Fundamenta Mathematicae, 51 (1962), 45–64.
11. D. Rose, R.E. Tarjan and G. Lueker, Algorithmic aspects on vertex elimination on graphs, SIAM J. on Computing, 5 (1976), 266–283.
12. R. Seidel, On the all-pair-shortest-path problem, Proceedings of the 24th ACM Symposium on Theory of Computing, 1992, 745–749.
13. Uri Zwick, Exact and Approximate Distances in Graphs - A survey, "Algorithms - ESA'01 " 9th Annual European Symposium, Aarhus, Denmark, August 2001, Springer, LNCS 2161, (2001), 33–48.

k-pseudosnakes in Large Grids*

Martín Matamala[1], Erich Prisner[2], and Ivan Rapaport[1]

[1] Departamento de Ingeniería Matemática and Centro de Modelamiento Matemático UMR 2071-CNRS, Universidad de Chile, Santiago, Chile.
{irapapor,mmatamal}@dim.uchile.cl

[2] University of Maryland University College, Schwäbisch Gmünd, Germany.
erich.prisner@aiug.de

Abstract. We study the problem of finding maximum induced subgraphs of bounded maximum degree k—so-called "k-pseudosnakes"— in D-dimensional grids with all side lengths large. We prove several asymptotic upper bounds and give several lower bounds based on constructions. The constructions turn out to be asymptotically optimal for every D when $k = 0, 1, D, 2D-2, 2D-1$ and $2D$.

1 Introduction

Many combinatorial optimization problems can be reformulated as the searching of a maximum independent set of a graph G, i.e. a set S of vertices inducing an edgeless subgraph. What happens if this total isolation of the vertices is not required, but instead we ask that every vertex of S is adjacent to at most k vertices of S in G?

Following the terminology of [5], we call an induced subgraph $G[S]$ of a graph G a *k-pseudosnake* if its maximum degree is at most k. The maximum size of the vertex set of a k-pseudosnake in a graph G is denote by $\alpha_k(G)$. In [7] it is shown that $\alpha_k(G) \geq \sum_{v \in G}(d_G(v) + \frac{1}{k})^{-1}$. Some authors have studied the relation of k-pseudosnakes with other properties such as the chromatic number [1] and the domination number [8]. For the particular graph consisting on the product of D complete graphs on n vertices K_n^D, good bounds have been obtained in [3]: $\alpha_k(K_n^D) \leq (1 + \frac{1}{D-1})n^{D-1}$ for $n \geq 2$ and $D \geq 2$, while $\alpha_k(K_n^D) \geq n^{D-1} + n^{\frac{D}{2}}$ for $n \geq 3$ and $D \geq 2$.

Nevertheless, most of the work has been devoted to the problem of searching 2-pseudosnakes in highly regular graphs (and in fact many results have been obtained for the even more restricted case of 2-pseudosnakes with induced degree exactly 2 [2,6,13,4,11,14]). A basic sharp result is known for the D-dimensional hypercube Q^D. In [5] it is proved that $\alpha_2(Q^3) = 6$, $\alpha_2(Q^4) = 9$ and $\alpha_2(Q^D) = 2^{D-1}$for $D \geq 5$.

Since the largest 2-pseudosnake in the D-dimensional hypercube contains half of the vertices, the following is a natural question: How much the density

* Partially supported by FONDAP on Applied Mathematics, Fondecyt 1990616 (I. R.) and Fondecyt 1010442 (M. M.).

S. Rajsbaum (Ed.): LATIN 2002, LNCS 2286, pp. 224–235, 2002.

of the 2-pseudosnake can increase if we consider grids instead of hypercubes? Notice that hypercubes are special kind of grids. Indeed, the grid P_n^D is the Cartesian product of D paths P_n on n vertices, and the hypercube Q^D is P_2^D.

Let us define the asymptotic density λ_k^D of k-pseudosnakes in D-dimensional grids as follows

$$\lambda_k^D = \overline{\lim}_{n\to\infty} \frac{\alpha_k(P_n^D)}{n^D}$$

The goal of this paper is to find sharp bounds for λ_k^D. In Section 2 we study upper bounds (we include some results valid for arbitrary graphs). In particular, we prove that $\lambda_2^3 \le 0.5373$ and $\lambda_2^4 \le 0.5128$. In Section 3 we exhibit constructions that yield lower bounds. In particular, we prove that $\lambda_2^3 \ge 0.5092$ and $\lambda_2^4 \ge 0.5008$.

2 Upper Bounds

Before concentrating on grids, let us establish some inequalities that hold for general graphs.

Proposition 1. *For every graph G and every integer $k \ge 1$*

$$\alpha_{k+1}(G) \ge \alpha_k(G) \ge \begin{cases} \frac{3}{5}\alpha_{k+1}(G) - 3 & \text{for } k \le 5 \\ \frac{k+1-\ln(k+2)}{k+2}\alpha_{k+1}(G) - 1 - \ln(k+2) & \text{for } k > 5 \end{cases}$$

Proof: The first inequality is obvious since every k-pseudosnake is also a $k+1$-pseudosnake. For the second one, let S be a $(k+1)$-pseudosnake in G. We add some complete graph K_{k+2} to $G[S]$, and join vertices of S to sufficiently many vertices of K_{k+2} such that each vertex of S has degree $k+1$ in the resulting graph G'. Certainly this graph has minimum degree $k+1$, so we can apply theorems in [9] and [10], assuring that G' has a dominating set U of cardinality at most $\frac{2(k+2+|S|)}{5}$ for $k \le 5$ and $\frac{(k+2+|S|)(1+\ln(k+2))}{k+2}$ for $k > 5$. By the construction of G', deleting U in $G[S]$ results in a graph of maximum degree at most k. This graph has at least $\frac{3|S|-2k-4}{5}$ vertices for $k \le 5$ and at least $\frac{|S|(k+1-\ln(k+2))-(k+2)(1+\ln(k+2))}{k+2} \ge \frac{k+1-\ln(k+2)}{k+2}\alpha_{k+1} - 1 - \ln(k+2)$ vertices for $k > 5$. □

Proposition 2. *Let G be a graph where $\alpha|V|$ of the vertices have degree smaller than $\Delta(G)$. Then*

$$\alpha_k(G) \le \frac{\Delta(G) + \alpha(\Delta(G) - \delta(G))}{2\Delta(G) - k}|V|.$$

Proof: Let S be a k-pseudosnake in G. Let m denote the number of edges between S and $V \setminus S$. Let $S' = \{x \in S : d_G(x) < \Delta(G)\}$. Since each $x \in S$ has at least $d_G(x) - k$ neighbors in $V \setminus S$, we get

$$\begin{aligned} m &\geq \sum_{x\in S}(d_G(x)-k) \\ &\geq |S'|(\delta(G)-k)+|S\setminus S'|(\Delta(G)-k) \\ &= |S|(\Delta(G)-k)+|S'|(\delta(G)-\Delta(G)) \\ &\geq |S|(\Delta(G)-k)+\alpha|V|(\delta(G)-\Delta(G)). \end{aligned}$$

On the other hand, each vertex y in $V\setminus S$ has at most $d_G(y)$ neighbors in S. Thus $m \leq (|V|-|S|)\Delta(G)$. By putting both inequalities together, we get $|S| \leq \frac{\Delta(G)+\alpha(\Delta(G)-\delta(G))}{2\Delta(G)-k}|V|$. □

Previous properties yield the following results for grids.

Proposition 3. *For every integers $D \geq 1$ and $1 \leq k \leq 2D$,*

1. $\lambda^D_{k+1} \geq \lambda^D_k$
2. $\lambda^D_k \geq \begin{cases} \frac{3}{5}\lambda^D_{k+1} & \text{for } k \leq 5 \\ \frac{k+1-\ln(k+2)}{k+2}\lambda^D_{k+1} & \text{for } k > 5 \end{cases}$
3. $\lambda^D_k \geq \lambda^{D+1}_k$

Proof: (1) and (2) follow from Proposition 1. For (3), take some maximum k-pseudosnake S in some grid P_n^{D+1} By the pigeonhole principle, some of the sets $S_i := \{(x_1, x_2, \ldots, x_{D+1}) \in S : x_{D+1} = i\}$ has at least $\frac{|S|}{n}$ elements and it induces a k-pseudosnake in P_n^D. □

It is easy to see that $\lambda^D_k = \overline{\lim}_{n\to\infty}\alpha_k(C_n^D)$ where C_n is the cycle with n vertices. Therefore, we can assume that our graph is not a grid but a torus. In particular, from Proposition 2 we have the following.

Corollary 1. *For all integers $D \geq 1$ and $1 \leq k \leq 2D$, $\lambda^D_k \leq \frac{2D}{4D-k} = \frac{1}{2-\frac{k}{2D}}$.*

By making use of the special structure of our graphs, the previous bound can be improved for all cases $k < D$. We need the following terminology: Any given k-pseudosnake S in a graph G partitions the set of edges into the set C of those edges contained in $G[S]$, the set T of those edges touching exactly one vertex in S and the set N of those edges with no end in S.

In our graphs we call two edges *parallel* if they are opposite edges of some 4-cycle.

Theorem 1. *For every $k < D$ we have that $\lambda^D_k \leq \frac{2(D-1)D}{4D(D-1)-k(k-1)} = \frac{1}{2-\frac{k}{2D}\frac{k-1}{D-1}}$.*

Proof: Let S be a k-pseudosnake in the torus $G = C_n^D$. The graph G is $2D$-regular and the incident edges with a vertex $x \in S$ are either contained in S ($e \in C$) or touched by S ($e \in T$). More precisely,

$$2D|S| = \sum_{x\in S} d_G(x) = 2|C| + |T|.$$

Since N, T, C is a partition of the edges in V we have that $D|V| = |N|+|T|+|C|$. Therefore

$$\frac{|S|}{|V|} = \frac{2|C|+|T|}{2|N|+2|T|+2|C|} = \frac{1}{1+\frac{2|N|+|T|}{2|C|+|T|}}$$

Since S induces a k-pseudosnake in G, every vertex x of S meet at most k members of C. Since members of C are counted twice in this way, we obtain $2|C| \leq k|S|$. Moreover, every vertex x of S meet at least $2D-k$ members of T which implies that $(2D-k)|S| \leq |T|$. Both inequalities combined yield

$$2a|C| \leq |T|$$

where $a = \frac{2D}{k} - 1$.

Let xy be an edge in C, i.e. $x, y \in S$. If t denotes the number of parallel edges of xy in T, and c the number of parallel edges of xy in C, then $2k \geq d_{G[S]}(x) + d_{G[S]}(y) \geq 2+t+2c$. Since each edge of G has $2(D-1)$ parallel edges we have that every edge in C has at least $2(D-1)-t-c \geq 2(D-1)-(2k-2) = 2(D-k)$ parallel edges in N. Therefore $2(D-k)|C| \leq 2(D-1)|N|$ and

$$b|C| \leq |N|$$

where $b = \frac{D-k}{D-1} = 1 - \frac{k-1}{D-1} < 1$.

Let $e = \frac{a+b}{a+1}$. Then $e < 1$ and it holds that $e = b + a(1-e)$. We get the following bounds

$$2|N|+|T| = 2|N|+(1-e)|T|+e|T| \geq 2b|C|+2a(1-e)|C|+e|T| = e(2|C|+|T|)$$

Therefore $\frac{2|N|+|T|}{2|C|+|T|} \geq e$ which implies $\frac{|S|}{|V|} \leq \frac{1}{1+e} = \frac{1}{2-(1-e)}$. Finally

$$\begin{aligned}\frac{|S|}{|V|} &\leq \frac{1}{2-\frac{1}{a+1}(1-b)} \\ &= \frac{1}{2-\frac{k}{2D}\frac{k-1}{D-1}} \\ &= \frac{2D(D-1)}{4D(D-1)-k(k-1)}\end{aligned}$$

□

The main idea in the proof of Theorem 1 was to concentrate on the N-parallels of a given edge f. Now we are going to refine this method by having a closer look on the parallels in C for those edges in N parallel to f.

For instance, for $D = 3$ and $k = 2$, although every edge in N may have up to 4 parallels in C, this is impossible for both N-parallels of $f \in C$ in the situation given in Figure 1. (In the other situation where f has just two N-parallels shown in Figure 1, both these parallels may have 4 C-parallels each, so we need to look into this a little more carefully, see the proof of Theorem 3.)

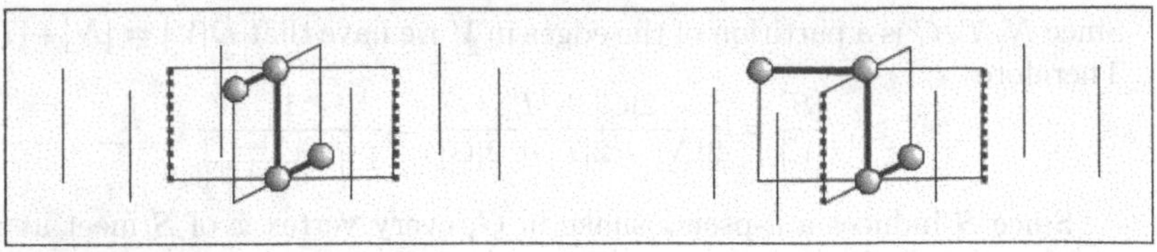

Fig. 1. a) The two N-parallels to f can not have four C-parallels. b) The two N-parallels to f can have four C-parallel.

The following Theorem improves the bound of Theorem 1 for the cases $k \leq \frac{2}{3}D$.

Theorem 2. *Let $k < D$. Then*

$$\lambda_k^D \leq \frac{1}{2 - \frac{k}{2D}\varphi(D,k)}.$$

where $\varphi(D,k) = \frac{k^2}{(2(D-1)-k)^2+k^2}$.

Proof: The bound is similar to those already obtained in Corollary 1 and Theorem 1. The reader should remark that we only have to prove that

$$|N| \geq \left(1 - \frac{k(k-1)}{2(D-k)^2 + k(k-1)}\right)|C|$$

For an edge f let us denote P_f the set of all parallel edges to f. We use the abbreviations $N_f := P_f \cap N$, $C_f := P_f \cap C$, $T_f := P_f \cap T$, $\lambda := |N_f|$ and $\mu := 2(D-1)$.

Since for every $f \in C$ and $h \in N$ we have that $h \in N_f$ if and only if $f \in C_h$, we deduce that

$$|N| = \sum_{h,h\in N}\sum_{f,f\in C_h}\frac{1}{|C_h|} = \sum_{f,f\in C}\sum_{h,h\in N_f}\frac{1}{|C_h|}$$

If for all $f \in C$ we have that $\sum_{h,h\in N_f}\frac{1}{|C_h|} \geq b$, where b does not depend on f, we would obtain $|N| \geq b|C|$.

Since for every set of n positive numbers we have that $(\sum_{i=1}^{n} x_i)(\sum_{i=1}^{n} x_i^{-1}) \geq n^2$, in order to find a lower bound for $\sum_{h,h\in N_f}\frac{1}{|C_h|}$ we have to find an upper bound for $\sum_{h,h\in N_f}|C_h|$. This quantity is the number of edges in C parallel to some parallel edge of f.

Since an edge g in T_f has at most $k-1$ parallels in C, among all edges parallel to g and to some h in N_f, at least $\lambda - 1 - (k-1)$ are not in C. Since an edge in C_f has at most $k-2$ parallels in C, at least $\lambda - 1 - (k-2)$ edges parallel to g and to some $h \in N_f$ are not in C. Clearly, these quantities are positive for $k \leq \frac{2}{3}(D+1)$ since $\lambda \geq \mu - 2(k-1)$. Moreover, we have $|T_f| + |C_f| = \mu - \lambda$. Therefore

$$\sum_{h, h \in N_f} |C_h| \leq \mu\lambda - |T_f|(\lambda - 1 - (k-1)) - |C_f|(\lambda - 1 - (k-2))$$
$$\leq \mu\lambda - (\mu - \lambda)(\lambda - k)$$
$$= \lambda^2 + k(\mu - \lambda)$$

Therefore we get

$$\sum_{h, h \in N_f} \frac{1}{|C_h|} \geq \frac{\lambda^2}{\lambda^2 + k(\mu - \lambda)}$$

The right hand side term is monotonically increasing for $\lambda = \mu - 2(k-1), \ldots, \mu$. Whence

$$|N| \geq \frac{(\mu - 2(k-1))^2}{(\mu - 2(k-1))^2 + 2k(k-1)} |C| = \left(1 - \frac{k(k-1)}{2(D-k)^2 + k(k-1)}\right)|C|$$

□

For $D = 3, k = 2$, we can even improve the bound of $6/11 \approx 0.545$ obtained in Theorem 1 and Theorem 2 slightly:

Theorem 3. $\lambda_2^3 \leq 36/67 \approx 0.537$.

Proof: We will show that for every two incident edges f and f' in C we have

$$s(f, f') := \sum_{h, h \in N_f} \frac{1}{|C_h|} + \sum_{h, h \in N_{f'}} \frac{1}{|C_h|} \geq \frac{7}{6}$$

If $|N_f| + |N_{f'}| \geq 5$ then $s(f, f') \geq \frac{5}{4} \geq \frac{7}{6}$, since $|C_h| \leq 4$. Otherwise $|N_f| + |N_{f'}| = 4$. In this situation f and f' are edges of some 4-cycle. If there are two adjacent edges h and h' with $h \in N_f$ and $h' \in N_{f'}$ then $|C_h| \leq 3$ and $|C_{h'}| \leq 3$ which gives $s(f, f') \geq \frac{7}{6}$. So we can assume that h and h' are not adjacent for any pair h, h' with $h \in N_f$ and $h' \in N_{f'}$.

To study this situation we need some notations. Let us denote by h_1, h_2, h_3, h_4 the parallel edges to f and by h'_1, h'_2, h'_3, h'_4 those parallel to f'. Moreover, let us assume that $h_1, h_2 \in N_f$, $h'_1, h'_2 \in N_{f'}$, h_3 is the edge adjacent to f' and h'_3 is the edge adjacent to f. Additionally, let us assume that the edge h_1 belongs to the plane defined by f and h_3 and that h'_1 belongs to the plane defined by f' and h'_3. Remember that we are assuming that h_2 and h'_2 are not adjacent. Let g be the edge different from f in $P_{h_2} \cap P_{h_3}$. If $g \in C$ then the edge r' different from f' in $P_{h'_1} \cap P_{h'_4}$ can not belong to C, due to the degree constrain in the vertex incident with g and h'_4. Therefore $|C_{h_2}| + |C_{h'_1}| \leq 7$. Similarly we can show that $|C_{h'_1}| + |C_{h_2}| \leq 7$ which implies that $s(f, f') \geq \frac{7}{6}$.

Finally we have that $2|N| = \sum_{f,f' \in C} s(f,f') \geq \frac{7}{6}|C|$. Thus

$$|N| \geq \frac{7}{12}|C|$$

and we obtain the bound $\lambda_2^3 \leq \frac{1}{2-\frac{2}{6}\frac{5}{12}} = \frac{36}{67}$ as indicated in the proof of Theorem 1. □

3 Lower Bounds

We start this section by showing some fairly simple constructions which imply that, at least for certain cases, the previously given bounds are sharp.

Let us label the vertices of the torus $G = C_n^D$ by $(x_1, x_2, \ldots, x_D)$ with $0 \leq x_i \leq n-1$ and $i = 1, \ldots, D$. Three canonical induced subgraphs of the D-dimensional torus will turn out to be optimal solutions for certain k's. The first graph is G itself and we have $\lambda_{2D}^D = 1$. The second graph is the checkerboard subgraph. Let S be the set of those vertices $(x_1, x_2, \ldots, x_D)$ of G for which $\sum_{i=1}^D x_i$ is odd. This set is independent and contains asymptotically half of the vertices of the torus. The optimality of this construction for $k = 0$ and $k = 1$ follows from Theorem 1. More precisely,

Corollary 2. *$\lambda_0^D = \lambda_1^D = 1/2$ for $D \geq 2$.*

The third subgraph, $G[S]$, is induced by the set S of all the vertices $x \in C_n^D$ for which the sum $\sum_{i=1}^D x_i$ of its coordinates equals 0 or 1 modulo 3. Surely $|S|/|V| \to 2/3$ for $n \to \infty$. Now consider $x = (x_1, x_2, \ldots, x_D) \in S$. If $\sum_{i=1}^D x_i = 0 \pmod 3$, then each neighbor $(x_1, \ldots, x_{i-1}, x_i + 1, x_{i+1}, \ldots, x_D)$ is in S, but each neighbor $(x_1, \ldots, x_{i-1}, x_i - 1, x_{i+1}, \ldots, x_D)$ is not. Thus x has exactly one neighbor in S in each dimension. The case $\sum_{i=1}^D x_i = 1 \pmod 3$ is analogous. Using Corollary 1 we get

Corollary 3. *For every D we have $\lambda_D^D = \frac{2}{3}$.*

For getting more general bounds let us fix $p > D$ and $a = (1, 2, \ldots, D)$. Let us define the linear function $f(x) = xa^t = \sum_{i=1}^D ix_i$, where $x = (x_1, x_2, \ldots, x_D) \in G = C_n^D$ and all the calculations are taken modulo p. For each $0 \leq i \leq p-1$, we define S_i as the set of vertices corresponding to $f^{-1}(\{i\})$. Since $p > D$, every S_i is independent. On the other hand, all the sets of the partition $S_0, S_1, \ldots, S_{p-1}$ have "almost" the same cardinality. Moreover, for the n's which are multiple of p, all the sets satisfy $|S_i| = \frac{n^D}{p}$. To see this, notice that if we fix $x_2, \ldots, x_D$ then the value of x_1 will determine to which class the vertex $x = (x_1, x_2, \ldots, x_D)$ belongs. We will consider the particular cases for which $p = 2D+1$ and $p = D+1$.

Theorem 4.

1. *For every integer k, $1 \leq k \leq 2D$, $\lambda_k^D \geq \frac{k+1}{2D+1}$.*
2. *For every integer k, $1 \leq k \leq D$, $\lambda_{2k}^D \geq \frac{k+1}{D+1}$.*

Proof: For the first inequality let $p = 2D + 1$. Let n be a multiple of $2D + 1$ and let $G = C_n^D$. Let us consider the graph $G[\bigcup_{i=0}^k S_i]$. Its maximum degree is at most k. In fact, let $x \in S_i$ and let $i' \neq i$. In order to find a vertex $x' \in S_{i'}$ such that $xx' \in E(G)$, we have to solve the equation $i' = i + y \pmod{2D+1}$, where $y \in \{1, \cdots, D\} \cup \{-1, \cdots, -D\}$. It's easy to notice that this equation admits only one solution. It follows that the number of vertices is $(k+1)n^D/2D+1$ and the first inequality is concluded.

We can improve the bound when the pseudosnake degree is bounded by an even number by taking $p = D + 1$. In fact, in this case the maximum degree of the graph $G[\bigcup_{i=0}^k S_i]$ is at most $2k$: The equation $i' = i + y \pmod{D+1}$, where $y \in \{1, \cdots, D\} \cup \{-1, \cdots, -D\}$, admits two solutions. It follows that the number of vertices is $(k+1)n^D/D+1$ and the second inequality is concluded. □

Using again Corollary 1 we have

Corollary 4. *$\lambda_{2D-1}^D = \frac{2D}{2D+1}$, and $\lambda_{2D-2}^D = \frac{D}{D+1}$.*

We know from Theorem 4 that, for every $1 \leq k \leq D$, $\lambda_{2k}^D \geq \frac{k+1}{D+1}$. But we can remove one third of the vertices of the pseudosnake in such a way that the degree is reduced to half the original, i.e:

Theorem 5. *For every integer $1 \leq k \leq D$, $\lambda_k^D \geq \frac{2}{3}(\frac{k+1}{D+1})$.*

Proof: We are in the context of the second construction of Theorem 4. Let us consider the sets $A_l \subseteq \bigcup_{i=0}^k S_i$ of those x's for which $\sum_{i=1}^D x_i = l \pmod 3$ where $l = 0, 1, 2$. Without loss of generality we can assume that $|A_0| \leq (\frac{1}{3})n^D$ (otherwise we choose the sum to be 1 or 2 instead of 0). All we need to prove is that for any vertex belonging to either A_1 or A_2, at least half of its neighbors belong to A_0. Therefore, by removing A_0, we reduce in one half the degree of the graph. Let $x = (x_1, x_2, x_3, \ldots, x_D) \in A_1 \cap S_r$. It follows that, for each $r' \neq r$, x has two neighbors in $S_{r'}$: $(x_1, \ldots, x_{r'-r} + 1, \ldots x_D) \in A_2$ and $(x_1, \ldots, x_{r'+r} - 1, \ldots x_D) \in A_0$. □

Since $\frac{2}{3}(\frac{k+1}{D+1}) \geq \frac{k+1}{2D+1}$ for all $D \geq 1$ previous result is an improvement of our bound in Theorem 4 for all $k \leq D$.

We will now concentrate on torus-hypercube constructions. The idea is very simple: k-pseudosnakes for tori C_n^D yield k-pseudosnakes for tori $C_{\alpha n}^D$. Let us recall that Q^D denotes the D-dimensional hypercube.

Proposition 4. *Let G be a fixed graph. Let $D \geq 1$. Assume that S as well as its complement $\overline{S}$ induce k-pseudosnakes in $Q^D \times G$. Then there exist complementary k-pseudosnakes S' and $\overline{S'}$ in $C_6^D \times G$ such that $|S'| \geq |S| + (3^D - 1)2^{D-1}|V(G)|$.*

Proof: By induction on the dimension D. Let D=1. We know that both $S \subseteq \{0,1\} \times V(G)$ and its complement $\overline{S} = (\{0,1\} \times V(G)) \setminus S$ induce k-pseudosnakes in $Q^1 \times G$. It is useful to see $Q^1 \times G$ as "two layers of G". Let S_0 and S_1 be the part of S in each layer. In other words, $S_0 = \{x \in V(G) : (0, x) \in S\}$ and $S_1 = \{x \in V(G) : (1, x) \in S\}$. Notice that $|S_0| + |S_1| = |S|$. Since $C_6^D \times G$

is a six layers graph, the k-pseudosnake S' will be constructed as a six layers pseudosnake. More precisely, $S'_0 = S_0$, $S'_1 = S_1$, $S'_2 = \overline{S_1}$, $S'_3 = \overline{S_0}$, $S'_4 = \overline{S_1}$, $S'_5 = S_1$. It is easy to notice that both S' and $\overline{S'}$ are k-pseudosnakes. On the other hand, $|S'| = |S_0| + |S_1| + |\overline{S_1}| + |\overline{S_0}| + |\overline{S_1}| + |S_1| = 2|V(G)| + |S|$.

Let us assume the proposition true for D. Let S induce a k-pseudosnake in $Q^{D+1} \times G$ such that $\overline{S}$ is also a k-pseudosnake. Since $Q^{D+1} \times G = Q^D \times (Q^1 \times G)$ then, by the induction hypothesis, we know the existence of a k-pseudosnake S' in $C_6^D \times (Q^1 \times G)$ such that $|S'| \geq |S| + (3^D - 1)2^{D-1}|V(Q^1 \times G)|$ and with $\overline{S'}$ in $C_6^D \times (Q^1 \times G)$ also being a k-pseudosnake. Since $C_6^D \times (Q^1 \times G) = Q^1 \times (C_6^D \times G)$ we can apply the result for the case $D = 1$ and deduce the existence of a k-pseudosnake S'' in $C_6^{D+1} \times G$ with $\overline{S''}$ also being a k-pseudosnake and such that

$$\begin{aligned}
|S''| &\geq |S'| + 2|V(C_6^D \times G)| \\
&\geq |S| + (3^D - 1)2^{D-1}|V(Q^1 \times G)| + 2|V(C_6^D \times G)| \\
&= |S| + (3^D - 1)2^{D-1}2|V(G)| + 2 \cdot 6^D|V(G)| \\
&= |S| + (3^{D+1} - 1)2^{D+1-1}|V(G)|.
\end{aligned}$$

□

Therefore, lower bounds for k-pseudosnakes in Q^D would be highly desirable. For $k = 2$, everything is known:

Theorem 6 (Danzer/Klee [5]). $\alpha_2(Q^3) = 6$, $\alpha_2(Q^4) = 9$. *For every* $D \geq 5$, $\alpha_2(Q^D) = 2^{D-1}$.

Corollary 5. $\alpha_2(C_6 \times C_6 \times C_6) \geq 110$ *and* $\alpha_2(C_6 \times C_6 \times C_6 \times C_6) \geq 649$.

Proof: We apply Proposition 4. For the first equation we use the 6-vertex 2-pseudosnake in Q^3, whose complement turns out to be also a 2-pseudosnake. For the second equation we take the 9-vertex 2-pseudosnake in Q^4, shown in Figure 2. □

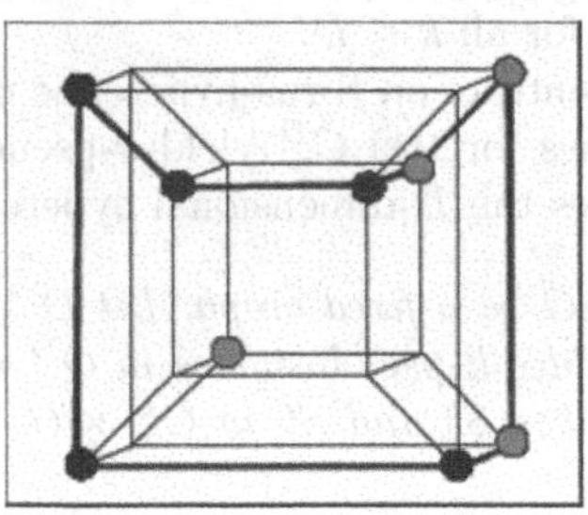

Fig. 2. A 9-vertex pseudosnake in the cube Q^4.

Corollary 6. $\lambda_2^3 \geq 110/216 \approx 0.509$, *and* $\lambda_2^4 \geq 649/1296 \approx 0.5008$.

For arbitrary $k > 2$, k-pseudosnakes in hypercubes have not yet been investigated. But for every k-pseudosnake S in Q^D, the set $S' = \{0,1\} \times S$ is a $(k+1)$-pseudosnake in Q^{D+1} and $|S'| = 2|S|$. Therefore each useful lower bound for $\alpha_k(Q^D)$ yields lower bounds for all $\alpha_{k+t}(Q^{D+t})$. Whence $\alpha_{2+t}(Q^{3+t}) \geq 2^t \cdot 6$ and $\alpha_{2+t}(Q^{4+t}) \geq 2^t \cdot 9$, for each $t \geq 0$. The second family of inequalities can be improved. In fact, the construction of Figure 3 gives the lower bounds $\alpha_{3+t}(Q^{5+t}) \geq 2^t \cdot 20$ for each integer $t \geq 0$.

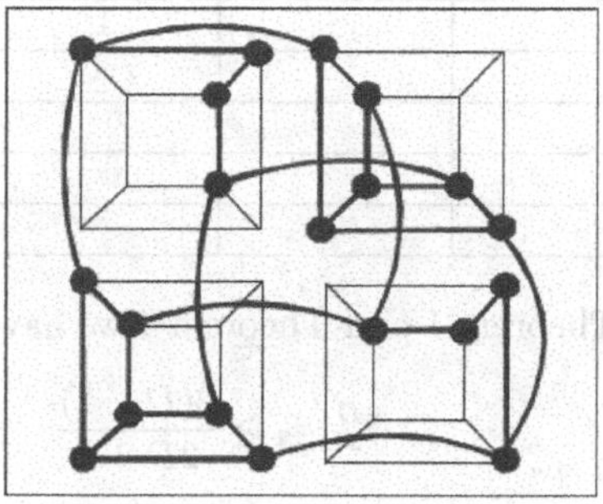

Fig. 3. A 20-vertex 3-pseudosnake in Q^5.

Corollary 7. *For every* $t \geq 0$, $\lambda_{2+t}^{3+t} \geq \frac{2^{t+1}+3*6^{2+t}}{6^{3+t}}$ *and* $\lambda_{3+t}^{5+t} \geq \frac{2^{t+2}+3*6^{4+t}}{6^{5+t}}$.

The first bounds are much weaker than that in Theorem 5. The second bounds are very close to 0.5 but for three values still the best we have at the moment. We get $\lambda_3^5 \geq 3892/7776 \approx 0.5005$, $\lambda_4^6 \geq 23336/46656 \approx 0.50017$, $\lambda_5^7 \geq 139984/279936 \approx 0.500057$, and the others are again much weaker than the bound in Theorem 5.

4 Conclusion

The current lower and upper bounds for λ_k^D for $D \leq 7$ can be seen in the following table.

k	$D=1$	$D=2$	$D=3$	$D=4$	$D=5$	$D=6$	$D=7$
0	.5	.5	.5	.5	.5	.5	.5
1	.667	.5	.5	.5	.5	.5	.5
2	1	.667	.509 − .537	$.500_{8} - .512_{8}$	.5 − .505	.5 − .502	.5 − .501
3		.8	.667	.533 − .571	$.500_{5} - .534_{4}$	.5 − .516	.5 − .509
4		1	.75	.667	.556 − .588	$.500_{17} - .556$	.5 − .530
5			.857	.667 − .727	.667	.571 − .6	$.500_{057} - .568$
6			1	.8	.667 − .714	.667	.583 − .609
7				.889	.727 − .769	.667 − .706	.667
8				1	.833	.714 − .75	.667 − .7
9					.909	.769 − .8	.667 − .737
10					1	.857	.75 − .778
11						.923	.8 − .824
12						1	.875
13							.933
14							1

Notice that from Theorem 1 and Theorem 4 we have

$$\lambda_{2D-3}^{D} \geq \frac{2(D-1)}{2D+1}$$

and

$$\lambda_{2(D-1)-3}^{D-1} \leq \frac{2(D-1)}{4(D-1)-2(D-1)+3} = \frac{2(D-1)}{2D+1}$$

which explains some regularities in previous table.

References

1. Andrews, J., Jacobson M.: On a generalization of chromatic number and two kinds of Ramsey numbers. Ars Comb. 23 (1987) 97-102.
2. Abbott, H.L., Dierker, P.F.: Snakes in powers of complete graphs. SIAM J. Appl. Math. 32(2) (1977) 347-355.
3. Abbott H.L., Katchalski, M.: Snakes and pseudo-snakes in powers of complete graphs. Discrete Math. 68 (1988) 1-8.
4. Abbott H.L., Katchalski, M.: Further results on snakes in powers of complete graphs. Discrete Math. 91 (1991) 111-120.
5. Danzer, L., Klee, V.: Lengths of snakes in boxes. J. Comb. Th. 2 (1967) 258-265.
6. Deimer, K.: A new upper bound for the length of snakes. Combinatorica 5(2) (1985) 109-120.
7. Favaron, O.: k-Domination and k-independence in graphs. Eleventh British Combinatorial Conference (London, 1987). Ars Combin. 25 (1988) C 159-167.
8. Fink, J., Jacobson, M.: On n-domination, n-dependence and forbidden subgraphs. Graph Theory with Applications to Algorithms and Computer Science, Proceedings 5th Int. Conf., Kalamazoo/MI (1985) 301-311.
9. McCuaig, W., Shepherd, B.: Domination in graphs with minimum degree two. J. Graph Theory 13 (1989) 749-762.

10. Payan, C.: Sur le nombre d'absorption d'un graphe simple. Cah. Centre Etud. Rech. Oper. 17 (1975) 307-317.
11. Snevily, H.S.: The snake-in-the-box problem: A new upper bound. Discrete Math. 133 (1994) 307-314.
12. Solov'jeva, F.I.: An upper bound for the length of a cycle in an n-dimensional unit cube. Diskretnyi Analiz 45 (1987).
13. Wojciechowski, J.: A new lower bound for snake-in-the-box codes. Combinatorica 9(1) (1989) 91-99.
14. Zémor, G.: An upper bound on the size of snake-in-the-box. Combinatorica 17(2) (1997) 287-298.

$L(2,1)$-Coloring Matrogenic Graphs
(Extended Abstract)

Tiziana Calamoneri and Rossella Petreschi

Department of Computer Science, University of Rome "La Sapienza" - Italy
via Salaria 113, 00198 Roma, Italy.
{calamo, petreschi}@dsi.uniroma1.it

Abstract. This paper investigates a variant of the general problem of assigning channels to the stations of a wireless network when the graph representing the possible interferences is a matrogenic graph. In this problem, channels assigned to adjacent vertices must be at least two apart, while the same channel can be reused for vertices whose distance is at least three. Linear time algorithms are provided for matrogenic graphs and, in particular, for two specific subclasses: threshold graphs and split matrogenic graphs. For the first one of these classes the algorithm is exact, while for the other ones it approximates the optimal solution. Consequently, improvements on previously known results concerning subclasses of cographs, split graphs and graphs with diameter two are achieved.

keywords: $L(2,1)$-Coloring, Threshold Graphs, Matrogenic Graphs, Split Graphs, Cographs.

1 Introduction

The $L(2,1)$-*coloring problem* consists in an assignment of colors from the integer set $0, \ldots, \lambda_{2,1}$ to the vertices of a graph G such that vertices at distance at most two get different colors and adjacent vertices get colors which are at least two apart. The aim is to minimize $\lambda_{2,1}$. For some special classes of graphs – such as paths, cycles, wheels, tiling and k-partite graphs – tight bounds for the number of colors necessary for a $L(2,1)$-coloring are known and such a coloring can be computed efficiently [2,4,6,7,11]. Nevertheless, in general, both determining the minimum number of necessary colors [11] and deciding if this number is $< k$ for every fixed $k \geq 4$ [9] is NP-complete. Therefore, for many classes of graphs – such as chordal graphs [6,15], interval graphs [7], split graphs [1], outerplanar and planar graphs [1,3] – approximate bounds have been looked for.

In this paper we consider the $L(2,1)$-coloring problem restricted to *matrogenic graphs*, proved to be characterized by their own degree sequence up to isomorphism [14].

Matrogenic graphs can be considered as a superclass including *split matrogenic graphs* and *threshold graphs* (see Sect. 2 for definitions).

S. Rajsbaum (Ed.): LATIN 2002, LNCS 2286, pp. 236–247, 2002.

In the following we present linear time algorithms to $L(2,1)$-color these graphs, taking advantage from the degree sequence's analysis. Namely, first an algorithm for threshold graphs is provided, and then it is modified for split matrogenic and matrogenic graphs.

For the first one of these classes the algorithm is exact, while for the other one it approximates the optimal solution. These algorithms improve results presented in the literature.

In particular, threshold graphs are a subclass of *cographs*, i.e. graphs not containing P_4 as subgraph. For cographs Chang and Kuo [7] proved that the $L(2,1)$-coloring problem is polynomially solvable, but they provide a theoretical result, while our algorithm constructively finds an optimal solution.

Moreover, we prove some upper bounds for $\lambda_{2,1}$ that are linear in Δ, the maximum degree of the graph. Then the upper bound $\lambda_{2,1} \leq \Delta^{1.5} + 2\Delta + 2$, showed by Bodlaender et al. in [1] for split graphs, is improved when the problem is restricted both to threshold and to split matrogenic graphs.

Finally, Griggs and Yeh [11] showed that for graphs with diameter 2 it holds $\lambda_{2,1} \leq \Delta^2$; threshold graphs have diameter 2, nevertheless for them our upper bound on $\lambda_{2,1}$ is linear in Δ.

This paper is organized as follows. In the next section we give definitions of the classes of graphs considered in the present work; then we present some properties useful in the successive sections.

In Sect. 3 a linear time algorithm for $L(2,1)$-coloring a threshold graph is presented; both its correctness and its optimality are claimed; indeed, in this extended abstract all the proofs are omitted for the sake of brevity; they can be found in [5]: the interested reader can contact one of the authors. Sect. 4 presents the algorithm for split matrogenic graphs and provides its performances. Finally, Sect. 5 is devoted to $L(2,1)$-color matrogenic graphs.

2 Graph Theory Preliminaries

Given a finite, simple, loopless graph $G = (V,E)$, a vertex $x \in V$ is called *universal* (*isolated*) if it is adjacent to all other vertices of V (no one vertex in V); if x is an universal (isolated) vertex, then its degree is $d(x) = n-1$ ($d(x) = 0$).

A graph $I = (V_I, E_I)$, where $V_I \subseteq V$ and $E_I = E \cap (V_I \times V_I)$ is said to be *induced* by V_I.

Let $DS(G) = \delta_1, \delta_2, \ldots, \delta_n$ be the degree sequence of a graph G sorted by non increasing values: $\delta_1 \geq \delta_2 \geq \ldots \geq \delta_n \geq 0$. Let m_i be the number of vertices with degree $\delta_i, 1 \leq m_i \leq n$. We call *boxes* the equivalence classes of vertices in G under equality of degree. In terms of boxes the degree sequence can be compressed in $d_1^{m_1}, d_2^{m_2}, \ldots d_r^{m_r}, d_1 > d_2 > \ldots > d_r \geq 0$, where d_i is the degree of the m_i vertices contained in box $B_i(G)$. We will say *isolated (universal)* a box containing only isolated (universal) vertices.

A graph I induced by subset $V_I \subseteq V$ is called

- *complete* or *clique* if any two distinct vertices in V_I are adjacent in G
- *stable* or *independent* if no two vertices in V_I are adjacent in G.

A graph G is said to be *split* if there is a partition $V = V_K \cup V_S$ of its vertices such that the induced subgraphs K and S are complete and stable, respectively.

For any graph G, let $N(x)$ be the set of x's adjacent vertices. Then, we define the *vicinal preorder* $\preceq$ on V as follows: $x \preceq y$ iff $N(x) - y \subseteq N(y) - x$.

2.1 Threshold Graphs

Definition 1. *[8,12] A graph $G = (V, E)$ is a threshold graph if and only if G is a split graph and the vicinal preorder on V is total, i.e. for any pair $x, y \in V$, either $x \preceq y$ or $y \preceq x$.*

Threshold graphs are very rich in structure (for details, see the book by Mahedev and Peled [13]). Among all their properties, we highlight that threshold graphs are a subclass of cographs, have diameter 2 and each induced subgraph of a threshold graph is threshold, i.e. thresholdness is a hereditary property.

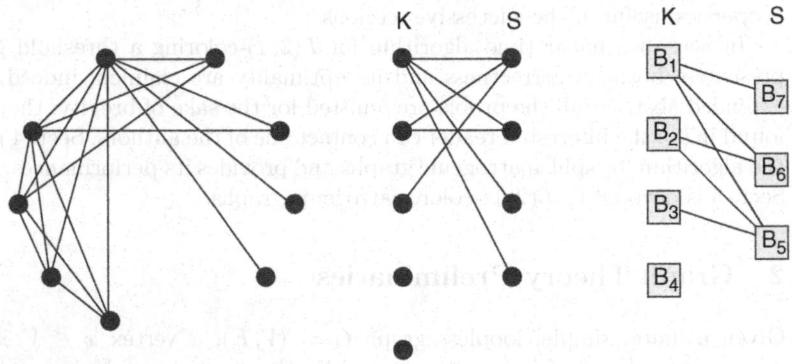

Fig. 1. A threshold graph.

In Fig. 1.a an example of a threshold graph with 9 vertices is depicted. Its degree sequence is 8, 6, 5, 4, 4, 3, 2, 1, 1. In Fig. 1.b the same graph is represented highlighting vertices in the clique K and vertices in the stable S, and the adjacencies in K are represented by a rectangle for the sake of clearness. Finally, in Fig. 1.c the same graph is represented in terms of boxes with compressed degree sequence $8^1, 6^1, 5^1, 4^2, 3^1, 2^1, 1^2$. Observe that if G is connected, the box of maximum degree contains all the universal vertices of G.

From now on we will consider only connected threshold graphs. Indeed this limitation is not restrictive because a not connected threshold graph is constituted by a connected one plus an isolated box. Analogous considerations hold for the classes of matrogenic graphs; hence, from now on we will speak about connected graphs, even if we will not explicitly underline this fact.

2.2 Matrogenic Graphs

Before introducing matrogenic graphs, let us recall some definitions.

A set M of edges is a *perfect matching of dimension* h of X onto Y if and only if X and Y are disjoint subsets with the same cardinality h and each edge is incident to exactly one vertex $x \in X$ and to one vertex $y \in Y$, and different edges must be incident to different vertices. We say x and y *dually correlated* (see Fig. 2.a).

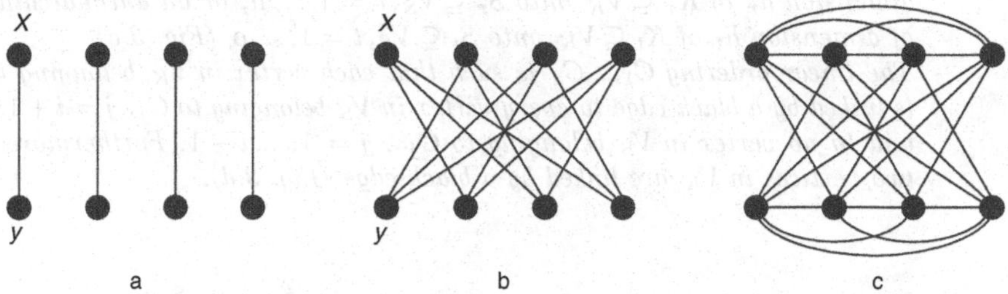

Fig. 2. a. a matching; b. an antimatching; c. a hyperoctahedron.

An *antimatching of dimension* h of X onto Y is a set A of edges such that $M(A) = X \times Y - A$ is a perfect matching of dimension h of X onto Y. $M(A)$ is said an *uncorrelated matching* and *dually uncorrelated* are two vertices endpoints of any edge in $M(A)$ (see Fig. 2.b). Observe that, by definition, a non trivial antimatching of dimension h must have $h \geq 3$, indeed $h = 1$ implies that the unique vertex in X (Y) is isolated, while $h = 2$ implies that the antimatching coincides with a matching.

A graph $G = (\{v_1, v_2, \ldots v_k\}, \emptyset)$ is a *null graph* if its edge set is empty, independently by the dimension of the vertex set.

Definition 2. *[10] A graph $G = (V, E)$ is* matrogenic *if and only if its vertex set V can be partitioned into three disjoint sets V_K, V_S, and V_C such that:*

1. *$V_K \cup V_S$ induces a split matrogenic graph in which K is the clique and S the stable;*
2. *V_C induces a* crown, *i.e. either a* perfect matching *or a* hyperoctahedron *(the complement of a perfect matching - see Fig. 2.c) or a* chordless C_5*;*
3. *each vertex in V_C is adjacent to each vertex in V_K and to no vertex in V_S.*

Matrogenicity is a hereditary property.

Now we have to characterize the class of *split matrogenic graphs*, i.e. matrogenic graphs in which $V_C = \emptyset$. It has been proved in [14] that a split matrogenic graph can be obtained by a superposition of a *black graph*, B, and a *red graph*, R. B is a threshold graph and R is the union of disjoint perfect matchings, antimatchings and null graphs. In Fig. 3.a a split matrogenic graph having degree sequence 11, 10, 10, 10, 7, 7, 6, 5, 5, 3, 3, 3 is presented and in Fig. 3.b its representation in terms of boxes with compact degree sequence $11^1, 10^3, 7^2, 6^1, 5^2, 3^3$ is depicted. Formally:

Theorem 1. *[14] A split graph G with clique K and stable S is matrogenic if and only if the edges of G can be colored red and black so that:*
- *The red partial graph is the union of vertex-disjoint pieces, $C_i, i = 1, ..., z$. Each piece is either a null graph N_j, $\sum_j |V_{N_j}| = \nu_K + \nu_S$, belonging either to K (exactly ν_K vertices) or to S (exactly ν_S vertices), or a matching M_r of dimension h_r of $K_r \subseteq V_K$ onto $S_r \subseteq V_S, r = 1, \ldots \mu$, or an antimatching A_t of dimension h_t of $K_t \subseteq V_K$ onto $S_t \subseteq V_S, t = 1, \ldots \alpha$ (Fig. 3.c).*
- *The linear ordering $C_1, ... C_z$ is such that each vertex in V_K belonging to C_i is linked by a black edge to every vertex in V_S belonging to C_j, $j = i+1, ..., z$ and to no vertex in V_S belonging to C_j , $j = 1, ..., i-1$. Furthermore, any two vertices in V_K are linked by a black edge (Fig. 3.d).*

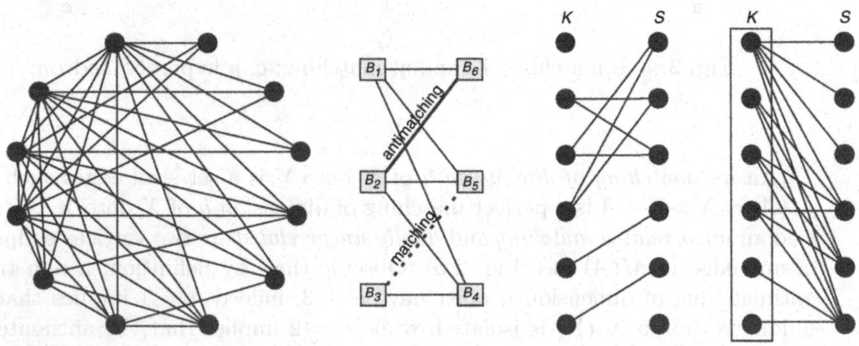

a b c d

Fig. 3. A split matrogenic graph (a) and its representation in terms of boxes (b); c. its red graph; d. its black graph.

The definition of perfect matching and Theorem 1 make not restrictive to suppose that any matching M_r has dimension at least 2; indeed, if M_r of $K_r = \{v_r\}$ onto $S_r = \{w_r\}$ has dimension 1, it is possible to color edge (v_r, w_r) black instead of red, and to transform v_r and w_r into null graphs.

We can characterize split matrogenic graphs also considering their representation in terms of boxes in the following way:

Theorem 2. *[14] If G is a split matrogenic graph, then one of the following four cases must occur:*
a. $B_r(G)$ consists of vertices of degree 0;
b. $B_1(G)$ consists of vertices of degree $n-1$;
c. $B_r(G)$ consists of t vertices of degree 1 and $B_1(G)$ consists of t vertices of degree $n-t$, $B_1(G) \cup B_r(G)$ induces a perfect matching of dimension t;
d. $B_r(G)$ consists of t vertices of degree ≥ 2 and $B_1(G)$ consists of t vertices of degree $n-2$, $B_1(G) \cup B_r(G)$ induces an antimatching of dimension t.

From the previous theorem, the following definition naturally arises: two boxes $B_i(G)$ and $B_j(G)$, $i \neq j$, of a matrogenic graph are *partially connected* if the graph induced by $B_i(G) \cup B_j(G)$ in the red graph is either a perfect matching or an antimatching. whenever it is neither partially connected

3 $L(2,1)$-Coloring Threshold Graphs

In the following we will deal with the representation of a graph $G=(V,E)$ in terms of boxes with degree sequence $d_1^{m_1}, d_2^{m_2}, \ldots, d_r^{m_r}, d_1 > d_2 > \ldots > d_r > 0$.

Note 1. [13] Any threshold graph can be derived from a one-vertex graph by repeatedly adding either an isolated vertex or an universal one. In other words, if $G_1 = \{v_1, \emptyset\}$, G_k is the threshold graph obtained by adding vertex v_k (either isolated or universal) to threshold graph G_{k-1}.

This note implies that it is possible to identify the structure of a threshold graph analyzing only its degree sequence. In particular, we exploit the decomposition of threshold graphs in boxes. The $L(2,1)$-coloring algorithm we are going to present proceeds by labeling the boxes according to their structure. Namely, it works on the current graph (G at the beginning) and colors vertices belonging to the extremal boxes $B_{imax}(G)$ and $B_{imin}(G)$ (at the beginning, $B_1(G)$ and $B_r(G)$, respectively). Notice that, at the first step, for the connectivity hypothesis, $B_{imin}(G)$ cannot be an isolated box. The derived *pruned graph* G_p is the subgraph induced by:

- $V - B_{imax}(G)$ if $B_{imax}(G)$ is universal;
- $V - B_{imin}(G)$ if $B_{imin}(G)$ is isolated.

Operatively, the pruning procedure works on the compressed degree sequence and transforms the degree sequence of G into the degree sequence of G_p by eliminating either $d_{imax}^{m_{imax}}$ or $d_{imin}^{m_{imin}}$. When $d_{imax}^{m_{imax}}$ is eliminated the degree of each other box B_i is decreased by m_{imax}. In the algorithm this procedure is called `Prune`$(G, imax, imin)$, where $imax$ and $imin$ indicate at each step the indices of the boxes to be considered. From the definition of pruned graph, observe that the third argument is equal to 0 if $B_{imax}(G)$ is an universal box while the second one is 0 when $B_{imin}(G)$ is an isolated box.

Before describing the algorithm for $L(2,1)$-coloring threshold graphs, let us partition colors $0, \ldots, \lambda_{2,1}$ into *even* and *odd* in trivial way. Furthermore, we say that color k is the *first odd (even) available* color if we have already used all odd (even) colors from 0 to $k-1$. Finally, we say that color k is *thrown out* if we decide not to use it; after k has been thrown out it is not available anymore.

The $L(2,1)$-coloring algorithm is the following:

ALGORITHM `Color-Threshold-Graphs`
INPUT: a connected threshold graph G by means of its degree sequence $d_1^{m_1}, \ldots d_r^{m_r}$
OUTPUT: a $L(2,1)$-coloring for G.

`Inizialize-Queue`: $Q = \emptyset$; $G \leftarrow G_p$; $imax \leftarrow 1$; $imin \leftarrow r$;
REPEAT
  Consider the boxes $B_{imax}(G_p)$ and $B_{imin}(G_p)$;
Step 1 ($B_{imax}(G_p)$ universal)
  IF $d_{imax} = (\sum_{j=imax}^{imin} m_i) - 1$
    THEN color the m_{imax} vertices with the first m_{imax} available even colors;
      FOR each $i = 1, \ldots m_{imax}$ DO
        IF `Queue-is-empty`
          THEN throw the first available odd color out
          ELSE $v \leftarrow$ `Extract-from-Queue`
            Color v with the first available odd color;
      $G_p \leftarrow$ `Prune`$(G_p, imax, 0)$; $imax \leftarrow imax + 1$;
    ELSE
Step 2 ($B_{imin}(G_p)$ isolated)
    IF $d_{imin} = 0$
      THEN `Enqueue`$(Q, B_{imin}(G_p))$;
        $G_p \leftarrow$ `Prune`$(G_p, 0, imin)$; $imin \leftarrow imin - 1$;
UNTIL $imax > imin$ (i.e. G_p is empty);
Step 3
    IF `Queue-is-not-empty`
      THEN color the k vertices in Q with the first k available
        consecutive colors (both odd and even).

Theorem 3. *Algorithm* `Color-Threshold-Graphs` *correctly* $L(2,1)$*-colors a connected threshold graph* G *with at most* $2\Delta + 1$ *colors.*

Theorem 4. *Algorithm* `Color-Threshold-Graphs` $L(2,1)$*-colors a connected threshold graph* G *with the minimum number of colors.*

Theorems 3.2 and 3.3 improve a set of previously known results: algorithm `Color-Threshold-Graphs` constructively finds an optimal solution for threshold graphs, a subclass of cographs, for which it was only known that the problem is in P [7].

Moreover, the upper bounds $\lambda_{2,1} \leq \Delta^{1.5} + 2\Delta + 2$ for split graphs [1] and $\lambda_{2,1} \leq \Delta^2$ for graphs with diameter 2 [11], are improved to $\lambda_{2,1} \leq 2\Delta + 1$ when the problem is restricted to threshold graphs.

4 $L(2,1)$-Coloring Split Matrogenic Graphs

When we $L(2,1)$-color split matrogenic graphs, in view of Thm. 2, an additive step dealing with perfect matchings and antimatchings must be added to Algorithm `Color-Threshold-Graphs`. In the following we will detail how these structures can be colored using no more than $3\Delta + 1$ colors for $L(2,1)$-coloring split matrogenic graphs, improving for this class the general result for split graphs due to Bodlaender et al [1].

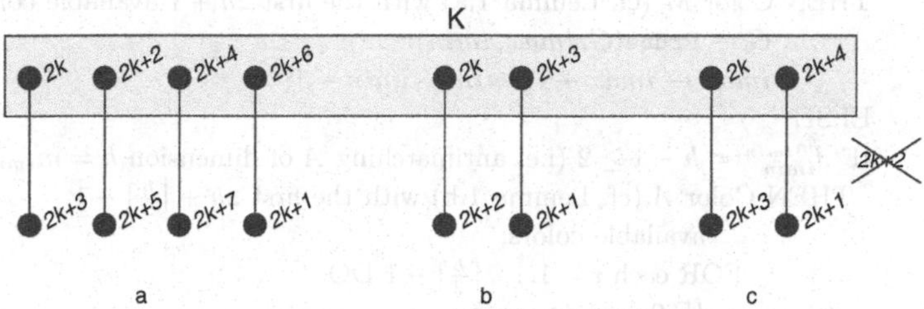

Fig. 4. Some colored matchings.

Lemma 1. *Let $G = (V_K \cup V_S, E)$ be a split matrogenic graph and let $K_r \subseteq V_K$ and $S_r \subseteq V_S$, $K_t \subseteq V_K$ and $S_t \subseteq V_S$ induce a perfect matching M_r and an antimatching A_t, respectively. Then:*

a. M_r of dimension $h_r \geq 2$ can be $L(2,1)$-colored with $2h_r+1$ consecutive colors, if all the vertices in K_r are colored with even colors and all the vertices in S_r are colored with odd colors (see Fig. 4);

b. A_t of dimension $h_t \geq 3$ can be $L(2,1)$-colored with at most $2h_t + \lceil \frac{h_t}{2} \rceil - 1$ consecutive colors, if the leftmost and rightmost vertices in K_t are colored with even colors, and the leftmost and rightmost vertices in S_t are colored with odd colors (see Fig. 5).

Now we are able to modify the algorithm for labeling threshold graphs into the next algortihm specific for split matrogenic graphs: we have to insert a new step between Step 2 and Step 3, call it Step 2bis. This new step is based on the fact that a split matrogenic graph can be pruned also by considering the subgraph induced by $V - B_1(G) - B_r(G)$, if $B_1(G)$ and $B_r(G)$ are partially connected. In this case, procedure `Prune`$(G, imax, imin)$ eliminates from the degree sequence of G both $B_{imax}(G)$ and $B_{imin}(G)$.

ALGORITHM `Color-Split-Matrogenic-Graphs`

......

REPEAT

Consider the boxes $B_{imax}(G_p)$ and $B_{imin}(G_p)$;

Step 1 ($B_{imax}(G_p)$ universal)

IF $d_{imax} = (\sum_{j=imax}^{imin} m_i) - 1$ THEN ...

ELSE

Step 2 ($B_{imin}(G_p)$ isolated)

IF $d_{imin} = 0$ THEN ...

ELSE

Step 2bis ($B_{imax}(G_p) \cup B_{imin}(G_p)$ induces either a matching or an antimatching)

IF $imax \neq imin$ THEN

IF $d_{imin}^{m_{imin}} = 1$ (i.e. matching M of dimension $h = m_{imin}$)

THEN Color M (cf. Lemma 1.a) with the first $2h+1$ available colors;

$G \leftarrow$ `Prune`$(G, imax, imin)$;

$imax \leftarrow imax + 1; imin \leftarrow imin - 1$;

ELSE

IF $d_{imin}^{m_{imin}} = h - 1 \geq 2$ (i.e. antimatching A of dimension $h = m_{imin}$)

THEN Color A (cf. Lemma 1.b) with the first $2h + \lceil \frac{h}{2} \rceil - 1$ available colors;

FOR each $i = 1, \ldots \lceil \frac{h}{2} \rceil - 1$ DO

IF `Queue-is-empty`

THEN throw the first available color out

ELSE $v \leftarrow$ `Extract-from-Queue`

Color v with the first available color;

$G \leftarrow$ `Prune`$(G, imax, imin)$;

$imax \leftarrow imax + 1; imin \leftarrow imin - 1$;

UNTIL $imax > imin$ (i.e. G_p is empty);

Step 3

......

Next theorem bounds the number of used colors and states that this number is linear in Δ.

Theorem 5. *Algorithm* `Color-Unigraphs` $L(2,1)$*-colors a split matrogenic graph* G *with at most* $3\Delta + 1$ *colors.*

5 $L(2,1)$-Coloring Matrogenic Graphs

Observe that matrogenic graphs differ from split matrogenic graphs only in the fact that they have a non empty crown induced by vertices in V_C. Hence, we have to furtherly specialize Step 2bis in order to guarantee that the crown is handled.

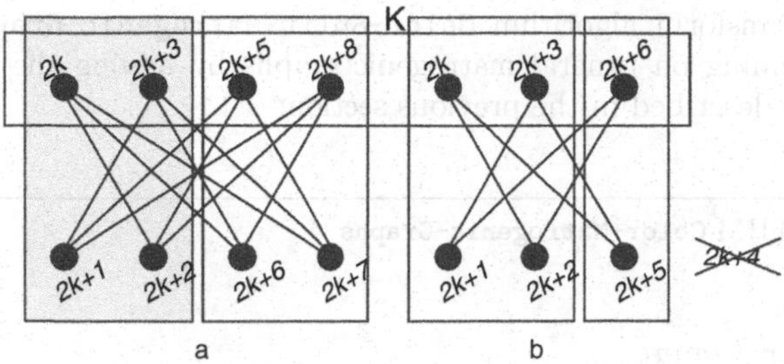

Fig. 5. Some colored antimatchings.

No color already used to $L(2,1)$-color $V_K \cup V_S$ can be reused for the crown, since each vertex in V_C is adjacent to each vertex of V_K then it is connected by a 2-length path to each vertex of V_S.

A crown that is a perfect matching of dimension h can be $L(2,1)$-colored with $2h$ consecutive colors in such a way that dually correlated vertices receive 'far' colors (see Fig. 6.a and Lemma 1.a).

A crown that is a chordless C_5 can be trivially $L(2,1)$-colored with 5 consecutive colors (see Fig. 6.b).

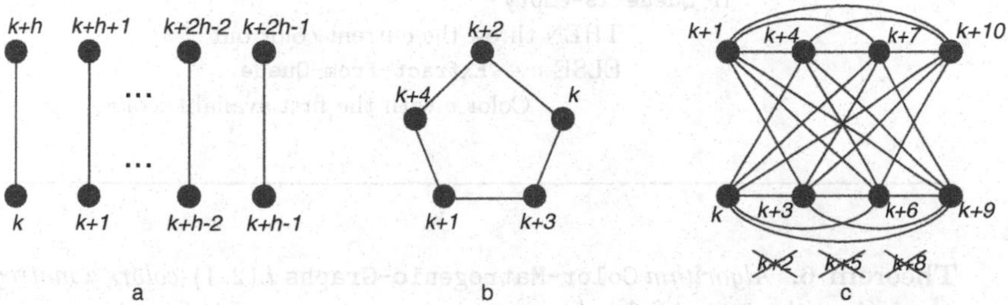

Fig. 6. All the possible crowns with a feasible $L(2,1)$-coloring: a. a perfect matching, b. a chordless C_5, c. a hyperoctahedron.

Now it remains to consider only the case in which the crown is a hyperoctahedron. Notice that a hyperoctahedron H of dimension $h, h \geq 2$, of X onto Y can be obtained from the clique induced by $X \cup Y$ eliminating a perfect matching of dimension h of X onto Y.

Lemma 2. *Let $G = (V_K \cup V_S \cup V_C, E)$ be a matrogenic graph and let V_C induce a hyperoctahedron H of dimension $h \geq 2$. Then H can be $L(2,1)$-colored with $3h-1$ consecutive colors (see Fig. 6.c), and this number is the minimum possible.*

We transform algorithm `Color-Split-Matrogenic-Graphs` into an algorithm running on general matrogenic graphs by adding the following code to Step 2bis described in the previous section:

ALGORITHM **Color-Matrogenic-Graphs**

......

Step 2bis

IF $imax \neq imin$

THEN

...

ELSE ($imin = imax = c$, i.e. the crown is reached)

IF $d_c^{m_c} = 1$ (i.e. crown = matching M of dimension $h = m_c$)

THEN Color M with the first $2h$ available colors;

ELSE

IF $d_c^{m_c} = 2$ (and $m_c = 5$) (i.e. crown = chordless P_5)

THEN Color P_5 with the first 5 available colors;

ELSE

IF $d_c^{m_c} = m_c - 2$ (i.e. crown = hyperoctahedron H of dimension $h = \frac{m_c}{2}$)

THEN Color H (cf. Lemma 2) with the first $3h - 1$ available colors;

FOR each not used color DO

IF `Queue-is-empty`

THEN throw the current color out

ELSE $v \leftarrow$ `Extract-from-Queue`

Color v with the first available color;

......

Theorem 6. *Algorithm* `Color-Matrogenic-Graphs` $L(2,1)$*-colors a matrogenic graph* G*, with at most* 3Δ *colors.*

References

1. Bodlaender, H.L., Kloks, T.,Tan, R.B.,van Leeuwen,J.: λ-Coloring of Graphs. Proc. of 17th Int.l Symp. on Theoretical Aspects of Computer Science (STACS 2000) LNCS 1770 (2000) 395–406
2. Bertossi,A.A., Pinotti, C., Tan, R.: $L(2,1)$- and $L(2,1,1)$-Labeling of Graphs with Application to Channel Assignment in Wireless Networks. Proc. of the 4th ACM Int.l Workshop on Disc. Alg. and Methods for Mobile Compu. and Comm. (DIAL M) (2000)
3. Calamoneri, T., Petreschi, R.: The $L(2,1)$-Labeling of Planar Graphs. Proc. of the 5th ACM Int.l Workshop on Disc. Alg. and Methods for Mobile Compu. and Comm. (DIAL M) (2001) 28–33

4. Calamoneri, T., Petreschi, R.: $L(2,1)$-Coloring of Regular Tiling. 1st Cologne-Twente Workshop (CTW '01) (2001).
5. Calamoneri, T., Petreschi, R.: λ-Coloring Unigraphs. Manuscript (2001).
6. Chang, G.J., Ke, W., Kuo, D., Liu, D., Yeh, R.: On $L(d,1)$-Labeling of Graphs. Disc. Math. **220** (2000) 57–66
7. Chang, G.J., Kuo, D.: The $L(2,1)$-labeling Problem on Graphs. SIAM J. Disc. Math. **9** (1996) 309–316
8. Chvatal, V., Hammer, P.: Aggregation of inequalities integer programming. Ann Discrete Math **1** (1977) 145–162
9. Fiala, J., Kloks, T., Kratochvíl, J.: Fixed-parameter Complexity of λ-Labelings. Proc. Graph-Theoretic Concepts of Compu. Sci. (WG99) LNCS 1665 (1999) 350–363
10. Foldes, S., Hammer, P.: On a class of matroid producing graphs. Colloq. Math.Soc.J. Bolyai (Combinatorics), **18** (1978) 331–352
11. Griggs, J.R., Yeh, R.K.: Labeling graphs with a Condition at Distance 2. SIAM J. Disc. Math **5** (1992) 586–595
12. Henderson, P.H., Zalcstein, Y.: A graph-theoretic characterization of the PV-chunk class of syncronizing primitives. SIAM J. Comput. **6** (1977) 88–108
13. Mahadev, N.V.R., Peled, U.N.: Threshold Graphs and Related Topics. Ann. Discrete Math. 56, North-Holland, Amsterdam (1995)
14. Marchioro, P., Morgana, A., Petreschi, R., Simeone, B.: Degree sequences of matrogenic graphs. Discrete Math. **51** (1984) 47–61
15. Sakai D.: Labeling Chordal Graphs: Distance Two Condition. SIAM J. Disc. Math **7** (1994) 133–140

Pipeline Transportation of Petroleum Products with No Due Dates*

Ruy Luiz Milidiú, Artur Alves Pessoa, and Eduardo Sany Laber

PUC-Rio, Informatics Department, Rua Marquês de São Vicente 225, RDC, 4º andar, CEP 22453-900, Rio de Janeiro - RJ, Brazil
{milidiu,artur,laber}@inf.puc-rio.br

Abstract. We introduce a new model for pipeline transportation of petroleum products with no due dates. We use a directed graph G with n nodes, where arcs represent pipes and nodes represent locations. We also define a set L of r transportation orders and a subset $F \subset L$ of further orders. A feasible solution to our model is a pumping sequence that delivers the products corresponding to all orders in $L - F$. We prove that the problem of finding such a solution is $\mathcal{NP}$-hard, even if G is acyclic. For the special case where the products corresponding to orders in F are initially stored at nodes, we propose the BPA algorithm. This algorithm finds a feasible solution in $O(r^2 \log r + s^2(rn + \log s))$ time, where s is the total volume in the arcs of G. We point out that the input size is $\Omega(s)$. If G is acyclic, then BPA finds a minimum cost solution.

1 Introduction

Petroleum products are typically transported through pipelines. Pipelines are different from all other transportation methods since they use stationary carriers whose cargo moves rather than moving carriers of stationary cargo. An important characteristic of pipelines is that they must be always full. Hence, assuming incompressible fluids, an elementary pipeline operation is the following: pump an amount of product into the pipeline and remove the same amount of product from the opposite side. Typically, each pipeline is a few inches wide and several miles long. As a result, reasonable amounts of distinct products can be transported through the same pipeline with a very small loss due to mixing at liquid boundaries.

Optimizing the transportation through pipelines is a problem of high relevance, since a non negligible component of a petroleum derivative's price depends on its transportation cost. Nevertheless, as far as we know, just a few authors have specifically addressed this problem [6, 2, 7, 10, 8]. Let us define an *order* as a requirement to transport a given amount of some product from one location to another. In [6], Hane and Ratliff present a model that assumes cyclic orders. In this case, the same orders always repeat after the completion of a given period of time. Moreover, the flow inside each pipeline is unidirectional. The authors

* Sponsored by CTPetro/FINEP, CENPES/Petrobras, and FAPERJ

S. Rajsbaum (Ed.): LATIN 2002, LNCS 2286, pp. 248–262, 2002.

also introduce the concept of pipeline *restarts*, which occur when the pipeline flow rate changes from zero to non-zero. In Hane's model, the objective is to minimize the total number of pipeline restarts.

In this paper, we propose a new model for pipeline transportation of petroleum products with non-cyclic orders [10, 9]. Let us refer to this model as the PTOP (Pipeline Transportation Optimization Problem) model. PTOP models a pipeline system through a directed graph G, where each node represents a location and each directed arc represents a pipeline, with a corresponding flow direction. In this sense, PTOP is more general than Hane's model, where the pipeline system must be represented by a directed tree. As in Hane's model, the flow inside each pipeline is assumed to be unidirectional. Furthermore, we also assume that both ordered volumes and pipeline capacities are integers. For the sake of simplicity, we represent each order with volume v as v unitary volume orders. Throughout this paper, we use the term *batch* to denote the amount of product that corresponds to a given unitary volume order. We assume that these batches cannot be split during transportation. Moreover, each batch is defined by both its initial position and its associated destination node. The initial position of a batch may be either a node or a pipeline. In the Petroleum industry terminology, batches with fixed destination nodes are usually called *proprietary batches*. PTOP assumes that all batches are proprietary.

A transportation order is defined by its source node or arc, destination node and product volume to be transported. Let L be a set of r transportation orders. An order $o \in L$ is satisfied by a sequence of elementary pipeline operations if each batch associated to the order is delivered to its corresponding destination. Since pipelines must always be full, some of the orders cannot be delivered. In fact, they are only used to move the pipeline contents. Hence, we define a subset $F \subset L$ of *further* orders that are not necessarily satisfied at the end of a feasible pumping sequence. As result, a feasible solution is a pumping sequence that delivers all batches corresponding to the *non-further* orders in $L - F$.

The problem of finding a feasible solution to PTOP is referred to as PTP. In this paper, we prove that PTP is $\mathcal{NP}$-hard, even if G is acyclic. Moreover, we consider the SPTOP, a special case of PTOP where all orders in F are initially stored at nodes (not at arcs). In this case, we give a necessary and sufficient condition for SPTOP to be feasible. Roughly speaking, we reduce the problem of finding a feasible solution to SPTOP to the problem of finding batch routes with some properties that we define later. Given such routes, we use them to construct a dependence graph H whose nodes are the arcs of G. A feasible solution to SPTOP is obtained through a decomposition of G that is associated to the strongly connected components of H.

Furthermore, we propose the Batch-to-Pipe Assignment (BPA) algorithm. This algorithm tests first the feasibility condition. When this condition is satisfied, the algorithm also generates a feasible pumping sequence in $O(r^2 \log r + s^2(rn + \log s))$ time, where s is the total volume capacity in the arcs of G and n is the number of vertices of G. We point out that BPA runs in a polynomial time since the input size is $\Omega(s)$. This size is required to represent the initial state of the pipeline system. If G is acyclic, then BPA finds a minimum pumping cost solution. Otherwise, if the graph has one or more cycles, then the obtained solution may be suboptimal. In this case, BPA also gives a lower bound on the minimum pumping cost solution.

For the sake of simplicity, we start describing a simplified version of BPA, called the *Unitary BPA algorithm* (UBPA), that deals with unitary volume orders. Since UBPA represents each order as several unitary volume orders, it runs in a pseudo-polynomial time. Later, we introduce the BPA algorithm as an enhancement of UBPA, having an improved (polynomial) time complexity.

This paper is organized as follows. In Section 2, we formalize our model. In Section 3, we prove that the problem of finding a feasible solution to our model is $\mathcal{NP}$-hard. In Section 4, we describe the UBPA algorithm for the special case where all orders in F are initially stored at nodes. In Section 5, we introduce the BPA algorithm and analyze its time complexity.

2 The PTOP Model

Our model for pipeline transportation of petroleum products has the following simplifying assumptions: fluids are incompressible, location storages are unlimited, all batches are proprietary, batch volumes are unitary and batches cannot be split.

The Pipeline System:

Let $G = (N, A)$ be a directed graph, where N is the set of nodes and A is the set of arcs. Given an arc $a = (i, j) \in A$, we say that i is the start node of a and j is the end node of a. Arcs represent pipes and nodes represent locations. Each arc $a \in A$ has an associated integer capacity $v(a)$. Moreover, we define the set of pipeline positions $A^* = \{(a, l) | a \in A \text{ and } l \in \{1, \ldots, v(a)\}\}$.

The Orders:

Let L be a set of orders, where each order $k \in L$ determines that $u(k)$ batches with weight $w(k)$ shall be transported to $d_k \in N$. Let also $F \subset L$ be a subset of the orders that are not necessarily satisfied at the end of a feasible pumping sequence. If $k \in F$, then k is a *further* order. Otherwise, if $k \in L - F$, then k is a *non-further* order.

Now, let us define two corresponding sets of unitary volume orders. For each order $k \in L$, let $b_{l'}(k)$ be the l'th unitary volume order corresponding to k, for $l' = 1, 2, \ldots, u(k)$. Hence, we have $w(b_{l'}(k)) = w(k)$ and $d_{b_{l'}(k)} = d_k$. Let us define $L^1 = \{b_1(k), \ldots, b_{u(k)}(k) | k \in L\}$ and $F^1 = \{b_1(k), \ldots, b_{u(k)}(k) | k \in F\}$.

Furthermore, let us also use b to denote the batch that corresponds to the unitary volume order b, for $b \in L^1$. If $b \in F^1$, then b is a *further* batch. Otherwise, if $b \in L^1 - F^1$, then b is a *non-further* batch.

Pipeline Contents:

Now, let us define our representation of the pipeline contents at a given state. Here, we only consider the states where each arc $a \in A$ contains exactly $v(a)$

integral batches. As a result, any solution to this model generates a discrete sequence of states, where the positions of all batches are well-defined.

Let us use $p_t(b)$ to denote the position of batch b at state t. If $p_t(b) = (a, l) \in A^*$, then batch b is located at the lth position of arc a at state t. Otherwise, if $p_t(b) = i \in N$, then batch b is stored at node i. Furthermore, the content of a given arc a at a given state t is represented by a list of batches $[b_1, b_2, \ldots, b_{v(a)}]$, where b_l is the batch such that $p_t(b_l) = (a, l)$, for $l = 1, 2, \ldots, v(a)$.

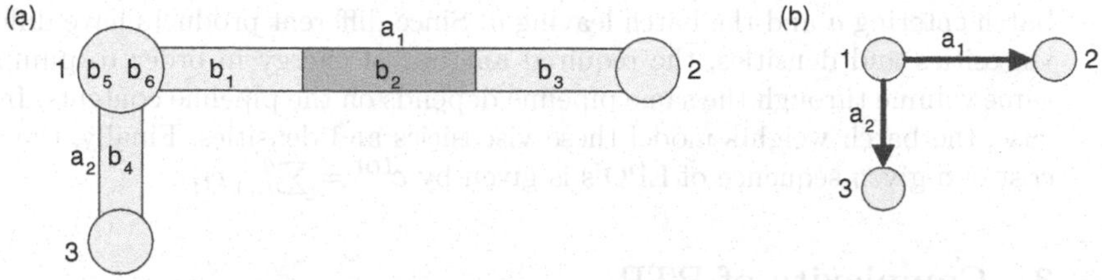

Fig. 1. (a) The contents of a pipeline system; (b) the corresponding graph.

As an example, Figure 1.(a) represents the pipeline contents corresponding to the graph of Figure 1.(b). Observe that the system has two pipelines $a_1 = (1, 2)$ and $a_2 = (1, 3)$, whose flow direction is indicated by the corresponding arcs. The capacities of a_1 and a_2 are $v(a_1) = 3$ and $v(a_2) = 1$, respectively. Let us assume that Figure 1.(a) corresponds to state t. In this case, we have $p_t(b_1) = (a_1, 1)$, $p_t(b_2) = (a_1, 2)$, $p_t(b_3) = (a_1, 3)$, and $p_t(b_4) = (a_2, 1)$, since the contents of a_1 and a_2 are respectively represented by the lists $[b_1, b_2, b_3]$ and $[b_4]$. Furthermore, we have $p_t(b_5) = p_t(b_6) = 1$ since both b_5 and b_6 are stored at node 1.

The positions of all batches at the initial state (state 0) are given. Without loss of generality, we assume that two batches that correspond to the same order always have the same position at the initial state.

Operations:

A solution for the model is a sequence of elementary pipeline operations (EPO), defined as follows. Let $a = (i, j)$ be an arc of G, whose contents at a given state t are given by the list $[b_1, b_2, \ldots, b_{v(a)}]$. Moreover, let b be a batch stored at node i at this moment. An EPO (b, a) is to pump b into a. As a result of this operation, the contents of a at state $t + 1$ are given by the list $[b, b_1, b_2, \ldots, b_{v(a)-1}]$ and $b_{v(a)}$ is stored at the node j. Observe that a sequence of q EPO's generates states $1, 2, \ldots, q$. Such a sequence is feasible when the following two conditions hold:

1. every batch $b \in L^1 - F^1$ is stored in node d_b, when the state is q;
2. for every batch $b \in F^1$ there is a path in G containing $p_q(b)$ and terminating at node d_b.

The Objective Function:

First, let us define the pumping cost of an EPO. Let a be an arc of G, whose contents at a given time are given by the list $[b_1, b_2, \ldots, b_{v(a)}]$. If the tth EPO is (b, a), then the pumping cost due to this EPO is given by $c_t = w(b)/2 + w(b_{v(a)})/2 + \sum_{\ell=1}^{v(a)-1} w(b_\ell)$. Observe that each term of the previous sum is due to one of the batches that are contained in a during the operation. The first two terms are divided by two because they respectively correspond to the batch entering a and the batch leaving a. Since different products have different viscosities and densities, the required amount of energy in order to pump the same volume through the same pipeline depends on the pipeline contents. In this case, the batch weights model these viscosities and densities. Finally, the total cost of a given sequence of EPO's is given by $c^{\text{tot}} = \sum_{t=1}^{q} c_t$.

3 Complexity of PTP

In this section, we prove that PTP is an $\mathcal{NP}$-hard problem by showing a polynomial reduction from the well-known Vertex Cover Problem [5] to PTP.

Given an undirected graph $G' = (V', E')$ and a positive integer k', the Vertex Cover Problem is to find a subset $S' \subset V'$ of vertices with $|S'| \leq k'$ such that, for all $e = (i, j) \in E'$, either $i \in S'$ or $j \in S'$ (or both). Then, we have the following theorem.

Theorem 1. *PTP is an $\mathcal{NP}$-hard problem.*

Sketch of the Proof: Let $G' = (V', E')$ be an undirected graph and k' a positive integer. We construct a corresponding instance of PTP as follows:

1. add the two nodes 0 and $|V'| + |E'| + 1$ to N;
2. add a further batch b_l, for $l = 1, \ldots, k'$, to F^1, with $d_{b_l} = |V'| + |E'| + 1$, and $p_O(b_l) = 0$;
3. for each vertex $i \in V' = \{1, 2, \ldots, |V'|\}$, add:
   (a) a corresponding node i to N and an arc $(0, i)$ to A;
   (b) a corresponding further batch b'_i to F^1, with $d_{b'_i} = i$ and $p_O(b'_i) = ((0, i), 1)$;
4. for each edge $e = (i, j) \in E' = \{1, 2, \ldots, |E'|\}$, add:
   (a) a corresponding node $|V'| + e$ to N;
   (b) three arcs $(i, |V'| + e)$, $(j, |V'| + e)$, and $(|V'| + e, |V'| + |E'| + 1)$ to A;
   (c) a corresponding further batch b^l_e, for $l = 1, \ldots, 5$, to F^1, with $d_{b^l_e} = |V'| + |E'| + 1$, $p_O(b^1_e) = i$, $p_O(b^2_e) = j$, $p_O(b^3_e) = ((i, |V'| + e), 1)$, $p_O(b^4_e) = ((j, |V'| + e), 1)$, and $p_O(b^5_e) = ((|V'| + e, |V'| + |E'| + 1), 1)$;
   (d) a corresponding non-further batch b^*_e to L^1 (not to F^1), with $d_{b^*_e} = |V'| + e$, and $p_O(b^*_e) = 0$;
5. for every arc $a \in A$, $v(a) = 1$.

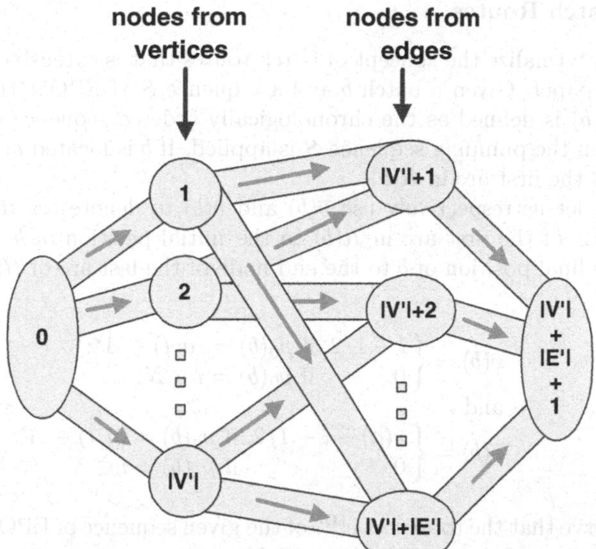

Fig. 2. A generic representation of a graph G obtained through a reduction from an instance of the Vertex Cover Problem.

Figure 2 shows a generic representation of a graph G obtained through a reduction from an instance of the Vertex Cover Problem. In this figure, node numbers are represented inside the corresponding circles. Moreover, arrows inside pipelines indicate the corresponding flow directions.

Observe that, for each edge $e = (i, j) \in E'$, there is a non-further batch $b_e^* \in L^1 - F^1$. Furthermore, b_e^* must either cross the arc $(0, i)$ or the arc $(0, j)$. Our strategy is to show that the obtained instance of PTP is feasible if and only if we can find routes for all non-further batches using at most k' arcs from $(0, 1), (0, 2), \ldots, (0, |V'|)$. This is true because only k' further batches $(b_1, b_2, \ldots, b_{k'})$ can be pumped into these arcs. If such routes exist, then the vertices corresponding to the selected arcs give a feasible solution to the original Vertex Cover Problem. ■

4 The Unitary Batch-to-Pipe Assignment Algorithm

In this section, we consider the special case of PTOP where all orders in F are initially stored at nodes. Let us refer to this special case as the Synchronous PTOP (SPTOP). In this case, we present a necessary and sufficient condition for an instance I of the SPTOP to be feasible. Furthermore, we describe and analyze the Unitary Batch-to-Pipe Assignment (UBPA) algorithm.

4.1 Batch Routes

Here, we formalize the concept of batch routes that is extensively used throughout this paper. Given a batch b and a sequence S of EPO's, the corresponding route $R(b)$ is defined as the chronologically ordered sequence of arcs traversed by b when the pumping sequence S is applied. If b is located at arc a at state 0, then a is the first arc in $R(b)$.

Now, let us respectively use $\bar{v}(b)$ and $\hat{v}(b)$ to denote the distance from the start node of the first arc in $R(b)$ to the initial position of b and the distance from the final position of b to the end node of the last arc of $R(b)$. Formally, we have

$$\bar{v}(b) = \begin{cases} l - 1/2, & \text{if } p_0(b) = (a, l) \in A^*; \\ 0, & \text{if } p_0(b) = i \in N; \end{cases}$$

and

$$\hat{v}(b) = \begin{cases} v(a) - l + 1/2, & \text{if } p_q(b) = (a, l) \in A^*; \\ 0, & \text{if } p_q(b) = i \in N. \end{cases}$$

Observe that the total cost c^{tot} of the given sequence of EPO's can be written as a function of the routes, the initial positions and the final positions of all batches as follows:

$$c^{\text{tot}} = \sum_{b \in L^1} w(b) \left(\sum_{a \in R(b)} v(a) - \bar{v}(b) - \hat{v}(b) \right).$$

Next, we relate batch routes and the final state of the pipeline system. Let us use the terminology *source (tail) node* of $p_t(b)$ to denote:

1. the start (end) node of a if $p_t(b) = (a, l) \in A^*$;
2. the node i, if $p_t(b) = i \in N$.

We say the $p_q(b)$ is a *valid final position* for $b \in F^1$ when there is a path connecting the tail node of $p_0(b)$ to the source node of $p_q(b)$ and another path connecting the tail node of $p_q(b)$ to d_b. If $b \in L^1 - F^1$ then d_b is the only *valid final position* for b. Furthermore, $R(b)$ is a *valid route* for $b \in L^1$ when it satisfies the following four conditions:

1. the first arc of $R(b)$ starts at the source node of $p_0(b)$;
2. if $p_0(b) = (a, l) \in A^*$ then a is the first arc of $R(b)$;
3. the last arc of $R(b)$ ends at the tail node of a valid final position $p_q(b)$ for b;
4. if $p_q(b) = (a, l) \in A^*$ then a is the last arc of $R(b)$.

4.2 Feasibility Conditions

Our first result is a simple necessary condition for an instance of SPTOP to be feasible. Since the pipeline system must contain only batches of F^1 at the final state, any feasible solution to an instance of SPTOP assigns a batch of F^1 to each pipeline position $(a,l) \in A^*$. This assignment determines the position of each batch at the final state. Therefore, each batch of F^1 is assigned to at most one pipeline position. Moreover, a batch b can be assigned to a pipeline position (a,l) only if (a,l) is a valid final position for b. Our necessary condition is that such an assignment exists.

Proposition 1. *If an instance I of the SPTOP is feasible then there exists an assignment from F^1 to A^* with the following three properties:*

1. *to each pipeline position $(a,l) \in A^*$, it is assigned a batch of F^1;*
2. *each batch $b \in F^1$ is assigned to at most one pipeline position;*
3. *for every batch b assigned to $(a,l) \in A^*$, (a,l) is a valid final position for b.*

Given a valid final position $p_q(b)$ for each batch $b \in L^1$, we say that these final positions describe a *valid final state* when they have the three properties of Proposition 1.

Now, we define the concept of a dependence graph. Suppose we are given

(a) an instance of the SPTOP with a corresponding graph G;
(b) a valid final state represented by $p_q(b), \forall b \in L^1$;
(c) a valid route $R(b)$ for each batch $b \in L^1$.

We remark that the routes described in Item (c) must be consistent with the given final state, that is, the last arc of each route must contain the final position of the corresponding batch. Then, the corresponding dependence graph $H = (A, D)$ is a directed graph defined as follows:

1. each node of H corresponds to an arc $a \in A$;
2. there is an arc $d = (a, a') \in D$ if and only if there is a batch $b \in L^1$ with $R(b) = [a_1, a_2, \ldots, a_{|R(b)|}]$ such that $a_j = a$ and $a_{j+1} = a'$ for some $j \in \{1, 2, \ldots, |R(b)| - 1\}$.

Observe that, if there is an arc in H from a to a', then there is at least one batch that leaves arc a and enters arc a', during a sequence of EPO's for the given routes and final state.

Theorem 2. *Suppose we are given an instance I of SPTOP, a valid final state and a valid route for each batch. If the resulting dependence graph H is acyclic, then there is a feasible solution for I that leads to the given final state using the given routes.*

Proof: We prove this theorem by showing how to construct such a feasible solution for I. Since H is acyclic, it has a source node. Let $a = (i, j)$ be an arc of G that corresponds to a source node in H. Every batch b that must pass by the arc a, according to $R(b)$, is stored at node i at the current state, otherwise a would have an incoming arc in H. Hence, we start our feasible solution by a sequence of EPO's that pump these batches into the arc a. We point out that the tail of this sequence must correspond to the reversal of the list of batches that determines the content of the arc a at the final state. After this sequence is executed, every batch b that must pass by the arc $a = (i, j)$ is already stored at the node j, including those having $p_q(b) = j$. Moreover, the current contents of the arc a are already equal to the required contents at the given final state. Therefore, we may remove node a (and all its outgoing arcs) from H. By repeating this procedure until the dependence graph H is empty, we assure that the contents of every arc and every node correspond to the given final state after the proposed solution is executed. ■

4.3 Algorithm Description

Let $n = |N|$, $m = |A|$, and $s = |A^*|$. Now, we describe the Unitary Batch-to-Pipe Assignment (UBPA) algorithm for SPTOP. UBPA performs six main steps, summarized in the pseudo-code of Figure 3. In this figure, $c()$ denotes a cost function and *Sequencing* is a procedure described in section 4.5.

The UBPA Algorithm

Step 1: **For** each arc $a \in A$, let $v(a)$ be the weight of a;
  For every pair of nodes $i, j \in N$,
    find the weighted shortest path $S(i,j)$ from i to j in G;
    let $v(i,j)$ be the weighted length of $S(i,j)$;
Step 2: construct a weighted bipartite graph $B = (N^1 \cup N^2, A^B)$ as follows:
  For each batch $b \in F^1$, $N^1 \leftarrow N^1 \cup n(b)$;
  For each pipeline position $(a,l) \in A^*$, $N^2 \leftarrow N^2 \cup n(a,l)$;
  For $l = 1, 2, \ldots, |F^1| - s$, $N^2 \leftarrow N^2 \cup n(\cdot, l)$;
  For each ordered pair $(n(b), n(a,l)) \in N^1 \times N^2$,
    $A^B \leftarrow A^B \cup (n(b), n(a,l))$;
    $c(n(b), n(a,l)) \leftarrow$ the minimum cost of transporting b through a valid route such that $p_q(b) = (a,l)$;
  For each ordered pair $(n(b), n(\cdot, l)) \in N^1 \times N^2$,
    $A^B \leftarrow A^B \cup (n(b), n(\cdot, l))$; $c(n(b), n(\cdot, l)) \leftarrow 0$;
Step 3: let X be a minimum cost assignment on B;
  If $c(X) = \infty$ **then** I is infeasible; stop;
Step 4: **For** every batch $b \in F^1$,
    let $(n(b), k)$ be the only outgoing arc from $n(b)$ in X;
    If $k = n(a,l)$ for some $(a,l) \in A^*$ **then**
      $p_q(b) \leftarrow (a,l)$;
      $R(b) \leftarrow$ a minimum cost valid route for b such that $p_q(b) = (a,l)$;
    Else $p_q(b) \leftarrow p_0(b)$; $R(b) \leftarrow$ an empty route;
  For every batch $b \in L^1 - F^1$,
    $R(b) \leftarrow$ a minimum cost valid route for b;
Step 5: construct the corresponding dependence graph H;
Step 6: call *Sequencing* to construct a feasible solution from both X and H.

Fig. 3. A pseudo-code for the UBPA algorithm.

At Step 1, if there is no path from i to j in G, then $v(i,j)$ is set equal to a sufficiently large value (say ∞). At Step 2, the minimum cost of transporting a batch b through a valid route such that b terminates at a pipeline position (a,l) with $a=(i,j)$ is given by $w(b)(v(p_0(b),i)+l-1/2)$. Observe that the value of $v(p_0(b),i)$ is computed at Step 1. At Step 4, for each batch $b \in F^1$, if $p_q(b)$ is set equal to $(a,l) \in A^*$ with $a=(i,j)$, then a corresponding minimum cost valid route for b is given by the concatenation of $S(p_0(b),i)$ with $[a]$. Moreover, for each batch $b \in L^1 - F^1$, if $p_0(b)=(a,l) \in A^*$ with $a=(i,j)$, then a minimum cost valid route for b is given by the concatenation of $[a]$ with $S(j,d_b)$. Otherwise, if $p_0(b) \in N$, then this route is given by $S(p_0(b),d_b)$. If $c(X)=\infty$, then the problem is infeasible since the condition of Proposition 1 cannot be satisfied. Otherwise, we have the following proposition.

Proposition 2. *For all $b \in L^1$, let $R(b)$ be the route assigned to the batch b by UBPA. If $c(X) \neq \infty$, then*

$$c(X) + \sum_{b \in L^1 - F^1} w(b)\left(\left(\sum_{a \in R(b)} v(a)\right) - \bar{v}(b)\right) \tag{1}$$

is a lower bound on the cost of any feasible solution for I.

Proof: First, we observe that $c(X)$ is the minimum cost of transporting the batches of F^1 through valid routes that terminate at a valid final state. Since, by Proposition 1, any feasible solution for I must terminate at a valid final state, $c(X)$ is a lower bound on the cost of transporting the batches of F^1 in any such solution. Since UBPA assigns the minimum cost valid route to each batch $b \in L^1 - F^1$, (1) gives a lower bound on the cost of any feasible solution for I. ■

At Step 5, UBPA constructs the dependence graph H corresponding to X. By Theorem 2, if H is acyclic, then we can construct a feasible solution for I with cost equal to the lower bound given by (1). Next, we describe the *Sequencing procedure*, called at Step 6. If H is acyclic then Sequencing gives an optimal solution for I. Otherwise, the obtained solution is feasible but not necessarily optimal.

4.4 Source Components

First, let us define the concept of source components in a directed graph.

Definition 1. *Given a directed graph $H=(A,D)$, a source component is a strongly connected component $C \subseteq A$ such that there is no arc $(i,j) \in D$ with $j \in C$ and $i \in A - C$.*

Observe that every graph has at least one source component [3, Exercise 23.5-4].

Now, let $((i,j),(i',j'))$ be an arc of H. By the definition of H, there is a route that passes by (i',j') immediately after passing by (i,j). Hence, we must have $i' = j$. In this case, we say that the arc $((i,j),(j,j'))$ of H *corresponds* to the node j of G. In fact, we may have many arcs in H corresponding to the same node in G. Given a strongly connected component T of H, with $|T| > 1$, we define T' as the induced subgraph of G with the nodes corresponding to the arcs of T.

Now, let us consider the following proposition.

Proposition 3. *For every strongly connected component T with $|T| > 1$ in H, the corresponding induced subgraph T' in G is also strongly connected.*

4.5 The Sequencing Procedure

Here, we describe *Sequencing*, a key procedure used by UBPA. Recall that this procedure receives the following additional input data:

1. a valid final state represented by $p_q(b), \forall b \in L^1$;
2. a valid route $R(b)$ for each batch $b \in L^1$, consistent with the given final state;
3. the corresponding dependence graph H.

We point out that, for each batch $b \in L^1$, the route $R(b)$ and final position $p_q(b)$ may be changed by *Sequencing*. Hence, we use $R^*(b)$ and $p_q^*(b)$ to respectively denote the original route and final position assigned to b at Step 4.

If H is acyclic, then *Sequencing* does exactly the same as the method described in the proof of Theorem 2. Recall that, at each iteration, this method selects a source node a from H. Then, it pumps into a every batch whose route contains a. After that, a is removed from H and no longer used. Similarly to the previous method, the sequencing procedure selects a source component T from H, at each iteration. Let T' be the corresponding induced subgraph in G. Then, *Sequencing* obtains a sequence of EPO's for T' that leads to a state with the following properties:

1. if b is a batch such that $p_q^*(b) = (a,l) \in A^*$, for $a \in T'$, then b is contained in an arc of T' (not necessarily in a);
2. if $b \in L^1 - F^1$ is a batch such that $p_q^*(b)$ is a node of T', then b is delivered;
3. if b is a batch not included in the previous items such that $R^*(b)$ contains at least one arc of T', then b is stored at the tail node of the last arc of T' in $R^*(b)$;

After that, it removes T from H and performs the next iteration.

A pseudo-code for *Sequencing* is given by Figure 4. In this figure, *Cycling* is a procedure described later. Given a selected source component T of H, this procedure checks whether $|T| = 1$ or not. If we have $|T| = 1$, then it pumps the corresponding batches into the only arc associated to T. In this case, *Sequencing* does exactly the same as the method described in the proof of Theorem 2. On the other hand, if we have $|T| > 1$, then *Sequencing* finds the induced subgraph T' of G corresponding to T. After that, it moves every batch in this subgraph to the appropriate position as follows. First, it fills every arc of T' with batches

```
                      The Sequencing Procedure
While H is not empty,
    T ← a source component of H;
    If |T| = 1 then
        a ← the only arc associated to T;
        pump the corresponding batches into a;
    Else
        T' ← the induced subgraph of G corresponding to T;
        For every batch b ∈ F^1 with p*_q(b) = ((i,j),k) ∈ A*, for (i,j) ∈ T',
            If b is stored at a node l ≠ i then
                call Cycling to move b from l to i, in T';
        For every batch b ∈ F^1 with p*_q(b) = (a,k) ∈ A*, for a ∈ T',
            pump b into a;
        For every batch b such that R*(b) contains at least one arc of T',
            i ← the tail node of the last arc of T' in R*(b);
            If b is stored at a node l ≠ i then
                call Cycling to move b from l to i, in T';
        remove T from H;
For every batch b ∈ L^1
    update both R(b) and p_q(b)
        according to the obtained pumping sequence;
```

Fig. 4. A pseudo-code for *Sequencing*

of F^1 whose final positions are arcs of T'. This is performed by the first two "for" loops in the pseudo-code of Figure 4. Next, in the last "for" loop, the remaining batches are moved to the appropriate nodes in T', following their corresponding routes. Observe that *Sequencing* uses the route $R^*(b)$ obtained at Step 4 of UBPA, for each batch b. Nevertheless, the final route $R(b)$ tracked by each batch b is not necessarily equal to $R^*(b)$.

Now, let us explain *Cycling*, a key procedure used by *Sequencing*. Given a batch b^* stored at a node $l \neq i$, with $i, l \in T'$, *Cycling* moves b^* from l to i. This is performed in such a way that no batch other than b^* is removed from a node. For that, *Cycling* finds the smallest cycle C' of T' that contains both l and i. Observe that such a cycle always exists because, by Proposition 3, T' is strongly connected. Figure 5 shows a pseudo-code for this procedure.

Let $k = |C'|$ and $s' = \sum_{j=1}^{k} v(a_j)$. We have the following proposition.

Proposition 4. *If the cycle C' is given, then Cycling runs in $O(ks')$ time.*

Proof: Clearly, for each k iterations of *Cycling* the batch b^* is moved exactly once. Since b^* always reaches its destination node i after at most $k + s' - 2$ movements, *Cycling* runs in $O(ks')$ time. ■

The Cycling Procedure
$C' \leftarrow$ the smallest cycle of T' that contains both l and i;
$K \leftarrow \{b \in L^1 \| b$ is currently contained in an arc of $C'\}$;
While b^* is not stored in i,
$b \leftarrow$ the only batch in $K \cup \{b^*\}$ that is stored at a node;
$a \leftarrow$ the only outgoing arc from this node in C';
pump b into a;

Fig. 5. A pseudo-code for *Cycling*

4.6 Analysis

Now, we analyze UBPA. The following theorem is a consequence of this algorithm.

Theorem 3. *Given an instance I of SPTOP, I is feasible if and only if the conditions of Proposition 1 hold.*

Proof: By Proposition 1, if I is feasible, then the given conditions hold. Conversely, we observe that UBPA tests these conditions at Step 3. If they hold, then UBPA always finds a feasible solution for I. ■

Let us consider the pseudo-code of UBPA described in Figure 3. We have the following theorem.

Theorem 4. *If the constructed dependence graph H is acyclic, then the feasible solution obtained by the UBPA algorithm is optimal for the SPTOP.*

Observe that, if G is acyclic, then any corresponding dependence graph is also acyclic. As a result, we obtain the following corollary.

Corollary 1. *If the graph G is acyclic, then the UBPA algorithm always gives an optimal solution for the SPTOP.*

5 The Batch-to-Pipe Assignment Algorithm

In Section 4, we describe a pseudo-polynomial time algorithm for the SPTOP called the *Unitary BPA algorithm* (UBPA). In this section, we introduce the BPA algorithm as an enhancement of UBPA. Let $r = |L|$. BPA runs in $O(r^2 \log r + s^2(rn + \log s))$ time. Recall that the input size for BPA is $\Omega(s)$. Furthermore, both Theorem 4 and Corollary 1 hold for BPA.

5.1 Description

In order to develop a polynomial time algorithm for the SPTOP, we must be able to represent a feasible solution in a more compact way. Hence, we extend the definition of EPO to allow for pumping more than one batch corresponding to the same order. Recall that we use $b_l(k)$ to denote the lth unitary order (batch) corresponding to k, for $l = 1, 2, \ldots, u(k)$. Let $a = (i, j)$ be an arc of G, whose content at a given state t is given by the list $[b_1, b_2, \ldots, b_{v(a)}]$. Moreover, let $k \in L$ be an order such that the batches $b_l(k), b_{l+1}(k), \ldots, b_{l'}(k)$ are stored at node i in this moment. An extended pipeline operation (XPO) is to pump the batches $b_l(k), b_{l+1}(k), \ldots, b_{l'}(k)$ into a. Observe that an XPO represents $l' - l + 1$ EPO's, using $O(1)$ space.

Now, we are ready to describe the BPA algorithm. For $k \in L$, let us use $p_0(k)$ (not $p_0(b_1(k))$) to denote the initial position of the batches of k. A pseudo-code for BPA can be obtained from the pseudo-code of UBPA, shown in Figure 3, through the following two updates:

1. insert an additional step (say Step $1\frac{1}{2}$) after Step 1;
2. slightly modify Steps 2 and 3 in order to construct a smaller graph.

At Step $1\frac{1}{2}$, BPA processes the batches of $L^1 - F^1$ that are initially stored at the nodes. A pseudo-code for this step is shown in Figure 6.

Step $1\frac{1}{2}$ of BPA

For each order $k \in L - F$ such that $p_0(k) \in N$,
    $l \leftarrow u(k)$; $j \leftarrow 1$;
    While $j \leq \lvert S(p_0(k), d_k)\rvert$ and $l \geq 1$,
        $a \leftarrow$ the jth arc of $S(p_0(k), d_k)$;
        pump $b_1(k), b_2(k), \ldots, b_l(k)$ into a;
        $l \leftarrow l - v(a)$; $j \leftarrow j + 1$;
remove all delivered batches from L^1;
consider the current state as the initial state;

Fig. 6. A pseudo-code for the Step $1\frac{1}{2}$ of the BPA algorithm

We point out that, after executing Step $1\frac{1}{2}$, for any order $k \in L - F$ with $p_0(k) \in N$, the number of corresponding batches that are still not delivered is not greater than $v(p_0(k), d_k) \leq s$. As a result, we have $|L^1 - F^1| = O(s|L - F|)$ after this step.

At Step 2, BPA constructs a bipartite B' graph smaller than B. Next, we define B' by describing how to obtain it from B. First, for each further order k, all the nodes from the first partition of B that correspond to a batch b of k are shrunk into the same node of B'. Moreover, we assign to each node from the first partition of B', an *excess* equal to the number of batches that it represents. Observe that we have $O(|F|)$ nodes at the first partition of B'. Furthermore, the nodes $n(\cdot, 1), n(\cdot, 2), \ldots, n(\cdot, |F^1| - s)$ from the second partition of B are also shrunk into the same node of B', with a *demand* equal to $|F^1| - s$. All other nodes from the second partition of B' have a *demand* equal to 1. Observe that we have exactly $s + 1$ nodes at the second partition of B'. We point out that such a reduction is possible because we do not need to distinguish two batches of the same order.

At Step 3, BPA solves a Minimum Cost Flow Problem on B' [1]. In this case, we consider that all arc capacities are infinity. The interpretation of the obtained optimal solution is omitted since it is analogous to that of UBPA.

5.2 Analysis

Next, we analyze the running time of BPA. For that, we assume that every arc in G has a positive volume and that every node in G has at least one adjacent arc. As a result, we obtain that $s \geq m \geq n/2$. Then, we have the following theorem.

Theorem 5. *The BPA algorithm runs in* $O(r^2 \log r + s^2(rn + \log s))$ *time.*

References

[1] R. Ahuja, T. Magnanti, and J. Orlin. *Network Flows: Theory, Algorithms and Applications.* Prentice Hall, 1993.

[2] Eduardo Camponogara. A-teams para um problema de transporte de derivados de petróleo. Master's thesis, Departamento de Ciência da Computação, IMECC - UNICAMP, December 1995.

[3] T. H. Cormen, C. E. Leiserson, and R. L. Rivest. *Introduction to algorithms.* MIT Press and McGraw-Hill Book Company, 6th edition, 1992.

[4] R. W. Floyd. Algorithm 97: Shortest path. *Communications of ACM*, 5:345, 1962.

[5] M. R. Garey and D. S. Johnson. *Computers and Intractability: A Guide to the Theory of NP-Completeness.* W. H. Freeman and Company, 1979.

[6] Christopher A. Hane and H. Donald Ratliff. Sequencing inputs to multi-commodity pipelines. *Annals of Operations Research*, 57, 1995. Mathematics of Industrial Systems I.

[7] Ruy L. Milidiú, Eduardo S. Laber, Artur A. Pessoa, and Pablo A. Rey. Petroleum products scheduling in pipelines. In *The International Workshop on Harbour, Maritime & Industrial Logistics Modeling and Simulation*, september 1999.

[8] Ruy L. Milidiú, Artur A. Pessoa, Viviane Braconi, Eduardo S. Laber, and Pablo A. Rey. Um algoritmo grasp para o problema de transporte de derivados de petróleo em oleodutos (to appear). In *Anais do XXXIII SBPO*, Campos do Jordão, SP, Brazil, november 2001.

[9] Ruy L. Milidiú, Artur A. Pessoa, and Eduardo S. Laber. Pipeline transportation of petroleum products. Technical Report 37, Departamento de Informática, PUC-RJ, Rio de Janeiro, Brasil, September 2000.

[10] Ruy L. Milidiú, Artur A. Pessoa, and Eduardo S. Laber. Transporting petroleum products in pipelines (abstract). In *ISMP 2000 – 17th International Symposium on Mathematical Programming*, pages 134–135, Atlanta, Georgia, USA, August 2000.

[11] R. E. Tarjan. Depth-first search and linear graph algorithms. *SIAM Journal on Computing*, 1(2):146–160, 1972.

Ancestor Problems on Pure Pointer Machines*

Enrico Pontelli and Desh Ranjan

Dept. Computer Science
New Mexico State University
{epontell,dranjan}@cs.nmsu.edu

Abstract. We study several problems related to computing ancestors in dynamic trees on *pure* pointer machines, i.e., pointer machines with *no* arithmetic capabilities. The problems are motivated by those that arise in implementation of declarative and search-based programming languages. We provide a data structure that allows us to solve many of these problems including the computation of the nearest common ancestor, determination of precedence in the in-order traversal of the tree and membership of two nodes in the same path in worst-case $O(\lg h)$ time per operation where h is the height of the tree. Our solutions work for the fully dynamic case (no preprocessing) and do not use any arithmetic.

1 Introduction

We study a number of problems related to computing ancestors in dynamic trees on pure pointer machines, i.e., pointer machines with no arithmetic capabilities. Ancestor problems arise in many contexts (e.g. computing maximum weight matchings [7], parallel data-structure design [4], suffix trees for strings [9]) and have been studied before. In particular, a number of results have been obtained for solving the nearest common ancestor (nca) problem – the problem of determining the deepest node z that is a common ancestor of two input nodes x and y. A summary of these results is provided in Section 1.2.

Our motivation to study the ancestor problems derives from the implementation of declarative and search-based programming languages (e.g., Prolog) [12]. A Prolog program execution gives rise to a dynamic execution tree where the nodes of the trees can be thought of tasks and the children of nodes as subtasks of the task at the parent node. Developing an efficient system for execution of the program maintaining the sequential semantics of Prolog entails solving many data structure problems pertaining to supporting specific operations on this dynamic execution tree. Efficient solution of ancestor problems in dynamic trees are of relevance to many of these problems. One such important problem is the is_left problem – the problem of determining if a given node x is to the left of another node y in an in-order traversal of the dynamic tree. Another such problem is the is_on_left_branch problem which is the problem of determining if a

* This work is partially supported by NSF grants CCR9875279, CCR9900320, CDA9729848, EIA9810732, EIA0130887,CCR9820852, and HRD9906130.

S. Rajsbaum (Ed.): LATIN 2002, LNCS 2286, pp. 263–277, 2002.

leaf node x is on the leftmost branch of the sub-tree rooted at node y. For more details the reader is referred to [8,12].

The key contribution of this paper is an algorithm to solve the nearest common ancestor problem in worst-case $O(\lg h)$ time on the pure pointer machines in the totally dynamic case (no preprocessing), where h is the height of the dynamic tree. The algorithm uses a data structure similar to the one presented in [16] but, crucially, our solution does not use any arithmetic. The data structure is useful for other dynamic tree problems on pointer machines. It is worth noting that all the previous results obtained for the ancestor problems are either for the static case (with $\Theta(n)$ preprocessing [10,15,3], where n is the number of nodes), or for the RAM models and make substantial use of arithmetic capabilities [5], or for the pointer machines with arithmetic capabilities [16,1]. We formally define several other problems that arise in the context of execution of declarative languages and that are related to the ancestor problems. We establish relationships between these problems. This, together with our solution of the nca problem, provides efficient solutions to some of these problems. E.g., we show that the is_left problem can be solved in time $O(\lg h)$ on pointer machines in the dynamic case. We also show that if the ancestor problem can be solved in time $O(f)$ then the problem is_on_left_branch(x, y) can be solved in time $f + O(\lg \lg h)$. This gives us an $O(\lg h)$ solution for this problem. Furthermore, we show that the problem can be solved in $O(\lg \lg h)$ time if y is known to be an ancestor of x.

1.1 The Pure Pointer Machine Model

A *Pure Pointer Machine (PPM)* consists of a finite but expandable collection R of *records*, called *global memory*, and a finite collection of *registers*. Each record is uniquely identified through an address. A special address *nil* is used to denote an invalid address. Each record is a finite collection of named fields. All the records in the global memory have the same structure, i.e., they all contains the same number of fields. Each field may contain either a *data* or an address. The PPM is also supplied with two finite collections of registers, $d_1, d_2, \ldots$ (data registers) and $r_1, r_2, \ldots$ (pointer registers). Each register d_i can contain a data element, while each register r_i can contain an address. The machine can execute *programs*. A program P is a finite, numbered sequence of instructions. The instructions allow to move addresses and data between registers and between registers and records' fields. The only "constant" which can be explicitly assigned to a register is *nil*. Special instructions are used to create a new record and to perform conditional jumps. The only conditions allowed in the jumps are equality comparisons between pointer registers. Observe that the content of the data fields will never affect the behavior of the computation. In terms of analysis of complexity, it is assumed that each instruction has a unit cost. Further details on the structure pure pointer machines can be found in [2].

Even though RAM is the most commonly used reference model in studies of complexity of algorithms, the PPM model has received increasing attention as a viable model to study complexity of algorithms. The PPM model is simpler, thus making it more suitable for analysis of lower bounds of time complexity [2].

Furthermore, RAM commonly hides the actual cost of arithmetic operations, by allowing operations on numbers of size up to $\lg n$ (n being the size of the input) to be treated as constant time operations. PPMs instead make these costs explicit. Indeed, many lower bound results have been developed using the Pointer Machine model. PPMs provide a good base for modeling implementation of linked data structures, like trees and lists. But they fail to capture some computational aspects of a "real" computer architecture. One of the major characteristics of PPMs is the lack of arithmetic. Numbers have to be explicitly represented, e.g., as linked lists of bits. This is realistic in the sense that, for arbitrarily large problems requiring arithmetic operations, the time to compute such operations will be a function of the size of the problem. Note that the PPM model differs from the pointer machine model used in [1] to obtain seemingly better results. Note also that the PPM model is similar to the Turing Machine model with respect to the fact that the complexity of the arithmetic operations has to be accounted for while analyzing the complexity of an algorithm.

1.2 Notation, Definitions, and Previous Work

All the problems considered in this paper relate to dynamic binary trees. The following operations are used to manipulate the structure of a dynamic tree:

- create_tree(v) creates a new tree containing a single vertex (the root) labeled v; the operation return as result the node representing the root of the tree.
- expand(x, v_1, v_2) assumes that x is a leaf in the tree, and it expands the tree by creating two children of x, respectively labeled v_1 and v_2;
- remove(x) is used to remove the node x, which is assumed to be a leaf.

A binary tree induces a natural partial ordering on the set of nodes in the tree. Given two nodes v and w, we write $v \preceq w$ if v is an ancestor of w. $v \prec w$ additionally says that $v \neq w$. We will assume $\preceq$ to be a reflexive relation. Observe that $\preceq$ is a partial but not a total order. Let us indicate with *left*(v) (*right*(v)) the left (right) child of the node v—in particular *left*(v) $= \perp$ if v does not have a left child. Given a node v, we can recursively define the set of nodes *left_branch*(v):

- if *left*(v) $= \perp$ and *right*(v) $= \perp$ then *left_branch*(v) $= \{v\}$
- if *left*(v) $= \perp$ and *right*(v) $= w$ then *left_branch*(v) $= \{v\} \cup$ *left_branch*(w)
- if *left*(v) $= w$ then *left_branch*(v) $= \{v\} \cup$ *left_branch*(w)

Observe that, given a node v, the elements of *left_branch*(v) constitute a chain w.r.t. $\preceq$. The notion of left branch allows us to define another relation between nodes, indicated by $\trianglelefteq$; given two nodes v, w we have that $v \trianglelefteq w$ if w is a node in the left branch of the subtree rooted at v: $v \trianglelefteq w \Leftrightarrow w \in$ *left_branch*(v). Observe that $\trianglelefteq$ is a partial order. Given a node v, let $\mu(v)$ indicate the node $\mu(v) = min_{\preceq}\{w | w \trianglelefteq v\}$; $\mu(v)$ indicates the highest node w (i.e., closest to the root) such that v is in the leftmost branch of the subtree rooted at w. $\mu(v)$ is called the *subroot* of v.

The initial problem considered in this context is the problem of determining nearest common ancestors in a dynamic tree. This problem introduces an additional operation to manipulate dynamic trees, called nca(x, y), where x and y

are two nodes in the tree. Given two nodes x, y, the operation $\mathsf{nca}(x, y)$ returns a pointer to the nearest common ancestor of the nodes x and y. More formally, $\mathsf{nca}(x, y) = z \Leftrightarrow \left(z \preceq x \wedge z \preceq y \wedge \forall w(w \preceq x \wedge w \preceq y \rightarrow w \preceq z)\right)$

We will indicate with $\mathcal{NCA}$ the problem of executing an arbitrary, correct, on-line sequence of **create_tree**, **expand**, **remove**, and **nca** operations.

Related Work

The problem of computing the nearest common ancestor (also called least common ancestor in the literature) has been extensively studied. For the *static case*, the problem of computing the nca has been pretty much satisfactorily solved. The work by Harel and Tarjan [10] provides a constant-time algorithm for computing $\mathsf{nca}(x, y)$ after a linear-time preprocessing of the input tree. This result was later simplified and parallelized by Schieber and Vishkin [15]. A nice exposition of this result can be found in [9]. Later, a nice piece of work by Bender and Farach-Colton [3] provides a simple and effectively implementable algorithm which provides constant time execution of $\mathsf{nca}(x, y)$ with linear-time preprocessing. In all the three algorithms above, complexity analysis is done assuming the RAM model. Also, the algorithms are for static trees, i.e., the tree is assumed to be known in advance (allowing for preprocessing).

For the dynamic case, a work by Tsakalidis [16] provides algorithms with $O(\lg h)$ worst-case time for the nca operation and amortized almost constant time for insertion and deletions in a dynamic tree. The algorithm is developed for a pointer machine model with uniform cost measure assumption (constant time arithmetic for $\Theta(\lg n)$ size integers). This result has been recently improved in [1], where the problem is solved in $O(n+m \lg\lg n)$ time where n is the number of link operations and m is the number of nca queries. Once again, the pointer machine model with uniform cost measure for arithmetic is assumed. The algorithms of Cole and Hariharan [5] provide the ability to insert (leaves and internal nodes) and delete nodes (leaves and internal nodes with one child) in a tree and execute the $\mathsf{nca}(x, y)$ operation in constant time. The method relies upon assigning codes to nodes such that nca calculation is constant-time for nodes on the same centroid path. Reorganizing the tree to maintain centroid paths and codes in constant worst-case times is non-trivial. However, with clever amortization over future insertions and deletions a worst-case $O(1)$ bound is obtained. In this case, the RAM model with the ability to perform arithmetic operations at unit cost, is assumed for the complexity analysis. Both methods make extensive and crucial use of arithmetic capabilities of the RAM.

To compare and contrast these results with ours, note that we deal with the totally dynamic case which rules out preprocessing. This is natural because the execution trees that might arise from the execution of declarative languages are completely dynamic. Also, we devise our algorithms for the pointer machines, which do not allow for arithmetic at unit cost. All our running times are worst-case and amortization is not used. Note that our problem is somewhat simpler because we allow deletion/insertion only at the leaves of the dynamic tree. This is also motivated by how declarative program execution trees change dynamically.

2 Solving $\mathcal{NCA}$ Problem on PPMs

We provide a worst-case $O(\lg h)$ solution for the $\mathcal{NCA}$ Problem on PPMs, where h is the height of the dynamic tree. It is worth noting that a worst-case constant-time solution is not possible for the $\mathcal{NCA}$ Problem on pure pointer machines, even in the static case. A lower bound of $\Omega(\lg \lg n)$ (n being the number of nodes in the tree) is provided at the end of this section (see also [10]).

The basic idea behind our solution is to maintain the depth of the tree nodes. For any vertex x in the tree, let us denote with $anc(x, d)$ the ancestor of vertex x lying at depth d in the tree. Thus, if we have two nodes x and y and $anc(x, d) = anc(y, d)$, then we can infer that $\mathsf{nca}(x, y)$ is at a depth at least d in the tree. Otherwise: $\mathsf{nca}(x, y) = \mathsf{nca}(anc(x, d), anc(y, d))$.

In our solution, with each node in the dynamic tree we store the *depth* of the node. The depth is encoded as a binary number—by using a list of records containing *nil* and *non-nil* pointers to represents zeros and ones respectively. Each time a new node is created (expand operation) we can calculate its depth by adding one to the depth of the parent node. All this can be accomplished in worst-case time complexity $O(\lg h)$, where h is the depth of the node. In addition, we maintain for each node in the tree a list of pointers to selected ancestor nodes (the *predecessors list* described in the next subsection). These pointers are used to perform a binary search leading to the identification of the nearest common ancestor. The resulting data structure, discovered independently, resembles the one used in [16], but does not assume constant time arithmetic capabilities.

The rest of this section describes how to realize these ideas. We start by describing the data structures needed to efficiently solve the $\mathcal{NCA}$ problem.

2.1 Predecessors List

To support the execution of the nca operation we need an additional data structure super-imposed on the dynamic tree: a *predecessors list (p-list)* attached to each node of the tree. Each element of the p-list is a record with two fields called *data* and *direction (dir)* (in addition to the fields to maintain the linked list).

The p-list of node x contains $\lfloor \lg h \rfloor + 1$ elements, where h is the depth of node x in the tree. The p-list of x is designed to contain pointers to ancestor nodes of x. In particular the p-list of x points to the ancestors of x which have distance $1, 2, 4, 8, \ldots, 2^{\lfloor \lg h \rfloor}$ from node x. Let us denote with $jump(x, k)$ the ancestor of node x which is at a distance k from x in the tree. The data field of the ith element of the p-list of x contains a pointer to the node $jump(x, 2^i)$, for $i = 0, 1, \ldots, \lfloor \lg depth(x) \rfloor$. The direction field of the ith element of the p-list of x contains the value L (which stands for "left") if x belongs to the left subtree of the node $jump(x, 2^i)$, and the value R otherwise.

The management of the p-lists is somewhat involved. Nothing special needs to be done whenever create_tree and remove are performed—observe that by definition the remove operation is applied only to leaves of the tree, and it will not affect the p-lists of other nodes. Let us show how we can create the p-list for each new node created when an expand operation is performed.

Creating the p-lists: Let us consider the execution of expand(x, y, z). Let us focus on the creation of y (the creation of z is basically identical). The data fields of the p-list of node y can be calculated from the p-list of node x: if the data field of the ith element of the p-list of x is $jump(x, 2^i)$, then the data field of the ith element of the p-list of y is a child of $jump(x, 2^i)$. If the direction field associated to $jump(x, 2^i)$ in the p-list of x is L (R), then the data field of the ith element in the p-list of y is the *left* (*right*) child of $jump(x, 2^i)$. Thus, we can create the data fields of the p-list of y by scanning the p-list of x and picking the appropriate child of the data field of each element. Since the p-list of x contains $O(\lg h)$ elements, and determining the corresponding element in the p-list of y can be done in constant time, this process requires $O(\lg h)$ time (h height of the tree). The only exception to this rule occurs when *depth*(y) is an exact power of two—in which case we need to extend the p-list by adding one element. The new element introduced will point to the root of the tree. The only problem is *how* to determine when this case occurs. This can be achieved in $O(\lg h)$ by scanning the list of records representing the *depth* of the parent node and ensuring that it is composed only by $non - nil$ pointers (i.e., the binary number representing the depth of the parent node is composed only by ones).

Updating the Directional Fields: In order to efficiently maintain the directional fields in the p-list, we need to modify the current picture. We introduce in each element of the p-list a field called *right-p-link (rplink)*, that points to an element belonging to the p-list of another node of the tree. In particular, the ith element of the p-list of node x contains a pointer to the $(i+1)^{th}$ element of the p-list of the node $jump(x, 2^i)$, if it exists (it is *nil* otherwise)—see Fig. 1.

Let us introduce some simplified notation. Given a node x in the tree, we will denote with *last(p-list*(x)) ($first$(*p-list*(x))) a pointer to the last (first) record in the p-list of node x. If s is a pointer to a p-list element, then *data*(s), *dir*(s), *lplink*(s) represent respectively the value of the *data*, *dir*, and *lplink* fields stored in the record pointed to by s. If s points to a p-list element, then *prev*(s) (*next*(s)) is a pointer to the p-list element preceding (following) s. The interesting property of this organization is that the creation of a new p-list can be obtained by simply following the appropriate right-p-links and copying the p-lists nodes encountered. Let y be a new node which is inserted in the tree as the left (right) child of node x. The creation of the p-list for y can be performed as follows. First of all, the first element of the p-list for y contains *(i)* a pointer to the node x in the data field, *(ii)* the direction L (R) in the direction field, and *(iii)* a pointer to the second node in the p-list of x in the right-p-link field. All the other elements of the p-list for y can be obtained essentially by following the (right-p-link) list of p-list elements starting with the first element

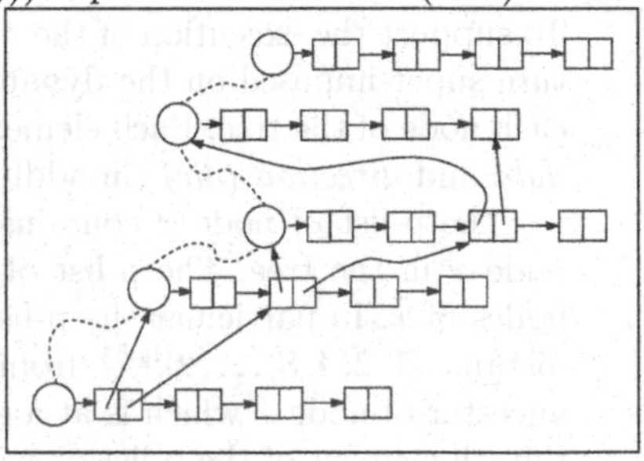

Fig. 1: Right-p-links in the p-lists

of the p-list of x and copying each one of them. The appropriate right-p-links for this new list can be calculated simultaneously as follows:

```
scan := first(p-list)(x) ; // first element of p-list of x
p-list(y) := new (); // create first element p-list of y
data(p-list(y)) := x;
rplink(p-list(y)) := next(p-list(x));
if (y = left(x)) then
     dir(p-list(y)) := L;
else
     dir(p-list(y)) := R;
endif
prev := p-list(y) ;
while (scan ≠ nil) do
     next(prev) := copy(scan); // create copy of element scan
     rplink(prev) := next(scan);
     scan := rplink(scan);
     prev := next(prev);
```

The new p-list may require an additional element if the depth of the node is a power of 2 (this can be done as described in the previous subsection).

As an example, consider the tree in Fig. 2[1] and let us assume that a new node (10) is inserted as right child of 9. The Figure shows the various p-lists (the notation m/L means that the p-list element points to the node m and the direction is L). The Figure shows the p-list for 10. The first element of the p-list contains a pointer to parent 9; the rest of the p-list is a copy of the list made by right-p-links and starting from the first element of the p-list of 9 (shown using solid lines). The right-p-links of the new elements are shown as dashed lines.

Note that the creation of the new p-list requires $\Theta(\lg h)$ time, where h is the depth of the new node. Hence, the **expand** operation takes time $\Theta(\lg h)$.

The correctness of the algorithm follows by observing that the data and directional fields of $(i+1)^{th}$ element of p-list(y) are copied from the ith element of the p-list of the node pointed to by the data field of the ith element of p-list(y). Noting that $jump(jump(y, 2^i), 2^i) = jump(y, 2^{i+1})$, it is easy to establish inductively that the data and directional fields of p-list(y) are correctly computed. The fact that the right-p-links are correct is also straightforward to establish.

The right-p-links are essential to guarantee a fast construction of the p-lists each time an **expand** operation is performed. To support the execution of the **nca** operation we will also require two additional fields in each element of the p-list. The first field is called *left-p-link (lplink)*. The left-p-link in the ith p-list element of the tree node x contains a pointer to the $(i-1)$th p-list element of the node $jump(x, 2^i)$. The second field is called *middle-p-link (mplink)*. The middle-p-link in the ith p-list element of the tree node x contains a pointer to the ith p-list element of the node $jump(x, 2^i)$. The left-p-links and middle-p-links can be maintained in a fashion similar to the right-p-links. Observe that it is not

[1] The Figure shows only the nodes on the path from the root to the new node.

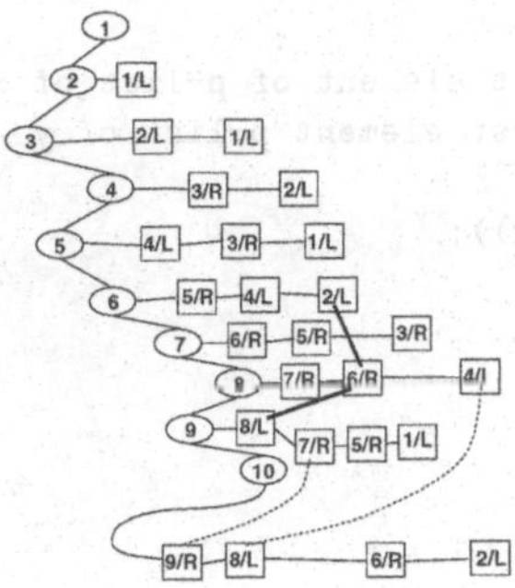

Fig. 2. Creation of p-list: example

```
nca(x,y):
1: repeat
2:   stepx := last(p-list(x))
3:   stepy := last(p-list(y))
4:   if (data(stepx) ≠ data(step(y)))
5:      x = data(stepx);
6:      y = data(stepy);
7:   endif
8: until (data(stepx) = data(stepy));
9: while ( true ) do
10: while (data(stepx) = data(stepy))
11:     stepx := prev(stepx);
12:     stepy := prev(stepy);
13: endwhile;
14: if (dir(next(stepx))≠dir(next(stepy)))
15:     then return (data(next(stepx)));
16: else
17:     x := data(stepx);
18:     y := data(stepy);
19:     stepx := lplink(stepx);
20:     stepy := lplink(stepy);
21: endif;
22: endwhile;
```

Fig. 3. Computation of the nca

necessary to maintain all the p-links; *e.g.*, the right-p-link can be determined from the middle-p-link. We use all the three p-links only for the sake of clarity.

2.2 Computing the nca

We subdivide the process of computing $\mathsf{nca}(x, y)$ into two subproblems: *(i)* determine the nearest common ancestor under the assumption that x and y are at the same depth; *(ii)* given x and y such that $depth(x) < depth(y)$ determine the ancestor y' of y such that $depth(y') = depth(x)$. It is clear that the ability to solve these two subproblems will provide a general solution to the task of determining $\mathsf{nca}(x, y)$ for arbitrary nodes x, y. Below we provide a $O(\lg h)$ solution for both these subproblems. This gives us a $O(\lg h)$ solution for the $\mathcal{NCA}$ problem.

Determining nca for Same Depth Nodes: We will make use of the elements of the p-list. Let us assume x and y to be two nodes that are at the same depth (h) in the tree. Then, the p-lists of x and y each contain $\lfloor \lg h \rfloor + 1$ elements. The computation of $\mathsf{nca}(x, y)$ is performed through the algorithm in Fig. 3.

The idea behind the algorithm is to locate the nearest common ancestor of two nodes by performing successive "jumps" in the tree, making use of the pointers stored in the p-list of the two nodes. The first loop (lines 1-8) is used to deal with the special case where the nca lies in the highest part of the tree (above the highest node pointed by the p-list). The loop in lines 10-13 compares ancestors of the two nodes, starting from the ancestor in the p-list which is farther away from the nodes x and y. During the successive iteration of this first loop we compare the nodes $jump(x, 2^i)$ and $jump(y, 2^i)$ for successively decreasing values of i (to avoid the use of arithmetic, we are simply moving pointers in the p-lists of x and y). The loop continues as long as the nodes reached are equal.

All these are common ancestors for x and y. As soon as we reach two ancestors $jump(x, 2^i)$ and $jump(y, 2^i)$ that are different (see Figure 4 on the left) we can verify whether the nearest common ancestor has been reached or not at the previous step ($jump(x, 2^{i+1})$). This test can be easily performed by making use of the directional information stored in the p-list (line 14).

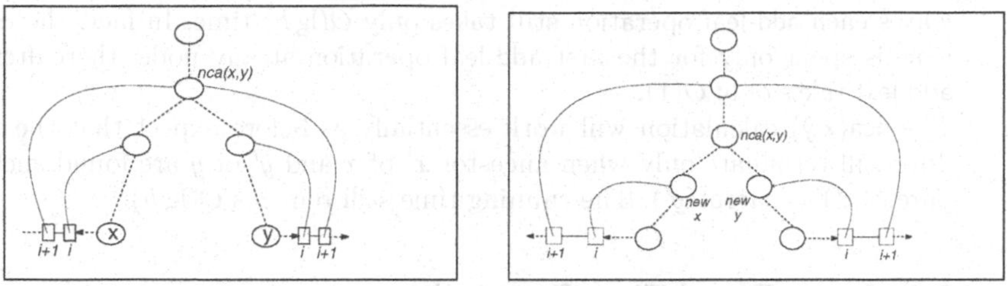

Fig. 4. Searching for Nearest Common Ancestor

If the ancestor $jump(x, 2^{i+1})$ is not the nearest common ancestor (see Figure 4 on the right), then the algorithm repeats the computation by replacing the nodes x and y respectively with $jump(x, 2^i)$ and $jump(y, 2^i)$. From the tests made earlier, we can limit the search to the ancestors of distance up to 2^i—and this is accomplished by starting a new iteration of the loop in line 9 not by taking the last element of the p-lists of the new x and y but by taking the p-list elements pointed by the left-p-links (i.e., maintain the same jump distance). Considering that the number of iterations performed is limited by the length of the longest p-list in the tree (that is $O(\lg h)$ for a tree of height h), the we can conclude that the algorithm has a complexity of $O(\lg h)$.

Determining Equal Height Ancestor: Fig. 5 provides a procedure that, given nodes x, y such that $depth(x) < depth(y)$, returns a node y' such that $y' \preceq y$ and $depth(y') = depth(x)$. The method uses the depth information stored in the nodes to determine the jump necessary to find the ancestor y', and uses the p-lists to jump to the correct ancestor. The subtraction in line 1 creates a list of records representing the binary number obtained from the subtraction of the depths of x and y. This operation can be done in time $O(\lg h)$ ($h = depth(y)$). In Fig. 5, a jump of size 2^ℓ is performed from the current ancestor of y if digit j_ℓ is one. All jumps together add to a total jump of j from y. The overall complexity is $O(\lg h)$.

```
1: j := depth(y) - depth(x);
2: let j = jt jt-1 ... j0; // digits of j
3: step := p-list(y);
4: current := y;
5: for i := 0 to t do
6:        if ( ji = 1 ) then
7             current := data(step);
8:            step := mplink(step);
9:        step := next(step);
10: return current;
```

Fig. 5: Determining Equal Height Ancestor

Comment: It is possible to work with p-lists where the directional information is not stored. In this case the p-lists of the nodes that are siblings are identical. These two facts can be used to save space and time as well as to extend the pro-

cedure to trees with unbounded arity. For trees with unbounded degree, when an operation $\mathsf{expand}(x, a_1, a_2, \ldots a_r)$ is performed only one p-list is created. Each of $a_1, a_2, \ldots a_r$ has a pointer to this p-list. Of course each of $a_1, a_2, \ldots a_r$ has a parent pointer to x. The time for this operation then clearly is $O(r + \lg h)$. Note that since the p-list is created only once this avoids the $\Omega(r \lg h)$ cost that the naive method will have to incur. If we think of expand operation as adding leaves each add-leaf operation still takes only $O(\lg h)$ time. In fact, the $\Theta(\lg h)$ time is spent only for the first add-leaf operation at any node, thereafter each add-leaf takes only $O(1)$.
The nca(x,y) calculation will work essentially as before expect that the procedure will terminate only when ancestor x' of x and y' of y are found such that $\text{parent}(x') = \text{parent}(y')$. The running time still remains $O(\lg h)$.

2.3 Lower Bound Time Complexity

We can prove a lower bound time complexity for this problem on PPMs as follows. Let us assume that we have a solution to the $\mathcal{NCA}$ problem and let us show how we can use it to build a solution for the *Temporal Precedence* ($\mathcal{TP}$) problem. The $\mathcal{TP}$ problem is the problem of performing an on-line sequence of operations, and the only two allowed operations are $\mathsf{insert}(x)$—that adds the element x to a collection of elements—and $\mathsf{precedes}(x, y)$—that verifies whether the element x has been inserted before the element y. In [14] we have proved that the $\mathcal{TP}$ problem has a lower bound time complexity of $\Omega(\lg \lg n)$ per operation, where n is the number of elements inserted.

Let us implement the two operations of the $\mathcal{TP}$ problem using the operations available in the $\mathcal{NCA}$ problem. The overall idea is to insert elements in a tree which is composed only by a right spine; each element inserted is a new leaf created down the right spine. Each precedes operation can be realized as follows: $(\mathsf{precedes}(x, y) \Leftrightarrow \mathsf{nca}(x, y) = x)$. This allows us to conclude that the $\mathcal{NCA}$ problem has a lower bound time complexity of $\Omega(\lg \lg n)$ as well (where n is the number of nodes in the tree).

3 Related Problems

In this section we formally define a number of problems pertaining to ancestor relationships in dynamic trees. The motivation to do this is two-fold. Several of these problems arise in concrete applications, such as implementation of declarative languages [12]. Secondly, we hope that the study of these problems will provide greater insight into the inherent complexity of the $\mathcal{NCA}$ problem. In particular, they might be of some help in closing the gap between the upper bound ($O(\lg n)$) and the lower bound ($\Omega(\lg \lg n)$) for the $\mathcal{NCA}$ problem on PPMs.

Ancestor Test: this problem introduces an additional operation on dynamic trees, called $\mathsf{is\_ancestor}(x, y)$, where x and y are two nodes in the tree. Given two nodes x, y in a dynamic tree, the operation determines whether x is an ancestor of y. More formally, $\mathsf{is\_ancestor}(x,y) = \begin{cases} \mathit{true} & x \preceq y \\ \mathit{false} & x \not\preceq y \end{cases}$ We will indicate with $\mathcal{ANC}$? the problem of executing an arbitrary correct, on-line sequence of create_tree, expand, remove, and is_ancestor operations. Note that this problem can be solved in worst-case constant time on pointer machines *with* uniform cost measure for arithmetic [16].

Nearest Common Ancestor Test: this problem introduces an additional operation on dynamic trees, called $\mathsf{is\_nca}(x,y,z)$, where x, y, and z are nodes in the tree. The operation returns *true* if z is the nearest common ancestor of x, y, and *false* otherwise. We will indicate with $\mathcal{IS\_NCA}$? the problem of executing a correct, on-line sequence of create_tree, expand, remove, and is_nca operations.

Left Test: this problem introduces an additional operation to manipulate dynamic trees, called $\mathsf{is\_left}(x,y)$, where x and y are arbitrary nodes in the tree. The operation returns *true* if x lies to the left of y in an in-order traversal of the tree, *false* otherwise. We will indicate with $\mathcal{LT}$? the problem of executing an arbitrary, correct, on-line sequence of create_tree, expand, remove, and is_left operations. Observe that the ability of performing arithmetic operation in constant time allows one to solve this problem in worst-case constant time [16].

Left Branch Test: this problem introduces an additional operation on dynamic trees, called $\mathsf{is\_on\_left\_branch}(x,y)$, where x and y are nodes in the tree and x is a leaf. The operation returns *true* if x lies in the leftmost branch of the tree rooted in y and *false* otherwise. More formally, $\mathsf{is\_on\_left\_branch}(x,y)$ returns *true* iff $y \trianglelefteq x$. We will indicate with $\mathcal{LB}$? the problem of executing a correct, on-line sequence of create_tree, expand, remove, and is_on_left_branch operations.

Same Branch Test: this problem introduces an additional operation to manipulate dynamic trees, called $\mathsf{same\_branch}(x,y)$, where x and y are nodes in the tree. The operation returns *true* if x and y lie in the same branch of the tree and *false* otherwise. More formally, $\mathsf{same\_branch}(x,y) \Leftrightarrow (x \preceq y \vee y \preceq x)$. We will indicate with $\mathcal{SB}$? the problem of executing a correct, on-line sequence of create_tree, expand, remove, and same_branch operations. Since the ancestor test can be performed in $O(1)$ time on pointer machines with arithmetic, the $\mathcal{SB}$? problem can be also solved in worst-case constant time on such machines.

In the successive subsections we consider some of the relationships between these problems and use them to build efficient solutions for them.

$\mathcal{ANC}$? and $\mathcal{NCA}$ Problems: The ability to solve the $\mathcal{NCA}$ problem provides a simple solution to the $\mathcal{ANC}$? problem as well. In fact, $\mathsf{ancestor}(x,y) \leftrightarrow \mathsf{nca}(x,y) = x$. Thus, a $f(n)$-time solution of the $\mathcal{NCA}$ problem automatically provides a $f(n)$-time solution for the $\mathcal{ANC}$? problem. The extent to which a solution of the $\mathcal{ANC}$? problem helps in solving the $\mathcal{NCA}$ problem is unclear.

$\mathcal{NCA}$ and $\mathcal{IS\_NCA}$? Problems: The ability to solve the $\mathcal{NCA}$ problem immediately provides a solution to the $\mathcal{IS\_NCA}$? problem—since $\mathsf{is\_nca}(x,y,z) \leftrightarrow \mathsf{nca}(x,y) = z$. We can also extend the lower bound proof proposed for the $\mathcal{NCA}$ problem to provide a lower bound time complexity for the $\mathcal{IS\_NCA}$? problem. The lower bound time complexity of this problem can be shown to be $\Omega(\lg\lg n)$

per operation. The proof is analogous to the one used in Sect. 2.3. In this case we can observe that precedes$(x, y) \Leftrightarrow$ is_nca(x, y, x). The question of whether a solution of $\mathcal{IS}\mathcal{NCA}$? helps in solving $\mathcal{NCA}$ problem is an open issue.

$\mathcal{NCA}$ and $\mathcal{LT}$? Problems: The ability to solve the $\mathcal{NCA}$ problem provides a solution to the $\mathcal{LT}$? problem. As illustrated in Figure 7, if z is the result of nca(x, y), then we have that is_left$(x, y) \leftrightarrow$ ancestor(left$(z), x)$, i.e., x is on the left of y if the left child of z is an ancestor of x. Note that the ancestor operation can be directly implemented using the nca operation, as seen earlier. The implementation of is_left using nca and ancestor is sketched in Fig. 6. Observe that, given an $f(n)$-time solution for the $\mathcal{NCA}$ problem, we can obtain an $f(n)$-time solution for the $\mathcal{LT}$? problem.

$\mathcal{NCA}$ and $\mathcal{SB}$? Problems: The ability to solve the the $\mathcal{NCA}$ problem provides an immediate solution to the $\mathcal{SB}$? problem. In fact, same_branch$(x, y) \Leftrightarrow \big(nca(x, y) = x \vee$ nca$(x, y) = y\big)$. It is not clear whether a solution of the $\mathcal{SB}$? problem is helpful to solve the $\mathcal{NCA}$.

```
is_left(x,y):
  if (ancestor(x,y)) then
     if (ancestor(left(x),y)) then
          return False;
     else
          return True;
  if (ancestor(y,x)) then
     return (ancestor(left(y),x));
  z := nca (x,y); // nca of the two nodes
  w := left (z); // left child of nca
return (ancestor(w,x));
```

Fig. 6. Implementation of is_left

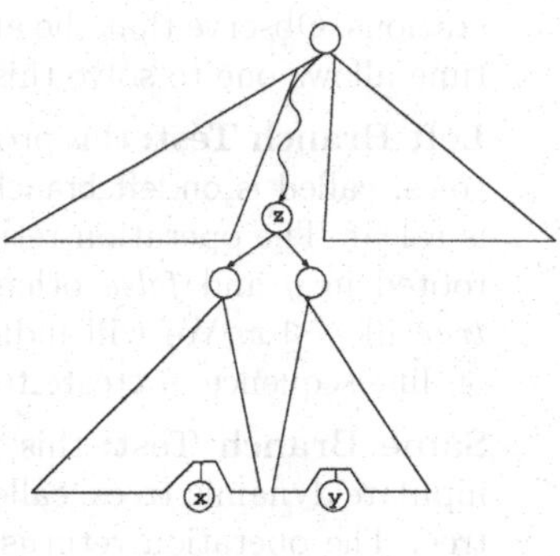

Fig. 7. is_left operation

$\mathcal{ANC}$? and $\mathcal{SB}$? Problems: A solution for the $\mathcal{ANC}$? problem can be used to provide a solution for the $\mathcal{SB}$? problem. In fact, same_branch$(x, y) \Leftrightarrow$ is_ancestor$(x, y) \vee$ is_ancestor(y, x). Vice versa, if we are given a solution for the $\mathcal{SB}$? problem, this can be used to provide a solution for the $\mathcal{ANC}$? problem. In fact

$$\text{is\_ancestor}(x, y) \Leftrightarrow \big(\text{same\_branch}(x, y) \wedge depth(x) \leq depth(y)\big)$$

This second property indicates that if we can provide a solution for $\mathcal{SB}$? which takes $f(n)$ per operation, then we can solve the $\mathcal{ANC}$? problem in time $f(n) + O(\lg\lg n)$ per operation—since the test on the depths of the nodes can be realized in $O(\lg\lg n)$ time using the solution to the $\mathcal{TP}$ problem.

Observe also that any solution to the $\mathcal{ANC}$ problem can be used to construct a solution to the $\mathcal{TP}$ problem (in a fashion similar to what used in Sect. 2.3, simply creating only the right spine of the tree at each insert and directly using is_ancestor to test precedes). This shows that the $\mathcal{ANC}$? problem has a worst-case time complexity $\Omega(\lg\lg n)$ per operation. The previous considerations also show that the $\mathcal{SB}$? problem has a time complexity of $\Omega(\lg\lg n)$ per operation.

Combining all these we can conclude that the problems $\mathcal{ANC}$? and $\mathcal{SB}$? are equivalent from the computational complexity aspects.

$\mathcal{NCA}$ and $\mathcal{LB}$? Problems: The $\mathcal{LB}$? problem introduces an additional operation to the manipulation of dynamic trees: is_on_left_branch(x, y), which returns *true* if the leaf x lies in the leftmost branch of the subtree rooted in y. In order to solve this problem we first need to introduce an additional data structure to the representation presented earlier. Below we present this additional data structure and we show how it can be used in the context of the $\mathcal{LB}$? problem.

Maintaining the Frontier and the Subroots: The problem we would like to solve is the following: we want to efficiently determine the subroot node of each leaf in the tree. The subroot node is defined as follows: the subroot node of a leaf ℓ is the highest node n in the tree (i.e., closest to the root) such that ℓ lies in the leftmost branch of n. Using the notation introduced earlier, $n = \mu(\ell)$.

Identification of the Frontier: Let us show how to keep an explicit representation of the frontier of a tree built using the expand and remove operations. Let us assume that the tree is implemented in the straightforward way on a PPM—each record contains pointers to the parent node and to the left and right children. Each operation to manage the tree has complexity $\Theta(1)$ [13]. The tree representation can be extended to maintain an explicit representation of the frontier of the tree at each step of execution. The nodes in the frontier are connected (from left to right) in a doubly-linked list. Each record representing a node is extended with two fields, pred and succ, used to link the node to its predecessor and successor in the frontier. These pointers are defined only for nodes that are leaves in the tree. During an *expansion*, a leaf n is expanded adding two new nodes n_1 and n_2. The new frontier is obtained by removing n from the list and replacing it with n_1, n_2. This can be achieved by modifying a constant number of pointers. During a *deletion*, a leaf n is removed from the frontier. If n is the only child of its parent (m), then we need to replace n with m in the frontier. If n is the left (right) child of its parent (m), and m has also a right (left) child, then the deletion involves eliminating one element from the frontier. Thus, the predecessor of n in the list is linked directly to the successor of n. In any case, at most 3 records need to be modified, and all of them can be obtained from the node to be removed. Thus, it is possible to prove [13] that the frontier of a tree created via a valid sequence of operations create_tree, expand, and remove, can be maintained as a doubly-linked list with $\Theta(1)$ time per operation.

Subroot Nodes: We tackle now the problem of maintaining (in constant time) the subroot node information for each leaf. The result is based on the following lemma [13]: If n and m are two leaves of a binary tree, $\mu(m)$ changes when n is removed iff n is the left child of a node k and $\mu(m)$ is the right child of k. In addition, if n and m are two distinct leaves, then $\mu(n) \neq \mu(m)$. These results imply that the removal of a leaf from a tree will affect the μ of at most one leaf.

The record structure is modified by adding one field (sub). sub contains a meaningful value only in the records of the leaves. If n is a node in the tree and r is its record, then *sub*(r) contains the address of the record representing $\mu(n)$. During the tree creation, the *sub* field of the root of the tree points to the node itself. When a node n is expanded, the subroot nodes for the new leaves can be determined as follows: the *sub* field of the left child is a copy of the *sub* field of n, while the *sub* field of the right child is set to point to the right child itself. During the removal of a leaf n, three cases may occur. If n is the only child of its parent, then the removal of the node will not affect the μ of any other leaf. Thus the only operation required is to copy the value of sub of the leaf to the sub field of the parent. If n is the right child of a node m, which has also a left child, then the removal of n will not modify any μ. In fact, from the previous lemma we gain that only the removal of a left child of a node can affect some μ. If n is the left child of k, and k has a right branch, then the removal of n can affect the μ of at most one leaf. Let m be the leaf which follows n in the frontier. From how the tree is built, m must be in the leftmost branch of the tree rooted in the right child of k. Thus, when n is removed, $\mu(m)$ should be given the value $\mu(n)$. It is easy to see that the cost of maintaining the sub field is $O(1)$ [13].

A Solution for the $\mathcal{LB}$? Problem: Once we have determined how to maintain the subroot node for each element in the frontier of the tree, we can proceed and provide a solution to the $\mathcal{LB}$? problem. Let us consider the code in Fig. 8.

If the node y is not an ancestor of x (first test) then clearly x cannot be in the leftmost branch of y. If y is an ancestor of x, then is_on_left_branch should return *true* only if the subroot node of x is an ancestor of node y. Since we know that both y and the subroot node of x are on the same path, this last test can be performed by simply comparing the depths of these two nodes (which can be efficiently performed in $O(\lg\lg n)$—by maintaining depths of the nodes in the tree using a temporal precedence list, as in the $\mathcal{TP}$ problem mentioned earlier). Since the management of the subroot nodes for the frontier of the tree can be performed with time complexity $O(1)$, then the cost of this implementation of is_on_left_branch is $O(\lg h) + O(\lg\lg n)$.

```
is_in_left_branch(x,y):
    if (not ancestor(y,x)) then
        return False;
    z = subroot (x);
    if (depth(z) ≤ depth(y)) then
        return True;
    else
        return False;
```

Fig. 8: Solving $\mathcal{LB}$?

If we simplify the problem, assuming that we perform the test only for cases where y is an ancestor of x, then the cost of the operation is sipmly the cost of a depth comparison ($O(\lg\lg n)$). Note that this last condition is met in some of the practical applications of this test [8]. Note that the ability to perform constant time arithmetic provides a worst-case constant time solution to this problem.

4 Remarks and Future Work

In the previous sections we have provided a novel data structure which can be used to efficiently compute nearest common ancestors in a dynamic tree on pure pointer machines. We have also shown how this solution can be employed to provide efficient solutions to a range of other problems on dynamic trees, that frequently arise in the context of sequential and parallel implementation of declarative and search-based programming languages [12]. The data structures have been described in the context of dynamic *binary* trees; nevertheless, it is straightforward to extend the data structures to deal with trees of *arbitrary* arity, without affecting the asymptotic complexity.

There are a number of open problems that still have to be explored. First of all, we only have an $\Omega(\lg\lg n)$ lower bound to the complexity of the $\mathcal{NCA}$ problem; a tighter result is needed. We have recently proposed [6] an extension of the scheme described in this paper to provide an $O(\lg\lg n)$ upper bound for the $\mathcal{NCA}$ problem, thus providing matching lower and upper bounds. Various relationships between the different problems have not been explored, and they may provide better insights on the actual complexity of these problems. We are also considering various relaxations of the problems described—e.g., by removing the requirements that x is a leaf in the $\mathsf{is\_left}(x, y)$ operation.

References

1. S. Alstrup and M. Thorup. Optimal Pointer Algorithms for Finding Nearest Common Ancestors in Dynamic Trees. *Journal of Algorithms*, 35:169–188, 2000.
2. A.M. Ben-Amram. What is a Pointer Machine? Tech. Rep., U. Copenhagen, 1995.
3. M.A. Bender and M. Farach-Colton. The LCA Problem Revisited. In *LATIN 2000*, Springer Verlag, 2000.
4. O. Berkman and U. Vishkin. Recursive $*$-tree Parallel Data-structure. In *FOCS*, IEEE Computer Society, 1989.
5. R. Cole and R. Hariharan. Dynamic LCA Queries on Trees. In *Proceedings of the Symposium on Discrete Algorithms (SODA)*, pages 235–244. ACM/SIAM, 1999.
6. A. Dal Palú, E. Pontelli, D. Ranjan. An Optimal Algorithm for Finding NCA on Pure Pointer Machines. Tech. Rep. NMSU-TR-CS-007/2001, 2001.
7. H.N. Gabow. Data structures for weighted matching and nearest common ancestor. In *ACM Symp. on Discrete Algorithms*, pages 434–443, 1990.
8. G. Gupta, E. Pontelli, et al. Parallel Execution of Prolog Programs: a Survey. *ACM TOPLAS*, 2002. (to appear).
9. D. Gusfield. *Algorithms on Strings, Trees, and Sequences.* Cambridge University Press, 1999.
10. D. Harel and R.E. Tarjan. Fast Algorithms for Finding Nearest Common Ancestor. *SIAM Journal of Computing*, 13(2):338–355, 1984.
11. D.E. Knuth. *The Art of Computer Programming*, volume 1. Addison-Wesley, 1968.
12. E. Pontelli, D. Ranjan, and G. Gupta. On the Complexity of Parallel Implementation of Logic Programs. In *FSTTCS*, 1997. Springer Verlag.
13. D. Ranjan, E. Pontelli, and G. Gupta. Data Structures for Order-Sensitive Predicates in Parallel Nondetermistic Systems. *ACTA Informatica*, 37(1):21–43, 2000.
14. D. Ranjan, E. Pontelli, L. Longpre, and G. Gupta. The Temporal Precedence Problem. *Algorithmica*, 28, 2000.
15. B. Schieber and U. Vishkin. On Finding Lowest Common Ancestors. *SIAM J. Computing*, 17:1253–1262, 1988.
16. A.K. Tsakalidis. The Nearest Common Ancestor in a Dynamic Tree. *ACTA Informatica*, 25:37–54, 1988.

Searching in Random Partially Ordered Sets
(Extended Abstract)

Renato Carmo[1,2]*, Jair Donadelli[2]**, Yoshiharu Kohayakawa[2]***, and Eduardo Laber[3]

[1] Departamento de Informática
Universidade Federal do Paraná
Centro Politécnico da UFPR, 81531–990, Curitiba PR, Brazil
http://www.inf.ufpr.br
renato@inf.ufpr.br

[2] Instituto de Matemática e Estatística
Universidade de São Paulo
Rua do Matão 1010, 05508–900 São Paulo SP, Brazil
http://www.ime.usp.br/mac
{renato,jair,yoshi}@ime.usp.br

[3] Departamento de Informática
Pontifícia Universidade Católica do Rio de Janeiro
R. Marquês de São Vicente 225, Rio de Janeiro RJ, Brazil
http://www.inf.puc-rio.br/
laber@info.puc-rio.br

Abstract. We consider the problem of searching for a given element in a partially ordered set. More precisely, we address the problem of computing efficiently near-optimal search strategies for typical partial orders. We consider two classical models for random partial orders, the *random graph model* and the *uniform model*.

We shall show that certain simple, fast algorithms are able to produce nearly-optimal search strategies for typical partial orders under the two models of random partial orders that we consider. For instance, our algorithm for the random graph model produces, in linear time, a search strategy that makes $O\left((\log n)^{1/2} \log\log n\right)$ more queries than the optimal strategy, for almost all partial orders on n elements. Since we need to make at least $\lg n = \log_2 n$ queries for *any* n-element partial order, our result tells us that one may efficiently devise near-optimal search strategies for almost all partial orders in this model (the problem of determining an optimal strategy is NP-hard, as proved recently in [1]).

1 Introduction

A fundamental problem in data structures is the problem of representing a dynamic set S in such a way that we may efficiently search for arbitrary elements u

* Partially supported by PICDT/CAPES.
** Supported by a CNPq doctorate Studentship (Proc. 141633/1998-0).
*** Partially supported by MCT/CNPq through ProNEx Programme (Proc. CNPq 664107/1997–4) and by CNPq (Proc. 300334/93–1 and 468516/2000–0).

S. Rajsbaum (Ed.): LATIN 2002, LNCS 2286, pp. 278–292, 2002.

in S. Perhaps the most common assumption about S is that its elements belong to some *totally ordered set* U. Here, we are interested in examining a certain variant of this problem, where the 'universe' U from which our elements are drawn is a *partially ordered set.* In fact, we address the basic problem of efficiently computing near-optimal search strategies for typical partial orders under two classical models for random partial orders: the *random graph model* and the *uniform model.*

The problem we consider is, intuitively speaking, as follows: suppose we are given a partial order $S = (S, \prec)$, where $S \subset U$ and $\prec$ is a partial order on U. We wish to construct a search strategy T that, given an arbitrary $u \in U$, determines whether or not $u \in S$. We suppose that T has access to an oracle that, given $s \in S$, replies whether or not $s = u$, and if this is not the case then tells T whether or not $u \prec s$. We measure the efficiency of T by the number of queries that it needs to send to the oracle in the worst case (we maximize over all $u \in U$). Our problem may then be summarized as follows:

1. an instance is a partial order $(S, \prec)$, its size given by the size of a directed acyclic graph representing its Hasse diagram;
2. a solution is a search strategy T;
3. the aim is to minimize number of queries in the worst-case performance of T, and
4. the number of steps needed to compute T is to be kept as low as possible.

We shall show that certain simple, fast algorithms are able to produce nearly-optimal search strategies for typical instances under the two models of random partial orders.

1. *Random graph model.* In the *random graph model*, we randomly select a graph on the vertex set $[n] = \{1, \ldots, n\}$ and say that $i \prec j$ in our partial order if $i < j$ and one may reach j from i along an 'increasing path' in the graph. (We in fact consider random graphs of density p, for constant p.) We present an algorithm that, *in time linear in the size of the instance* $(S, \prec)$, produces a search strategy that makes at most
$$\lg n + O\left((\log n)^{1/2} \log \log n\right)$$
queries in the worst case, for almost all n-element partial orders in this model (that is, with probability tending to 1 as $n \to \infty$).
2. *Uniform model.* In the *uniform model*, we simply consider all partial orders on $[n]$ equiprobable. The situation here is somewhat different: it is easy to show that almost all n-element partial orders are such that any search strategy makes at least
$$\left(\frac{1}{4} + o(1)\right) n$$
queries in the worst case. However, if we consider a slightly more 'generous' oracle, then almost all partial orders admit search strategies that require only
$$O(\log n)$$

queries. Moreover, this search strategy may be computed in time $O(n^2)$. For this result, we suppose that, presented with the query $s \in S$, the oracle replies whether or not $s = u$, and if this is not the case then it tells us whether $u \prec s$, $s \prec u$, or u is not comparable with s. Thus, our more generous oracle tells us exactly which of the three possibilities holds when $u \neq s$, whereas the previous oracle only told us whether or not $u \prec s$ holds.

Since we need to make at least $\lg n$ queries for *any* n-element partial order, our result tells us that one may efficiently devise near-optimal search strategies for almost all partial orders in the first model, and we may devise search strategies for almost all partial orders in the second model that are only worse than the optimal by a constant factor. We remark that the constants hidden in the big-O notation and in the discussion above are not at all large.

Summarizing, (*i*) one may compute an essentially best possible search strategy for almost all random graph orders and (*ii*) one may compute a constant factor approximation for almost all partial orders. This is in pleasant constrast to the fact that determining an optimal strategy is NP-hard, as recently proved in [1]. Finally, as the reader will see, it will be quite surprisingly simple to prove our results.

1.1 Definitions and Notation

A *partial order* is a pair $(U, \prec)$ where U is a set and $\prec$ a binary relation on U which is anti-symmetric, transitive and irreflexive. If $x, y \in U$ then $x \prec y$ stands for $(x, y) \in \prec$. The elements $u, v \in U$ are said to be *comparable* in $(U, \prec)$ if $u \prec v$ or $v \prec u$, being otherwise said to be *incomparable*; a *chain* in $(U, \prec)$ is a set $X \subseteq U$ of pairwise comparable elements; an *antichain* of $(U, \prec)$ is a set $X \subseteq U$ of pairwise incomparable elements; the *height* of $(U, \prec)$ is the size of a maximum chain and the *width* is the size of a maximum antichain. An element $u \in U$ is maximal in $(U, \prec)$ if there is no $x \in U$ such that $u \prec x$.

An *ideal* of $(U, \prec)$ is a set $I \subseteq U$ such that $u \prec i \Rightarrow u \in I$ for all $i \in I$ and a *filter* of $(U, \prec)$ is a set $F \subseteq U$ such that $f \prec u \Rightarrow u \in F$ for all $f \in F$; if $X \in U$, we denote by $I(X)$ (resp. $F(X)$) the minimal ideal I (resp. filter F) of $(U, \prec)$ such $X \subseteq I$ (resp. $X \subseteq F$).

Thoughout the text, we denote a set of integers $\{z \in \mathbb{N}\colon x \leq z \leq y\}$ by $[x, y]$ and define $[n] = [1, n]$ for any integer $n \geq 1$. Also, ω will denote a function $\omega : \mathbb{N} \to \mathbb{R}$ satisfying $\lim_{n\to\infty} \omega(n) = \infty$.

The problem of searching through a partially ordered set can be stated as follows. We are given a partially ordered set $(U, \prec)$ and a finite set $S \subseteq U$ with the partial order induced by $\prec$. Our goal is to determine whether a given $u \in U$ is an element of S and we are allowed to pose *queries* about u to the elements of S.

A *query about* $u \in U$ *to* $s \in S$ has three possible outcomes, namely, hit, smaller and no, meaning $u = s$, $u \prec s$ and $u \not\prec s$, respectively. A *search for* $u \in U$ *through* S is a sequence of queries allowing to decide whether $u \in S$ or not. The goal is to devise a strategy for querying the elements of S in such a way that the longest search through S poses the smallest number of queries.

Such a strategy can be conveniently thought of as a binary decision tree whose internal nodes are labelled with elements of S, whose external nodes are labelled with pairs $(s_{\min}, s_{\max}) \in (S \cup \{-\infty\}) \times (S \cup \{+\infty\})$ and whose edges are labelled with $\prec$ and $\not\prec$.

A path in the decision tree from its root to an internal node labelled s represents a successful search whose outcome is $u = s$, while a path from the root to an external node labelled $(s_{\min}, s_{\max})$ represents an unsuccessful search whose outcome is $u \notin S$ with the additional information that $s_{\min} \prec u \prec s_{\max}$. We define the *height of a binary decision tree* as the length of a longest path from its root to an external node.

We can then restate our problem as follows: *given a partial order* $(S, \prec)$, *compute a binary decision tree of minimum height for* S.

We will denote by n the number of elements of S and will assume that the partial order $(S, \prec)$ is given as a directed acyclic graph $G_\prec$ of m edges and n vertices , representing its Hasse diagram. Then, given an algorithm $\mathcal{A}$ for computing such a decision tree, we focus on the height $H_{\mathcal{A}}(n, m)$ of the tree that we construct and on the number of steps $T_{\mathcal{A}}(n, m)$ required by $\mathcal{A}$ to construct such tree.

We note that when $\prec$ is a total order on S, the optimum decision tree for S is the usual binary search tree for S. In this case we have $m = n - 1$, $H(n, m) = \lceil \lg n \rceil + 1$ and $T(n, m) = O(n)$, these being the best possible values for H and T.

On the other extreme, if the whole of S is an antichain, then we have $m = 0$, $H(n, m) = n$ and $T(n, m) = O(n)$.

1.2 Organization

This extended abstract is organized as follows. In Section 2 we mention some relevant results in the context of searching in partially ordered sets. In Section 3 we present a linear time algorithm for building a decision tree which has height almost surely bounded by $\lg n + O\left((\log n)^{1/2} \log\log n\right)$ under the random graph model for n-element partial orders.

In Section 4 we present a $O(n^2 \log n)$ time algorithm for building a decision tree which has height almost surely bounded by $O(\log n)$ under the uniform model for n-element partial orders (assuming the search model with the 'generous' oracle).

In Section 5 we make some general ramarks and briefly discuss some connections among our results and a related problem proposed in [2].

2 Related Work

It is shown in [1] that the problem of searching through a partially ordered set is NP-hard, even when restricted to the case in which $(S, \prec)$ has a maximum element. The more restricted case in which the given partial order has a maximum element and $G_{\prec}$ is a tree (the so called *rooted tree* variant) can be solved in polynomial time as shown in [3], where an algorithm for computing a minimum height decision tree is presented with $T(n,m) = O(n^4(\log n)^3)$. Their algorithm does not yield an easy way to estimate $H(n,m)$, except in a few cases: for instance, when $G_{\prec}$ is a complete binary tree, the decision tree built by their algorithm has $H(n,m) = \lg n + \lg^* n + \Theta(1)$.

The work in [1] presents a much simpler algorithm for the case in which $G_{\prec}$ is a rooted tree, which computes in time $T(n,m) = O(n \log n)$ a decision tree whose height is at most $\lg n$ greater than the height of the optimum decision tree. Since the optimum decision tree must have height at least $\lg n$, their algorithm constitutes a 2-approximation algorithm for the problem.

The case in which $(S, \prec)$ has a maximum element and the maximum degree of $G_{\prec}$ is d is also studied in [1], where it is shown that $d \log_d n$ is an upper bound for the height of an optimum decision tree, which improves to the best possible a previous bound from [4].

Lipman and Abrahams [5] present optimized exponential time algorithms for building decision trees for searching in general partially ordered sets. Nevertheless, they considered the minimization of the expected path length of the decision tree.

Linial and Saks consider in [2] a different although related problem, motivated by the following setting: suppose we are given an $m \times n$ real matrix M whose (say, distinct) entries are known to be increasing along the rows and along the columns and suppose we wish to decide whether a given real number x occurs in M. The goal is to devise a search strategy which minimizes the number of inspections of entries of M in the worst case.

If one looks at the matrix as the product of two chains of length m and n, the problem may be thought of as a problem of searching in a partially ordered set. However, the underlying assumption that the entries of M come from a totally ordered set actually turns it into a different problem, which we discuss in Section 5.

Linial and Saks (see [2] and [6]) have determined bounds for arbitrary orders $\prec$ and have studied in detail the case in which $\prec$ is a product of chains and the case in which $G_{\prec}$ is a rooted tree.

3 The Random Graph Model

Let $(S, \prec)$ be a partial order and denote by $\max_{\prec} X$ the set of maximal elements of $X \subseteq S$. Consider the following recursive decomposition of S into antichains: let $L_1 = \max_{\prec} S$, and let $L_i = \max_{\prec}\left(S - \bigcup_{j=1}^{i-1} L_j\right)$ for $i > 1$. Let h be the height of $(S, \prec)$ and let $w = \max_{1 \le i \le h} |L_i|$. We will call each set L_i the *layer* i of S.

A possible adaptation of the usual binary search strategy to the case of partial orders may be described as follows.

Algorithm $\mathcal{B}$

1. Let m be the index of the layer which divides S in two parts, each of them with less than $|S|/2$ elements, that is, m is such that $|\bigcup_{i=1}^{m-1} L_i| < |S|/2$ and $|\bigcup_{i=m+1}^{h} L_i| < |S|/2$. Denote these halves of S by $L_\downarrow$ and $L_\uparrow$, respectively.
2. Perform a query about u to each $s \in L_m$:
   (a) if the outcome of one of these queries is hit, the search is over;
   (b) if the outcome of one of these queries is smaller, restart the search, restricted to $L_\downarrow$;
   (c) if the outcome to all these queries is no, restart the search, restricted to $L_\uparrow$.

Let us call the above strategy an *extended binary search*, and denote the algorithm which computes the respective decision tree by $\mathcal{B}(S, \prec)$. We clearly have $H_\mathcal{B}(n,m) \leq w \lceil \lg h \rceil$; moreover, as the layering of S can be produced in time $O(m)$ and the building of the tree takes one step per element of S, we have $T_\mathcal{B}(n,m) = O(n+m)$.

We now turn our attention to another strategy: suppose we have $\{d_i\}_{i=1}^k \subseteq S$ such that $\{d_1\} = I(d_1) \subset I(d_2) \subset I(d_3) \subset \ldots \subset I(d_k) \subseteq S$. In this case we define the *segments* S_i $(0 \leq i \leq k)$ of $(S, \prec)$ by

$$S_i = \begin{cases} \{d_1\} & \text{if } i = 0, \\ I(d_{i+1}) - I(d_i) & \text{if } 0 < i < k, \\ S - I(d_k) & \text{if } i = k. \end{cases}$$

We can then formulate the following algorithm which takes advantage of the above structure.

Algorithm $\mathcal{A}$

1. Perform a (usual) binary search of u through $\{d_i\}_{i=1}^k$; if u is found stop, otherwise let i be the minimum index such that $u \prec d_i$ (or k, if there is no such index).
2. Perform an extended binary search through the segment S_i.

We have

$$H_\mathcal{A}(n,m) \leq \lceil \lg k \rceil + \max_{1 \leq i \leq k} H_\mathcal{B}(|S_i|, m) \leq \lg n + w \lg \left(\max_{1 \leq i \leq k} |S_i| \right),$$

and this tree can be built in $T_\mathcal{B}(n,m) = O(n+m)$ steps.

In the following sections we introduce the *random graph order* model for partial orders and show that, in this model, we almost surely have

$$H_\mathcal{A}(n,m) = \lg n + O(\sqrt{\log n} \log \log n).$$

3.1 The Random Graph Model

The *random graph order* probability space, denoted $\mathcal{P}_{n,p}$, is the probability space of all partial orders $([n], \prec)$ obtained by independently choosing each pair of $\{(i,j) \in [n]^2 : i < j\}$ with probability p and taking the transitive closure of the resulting relation. We denote a generic partial order $([n], \prec)$ in $\mathcal{P}_{n,p}$ by $P_{n,p}$.

We say that $d \in [n]$ is *k-dominating* if $[d-k, d-1] \subseteq I(d)$ and call d a *dominating* element of $[n]$ if d is $(d-1)$-dominating (that is, if $x \prec d$ for all $1 \leq x < d$). The conditional probability that d should be k-dominating, given that d is $(k-1)$-dominating, is $1-(1-p)^k$, which leads us to

$$\mathbb{P}(d \text{ is } k\text{-dominating}) = (1-(1-p)^k)\mathbb{P}(d \text{ is } (k-1)\text{-dominating}).$$

By induction on k, we have

$$\mathbb{P}(d \text{ is } k\text{-dominating}) = \eta(k,p),$$

where $\eta(k,p) = \prod_{i=1}^{k}(1-(1-p)^i)$. We note that η is a strictly decreasing function of k and define $\eta(p) = \lim_{k\to\infty} \eta(k,p)$; then, for all $k \geq 1$, we have $\eta(p) < \eta(k,p) \leq p$.

Lemma 1. *The probability that $d \in [n]$ is dominating is $\eta(d-1,p)$.*

As a consequence, we note that the expected number of dominating elements in $P_{n,p}$ is greater than $n\eta(n,p)$. By setting $\{d_i\}_{i=1}^k$ to be the dominating elements in $P_{n,p}$, we meet the necessary conditions for applying algorithm $\mathcal{A}$ on $P_{n,p}$.

In what follows, we will show that the decision tree built by $\mathcal{A}(P_{n,p})$ almost surely has "small height", by showing that both the size of each segment S_i and the size of each layer L_j are suitably small.

To show that the size of each segment is not too large we show that we have no large intervals of $[n]$ free of dominating elements in $P_{n,p}$; to show that the size of each layer is not too large, we use that we cannot have large antichains in $P_{n,p}$.

On the Size of the Segments. Let us consider the case in which d is not dominating. In this case, there must be a minimum b such that d is $(b-1)$-dominating but it is not b-dominating. If we call such b a *barrier* for (the domination of) d, then we have, for each $0 < b < d-1$,

$$\begin{aligned}\mathbb{P}(b \text{ is a barrier for } d) &= \mathbb{P}(d-b \not\prec d \textbf{ and } d \text{ is } (b-1)\text{-dominating}) \\ &= \mathbb{P}(d-b \not\prec d \mid d \text{ is } (b-1)\text{-dominating}) \\ &\qquad \times \mathbb{P}(d \text{ is } (b-1)\text{-dominating}) \\ &= (1-p)^b \eta(b-1,p).\end{aligned}$$

Thus, for any $0 < s < d-1$, we have that the probability that d is not dominating but d is s-dominating is

$$\mathbb{P}\left(\bigvee_{b=s+1}^{d-2}\left\{b \text{ is a barrier for } d\right\}\right) = \sum_{b=s+1}^{d-2} \mathbb{P}(b \text{ is a barrier for } d)$$

$$= \sum_{j=0}^{d-s-3} \mathbb{P}(j+s+1 \text{ is a barrier for } d) = \sum_{j=0}^{d-s-3} (1-p)^{j+s+1}\eta(j+s,p)$$

$$\leq \eta(s,p)(1-p)^{s+1}\sum_{j\geq 0}(1-p)^j = \frac{\eta(s,p)}{p}(1-p)^{s+1} \leq (1-p)^{s+1}.$$

Now, if $M \subseteq [n]$ and $d = \min\{|m-m'|\colon m, m' \in M\}$, then the events "$m$ is d-dominating" are mutually independent for all $m \in M$. For convenience, let D and D_d denote the set of dominating and d-dominating elements in $P_{n,p}$. We have

$$\begin{aligned}\mathbb{P}(D \cap M = \emptyset) &= \mathbb{P}(D \cap M = \emptyset \mid D_d \cap M = \emptyset)\mathbb{P}(D_d \cap M = \emptyset)\\ &\qquad + \mathbb{P}(D \cap M = \emptyset \text{ and } D_d \cap M \neq \emptyset)\\ &\leq \mathbb{P}\left(\bigwedge_{m\in M}\{m \notin D_d\}\right) + \mathbb{P}\left(\bigvee_{m\in M}\{m \notin D \text{ and } m \in D_d\}\right)\\ &\leq (1-\eta(d,p))^{|M|} + \sum_{m\in M}(1-p)^{d+1}\\ &= (1-\eta(d,p))^{|M|} + |M|(1-p)^{d+1}.\end{aligned} \tag{1}$$

Let us put

$$\beta(p) = \lg\frac{1}{1-p}, \qquad \gamma(p) = \lg\frac{1}{1-\eta(p)}, \qquad \alpha(p) = \frac{2}{\beta(p)\gamma(p)}.$$

Theorem 2. *If $g \geq \lceil \alpha(p)(\lg n\omega(n))^2\rceil$ and $x < n-g$, the probability that the set $[x, x+g]$ has no dominating element is at most $1/n\omega(n)$.*

Proof. If

$$d \geq \frac{\gamma(p)m + \lg m}{\beta(p)} - 1, \tag{2}$$

then $m(1-p)^{d+1} \leq (1-\eta(p))^m$, so that if $|M| = m$ in (1) then the probability that there is no dominating element in M is $2(1-\eta(p))^m$. If

$$m \geq \frac{\lg 2n\omega(n)}{\gamma(p)}, \tag{3}$$

then the probability that there is no dominating elements in M is $\leq (n\omega(n))^{-1}$.

We can then set $M = \{x + id\}_{i=0}^{m-1}$, with d and m satisfying (2) and (3) so that we have $M \subseteq [x, x+g]$ with

$$g \geq dm \geq \frac{(\lg 2n\omega(n))^2 + (\lg 2n\omega(n))\left(\lg\lg 2n\omega(n) - \left(\frac{1}{\beta(p)} + \lg\gamma(p)\right)\right)}{\beta(p)\gamma(p)}$$

$$\geq \frac{2(\lg n\omega(n))^2}{\beta(p)\gamma(p)} = \alpha(p)(\lg n\omega(n))^2.$$

□

Corollary 3. *The probability that $P_{n,p}$ has a segment of size $\geq \alpha(p)(\lg n\omega(n))^2$ is at most $1/\omega(n)$.*

Proof. Let $\{S_j\}_{j=1}^k$ be the segments of $P_{n,p}$. Then

$$\mathbb{P}\left(\bigvee_{j=1}^{k}\left\{|S_j| > \alpha(p)(\lg n\omega(n))^2\right\}\right) \leq \sum_{j=1}^{k}\mathbb{P}(|S_j| > \alpha(p)(\lg n\omega(n))^2)$$

$$\leq \sum_{j=1}^{k}\frac{1}{n\omega(n)} = \frac{k}{n\omega(n)} \leq \frac{1}{\omega(n)}.$$

□

On the Size of the Layers. Consider a layer L_i. Since L_i is an antichain in $P_{n,p}$, we can make direct use of the following result from Barak and Erdős [7].

Theorem 4 (Barak and Erdős [7]). *The probability that $P_{n,p}$ has an antichain of size larger than*

$$K_n = \sqrt{\frac{2\lg n}{\beta(p)} + \frac{1}{4}} + \frac{1}{2} \tag{4}$$

tends to zero as $n \to \infty$.

The Decision Tree Is Not Too Tall. We are now in position to prove the main result of this section.

Theorem 5. *The decision tree built by algorithm $\mathcal{A}(P_{n,p})$ has height almost surely bounded by $\lg n + O(\sqrt{\log n}\log\log n)$.*

Proof. Let H be the random variable in $\mathcal{P}_{n,p}$ whose value is the height of the decision tree built by the algorithm on the input $P_{n,p}$. As has been noted, we have $H \leq \lg n + w \lg \max_{1\leq i\leq h}|S_i|$, where w is the size of the greatest layer of $P_{n,p}$, the S_i are its segments, and h is its height.

Corollary 3 tells us that

$$\mathbb{P}\left(\max_{1\le i\le h} |S_i| > \alpha(p)(\lg n\omega(n))^2\right) < \frac{1}{\omega(n)},$$

while Theorem 4 gives

$$\mathbb{P}\left(w > \frac{1}{2} + \sqrt{\frac{2\lg n}{\beta(p)} + \frac{1}{4}}\right) < \frac{1}{\omega'(n)},$$

for some function $\omega' : \mathbb{N} \to \mathbb{R}$ satisfying $\lim_{n\to\infty} \omega'(n) = \infty$. Therefore

$$\begin{aligned}
&\mathbb{P}\left(\max_{1\le i\le h} |S_i| \le \alpha(p)(\lg n\omega(n))^2 \textbf{ and } w \le \frac{1}{2} + \sqrt{\frac{2\lg n}{\beta(p)} + \frac{1}{4}}\right) \\
&= 1 - \mathbb{P}\left(\max_{1\le i\le h} |S_i| > \alpha(p)(\lg n\omega(n))^2 \textbf{ or } w > \frac{1}{2} + \sqrt{\frac{2\lg n}{\beta(p)} + \frac{1}{4}}\right) \\
&< 1 - \left(\frac{1}{\omega(n)} + \frac{1}{\omega'(n)}\right) = 1 - o(1).
\end{aligned}$$

We conclude that

$$\lim_{n\to\infty} \mathbb{P}\left(H \le \lg n + \left(\frac{1}{2} + \sqrt{\frac{2\lg n}{\beta(p)} + \frac{1}{4}}\right) \lg(\alpha(p)(\lg n)^2)\right) = 1.$$

□

4 The Uniform Model

In this section we study the problem of searching in a typical partial order according to the uniform model. We start by stating some definitions and a key auxiliary result.

Denote by $\mathcal{P}(n)$ the set of all partial orders on $[n]$. Taking $\mathcal{P}(n)$ with the uniform distribution, that is, making each partial order equally likely, we have the *uniform model* for random partial orders; a random element in this model will be denoted by U_n.

It is known that almost all U_n have a strong structural property, which we now describe. Let $\{X_1, X_2, X_3\}$ be a partition of $[n]$, and let $\mathcal{A}(X_1, X_2, X_3)$ be the set of partial orders $([n], \prec)$ satisfying the following conditions:

- whenever $x \prec y$, for $x \in X_i$ and $y \in X_j$, we have $i < j$,
- whenever $x \in X_1$ and $y \in X_3$ we have $x \prec y$.

The partial orders in $\mathcal{A}(X_1, X_2, X_3)$ are said to be *3-layered*.

Answering the question "what does a 'typical' partial order on $[n]$ look like?", in [8] Kleitman and Rothschild proved that, rather surprisingly, *almost all partially ordered sets are 3-layered.*

Theorem 6 (Kleitman and Rothschild [8]). *Suppose $\omega(n) \to \infty$ as $n \to \infty$. Almost every partial order on $[n]$ lies in $\mathcal{A}(X_1, X_2, X_3)$ for some partition $\{X_1, X_2, X_3\}$ of $[n]$ with $\big||X_2| - n/2\big| < \omega(n)$ and $\big||X_1| - n/4\big| < \omega(n)\sqrt{n}$.*

Theorem 6 makes the problem of searching in typical partial orders as posed in Section 1 rather uninteresting, since our search model makes it unavoidable to query each of the maximal elements of the given order. Theorem 6 tells us that almost all orders have $(1/4 + o(1))n$ such elements.

To make the problem more interesting, we now consider a variation of our search model where a query to s about u has four possible outcomes: smaller, greater, hit and no meaning, respectively, $u \prec s$, $s \prec u$, $s = u$ and s is not comparable to u; a strategy, accordingly, is redefined to be a decision tree as described previously, with the difference that each internal node has three children. In this section we shall prove that it is almost always possible, under the uniform model, to construct a ternary decision tree of height $O(\log n)$ in time $O(n^2 \log n)$.

Let us first make the following definition: given two disjoint sets X_1, X_2, the set of *2-layered orders* $\mathcal{A}(X_1, X_2)$ is the set of all partial orders $(X_1 \cup X_2, \prec)$ such that $\prec \subseteq X_1 \times X_2$.

Given a partition $\{X_1, X_2, X_3\}$ of $[n]$, there is a natural correspondence between $\mathcal{A}(X_1, X_2, X_3)$ and $\mathcal{A}(X_1, X_2) \times \mathcal{A}(X_2, X_3)$, so that we can devise a search strategy for a 3-layered order as a 'composition' of search strategies for 2-layered orders, by applying the latter on the suborders induced by $X_1 \cup X_2$ and $X_2 \cup X_3$.

Our strategy for searching for $u \in U$ in $P \in \mathcal{A}(X_1, X_2)$ is simple: we start by making the assumption $u \in X_1$, which we then verify by means of a series of queries to elements of X_2 in such a way that, at the end, we either have found u in X_1 or know that $u \notin X_1$. In the latter case, we restart the algorithm with the rôles of X_1 and X_2 exchanged, so that, after this is done, we either have found u in $X_1 \cup X_2$ or can conclude that $u \notin X_1 \cup X_2$. The algorithm to search for $u \in U$ in the layer X_1, say, starts by setting $S = X_1$ and proceeds in two phases.

Algorithm $\mathcal{C}$

querying phase: at each step,

1. choose $x \in X_2$ such that $\big||S \cap I(x)| - |S - I(x)|\big|$ is minimal.
2. query x about u:
   (a) if the answer is hit, the search is over;
   (b) if the answer is smaller, replace S with $|S \cap I(x)|$ and restart;
   (c) if the answer is no, replace S with $|S - I(x)|$ and restart.

This procedure is repeated until we reach the point where $S \cap I(x) = \emptyset$ or $S - I(x) = \emptyset$ for all $x \in X_2$, at which point we go to the next phase.

sweeping phase: for each $s \in S$:

1. query s about u: if the answer is hit, the search is over, otherwise, proceed.

If the search is not over after this, we can conclude that $u \notin X_1$.

Let us call $s_i(X_1, X_2)$ the size of the set S at the beginning of the ith step of the querying phase, $s(X_1, X_2)$ the size of the set S at the beginning of the sweeping phase and $q(X_1, X_2)$ the number of queries made at the querying phase. It is clear that the height of the decision tree corresponding to the above procedure is bounded by $q(X_1, X_2) + s(X_1, X_2)$; moreover, we note that the choice of x at step i of the querying phase can be accomplished in $s_i(X_1, X_2)|X_2|$ steps.

From this and the above discussion, we can conclude that it is possible to construct a decision tree of height

$$H_C \leq \sum_{(i,j)\in J} (q(X_i, X_j) + s(X_i, X_j)), \tag{5}$$

and this can be done in time

$$T_C \leq \sum_{(i,j)\in J} \left(s(X_i, X_j) + |X_j| \sum_{k=1}^{q(X_i,X_j)} s_k(X_i, X_j) \right), \tag{6}$$

where $J = \{(1,2), (3,2), (2,1)\}$.

At this point we define the *binomial probability model* $\mathcal{A}_{1/2}(X_1, X_2)$ for 2-layered orders, given by independently choosing each $(x_1, x_2) \in X_1 \times X_2$ with probability 1/2. We shall show that, in this model, we almost always have

1. $s_i(X_1, X_2) \leq (2/3)^i |X_1|$,
2. $q(X_1, X_2) \leq \log_{3/2} |X_1|$,
3. $s(X_1, X_2) < 40$,

which, along with (5), (6) and the sizes of X_1, X_2 and X_3 in Theorem 6 show that

$$H_C \leq 3 \log_{3/2} n + 120 \qquad \text{and} \qquad T_C \leq \frac{3}{2} n|X_2| + 120.$$

There is a natural correspondence between the product space $\mathcal{A}_{1/2}(X_1, X_2) \times \mathcal{A}_{1/2}(X_2, X_3)$ and the uniform probability space $\mathcal{A}(X_1, X_2, X_3)$. On the other hand, Theorem 6 and a standard argument give that if an event is almost certain in the space $\mathcal{A}(X_1, X_2, X_3) = \mathcal{A}_{1/2}(X_1, X_2) \times \mathcal{A}_{1/2}(X_2, X_3)$ for *any* fixed partition (X_1, X_2, X_3) as in Theorem 6, then this event must also be an almost certain event in $\mathcal{U}_n$ (details are given in the final version [9] of this extended abstract).

Therefore, in the remaining of this section, we concentrate in proving the relevant results for $\mathcal{A}_{1/2}(X_1, X_2) \times \mathcal{A}_{1/2}(X_2, X_3)$, where the X_i are as in the Kleitman–Rothschild theorem.

In what follows, we assume $S \subseteq X_1$, $x \in X_2$ and let $d_S(x) = |I(x) \cap S|$ be the number of elements in S smaller than x. The reader may notice that it is easy to prove analogous versions of the proposition and corollary below with X_1 and X_2 interchanged. Moreover, both, proposition and corollary, obviously remain true if we consider X_3 instead of X_1.

Proposition 7. *Almost surely, for a fixed $S \subseteq X_1$ with $|S| \geq 40$, the probability that there is no vertex $x \in X_2$ with $|S|/3 \leq d_S(x) \leq 2|S|/3$ is*

$$\leq \left(2\exp\{-|S|/40\}\right)^{|X_2|}.$$

Proof. Fix $S \subseteq X_1$ and $x \in X_2$. Then, by Chernoff's inequality, we have

$$\mathbb{P}\left(|d_S(x) - \mu| > |S|/6\right) < 2\exp\{-\mu/20\},$$

where $\mu = |S|/2$ is the expected value of $d_S(x)$. Thus, we have

$$\mathbb{P}\left(\nexists x \in X_2 \colon |S|/3 \leq d_S(x) \leq 2|S|/3\right) \leq \left(2\exp\{-|S|/40\}\right)^{|X_2|},$$

as required. □

Corollary 8. *For X_1, X_2 as above, and q, s_i, s be as defined in the preceding discussion, the following almost surely holds:*

1. *$s(X_1, X_2) \leq 40$,*
2. *$s_i(X_1, X_2) \leq (2/3)^i |X_1|$, and so,*
3. *$q(X_1, X_2) \leq \log_{3/2}|X_1|$.*

Proof. To prove (1) we observe that the probability of $s(S, X_2) > 40$, for some $S \subset X_1$ with $|S| > 40$, is given by

$$\begin{aligned}
&\mathbb{P}\left(\exists S \subseteq X_1\, \forall x \in X_2 \text{ we have } \left|d_S(x) - |S|/2\right| > |S|/6\right) \\
&\leq \sum_{S \subseteq X_1,\, |S| \geq 40} \mathbb{P}\left(\forall x \in X_2 \text{ we have } \left|d_S(x) - |S|/2\right| > |S|/6\right) \\
&\leq \sum_{s=40}^{|X_1|} \binom{|X_1|}{s} \exp\{-(s/40 - \ln 2)|X_2|\} \\
&\leq \sum_{s=40}^{|X_1|} \exp\{-(s/40 - \ln 2)|X_2| + s \ln |X_1|\} = o(1).
\end{aligned}$$

To prove (2) and (3), it suffices to notice the following. The probability that, at some point in step 1 of the querying phase, we cannot choose $x \in X_2$ such that $|S|/3 \leq |S \cap I(x)| \leq 2|S|/3$ is at most

$$(\log_{3/2}|X_1|)\mathbb{P}\left(\forall x \colon \left|d_S(x) - |S|/2\right| > |S|/6\right) \leq n\left(\frac{2}{e}\right)^{|X_2|} = o(1).$$

Thus in both cases (b) and (c) of step 2 of algorithm $\mathcal{C}$, we have reduced the search space by a factor $\leq 2/3$. Therefore we have $q(X_1, X_2) \leq \log_{3/2}|X_1|$. □

Theorem 9. *Almost every U_n admits a ternary search tree of height $O(\log n)$ which can be constructed in time $O(n^2)$.*

5 Concluding Remarks

We observe that one may also study the uniform model conditioning on having sparser partial orders. To carry out this investigation, the recent results in [10] and [11] are crucial. We shall come back to this in [9].

The arguments in the proofs of Corollary 3 and preceding lemmas and theorems are adapted rewritings of arguments found in [12] (Lemma 2.3 and Theorems 2.10 and 2.11). The results in that paper, on the structure of random graph orders, were what first suggested to the authors the idea of the algorithm.

In the present work we consider only the case in which p does not depend on n. It should be noted, however, that the results in [12] imply that ours remain valid even in the case in which $p = p(n)$ is a decreasing function of n, as long as $p \lg n \to \infty$.

It is worth noting that Theorem 4 is part of a deeper investigation on the width of random graph orders found in [7], where a much stronger result is proven, namely, that the width of a random graph order is an almost determined random variable whose value, rather surprisingly, is almost surely $\lfloor K_n \rfloor$ or $\lceil K_n \rceil$, where K_n is given in (4).

As mentioned in Section 2, Linial and Saks [2] consider a different although related problem where the set U is assumed to be totally ordered, the given relation $(S, \prec)$ is a partial order compatible with this total order and where each query about $u \in U$ to $s \in S$ is made with respect to the total order induced on S and not with respect to the given relation $\prec$ as is our case.

To see why this turns out to define a different problem, consider what information is gained in a search through S for $u \in U$ when we query $s \in S$ about u and the outcome is smaller: in our problem, such an outcome is enough to confine the remaining of the search to $S \cap I(s)$; in their problem, however, this is not the case: as the query is made with respect to the underlying total order in S, the outcome smaller leaves all elements in $S - F(s)$ as valid candidates.

While not presenting explicitly an algorithm to compute an optimal decision tree for the problem, it is a consequence of the results in [2] and [6] that the height H of an optimal decision tree for their problem satisfies

$$\lg \iota(S,\prec) \leq H \leq k \lg \iota(S,\prec), \tag{7}$$

where $\iota(S,\prec)$ is the number of ideals in $(S,\prec)$ and $k = (2 - \lg(1 + \lg 5))^{-1} \approx 3.73$.

We note that this fact alone can lead us to similar results to those given in Section 3 when we consider *their* problem in the random graph order model. Let us briefly discuss this. Redefine an element of S to be *dominating* if it is comparable to every other element in $(S,\prec)$, and a *segment* to be an interval free of dominating elements.

A possible search strategy, then, would be isolating one segment by means of a binary search restricted to the dominant elements of the order and then searching through this segment.

An ideal in a partial order is uniquely determined by the antichain of its maximal elements. Therefore, the number of ideals in a segment R is bounded by the number of antichains it contains, and hence if w is the width of the order, then

$$\iota(R, \prec) \leq \sum_{i=1}^{w} \binom{|R|}{i} \leq |R|^w.$$

Together with the bounds in (7) and Theorem 4, this allows us to conclude that there is a decision tree for R of height at most

$$H \leq k \lg \iota(R, \prec) \leq kw \lg |R|.$$

Therefore we can conclude that, for any partial order, there is a decision tree of height $H \leq \lg n + w \lg s$, where s is the size of the largest segment in the order.

Now, the argument leading to Theorem 2 proves a result for dominating elements as defined now (with another value for $\alpha(p)$, of course, say $\alpha'(p)$), which, together with Theorem 4 gives us

$$\lim_{n\to\infty} \mathbb{P}\left(H \leq \lg n + k \left(\frac{1}{2} + \sqrt{\frac{2 \lg n}{\beta(p)} + \frac{1}{4}} \right) \lg(\alpha'(p)(\lg n)^2) \right) = 1.$$

References

1. Laber, E.S., Nogueira, L.T.: Binary searching in posets. Submitted for publication (2001)
2. Linial, N., Saks, M.: Searching ordered structures. Journal of Algorithms **6** (1985) 86–103
3. Ben-Asher, Y., Farchi, E., Newman, I.: Optimal search in trees. SIAM Journal on Computing **28** (1999) 2090–2102
4. Ben-Asher, Y., Farchi, E.: The cost of searching in general trees versus complete binary trees. Technical Report 31905, Technical Report Research Center (1997)
5. Lipman, M.J., Abrahams, J.: Minimum average cost testing for partially order. IEEE Transactions on Information Theory **41** (1995) 287–291
6. Linial, N., Saks, M.: Every poset has a central element. Journal of Combinatorial Theory, Series A **40** (1985) 195–210
7. Barak, A.B., Erdős, P.: On the maximal number of strongly independent vertices in a random acyclic directed graph. SIAM Journal on Algebraic and Discrete Methods **5** (1984) 508–514
8. Kleitman, D.J., Rothschild, B.L.: Asymptotic enumeration of partial orders on a finite set. Trans. Amer. Math. Soc. **205** (1975) 205–220
9. Carmo, R., Donadelli, J., Laber, E., Kohayakawa, Y.: Searching in random partially ordered sets. In preparation (2002)
10. Prömel, H.J., Steger, A., Taraz, A.: Counting partial orders with a fixed number of comparable pairs. Combin. Probab. Comput. **10** (2001) 159–177
11. Prömel, H.J., Steger, A., Taraz, A.: Phase transitions in the evolution of partial orders. J. Combin. Theory Ser. A **94** (2001) 230–275
12. Bollobás, B., Brightwell, G.: The structure of random graph orders. SIAM Journal on Discrete Mathematics **10** (1997) 318–335

Packing Arrays

Brett Stevens[1] and Eric Mendelsohn[2]

[1] School of Mathematics and Statistics
Carleton University
1125 Colonel By Dr.
Ottawa ON K1S 5B6
brett@math.carleton.ca
http://mathstat.carleton.ca/~brett/
[2] Department of Mathematics
University of Toronto
100 St. George St.
Toronto ON M6G 3G3
mendelso@math.toronto.edu,
http://www.math.toronto.edu/mendelso/

Abstract. A packing array is a $b \times k$ array of values from a g-ary alphabet such that given any two columns, i and j, and for all ordered pairs of elements from the g-ary alphabet, (g_1, g_2), there is at most one row, r, such that $a_{r,i} = g_1$ and $a_{r,j} = g_2$. Further, there is a set of at least n rows that pairwise differ in each column: they are disjoint. A central question is to determine, for given g and k, the maximum possible b. We develop general direct and recursive constructions and upper bounds on the sizes of packing arrays. We also show the equivalence of the problem to a matching problem on graphs and a class of resolvable pairwise balanced designs. We provide tables of the best known upper and lower bounds.

1 Introduction

A $g^2 \times k$ array filled with elements from a g-ary alphabet such that each ordered pair from the alphabet occurs exactly once in each pair of columns is called an *orthogonal array.* By having the first column index the rows of a $g \times g$ square, the second column index its columns, and the remaining $k - 2$ columns give the entries of $k - 2$ different squares, Orthogonal arrays are transformed into a set of $k - 2$ *Mutually Orthogonal Latin Squares* (MOLS). Both are known never to exist for $k > g + 1$ [6]. It is natural to ask for structures that have similarly useful properties as orthogonal arrays for larger k. One generalization is to require that all pairwise interactions be covered at least once. These objects are known as covering arrays or transversal covers and have been extensively studied, see [13,14,15,17,18] and their references. The other natural generalization of orthogonal arrays is the packing array.

Definition 1. *A packing array, $PA(k, g : n)$, is a $b \times k$ array of values from a g-ary alphabet such that given any two columns, i and j, and for all ordered*

S. Rajsbaum (Ed.): LATIN 2002, LNCS 2286, pp. 293–305, 2002.

pairs of elements from the g-ary alphabet, (g_1, g_2), there is at most one row, r, such that $a_{r,i} = g_1$ and $a_{r,j} = g_2$. Further, there is a set of at least n rows that pairwise differ in each column: they are called disjoint. The largest number of rows possible in a PA(k, g : n) is denoted by pa(k, g : n).

The introduction of the set of disjoint rows is novel. It is introduced as a necessary substructure for several constructions to work, most notable Theorems 5 and 6. If this set does not occur inside the ingredient arrays of these recursive constructions then the arrays formed will have many repeated pairs which is forbidden in a packing array. The same constructions motivated the inclusion of the study of sets of disjoint rows in covering arrays [15,17]. In that case these sets allowed the optimization of the construct, here they are absolutely necessary.

Row and column permutations, as well as permuting symbols within each column, leave the packing conditions intact.

Example 1. A PA(5, 3 : 1) with six rows:

```
0 0 0 0 0
0 1 2 2 1
1 0 1 2 2
2 1 0 1 2
2 2 1 0 1
1 2 2 1 0
```

We will often and without loss of generality use $\mathbb{Z}_g$ as the symbol set on each column and let $r_{i,j}$ equal the number of times symbol $i \in \mathbb{Z}_g$ appears in column j.

Packing arrays are also called *transversal packings* [14] and *mutually orthogonal partial latin squares* [1,2]. Observing that any two rows of the packing array must have Hamming distance at least $k - 1$ we see that packing arrays are also error correcting codes, specifically, Maximal Distance Separating (MDS) codes or Partial MDS codes.

Abdel-Ghaffar and Abbadi [2] use MDS codes to allocate large database files to multiple hard disk systems so that the retrieval time is optimal. For storage on multiple disks, each attribute space D_i, $1 \le i \le k$ of a data base file with n records and k attributes is broken up into g parts, D_{ij}, $1 \le i \le k$ and $1 \le j \le g$. The database file can be seen as a subset of $D_1 \times D_2 \times \cdots \times D_k$. The set of all records with entries in $D_{1j_1} \times D_{2j_2} \times \cdots \times D_{kj_k}$ for $1 \le j_i \le g$ and $1 \le i \le k$ is called a *bucket*, which can be made equivalent to an element of $\mathbb{Z}_g^k$. Abdel-Ghaffar and Abbadi begin by first dividing up the database into all its buckets and partitioning the set of all buckets into m pieces each of which will be stored on a single disk in an array of m disks. A *partial match query* is a request for all records in the database file that match a choice of from zero to k attribute values. A search will begin by retrieving from disk all the buckets that could contain records matching the request. After retrieval, this set of buckets will be searched for the matching records therein. Abdel-Ghaffar and Abbadi concern themselves with minimizing the time required to retrieve the buckets from the disk array. If a request matches b_i buckets from disk i, $1 \le i \le m$ then they define the retrieval time to be $\max\{b_1, b_2, \ldots, b_m\}$ and ask how they can optimize this time. A response time is *strictly optimal* if

$$\max\{b_1, b_2, \ldots, b_m\} = \left\lceil \frac{\sum_{i=1}^{m} b_i}{m} \right\rceil$$

The conditions necessary for strict optimality are very tight and cannot be achieved in a large number of reasonable situations. Abdel-Ghaffar and Abbadi discuss best possible response times when strict optimality is not possible, and show that if the set of buckets on each disk is an MDS error correcting code then the response time, although not strictly optimal, can be shown to be better than all other allocations schemes of the records from this database across m disks.

They establish bounds on the size of packing arrays, the smallest form of MDS codes, to yield extreme bounds on the efficiency of their system. The MDS or Partial MDS codes that correspond to packing arrays would apply to optimal disk allocation over a large number of disks, specifically at least g^{k-2}. If packing arrays were used for optimal disk allocation in their model, any search with at least two specified attributes would yield search time of one. Their method guarantees that any partial match query with at least two specified attributes matches at most one bucket on each disk. This is the best possible response time to retrieve the buckets from the disk array. They briefly but not completely consider the question of decomposing $\mathbb{Z}_g^k$ into m MDS codes. They do not offer a complete solution to this problem; nor do we. This is a very interesting open direction for research.

Abdel-Ghaffar and Abbadi [2] showed that

$$pa(k, g : 1) \geq g + 1 \Rightarrow k \leq \frac{g^2 + g}{2},$$

which together with $\text{pa}(k, g : 1) \geq g$ implies

$$pa(\frac{g^2 + g + 2}{2}, g : 1) = g.$$

and

$$pa(\frac{g^2 + g}{2}, g : 1) \geq g.$$

Abdel-Ghaffar modified the Plotkin bound for packing arrays with $n = 1$ and showed that this bound is met when $k \geq 2g - 1$ [1]. We will extend this result to include all n. He also solved completely the case where $n = 1$ and $g = 3, 4$ and showed that a packing array with $g^2 - 1$ rows can be completed to one with g^2 rows.

In general, the codes corresponding to packing arrays could be useful for any of the standard applications for codes: error detection, error correction. One would imagine that packing arrays would be most useful in a situation where the error rate in the information channel was exceedingly high, requiring that the minimum distance between codewords be extremely large.

In Section 2, two sets of upper bounds on packing arrays are derived. The first set of bounds consists of our modification of the Plotkin bound to account for sets of disjoint rows and the implications of non-integrality. The second set comes from an observation on the constraints that sets of disjoint rows place on these structures.

In Section 3, we discuss constructions which yield lower bounds. We review some constructive techniques motivated by design theory that are also useful to construct transversal covers [17]. We also present a direct construction using matchings in graphs. Finally, we present a set of recursive constructions based on this matching problem. Both these constructions extend Abdel-Ghaffar's result to packing arrays with sets of disjoint rows.

2 Packing Arrays: Upper Bounds

2.1 Modification of the Plotkin Bound

The rows of a PA$(k, g : n)$ form a code on a g-ary alphabet with word length k and Hamming distance at least $k-1$. The set of n disjoint rows form a set of n codewords with mutual Hamming distance k. If the PA$(k, g : n)$ has pa$(k, g : n)$ rows then the number of codewords in the code is maximum among the codes containing the specified set of n codewords.

One of the most potent coding theory bounds is the Plotkin bound [3]. Abdel-Ghaffar derived a generalization of the Plotkin bound for packing arrays. We will adapt this bound further to include consideration of sets of disjoint rows and to extend its utility to parameter sets where the number of rows is not a multiple of g.

Theorem 1 (Plotkin Bound Modified for Packing Arrays). *A $b \times k$ packing array must satisfy the following bound for all $\beta \leq b$, $\beta = ug + v$ where $0 \leq v < g$:*

$$k((g-v)u^2 + v(u+1)^2) \leq \beta^2 - \beta - n^2 + n + k\beta.$$

Although this generalization of the Plotkin bound is generated by consideration of a structure with two replication numbers, this bound applies in general to packing arrays that may have more than two replication numbers. In the proof of Theorem 1, we used the fact that $\sum_{i-1}^{g} r_{i,j}$ is minimized when the replication numbers are either $\lceil \frac{b}{g} \rceil$ or $\lfloor \frac{b}{g} \rfloor$. If the replication numbers are more varied then the bound derivable from this method will be tighter.

When equality is reached in Theorem 1, strong implications on the structure of the code arise [3]. In this case, the bound must be an integer (although the standard Plotkin bound might not be integral in many cases); any two code words, except those in the set of disjoint rows must intersect; and each symbol must appear nearly equally often in each column (either u or $u+1$ times). Let us examine these consequences for packing arrays.

We will dualize the packing array into a block design.

Definition 2. *A* Pairwise Balanced Block Design *(PBD(v, K, λ)) is a base set V of v elements, and a collection $\mathcal{B}$ of subsets $B \subseteq V$, and $|B| \in K$, called* blocks, *such that every pair of points from V occur in exactly λ blocks. Furthermore if the collection of blocks can be partitioned so that each part contains every element of V exactly once, we say that the PBD is* resolvable. *A generalization of PBDs admits holes. A subset, $H \subseteq V$ is called a* hole *if every pair of elements of H occurs in no blocks.*

The base set of the design of interest will be the rows of the array. The blocks of this design will be any maximal collection of rows that have the same symbol in some column. Each column, therefore, defines a spanning subset of blocks, or resolution class. This dual structure will be a resolvable PBD$(v, \{u, u+1\}, 1)$ with a hole of order n. Each resolution class must have each block size occurring the same number of times as every other resolution class. If b is a multiple of g then there will only be one block size in the resolvable PBD.

In fact, for $n = 1$, these structures are a particular class of design: A *Class-Uniformly Resolvable Design* (CURD). This recently studied class of designs provide insight into resolvable structures and have important applications [7,8,9,10,11,20].

If $b \leq 2g$, consideration of Theorem 1 implies that this bound will be tight when the only two replication numbers in the array are two and one. In this case, there will be $b - g$ symbols in each column appearing twice and $2g - b$ symbols appearing only once. The dual of this structure is a packing of pairwise edge disjoint $(b-g)$-matchings into $K_b - K_n$. We will discuss this specific case when we discuss constructions, in Subsection 3.2 because the existence can be constructively solved for a large range of parameters.

2.2 Disjoint Row Bound

If we have a symbol appearing $m > n$ times in one column of a PA$(k, g : n)$ then deleting this column yields a PA$(k-1, g : m)$. Therefore, bounds on the sizes of packing arrays can be translated into bounds on the admissible replication numbers for packing arrays with one more column. To this end, we calculate the maximum number of columns possible in a packing arrays with n disjoint rows and at least $g+1$ rows.

Theorem 2. *The maximum number of columns in a packing arrays with at least $g+1$ rows is*

$$\frac{g(g+1)}{2} - \frac{n(n-1)}{2}.$$

Theorem 3. *If, in a PA$(k, g : n)$ with more than g rows, we define $r_{\max} = \max\{r_{ij} : i \in \mathbb{Z}_g,\ 1 \leq j \leq k\}$, then*

$$k - 1 \leq \frac{g(g+1)}{2} - \frac{r_{\max}(r_{\max}-1)}{2}.$$

Proof. Remove any column which has a point achieving $r_{\max}$. This yields a $\mathrm{PA}(k-1, g : \max(n, r_{\max}))$. □

Observing that if a $\mathrm{PA}(k, g : n)$ has more than mg rows, it must have a point with replication number at least $m+1$, we have the following Corollary.

Corollary 1. *The maximum number of columns in a packing array with more than mg rows is*

$$\frac{g(g+1)}{2} - \frac{m(m-1)}{2} + 1,$$

or alternatively, noting that $pa(k, g : n)$ is non-increasing in k, we have

$$pa\left(\frac{g(g+1)}{2} - \frac{m(m-1)}{2} + 2, g : n\right) \leq mg.$$

One notable consequence of this is the fact that

$$pa(g+2, g : n) \leq g^2 - g.$$

3 Packing Arrays: Constructions and Lower Bounds

3.1 Constructions from Design Theory

Definition 3. *An* incomplete orthogonal array *is the same as an orthogonal array except there are s mutually disjoint subsets of the alphabet, H_i, called* holes *with cardinalities b_i, such that every ordered pair, both from H_i $1 \leq i \leq s$, never occurs in any pair of columns.*

They are also called incomplete transversal designs ($\mathrm{ITD}(k, g : b_1, b_2, \ldots, b_s)$) and have been used to construct transversal covers [17]. This construction can be similarly formulated for packing arrays to yield:

Theorem 4. *If there exists an $ITD(k, g; b_1, b_2, \ldots, b_s)$ then*

$$pa(k, g : i) \geq \max_{\substack{i_i + i_2 + \cdots + i_s = i \\ i_j \leq b_j}} \left(g^2 - \sum_{j=1}^{s} (b_j^2 - pa(k, b_j : i_j))\right).$$

Proof. Fill the holes of the ITD with the $\mathrm{PA}(k, b_j : i_j)$. These holes are disjoint, thus the union of the sets of disjoint rows from the $\mathrm{PA}(k, b_j : i_j)$ will also be a set of disjoint rows. □

Example 2. The existence of $ITD(4, 6; 2)$ and $ITD(6, 10; 2)$, yield $\mathrm{pa}(4, 6) = 34$, as also stated by Abdel-Ghaffar [1] and $\mathrm{pa}(6, 10) = 98$ or 100. $\mathrm{PA}(4, 6)$ is explicitly

$$\begin{pmatrix} 0&0&0&0&0&0&1&1&1&1&1&1&2&2&2&2&2&2&3&3&3&3&3&3&4&4&4&4&5&5&5&5&4&5 \\ 0&1&2&3&4&5&0&1&2&3&4&5&0&1&2&3&4&5&0&1&2&3&4&5&0&1&2&3&0&1&2&3&5&4 \\ 4&5&2&3&0&1&1&0&5&4&2&3&5&4&0&1&3&2&3&2&4&5&1&0&0&3&1&2&2&1&3&0&4&5 \\ 0&1&4&5&2&3&5&4&0&1&3&2&3&2&5&4&0&1&4&5&3&2&1&0&1&3&2&0&2&0&1&3&4&5 \end{pmatrix}^T$$

where the last two rows are the filled hole. We observe that this is actually a $\mathrm{PA}(4, 6 : 5)$: rows 1, 8, 18, 27 and 34 are a set of disjoint rows.

See [17] for more details and for restrictions on and discussion of this method.

Generalizing Wilson's Construction to Packings

Theorem 5. *Let C be a PA$(k+l,t)$ with columns $G_1, G_2, \ldots, G_k$ and, $H_1, H_2, \ldots, H_l$. Let $\mathcal{S}$ be any set of choices of a subset of each symbol set on $H_1, \ldots, H_l$, let h_i be the number of symbols chosen from column H_i, $\sum_{i=1}^{l} h_i = u$, and m be any nonnegative integer. For any row A of C let $u_A = |\mathcal{S} \cap A|$, the number of times the symbol in the intersection of column H_i with row A is in the set of symbols chosen for H_i summed from H_1 to H_l. Then*

$$pa(k, mt+u) \geq \sum_{A} (pa(k, m+u_A : u_A) - u_A) + \sum_{i=1}^{l} pa(k, h_i).$$

Proof. The proof is a straightforward generalization of Wilson's construction for transversal designs [21] .

□

If $l = 0$, we get the obvious generalization of MacNeish's theorem

Corollary 2.

$$pa(k, g:n) \geq \max_{\substack{2 \leq i \leq \lfloor g/2 \rfloor \\ \max(1, n/\lfloor g/i \rfloor) \leq j \leq \min(n,i)}} ((pa(k, i:j)\, pa(k, \lfloor g/i \rfloor : \lceil n/j \rceil)).$$

When $l = 1$, using some probabilistic analysis, we can remove all structural information and state the theorem purely as a recursion on the packing numbers alone.

Corollary 3.

$$\begin{aligned} pa(k, mt+u:n) \geq \max_{(i,j,\ell) \in A_{n,u,t,m}} & ((t-u)/t\; pa(k+1, t:i) pa(k, m:j) \\ & + u/t\; pa(k+1, t:i)(pa(k, m+1:j+1) - 1) \\ & + pa(k, u:\ell)), \end{aligned} \tag{1}$$

where

$$A_{n,u,t,m} = \{(i,j,\ell) \in \mathbb{N}^3 : ij + \ell \geq n,\ 1 \leq \ell \leq u,\ 1 \leq i \leq t, 1 \leq j \leq m\}.$$

By setting $m = 0$, we get another generalization of Theorem 5. For a full explanation of the motivation behind this PBD construction and detailed proofs see [17].

Theorem 6. *Given a $(v, \{2, 3, \ldots, g-1\}, 1)$-design, and for each point x, a chosen block to represent x, B_x, with $x \in B_x$, we can construct a PA(k, g). For each block, B, of the design, we define u_B to be the number of points on this block not represented by it. Then*

$$pa(k, g) \geq \sum_{B} pa(k, |B| : u_B) - u_B.$$

See [17] for results concerning the optimization of this type of construction and calculation the number of disjoint rows it yields.

3.2 Direct Construction for $b \leq 2g$: Matching Packings

The equivalence , given in Subsection 2.1 between packing arrays which meet the bound of Theorem 1 and packings of matchings into $K_b - K_n$ when $b \leq 2g$ also yields a number of constructive results.

Graph Theoretical Results

Lemma 1. *$K_v - K_n$ is $(v-1)$ edge colourable if and only if v is even or $n^2 - n \geq v - 1$.*

Theorem 7. *The maximum number, k, of edge disjoint copies of an m-matching that can be packed into a $K_v - K_n$ is subject to the following conditions:*

$$2km \leq v(v-1) - n(n-1)$$
$$k(2m - v + n) \leq n(v-n)$$

and a packing achieving the largest integer k subject to these conditions always exists.

Theorem 8. *The maximum number, k, of edge disjoint copies of an m-matching that can be packed into a $K_{v_1,v_2} - K_{n_1,n_2}$ (assume without loss of generality that $v_1 \geq v_2$) is subject to the following constraints:*

$$k(m - v_1 + n_1) \leq n_1(v_2 - n_2)$$
$$k(m - v_2 + n_2) \leq n_2(v_1 - n_1)$$
$$km \leq v_1 v_2 - n_1 n_2$$

and the largest integer k satisfying all these is always achievable.

Packing Array Constructions A direct application of Theorem 7 to packing arrays gives

Corollary 4.

$$pa(k, g : n) \geq \min\left(2g, \frac{2kg - kn - n^2}{k - n}, \frac{2k + 1 - \sqrt{(1 + 2k)^2 + 4(n^2 - n - 2kg)}}{2}\right)$$

If $k \leq n \leq g$ then a PA$(k, g : n)$ with $2g$ rows is easily constructed using Theorem 7 and so we can ignore that bound if the denominator is zero or negative. Similarly if the discriminant in the third bound is negative then the corresponding bound from Theorem 7 is always satisfied, so it may also be ignored in that case. In particular, when $k \geq 2g - 1$, Corollary 4 give a packing array that meets the bound in Theorem 1 and is optimal.

Abdel-Ghaffar and Abbadi [2] asked and answered the question: when does a packing array have more than g rows? We have extended this definite result to include sets of disjoint rows in Theorem 2. Theorem 7 and Theorem 1 also allow us to derive several exact packing arrays numbers:

Corollary 5.

$$pa(\frac{g^2+3g+2-n^2+n}{4}, g:n) = g+2.$$

$$pa(\frac{g^2+3g+2-n^2+n}{4}+1, g:n) = g+1.$$

When $k \geq 2g-1$ and $n > 1$, then

$$pa(k,g:n) = \left\lfloor \frac{2k+1-\sqrt{4k^2+4k+1-8kg+4n^2-4n}}{2} \right\rfloor.$$

And finally

$$pa(2g-2, g:2) = 2$$

We also have recursive constructions.

Theorem 9. *Within a PA$(k,g:n)$ fix n' rows which contain no members of the set of n disjoint rows. Let $q_{i,j}$ be the number of times that symbol j appears in i^{th} column of a row in the union of this set of rows and the original n disjoint rows. Let m_i, and $h_{i,j}$, $1 \leq i \leq k, 1 \leq j \leq g$ be a positive integers such that $pa(m_i, h_{i,j} : q_{i,j}) \geq r_{i,j}$. Additionally, let $h = \max_i\{\sum_{j=1}^{g} h_{i,j}\}$. Then*

$$pa\left(\left(\sum_{i=1}^{k} m_i\right), h : n + n'\right) \geq pa(k,g:n).$$

Proof. This is the Dual of Theorem 2.7 from [10]. □

We can use Theorem 8 to facilitate a recursive construction for packing arrays.

Theorem 10. *Suppose that there exists a PA$(k,g_1:n_1)$ with b_1 rows and a PA$(k,g_2:n_2)$ with b_2 rows and assume $b_1 \geq b_2$. If $b_1 \leq g_1+g_2$ then there exists a PA$(k+k', g_1+g_2 : n_1+n_2)$ with b_1+b_2 rows where k' is the largest integer satisfying*

$$k'(b_1+b_2-g_1-g_2) \leq b_1b_2 - n_1n_2$$
$$k'(b_2-g_1-g_2+n_1) \leq n_1(b_2-n_2)$$
$$k'(b_1-g_1-g_2+n_2) \leq n_2(b_1-n_1)$$

And there exists a PA$(k+k'', g_1+g_2 : \max(n_1,n_2))$ with b_1+b_2 rows where

$$k'' = \left\lfloor \frac{b_1b_2}{b_1+b_2-g_1-g_2} \right\rfloor$$

Example 3. An example of Theorem 10 is given here. We start with two copies of a PA(4, 3 : 2), one on symbol set $\{0, 1, 2\}$ and the other on symbol set $\{3, 4, 5\}$, vertically concatenated.

0	0	0	0
1	1	1	1
0	1	2	2
2	2	0	1
2	0	1	2
1	2	2	0
3	3	3	3
4	4	4	4
3	4	5	5
5	5	3	4
5	3	4	5
4	5	5	3

Each one represents a resolvable packing in $K_6 - K_2$. We find the maximum number of 6-matchings in the bipartite graph $K_{6,6}$ of edges between the two K_6. Each additional matching allows us to add one more column to the packing array:

0	0	0	0	0	0	0	0	0	0
1	1	1	1	1	1	1	1	1	1
0	1	2	2	2	2	2	2	2	2
2	2	0	1	3	3	3	3	3	3
2	0	1	2	4	4	4	4	4	4
1	2	2	0	5	5	5	5	5	5
3	3	3	3	0	5	4	3	2	1
4	4	4	4	1	0	5	4	3	2
3	4	5	5	2	1	0	5	4	3
5	5	3	4	3	2	1	0	5	4
5	3	4	5	4	3	2	1	0	5
4	5	5	3	5	4	3	2	1	0

Thus we get that $\mathrm{pa}(10, 6 : 2) \geq 12$ which is the best possible by Theorem 1.

3.3 Implementation

We implemented on computer all the packing array upper and lower bounds contained in this paper than can be expressed directly or as a recursion on the packing array numbers, without structural information. We only considered $3 \leq g \leq 9$, although the PBD bound was only implemented for $g \leq 7$ since it requires enumeration of all PBDs on g points along with all possible assignments of blocks to the points.

Except for the optimum existence results (Corollary 5, Theorem 2, and the existence of *MOLS*), the relative utility of these constructions is not obvious. Corollary 5 and Theorem 2 dominate the tables; but these existence results are for a restricted class of packing arrays. The other constructions are more interesting. The constructions from Theorem 7 produce a large number of the best currently known lower bounds. The constructions most useful for constructing packing arrays with more than $2g$ rows seem to be Wilson's construction and packings arrays obtained from deleting rows or columns. Abdel-Ghaffar has previously solved the packing array problem for $g = 3, 4$ and $n = 1$. He found $\mathrm{pa}(6, 4 : 1) = 9$ [1], which is also derivable as the dual of a CURD [10].

Table 1 contains the best upper and lower bounds. Since $k \geq 2g - 1$ is completely known we do not proceed farther than $k = 2g$. The values that are known to be optimal are emphasized.

k	4	5	6
pa(k, 3 : 1)	***9***	***6***	***4***
pa(k, 3 : 2)	***6***	***4***	***3***
pa(k, 3 : 3)	***3***	***3***	***3***

k	5	6	7	8
pa(k, 4 : 1)	***16***	***9***	***8***	***5***
pa(k, 4 : 2)	14/9	***8***	***6***	***5***
pa(k, 4 : 3)	13/9	***7***	***5***	***4***
pa(k, 4 : 4)	***4***	***4***	***4***	***4***

k	6	7	8	9	10
pa(k, 5 : 1)	***25***	***15***	***10***	***10***	***7***
pa(k, 5 : 2)	23/15	13/11	***10***	***8***	***7***
pa(k, 5 : 3)	22/15	12/11	10/9	***7***	***6***
pa(k, 5 : 4)	19/11	***11***	***6***	***6***	***5***
pa(k, 5 : 5)	***5***	***5***	***5***	***5***	***5***

k	3	4	5	6	7	8	9	10	11	12
pa(k, 6 : 1)	***36***	***34***	34/26	34/26	34/16	19/12	14/12	***12***	***12***	***9***
pa(k, 6 : 2)	***36***	***34***	34/26	34/26	34/16	19/12	14/12	***12***	***10***	***8***
pa(k, 6 : 3)	***36***	***34***	34/26	34/16	33/12	19/12	13/12	12/11	***9***	***8***
pa(k, 6 : 4)	***36***	***34***	34/26	34/16	32/12	18/12	13/11	12/9	***8***	***7***
pa(k, 6 : 5)	***36***	***34***	34/26	34/12	26/12	13/10	***8***	***7***	***7***	***6***
pa(k, 6 : 6)	***36***	34/26	34/26	34/12	***6***	***6***	***6***	***6***	***6***	***6***

k	8	9	10	11	12	13	14
pa(k, 7 : 1)	***49***	***28***	***21***	15/14	***14***	***14***	***10***
pa(k, 7 : 2)	47/28	26/21	19/14	15/14	***14***	***12***	***10***
pa(k, 7 : 3)	46/28	25/21	18/14	15/14	14/13	***11***	***10***
pa(k, 7 : 4)	45/28	24/15	17/14	15/13	14/13	***10***	***9***
pa(k, 7 : 5)	44/15	23/15	16/13	14/12	***9***	***9***	***8***
pa(k, 7 : 6)	34/15	18/15	***11***	***8***	***8***	***8***	***7***
pa(k, 7 : 7)	***7***	***7***	***7***	***7***	***7***	***7***	***7***

k	9	10	11	12	13	14	15	16
pa(k, 8 : 1)	***64***	34/22	25/16	19/16	17/16	***16***	***16***	***12***
pa(k, 8 : 2)	62/29	34/22	25/16	19/16	17/16	***16***	***14***	***12***
pa(k, 8 : 3)	61/22	33/16	25/16	19/16	17/16	16/15	***13***	***11***
pa(k, 8 : 4)	60/22	33/16	24/16	18/16	17/15	16/15	***12***	***11***
pa(k, 8 : 5)	59/16	32/16	24/16	17/15	16/14	***12***	***11***	***10***
pa(k, 8 : 6)	58/16	32/16	19/14	17/14	***11***	***10***	***10***	***9***
pa(k, 8 : 7)	43/16	20/13	13/12	***10***	***9***	***9***	***9***	***8***
pa(k, 8 : 8)	***8***	***8***	***8***	***8***	***8***	***8***	***8***	***8***

k	10	11	12	13	14	15	16	17	18
pa(k, 9 : 1)	***81***	45/27	30/27	***27***	21/18	19/18	***18***	***18***	***14***
pa(k, 9 : 2)	79/27	43/27	30/27	25/18	20/18	19/18	***18***	***16***	***14***
pa(k, 9 : 3)	78/27	42/27	30/27	24/18	20/18	19/18	18/17	***15***	***13***
pa(k, 9 : 4)	77/18	41/18	29/18	23/18	20/18	29/17	18/17	***14***	***13***
pa(k, 9 : 5)	76/18	40/18	29/18	22/18	19/17	18/17	***14***	***13***	***12***
pa(k, 9 : 6)	75/18	39/18	28/18	21/17	19/16	18/16	***12***	***12***	***11***
pa(k, 9 : 7)	74/18	38/18	27/16	20/15	18/15	***12***	***11***	***11***	***10***
pa(k, 9 : 8)	53/18	25/15	17/14	***11***	***10***	***10***	***10***	***10***	***9***
pa(k, 9 : 9)	***9***	***9***	***9***	***9***	***9***	***9***	***9***	***9***	***9***

Table 1. Upper and lower bounds for packing arrays.

4 Conclusion

A number of upper and lower bounds for packing arrays were presented. The two upper bounds were the generalization of the Plotkin bound (Theorem 1) and the bound derived from the consideration of sets of disjoint rows (Theorem 2). Graph decomposition results provide many constructive techniques and for the same range of parameters, $k \geq 2g-1$, that Abdel-Ghaffar solved when $n=1$, we completed the solution for all $n>1$. We have restricted the unsolved range of packing arrays to $k \leq 2g-2$ and k larger than the largest transversal design known. This is a small range in which $\text{pa}(k, g: n)$ drops from g^2 to $2g$. Abdel-Ghaffar has shown that the drop is initially very quick [1].

Consideration of sets of disjoint rows is crucial for the use of Wilson's construction. Removing a group from a packing array produces a smaller structure with a large set of disjoint rows. Bounds on the values of packing arrays with disjoint rows can be pulled back through this operation to restrict the existence of arrays with $n=1$.

An interesting graph theory problem that was solved in the pursuit of packing arrays is the maximum number of edge disjoint m-matchings that can be packed into a graph G. If $g \leq b \leq 2g$, $n \leq g$, $m=b-g$, and the graph G is K_b with the edges of a K_n removed, then the dual is a packing array. Lastly, for packing arrays, we would like to extend the graph constructions beyond the restriction that $b \leq 2g$. Since we have exhausted the use of matching packings towards constructing packing arrays here, the next step is to consider similar packings with triangles.

5 Acknowledgements

Brett Stevens was supported by NSF Graduate Fellowship GER9452870. He would also like to thank SFU, PIMS, MITACS, and IBM Watson Research. Eric Mendelsohn was supported by NSERC of Canada Grant No. OGP007681.

References

1. K. A. S. Abdel-Ghaffar. On the number of mutually orthogonal partial latin squares. *Ars Combin.*, 42:259–286, 1996.
2. K. A. S. Abdel-Ghaffar and A. E. Abbadi. Optimal disk allocation for partial match queries. *ACM Transactions on Database Systems*, 18(1):132–156, Mar. 1993.
3. P. J. Cameron. *Combinatorics: Topics, Techniques, Algorithms.* Cambridge University Press, Cambridge, 1994.
4. C. Colbourn and D. Kreher. Concerning difference methods. *Des. Codes Cryptogr.*, 9:61–70, 1996.
5. C. Colbourn and S. Zhao. Maximum Kirkman signal sets for synchronous uni-polar multi-user communication systems. *Des. Codes Cryptogr.*, 20:219–227, 2000.
6. C. J. Colbourn and J. H. Dinitz, editors. *The CRC Handbook of Combinatorial Designs.* CRC Press, Boca Raton, 1996.

7. P. Danziger, M. Greig, and B. Stevens. Geometrical constructions of class-uniformly resolvable structures. Unpublished Manuscript.
8. P. Danziger and B. Stevens. On class-uniformly resolvable frames. Unpublished Manuscript.
9. P. Danziger and B. Stevens. On class-uniformly resolvable group divisible designs. *Discrete Math.* (Submitted).
10. P. Danziger and B. Stevens. Class-uniformly resolvable designs. *J. Combin. Des.*, 9:79–99, 2001.
11. E. Lamken, R. Rees, and S. Vanstone. Class-uniformly resolvable pairwise balanced designs with block sizes two and three. *Discrete Math.*, 92:197–209, 1991.
12. E. Lucas. *Récréations Mathématiques*, volume 2. Gauthier-Villars, Paris, 1883.
13. N. J. A. Sloane. Covering arrays and intersecting codes. *J. Combin. Des.*, 1:51–63, 1993.
14. B. Stevens. *Transversal Covers and Packings.* Ph.D. thesis, University of Toronto, Toronto, 1998.
15. B. Stevens, A. Ling, and E. Mendelsohn. A direct construction of transversal covers using group divisible designs. *Ars Combin.* (To Appear).
16. B. Stevens and E. Mendelsohn. Packing arrays and packing designs. *Des. Codes Cryptogr.* (To Appear).
17. B. Stevens and E. Mendelsohn. New recursive methods for transversal covers. *J. Combin. Des.*, 7(3):185–203, 1999.
18. B. Stevens, L. Moura, and E. Mendelsohn. Lower bounds for transversal covers. *Des. Codes Cryptogr.*, 15(3):279–299, 1998.
19. W. Wallis, editor. *Computational and Constructive Design Theory*, volume 368 of *Mathematics and Its Applications.* Kluwer Academic Publishers, Dordrecht, 1996.
20. D. Wevrick and S. A. Vanstone. Class-uniformly resolvable designs with block sizes 2 and 3. *J. Combin. Des.*, 4:177–202, 1996.
21. Richard M. Wilson. Concerning the number of mutually orthogonal Latin squares. *Discrete Math.*, 9:181–198, 1974.

Generalized Shannon Code Minimizes the Maximal Redundancy*

Michael Drmota[1] and Wojciech Szpankowski[2]

[1] Institut für Geometrie, TU Wien, A-1040 Wien, Austria
[2] Dept. Computer Science, Purdue University, W. Lafayette, IN 47907, USA

Abstract. Source coding, also known as data compression, is an area of information theory that deals with the design and performance evaluation of optimal codes for data compression. In 1952 Huffman constructed his optimal code that minimizes the *average* code length among all prefix codes for known sources. Actually, Huffman codes minimizes the average *redundancy* defined as the difference between the code length and the entropy of the source. Interestingly enough, no optimal code is known for other popular optimization criterion such as the *maximal redundancy* defined as the maximum of the pointwise redundancy over all source sequences. We first prove that a generalized Shannon code minimizes the maximal redundancy among all prefix codes, and present an efficient implementation of the optimal code. Then we compute precisely its redundancy for memoryless sources. Finally, we study universal codes for unknown source distributions. We adopt the minimax approach and search for the best code for the worst source. We establish that such redundancy is a sum of the likelihood estimator and the redundancy of the generalize code computed for the maximum likelihood distribution. This replaces Shtarkov's bound by an exact formula. We also compute precisely the maximal minimax redundancy for a class of memoryless sources. The main findings of this paper are established by techniques that belong to the toolkit of the "analytic analysis of algorithms" such as theory of distribution of sequences modulo 1 and Fourier series. These methods have already found applications in other problems of information theory, and they constitute the so called *analytic information theory.*

1 Introduction

The celebrated Huffman code minimizes the average code length among all prefix codes (i.e., satisfying the Kraft inequality [3]), provided the probability distribution is known. As a matter of fact, the Huffman code minimizes the average *redundancy* that is defined as the difference between the code length and the entropy for the source. But other than the average redundancy optimization criteria have been also considered in information theory. One of the most popular is the maximal redundancy defined as the maximum over all source sequences

* This work was supported by NSF Grant CCR-9804760 and contract 1419991431A from sponsors of CERIAS at Purdue.

S. Rajsbaum (Ed.): LATIN 2002, LNCS 2286, pp. 306–318, 2002.

of the sum of the code length and the logarithm of the probability of source sequences (cf. Shtarkov [14]). A seemingly innocent, and still open, problem is what code minimizes the maximal redundancy. To make it more precise we need to plunge a little into source coding, better known as data compression.

We start with a quick introduction of the *redundancy problem.* A code $C_n : \mathcal{A}^n \to \{0,1\}^*$ is defined as a mapping from the set $\mathcal{A}^n$ of all sequences of length n over the finite alphabet $\mathcal{A}$ to the set $\{0,1\}^*$ of all binary sequences. A message of length n with letters indexed from 1 to n is denoted by x_1^n, so that $x_1^n \in \mathcal{A}^n$. We write X_1^n to denote a random variable representing a message of length n. Given a probabilistic source model, we let $P(x_1^n)$ be the probability of the message x_1^n; given a code C_n, we let $L(C_n, x_1^n)$ be the code length for x_1^n. Information-theoretic quantities are expressed in binary logarithms written $\lg := \log_2$.

From Shannon's work we know that the entropy $H_n(P) = -\sum_{x_1^n} P(x_1^n) \lg P(x_1^n)$ is the absolute lower bound on the expected code length. Hence $-\lg P(x_1^n)$ can be viewed as the "ideal" code length. The next natural question to ask is by how much the code length $L(C_n, x_1^n)$ differs from the ideal code length, either for individual sequences or on average. The *pointwise redundancy* $R_n(C_n, P; x_1^n)$ and the *average redundancy* $\overline{R}_n(C_n, P)$ are defined as

$$R_n(C_n, P; x_1^n) = L(C_n, x_1^n) + \lg P(x_1^n),$$
$$\overline{R}_n(C_n, P) = \mathbf{E}_P[R_n(C_n, P; X_1^n)] = \mathbf{E}[L(C_n, X_1^n)] - H_n(P),$$

where the underlying probability measure P represents a particular source model and $\mathbf{E}$ denotes the expectation. Another natural measure of code performance is the *maximal* redundancy defined as

$$R_n^*(C_n, P) = \max_{x_1^n}[L(C_n, x_1^n) + \lg P(x_1^n)].$$

While the pointwise redundancy can be negative, maximal and average redundancies cannot, by Kraft's inequality and Shannon's source coding theorem, respectively (cf. [3]).

Source coding is an area of information theory that searches for optimal codes under various optimization criteria. It has been known from the inception of the Huffman code that its average redundancy is bounded from above by 1 (cf. [3]), but its precise characterization for memoryless sources was proposed only recently in [16]. In [4,8,13] conditions for optimality of the Huffman code were given for a class of weight function and cost criteria. Surprisingly enough, to the best of our knowledge, no one was looking at another natural question: What code minimizes the maximal redundancy? More precisely, we seek a prefix code C_n such that

$$\min_{C_n} \max_{x_1^n}[L(C_n, x_1^n) + \lg P(x_1^n)].$$

We shall prove in this paper, that a generalized Shannon code[1] is the optimal code in this case, and propose an efficient algorithm to construct such a code.

[1] Shannon's code assigns length $\lceil -\lg P(x_1^n) \rceil$ to the source sequence x_1^n for known source distribution P.

Our algorithm runs in $O(N \log N)$ steps if source probabilities are not sorted and in $O(N)$ steps if the probabilities are sorted, where N is the number of source sequences. We also compute precisely the maximal redundancy of the optimal generalized Shannon code. In passing we observe that Shannon codes, in one form or another, are often used in practice; e.g., in arithmetic coder.

It must be said, however, that in practice probability distribution (i.e., source) P is unknown. So the next natural question is to find optimal codes for sources with unknown probabilities. In information theory this is handled by the so called *minimax* redundancy introduced next. In fact, for unknown probabilities, the redundancy rate can be also viewed as the penalty paid for estimating the underlying probability measure. More precisely, *universal codes* are those for which the redundancy is $o(n)$ for all $P \in \mathcal{S}$ where $\mathcal{S}$ is a class of source models (distributions). The (asymptotic) *redundancy-rate problem* consists in determining for a class $\mathcal{S}$ the rate of growth of the minimax quantities as $n \to \infty$ either on average

$$\overline{R}_n(\mathcal{S}) = \min_{C_n \in \mathcal{C}} \max_{P \in \mathcal{S}} [\overline{R}_n(C_n, P)], \tag{1}$$

or in the worst case

$$R_n^*(\mathcal{S}) = \min_{C_n \in \mathcal{C}} \max_{P \in \mathcal{S}} [R_n^*(C_n, P)], \tag{2}$$

where $\mathcal{C}$ denotes the set of all codes satisfying the Kraft inequality.

In this paper we deal with the maximal *minimax redundancy* $R_n^*(\mathcal{S})$ defined by (2). Shtarkov [14] proved that

$$\lg \left(\sum_{x_1^n} \sup_{P \in \mathcal{S}} P(x_1^n) \right) \le R_n^*(\mathcal{S}) \le \lg \left(\sum_{x_1^n} \sup_{P \in \mathcal{S}} P(x_1^n) \right) + 1. \tag{3}$$

We replace the inequalities in the above by an exact formula. Namely, we shall prove that

$$R_n^*(\mathcal{S}) = \lg \left(\sum_{x_1^n} \sup_{P \in \mathcal{S}} P(x_1^n) \right) + R^{GS}(Q^*)$$

where $R^{GS}(Q^*)$ is the maximal redundancy of the generalized Shannon code for the (known) distribution $Q^*(x_1^n) = \sup_P P(x_1^n) / \sum_{x_1^n} \sup_P P(x_1^n)$. For a class of memoryless sources we derive an asymptotic expansion for the maximal minimax redundancy $R_n^*(\mathcal{S})$.

2 Main Result

We first consider sources with known distribution P and find an optimal code that minimizes the maximal redundancy, that is, we compute

$$R_n^*(P) = \min_{C_n \in \mathcal{C}} \max_{x_1^n} [L(C_n, x_1^n) + \log_2 P(x_1^n)]. \tag{4}$$

We recall that Shannon code C_n^S assigns length $L(C_n^S, x_1^n) = \lceil -\lg P(x_1^n)\rceil$ to the source sequence x_1^n. We define a *generalized Shannon* code C_n^{GS} as

$$L(C_n^{GS}, x_1^n) = \begin{cases} \lfloor \lg 1/P(x_1^n)\rfloor & \text{if } x_1^n \in \mathcal{L} \\ \lceil \lg 1/P(x_1^n)\rceil & \text{if } x_1^n \in \mathcal{A}^n \setminus \mathcal{L} \end{cases}$$

where $\mathcal{L} \subset \mathcal{A}^n$, and the Kraft inequality holds (cf. [3]).

Our first main result proves that a generalized Shannon code is an optimal code with respect to the maximal redundancy.

Theorem 1. *If the probability distribution P is dyadic, i.e. $\lg P(x_1^n) \in \mathbf{Z}$ ($\mathbf{Z}$ is the set of integers) for all $x_1^n \in \mathcal{A}^n$, then $R_n^*(P)=0$. Otherwise, let $p_1, p_2, \dots, p_{|\mathcal{A}|^n}$ be the probabilities $P(x_1^n)$, $x_1^n \in \mathcal{A}^n$, ordered in a nondecreasing manner, that is,*

$$0 \le \langle -\lg p_1\rangle \le \langle -\lg p_2\rangle \le \cdots \le \langle -\lg p_{|\mathcal{A}|^n}\rangle \le 1,$$

where $\langle x\rangle = x - \lfloor x\rfloor$ is the fractional part of x. Let now j_0 be the maximal j such that

$$\sum_{i=1}^{j-1} p_i 2^{\langle -\lg p_i\rangle} + \frac{1}{2}\sum_{i=j}^{|\mathcal{A}|^n} p_i 2^{\langle -\lg p_i\rangle} \le 1, \tag{5}$$

that is, the Kraft inequality holds for a generalized Shannon code. Then

$$R_n^*(P) = 1 - \langle -\lg p_{j_0}\rangle. \tag{6}$$

Proof. First we want to recall that we are only considering codes satisfying Kraft's inequality

$$\sum_{x_1^n} 2^{-L(C_n, x_1^n)} \le 1.$$

Especially we will use the fact that for any choice of positive integers $l_1, l_2, \dots, l_{|\mathcal{A}|^n}$ with

$$\sum_{i=1}^{|\mathcal{A}|^n} 2^{-l_i} \le 1$$

there exists a (prefix) code C_n with code lengths l_i, $1 \le i \le |\mathcal{A}|^n$.

If P is dyadic then the numbers $l(x_1^n) := -\lg P(x_1^n)$ are positive integers satisfying

$$\sum_{x_1^n} 2^{-l(x_1^n)} = 1.$$

Thus, Kraft's inequality is satisfied and consequently there exists a (prefix) code C_n with $L(C_n, x_1^n) = l(x_1^n) = -\lg P(x_1^n)$. Of course, this implies $R_n^*(P) = 0$.

Now assume that P is not dyadic and let C_n^* denote the set of optimal codes, i.e.

$$\mathcal{C}^* = \{C_n \in \mathcal{C} : R_n^*(C_n, P) = R_n^*(P)\}.$$

The idea of the proof is to find some properties of the optimal code. Especially we will show that there exists an optimal code $C_n^* \in \mathcal{C}^*$ with

(i)
$$\lfloor -\lg P(x_1^n) \rfloor \le L(C_n^*, x_1^n) \le \lceil -\lg P(x_1^n) \rceil \quad (7)$$

(ii) There exists $s_0 \in [0,1]$ such that
$$L(C_n^*, x_1^n) = \lfloor \lg 1/P(x_1^n) \rfloor \quad \text{if} \quad \langle \lg 1/P(x_1^n) \rangle < s_0 \quad (8)$$
and
$$L(C_n^*, x_1^n) = \lceil \lg 1/P(x_1^n) \rceil \quad \text{if} \quad \langle \lg 1/P(x_1^n) \rangle \ge s_0, \quad (9)$$

that is, C_n^* is a generalized Shannon code. Observe that w.l.o.g. we may assume that $s_0 = 1 - R_n^*(P)$. Thus, in order to compute $R_n^*(P)$ we just have to consider codes satisfying (8) and (9). It is clear that (5) is just Kraft's inequality for codes of that kind. The optimal choice is $j = j_0$ and consequently $R_n^*(P) = 1 - \langle -\lg p_{j_0} \rangle$.

It remains to prove the above properties (i) and (ii). Assume that C_n^* is an optimal code. First of all, the upper bound in (7) is obviously satisfied for C_n^*. Otherwise we would have
$$\max_{x_1^n} [L(C_n^*, x_1^n) + \log_2 P(x_1^n)] > 1$$
which contradicts Shtarkov's bound (3). Second, if there exists x_1^n such that $L(C_n^*, x_1^n) < \lfloor \lg 1/P(x_1^n) \rfloor$ then (in view of Kraft's inequality) we can modify this code to a code $\widetilde{C}_n^*$ with
$$L(\widetilde{C}_n^*, x_1^n) = \lceil \lg 1/P(x_1^n) \rceil \quad \text{if } L(C_n^*, x_1^n) = \lceil \lg 1/P(x_1^n) \rceil,$$
$$L(\widetilde{C}_n^*, x_1^n) = \lfloor \lg 1/P(x_1^n) \rfloor \quad \text{if } L(C_n^*, x_1^n) \le \lfloor \lg 1/P(x_1^n) \rfloor.$$

By construction $R_n^*(\widetilde{C}_n^*, P) = R_n^*(C_n^*, P)$. Thus, $\widetilde{C}_n^*$ is optimal, too. This proves (i).

Now consider an optimal code C_n^* satisfying (7) and let x_1^{n*} be a sequence with $R_n^*(P) = 1 - \langle -\lg P(x_1^{n*}) \rangle$. Thus, $L(C_n^*, x_1^n) = \lfloor \lg 1/P(x_1^n) \rfloor$ for all x_1^n with $\langle -\lg P(x_1^n) \rangle < \langle -\lg P(x_1^{n*}) \rangle$. This proves (8) with $s_0 = \langle -\lg P(x_1^{n*}) \rangle$. Finally, if (9) is not satisfied then (in view of Kraft's inequality) we can modify this code to a code $\widetilde{C}_n^*$ with
$$L(\widetilde{C}_n^*, x_1^n) = \lceil \lg 1/P(x_1^n) \rceil \quad \text{if } \langle \lg 1/P(x_1^n) \rangle \ge s_0,$$
$$L(\widetilde{C}_n^*, x_1^n) = \lfloor \lg 1/P(x_1^n) \rfloor \quad \text{if } \langle \lg 1/P(x_1^n) \rangle < s_0.$$

By construction $R_n^*(\widetilde{C}_n^*, P) = R_n^*(C_n^*, P)$. Thus, $\widetilde{C}_n^*$ is optimal, too. This proves (ii). ∎

Thus, we proved that the following generalized Shannon code code is the desired optimal code and it satisfies
$$L(C_n^{GS}, x_1^n) = \begin{cases} \lfloor \lg 1/P(x_1^n) \rfloor & \text{if } x_1^n \in \mathcal{L}_{s_0} \\ \lceil \lg 1/P(x_1^n) \rceil & \text{if } x_1^n \in \mathcal{A}^n \setminus \mathcal{L}_{s_0}, \end{cases}$$

where

$$\mathcal{L}_t := \{x_1^n \in \mathcal{A}^n : \langle -\lg P(x_1^n)\rangle < t\}$$

and $s_0 = \langle -\lg p_{j_0}\rangle$ is defined in (5).

The next question is how to construct efficiently the optimal generalized Shannon code? This turns out to be quite simple due to property (ii) (cf. (8) and (9)). The algorithm is presented below.

Algorithm GS–CODE

Input: Probabilities $P(x_1^n)$.
Output: Optimal generalized Shannon code.
1. Let $s_i = \langle -\lg P(x_1^n)\rangle$ for $i = 1, 2, \ldots, N$, where $N \le |\mathcal{A}^n|$.
2. Sort $s_1, \ldots, s_N$.
3. Use *binary search* to find the largest j_0 such that (5) holds, and set $s_0 = 1 - s_{j_0} = 1 - \langle -\lg p_{j_0}\rangle$.
5. Set code length $l_i = \lfloor -\lg p_i \rfloor$ for $i \le j_0$, otherwise $l_i = \lceil -\lg p_i \rceil$.
end

Observe that property (ii) above was crucial to justify the application of the binary search in Step 3 of the algorithm. Obviously, Step 2 requires $O(N \log N)$ operations which determines the complexity of the algorithm. If probabilities are sorted, then the complexity is determined by Step 5 and it is equal to $O(N)$, as for the Huffman code construction (cf. [9]).

Now, we turn our attention to universal codes for which the probability distribution P is unknown. We assume that P belongs to a set $\mathcal{S}$ (e.g., class of memoryless sources with unknown parameters). The following result summarizes our next finding. It transforms the Shtarkov bound (3) into an equality.

Theorem 2. *Suppose that $\mathcal{S}$ is a system of probability distributions P on $\mathcal{A}^n$ and set*

$$Q^*(x_1^n) := \frac{\sup_{P\in\mathcal{S}} P(x_1^n)}{\sum_{y_1^n \in \mathcal{A}^n} \sup_{P\in\mathcal{S}} P(y_1^n)}.$$

If the probability distribution Q^* is dyadic, i.e. $\lg Q^*(x_1^n) \in \mathbf{Z}$ for all $x_1^n \in \mathcal{A}^n$, then

$$R_n^*(\mathcal{S}) = \lg\left(\sum_{x_1^n\in\mathcal{A}^n} \sup_{P\in\mathcal{S}} P(x_1^n)\right). \tag{10}$$

Otherwise, let $q_1, q_2, \ldots, q_{|\mathcal{A}|^n}$ be the probabilities $Q^*(x_1^n)$, $x_1^n \in \mathcal{A}^n$, ordered in such a way that

$$0 \le \langle -\lg q_1\rangle \le \langle -\lg q_2\rangle \le \cdots \le \langle -\lg q_{|\mathcal{A}|^n}\rangle \le 1,$$

and let j_0 be the maximal j such that

$$\sum_{i=1}^{j-1} q_i 2^{\langle -\lg q_i\rangle} + \frac{1}{2}\sum_{i=j}^{|\mathcal{A}|^n} q_i 2^{\langle -\lg q_i\rangle} \le 1. \tag{11}$$

Then

$$R_n^*(\mathcal{S}) = \lg\left(\sum_{x_1^n\in\mathcal{A}^n} \sup_{P\in\mathcal{S}} P(x_1^n)\right) + R_n^*(Q^*), \tag{12}$$

where $R_n^*(Q^*) = 1 - \langle -\lg q_{j_0}\rangle$ *is the maximal redundancy of the optimal generalized Shannon code designed for the distribution* Q^*.

Proof. By definition we have

$$\begin{aligned}
R_n^*(\mathcal{S}) &= \min_{C_n\in\mathcal{C}} \sup_{P\in\mathcal{S}} \max_{x_1^n}(L(C_n,x_1^n) + \lg P(x_1^n))\\
&= \min_{C_n\in\mathcal{C}} \max_{x_1^n}\left(L(C_n,x_1^n) + \sup_{P\in\mathcal{S}} \lg P(x_1^n)\right)\\
&= \min_{C_n\in\mathcal{C}} \max_{x_1^n}\left(L(C_n,x_1^n) + \lg Q^*(x_1^n) + \lg\left(\sum_{y_1^n\in\mathcal{A}^n} \sup_{P\in\mathcal{S}} P(y_1^n)\right)\right)\\
&= R_n^*(Q^*) + \lg\left(\sum_{y_1^n\in\mathcal{A}^n} \sup_{P\in\mathcal{S}} P(y_1^n)\right),
\end{aligned}$$

where $R_n^*(Q^*) = 1 - \langle -\lg q_{j_0}\rangle$, and by Theorem 1 it can be interpreted as the maximal redundancy of the optimal generalized Shannon code designed for the distribution Q^*. Theorem 2 is proved. ■

3 Memoryless Sources

Let us consider a binary memoryless source with $P_p(x_1^n) = p^k(1-p)^{n-k}$ where k is the number of "0" in x_1^n and p is the probability of generating a "0". In the next theorem we compute the maximal redundancy $R_n^*(P_p)$ of the optimal generalized Shannon code assuming p is *known*.

Theorem 3. *Suppose that* $\lg\frac{1-p}{p}$ *is irrational. Then as* $n\to\infty$

$$R_n^*(P_p) = -\frac{\log\log 2}{\log 2} + o(1) = 0.5287\ldots + o(1).$$

If $\lg\frac{1-p}{p} = \frac{N}{M}$ *is rational and non-zero then as* $n\to\infty$

$$R_n^*(P_p) = -\frac{\lfloor M\lg(M(2^{1/M}-1)) - \langle Mn\lg 1/(1-p)\rangle\rfloor + \langle Mn\lg 1/(1-p)\rangle}{M} + o(1).$$

Finally, if $\lg\frac{1-p}{p} = 0$ *then* $p=\frac{1}{2}$ *and* $R_n^*(P_{1/2}) = 0$.

Proof. Set

$$\alpha_p = \lg\frac{1-p}{p},$$
$$\beta_p = \lg\frac{1}{1-p}.$$

Then

$$-\lg(p^k(1-p)^{n-k}) = \alpha_p k + \beta_p n.$$

First we assume that α_p is irrational. We know from [16] that for every Riemann integrable function $f : [0,1] \to \mathbf{R}$ we have

$$\lim_{n\to\infty} \sum_{k=0}^{n} \binom{n}{k} p^k (1-p)^{n-k} f(\langle \alpha_p k + \beta_p n \rangle) = \int_0^1 f(x)\, dx. \tag{13}$$

Now set $f_{s_0}(x) = 2^x$ for $0 \le x < s_0$ and $f_{s_0}(x) = 2^{x-1}$ for $s_0 \le x \le 1$. We obtain

$$\lim_{n\to\infty} \sum_{k=0}^{n} \binom{n}{k} p^k (1-p)^{n-k} f_{s_0}(\langle \alpha k + \beta n \rangle) = \frac{2^{s_0-1}}{\log 2}.$$

In particular, for

$$s_0 = 1 + \frac{\log\log 2}{\log 2} = 0.4712\ldots$$

we get $\int_0^1 f(x)\, dx = 1$ so that (5) holds. This implies that

$$\lim_{n\to\infty} R_n^*(P_p) = 1 - s_0 = 0.5287\ldots$$

If $\alpha_p = \frac{N}{M}$ is rational and non-zero then we have (cf. [16] or [17] Chap. 8)

$$\begin{aligned}\lim_{n\to\infty} \sum_{k=0}^{n} \binom{n}{k} p^k (1-p)^{n-k} f(\langle \alpha_p k + \beta_p n \rangle) &= \frac{1}{M} \sum_{m=0}^{M-1} f\left(\left\langle \frac{mN}{M} + \beta_p n \right\rangle \right) \\ &= \frac{1}{M} \sum_{m=0}^{M-1} f\left(\frac{m + \langle M\beta_p n \rangle}{M} \right).\end{aligned}$$

Of course, we have to use $f_{s_0}(x)$, where s_0 is of the form

$$s_0 = \frac{m_0 + \langle M\beta_p n \rangle}{M},$$

and choose maximal m_0 such that

$$\begin{aligned}\frac{1}{M} \sum_{m=0}^{M-1} f_{s_0}\left(\frac{m + \langle M\beta_p n \rangle}{M} \right) &= \frac{2^{\langle M\beta_p n \rangle}/M}{M} \left(\sum_{m=0}^{m_0-1} 2^{m/M} + \sum_{m=m_0}^{M-1} 2^{m/M-1} \right) \\ &= \frac{2^{(\langle M\beta_p n \rangle + m_0)/M-1}}{M(2^{1/M}-1)} \\ &\le 1.\end{aligned}$$

Thus,

$$m_0 = M + \lfloor M \lg(M(2^{1/M}-1)) - \langle Mn \lg 1/(1-p) \rangle \rfloor$$

and consequently

$$
\begin{aligned}
R_n^*(P_p) &= 1 - s_0 + o(1) \\
&= 1 - \frac{m_0 + \langle M\beta_p n\rangle}{M} + o(1) \\
&= -\frac{\lfloor M \lg(M(2^{1/M}-1)) - \langle Mn \lg 1/(1-p)\rangle \rfloor + \langle Mn\beta_p\rangle}{M} + o(1).
\end{aligned}
$$

This completes the proof of the theorem. ■

The next step is to consider memoryless sources P_p such that p is *unknown* and say contained in an interval $[a,b]$, i.e. we restrict $\mathcal{S}_{ab}$ to the class of memoryless sources with $p \in [a,b]$. Here, the result reads as follows.

Theorem 4. *Let* $0 \le a < b \le 1$ *be given and let* $\mathcal{S}_{a,b} = \{P_p : a \le p \le b\}$. *Then as* $n \to \infty$

$$
R_n^*(\mathcal{S}_{a,b}) = \frac{1}{2}\lg n + \lg C_{a,b} - \frac{\log\log 2}{\log 2} + o(1), \tag{14}
$$

where

$$
C_{a,b} = \frac{1}{\sqrt{2\pi}}\int_a^b \frac{dx}{\sqrt{x(1-x)}} = \sqrt{\frac{2}{\pi}}(\arcsin\sqrt{b} - \arcsin\sqrt{a}).
$$

Proof. First observe that

$$
\sup_{p\in[a,b]} p^k(1-p)^{n-k} = \begin{cases} a^k(1-a)^{n-k} & \text{for } 0 \le k < na, \\ \left(\frac{k}{n}\right)^k\left(1-\frac{k}{n}\right)^{n-k} & \text{for } na \le k \le nb, \\ b^k(1-b)^{n-k} & \text{for } nb < k \le n. \end{cases}
$$

By Theorem 2 we must evaluate $T_n = \sum_{x_1^n} \sup_{P\in\mathcal{S}_{ab}} P(x_1^n)$, which becomes

$$
T_n := \sum_{k<na}\binom{n}{k}a^k(1-a)^{n-k} + \sum_{na\le k\le nb}\binom{n}{k}\left(\frac{k}{n}\right)^k\left(1-\frac{k}{n}\right)^{n-k} + \sum_{k>nb}\binom{n}{k}b^k(1-b)^{n-k}.
$$

It is easy to show that

$$
\sum_{k<na}\binom{n}{k}a^k(1-a)^{n-k} = \frac{1}{2} + O(n^{-1/2})
$$

and

$$
\sum_{k>nb}\binom{n}{k}b^k(1-b)^{n-k} = \frac{1}{2} + O(n^{-1/2}).
$$

Furthermore, we have (uniformly for $an \le k \le bn$)

$$
\binom{n}{k}\left(\frac{k}{n}\right)^k\left(1-\frac{k}{n}\right)^{n-k} = \frac{1}{\sqrt{2\pi}}\sqrt{\frac{n}{k(n-k)}} + O(n^{-3/2}).
$$

Consequently

$$\sum_{na\le k\le nb}\binom{n}{k}\left(\frac{k}{n}\right)^k\left(1-\frac{k}{n}\right)^{n-k}=\sqrt{\frac{n}{2\pi}}\int_a^b\frac{dx}{\sqrt{x(1-x)}}+O(n^{-1/2})$$

$$=2\sqrt{\frac{n}{2\pi}}(\arcsin\sqrt{b}-\arcsin\sqrt{a})+O(n^{-1/2})$$

which gives

$$T_n=C_{a,b}\sqrt{n}+1+O(n^{-1/2})$$

and

$$\lg T_n=\frac{1}{2}\lg n+\lg C_{a,b}+O(n^{-1/2}).$$

To complete the proof we must evaluate the redundancy $R_n^*(Q^*)$ of the optimal generalized Shannon code designed for the maximum likelihood distribution Q^*. We proceed as in the proof of Theorem 3, and define a function $f_{s_0}(x)=2^x$ for $x\le s_0$ and otherwise $f_{s_0}(x)=2^{x-1}$. In short, $f_{s_0}(x)=2^{-\langle s_0-x\rangle+s_0}$ (now considered as a periodic function with period 1). The problem is to evaluate the following sum (cf. (11))

$$\sum_{k=0}^{n}\binom{n}{k}\frac{\sup_{p\in[a,b]}p^k(1-p)^{n-k}}{T_n}f_{s_0}\left(-\lg\left(\sup_{p\in[a,b]}p^k(1-p)^{n-k}\right)+\lg T_n\right)$$

$$=\frac{1}{T_n}\sum_{k<an}\binom{n}{k}a^k(1-a)^{n-k}f_{s_0}(-\lg(a^k(1-a)^{n-k})+\lg T_n)$$

$$+\frac{1}{T_n}\sum_{an\le k\le bn}\binom{n}{k}\left(\frac{k}{n}\right)^k\left(1-\frac{k}{n}\right)^{n-k}f_{s_0}\left(-\lg\left(\left(\frac{k}{n}\right)^k\left(1-\frac{k}{n}\right)^{n-k}\right)+\lg T_n\right)$$

$$+\frac{1}{T_n}\sum_{k>bn}\binom{n}{k}b^k(1-b)^{n-k}f_{s_0}(-\lg(b^k(1-b)^{n-k})+\lg T_n)$$

$$=S_1+S_2+S_3.$$

Obviously, the first and third sum can be estimated as

$$S_1=O(n^{-1/2})\quad\text{and}\quad S_3=O(n^{-1/2}).$$

Thus, is remains to consider S_2.

We will use the property that for every (Riemann integrable) function $f:[0,1]\to\mathbf{C}$ and for every sequence $x_{n,k}$ of the following form

$$x_{n,k}=k\lg k+(n-k)\lg(n-k)+c_n,\qquad an\le k\le bn,$$

where c_n is an arbitrary sequence, we have

$$\lim_{n\to\infty}\frac{1}{T_n}\sum_{an\le k\le bn}\binom{n}{k}\left(\frac{k}{n}\right)^k\left(1-\frac{k}{n}\right)^{n-k}f(\langle x_{n,k}\rangle)=\int_0^1 f(x)\,dx. \tag{15}$$

Note that we are now in a similar situation as in the proof of Theorem 3. We apply (15) with $f_{s_0}(x)$ for $s_0 = -\log\log 2/\log 2$, and (14) follows.

For the proof of (15), we verify the Weyl criteria (cf. [5,17]), that is, we first consider the following exponential sums

$$S := \sum_{an \le k \le cn} e(h(k \lg k + (n-k)\lg(n-k)),$$

where $e(x) = e^{2\pi i x}$, $c \in [a,b]$, and h is an arbitrary non-zero integer. By Van-der-Corput's method (see [7, p. 31]) we know that

$$|S| \ll \frac{|F'(cn) - F'(an)| + 1}{\sqrt{\lambda}},$$

where $\lambda = \min_{an \le y \le cn} |F''(y)| > 0$ and

$$F(y) = h(y \lg y + (n-y)\lg(n-y)).$$

Since $|F'(y)| \ll h \log n$, and $|F''(y)| \gg h/n$ (uniformly for $an \le y \le cn$) we conclude

$$|S| \ll \log n \sqrt{hn}$$

and consequently

$$\left| \sum_{an \le k \le cn} e(hx_{nk}) \right| \ll \log n \sqrt{hn}.$$

Note that all these estimates are uniform for $c \in [a,b]$. Next we consider exponential sums

$$\widetilde{S} := \sum_{an \le k \le bn} a_{n,k} e(hx_{nk}),$$

where

$$a_{n,k} = \binom{n}{k} \left(\frac{k}{n}\right)^k \left(1 - \frac{k}{n}\right)^{n-k}.$$

By elementary calculations we get (uniformly for $an \le k \le bn$) $a_{n,k} \ll n^{-1/2}$ and

$$|a_{n,k+1} - a_{n,k}| \ll n^{-3/2}.$$

Thus, by Abel's partial summation (cf. [17])

$$\begin{aligned}
|\widetilde{S}| &\le a_{n,bn} \left| \sum_{an \le k \le bn} e(hx_{n,k}) \right| \\
&\quad + \sum_{an \le k < bn} |a_{n,k+1} - a_{n,k}| \left| \sum_{an \le \ell \le k} e(hx_{n,\ell}) \right| \\
&\ll n^{-1/2} \log n \sqrt{hn} + n n^{-3/2} \log n \sqrt{hn} \\
&\ll \sqrt{h} \log n.
\end{aligned}$$

This means that for every non-zero integer h we have

$$\lim_{n\to\infty} \frac{1}{T_n} \sum_{an\le k\le bn} a_{n,k} e(hx_{n,k}) = 0. \tag{16}$$

Consequently, by standard tools in Fourier analysis (16) implies (for every Riemann integrable function $f : [0,1] \to \mathbf{C}$)

$$\lim_{n\to\infty} \frac{1}{T_n} \sum_{an\le k\le bn} a_{n,k} f(\langle x_{n,k}\rangle) = A_0(f),$$

where A_0 is the zero-th Fourier coefficient

$$A_0 = \int_0^1 f(x)\, dx.$$

This means that we have proved (15). ■

Remark 1. We can derive a full asymptotic expansion for the maximal minimax redundancy $R_n^*(\mathcal{S})$ for memoryless sources. Indeed, for a change we consider an m–ary alphabet $\mathcal{A}$ ($m \ge 2$). Following the footsteps of the above derivation, and using the approach from [15] for $p \in (0,1)$, we arrive at

$$R_n^*(\mathcal{S}) = \frac{m-1}{2}\log\left(\frac{n}{2}\right) - \frac{\ln\frac{1}{m-1}\ln m}{\ln m} + \log\left(\frac{\sqrt{\pi}}{\Gamma(\frac{m}{2})}\right) + \frac{\Gamma(\frac{m}{2})m}{3\Gamma(\frac{m}{2}-\frac{1}{2})}\cdot\frac{\sqrt{2}}{\sqrt{n}}$$
$$+\left(\frac{3+m(m-2)(2m+1)}{36} - \frac{\Gamma^2(\frac{m}{2})m^2}{9\Gamma^2(\frac{m}{2}-\frac{1}{2})}\right)\cdot\frac{1}{n} + O\left(\frac{1}{n^{3/2}}\right)$$

for large n. To the best of our knowledge, the above formula is the first asymptotic expansion with the correct constant term (i.e., containing the term $R_n^*(Q^*)$). This is of some importance since some authors (cf. [18]) propose to design optimal codes that optimize the constant term.

Remark 2. Parker [13] (cf. also [4,8]) investigated other than average cost functions but such for which the Huffman construction still produces optimal code. For example, Campbell [4] has shown that Huffman's code is optimal if the average code length is replaced by

$$W(r) = \frac{1}{r}\log_m\left(\sum_{x_1^n} P(x_1^n) m^{rL(x_1^n)}\right)$$

where $m = |\mathcal{A}|$, $r > 0$ is any positive number, and $L(x_1^n)$ is the code length. Observe that $\lim_{r\to 0} W(r) = \mathbf{E}[L(C_n, X_1^n)]$, while $\lim_{r\to\infty} W(r) = \max_{x_1^n} L(C_n, x_1^n)$ ($= \lceil \log_m N\rceil$). In this paper, we proved that when the $\max_{x_1^n} L(C_n, x_1^n)$ is replaced by the maximal redundancy $R_n^* = \max_{x_1^n}[L(C_n, x_1^n) + \log P(x_1^n)]$, then the Huffman code is not any more optimal. In general, let us define the r-th redundancy $R_n^{[r]}$ $(r > 0)$ as

$$R_n^{[r]} = \left(\sum_{x_1^n} P(x_1^n) \left[L(C_n, x_1^n) + \log P(x_1^n) \right]^r \right)^{1/r} .$$

Observe that the average redundancy is $\overline{R}_n = R_n^{[1]}$, while the maximal redundancy is $R_n^* = R_n^{[\infty]}$. The open question is what code minimizes the r-th redundancy $R_n^{[r]}$?

References

1. J. Abrahams, "Code and Parse Trees for Lossless Source Encoding, *Proc. of Compression and Complexity of SEQUENCE'97*, Positano, IEEE Press, 145–171, 1998.
2. A. Barron, J. Rissanen, and B. Yu, The Minimum Description Length Principle in Coding and Modeling, *IEEE Trans. Information Theory*, 44, 2743-2760, 1998.
3. T. Cover and J.A. Thomas, *Elements of Information Theory*, John Wiley & Sons, New York 1991.
4. L. Campbell, A Coding Theorem and Rényi's Entropy, *Information and Control*, 8, 423–429, 1965.
5. M. Drmota and R. Tichy, *Sequences, Discrepancies, and Applications*, Springer Verlag, Berlin Heidelberg, 1997.
6. D. E. Knuth, Dynamic Huffman Coding, *J. Algorithms*, 6, 163-180, 1985.
7. E. Krätzel, Lattice Points, Kluwer, Dordrecht, 1988.
8. P. Nath, On a Coding Theorem Connected with Rényi's Entropy, *Information and Control*, 29, 234–242, 1975.
9. J. van Leeuwen, On the Construction of the Huffman Trees, *Proc. ICALP'76*, 382–410, 1976.
10. J. Rissanen, Complexity of Strings in the Class of Markov Sources, *IEEE Trans. Information Theory*, 30, 526–532, 1984.
11. J. Rissanen, Fisher Information and Stochastic Complexity, *IEEE Trans. Information Theory*, 42, 40–47, 1996.
12. P. Shields, Universal Redundancy Rates Do Not Exist, *IEEE Trans. Information Theory*, 39, 520-524, 1993.
13. D. S. Parker, Conditions for Optimiality of the Huffman Algorithm, *SIAM J. Compt.*, 9, 470–489, 1980.
14. Y. Shtarkov, Universal Sequential Coding of Single Messages, *Problems of Information Transmission*, 23, 175–186, 1987.
15. W. Szpankowski, On Asymptotics of Certain Recurrences Arising in Universal Coding, *Problems of Information Transmission*, 34, No.2, 55-61, 1998.
16. W. Szpankowski, Asymptotic Redundancy of Huffman (and Other) Block Codes, *IEEE Trans. Information Theory*, 46, 2434-2443, 2000.
17. W. Szpankowski, *Average Case Analysis of Algorithms on Sequences*, Wiley, New York, 2001.
18. Q. Xie, A. Barron, Minimax Redundancy for the Class of Memoryless Sources, *IEEE Trans. Information Theory*, 43, 647-657, 1997.
19. Q. Xie, A. Barron, Asymptotic Minimax Regret for Data Compression, Gambling, and Prediction, *IEEE Trans. Information Theory*, 46, 431-445, 2000.

An Improved Algorithm for Sequence Comparison with Block Reversals*

S. Muthukrishnan[1] and S. Cenk Şahinalp[2]

[1] AT& T Labs – Research, 180 Park Avenue, Florham Park, NJ 07932;
muthu@research.att.com
[2] Dept of EECS, Dept of Genetics, and Cntr for Computational Genomics,
Case Western Reserve University, Cleveland, OH 44106;
cenk@cwru.edu

Abstract. Given two sequences X and Y that are strings over some alphabet set, we consider the distance $d(X,Y)$ between them defined to be minimum number of character replacements and block (substring) reversals needed to transform X to Y (or vice versa). This is the "simplest" sequence comparison problem we know of that allows natural block edit operations. Block reversals arise naturally in genomic sequence comparison; they are also of interest in matching music data. We present an improved algorithm for exactly computing the distance $d(X,Y)$; it takes time $O(|X|\log^2|X|)$, and hence, is near-linear. Trivial approach takes quadratic time and the best known previous algorithm for this problem takes time $\Omega(|X|\log^3|X|)$.

1 Introduction

Computing the distance between two sequences under edit operations is a central problem in Combinatorial Pattern Matching. The distance $d(X,Y)$ between sequences X and Y can be defined as the minimum number of permitted edit operations needed to transform one to another (all edit operations of interest are reversible so that $d(X,Y) = d(Y,X)$ for any two sequences X and Y). The goal is to compute $d(X,Y)$ for given X and Y as efficiently as possible. The nature of the distance computation problem depends on the edit operation that is permitted, which corresponds to the notion of similarity between sequences that we wish to capture for the application in question.

Natural edit operations include character edits and block edits.

Character edits include

1. inserting a single character to any specified location,
2. deleting any given character, and
3. replacing a single character by another.

* Supported in part by a grant from Charles B. Wang foundation.

S. Rajsbaum (Ed.): LATIN 2002, LNCS 2286, pp. 319–325, 2002.

Among these edit operations, character replacements are particularly common in computational genomics, in the form of mutations in biomolecular sequences. Sequence distances involving character replacements are used to estimate the evolutionary distances between DNA, RNA or protein sequences [Se80], construct phylogenetic trees, search for common motifs in the genome with functional homology, etc. [JKL96]). Insert and delete operations are also quite commonly observed especially as sequencing errors in computational genomics [GD91], transcription errors in text, transmission errors in communication channels, etc. These are of classical interest, having been studied at least as early as 60's [Lev66].

Block (substring) edits include

1. reversing a block (commonly observed in genomic sequences, usually as a consequence of large scale inter or intra chromosomal genomic duplications; also common in multimedia data such as music scales [SK83]),
2. copying a block from one location to another (another genomic phenomena commonly observed in the form of genomic duplications and rearrangements [JKL96]),
3. deleting a copy of a block that exists elsewhere (motivated by general macro compression schemes [St88], especially in the form of Ziv-Lempel type data compression [ZL77]),
4. moving a block from one location to another (due to moving a paragraph of a text to another location [T84], or moving objects around in pen computing [LT97]).[1]

Although these edit operations are motivated from many applications, we do not consider the specifics of any application. Rather, we focus on a central computational complexity question, namely, how to compute the various character and block edit distances efficiently.

The main result of this paper is on the basic problem of computing the distance between two sequences with both character and block edit operations. If the distance $d(X,Y)$ between two sequences X and Y is defined so as to allow any block copy, delete or move operations the problem of computing $d(X,Y)$ becomes NP-hard [LT97]. Thus, among the block edit operations described above, we allow only block reversals. Since it is not possible to transform a given sequence X (e.g., all 0's) to every other sequence Y (e.g., all 1's) by block reversals alone, $d(X,Y)$ is not well defined when only block reversals are allowed. Therefore, the "simplest" well-defined distance with block reversals additionally allows character replacements. Henceforth, we will focus on this distance $d(X,Y)$ between two sequences X and Y formally defined to be the minimum number of

[1] There are other block edit operations of interest, such as *block linear transformations* which involves changing each position of $Q[i]$ in a block Q, to $\phi Q[i]+\kappa$, where ϕ and κ are respective multiplicative and additive scaling constants. Such edit operations are quite common in financial data analysis where one allows scaling effects in tracking similar trends [AL+95].

character replacements and block reversals needed to convert X to Y (equivalently, vice versa). In particular, observe that $d(X,Y)$ is only defined when $|X| = |Y|$.

The distance $d()$ of our interest combines the most commonly observed character edit operations in genomic sequences with common block edit operations. Using dynamic programming, we can compute $d(X,Y)$ exactly in time $\Theta(|X|^2)$; the solution is straightforward. Of course it is trivial to compute the distance in linear time if only character replacements are considered and block reversal are *not* allowed. The first nontrivial algorithm for solving our problem was presented in [MS00] which took time $O(|X| \cdot \log^3 |X|)$. In this paper we present an improved, $O(|X| \cdot \log^2 |X|)$ time deterministic algorithm for this problem. Subquadratic algorithms are in general of interest in computing various edit distances since the basic problem of computing character edit distance is not known to have a near-linear time solution at present.

2 Preliminaries

In the rest of the paper we denote sequences by P, Q, R, S.., integers and binary numbers by i, j, k.. and constants by α, β, γ... All sequences have characters drawn from alphabet σ. Given a sequence S, let $|S|$ denote its length and $S[i]$ denote its ith character. A *block* of S will be any subsequence $S[i:j]$ extending from $S[i]$ to $S[j]$. We denote by $S||Q$ the concatenation of two sequences S and Q, and by Q^l, a sequence obtained by concatenating l copies of Q. The reverse of S is the sequence $S[|S|], \ldots, S[2], S[1]$, and is denoted by $S^{\leftarrow}$. A sequence S is called *periodic* with $l > 1$ repetitions if there exists a block Q such that $S = Q^l||Q^-$ where Q^- is a prefix of Q; Q is called a *period* of S. The shortest such Q is called *the* period of S; the period of S is the period of all periods of S.

3 The Main Result

Let X and Y be two size ℓ sequences. Recall that $d(X,Y)$ is the minimum number of character replacements and block reversals needed to transform X into Y. Below we describe an algorithm for computing $d(X,Y)$ in time $O(\ell \log^2 \ell)$. The procedure relies on combinatorial properties of reversals in sequences.

We say that block $X[i:j]$ is *reversible* (with respect to Y) if $X[i:j] = Y[i:j]^{\leftarrow}$.

The algorithm. We compute $d(X,Y)$ in ℓ iterations as follows. In iteration i, we compute $d(i) = d(X[1:i], Y[1:i])$. Let $X[i_1:i]$ be the longest suffix of $X[1:i]$ that is reversible and let p_1 be the length of the period of $X[i_1:i]$. For $h \geq 1$, we inductively define

$$\#_h = \left\lceil \frac{i - i_h + 1}{p_h} \right\rceil - 1$$

$$i_{h+1} = i_h + \#_h \cdot p_h + 1.$$

We also define p_{h+1} to be the length of the period of $X[i_{h+1} : i]$ and define H the minimum h such that $\#_h = 0$; notice that $H \leq \log i$.

The algorithm computes $d(i)$ as follows. For $1 \leq h \leq H$, we set

$$d_h(i) = d(i_h - 1) + 1 \text{ if } \#_h \leq 1$$

$$d_h(i) = d_h(i - p_h) \text{ if } \#_h > 1.$$

We also set $d_0(i) = d(i-1) + d(X[i], Y[i])$. Then we compute $d(i) = \min\{d_h(i)\}$ which completes the description of the algorithm. □

We prove the correctness of the algorithm through the following lemma and its corollaries.

Lemma 1. *If $X[j : i]$ is reversible and p is the length of the period of $X[j : i]$, then for any $k \leq i - j - p + 1$, the substring $X[j + k : i]$ is reversible if and only if $k = p \cdot h$ for $0 \leq h \leq \lceil (k+1)/p \rceil - 1$.*

Proof. (1) If $X[j : i]$ is reversible, then $X[j : i] = Y[j : i]^{\leftarrow}$, and hence $X[j+k] = Y[i-k]$ for all $k \leq j - i + 1$. Because p is the length of the period of $X[j : i]$, $X[j+k] = X[j+p \cdot h+k] = Y[i-k]$ for all $k < i - j - (p \cdot h) + 1$, which implies $X[j+p \cdot h : i] = Y[j+p \cdot h : i]^{\leftarrow}$; therefore $X[j + p \cdot h : i]$ is reversible.
(2) If both $X[j : i]$ and $X[j + k : i]$ are reversible, then $X[j + h] = Y[i - h] = X[j + k + h]$ for all $h \leq i - j - k$. Therefore k must be the length of a period of $X[j : i]$, which implies that k must be a multiple of p. □

The statements below follow.

Corollary 1. *(i) The only reversible suffixes of $X[1 : i]$ are of the form $X[i_h + p_h \cdot j : i]$ for all $0 \leq j \leq \#_h$ and $1 \leq h \leq H$.*
(ii) If $X[j : i]$ is reversible and p is the length of the period of $X[j : i]$, then for any $k \geq j + p$, the substring $X[j : k]$ is reversible if and only if $k = j + p \cdot h$ for $0 \leq h \leq \lceil (k - j + 1)/p \rceil - 1$.
(iii) Let $X[j : i]$ be a reversible string whose period is of length p and let $X[h : i - k \cdot p]$ be a substring of $X[j : i]$ such that $h + p < i - k \cdot p$. If $X[h : i - k \cdot p]$ is reversible then so is $X[h : i]$.

Thus the proof of the lemma below follows, immediately giving the correctness of the entire algorithm.

Lemma 2. *Consider any i and let $p_0 = i - i_1$ where i_1 is as defined in the algorithm. For $1 \leq h \leq H$, $d_h(i)$ is equal to the distance between $X[1 : i]$ and $Y[1 : i]$, provided a suffix of length k for which $p_{h-1} \geq k > p_h$ is reversed (while transforming $X[1 : i]$ to $Y[1 : i]$).*

It remains for us to show how to implement the algorithm, and determine its running time.

We first compute i_1 for all i, $1 \le i \le \ell$ in $O(\ell)$ time as follows.

1. First we set $X' = X[1], \$, X[2], \$, \ldots, \$, X[\ell]$ and $Y' = Y[1], \$, Y[2], \$ \ldots, \$, Y[\ell]$ where \$ is a special character which is not in the original alphabet and used for treating reversals whose center is between two characters and those whose center is a character uniformly. Notice that for $X[i:j]$ is reversible with respect to Y if and only if $X'[2i-1:2j-1]$ is reversible with respect to Y'.
2. For all $1 \le j \le 2\ell - 1$ we compute the largest k for which $X'[j-k:j+k]$ is reversible with respect to Y'; let this be k_j. This can be done by finding the longest common prefix between $X'[j:2\ell-1]$ and $Y'[1:j]^{\leftarrow}$, the longest common suffix of $X'[1:j]$ and $Y'[j:2\ell-1]^{\leftarrow}$, and picking the shortest of the two. For finding either of the two longest common items, it suffices to build the joint suffix trees of X and $Y'^{\leftarrow}$; this takes $O(\ell)$ time to construct. [W73] This joint suffix tree can then be preprocessed in $O(\ell)$ time in order to answer the lowest common ancestor queries in time $O(1)$ each. [HT84]
3. For all $i = 1, \ldots, 2\ell - 1$, notice that i_1 is the equal to the smallest j for which $j + k_j \ge i$. We compute i_1 for all i in $O(\ell)$ time due to the observation that for $i < i'$, $i_1 \le i'_1$. Thus for each i, we only consider $j = (i-1)_1, (i-1)_1 + 1, \ldots$ until we find $j = i_1$.

Computing the periods of substrings. Next we focus on computing the period of any substring $S = X[j:i]$, which is of independent interest. We show that one can preprocess X in $O(\ell \log \ell)$ time to be able to compute the period of any given substring $S = X[j:i]$ in $O(\log |S|)$ time.

1. **Preprocessing.** We simply assign labels to all substrings of X of size 2^k, $0 \le k \le \lfloor \log \ell \rfloor$, such that (i) each distinct substring gets a distinct label, and (ii) each substring has a pointer to the nearest substring to its left which has the same label. This can be done in $O(\ell \log \ell)$ time using [KMR72] technique.
   We use these labels and some common facts about periods of strings to answer two types of queries:
   (a) Checking whether any two substrings R and Q (of any given length i) are identical:
   Q and R are identical if and only if their length $2^{\lfloor \log i \rfloor}$ prefixes are identical and their length $2^{\lfloor \log i \rfloor}$ suffixes are identical. Thus to check whether Q and R are identical, all we need to do is to check whether these suffixes and prefixes are identical by simply comparing their respective labels in $O(1)$ time.
   (b) Checking whether S has a period of length j for any specified length $j \le |S|/2$:

S has a period of length j if and only if $S[1 : |S| - j]$ is identical to $S[j+1 : |S|]$. This can be done by the simple method described above in $O(1)$ time.

2. **Querying.** Using the labels obtained in the preprocessing step we compute the period of any given substring $S = X[j : i]$ in $\log |S|$ iterations. In the kth iteration ($k = 0, 1, \ldots, \lceil \log |S| \rceil - 1$) we decide either (i) S does not have a period of length in the range $2^{k-1} + 1, \ldots 2^k$ and thus S does not have a period shorter than 2^k), or (ii) S has a period of length in the range $2^k, \ldots 2^{k-1} + 1$, and explicitly compute the length of the period, or (iii) S is non-periodic. This is done in $O(1)$ time per iteration as follows.
   We first check if there exists 0, 1 or more *candidate* period lengths l in the range $2^{k-1} + 1, \ldots, 2^k$: a candidate period length l is one which satisfies the condition that $S[|S| - l - 2^{\lceil \log |S| \rceil - 1} + 1 : |S| - l]$ is identical to $S[|S| - 2^{\lceil \log |S| \rceil - 1} + 1 : |S|]$. This can be done in $O(1)$ time by (i) following the pointer of $X[i - 2^{\lceil \log |S| \rceil - 1} + 1 : i]$ to the nearest substring to its left with an identical label, (ii) checking if the rightmost position of this substring is in the range $i - 2^k, \ldots, i - 2^{k-1}$ (if yes, there is at least one candidate substring), and (iii) following the pointer of this substring to a new substring and checking whether this new one has its rightmost position in the same range (if yes, there are at least two candidates).
   Notice that if l is a period length of S then $S[1 : |S| - l]$ and $S[l + 1 : |S|]$ should be identical, and so as their equal length prefixes. Thus if l is the length of a period of S then it should be identified as a candidate period length according to our definition.
   (a) If no such candidate period length exists, then S can not have a period shorter than $2^k + 1$ and thus we move to the next iteration.
   (b) If only one candidate period length l exists, then l is the only possible period length for S in the range $2^{k-1}, \ldots, 2^k$. We simply check in $O(1)$ time whether l is a period length of S via the simple test described in the preprocessing stage.
   (c) If two (or more) such candidate period lengths exist, then we conclude that S is non-periodic due to the following observation. If the length of the period of S is p_S and if for any of its substrings, R, the length of its period is p_R, then either $p_S = p_R$ or $p_S \geq |R|$.
   Now consider two candidate period lengths $l_1 \leq l_2$. We know that $R_1 = S[|S| - l_1 - 2^{\lceil \log |S| \rceil - 1} + 1 : |S| - l_1]$ is identical to $S[|S| - 2^{\lceil \log |S| \rceil - 1} + 1 : |S|]$ which in turn is identical to $R_2 = S[|S| - l_2 - 2^{\lceil \log |S| \rceil - 1} + 1 : |S| - l_2]$; thus the length of the period of $R = S[|S| - l_2 - 2^{\lceil \log |S| \rceil - 1} + 1 : |S| - l_1]$ (the concatenation of R_1 and R_2) is $p_R = l_2 - l_1 \leq 2^{k-1}$. We inductively know that $p_S > 2^{k-1} \geq p_R$, thus it must be the case that $p_S \geq |R| > 2^{\lceil \log |S| \rceil - 1}$ which means that S is non-periodic.

This combined with the iteration for computing $d(i)$'s gives,

Theorem 1. *Given two sequences X and Y of length ℓ, there exists an $O(\ell \log^2 \ell)$ time algorithm to compute $d(X, Y)$.*

4 Discussion

We have given a near linear time algorithm to compute $d(X, Y)$ which is the minimum number of character replacements and block reversals needed to convert X and Y (and vice versa). The function $d()$ is the "simplest" distance involving block edit operations, and this paper gives an improved subquadratic algorithm for computing any block edit distance. It will be of interest to design efficient algorithms for simple sequence comparison problems with other block operations [LT97].

References

[AL+95] R. Agarwal, K. Lin, H. Sawhney and K. Shim. Fast similarity search in the presence of noise, scaling and translation in time-series databases. *Proc. 21st VLDB conf*, 1995.

[GD91] M. Gribskov and J. Devereux *Sequence Analysis Primer*, Stockton Press, 1991.

[HT84] D. Harel and R. Tarjan. Fast Algorithms for Finding Nearest Common Ancestors. *SIAM J. Comput.*, 13(2): 338–355, 1984.

[JKL96] M. Jackson, T. Strachan and G. Dover. Human Genome Evolution, *Bios Scientific Publishers*, 1996.

[KMR72] R. Karp, R. Miller and A. Rosenberg, Rapid Identification of Repeated Patterns in Strings, Trees, and Arrays, Proceedings of *ACM Symposium on Theory of Computing*, (1972).

[LT97] D. Lopresti and A. Tomkins. Block Edit Models for Approximate String Matching. *Theoretical Computer Science*, 181(1): 159-179, 1997.

[SK83] D. Sankoff and J. Kruskal, Time Warps, String Edits, and Macromolecules: The Theory and Practice of Sequence Comparison, *Addison-Wesley*, Reading, Mass., 1983.

[Lev66] V. I. Levenshtein, Binary Codes Capable of Correcting Deletions, Insertions and Reversals, *Cybernetics and Control Theory*, 10(8):707-710, 1966.

[MS00] S. Muthukrishnan and S. C. Sahinalp, Approximate Nearest Neighbors and Sequence Comparison with Block Operations, Proceedings of *ACM Symposium on Theory of Computing*, 2000.

[Se80] P. Sellers, The Theory and Computation of Evolutionary Distances: Pattern Recognition. *Journal of Algorithms*, 1, (1980):359-373.

[St88] J. A. Storer, Data Compression, Methods and Theory. *Computer Science Press*, 1988.

[T84] W. F. Tichy, The String-to-String Correction Problem with Block Moves. *ACM Trans. on Computer Systems*, 2(4): 309-321, 1984.

[ZL77] J. Ziv and A. Lempel, A Universal Algorithm for Sequential Data Compression IEEE Trans. on Information Theory, 337–343, 1977.

[W73] P. Weiner Linear Pattern Matching Algorithms. *Proc. IEEE Foundations of Computer Science (FOCS)*, 1–11, 1973.

Pattern Matching and Membership for Hierarchical Message Sequence Charts

Blaise Genest and Anca Muscholl

LIAFA, Université Paris VII
2, pl. Jussieu, case 7014
F-75251 Paris cedex 05
genest@crans.org, muscholl@liafa.jussieu.fr

Abstract. Several formalisms and tools for software development use hierarchy for system design, for instance statecharts and diagrams in UML. Message sequence charts are an ITU standardized notation for asynchronously communicating processes. The standard Z.120 allows (high-level) MSC-references that correspond to the use of macros. We consider in this paper two basic verification tasks for hierarchical MSCs (nested high-level MSCs, nHMSC), the membership and the pattern matching problem. We show that the membership problem for nHMSCs is PSPACE-complete, even using a weaker semantics for nMSCs than the partial-order semantics. For pattern matching nMSCs M, N we exhibit a polynomial algorithm of time $O(|M|^2 \cdot |N|^2)$. We use here techniques stemming from algorithms on compressed texts.

1 Introduction

It is common to use macros to write a program or to specify the behavior of a system. Macros or hierarchical specifications allow a modular design of complex systems and have the advantage of being more succinct and user-friendly. Several formalisms and tools for software development use hierarchy for system design. One of the most prominent examples is the formalism of statecharts [8], which is a component of several object-oriented notations, such as the Unified Modeling Language (UML). Besides statecharts, UML widely uses several kinds of diagrams (activity, interaction diagrams etc), all based on the ITU standard Z.120 of message sequence charts (MSCs). While statecharts extend finite state machines by hierarchy and communication mechanisms, MSCs are a visual notation for asynchronously communicating processes. The usual application of MSCs in telecommunication is for capturing requirements of communication protocols in form of scenarios in early design stages. MSCs usually represent incomplete specifications, obtained from a preliminary view of the system that abstracts away several details such as variables or message contents. High-level MSCs (HMSCs) combine basic MSCs using choice and iteration, thus describing possibly infinite collections of scenarios. For abstract specifications as with HMSCs, hierarchy is of primary importance. Since a scenario corresponds to a specification level

S. Rajsbaum (Ed.): LATIN 2002, LNCS 2286, pp. 326–340, 2002.

which can be very abstract, a designer should be able to merge different specification cases yielding the same abstract scenario and to use this scenario as a macro. By using macros designers may identify sub scenarios which have to be refined at a later stage.

In this paper we consider two fundamental verification problems for *hierarchical* (or *nested high-level MSCs*, nHMSCs), the membership problem and pattern matching. The membership problem is a basic question, asking for instance whether a negative scenario occurs in a system specification, or asking whether a positive scenario is redundant, since already covered by the specification. Without hierarchy, the membership problem for HMSCs has been recently shown to be NP-complete, [1]. The reason for this complexity blow-up (compared to finite-state machines) is that executions of MSCs involve concurrency of events. We show that hierarchy yields an additional increase in complexity, precisely we show that the membership problem for nHMSCs is PSPACE-complete. Surprisingly, hierarchy alone is the source of the complexity. We show namely that the membership problem for hierarchical automata is already PSPACE-complete. This result can be compared with [3], that shows e.g. that reachability of communicating hierarchical automata is already EXPSPACE-complete, whereas without communication it is PSPACE-complete.

The second problem considered in this paper is pattern matching for nMSCs. Given two nMSCs M, N, we want to know whether M occurs as a pattern (factor graph) of N. A polynomial solution seems out of reach for this problem, if one follows the naive approach of flattening M and N. However, by exploiting some combinatorial arguments used in pattern matching of compressed texts we obtain an algorithm of time $O(|M|^2 \cdot |N|^2)$.

Related work. Regarding the complexity of extended finite state machines, [9] considers the reachability and trace equivalence problems for communicating FSMs. Model-checking hierarchical FSMs against LTL and CTL is done in [4]. The paper [3] combines hierarchy and concurrency, analyzing the complexity of several problems (reachability, equivalence etc.) for communicating, hierarchical FSMs. There is also a rich body of theoretical results concerning MSCs. Several verification problems have been considered recently, e.g. detecting races [2,12], model-checking [5], pattern matching with gaps [13], inference [1], model-checking against partial-order logics [11]. Increasing the expressivity of this notation is also a central topic, [7].

For lack of space several proofs are omitted and can be found in the full version of the paper.

2 Syntax and Semantics of nMSCs

We adopt the definition of message sequence charts (MSC) as described in the ITU-standard [10].

Message Sequence Charts. An MSC M is given as a tuple $\langle E, <, P, l, t, m \rangle$ where:

– E is a finite set of events.
– $<$ is an acyclic relation on events.
– P is a finite set of processes.
– $l : E \longrightarrow P$ associates each event with a process (localization function).
– $t : E \longrightarrow \{s, r\}$ associates each event with its type, where s holds for send and r for receive (type function).
– $m : t^{-1}(s) \longrightarrow t^{-1}(r)$ associates bijectively each send event with a receive event (message function). When $e, f \in E$ are such that $m(e) = f$, then we say that e, f are matching events and that (e, f) is a message.

The relation $<$ is defined by two conditions. First, if (e, f) are matching events, $m(e) = f$, then $e < f$. Second, if e and f are located on the same process, $l(e) = l(f)$, then either $e < f$ or $f < e$.

For communication protocols it is very natural to assume that each communication channel $(i, j) \in P^2$ delivers messages first-in-first-out (*FIFO*). We assume the FIFO condition throughout the paper. That is, for all messages (e_i, f_i), $i = 1, 2$, such that $l(e_1) = l(e_2)$ and $l(f_1) = l(f_2)$ we require that $e_1 < e_2$ iff $f_1 < f_2$. The reflexive-transitive closure $\leq$ of the acyclic relation $<$ is a partial order called *visual order*. Every total order on the set of events E that extends $\leq$ is called *linearization* of M.

Remark 1 Let M and N be two MSCs. Due to the FIFO order, any total order on events defines a unique MSC, if any. So, in order to check whether $M = N$ (up to renaming events), one can choose any linearization α of M and check whether α is a linearization of N, too. Alternatively, one can check that on each process $i \in P$, the *projection* $M|_i$ of M on the set $l^{-1}(i)$ of events located on i, is equal to the projection $N|_i$ of N on that process.

We follow the ITU-standard and define nested MSCs (nMSC) by allowing the reuse of an already defined MSC in a definition. Our definition of nMSCs below formalizes the visual character of MSCs.

Definition 1 *(***Nested MSC***) A nested MSC (nMSC)* $M = (M_q)_{q=1}^n$ *is a finite sequence of modules. Each module* $M_q = \langle E_q, P_q, B_q, \varphi_q, l_q, t_q, m_q, <_q \rangle$ *is defined as an MSC, containing in addition a finite set* B_q *of references (boxes), together with a function* φ_q *that associates each reference* b *of* B_q *with an index* $q < \varphi_q(b) \leq n$*. The meaning is that the reference* b *represents the module* $M_{\varphi_q(b)}$*. Hence,* $B_n = \emptyset$*.*

The acyclic relation $<_q$ *is defined over the set of events and references* $E_q \cup B_q$ *by the following conditions:*

1. *For each process* $k \in P_q$*, the relation* $<_q$ *is total over the events located on* k *and over the references* b *involving process* k*. That is,* $<_q$ *is total over* $l_q^{-1}(k) \cup \{b \in B_q \mid k \in P_{\varphi_q(b)}\}$*.*
2. *For each matching pair* $m_q(e) = f$*, we require* $e <_q f$*.*

The semantics of an nMSC M is the MSC defined by replacing recursively each reference of M by the corresponding MSC. We define the *visual order* of an nMSC M as the visual order of the MSC represented by M. The *nesting depth* of M is the maximal number of nested references of M.

Remark 2 We required that the order $<_q$ of M_q is total over each process, including the references using that process. However, this does not mean that *all* events of a reference $b \in B_q$ have to occur after e.g. the last event preceding b on some process of $\varphi_q(b)$. The ordering of a reference $b \in B_q$ is just the ordering needed for the visual representation of M_q. Of course, the position of b on each process of $\varphi_q(b)$ within M_q must represent a cut of M_q (pairwise incomparable positions, one for each process).

Moreover, a syntactically correct nMSC M might not yield an MSC because of the FIFO order. However, we can check in polynomial time whether an nMSC satisfies FIFO.

Straight-line program. A straight-line program (SLP) over the alphabet A is a set $V = \{X_1, \dots, X_k\}$ of variables together with a set of rules $V \times (V \cup A)^+$ such that there is exactly one rule for each left-hand side and if $X_i \longrightarrow \alpha$ is a rule, then each X_j in α is such that $j > i$.

A variable X_i with $X_i \longrightarrow \alpha$ and $\alpha \in A^*$ is denoted as a variable on the lowest level of the hierarchy. Any variable X_i generates a unique word $w(X_i)$. The length of a variable X_i represents the length of the word $w(X_i)$ and is denoted as $||X_i||$. Note that $||X_i||$ can be exponential in the size of the SLP.

In the same way as we can describe any MSC unambiguously by its projections, we associate each nMSC M over process set P with $|P|$ straight-line programs (SLP) $L^1, \dots, L^p$. The SLP L^i corresponds to the projection of M on the set of events of process $i \in P$. We will denote the variables used by L^i as $X|_i$, where X is a reference of the nMSC M.

Size of nMSC. The size of an nMSC M is denoted as $|M|$, and it represents the size of the syntactical description of M. Here, the size of event is 1 and the size of a reference is the number of processes it involves.

3 Syntax and Semantics of nHMSC

An MSC can only specify a finite scenario. For describing more complex behaviors such as infinite sets of scenarios, the simplest way is to compose MSCs in form of MSC-graphs (also called high-level MSC, HMSC), by allowing choice and iteration.

MSC-graph. An MSC-graph G is given as a tuple $\langle V, E, s, f, \varphi\rangle$, where:

- (V, E) is a directed graph with starting vertex $s \in V$ and final vertex $f \in V$.
- Each vertex v is labeled by an MSC $\varphi(v)$.

In the same way as we defined nested MSCs from (flat) MSCs we can generalize MSC-graphs to *hierarchical MSCs* (or *nested high-level MSCs*, nHMSC for short).

Nested high-level MSC. An nHMSC is a finite sequence $G = (G_q)_{q=1}^n$, where each (reference) G_q is either a labeled graph or an extended nMSC.

A labeled graph G_q is a tuple $\langle V_q, E_q, \varphi_q, s_q, f_q \rangle$ consisting of:

- A directed graph (V_q, E_q) with starting vertex s_q and final vertex f_q.
- A function φ_q that associates each vertex v with a reference $q < \varphi_q(v) \leq n$, representing $G_{\varphi_q(v)}$.

An *extended nMSC* G_q is an nMSC where a reference b represents $G_{\varphi_q(b)}$ with $q < \varphi_q(b) \leq n$.

Semantics (executions) of nHMSC. The semantics of an nHMSC $G = (G_q)_{q=1}^n$ is defined recursively. Let us consider first a vertex $v \in V_q$. If $\varphi_q(v)$ is an nMSC, then it represents the execution of that vertex. Else, let $\varphi_q(v)$ be an extended nMSC G_r. Then the set of executions of v is obtained by replacing each reference b of G_r by some nMSC execution of the nHMSC $G_{\varphi_r(b)}$. Finally, let $\varphi_q(v)$ be an nHMSC. Then the executions of v are exactly the executions of $G_{\varphi_q(v)}$.

With a path $v_1, \ldots, v_s$ of G_q we associate the set of executions of the form $M_1 \circ \cdots \circ M_s$, where each M_k is an execution of v_k. Finally, the set of executions $L(G_q)$ of the nHMSC G_q consists of the executions of all paths from the starting vertex s_q to the final vertex f_q. The set of executions of the nHMSC G is defined as $L(G_1)$.

It remains to define the sequential composition of two nMSCs N_1, N_2. For $q = 1, 2$ and $N_q = \langle E_q, P_q, B_q, \varphi_q, l_q, t_q, m_q, <_q \rangle$ let $N_1 \circ N_2 = \langle E, B, \varphi, l, t, m, < \rangle$ with $E, P, B, \varphi, l, t, m$ defined as the disjoint union of the corresponding components, $P = P_1 \cup P_2$ and

$$< = <_1 \cup <_2 \cup \bigcup_{i \in P} l_1^{-1}(i) \times l_2^{-1}(i).$$

As usual, we just require ordering on each process from one nMSC to the next one (weak sequencing). Of course, a process can execute an event of N_1 while another process already executes an event of N_2.

Notice that the execution of every path of an nHMSC is a well-defined nMSC, in contrast to compositional HMSCs [7], where matching sends and receives is done along paths in the graph.

Remark 3 Each MSC-graph can be seen as an nHMSC where nodes are mapped to MSCs. However, nHMSCs are more expressive than MSC-graphs, since in extended nMSCs references can be graphs. For instance, the nHMSC of Figure 1 cannot be represented by a finite MSC-graph, since there can be arbitrary many events between the send-receive pair depicted in the initial node of the graph on the left (note that the reference G is a graph).

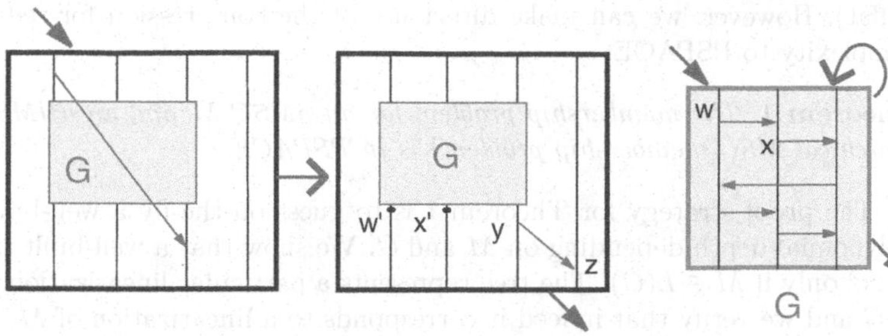

Fig. 1. An nHMSC that cannot be expressed as a *finite* MSC-graph.

According to [1] we will also consider a weaker semantics for nHMSCs, called the *weak closure.* This is a semantics based on taking the (unsynchronized) product of the sequential behaviors of single processes. Several algorithmic problems can be solved more efficiently for the weak closure of MSC-graphs. This makes it interesting to compare it with the usual semantics also in the setting of nHMSCs.

Weak closure of nHMSC. Let G be an nHMSC. Then $L^w(G)$ denotes the set of MSCs M such that for each process i there is some execution $N \in L(G)$ such that $M|_i$ is equal to $N|_i$.

4 MSC Membership Problem

The membership problem of an MSC M in an MSC-graph G is a basic question since it allows to check for instance if a bad scenario is possible in some HMSC specification. Another application is checking whether a good scenario is already covered by a specification. The MSC membership problem was recently considered in [1], together with the weak membership problem $M \overset{?}{\in} L^w(G)$. Recall that $L^w(G)$ is built from projections of possibly distinct paths in G. The results of [1] can be summarized as follows:

- The MSC membership problem is NP-complete.
- The MSC membership problem is solvable in time[1] $O(|G|^{O(1)} \cdot |M|^p)$, where p is the number of processes.
- The weak MSC membership problem is solvable in time $O(|G| \cdot |M|)$.

4.1 Hierarchical Membership Problem

The hierarchical membership problem seems *a priori* more difficult, since the naive approach of guessing a path of an nHMSC G and checking equality with an nMSC M is too expensive (the path may be of exponential size, even if M

[1] This is a slightly improved running time compared to the result stated in [1].

is flat). However, we can make direct use of the compression for reducing the complexity to PSPACE:

Theorem 1 *The membership problem for an nMSC M and an nHMSC G (hierarchical MSC membership problem) is in PSPACE.*

The proof strategy for Theorem 1 is to guess on-the-fly a well-built tree of polynomial depth depending on M and G. We show that a well-built tree exists if and only if $M \in L(G)$. The tree represents a particular linearization of a path in G and we verify that indeed it corresponds to a linearization of M. We show how to check on-the-fly in PSPACE that a tree is well-built according to M and G. This gives a (non-deterministic) polynomial space algorithm.

The depth of a well-built tree corresponds to the nesting depth of the definition of M. Any node of the tree is labeled by a claim (α, N, β) where N is a factor of M and α, β are positions of G. Moreover, there exists a path in G from α to β that is labeled by N. Formally, we use configurations for specifying positions in G:

Configurations. Let p be the number of processes. A configuration α of an nHMSC G is a tuple $\langle (A_1^1, \ldots, A_1^{\lambda_1}, a_1), \ldots, (A_p^1, \ldots, A_p^{\lambda_p}, a_p) \rangle$, where each A_i^k is a vertex of some reference used in $G = (G_q)_{q=1}^n$. Furthermore, A_i^1 is a vertex of G_1, A_i^k is a vertex of A_i^{k-1}, and a_i is an event appearing literally on the i-th process of $A_i^{\lambda_i}$. A configuration is called *strong*, if $A_i^k = A_j^k$ for all $1 \leq k \leq \min(\lambda_i, \lambda_j)$.

Remark 4 We need strong configurations for ensuring that we consider the same path in G for all processes. Without hierarchy, a strong configuration just corresponds to a single vertex, in the hierarchical case it corresponds to vertices that are comparable w.r.t. nesting. Clearly, each execution of G has at least one linearization where all prefixes are strong configurations.

The MSC factor N in a claim (α, N, β) is of the form $< (N^1, a_1, b_1), \ldots, (N^p, a_p, b_p) >$, where each N^i is a reference of $M = (M_q)_{q=1}^n$ and $a_i < b_i$ are events of the i-th process in the MSC defined by N^i.

M-defined nMSC. A tuple $< (N^1, a_1, b_1), \ldots, (N^p, a_p, b_p) >$ is called $(M^1, \ldots, M^p)$-*defined* if N^i is a reference used in the definition of M^i.

We will use only tuples $< (N^1, a_1, b_1), \ldots, (N^p, a_p, b_p) >$ forming MSCs, consisting of the sequence of events of N^i from a_i to b_i, for each projection i. Hence, the concatenation $\circ$ will be meant as MSC product.

Formally, a well-built tree is a finite tree where each vertex v is labeled by a claim (α, X, β) and every child of v is labeled by a claim $(\alpha_i, X_i, \alpha_{i+1})$, where α, β are strong configurations, $X =< (M^1, a_1, b_1), \ldots, (M^p, a_p, b_p) >$, each X_i is $(M^1, \ldots, M^p)$-defined, and such that the following conditions are satisfied:

1. $X = X_1 \circ \cdots \circ X_m$.
2. $\alpha_0 = \alpha$ and $\alpha_m = \beta$.
3. Each α_i is a strong configuration of G.

4. For each process i there exists $1 \leq j \leq n$ such that $X_j =< (N^1, c_1, d_1), \ldots, (N^p, c_p, d_p) >$ and d_i is the last event of $N^i|_i$.
5. If v is a leaf, then there exists a path from α to β that is labeled by X, which is a (non-hierarchical) MSC.
6. If v is the root of the tree, then α and β are the initial and final vertex, respectively, of G, and $X = M$.

Note that Condition 4 implies that every node has at most $p|M|$ children. For the complexity we note that guessing the claims labelling the children of a node can be done in polynomial time. Moreover, we just need to remember all children of one vertex by level, hence just a polynomial number of vertices at a time. Therefore we can guess and check on-the-fly whether a tree is well-built in PSPACE.

The next theorem shows the PSPACE lower bound for the hierarchical MSC membership problem.

Theorem 2 *The hierarchical MSC membership problem is PSPACE-hard, even if the graph is non-hierarchical.*

Proof. The problem we reduce from is a variant of TQBF, called (1-in-3)TQBF (true quantified one-in-three boolean formula). Let F be an instance of (1-in-3)TQBF of the form $F = (Q_n x_n) \ldots (Q_1 x_1)\varphi$, where $Q_i \in \{\exists, \forall\}$ and the formula φ is of the form $\wedge_{j=1\ldots m} R(\alpha_{j,1}, \alpha_{j,2}, \alpha_{j,3})$, with $\alpha_{j,k}$ literals. The function $R(x, y, z)$ is true iff exactly one of x, y, z is true. The PSPACE-hardness of this problem is shown in [17].

The idea is to let valuations of the variables to correspond to paths of G and to validate the valuations using the nMSC M. We define the graph G and the nMSC M by induction on $F = F_n$. Let $F_i = (Q_i x_i)\, F_{i-1}$, with $F_0 = \varphi$. Each F_i will determine the nHMSC G_i and the nMSC M_i.

The processes used in the construction are $\mathrm{SC}_1, \ldots, \mathrm{SC}_m$ and $\mathrm{RC}_1, \ldots, \mathrm{RC}_m$, plus $\mathrm{VY}_1, \ldots, \mathrm{VY}_n$ and $\mathrm{VN}_1, \ldots, \mathrm{VN}_n$. Here V means a variable and C a clause, S stands for "send", R for "receive", Y for "yes" and N for "no".

For all i, let MY_i be the MSC consisting of a message from VY_i to VN_i, then back from VN_i to VY_i, and a message from SC_j to RC_j for all j such that $x_i \in \{\alpha_{j,1}, \alpha_{j,2}, \alpha_{j,3}\}$. Symmetrically, let MN_i be the MSC consisting of a message from VN_i to VY_i, then back from VY_i to VN_i, and a message from SC_j to RC_j for all j such that $\neg x_i \in \{\alpha_{j,1}, \alpha_{j,2}, \alpha_{j,3}\}$.

M_0 is an MSC consisting of one message from SC_j to RC_j, for all j. The MSC-graph G_0 consists of $4n$ vertices, labeled by the MSCs MY_i, MN_i, or $\emptyset$. The graph allows to choose between MY_i (x_i true) and MN_i (x_i false) for all variables x_i:

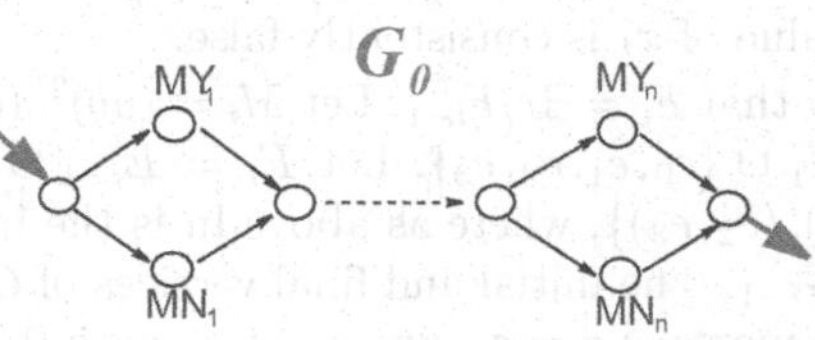

Note that all messages defined above commute, except for the messages between VY_i and VN_i. Let us denote by a_i the message from VY_i to VN_i, and by b_i the message from VN_i to VY_i. We use the ordering between a_i, b_i as follows: The sequence a_ib_i means that x_i is true, while b_ia_i means that x_i is false.

Assume now that G_{i-1}, M_{i-1} are already defined, and that there are f universal quantifiers in F_{i-1}. For simplicity, we denote below $a = a_i$ and $b = b_i$. Note that in a valuation tree for F showing that F is true, each value 0 or 1 assigned to the variable x_i is used by 2^f leaves. A valuation tree is defined as usual, by assigning each universally quantified variable two children labeled 0 and 1, respectively each existentially quantified variable one child labeled 0 or 1.

If $F_i = \forall x_i F_{i-1}$, then define $M_i = (ab)^{2^f} M_{i-1} S_i (ba)^{2^f} M_{i-1}$. The MSC S_i is used for synchronizing processes occurring in M_i. It contains a message between each (ordered) pair of processes of M_i (in some arbitrary order). Note that using the hierarchy we can describe $(ab)^{2^f}$, and thus M_i, by an expression of polynomial size in f.

Let $G_i = (V_i, E_i)$, where $V_i = V_{i-1} \cup \{e_0\}$ and $E_i = E_{i-1} \cup \{(\text{Fin}, e_0), (e_0, \text{In})\}$. The initial node In (the final node Fin, respectively) of G_i is the same as for G_{i-1}. The vertex e_0 is labeled by the synchronization MSC S_i.

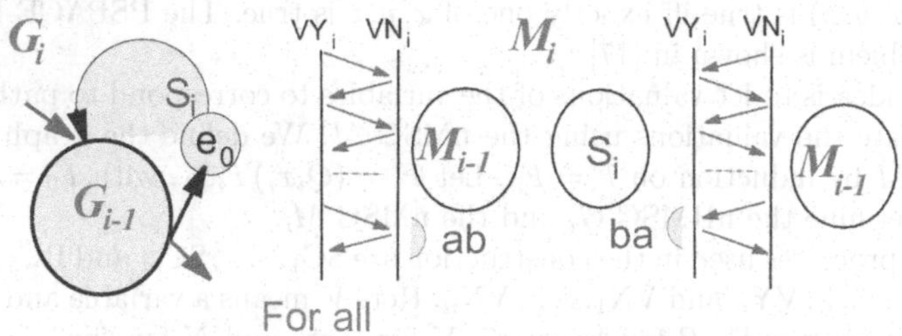

The definition of M_i, G_i can be explained intuitively as follows. Let ρ be a path of G_i labeled by M_i. Note that the MSC S_i occurring in M_i has to match the MSC S_i of e_0. Thus $\rho = \rho_1 e_0 \rho_2$, with ρ_1 an accepting path of G_{i-1} labeled by $(ab)^{2^f} M_{i-1}$ and ρ_2 an accepting path of G_{i-1} labeled by $(ba)^{2^f} M_{i-1}$. Each time ρ_1 goes through G_0 (which happens 2^f times), ρ_1 consumes either a message pair ab of MY_i or a pair ba of MN_i. Since ρ_1 must consume all occurrences of a, b in $(ab)^{2^f}$, all occurrences are of the form ab, which ensures that the valuation of x_i associated with ρ_1 is consistent (x_i is true). The same holds for the path ρ_2, where the value of x_i is consistently false.

Suppose now that $F_i = \exists x_i F_{i-1}$. Let $M_i = (ab)^{2^f}(a) M_{i-1}$, and $G_i = (V_i, E_i)$, where $V_i = V_{i-1} \cup \{e_0, e_1, e_2, e_3\}$. Let $E_i = E_{i-1} \cup \{(e_0, \text{In}), (\text{Fin}, e_3), (e_0, e_1), (e_1, \text{In}), (\text{Fin}, e_2), (e_2, e_3)\}$, where as above In is the initial vertex and Fin is the final vertex of G_{i-1}. The initial and final vertices of G_i are e_0 et e_3. We label e_1 and e_2 with the message $a = a_i$, and e_0 et e_3 with $\emptyset$.

The underlying idea in this case is that the additional occurrence of a in M_i must be matched by e_1 or e_2 (nowhere else there is an a). If it is e_1, every time the path ρ goes through G_0, it must choose the message pair ba, hence it goes through VN_i. The corresponding value for x_i is then forced to be false. If it is e_2, then ρ must choose the pair ab, hence it goes through VY_i.

□

4.2 Membership Problem for Automata

A natural restriction of the membership problem is to consider only one process. Several reasons explain this choice: if the problem is not PSPACE-hard anymore, maybe in the case where we fix the number of processes it is easier than PSPACE. Moreover, the weak membership problem [1] can be reduced to this question. At last, this is the natural problem of knowing whether an SLP-compressed word W is accepted by a hierarchical automaton A. We adopt the definition of hierarchical automaton of [4], where states can be themselves (hierarchical) automata with unique initial and final state, respectively. One good news is that in the case where the word W or A are flat, the membership problem is solvable in polynomial time. Both algorithms are straightforward and can be found in the full version.

Theorem 3 *1. Let W be an SLP-compressed word and let A be an NFA. Deciding whether $W \in L(A)$ can be done in time $O(|W| \cdot |A|^3)$.*
2. Let W be a word and let A be a hierarchical automaton (hNFA). Deciding whether $W \in L(A)$ can be done in time $O(|W|^3 \cdot |A|^3)$.

Surprisingly, the membership problem remains PSPACE-hard when both the word W is compressed and the automaton A is hierarchical.

Theorem 4 *Let W be an SLP-compressed word and A be a hierarchical automaton. Then the membership problem $W \in L(A)$ is PSPACE-complete. Moreover, the membership problem for unary alphabets is NP-complete.*

The NP-hardness result in the unary case follows also from [16]. We omit the proof for lack of space.

M	G	automaton	MSC
Flat	Flat	P	NP-complete
Flat	Nested	P	NP-complete
Nested	Flat	P	PSPACE-complete
Nested	Nested	PSPACE-complete	PSPACE-complete

Fig. 2. Complexity of membership problems.

5 Pattern Matching of nMSCs

Let M and N be two nMSCs. If we want to know whether M and N represent the same MSC then we only have to check if for all processes i, we have $M|_i = N|_i$. This suffices since we assume that the messages are in FIFO order. Nevertheless, the property "$M|_i$ is a pattern of $N|_i$, for all i" doesn't imply that M is a pattern of N.

From the definition of an nMSC M, we can easily derive for each process i a straight-line program generating the i-th projection $M|_i$ of M.

The strategy we will use is to compute a representation of *all* positions where projections of M are a pattern of projections of N. Then we will compute all positions where the projections of the pattern form an MSC-factor.

Let X and Y be two nMSCs (or two variables of an SLP). By $X \in Y$ we denote that X is used in the definition of Y or in the definition of Z with $Z \in Y$. By $X \subseteq Y$ we denote that the MSC (word, resp.) generated by X is a factor of the MSC (word, resp.) generated by Y. We also say that X is a pattern of Y.

The basis of our algorithm is a pattern matching algorithm for SLP-compressed words, described in [14,15]. We recall below briefly the main idea of [14].

Theorem 5 ([14]) *Let P be a SLP and let A, B be two variables of P. We can determine whether $A \subseteq B$ in time $O(|A|^2|B|^2)$.*

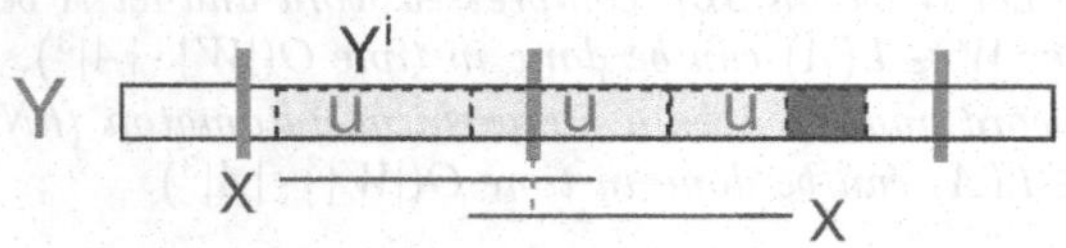

Occ(X,Y) as arithmetic progression*

Let X be a variable of A and suppose that $X \subseteq B$. Then X can occur as a pattern of B entirely in a variable Y on the lowest hierarchy level. Otherwise, X occurs in some variable Y such that $Y \longrightarrow Y^1 \cdots Y^k$, and i is such that Y^i is the first variable that X intersects, such that X ends beyond Y^i. Hence, for all Y, Y^i with Y^i used in the definition of Y, let $\mathrm{Occ}^*(X,Y,Y^i)$ denote the set of positions of Y at which an occurrence of X starts within Y^i and ends beyond Y^i. Let $\mathrm{Occ}^*(X,Y) = \bigcup_i \mathrm{Occ}^*(X,Y,Y^i)$.

Using a combinatorial argument (lemma of Fine and Wilf [6]), it is shown in [14] that $\mathrm{Occ}^*(X,Y,Y^i)$ is an arithmetic progression that is computable by dynamic programming in polynomial time. Note also that $\mathrm{Occ}^*(X,Y)$ contains at most $|Y|$ arithmetic progressions (at most one per Y^i).

Remark 5 Using Theorem 5 we can check the equality of two nMSCs M, N in time $O(|M|^2|N|^2)$.

We consider now the pattern matching problem for an nMSCs pattern M in an nMSC N. We will use arithmetic progressions representing sets of starting positions of patterns $M|_i$ within $N|_i$. Each arithmetic progression will be presented as a triple $\pi = (u, \lambda, v)$, where u, v are words with v a strict prefix of u and λ a positive integer. An arithmetic progression π represents a subword of the form $u^\lambda v$. If $\pi = \pi_i$ is associated with the i-th projection of the pattern M, then $M|_i$ has the form $u^{\lambda_0} v$ for some $\lambda_0 \le \lambda$. An **index** of an arithmetic progression will be the starting position of some u in $u^\lambda v$. For nMSCs M, Y we will denote by $\mathrm{Occ}^*(M, Y)$ the set of occurrences M^0 of M in Y such that M^0 does not occur in any $Z \in Y$. Occurrences of patterns will be denoted using superscripts.

Theorem 6 *Let M, N be two nMSCs where M is connected, i.e. M cannot be written as $M_1 M_2$, where M_1, M_2 are MSCs over disjoint sets of processes. Then checking $M \subseteq N$ is solvable in time $O(|M|^2|N|^2)$.*

Remark 6
1. We ensure that the transformation from nMSC to SLPs on the projections keeps the initial form of the nMSC w.r.t. messages, i.e. if (e, f) is a message then e and f appear literally in the same variable.
2. We note that for all occurrences $M|_h^0$ in $\mathrm{Occ}^*(M|_h, Y|_h)$, there is at most one tuple of occurrences $(M|_j^0)_{j \neq h}$ with $M|_j^0 \subseteq Y|_j$ for all j that can match $M|_h^0$. This follows from FIFO and because M is connected. (There can be more than one tuple $(M|_j^0)_{j \neq h}$ if $M|_j^0 \subseteq Z|_j$ with $Y \in Z$ for some j).
3. A simple argument shows that for M connected and for each occurrence M^0 in $\mathrm{Occ}^*(M, Y)$ there exists a process h such that $M^0|_h \in \mathrm{Occ}^*(M|_h, Y|_h)$. So $(M^0|_k)_{k \neq h}$ is fixed by $M^0|_h$. Thus, the variable $Y|_h$ associated with the process h is the *highest* variable for this occurrence.
4. Two occurrences $M|_i^1$ and $M|_j^2$ can be combined only if the first send (resp. receive) between the processes i and j on $M|_i^1$ matches the first receive (resp. send) on $M|_j^2$. We say in this case that $M|_i^1$ and $M|_j^2$ are *compatible*. By extending the notations we also say that the indices of an arithmetic progression corresponding to $M|_i^1, M|_j^2$ are compatible.
5. $(M|_i^0)_i$ is an occurrence of M iff $(M|_1^0, \ldots, M|_p^0)$ are compatible two by two.

For simplifying the presentation of the algorithm we will assume that every process in M sends at least one message to every other process. This is just a technical assumption and the general case is analogous (we use that M is connected). We denote for a word u and $i < j$ by $m_{i,j}(u)$ the number of sends of i to j in u, respectively by $m_{j,i}(u)$ the number of receives of j from i in u.

Pattern-Matching `(nMSC M , N)`

```
For each variable Y ∈ N in the lowest level of hierarchy:
    If M ⊆ Y at position pos then return (Y, pos);
For all variables Y,V of N where V ∈ Y:
    Compute Occ*(M|1,Y|1,V|1),...,Occ*(M|p,Y|p,V|p);
For every variable Y of N:
    For every process h: // special case first/last index
```

```
For every pos(h) as first or last index of some
    arithmetic progression in Occ*(M|_h, Y|_h):
  Let (M|_h)^{pos(h)} be the corresponding occurrence of M|_h:
  If there exists ((M|_k)^{pos(k)})_{k≠h} compatible with (M|_h)^{pos(h)}
    s.t. for all k, pos(k) ∈ Occ*(M^k|_k, Z^k|_k) with Z^k ∈ Y:
      return (Y, pos);
For each V ∈ Y s.t. ∀i, π_i = Occ*(M|_i, Y|_i, V|_i) ≠ ∅: //int.index
  For each i, let π_i = (u_i, λ_i, v_i);
  (π_1,...,π_p) := Reduce(π_1,...,π_p);
  Let u_i^0 be the first u_i of π_i;
  Let (t_1,...,t_p, e_1,...,e_p) = Periods(π_1,...,π_p);
  Let π'_i = (u_i^{e_i}, λ_i/e_i, v_i)) be the subarithmetic
          progression of π_i beginning at the t_i-th occurrence
          of u_i starting from u_i^0;
  If (π'_i)_i ≠ ∅ then return (Y, (π'_i)_i)
```

Let us give an intuition of the way our algorithm works. The main difficulty is to check the last condition of Remark 6. Suppose that M^0 is an occurrence of M in N. Hence, $M^0 \in \mathrm{Occ}^*(M, Y)$ for some Y. By Remark 6 (3), there exists a process h on the highest hierarchy level. That is, $M^0|_h \in \mathrm{Occ}^*(M|_h, Y|_h)$ and for all i, $M^0|_i \in \mathrm{Occ}^*(M|_i, Z^i|_i)$, where the variables Z^i are such that either $Z^i \in Y$ or $Z^i = Y$.

There are two cases to consider. The first case, where $M^0|_h$ is the first or last index of an arithmetic progression, is easy. It suffices to use Remark 6 (2) and to look directly for the (unique, if any) compatible occurrences of $M^0|_j$, $j \neq h$. This is easy, since M is connected.

The second case is when for all i such that $M^0|_i$ is the first or last index of an arithmetic progression, we have $Z^i \in Y$, i.e., i is not on the highest hierarchy level w.r.t. M^0. Equivalently, on the highest hierarchy level, all occurrences $M^0|_i$ are internal indices of the arithmetic progression $\mathrm{Occ}^*(M|_i, Y|_i)$.

The next proposition is crucial for the polynomial complexity of the algorithm. It shows that in this second case it suffices to consider simultaneously (non empty) arithmetic progressions of the form $\mathrm{Occ}^*(M|_i, Y|_i, V|_i)$, for all processes i. Notice that the arithmetic progressions refer to the same variables Y, V with $V \in Y$ (the important fact is that V is the *same* for all processes).

Proposition 1. *Let M^0 be an occurrence of M in N and let h be a process on the highest hierarchy level. Assume that $M^0|_h$ is an internal index of $Occ^*(M|_h, Y|_h, V|_h)$. Moreover, suppose that each $M^0|_i$ that is the first or last index of some $Occ^*(M|_i, Z^i|_i)$ satisfies $Z^i \in Y$. Then $M^0|_i$ is an internal index of $Occ^*(M|_i, Y|_i, V|_i)$, for all i.*

We consider now the second case where we look for compatible occurrences located in arithmetic progressions $\pi_i = \mathrm{Occ}^*(M|_i, Y|_i, V|_i) = (u_i, \lambda_i, v_i)$, for *fixed* $V \in Y$. Each pair of processes i, j will enforce restricting the arithmetic progressions π_i, π_j by intersecting with periodical sets of indices (however, the words u_i do not change during the algorithm). The algorithm finds an occurrence of M iff we can proceed restricting the π_i's, up to non empty subprogressions. The first restriction (`Reduce`) consists in modifying the first and/or last index of each arithmetic progression π_i in such a way that we keep exactly those indices u_i containing an event with matching event in π_j.

We turn now to the second part of the algorithm and we show that it returns only occurrences of M which are indeed factors of N. In this part we also describe the subroutine `Periods`.

Let $\pi_1, \dots, \pi_p$ be arithmetic progressions of occurrences of $M|_1, \dots, M|_p$, such that for each pair i, j there exists a message between i, j from each u_i in π_i to some index u_j of π_j, and vice-versa. That is, $\pi_1, \dots, \pi_p$ is the result of a call of `Reduce`. Let u_i^0 be the first index of each arithmetic progression.

We need all tuples of occurrences $((M|_1)^0, \dots, (M|_p)^0)$ that are two by two compatible. Actually it suffices to know all tuples $(u_1, \dots, u_p)$ corresponding to the beginning of the $((M|_1)^0, \dots, (M|_p)^0)$ that are two by two compatible. As we show later, such tuples appear periodically, hence we just have to keep some periods $(\mu_1, \dots, \mu_p)$ and the starting indices $(u_1^1, \dots, u_p^1)$, and to keep in mind that we are interested in the intersection with $(\pi_1, \dots, \pi_p)$. In the beginning, $\mu_i = 1$ and $u_i^1 = u_i^0$.

For all $i < j$ let $z_{i,j} < m_{i,j}$ be the number of send(i,j) in u_i^0 before the first one that reaches π_j. Let $z_{j,i} < m_{j,i}$ be the number of receive(i,j) in u_j^0 before the first that comes from π_i. Let $z_{i,j}^0$ be such that after reading the first $z_{i,j}^0$ messages from π_i to π_j we arrive at a message consisting of the first send(i,j) of some u_i and the first receive(i,j) of some u_j. So $z_{i,j}^0 + z_{i,j} \equiv 0 \pmod{m_{i,j}}$ and $z_{i,j}^0 + z_{j,i} \equiv 0 \pmod{m_{j,i}}$. Using the Chinese Remainder Theorem the subroutine `Periods` first computes the least solutions $z_{i,j}^0$ modulo $\mathrm{lcm}(m_{i,j}, m_{j,i})$ to the above equations in time $O(\min(|M|_i|, |M|_j|)^3)$. Hence we obtain some periods μ_i and starting indices u_i^1. We apply this argument for each pair of processes. Note that μ_i divides $\mathrm{lcm}\{m_{i,j} \mid i, j\}$.

The first send(i,j) of each u_i in the restricted arithmetic progression π_i' corresponds to the first receive(i,j) of some u_j of the unrestricted arithmetic progression π_j. The final step of `Periods` is to compute occurrences of M from $(\pi_i')_i$. Let $x_{i,j}$ be an integer denoting the number of u_j between u_j^1 and the reception of the first message from u_i^1. We want to compute all tuples $(u_i)_i$ such that the first send(i,j) of u_i matches the first receive(i,j) of u_j. That is, we need a solution $(t_i)_i$ of the system of linear equations $\mu_i m_{i,j} t_i = x_{i,j} m_{j,i} + \mu_j m_{j,i} t_j$. By combining equations we obtain an equivalent system of the form $\delta_{i,j} t_1 = y_{i,j} + \nu_{i,j} t_j$, with $\gcd(\delta_{i,j}, \nu_{i,j}) = 1$. Thus, the value of t_1 determines each t_i, modulo some value e_i depending on the $(m_{i,j})_{i,j}$.

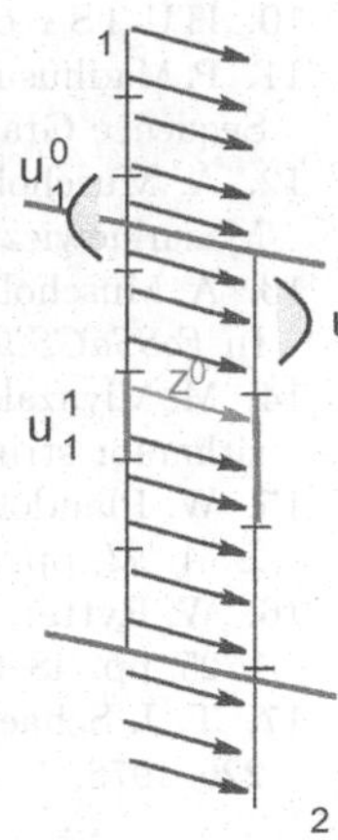

Let $\gamma_{i,j}$ be the inverse of $\delta_{i,j}$ modulo $(\nu_{i,j})$. Hence the equations are reduced to $p(p-1)$ trivial equations of the form $t_1 \equiv y_{i,j}\gamma_{i,j} \pmod{\nu_{i,j}}$. The subroutine `Periods` finally computes a solution $(t_1, \dots, t_p)$ using again the Chinese Remainder Theorem and returns $(t_i + u_i^1 - u_i^0, e_i)_i$. Since the intersection of a arithmetic progression with the periodic set is still a arithmetic progression, in the end we have arithmetic progressions of periods increased by a factor of e_i, that contains only compatible occurrences.

This algorithm is applied for each pair of processes, thus a call of `Periods` costs time $O(\sum_j \sum_i |M|_i||M|_j| \min(|M|_i, |M|_j)) \in O(|M|^3)$. This ends the description of the algorithm and the correctness proof.

Acknowledgment. We thank Markus Lohrey and in particular Marc Zeitoun for many discussions and useful comments on the presentation of this paper. We also thank an anonymous referee for his/her detailed remarks and suggestions.

References

1. R. Alur, K. Etessami, and M. Yannakakis. Realizability and verification of MSC graphs. In *ICALP'01*, LNCS 2076, pp. 797–808, 2001.
2. R. Alur, G. H. Holzmann, and D. A. Peled. An analyzer for message sequence charts. *Software Concepts and Tools*, 17(2):70–77, 1996.
3. R. Alur, S. Kannan, and M. Yannakakis. Communicating hierarchical state machines. In *ICALP'99*, LNCS 1644, pp. 169–178, 1999.
4. R. Alur and M. Yannakakis. Model checking of hierarchical state machines. In *SIGSOFT '98*, pp. 175–188, 1998.
5. R. Alur and M. Yannakakis. Model checking of message sequence charts. In *CONCUR'99*, LNCS 1664, pp. 114–129, 1999.
6. N.J. Fine and H.S. Wilf. Uniqueness theorems for periodic functions. *Proceedings of the American Mathematical Society*, 16, 1965.
7. E. Gunter, A. Muscholl, and D. Peled. Compositional message sequence charts. In *TACAS'01*, LNCS 2031, pp. 496–511, 2001.
8. D. Harel. Statecharts: A visual formulation for complex systems. *Science of Computer Programming*, 8(3):231–274, 1987.
9. M. Y. Vardi, D. Harel, O. Kupferman. On the complexity of verifying concurrent transition systems. In *CONCUR '97*, LNCS 1243, pp. 258–272, 1997.
10. ITU-TS recommendation Z.120, 1996.
11. P. Madhusudan. Reasoning about Sequential and Branching Behaviours of Message Sequence Graphs. In *ICALP'01*, LNCS 2076, pp. 809–820, 2001.
12. A. Muscholl and D. Peled. Message sequence graphs and decision problems on Mazurkiewicz traces. In *MFCS'99*, LNCS 1672, pp. 81–91, 1999.
13. A. Muscholl, D. Peled, and Z. Su. Deciding properties of message sequence charts. In *FoSSaCS'98*, LNCS 1378, pp. 226–242, 1998.
14. M. Miyazaki, A. Shinohara, and M. Takeda. An improved pattern matching algorithm for strings in terms of straight-line programs. In *CPM '97*, LNCS 1264.
15. W. Plandowski. Testing equivalence of morphisms on context-free languages. In *ESA '94*, pp. 460–470, 1994.
16. W. Rytter. Algorithms on compressed strings and arrays. In *SOFSEM '99*, LNCS 1725, pp. 48–65, 1999.
17. T. J. Schaefer. The complexity of satisfiability problems. In *STOC '78*, pp. 216–226, 1978.

Improved Exact Algorithms for MAX-SAT

Jianer Chen[1]* and Iyad A. Kanj[2]**

[1] Department of Computer Science, Texas A&M University, College Station TX 77843-3112, USA,
chen@cs.tamu.edu

[2] School of CTI, DePaul University, 243 S. Wabash Avenue, Chicago IL 60604, USA,
ikanj@cs.depaul.edu

Abstract. In this paper we present improved exact and parameterized algorithms for the maximum satisfiability problem. In particular, we give an algorithm that computes a truth assignment for a boolean formula F satisfying the maximum number of clauses in time $O(1.3247^m|F|)$, where m is the number of clauses in F, and $|F|$ is the sum of the number of literals appearing in each clause in F. Moreover, given a parameter k, we give an $O(1.3695^k k^2 + |F|)$ parameterized algorithm that decides whether a truth assignment for F satisfying at least k clauses exists. Both algorithms improve the previous best algorithms by Bansal and Raman for the problem.

Key words. maximum satisfiability, exact algorithms, parameterized algorithms.

1 Introduction

The maximum satisfiability problem, abbreviated MAX-SAT, is to compute for a given boolean formula F in conjunctive normal form, a truth assignment that satisfies the largest number of clauses in F. In the parameterized maximum satisfiability problem, we are given an additional parameter (positive integer) k, and we are asked to decide if F has a truth assignment that satisfies at least k clauses. It is easy to see that both MAX-SAT and parameterized MAX-SAT are NP-complete since they generalize the satisfiability problem [26]. According to the theory of NP-completeness [26], these problems cannot be solved in polynomial time unless P = NP.

The above fact does not diminish the need for solving these problems for their practical importance. Due to its clause relationship to the satisfiability problem (SAT), MAX-SAT has many applications in artificial intelligence, combinatorial optimization, expert-systems, and database-systems [6,9,25,28,29,32,35]. Many approaches have been employed in dealing with the NP-hardness of the MAX-SAT problem including approximation algorithms [4,5,10], heuristic algorithms [9,10,11,35], and exact and parameterized algorithms [8,14,27,31,33].

* This author was supported in part by NSF under the grant CCR-0000206.
** The Corresponding author.

S. Rajsbaum (Ed.): LATIN 2002, LNCS 2286, pp. 341–355, 2002.

In this paper we focus our attention on finding exact solutions for the Max-Sat problem. Acknowledging the apparent inevitability of an exponential time complexity for NP-hard problems, this line of research seeks designing efficient exponential time algorithms that improve significantly on the straightforward exhaustive search algorithms. Numerous NP-hard combinatorial problems were studied from this point of view. Examples of exponential time algorithms for NP-hard optimization problems include the Independent Set problem [12,34], Dantsin et al.'s $O(1.481^n)$ time deterministic algorithm for the 3-Sat problem [22], and Beigel and Eppstein's $O(1.3446^n)$ time algorithm for the 3-Coloring problem [13]. A closely related line of research to the above one that has been receiving a lot of attention recently, is the area of fixed-parameter tractability [20], which has found many applications in fields like databases, artificial intelligence, logic, and computational biology [21]. A famous problem that belongs to this category is the Vertex Cover problem. Recently, there has been extensive research done aiming at improving the complexity of parameterized algorithms that solve the Vertex Cover problem [7,18,19,24]. Other examples of parameterized problems include Planar Dominating Set [1,2], k-Dimensional Matching [16,20], and Constraint Bipartite Vertex Cover [17,24].

The Max-Sat problem has played a significant role in both these lines of research. Considerable efforts have been paid trying to lower the worst-case complexity for the problem. Cai and Chen [14] presented an algorithm of running time $O(2^{2ck}cm)$ for the Max-c-Sat problem. Mahajan and Raman [31] improved on that bound by presenting an algorithm of running time $O(cm + ck\phi^k)$ ($\phi = (1+\sqrt{5})/2 \approx 1.618$). Mahajan and Raman also gave an algorithm of running time $O(|F| + k^2\phi^k)$ for the parameterized Max-Sat problem. They described how their algorithm for parameterized Max-Sat implies an algorithm for the parameterized Max-Cut problem. Niedermeier and Rossmanith [33] presented an algorithm of running time $O(1.3995^k k^2 + |F|)$ for parameterized Max-Sat. For Max-Sat they gave $O(1.3803^m|F|)$ and $O(1.1272^{|F|}|F|)$ algorithms, where m and $|F|$ are as defined above. Bansal and Raman [8] improved Niedermeier and Rossmanith result and presented algorithms of running time $O(1.3802^k k^2 + |F|)$ for parameterized Max-Sat, and $O(1.341294^m|F|)$ and $O(1.105729^{|F|}|F|)$ for Max-Sat. Recently, Gramm et al. [27] considered the Max-2-Sat problem, a special case of the Max-Sat problem. They gave an $O(2^{m/5}|F|)$ algorithm for the problem. Their algorithm also implies better algorithms for the parameterized Max-Cut problem.

In this paper, we further improve on Bansal and Raman's algorithms. More specifically, we give an $O(1.3247^m|F|)$ algorithm for the Max-Sat problem , and $O(1.3695^k k^2 + |F|)$ algorithm for the parameterized Max-Sat problem. Our techniques are basically similar to the previous ones in the careful case-by-case analysis, which seems unavoidable when designing exact algorithms for NP-hard problems (for instance, see [7,8,12,13,17,18,19,27,33,24,34]). However, we add a number of nice observations, such as the **Create-Low-Literal** subroutine, to show how to make use of certain structures in the formula to yield more efficient algorithms and reduce the case-by-case combinatorial anlaysis.

The paper is organized as follows. In Section 2, we present the basic notations, definitions, and transformation rules. In Section 3 we present our algorithm for the parameterized MAX-SAT problem. In Section 4 we give an algorithm for the MAX-SAT problem based on the algorithm in Section 3 for parameterized MAX-SAT, and a new technique that we introduce in Section 4. We conclude the discussion in Section 5 with some remarks and further research directions.

2 Definitions and Background

2.1 Definitions and Notations

We assume familiarity with the basic concepts of propositional logic, and we use a notation that is similar to that of [33]. The formula F is assumed to be given as a set of clauses in conjunctive normal form (CNF). Each clause is a disjunction of literals, where each literal is a variable or its negation. A variable will be denoted by an alphabatical character (e.g., x), its negation by its name with bar on the top of it (e.g., $\bar{x}$), and a literal by the variable name with a tilde sign on the top of it (e.g., $\tilde{x}$). A literal $\tilde{x}$ occurs *positively* (resp. *negatively*) in a clause C if x (resp. $\bar{x}$) occurs in C. Two occurences of a literal $\tilde{x}$ have *opposite* signs, if in one of the occurences $\tilde{x}$ occurs positively, and negatively in the other; otherwise, the two occurences have the *same* sign. A *truth assignment* to F is a function that assigns every variable in F a boolean value *true* or *false*. For simplicity, we will denote from now on the boolean value *true* by 1 and *false* by 0. A *partial* assignment τ to F is an assignment to a subset of the variables in F. If τ is a partial assignment to the variables $\{x_1, \ldots, x_r\}$ in F, and if τ assigns the value 1 to the variables $\{x_1, \ldots, x_i\}$, and the value 0 to the variables $\{x_{i+1}, \ldots, x_r\}$, where $i \leq r$, we denote by F/τ, or $F[x_1, \ldots, x_i][x_{i+1}, \ldots, x_r]$, the formula resulting from F by replacing the variables in $\{x_1, \ldots, x_i\}$ in F by the value 1 and $x_{i+1}, \ldots, x_r$ by the value 0, and eliminating from F all clauses whose values have been determined upon that assignment. Note that these sets of variables might be empty.

The maximum number of simultaneously satisfiable clauses in F will be denoted by $maxsat(F)$. We call a partial assignment τ *safe*, if there is a truth assignment τ' to F such that τ and τ' agree on the variables $\{x_1, \ldots, x_i\}$ and τ' satisfies $maxsat(F)$ clauses. In such case we also call the formula F/τ a safe formula.

A *subformula* H of F is a subset of clauses of F. A subformula H is said to be *closed* if no literal in H appears outside H. The *length* of a clause C is simply the number of literals in C, denoted $|C|$. The length of F, denoted $|F|$, is $\sum_{C \in F} |C|$. If a clause C has length r then C is called an r-clause. A literal $\tilde{l}$ is called an (i, j) literal, if $\tilde{l}$ occurs exactly i times positively and j times negatively. Similarly, $\tilde{l}$ is an (i^+, j^+) literal if it occurs at least i times positively and at least j times negatively. We can define in a similar fashion (i^-, j^-), (i, j^+), (i, j^-), (i^+, j), (i^+, j^-), (i^-, j), (i^-, j^-) literals. A formula F is called *simple*, if negative literals occur only in 1-clauses, and each pair of variables occurs together in at most one clause.

2.2 Transformation and Reduction Rules

A transformation rule is a rule that is applied to a formula F to transform it into a simpler formula G such that a solution of G can be mapped back to a solution of F. We describe next some standard transformation rules that appear in the literature [8,31,33]. They can be carried out in linear time, and their correctness can be easily verified.

TransRule 1. *Pure literal*
If $\tilde{x}$ is an $(i, 0)$ (resp. $(0, i)$) variable in F, let G be the formula resulting from F by assigning x the value 1 (resp. 0) and eliminating all satisfied clauses. Work on G with $maxsat(G) = maxsat(F) - i$.

TransRule 2. *Dominating unit-clause*
If an (i, j) literal $\tilde{x}$ occurs positively (resp. negatively) i' times, where $j \leq i' \leq i$ (resp. $i \leq i' \leq j$) in unit clauses, assign x the value 1 (resp. the value 0), and eliminate all clauses containing $\tilde{x}$, and work on the remaining formula G with $maxsat(G) = maxsat(F) - i$ (resp. $maxsat(G) = maxsat(F) - j$).

TransRule 3. *Resolution*
If $F = (x \vee p_1 \vee \ldots \vee p_r) \wedge (\bar{x} \vee q_1 \vee \ldots \vee q_s) \wedge H$, where H does not contain $\tilde{x}$, then work on $G = (p_1 \vee \ldots \vee p_r \vee q_1 \vee \ldots \vee q_s) \wedge H$, with $maxsat(G) = maxsat(F) - 1$.

TransRule 4. *Reduction to problem kernel* [31]
Given a formula F and a positive integer k, then in linear time, we can compute a formula G and a positive integer $k' \leq k$ with $|G| \in O(k'^2)$, such that F has an assignment satisfying at least k clauses if and only if G has an assignment satisfying at least k' clauses. Morever, such an assignment for F is computable from an assignment for G in linear time.

3 An Algorithm for Parameterized Max-Sat

Recall that in the parameterized Max-Sat problem we are given a boolean formula F and a positive integer k, and we are asked to decide if there is a truth assignment for F satisfying at least k clauses. By **TransRule 4**, we can assume that $|F| = O(k^2)$. We will also assume that no clause contains more than one occurence of each literal. This assumption is justified since if a literal occurs in a clause C more than once, then either all occurences of the literal have the same sign, or at least two of them have opposite signs. In the former case all the occurences can be removed and replaced with a single occurence, and the resulting formula is equivalent to the original one. In the latter case the clause C will always be satisfied, so C can be removed, and the formula and the parameter can be updated accordingly.

The execution of our algorithm is recursive and is depicted by a search tree (branching tree). A node in the search tree represents a boolean formula and its children are the boolean formulas resulting from applying to the formula the first branching case that applies. By branching at a formula F in the search tree we mean replacing F with formulas of the form $F[x_1^1, \ldots, x_r^1][y_1^1, \ldots, y_s^1], \ldots, F[x_1^l, \ldots, x_r^l][y_1^l, \ldots, y_s^l]$, according to one of the branching rules of the algorithm, and then working on each of the formulas recursively. This technique is similar to the Davis-Putnam procedure [23]. If at any time in the algorithm F becomes empty while the parameter k is still positive, then we stop and report that no truth assignment for F satisfying k clauses exists. Also, at the beginning of each stage the algorithm applies the above transformation rules until they are no longer applicable.

Let $C(k)$ be the number of leaves in the search tree of our algorithm looking for an assignment satisfying k or more clauses in F. If we branch at a certain node x in the search tree by reducing the parameter k' by $k'_1, k'_2, \ldots, k'_r$, in each branch respectively, where $k'_1 \leq k'_2 \leq \ldots \leq k'_r$, then the following recurrence relation for the size of the search tree $C(k')$ rooted at x holds:

$$C(k') = \sum_{i=1}^{r} C(k' - k'_i)$$

It is well known that the solution to the above recurrence (see for instance [30]) is $C(k') = O(\alpha^{k'})$, where α is the unique positive root of the characteristic polynomial $p(x) = x^{k'_r} - x^{k'_r - k'_1} - \ldots - x^{k'_r - k'_{r-1}} - 1$. Now the size of the whole search tree will be $O(\alpha^k_{max})$, where α_{max} is the largest root among all roots of the characteristic polynomials resulting from the branching cases of the algorithm. The running time of the algorithm will be $O(\alpha^k_{max}k^2 + |F|)$, since an $O(|F|)$ time is needed to reduce $|F|$ to a boolean formula whose length is $O(k^2)$, and along each root-leaf path in the searching tree we spend $O(|F|) = O(k^2)$ time. The following general branching rules prove to be useful in our algorithm.

BranchRule 1. If $\tilde{x}$ is an $(i, 1)$ literal, and $p_1, \ldots, p_r, \bar{q}_1, \ldots, \bar{q}_s$ occur with $\bar{x}$ in a clause C, then branch as $F[x][]$ and $F[q_1, \ldots, q_s][x, p_1, \ldots, p_r]$.

To see why the above branching rule is correct, observe that if the formula $F[p_i][]$ $(1 \leq i \leq r)$ or $F[][q_j]$ $(1 \leq j \leq s)$ is safe, then so is $F[x, p_i][]$ or $F[x][q_j]$, which are subbranches of the branch $F[x][]$. Hence when we branch as $F[][x]$, we can assume that $p_i = 0$ and $q_j = 1$, and branch as $F[q_1, \ldots, q_s][x, p_1, \ldots, p_r]$.

BranchRule 2. Let τ_1 be a partial assignment to F satisfying i clauses such that $F/\tau_1 = C_1 \wedge \ldots \wedge C_r \wedge G$, and τ_2 be a partial assignment to F satisfying j clauses such that $F/\tau_2 = C'_1 \wedge \ldots \wedge C'_s \wedge G$, where $i \geq s + j$ and G is the maximal common subclause between F/τ_1 and F/τ_2. If there is a branching rule in which we branch as F/τ_1 and F/τ_2, then it suffices to branch as F/τ_1.

Proposition 1. BranchRule 2 *is correct.*

Proof. To prove the proposition it suffices to prove that if F/τ_2 is safe then so is F/τ_1. Suppose that F/τ_2 is safe and let τ be a truth assignment to F satisfying $maxsat(F)$ clauses that agrees with τ_2. Suppose that τ satisfies l clauses in G. Then τ satisfies at most $l+s+j$ clauses in F, and hence, $maxsat(F) \leq l+s+j$. Now cosider the truth assignment τ' to F that agrees with τ on G and agrees with τ_1 (it does not matter what τ' assigns to the other variables in F). Now τ' satisfies at least $l + i$ clauses. Since $i \geq j + s$, τ' satisfies at least $l + j + s \geq maxsat(F)$ clauses. It follows that τ' is a truth assignment to F satisfying $maxsat(F)$ clauses that agrees with τ_1, and τ_1 is safe.

Now we are ready to present the algorithm. For each branching case we will write the recurrence relation resulting from that case next to it.

Case 3.1. There is an (i,j) literal $\tilde{x}$ with $i+j \geq 6$, or $i+j=5$ and $i,j>1$. Branch as $F[x][]$ and $F[][x]$. This gives a worst case recurrence $C(k) \leq C(k-1)+C(k-5)$.

Case 3.2. There is a $(3,1)$ or $(4,1)$ literal $\tilde{x}$ such that $\bar{x}$ does not occur as a unit clause. Let $\tilde{y}$ be a literal occuring with $\bar{x}$, and suppose that y occurs with $\bar{x}$ (the analysis in the other case is the same). By **BranchRule 1**, branch as $F[x][]$ and $F[][x,y]$. $C(k) \leq C(k-3)+C(k-2)$

Excluding **Case 3.1** and **Case 3.2**, we can assume that the formula contains only $(2,2)$ literals and $(4-,1-)$ literals, and all $(4,1)$ and $(3,1)$ literals have their negations as unit clauses.

Case 3.3. There is a $(2,2)$ literal $\tilde{y}$ that occurs with a $(2,1)$ literal $\tilde{x}$. Assume that y occurs with x (the case is the same if y occurs with $\bar{x}$). Branch as $F[y][]$ and apply either **TransRule 1** or **TransRule 3** to $\tilde{x}$, and $F[][y]$. $C(k) \leq C(k-3)+C(k-2)$

Case 3.4. There are two $(2,1)$ literals $\tilde{x}$ and $\tilde{y}$ such that $\bar{y}$ occurs with $\bar{x}$. Branch as $F[x][]$ and $F[y][x]$. $C(k) \leq C(k-2)+C(k-3)$

Case 3.5. There is a $(2,1)$ literal $\tilde{x}$ such that $\bar{x}$ does not occur as a unit clause. Let C be the clause containing $\bar{x}$. We distinguish the following subcases.

Subcase 3.5.1 $|C| \geq 3$. If there is a $(3^+,1)$ literal $\tilde{y}$ that occurs in C, let $\tilde{t}$ be another literal that occurs in C. Assume that t occurs in C (the analysis is the same in the case $\bar{t}$ occurs in C). Since $\tilde{y}$ is a $(3^+,1)$ literal, by **Case 3.2**, y occurs with $\bar{x}$. By **BranchRule 1**, we can branch as $F[x][]$ and $F[][x,y,t]$. Noting that $\bar{y}$ occurs as a singleton, and hence, in a different clause than $\bar{t}$, we branch with recurrence $C(k) \leq C(k-2)+C(k-3)$. If there is no $(3^+,1)$ literal occuring in C, then all literals in C must be $(2,1)$ literals. Morever, since **Case 3.4** was eliminated, all $(2,1)$ literals except x occur positively in C. Let $\tilde{r}$ and $\tilde{s}$ be two $(2,1)$ literals such that r and s occur in C. By **BranchRule 1**, we can branch as $F[x][]$ and $F[][x,r,s]$. Since **Case 3.4** was eliminated, $\bar{r}$ and $\bar{s}$ occur in different clauses. $C(k) \leq C(k-2)+C(k-3)$

Subcase 3.5.2. C is a 2-clause. Let $\tilde{y}$ be the literal occuring with $\bar{x}$. Since all the previous cases do not apply, it must be the case that y occurs with $\bar{x}$. Observe that when $y=1$, x becomes a pure variable, and hence can be assigned the value 1. It follows that branch $F[y][]$ is equivalent to $F[y][x]$. Moreover, since $\tilde{y}$ is a $(2^+,1)$ literal, by **BranchRule 2**, the two branches $F[y,x][]$ and $F[x][y]$ reduce to $F[x,y][]$. Hence, we branch as $F[y,x][]$ and $F[][y,x]$. $C(k) \leq C(k-3)+C(k-2)$

Case 3.6. For every $(2,1)$ literal $\tilde{x}$, $\bar{x}$ ocurs as a unit clause. We distinguish the following subcases.

Subcase 3.6.1 There is a $(4,1)$ literal $\tilde{y}$ that occurs with x. Clearly y has to occur with x. Branch as $F[y][]$ and apply **TransRule 1** or **TransRule 3** to $\tilde{x}$, and $F[][y]$. $C(k) \leq C(k-5)+C(k-1)$

Subcase 3.6.2 There is a $(2^+,1)$ literal $\tilde{y}$ that occurs with x. Clearly, y occurs with x. Let C be a clause containing x and y. If C is a 2-clause, branch as $F[y][]$ and apply **TransRule 1** or **TransRule 3** to $\tilde{x}$, and $F[x][y]$. This is correct since by **BranchRule 2**, branching as $F[][x,y]$ and $F[x][y]$ reduces to branching as $F[x][y]$. Thus, we branch with the recurrence relation $C(k) \leq C(k-3)+C(k-3)$. Observe that since **TransRule 2** was applied, no clause containing x can be a unit clause. Thus, we can assume at this point that both clauses containing x have cardinality at least three. Now let t be another variable occuring with x in C and let z be another variable occuring with x in the other clause. Note that since all the previous cases do not apply, we must have $\bar{y}$, $\bar{t}$, and $\bar{z}$ occuring as unit clauses. Branch as $F[][x]$ and $F[x][s,t,y]$. To see why this is true, note that by **BranchRule 2**, branching as $F[x,l][]$ and $F[l][x]$, where $l \in \{y,t,z\}$, reduces to branching as $F[l][x]$ which is contained in the branch $F[][x]$. $C(k) \leq C(k-1)+C(k-5)$

Now we can assume that we do not have any $(2,1)$ literals. Note that in the previous cases when we had a $(2,1)$ literal, we were always able to branch with a worst case recurrence relation $C(k) \leq C(k-1)+C(k-5)$. Suppose that there is a $(4,1)$ literal $\tilde{x}$. Clearly, since **Case 3.2** was excluded, $\bar{x}$ occurs as a unit clause.

Case 3.7. There is a literal $\tilde{y}$ occuring with x such that either $\tilde{y}$ is a $(3,1)$ literal, or a $(4,1)$ literal that has multiple occurences with x, or a $(2,2)$ literal that is not dominated by x (i.e., not all occurences of $\tilde{y}$ are with x). Branch as $F[x][]$ and $F[][x]$. In the first branch four clauses are satisfied and y becomes a $(2^-,1^-)$ literal allowing a further branch with a recurrence relation $C(k) \leq C(k-1)+C(k-5)$ according to one of the cases **Case 3.3**–**Case 3.6**. In the second branch one clause is satisfied. $C(k) \leq C(k-5)+C(k-9)+C(k-1)$

Case 3.8. There is a $(2,2)$ literal y that is dominated by x. Branch as $F[y][]$ and $F[][y]$ and in both branches $\tilde{x}$ becomes a $(2,1)$ literal allowing a further branch with a recurrence $C(k) \leq C(k-1)+C(k-5)$. $C(k) \leq 2C(k-3)+2C(k-7)$

Excluding **Case 3.7** and **Case 3.8**, no $(2,2)$ or $(3,1)$ literal occurs with a $(4,1)$ literal, and every two $(4,1)$ literals can have at most one common occurence.

Case 3.9. All literals are $(4,1)$ literals and the formula is simple. Let $\tilde{x}$ be a $(4,1)$ literal and let C_1 be the clause with minimum cardinality containing x. Let $i = |C_1|$. We distinguish two subcases.

Subcase 3.9.1. $i \leq 5$. Let $y_1, \ldots, y_{i-1}$ be the variables occuring with x in C_1. Note that by **BranchRule 2**, $F[x][]$ and $F[][x, y_1, \ldots, y_{i-1}]$ reduce to $F[x][]$. Thus, we can branch as $F[x][]$ and $F[y_1][x]$ and $F[y_2][x, y_1]$ and ... and $F[y_{i-1}][x, y_1, \ldots, y_{i-2}]$. We get a worst case recurrence $C(k) \leq C(k-4)+C(k-5)+C(k-6)+C(k-7)+C(k-8)$.

Subcase 3.9.2. $i > 5$. Let C_1, ..., C_4 be the clauses containing x, and suppose that besides x, C_1 contains the variables $x_1^1, \ldots, x_1^p$, C_2 the variables $x_2^1, \ldots, x_2^q$, C_3 the variables $x_3^1, \ldots, x_3^r$, and C_4 the variables $x_4^1, \ldots, x_4^s$, where p, q, r, $s > 4$. Branch as $F[][x]$ and $F[x][x_1^1, \ldots, x_1^p]$ and $F[x][x_2^1, \ldots, x_2^q]$ and $F[x][x_3^1, \ldots, x_3^r, x_4^1, \ldots, x_4^s]$. To justify the above branch, note that if three variables other than x in three distinct clauses in $C_1, \ldots, C_4$ have the value 1, then we can safely set $x = 0$. Thus, either we branch as $F[][x]$, or $F[x][]$ and no three clauses among $C_1, \ldots, C_4$ have variables other than x set to 1. We get a worst case recurrence $C(k) \le C(k-1) + 2C(k-8) + C(k-12)$.

Now we can assume that the resulting formula does not contain $(2,1)$ and $(4,1)$ literals.

Case 3.10. There is a $(2,2)$ literal $\tilde{x}$. Clearly since **TransRule 2** was applied, not both occurences of x or $\bar{x}$ are in unit clauses. Branch as $F[x][]$ and $F[][x]$, and in both branches we have a $(2^-, 1^-)$ literal remaining allowing a further branch with a recurrence $C(k) \le C(k-1) + C(k-5)$ according to one of the cases **Case 3.3–Case3.6**. $C(k) \le 2C(k-3) + 2C(k-7)$

Now we can assume that all literals are $(3,1)$ literals.

Case 3.11. There are two literals $\tilde{x}$ and $\tilde{y}$ that occur three times together. Since by **BranchRule 2** $F[x][y]$ and $F[y][x]$ reduce to $F[x][y]$, we can branch as $F[x][]$ and apply **TransRule 1** to $\tilde{y}$, and $F[][x,y]$. $C(k) \le C(k-4) + C(k-2)$

Case 3.12. There are two literals $\tilde{x}$ and $\tilde{y}$ that occur twice together. Let t be a variable that occurs with y but not x. Since **TransRule 2** was applied, such a variable must exist. Branch as $F[][y]$ and $F[y][x,t]$. The reason is that by **BranchRule 2**, $F[y][]$ and $F[y][x]$ reduce to $F[y][x]$ and $F[t,x][y]$ and $F[y,t][x]$ reduce to $F[t,x][y]$ which is a subbranch of $F[][y]$. $C(k) \le C(k-1) + C(k-5)$

Case 3.13. F is a simple formula. Let $\tilde{x}$ be a $(3,1)$ literal and let C_1 be the clause with minimum cardinality containing x. Let $i = |C_1|$. As in **Case 3.9**, we distinguish two subcases.

Subcase 3.13.1. $i \le 3$. Let y_i, y_{i-1} be the variables occuring with x in C_1. Branch as $F[x][]$ and $F[y_{i-1}][x]$ and $F[y_i][x, y_{i-1}]$ (in case y_i exists). $C(k) \le C(k-3) + C(k-4) + C(k-5)$

Subcase 3.13.2. $i > 3$. Let C_1, C_2, C_3 be the clauses containing x, and suppose that besides x, C_1 contains the variables $x_1^1, \ldots, x_1^p$, C_2 the variables $x_2^1, \ldots, x_2^q$, and C_3 the variables $x_3^1, \ldots, x_3^r$, where p, q, $r > 2$. Branch as $F[][x]$ and $F[x][x_1^1, \ldots, x_1^p]$ and $F[x][x_2^1, \ldots, x_2^q, x_3^1, \ldots, x_3^r]$. $C(k) \le C(k-1) + C(k-6) + C(k-9)$

Theorem 1. *Given a boolean formula F and a parameter k, then in time $O(1.3695^k k^2 + |F|)$ we can either compute a truth assignment for F satisfying at least k clauses, or we can report that such assignment does not exist.*

Proof. *At each stage of the algorithm, either one of the above transformation rules is applicable to F, or one of the cases* **Case 3.1–Case 3.13**. *It takes linear time to apply one of the transformation rules or to update F after applying a branching rule. In all the above branching rules we branch with a recurrence relation not worse than $C(k) \le 2C(k-3) + 2C(k-7)$. Thus, the size of the branching tree is not larger than $O(\alpha^k)$ where $\alpha \approx 1.3695$ is the unique positive root of the characteristic polynomial $x^7 - 2x^4 - 2$ associated with the recurrence $C(k) \le 2C(k-3) + 2C(k-7)$. From the above discussion, the running time of the algorithm is $O(1.3695^k k^2 + |F|)$.*

Theorem 1 is an improvement on Bansal and Raman's $O(1.3802^k k^2 + |F|)$ algorithm for parameterized MAX-SAT.

4 An Algorithm for MAX-SAT

We say a literal has *low occurence* if it is a $(2^-,1^-)$ literal. From the above discussion we can observe the importance of having literals with low occurences. Basically, if we have a literal with low occurence, then either the literal is a $(1^-,1^-)$ literal, and hence we can apply one of the transformation rules to reduce the parameter directly (like **TransRule 1** or **TransRule 3**), or the literal is a $(2,1)$ literal, and we can always branch with a worst case recurrence $C(k) \leq C(k-1)+C(k-5)$. For the non-parameterized case (i.e., for the general MAX-SAT problem where the parameter is the total number of clauses), this recurrence becomes $C(m) \leq C(m-1)+C(m-5)$. Note that the difference between the parameterized MAX-SAT and MAX-SAT, is that in the parameterized MAX-SAT the parameter is reduced by i only when i clauses are satisfied. However, for the MAX-SAT problem, the parameter m is reduced by i when i clauses are eliminated (that is, their values have been determined). Hence, a branch of the form $C(k) \leq C(k-i)+C(k-j)$ for the parameterized case implies directly a branch of the form $C(m) \leq C(m-i)+C(m-j)$ for the general MAX-SAT. The idea then becomes to take advantage of literals of low occurences. Since the existence of a literal with low occurence is not always guaranteed, we try to enforce it after each branching case of our algorithm. To do that, we will use a subroutine called **Create-Low-Literal**. This subroutine assumes that the formula contains only $(3,1)$ and $(2,2)$ literals, with at least one $(2,2)$ literal, and it guarantees that a $(2^-,1^-)$ literal exists in F after its termination. We give this subroutine in Figure 1.

Create-Low-Literal

Precondition: F only contains $(2,2)$ and $(3,1)$ literals with at least one $(2,2)$ literal
Postcondition: F contains a $(2^-,1^-)$ literal

1. **while** the formula F contains a $(2,2)$ literal $\tilde{y}$ such that both occurences of y (resp. $\bar{y}$) are in unit clauses **do**
   $F = F[y][]$ (resp. $F = F[][y]$);
   $m = m - 2$;
2. let $\tilde{y}$ be a $(2,2)$ literal;
3. branch as $F[y][]$ and $F[][y]$;

Fig. 1. The subroutine: Create-Low-Literal

Proposition 2. *If the subroutine* **Create-Low-Literal** *is called on a formula F containing only* $(2,2)$ *and* $(3,1)$ *literals, then when the subroutine terminates,* *the invariant that F contains a $(2-,1-)$ literal is satisfied. Moreover, the subroutine branches with a worst case recurrence $C(k) \leq 2C(k-2)$.*

Proof. The subroutine **Create-Low-Literal** works as follows. First it eliminates all $(2,2)$ literals $\tilde{y}$, where both occurrences of y (or $\bar{y}$) are in unit clauses, deterministically by setting $F = F[y][]$ (resp. $F = F[][y]$). This step is correct by **TransRule 2**. The recurrence relation for Step 1 is $C(m) = C(m-2i)$, where $i \geq 0$ is the number of times the **while** loop in Step 1 is executed. In step 2 we pick a $(2,2)$ literal $\tilde{y}$. Since Step 1 is terminated, we know that both y and $\bar{y}$ occur with other literals say $\tilde{r}$ and $\tilde{s}$, respectively. Now when we branch as $F[y][]$ two clauses are satisfied and $\tilde{r}$ becomes a $(2^-,1^-)$ literal. Similarly, when we branch as $F[][y]$, $\tilde{s}$ becomes a $(2^-,1^-)$ literal. Thus, the invariant is always maintained at the end of the subroutine **Create-Low-Literal**. Now in Step 3 we branch with a recurrrence relation $C(m) \leq 2C(m-2)$. Combining Step 1 and Step 3 together, when **Create-Low-Literal** is called we branch with recurrence relation $C(m) \leq 2C(m-2i-2)$, $i \geq 0$. The worst case happens when $i = 0$, and we branch with recurrence relation $C(m) \leq 2C(m-2)$ with the invariant that the formula contains a $(2^-,1^-)$ literal is satisfied.

Our algorithm is divided into two phases. In the first phase we apply **Case 3.1** and **Case 3.2** of the algorithm in Section 3 to eliminate all literals (i,j) where $i+j \geq 6$ or $i+j=5$ with $i, j > 1$, and all $(3,1)$ and $(4,1)$ literals whose negations do not occur as unit clauses. **Case 3.1** and **Case 3.2** guarantee a recurrence relation $C(k) \leq C(k-1) + C(k-5)$ for the parameterized case, and hence a recurrence relation $C(m) \leq C(m-1) + C(m-5)$ for the non-parameterized case. Now if F contains a $(4,1)$ literal $\tilde{x}$, then $\bar{x}$ has to occur as a unit clause, and we branch as $F[x][]$ and $F[][x]$. In the first branch four clauses are satisfied and one is eliminated (the unit clause containing $\bar{x}$), and in the second branch one clause is satisfied. Hence, we branch with the recurrence relation $C(m) \leq C(m-1) + C(m-5)$. After this phase of the algorithm, we know that F does not contain any (i,j) literal where $i+j \geq 5$, and all $(3,1)$ literals have their negations occuring as unit clauses.

The next phase of the algorithm works in stages. The algorithm at each stage picks the first branching rule that applies and uses it. Also, at the beginning of each stage we will maintain the following invariant: The formula F has a $(2^-,1^-)$ literal. The way we keep this invariant is by guaranteeing that after each branching case, either a $(2^-,1^-)$ literal remains in the formula, or we can create one by applying the **Create-Low-Literal** subroutine. We can assume, without loss of generality, that such a literal exists in the first stage of the algorithm. Otherwise, we can introduce a new unit clause containing a new variable x and increase the parameter m by 1. It can be easily seen that the running time of the algorithm will not increase by more than a multiplicative constant. We also assume that in each stage of the algorithm F contains at least one $(2,2)$ literal. If this is not the case, then F contains only $(3^-,1^-)$ literals, so we stop and apply the following branching case:

Case 4.0. F does not contain any $(2,2)$ literal. In this case either a $(2,1)$ literal exists, or the formula consists of only $(3,1)$ literals. In the former case one of the cases **Case 3.4** through **Case 3.6** must apply, hence branching with a worst case recurrence $C(m) \leq C(m-1) + C(m-5)$. In the latter case one of cases **Case 3.11** through **Case 3.13** applies. Now observing that the recurrence relation in **Subcase 3.13.2** becomes $C(m) \leq C(m-1) + C(m-7) + C(m-10)$ for the non-parameterized case (since when $x = 1$ the unit clause containing $\bar{x}$ will be eliminated), we conclude that in these cases we can branch with a worst case recurrence relation $C(m) \leq C(m-1) + C(m-5)$.

Let $\tilde{y}$ be a $(2,2)$ literal and $\tilde{x}$ be a $(2,1)$ literal. At each stage of the algorithm we branch according to one of the following branching cases.

Case 4.1. Any of **TransRule 1** through **TransRule 3** is applicable. Apply the transformation rule and then **Create-Low-Literal** if necessary. $C(m) \leq 2C(m-3)$

Case 4.2. y or $\bar{y}$ occurs as a unit clause. Assume, without loss of generality, that y occurs as a unit clause. Branch as $F[y][]$ and $F[][y]$. In both branches at least one occurrence of the literal $\tilde{x}$ remains, and hence, the invariant is satisfied. $C(m) \leq C(m-2) + C(m-3)$

Now assume that $\tilde{y}$ occurs with $\tilde{x}$. Without loss of generality, suppose that y occurs with $\tilde{x}$.

Case 4.3. y occurs only with $\tilde{x}$. Branch as $F[y][]$ and apply **TransRule 1** to x, and then **Create-Low-Literal**, and $F[][y]$. Note that in the second branch a $(2^-, 1^-)$ literal remains (namely x). $C(m) \leq 2C(m-6) + C(m-2)$

Case 4.4. There is a literal $\tilde{r}$ that occurs with y such that $\tilde{r}$ occurs also outside y and $\tilde{x}$. Branch as $F[y][]$ and apply **TransRule 1** or **TransRule 3** to $\tilde{x}$, and $F[][y]$. In the first branch $\tilde{r}$ becomes a $(2^-, 1^-)$ literal, and in the second $\tilde{x}$. $C(m) \leq C(m-3) + C(m-2)$

Case 4.5. The other occurence of y is outside $\tilde{x}$. Let $\tilde{r}$ be a literal that occurs with y outside $\tilde{x}$. Sine **Case 4.4** was eliminated, all other occurences of $\tilde{r}$ have to be with $\tilde{x}$. Branch as $F[y][]$ and apply **TransRule 1** or **TransRule 3** to $\tilde{x}$ and $\tilde{r}$, and apply **Create-Low-Literal**, and $F[][y]$. In the second branch $\tilde{x}$ remains. $C(m) \leq 2C(m-6) + C(m-2)$

Now both occurrences of y are with $\tilde{x}$. Also, both occurences of $\bar{y}$ are outside $\tilde{x}$, otherwise, by symmetry, we can apply one of the above cases with y exchanged with $\bar{y}$.

Case 4.6. There is a literal $\tilde{r}$ that occurs simultaneously with $\tilde{x}$ and outside $\tilde{x}$. Branch as $F[y][]$ and apply **TransRule 1** or **TransRule 3** to $\tilde{x}$, and $F[][y]$. In the first $\tilde{r}$ becomes a $(2^-, 1^-)$ literal, and in the second $\tilde{x}$ remains. $C(m) \leq C(m-3) + C(m-2)$

Case 4.7. There is a literal $\tilde{r}$ distinct from y that occurs with $\tilde{x}$. Since the previous case was eliminated, $\tilde{r}$ can only occur with $\tilde{x}$, and hence it must be a $(2,1)$ literal. In this case a safe partial assignment that satisfies the three clauses containing $\tilde{x}$ and $\tilde{r}$ can be deterministically and easily computed, and we can apply **Create-Low-Literal** to create a $(2^-, 1^-)$ literal in the resulting formula. $C(m) \leq 2C(m-5)$

Case 4.8. y is the only variable that occurs with $\tilde{x}$ and $\bar{x}$ occurs as a unit clause. By **BranchRule 2**, branch $F[x][y]$ and $F[][y,x]$ reduce to $F[][y,x]$.

Thus, we branch as $F[y][]$ and apply **TransRule 1** or **TransRule 3** to $\tilde{x}$, then apply **Create-Low-Literal**, and $F[x][y]$ and apply **Create-Low-Literal**. $C(m) \leq 2C(m-5) + 2C(m-7)$

Now we can assume that no $(2,2)$ literal occurs with a $(2,1)$ literal.

Case 4.9. The $(2,1)$ literals form a claused subformula. Let H be the claused subformula consisting of the $(2,1)$ literals. If the number of clauses in H is less than a prespecified constant c, we can choose $c = 10$ for instance, then a truth assignment for H satisfying at least one clause can be computed in constant time (note that H has to contain at least one clause), and we can apply the **Create-Low-Literal** subroutine to create a $(2^-, 1^-)$ literal in the remaining formula. In this case we branch with recurrence relation $C(m) \leq 2C(m-3)$. If $|H| \geq 10$, then we can apply one of the cases: **Case 3.3** through **Case 3.6**. In this case we branch with recurrence relation $C(m) \leq C(m-1) + C(m-5)$, and the remaining formula of H will contain a $(2^-, 1^-)$ literal.

Now we must have a $(3,1)$ literal $\tilde{y}$ occuring with a $(2,1)$ literal $\tilde{x}$ or else the $(2,1)$ literals would form a closed subformula. Note that all $(3,1)$ literals have their negative occurences as unit clauses.

Case 4.10. $\tilde{x}$ is dominated by $\tilde{y}$. Now branch $F[][y]$ contains $F[x][y]$. By **BranchRule 2**, $F[x][y]$ and $F[y][x]$ reduce to $F[x][y]$. Also $F[x][y]$ and $F[x,y][]$ reduce to $F[x][y]$. Thus, in this case we deterministically branch as $F[][y]$ and $\tilde{x}$ remains in the resulting formula. We get recurrence relation $C(m) = C(m-1)$.

Case 4.11. y and $\tilde{x}$ occur twice together. Note that if we do not have a literal $\tilde{r}$ that occurs with $\tilde{x}$ or $\tilde{y}$ and that has at least one occurrence outside the clauses containing $\tilde{x}$ or $\tilde{y}$, then the clauses containing $\tilde{x}$ or $\tilde{y}$ would form a closed subformula of F of fewer than four clauses. In this case (as in **Case 4.9**) a partial safe assignment for this subformula satisfying at least two clauses could be computed easily and we can apply **Create-Low-Literal** to create a $(2^-, 1^-)$ literal in the resulting formula. Hence, the parameter m is reduced by at least two and we can create a $(2^-, 1^-)$ literal using **Create-Low-Literal** giving a recurrence relation $C(m) \leq 2C(m-4)$. Now we can assume that such a literal $\tilde{r}$ exists. Branch as $F[y][]$ and apply **TransRule 1** or **TransRule 3** to $\tilde{x}$, and $F[][y]$. In the first branch $\tilde{r}$ becomes a $(2^-, 1^-)$, and in the second $\tilde{x}$ remains. $C(m) \leq C(m-5) + C(m-1)$

Case 4.12. y occurs exactly once with $\tilde{x}$. We distinguish two subcases.

Subcase 4.12.1. y occurs with x. By **TransRule 2**, none of the occurrences of y can be in a unit clause. Now let $\tilde{r}$ be a literal that occurs with y. Since the previous two cases were eliminated, $\tilde{r}$ must occur outside y (if $\tilde{r}$ is a $(3,1)$ literal, this condition is automatically satisfied). Branch as $F[y][]$ and apply **TransRule 3** to $\tilde{x}$, and $F[][y]$. In the first branch $\tilde{r}$ becomes a $(2^-, 1^-)$ literal, and in the second $\tilde{x}$ remains. $C(m) \leq C(m-5) + C(m-1)$

Subcase 4.12.1. y occurs with $\bar{x}$. Branch as $F[y][]$ and **TransRule 1** to $\tilde{x}$ then apply **Create-Low-Literal**, and $F[][y]$. In the second branch $\tilde{x}$ remains. $C(m) \leq 2C(m-8) + C(m-1)$

Theorem 2. *Given a boolean formula F of m clauses, then in time $O(1.3247^m|F|)$, we can compute a truth assignment to F satisfying the largest number of clauses.*

Proof. *The above cases are comprehensive. The size of the branching tree is the largest when we branch with the recurrence relation $C(m) \leq C(m-1)+C(m-5)$. This gives a tree size of $O(\alpha^m)$ where $\alpha \approx 1.3247$ is the unique positive root of the polynomial $x^5 - x^4 - 1$. Along each root-leaf path in the branching tree we spend $O(|F|)$ time. Hence, the running time of the algorithm is $O(1.3247^m|F|)$.*

The above algorithm is an improvement over Bansal and Raman's $O(1.341294^m|F|)$ algorithm for the MAX-SAT problem.

5 Concluding Remarks

In this paper we presented two exact algorithms for the MAX-SAT problem. Both algorithms induce improvements on the previously best algorithms by Bansal and Raman for the problem. Basically the technique used in this paper is a variation of the search tree technique which has been widely employed in designing exact algorithms for NP-hard problems [7,8,12,13,17,18,19,27,33,24,34]. Although case-by-case analysis seems unavoidable for such problems, reducing the number of cases by introducing new techniques that either enable the classification of multiple cases into a general case, or exploit the structure of the combinatorial problem by looking more carefully at its nature, are always desirable. Such techniques like the "vertex folding" and the "iterative branching" introduced in [18], the "struction" in [12], and the **Create-Low-Literal** subroutine introduced in this paper, have reduced significantly the number of cases in the problems considered.

The general open question that is posed is to what extent we can keep imrpoving these upper bounds? Put it differently, is it possible to come up with an exact algorithm scheme (if we are to call it), that for any given $\epsilon > 0$, there is an algorithm of running time $O((1+\epsilon)^m)$ for the MAX-SAT problem, and similarly for the VERTEX COVER and the INDEPENDENT SET problems? In fact, there are some problems, like the PLANAR DOMINATING SET for instance, that has such an exact algorithm scheme. More specifically, the PLANAR DOMINATING SET can be solved in time $O(c^{\sqrt{k}}n)$ for some constant $c > 0$, which is assymptotically much better than $\Theta((1+\epsilon)^k n)$ [1]. Similarly for the PLANAR VERTEX COVER and PLANAR INDEPENDENT SET problems [3]. In a recent progress in proving lower bounds for NP-hard problems, Cai and Juedes were able to show that the above $O(c^{\sqrt{k}}n)$ upper bounds for PLANAR DOMINATING SET, PLANAR VERTEX COVER, and PLANAR INDEPENDENT SET, are tight unless 3-SAT $\in$ DTIME$(2^{o(n)})$ [15]. The fact that the above problems admit $O(c^{\sqrt{k}}n)$ algorithms seems highly dependent on the planarity of the graph. It would be interesting to identify more problems that enjoy this property naturally.

Finally, we note that even though the algorithms for such NP-hard problems tend to be based on case-by-case analysis, these cases are easy to distinguish, and hence can be implemented easily. The question remainining would be how well these algorithms can perform in practice in comparison with their theoretical upper bounds and other existing heuristics for the problems.

References

1. J. ALBER, H. L. BODLAENDER, H. FERNAU, AND R. NIEDERMEIER, Fixed parameter algorithms for dominating set and related problems on planar graphs, *Lecture Notes in Computer Science* **1851**, (2000), pp. 97–110. Accepted for publication in *Algorithmica*.
2. J. ALBER, H. FAN, M. R. FELLOWS, H. FERNAU, R. NIEDERMEIER, F. ROSAMOND, AND U. STEGE, Refined search tree techniques for Dominating Set on planar graphs, *Lecture Notes in Computer Science* **2136**, (2001), pp 111-122.
3. J. ALBER, H. FERNAU, AND R. NIEDERMEIER, Parameterized complexity: Exponential speed-up for planar graph problems, in *Proceedings of the 28th International Colloquium on Automata, Languages, and Programming (ICALP 2001), Lecture Notes in Computer Science* **2076**, (2001), pp 261-272.
4. T. ASANO, K. HORI, T. ONO, AND T. HIRATA, A theoretical framwork of hybrid approaches to MAX-SAT, *Lecture Notes in Computer Science* **1350**, (1997), pp. 153-162.
5. T. ASANO, AND D. P. WILLIAMSON, Improved Aproximation Algorithms for MAX-SAT, in *Proceedings of the 11th ACM-SIAM Symposium on Discrete Algorithms (SODA)*, (2000), pp. 96-105.
6. P. ASIRELLI, M. DE SANTIS, AND A. MARTELLI, Integrity constraints in logic databases, *Journal of Logic Programming* **3**, (1985), pp. 221-232.
7. R. BALASUBRAMANIAN, M. R. FELLOWS, AND V. RAMAN, An improved fixed parameter algorithm for vertex cover, *Information Processing Letters* **65**, (1998), pp. 163-168.
8. N. BANSAL AND V. RAMAN, Upper bounds for MAX-SAT further improved, *Lecture Notes in Computer Science* **1741**, (1999), pp. 247-258.
9. R. BATTITI AND M. PROTASI, Reactive research, a history base heuristic for MAX-SAT, *J. Exper. Algorithmics* **2**, No. 2, (1997).
10. R. BATTITI AND M. PROTASI, Approximate algorithms and heuristics for MAX-SAT, in *Handbook of Combinatorial Optimization* **1**, D. Z. Du and P. M. Pardalos Eds., (1998), pp. 77-148.
11. B. BORCHERS AND J. FURMAN, A two-phase exact algorithm for MAX-SAT and weighted MAX-SAT problems, *J. Combinatorial Optimization* **2**, (1999), pp. 465-474.
12. R. BEIGEL, Finding mximum independent sets in sparse and general graphs, in *Proceedings of the 10th ACM-SIAM Symposium on Discrete Algorithms* (SODA'99), (1999), pp. 856-857.
13. R. BEIGEL AND D. EPPSTEIN, 3-coloring in time $O(1.3446^n)$: a no-MIS algorithm, in *Proceedings of the 36th IEEE Symposium on Foundation of Computer Science* (FOCS'95), (1995), pp. 442-452.
14. L. CAI AND J. CHEN, On fixed-parameter tractability and approximation of NP-hard optimization problems, *Journal of Computer and System Sciences* **54**, (1997), pp. 465-474.
15. L. CAI AND D. JUEDES, On the existence of subexponential-time parameterized algorithms, available at `http://www.cs.uga.edu/ ~ cai/`.

16. J. CHEN, D. K. FRIESEN, W. JIA, AND I. A. KANJ, Using nondeterminism to design efficient deterministic algorithms, to appear in *proceedings of the 21st annual conference on Foundations of Software Technology and Theoreical Computer Science (FSTTCS'01)*, December 13-15, (2001), Indian Institute of Science, Bangalore, India.
17. J. CHEN AND I. A. KANJ, On constrained minimum vertex covers of bibartite graphs: Improved Algorithms, *Lecture Notes in Computer Science*,
18. J. CHEN, I. A. KANJ, AND W. JIA, Vertex cover, further observations and further improvements, *Lecture Notes in Computer Science* **1665**, (1999), pp. 313–324.
19. J. CHEN, L. LIU, AND W. JIA, Improvement on Vertex Cover for low-degree graphs, *Networks* **35**, (2000), pp. 253-259.
20. R. G. DOWNEY AND M. R. FELLOWS, *Parameterized Complexity*, New York, New York: Springer, (1999).
21. R. G. DOWNEY, M. R. FELLOWS, AND U. STEGE, Parameterized complexity: A framework for systematically confronting computational intractability, in *Contemporary Trends in Discrete Mathematics: From DIMACS and DIMATIA to the Future*, F. Roberts, J. Kratochvil, and J. Nesetril, eds., *AMS-DIMACS Proceedings Series* **49**, AMS, (1999), pp. 49-99.
22. E. DANTSIN, M. GOERDT, E. A. HIRSCH, AND U. SCHÖNING, Deterministic algorithms for k-SAT based on covering codes and local search, *Lecture Notes in Compuer Science* **1853**, (2000), pp. 236-247.
23. M. DAVIS AND H. PUTNAM, A computing procedure for quantification theory, *Journal of the ACM* **7**, (1960), pp. 201-215.
24. H. FERNAU AND R. NIEDERMEIER, An efficient exact algorithm for constraint bipartite vertex cover, *Journal of Algorithms* **38**, (2001), pp. 374-410.
25. H. GALLAIRE, J. MINKER, AND J. M. NICOLAS, Logic and databases: A deductive approach, *Computing Surveys* **16**, No. 2, (1984), pp. 153-185.
26. M. GAREY AND D. JOHNSON, *Computers and Intractability: A Guide to the Theory of NP-completeness*, Freeman, San Francisco, 1979.
27. J. GRAMM, E. A. HIRSCH, R. NIEDERMEIER, AND P. ROSSMANITH, New worst-case upper bounds for MAX-2-SAT with application to MAX-CUT, accepted for publication in *Discrete Applied Mathematics* (2001).
28. P. HANSEN AND B. JAUMARD, Algorithms for the maximum satisfiability problem, *Computing* **44**, (1990) pp. 279-303.
29. F. HAYES, D. A. WATERMAN, AND D. B. LENAT, *Building Expert Systems*, Reading Massachusetts: Addison Wesley, (1983).
30. O. KULLMAN AND H. LUCKHARDT, Deciding propositional tautologies: Algorithms and their complexity, submitted for publication, available at `http://cs-svr1.swan.ac.uk/ csoliver/papers.html`.
31. M. MAHAJAN AND V. RAMAN, Parameterizing above guaranteed values: MAX-SAT and MAX-CUT, *Journal of Algorithms* **31**, (1999), pp. 335-354.
32. T. A. NGUYEN, W. A. PERKINS, T. J. LAFFEY, AND D. PECORA, Checking an expert systems knowledge base for consistency and completeness, *IJCAI'85*, Arvind Joshi Ed., Los Altos, CA, (1983), pp. 375-378.
33. R. NIEDERMEIER AND P. ROSSMANITH, New upper bounds for maximum satisfiability, *Journal of Algorithms* **36**, (2000), pp. 63-88.
34. J. M. ROBSON, Algorithms for maximum independent set, *Journal of Algorithms* **6**, (1977), pp. 425-440.
35. R. J. WALLACE, Enhancing maximum satisfiability algorithms with pure literal strategies, *Lecture Notes on Artificial Intelligence* **1081**, (1996) pp. 388-401.

Characterising Strong Normalisation for Explicit Substitutions

Steffen van Bakel[1] and Mariangiola Dezani-Ciancaglini[2]

[1] Department of Computing, Imperial College,
180 Queen's Gate, London SW7 2BZ, UK,
svb@doc.ic.ac.uk

[2] Dipartimento di Informatica, Università di Torino,
Corso Svizzera 185, 10149 Torino, Italy,
dezani@di.unito.it

Abstract. We characterise the strongly normalising terms of a composition-free calculus of explicit substitutions (with or without garbage collection) by means of an intersection type assignment system. The main novelty is a cut-rule which allows to forget the context of the minor premise when the context of the main premise does not have an assumption for the cut variable.

Introduction

Intersection type disciplines originated in [8] to overcome the limitations of Curry's type assignment system and to provide a characterisation of the *strongly normalising terms* of the λ-calculus [22]. Since then, intersection types disciplines were used in a series of papers for characterising *evaluation properties* of λ-terms [5, 17, 16, 3, 4, 13, 2, 12, 10].

We are interested here in considering *calculi of explicit substitutions*, over terms Λx, originated in [1] for improving λ-calculus implementations.

In the literature there are many different proposals for explicit substitution calculi [7, 6, 15, 23], that are powerful tools for enlightening the relations between abstraction and application, as they decompose the evaluation rule of the λ-calculus into elementary steps. For this reason, it is crucial to characterize the computational behaviour of substitution calculi.

In a seminal paper [11], Dougherty and Lescanne show that intersection type systems can characterize normalising and head-normalising terms of a composition-free calculus of explicit substitutions (with or without garbage collection). Allowing composition between substitutions leads to the (unexpected) failure of termination of simply typed terms, as proved by Melliès [19]. Therefore, the choice of [11] and of the present paper is to consider calculi that are composition-free.

Characterisation of *strongly* normalising Λx-terms using intersection types has up to now been an open problem. In part, this problem has been addressed in [11], where the type assignment system $\mathcal{D}$ has the property that all typeable terms in Λx are strongly normalising; however, the converse of this property fails. The aim of the present paper is to recover from this failure with a very simple move: we add a new cut-rule to the

S. Rajsbaum (Ed.): LATIN 2002, LNCS 2286, pp. 356–370, 2002.

system $\mathcal{D}$. This rule essentially takes into account that, by putting a term of the shape $M\langle x=N\rangle$, where x does not occur free in M, in an arbitrary context, the free variables of N will never be replaced. Therefore, we can discharge the assumptions used to type N when we derive a type for $M\langle x=N\rangle$. Our main result is then:

a term in Λx is typeable if and only if it is strongly normalising.

In order to prove one direction of this result we devise an inductive characterisation of the set of strongly normalising terms in Λx inspired by that for pure λ-terms of [24]. This allows us to show that all strongly normalising terms have a type. Notice that, in general, only typeability (not types!) is preserved by subject expansion that does not leave the set of strongly normalising terms.

In order to prove the other direction, we use the set-theoretic semantics of intersection types and saturated sets, which is referred to as the *reducibility method*. This is a generally accepted method for proving the strong normalisation property of various type systems such as the simply typed lambda calculus in Tait [25], the polymorphic lambda calculus in Tait [26] and Girard [14]. All the above mentioned papers characterising evaluation properties of λ-terms and of terms in Λx by means of intersection types apply variants of this method.

Only after submitting the first version of the present paper we became aware of the fact that Dougherty, Lengrand and Lescanne solved the same problem [18] with a similar type assignment system. The present version is strongly influenced by [18] and by discussions with its authors.

1 The Calculus

Following [11], we consider the set of terms Λx which uses names rather than De Bruijn indices.

Definition 1 (Set of Terms Λx). The set of terms Λx is defined by the grammar:

$$M,N ::= x \mid (\lambda x.M) \mid (MN) \mid (M\langle x=N\rangle)$$

As usual we consider terms modulo renaming of bound variables.

In writing terms, we will use the standard conventions for removing brackets, and use the following abbreviations:

$$\overrightarrow{M} = M_1,\ldots,M_n\ (n\geq 0)$$
$$M\overrightarrow{M} = MM_1\ldots M_n\ (n\geq 0)$$
$$M\,\overrightarrow{\langle x=N\rangle} = M\langle x_1=N_1\rangle\ldots\langle x_n=N_n\rangle\ (n\geq 0)$$
$$|\overrightarrow{M}| = n,\ \text{where } n \text{ is the number of terms appearing in } \overrightarrow{M}.$$

Apart from defining the notion of *free variables* of a term, we need to single out the free variables which occur in a term without considering some explicit substitution.

Definition 2. The set *pfv*(M) of *proper free variables of* M is inductively defined by:

$$\begin{aligned} \mathit{pfv}(x) &= \{x\} \\ \mathit{pfv}(\lambda x.M) &= \mathit{pfv}(M)\backslash x \\ \mathit{pfv}(MN) &= \mathit{pfv}(M) \cup \mathit{pfv}(N) \\ \mathit{pfv}(M\langle x = N\rangle) &= \begin{cases} \mathit{pfv}(M)\backslash x \cup \mathit{pfv}(N) & \text{if } x \in \mathit{pfv}(M) \\ \mathit{pfv}(M) & \text{otherwise.} \end{cases} \end{aligned}$$

For example, $\mathit{pfv}(z\langle y = xx\rangle\langle z = t\rangle) = \{t\}$, while $\mathit{fv}(z\langle y = xx\rangle\langle z = t\rangle) = \{x, t\}$.

Notice that the set *fv*(M) of free variables of M can be defined in the same way, but for the last clause which then states

$$\mathit{fv}(M\langle x = N\rangle) = \mathit{fv}(M)\backslash x \cup \mathit{fv}(N).$$

Clearly we get $\mathit{pfv}(M) \subseteq \mathit{fv}(M)$ for all terms M.

We use the reduction relation $\lambda\mathsf{x}_{\mathsf{gc}}$ on $\Lambda\mathsf{x}$ as defined in [7].

Definition 3 (Reduction Relation). The reduction rules of $\lambda\mathsf{x}_{\mathsf{gc}}$ are:

$$\begin{array}{lrcl} (\mathsf{B}) & (\lambda x.M)N & \to & M\langle x = N\rangle \\ (\mathsf{App}) & (MP)\langle x = N\rangle & \to & (M\langle x = N\rangle)(P\langle x = N\rangle) \\ (\mathsf{Abs}) & (\lambda y.M)\langle x = N\rangle & \to & \lambda y.(M\langle x = N\rangle) \\ (\mathsf{VarI}) & x\langle x = N\rangle & \to & N \\ (\mathsf{gc}) & M\langle x = N\rangle & \to & M \text{ if } x \notin \mathit{fv}(M) \end{array}$$

As usual, the reduction relation we consider in this paper is the contextual, transitive closure of the relation generated by these rules, and we will write $M \to N$ if M reduces to N using this relation.

Inside $\Lambda\mathsf{x}$ we are interested in the set of strongly normalisable terms, i.e. those terms for which all reduction paths are of finite length.

Definition 4 ($\mathcal{SN}$). We define $\mathcal{SN} = \{M \in \Lambda\mathsf{x} \mid M \text{ is strongly normalisable}\}$.

As proved in [11], the set $\mathcal{SN}$ coincides with the set of strongly normalisable terms using the reduction relation obtained by removing the rule (gc) and by adding the following rule:

$$(\mathsf{VarK})\quad y\langle x = N\rangle \to y$$

To simplify proofs below, it is useful to consider a stronger version of the reduction rule (gc), which uses *proper* free variables instead of free variables, and the corresponding set of strongly normalisable terms.

Definition 5 ((gc_p), $\to_p$ and $\mathcal{SN}_p$). The reduction relation $\to_p$ is obtained by removing rule (gc) and adding the following rule:

$$(\mathsf{gc}_p)\quad M\langle x = N\rangle \to M \text{ if } x \notin \mathit{pfv}(N)$$

The set $\mathcal{SN}_p$ is the set of strongly normalising terms with respect to $\to_p$.

$$\frac{M \in \mathcal{A}}{\lambda x.M \in \mathcal{A}}\ (1) \qquad \frac{\overrightarrow{M} \in \mathcal{A}}{x\overrightarrow{M} \in \mathcal{A}}\ (2) \qquad \frac{M\langle x=N\rangle \overrightarrow{P} \in \mathcal{A}}{(\lambda x.M)N\overrightarrow{P} \in \mathcal{A}}\ (3)$$

$$\frac{(U\langle x=N\rangle)(V\langle x=N\rangle)\overrightarrow{\langle z=Q\rangle}\overrightarrow{P} \in \mathcal{A}}{(UV)\langle x=N\rangle \overrightarrow{\langle z=Q\rangle}\overrightarrow{P} \in \mathcal{A}}\ (4)$$

$$\frac{(\lambda y.M\langle x=N\rangle)\overrightarrow{\langle z=Q\rangle}\overrightarrow{P} \in \mathcal{A}}{(\lambda y.M)\langle x=N\rangle \overrightarrow{\langle z=Q\rangle}\overrightarrow{P} \in \mathcal{A}}\ (y \notin fv(M)) \quad (5)$$

$$\frac{N\overrightarrow{\langle z=Q\rangle}\overrightarrow{P} \in \mathcal{A}}{x\langle x=N\rangle \overrightarrow{\langle z=Q\rangle}\overrightarrow{P} \in \mathcal{A}}\ (6)$$

$$\frac{M\overrightarrow{\langle z=Q\rangle}\overrightarrow{P}, N \in \mathcal{SN}}{M\langle x=N\rangle \overrightarrow{\langle z=Q\rangle}\overrightarrow{P} \in \mathcal{SN}}\ (x \notin fv(M)) \quad (7)$$

$$\frac{M\overrightarrow{\langle z=Q\rangle}\overrightarrow{P}, N \in \mathcal{SN}_p}{M\langle x=N\rangle \overrightarrow{\langle z=Q\rangle}\overrightarrow{P} \in \mathcal{SN}_p}\ (x \notin pfv(M)) \quad (8)$$

Fig. 1. Rules generating $\mathcal{SN}$ and $\mathcal{SN}_p$

Notice that

$$M \to N \;\Rightarrow\; M \to_p N$$
$$\mathcal{SN}_p \subseteq \mathcal{SN}$$

but that, since $pfv(M) \subset fv(M)$ for some M, $M \to_p N \Rightarrow M \to N$ does not hold in general. However, we will prove that $\mathcal{SN}_p = \mathcal{SN}$ by means of inductive characterisations of the sets $\mathcal{SN}$ and $\mathcal{SN}_p$. The correctness of the characterisation of $\mathcal{SN}$ follows from Lem. 5 of [11]. It is easy to verify that the proof of Lem. 5 of [11] easily adapts to the set $\mathcal{SN}_p$ and so we get the correctness of the characterisation of $\mathcal{SN}_p$.

Lemma 6. *1. The set $\mathcal{SN}$ can be defined inductively through rules (1) to (6), where $\mathcal{A}$ is $\mathcal{SN}$, and (7) of Figure 1.*

2. The set $\mathcal{SN}_p$ can be defined inductively through rules (1) to (6), where $\mathcal{A}$ is $\mathcal{SN}_p$, and (8) of Figure 1.

3. $\mathcal{SN} = \mathcal{SN}_p$.

Proof. *1 and 2.* With $\mathcal{A} = \mathcal{SN}$, the first seven rules generate only terms in $\mathcal{SN}$: for the first two rules it is trivial and for the remaining it is proved in Lem. 5 of [11]. Since we can show Lem. 5 of [11] for $\mathcal{SN}_p$ (after replacing rule (8) to rule (7)) we can similarly show, with $\mathcal{A} = \mathcal{SN}_p$, that the first six rules and rule (8) generate only terms in $\mathcal{SN}_p$.

To see that these rules generate all strongly normalising terms, first notice that the terms in the conclusions of the given rules cover all possible shapes of terms in Λx, as observed also in [11] (Lem. 1). Moreover, if the term in the conclusion is strongly normalising, then also the terms in the premise must be strongly normalising: this can be proved by a double induction on the length of the longest derivation to normal form (using, respectively, $\rightarrow$ and $\rightarrow_p$) and on the structure of terms.

3. Immediate from 1 & 2 taking into account that $x \notin fv(M)$ implies $x \notin pfv(M)$ and that $\mathcal{SN}_p \subseteq \mathcal{SN}$. ■

Always from Lem. 5 of [11] we get that distribution of substitution preserves strong normalisation:

Lemma 7. If $(P\,\overrightarrow{\langle x=N\rangle}\,\langle y=Q\,\overrightarrow{\langle x=N\rangle}\rangle)\overrightarrow{M} \in \mathcal{SN}$ *then* $((P\,\langle y=Q\rangle)\,\overrightarrow{\langle x=N\rangle})\overrightarrow{M} \in \mathcal{SN}$.

2 The Type Assignment

We will consider intersection types as first defined in [8] with a pre-order which takes the idem-potence, commutativity and associativity of the intersection type constructor into account.

Definition 8 (Types).

1. The set of types considered in this paper is the set $\mathcal{T}$ of *intersection types*, defined by the following grammar:

$$\sigma, \tau ::= \varphi \mid (\sigma \rightarrow \tau) \mid (\sigma \cap \tau)$$

where φ ranges over a denumerable set of type atoms.

2. On $\mathcal{T}$, the type inclusion relation $\leq$ is inductively defined by: $\sigma \leq \sigma$, $\sigma\cap\tau \leq \sigma$, $\sigma\cap\tau \leq \tau$, $\sigma \leq \tau \;\&\; \sigma \leq \rho \Rightarrow \sigma \leq \tau\cap\rho$, and $\sigma \leq \tau \leq \rho \Rightarrow \sigma \leq \rho$.
3. $\sigma \sim \tau \Leftrightarrow \sigma \leq \tau \leq \sigma$.

In the notation of types, as usual, right-most outer-most brackets will be omitted, and, since the type constructor $\cap$ is associative and commutative, we will write $\sigma\cap\tau\cap\rho$ rather than $(\sigma\cap\tau)\cap\rho$, and denote $\sigma_1\cap\ldots\cap\sigma_n$ by $\cap_{\underline{n}}\sigma_i$, where $\underline{n} = \{1,\ldots,n\}$[1].

Before presenting the type assignment system we need a last definition.

Definition 9 (Statements and Contexts).

1. A *statement* is an expression of the form $M{:}\sigma$, where M is the *subject* and σ is the *predicate* of $M{:}\sigma$.
2. A *context* Γ is a partial mapping from variables to types, and can be seen as a set of statements with (distinct) variables as subjects.
3. The relations $\leq$ and $\sim$ are extended to contexts by:

$$\Gamma \leq \Gamma' \Leftrightarrow \forall x{:}\sigma' \in \Gamma'\; \exists x{:}\sigma \in \Gamma\; [\sigma \leq \sigma']$$
$$\Gamma \sim \Gamma' \Leftrightarrow \Gamma \leq \Gamma' \leq \Gamma.$$

[1] We allow $n = 1$ as an *abus de langage*.

We introduce the following notational conventions:

$$\begin{aligned}\Gamma\cap\Delta &= \{x{:}\sigma\cap\tau \mid x{:}\sigma\in\Gamma \;\&\; x{:}\tau\in\Delta\}\cup\{x{:}\sigma \mid x{:}\sigma\in\Gamma \;\&\; x\notin\Delta\}\cup\\ &\quad \{x{:}\tau \mid x\notin\Gamma \;\&\; x{:}\tau\in\Delta\}\\ \cap_{\underline{n}}\Gamma_i &= \Gamma_1\cap\ldots\cap\Gamma_n\\ \Gamma,x{:}\sigma &= \Gamma\backslash x\cup\{x{:}\sigma\}.\end{aligned}$$

For example $\{x{:}\sigma\}\cap\{x{:}\tau\}$ denotes $\{x{:}\sigma\cap\tau\}$, while $\{x{:}\sigma\},x{:}\tau$ denotes $\{x{:}\tau\}$.

As discussed in the introduction, the key of our type assignment is a cut-rule (*cut***K**) which allows to forget the context of the minor premise. We will call the standard cut-rule (*cut***I**).

Definition 10 (Type Assignment Rules). *Type assignment* for terms in Λx and *derivations* are defined by the following natural deduction system:

$$(Ax)\ \frac{}{\Gamma\vdash x{:}\sigma}\ (x{:}\sigma\in\Gamma)$$

$$(\to I)\ \frac{\Gamma,x{:}\sigma\vdash M{:}\tau}{\Gamma\vdash\lambda x.M{:}\sigma\to\tau}\qquad (\to E)\ \frac{\Gamma\vdash M{:}\sigma\to\tau\quad \Gamma\vdash N{:}\sigma}{\Gamma\vdash MN{:}\tau}$$

$$(\cap I)\ \frac{\Gamma\vdash M{:}\sigma\quad \Gamma\vdash M{:}\tau}{\Gamma\vdash M{:}\sigma\cap\tau}\qquad (\cap E)\ \frac{\Gamma\vdash M{:}\sigma_1\cap\sigma_2}{\Gamma\vdash M{:}\sigma_i}\ (i\in\underline{2})$$

$$(cut\mathbf{I})\ \frac{\Gamma,x{:}\sigma\vdash M{:}\tau\quad \Gamma\vdash N{:}\sigma}{\Gamma\vdash M\langle x=N\rangle{:}\tau}\qquad (cut\mathbf{K})\ \frac{\Gamma\vdash M{:}\tau\quad \Delta\vdash N{:}\sigma}{\Gamma\vdash M\langle x=N\rangle{:}\tau}\ (x\notin\Gamma)$$

We write $\Gamma\vdash M{:}\sigma$ if there exists a derivation, built using this rule, that has this as its conclusion.

The following example explains in detail the difference between the system of [11] and that of this paper, and shows that the counter-example to the characterisation of strongly normalisabilty of that paper is dealt with successfully here.

Example 11. Let $D\equiv\lambda a.aa$ and $M\equiv(\lambda x.(\lambda y.z)(xx))D$. First of all,

$$\dfrac{\dfrac{\dfrac{\{a{:}(\sigma\to\tau)\cap\sigma\}\vdash a{:}(\sigma\to\tau)\cap\sigma}{\{a{:}(\sigma\to\tau)\cap\sigma\}\vdash a{:}\sigma\to\tau}\quad \dfrac{\{a{:}(\sigma\to\tau)\cap\sigma\}\vdash a{:}(\sigma\to\tau)\cap\sigma}{\{a{:}(\sigma\to\tau)\cap\sigma\}\vdash a{:}\sigma}}{\{a{:}(\sigma\to\tau)\cap\sigma\}\vdash aa{:}\tau}}{\emptyset\vdash\lambda a.aa{:}((\sigma\to\tau)\cap\sigma)\to\tau}$$

Notice that $M\to z\langle y=DD\rangle\to z\langle y=DD\rangle\to\ldots$, so M is not strongly normalisable. Also

$$M\to(\lambda x.z\langle y=xx\rangle)D\to M'\equiv z\langle y=xx\rangle\langle x=D\rangle\to M''\equiv z\langle x=D\rangle.$$

Then both M' and M'' are strongly normalisable, but only the latter is typeable in the system of [11]. Instead, in the system presented here, M' is typeable[2] (where **D** is the derivation given above):

[2] By Subject Reduction (Thm. 14) M'' is typeable as well.

$$\dfrac{\dfrac{\overline{\{z{:}\mu\} \vdash z{:}\mu} \quad \dfrac{\dfrac{\overline{\{x{:}(\rho\to\nu)\cap\rho\} \vdash x{:}(\rho\to\nu)\cap\rho}}{\{x{:}(\rho\to\nu)\cap\rho\} \vdash x{:}\rho\to\nu} \quad \dfrac{\overline{\{x{:}(\rho\to\nu)\cap\rho\} \vdash x{:}(\rho\to\nu)\cap\rho}}{\{x{:}(\rho\to\nu)\cap\rho\} \vdash x{:}\rho}}{\{x{:}(\rho\to\nu)\cap\rho\} \vdash \langle y = xx\rangle :\nu}}{\{z{:}\mu\} \vdash z\,\langle y = xx\rangle :\mu} \quad \dfrac{\mathbf{D}}{\emptyset \vdash \lambda a.aa{:}(\sigma\to\tau)\cap\sigma\to\tau}}{\{z{:}\mu\} \vdash z\,\langle y = xx\rangle\,\langle x = \lambda a.aa\rangle :\mu}$$

This implies, of course, that, in contrast to the system of [11], the terms $(\lambda x.M)N$ and $M\,\langle x = N\rangle$ *no longer* have the same typing behaviour. In fact, when typing the term $(\lambda x.M)N$, the type used for x in the sub-derivation for $\Gamma \vdash \lambda x.M{:}\sigma$ *has* to be a type for N, even if x does not appear in M, whereas rule (*cut*$\mathbf{K}$) only needs that N is typeable, not that is has the type assumed for x; actually, in rule (*cut*$\mathbf{K}$), *no* type is assumed for x. This then solves the problem of [11], in that the type used for x to type xx has no relation at all to the type derived for $\lambda a.aa$.

As usual for type assignment systems, we have a Generation Lem. and a Subject Reduction Theorem. First, we formulate some general properties of our system. Let $M[y/x]$ denote standard substitution of y for free occurrences of x.

Lemma 12. *1. If $\Gamma \vdash M{:}\tau$ and $x, y \notin \Gamma$, then $\Gamma \vdash M[y/x]{:}\tau$.*
2. If $\Gamma \vdash M{:}\tau$ and $\Gamma' \leq \Gamma$ and $\tau \leq \tau'$, then $\Gamma' \vdash M{:}\tau'$.
3. If $\Gamma \vdash M{:}\sigma\to\tau$ and $\Delta \vdash N{:}\sigma$, then $\Gamma\cap\Delta \vdash MN{:}\tau$.
4. If $\Gamma \vdash M{:}\sigma$ and $x \notin \Gamma$, then $x \notin pfv(M)$.
5. Let $\Gamma \vdash M{:}\sigma$, and $\Gamma' = \{x{:}\tau \in \Gamma \mid x \in pfv(M)\}$, then $\Gamma' \vdash M{:}\sigma$.

Proof: All proofs are by induction on the structure of derivations, but part 3 which follows immediately from part 2 and rule $(\to E)$. The only interesting case is part 2 when the last applied rule is (*cut*$\mathbf{K}$)

$$\dfrac{\Gamma \vdash M{:}\tau \quad \Delta \vdash N{:}\sigma}{\Gamma \vdash M\,\langle x = N\rangle :\tau}\;(x \notin \Gamma)$$

and $x \in \Gamma'$. If y is a fresh variable, $M\,\langle x = N\rangle$ is alpha-convertible with $M[y/x]\,\langle y = N\rangle$. We can derive by part 1 and induction $\Gamma' \vdash M[y/x]{:}\tau'$ and conclude, using (*cut*$\mathbf{K}$)

$$\dfrac{\Gamma' \vdash M[y/x]{:}\tau' \quad \Delta \vdash N{:}\sigma}{\Gamma' \vdash M[y/x]\,\langle y = N\rangle :\tau'}\;(y \notin \Gamma)$$

■

Lemma 13 (Generation Lemma).

1. If $\Gamma \vdash x{:}\sigma$, then there exists $x{:}\tau \in \Gamma$ such that $\tau \leq \sigma$.
2. If $\Gamma \vdash MN{:}\sigma$, then there exist n, σ_i, ρ_i ($i \in \underline{n}$) such that $\sigma = \cap_{\underline{n}}\sigma_i$, and $\Gamma \vdash M{:}\rho_i\to\sigma_i$ and $\Gamma \vdash N{:}\rho_i$, for all $i \in \underline{n}$.

3. *If $\Gamma \vdash \lambda y.M : \sigma$, then there exist n, ρ_i, μ_i ($i \in \underline{n}$) such that $\sigma = \cap_{\underline{n}}(\rho_i \to \mu_i)$ and, for all $i \in \underline{n}$, $\Gamma, y{:}\rho_i \vdash M : \mu_i$.*
4. *If $\Gamma \vdash M \langle x = N \rangle : \sigma$, then either*
   (a) there exists τ such that $\Gamma, x{:}\tau \vdash M : \sigma$ and $\Gamma \vdash N : \tau$, or
   (b) $\Gamma \backslash x \vdash M : \sigma$ and there exist Δ, τ such that $\Delta \vdash N : \tau$.

Proof: Easy, using Lem. 12:2 and rule $(\cap I)$ for part 4. ■

A minimal requirement of our system is that it satisfies the subject reduction property (SR). We show SR for the reduction $\to_p$: this gives us SR for $\to$ for free.

Theorem 14 (Subject Reduction Theorem). If $M \to_p N$, and $\Gamma \vdash M : \sigma$, then $\Gamma \vdash N : \sigma$.

Proof: By induction on the definition of the reduction relation, '$\to$'. We only show the base cases.

(B) : Then $\Gamma \vdash (\lambda x.M)N : \sigma$, and, by Lem. 13:2, there exist n, σ_i, ρ_i ($i \in \underline{n}$) such that $\sigma = \cap_{\underline{n}}\sigma_i$, and, for all $i \in \underline{n}$, $\Gamma \vdash \lambda x.M : \rho_i \to \sigma_i$ and $\Gamma \vdash N : \rho_i$. We can assume none of the σ_i to be an intersection, so, by Lem. 13:3, for $i \in \underline{n}$, $\Gamma, x{:}\rho_i \vdash M : \sigma_i$, and therefore, by rule (*cut***I**), $\Gamma \vdash M \langle x = N \rangle : \sigma_i$. So, by rule $(\cap I)$, $\Gamma \vdash M \langle x = N \rangle : \sigma$.

(App) : Then $\Gamma \vdash (MN) \langle x = P \rangle : \sigma$. Let $\sigma = \cap_{\underline{n}}\sigma_i$ where none of the σ_i is an intersection. By Lem. 13:4, we have two cases:

(a) there exists τ such that $\Gamma, x{:}\tau \vdash MN : \sigma$ and $\Gamma \vdash P : \tau$. Then, by Lem. 13:2, for every $i \in \underline{n}$, there exists ρ_i such that $\Gamma, x{:}\tau \vdash M : \rho_i \to \sigma_i$ and $\Gamma, x{:}\tau \vdash N : \rho_i$. Then, by rule (*cut***I**), $\Gamma \vdash M \langle x = P \rangle : \rho_i \to \sigma_i$ and $\Gamma \vdash N \langle x = P \rangle : \rho_i$.

(b) $\Gamma \backslash x \vdash MN : \sigma$ and there exist Δ, τ such that $\Delta \vdash P : \tau$. As above, by Lem. 13:2, there exists ρ_i such that $\Gamma \backslash x \vdash M : \rho_i \to \sigma_i$ and $\Gamma \backslash x \vdash N : \rho_i$. Then, by rule (*cut***K**) and Lem. 12:2, $\Gamma \vdash M \langle x = P \rangle : \rho_i \to \sigma_i$ and $\Gamma \vdash N \langle x = P \rangle : \rho_i$.

In both cases, for $i \in \underline{n}$, by rule $(\to E)$, also $\Gamma \vdash (M \langle x = P \rangle)(N \langle x = P \rangle) : \sigma_i$, so by rule $(\cap I)$, $\Gamma \vdash (M \langle x = P \rangle)(N \langle x = P \rangle) : \sigma$.

(Abs) : Then $\Gamma \vdash (\lambda y.M) \langle x = N \rangle : \sigma$. Let $\sigma = \cap_{\underline{n}}\sigma_i$ where none of the σ_i is an intersection. By Lem. 13:4, we have two cases:

(a) $\Gamma, x{:}\tau \vdash \lambda y.M : \sigma$ and there exists τ such that $\Gamma \vdash N : \tau$. By Lem. 13:3, for $i \in \underline{n}$, there exist ρ_i, μ_i such that $\sigma_i = \rho_i \to \mu_i$ and $\Gamma, x{:}\tau, y{:}\rho_i \vdash M : \mu_i$. Then, by rule (*cut***I**), $\Gamma, y{:}\rho_i \vdash M \langle x = N \rangle : \mu_i$.

(b) $\Gamma \backslash x \vdash \lambda y.M : \sigma$ and there exist Δ, τ such that $\Delta \vdash N : \tau$. As above, there exist ρ_i, μ_i such that $\sigma_i = \rho_i \to \mu_i$ and $\Gamma \backslash x, y{:}\rho_i \vdash M : \mu_i$. Then, by rule (*cut***K**) and Lem. 12:2, $\Gamma, y{:}\rho_i \vdash M \langle x = N \rangle : \mu_i$.

In both cases, by rule $(\to I)$, $\Gamma \vdash \lambda y.(M \langle x = N \rangle) : \sigma_i$, and, by rule $(\cap I)$, $\Gamma \vdash \lambda y.(M \langle x = N \rangle) : \sigma$.

(VarI) : Then $\Gamma \vdash x \langle x = N \rangle : \sigma$, and, by Lem. 13:4 & 1, there exists τ such that $\Gamma, x{:}\tau \vdash x : \sigma$. and $\Gamma \vdash N : \tau$. Then, by Lem. 13:1, $\tau \leq \sigma$, and, by Lem. 12:2, $\Gamma \vdash N : \sigma$.

(gc_p) : Then $\Gamma \vdash M\langle x = N\rangle : \sigma$ and $x \notin pfv(M)$. Then, by Lem. 13:4,
(a) either there exists τ such that $\Gamma, x{:}\tau \vdash M : \sigma$, and, by Lem. 12:5, $\Gamma \vdash M : \sigma$.
(b) or $\Gamma\backslash x \vdash M : \sigma$, and $\Gamma \vdash M : \sigma$ follows by Lem. 12:2. ∎

3 All Strongly Normalisable Terms Are Typeable

The main result of this paper, that our system types exactly the strongly normalisable terms, comes in two parts. First we show that all terms in $\mathcal{SN}$ are typeable in our system (Theorem 17), and then that all typeable terms are in $\mathcal{SN}$ (Theorem 22).

First we show two technical lemmas which are useful for the proof of Theorem 17.

Lemma 15. *1. If $\Gamma \vdash M\langle x = N\rangle : \tau$, then there exists $\Gamma' \leq \Gamma$ such that $\Gamma' \vdash (\lambda x.M)N : \tau$.*
2. If $\Gamma \vdash (U\langle x = N\rangle)(V\langle x = N\rangle) : \tau$ then $\Gamma \vdash (UV)\langle x = N\rangle : \tau$.
3. If $\Gamma \vdash \lambda y.M\langle x = N\rangle : \tau$ and $y \notin fv(N)$, then $\Gamma \vdash (\lambda y.M)\langle x = N\rangle : \tau$.
4. If $\Gamma \vdash M : \tau$, $\Delta \vdash N : \sigma$ and $x \notin pfv(M)$, then $\Gamma \vdash M\langle x = N\rangle : \tau$.

Proof: 1. By Lem. 13:4, if $\Gamma \vdash M\langle x = N\rangle : \tau$ then:
(a) either there is ρ such that $\Gamma, x{:}\rho \vdash M : \tau$, and $\Gamma \vdash N : \rho$. But then, by rules ($\rightarrow I$) and ($\rightarrow E$), we get $\Gamma \vdash (\lambda x.M)N : \tau$.
(b) or $\Gamma\backslash x \vdash M : \tau$ and there are Δ, ρ such that $\Delta \vdash N : \rho$. But then, by Lem. 12:2, $\Gamma, x{:}\rho \vdash M : \tau$ so, by rule ($\rightarrow I$) and using Lem. 12:3, $\Gamma \cap \Delta \vdash (\lambda x.M)N : \tau$. Notice that $\Gamma \cap \Delta \leq \Gamma$.

2. By Lem. 13:2, if $\Gamma \vdash (U\langle x = N\rangle)(V\langle x = N\rangle) : \tau$ there are n, ρ_i, τ_i ($i \in \underline{n}$) such that $\tau = \cap_n \tau_i$, $\Gamma \vdash U\langle x = N\rangle : \rho_i \rightarrow \tau_i$ and $\Gamma \vdash V\langle x = N\rangle : \rho_i$. By Lem. 13:4, for each $i \in \underline{n}$ we can have four different cases:
(a) there are μ_i, ν_i such that $\Gamma, x{:}\mu_i \vdash U : \rho_i \rightarrow \tau_i$, $\Gamma \vdash N : \mu_i$, $\Gamma, x{:}\nu_i \vdash V : \rho_i$, and $\Gamma \vdash N : \nu_i$. But then, by Lem. 12:3 and rule ($\cap I$):

$$\Gamma, x{:}\mu_i \cap \nu_i \vdash UV : \tau_i, \text{ and } \Gamma \vdash N : \mu_i \cap \nu_i$$

so, using rule (*cut*I), we get $\Gamma \vdash (UV)\langle x = N\rangle : \tau_i$.
(b) $\Gamma\backslash x \vdash V : \rho_i$, and there are μ_i, Δ_i, ν_i such that $\Gamma, x{:}\mu_i \vdash U : \rho_i \rightarrow \tau_i$, $\Gamma \vdash N : \mu_i$, $\Delta_i \vdash N : \nu_i$. But then, by Lem. 12:3

$$\Gamma, x{:}\mu_i \vdash UV : \tau_i, \text{ and } \Gamma \vdash N : \mu_i$$

so, using rule (*cut*I), we get $\Gamma \vdash (UV)\langle x = N\rangle : \tau_i$.
(c) $\Gamma\backslash x \vdash U : \rho_i \rightarrow \tau_i$, and there are μ_i, Δ_i, ν_i such that $\Delta_i \vdash N : \mu_i$, $\Gamma, x{:}\nu_i \vdash V : \rho_i$, $\Gamma \vdash N : \nu_i$. The proof in this case is similar to that of the previous one.
(d) $\Gamma\backslash x \vdash U : \rho_i \rightarrow \tau_i$, $\Gamma\backslash x \vdash V : \rho_i$, and there are $\Delta_i, \mu_i, \Delta'_i, \nu_i$ such that $\Delta_i \vdash N : \mu_i$, $\Delta'_i \vdash N : \nu_i$. But then, by rule ($\rightarrow E$):

$$\Gamma\backslash x \vdash UV : \tau_i, \text{ and } \Delta_i \vdash N : \mu_i$$

so we get $\Gamma \vdash (UV)\langle x = N\rangle : \tau_i$, using rule (*cut***K**) and Lem. 12:2.

Finally, using rule $(\cap I)$ we conclude $\Gamma \vdash (UV)\langle x = N\rangle : \tau$.

3. By Lem. 13:3, if $\Gamma \vdash \lambda y.M\langle x = N\rangle : \tau$ then there are n, μ_i, ν_i $(i \in \underline{n})$ such that $\Gamma, y{:}\mu_i \vdash M\langle x = N\rangle : \nu_i$, and $\tau = \cap_{\underline{n}}\nu_i$. By Lem. 13:4, for each $i \in \underline{n}$, either:
   (a) there is ρ_i such that $\Gamma, y{:}\mu_i, x{:}\rho_i \vdash M : \nu_i$ and $\Gamma, y{:}\mu_i \vdash N : \rho_i$. Then, by rule $(\to I)$, we get $\Gamma, x{:}\rho_i \vdash \lambda y.M : \mu_i \to \nu_i$.
   (b) or $\Gamma \backslash x, y{:}\mu_i \vdash M : \nu_i$ and there are Δ_i, ρ_i such that $\Delta_i \vdash N : \rho_i$. Then, by rule $(\to I)$, we get $\Gamma \backslash x \vdash \lambda y.M : \mu_i \to \nu_i$.

   Let $\underline{m}$ be the subset of $\underline{n}$ which match the first alternative. If $\underline{m}$ is not empty then by Lem. 12:2 & 4 (notice that by hypothesis $y \notin fv(N)$) and rule $(\cap I)$:
   $$\Gamma, x{:}\cap_{\underline{m}}\rho_i \vdash \lambda y.M : \cap_{\underline{n}}(\mu_i \to \nu_i) \text{ and } \Gamma \vdash N : \cap_{\underline{m}}\rho_i.$$
   We conclude, using rule (*cut***I**):
   $$\Gamma \vdash (\lambda y.M)\langle x = N\rangle : \tau.$$
   Otherwise, if $\underline{m}$ is empty then, by Lem. 12:2 and rule $(\cap I)$:
   $$\Gamma \backslash x \vdash \lambda y.M : \cap_{\underline{n}}(\mu_i \to \nu_i) \text{ and } \Delta_i \vdash N : \rho_i.$$
   We conclude, choosing an arbitrary $\Delta_j \vdash N : \rho_j$ and using rule (*cut***K**) and Lem. 12:2:
   $$\Gamma \vdash (\lambda y.M)\langle x = N\rangle : \tau.$$
4. If $\Gamma \vdash M : \tau$ and $x \notin pfv(M)$, then, by Lem. 12:5, $\Gamma \backslash x \vdash M : \tau$. We conclude using rule (*cut***K**) and Lem. 12:2. ■

Lemma 16. 1. *Assume, for all Γ, that $\Gamma \vdash M : \sigma$ implies $\Gamma \vdash M' : \sigma$. Then, for all Δ, N, τ: $\Delta \vdash M\langle x = N\rangle : \tau$ implies $\Delta \vdash M'\langle x = N\rangle : \tau$;*
2. *Assume, for all Γ, that $\Gamma \vdash M : \sigma$ implies there exists $\Gamma' \leq \Gamma$ such that $\Gamma' \vdash M' : \sigma$. Then, for all Δ, N, τ: $\Delta \vdash MN : \tau$ implies that there exists $\Delta' \leq \Delta$ such that $\Delta' \vdash M'N : \tau$. Moreover, we get $\Delta' = \Delta$ whenever $\Gamma' = \Gamma$.*

Proof: 1. By Lem. 13:4, $\Delta \vdash M\langle x = N\rangle : \tau$ implies that either there is ρ such that $\Delta, x{:}\rho \vdash M : \tau$ and $\Delta \vdash N : \rho$, or $\Delta \backslash x \vdash M : \tau$ and there are Δ', ρ such that $\Delta' \vdash N : \rho$. In both cases the conclusion easily follows from the assumption using rules (*cut***I**) and (*cut***K**).
2. By Lem. 13:2, $\Delta \vdash MN : \tau$ implies that $\tau = \cap_{\underline{n}}\tau_i$ and there are ρ_i such that $\Delta \vdash M : \rho_i \to \tau_i$ and $\Delta \vdash N : \rho_i$ for all $i \in \underline{n}$. By assumption there are $\Delta'_i \leq \Delta$ such that $\Delta'_i \vdash M' : \rho_i \to \tau_i$. Let $\Delta' = \cap_{\underline{n}}\Delta_i$. So we can conclude using Lem. 12:2 to get $\Delta' \vdash M' : \rho_i \to \tau_i$, $\Delta' \vdash N : \rho_i$ and rules $(\to E)$, $(\cap I)$ to get $\Delta' \vdash M'N : \tau$. ■

The characterisation of $\mathcal{SN}$ given in Lem. 6:1 is crucial to prove that all strongly normalisable terms are typeable.

Theorem 17. If $M \in \mathcal{SN}$ then $\Gamma \vdash M : \sigma$ for some Γ, σ.

Proof: By induction on the rules generating $\mathcal{SN}$ (Lem. 6:1) using the Generation Lem. (Lem. 13).

(*1*) : By induction, $\Gamma \vdash M:\sigma$. We distinguish two cases:
(a) If $x{:}\tau \in \Gamma$, so, by rule $(\to I)$, $\Gamma \backslash x \vdash \lambda x.M:\tau\to\sigma$.
(b) If $x \notin \Gamma$ then, by Lem. 12:2, $\Gamma, x{:}\tau \vdash M:\sigma$, so, by rule $(\to I)$, $\Gamma \vdash \lambda x.M:\tau\to\sigma$.

(*2*) : Let $|\overrightarrow{M}| = n$, then, by induction, there are Γ_i, σ_i $(i \in \underline{n})$ such that $\Gamma_i \vdash M_i:\sigma_i$, for $i \in \underline{n}$. Then $(\cap_{\underline{n}}\Gamma_i) \cap \{x{:}\sigma_1\to\ldots\to\sigma_n\to\tau\} \vdash x\overrightarrow{M}:\tau$.

(*3*) : By induction, $\Gamma \vdash M\,\langle x=N\rangle\,\overrightarrow{P}:\sigma$, and, by Lem. 15:1 & 16:2, there exists $\Gamma' \leq \Gamma$ such that $\Gamma' \vdash (\lambda x.M)N\overrightarrow{P}:\sigma$.

(*4*) : By induction, $\Gamma \vdash (U\,\langle x=N\rangle)(V\,\langle x=N\rangle)\,\overrightarrow{\langle z=Q\rangle}\,\overrightarrow{P}:\sigma$, and, by Lem. 15:2 & 16, $\Gamma \vdash (UV)\langle x=N\rangle\,\overrightarrow{\langle z=Q\rangle}\,\overrightarrow{P}:\sigma$.

(*5*) : By induction, $\Gamma \vdash (\lambda y.M\,\langle x=N\rangle)\,\overrightarrow{\langle z=Q\rangle}\,\overrightarrow{P}:\sigma$, and, by Lem. 15:3 & 16, $\Gamma \vdash (\lambda y.M)\,\langle x=N\rangle\,\overrightarrow{\langle z=Q\rangle}\,\overrightarrow{P}:\sigma$.

(*6*) : This case follows immediately from Lem. 16 and rule (*cut***I**).

(*7*) : If $x \notin \mathit{fv}(M)$, then $x \notin \mathit{pfv}(M)$. By induction, $\Gamma \vdash M\,\overrightarrow{\langle z=Q\rangle}\,\overrightarrow{P}:\sigma$ and $\Delta \vdash N:\tau$, and, by Lem. 15:4 & 16, $\Gamma \vdash M\,\langle x=N\rangle\,\overrightarrow{\langle z=Q\rangle}\,\overrightarrow{P}:\sigma$. ■

Notice that, since subject expansions preserving strong normalisation do not necessarily preserve types, we can only assure that the expanded term is always typeable by the theorem above. For example, we can derive $\emptyset \vdash \lambda y.(\lambda z.z)\,\langle x=yy\rangle:\varphi\to\sigma\to\sigma$ where φ is an atom, but we cannot derive $\emptyset \vdash \lambda y.(\lambda xz.z)(yy):\varphi\to\sigma\to\sigma$. A typing for the last term is for example: $\emptyset \vdash \lambda y.(\lambda xz.z)(yy):\tau\cap(\tau\to\rho)\to\sigma\to\sigma$.

4 All Typeable Terms Are Strongly Normalisable

The general idea of the reducibility method is to interpret types by suitable sets (saturated and stable sets for Tait [25] and Krivine [16] and admissible relations for Mitchell [20, 21]) of terms (*reducible terms*) which satisfy the required property (e.g. strong normalisation) and then to develop semantics in order to obtain the soundness of the type assignment. A consequence of soundness, the fact that every term typeable by a type in the type system belongs to the interpretations of that type, leads to the fact that terms typeable in the type system satisfy the required property, since the type interpretations are built up in that way.

In order to develop the reducibility method we consider the applicative structure whose domain are the terms in Λx and where the application is just the application of terms.

Definition 18 (Reducible terms).

1. We define the collection of set of terms $\mathcal{R}^\rho$ inductively over types by:

$$\begin{aligned} \mathcal{R}^\varphi &= \mathcal{SN} \\ \mathcal{R}^{\sigma\to\tau} &= \{M \mid \forall N \in \mathcal{R}^\sigma\,[MN \in \mathcal{R}^\tau]\} \\ \mathcal{R}^{\sigma\cap\tau} &= \mathcal{R}^\sigma \cap \mathcal{R}^\tau. \end{aligned}$$

2. We define the set $\mathcal{R}$ of *reducible terms* by: $\mathcal{R} = \{M \mid \exists\rho\,[M \in \mathcal{R}^\rho]\} = \bigcup_{\rho\in\mathcal{T}} \mathcal{R}^\rho$.

Notice that, if $M \in \mathcal{R}^{\sigma}$, not necessarily there exists a Γ such that $\Gamma \vdash M:\sigma$. For example, if φ, φ' are two different type variables, then $\lambda x.x \in \mathcal{R}^{\varphi\to\varphi'}$, since $(\lambda x.x)M \in \mathcal{SN}$ whenever $M \in \mathcal{SN}$, but we cannot derive $\emptyset \vdash \lambda x.x:\varphi\to\varphi'$. Also, since $\lambda x.x \in \mathcal{SN}$, $\lambda x.x \in \mathcal{R}^{\varphi}$, but we cannot derive $\emptyset \vdash \lambda x.x:\varphi$.

We now show that reducibility implies strongly normalisability and that all term-variables are reducible. For the latter, we need to show that all typeable strongly normalisable terms that start with a term variable are reducible.

Lemma 19. *1.* $\mathcal{R} \subseteq \mathcal{SN}$.
2. $x\overrightarrow{N} \in \mathcal{SN} \Rightarrow \forall \rho\,[x\overrightarrow{N} \in \mathcal{R}^{\rho}]$.

Proof: By simultaneous induction on the structure of types.

1. (φ) : By Def. 18.
$(\sigma\to\tau)$: $M \in \mathcal{R}^{\sigma\to\tau} \Rightarrow (IH{:}2)\ M \in \mathcal{R}^{\sigma\to\tau}\ \&\ x \in \mathcal{R}^{\sigma} \Rightarrow (18)$
$Mx \in \mathcal{R}^{\tau} \Rightarrow (IH{:}1)\ Mx \in \mathcal{SN} \Rightarrow M \in \mathcal{SN}$.
$(\sigma\cap\tau)$: $M \in \mathcal{R}^{\sigma\cap\tau} \Rightarrow (18)\ M \in \mathcal{R}^{\sigma}\ \&\ M \in \mathcal{R}^{\tau} \Rightarrow (IH{:}1)\ M \in \mathcal{SN}$.
2. (φ) : $x\overrightarrow{N} \in \mathcal{SN} \Rightarrow (18)\ x\overrightarrow{N} \in \mathcal{R}^{\varphi}$.
$(\sigma\to\tau)$: $x\overrightarrow{N} \in \mathcal{SN} \Rightarrow (6{:}1)$
$\forall M \in \mathcal{SN}\,[x\overrightarrow{N}M \in \mathcal{SN}] \Rightarrow (IH{:}1)$
$\forall M \in \mathcal{R}^{\sigma}\,[x\overrightarrow{N}M \in \mathcal{SN}] \Rightarrow (IH{:}2)$
$\forall M \in \mathcal{R}^{\sigma}\,[x\overrightarrow{N}M \in \mathcal{R}^{\tau}] \Rightarrow (18)\quad x\overrightarrow{N} \in \mathcal{R}^{\sigma\to\tau}$
$(\sigma\cap\tau)$: $x\overrightarrow{N} \in \mathcal{SN} \Rightarrow (IH{:}2)\ x\overrightarrow{N} \in \mathcal{R}^{\sigma}\ \&\ x\overrightarrow{N} \in \mathcal{R}^{\tau} \Rightarrow (18)\ x\overrightarrow{N} \in \mathcal{R}^{\sigma\cap\tau}$. ■

We will now show that the reducibility predicate is closed for subject expansion – that preserves strong normalisation – with respect to the reduction rules (B), (App), (Abs), (VarI), (gc_p) and with respect to distribution of substitution. These results are needed in the proof of Theorem 22.

Lemma 20. *1. If* $M\,\overrightarrow{\langle x=N\rangle}\,\overrightarrow{Q} \in \mathcal{R}^{\mu}$, *then* $(\lambda x.M)N\overrightarrow{Q} \in \mathcal{R}^{\mu}$.
2. *If* $(M_1\,\overrightarrow{\langle x=N\rangle})(M_2\,\overrightarrow{\langle x=N\rangle})\,\overrightarrow{Q} \in \mathcal{R}^{\mu}$, *then* $(M_1M_2)\,\overrightarrow{\langle x=N\rangle}\,\overrightarrow{Q} \in \mathcal{R}^{\mu}$.
3. *If* $(\lambda y.M'\,\overrightarrow{\langle x=N\rangle})\,\overrightarrow{P} \in \mathcal{R}^{\mu}$ *and* $y \notin fv(\overrightarrow{N})$, *then* $(\lambda y.M')\,\overrightarrow{\langle x=N\rangle}\,\overrightarrow{P} \in \mathcal{R}^{\mu}$.
4. *If* $N\,\overrightarrow{\langle z=Q\rangle}\,\overrightarrow{P} \in \mathcal{R}^{\mu}$, *then* $x\,\langle x=N\rangle\,\overrightarrow{\langle z=Q\rangle}\,\overrightarrow{P} \in \mathcal{R}^{\mu}$.
5. *If* $M\,\overrightarrow{\langle z=Q\rangle}\,\overrightarrow{P} \in \mathcal{R}^{\mu}$, $N \in \mathcal{SN}$, $x \notin pfv(M)$, *then* $M\,\langle x=N\rangle\,\overrightarrow{\langle z=Q\rangle}\,\overrightarrow{P} \in \mathcal{R}^{\mu}$.
6. *If* $(P\,\overrightarrow{\langle x=N\rangle}\,\langle y=Q\,\overrightarrow{\langle x=N\rangle}\rangle)\,\overrightarrow{M} \in \mathcal{R}^{\mu}$, *then* $((P\,\langle y=Q\rangle)\,\overrightarrow{\langle x=N\rangle})\,\overrightarrow{M} \in \mathcal{R}^{\mu}$.

Proof: By induction on the structure of types.

(φ) : The proofs of these properties are all very similar, using Lem. 6:1 for the first four parts, Lem. 6:2 & 3 for part 5, and Lem. 7 for part 6. So it suffices to show this last case.

$$\begin{array}{ll}(P\,\overrightarrow{\langle x=N\rangle}\,\langle y=Q\,\overrightarrow{\langle x=N\rangle}\rangle)\overrightarrow{M} \in \mathcal{R}^{\varphi} & \Rightarrow (18)\\ (P\,\overrightarrow{\langle x=N\rangle}\,\langle y=Q\,\overrightarrow{\langle x=N\rangle}\rangle)\overrightarrow{M} \in \mathcal{SN} & \Rightarrow (\text{Lem. }7)\\ ((P\,\langle y=Q\rangle)\,\overrightarrow{\langle x=N\rangle})\overrightarrow{M} \in \mathcal{SN} & \Rightarrow (18)\\ ((P\,\langle y=Q\rangle)\,\overrightarrow{\langle x=N\rangle})\overrightarrow{M} \in \mathcal{R}^{\varphi}. & \end{array}$$

$(\sigma{\rightarrow}\tau)$: We consider again part 6, the other parts being similar.

$$\begin{array}{ll} (P\langle \overrightarrow{x=N}\rangle \langle y=Q\langle \overrightarrow{x=N}\rangle\rangle)\overrightarrow{M} \in \mathcal{R}^{\sigma\to\tau} & \Rightarrow (18)\\ \forall R\in\mathcal{R}^{\sigma}\, [(P\langle \overrightarrow{x=N}\rangle \langle y=Q\langle \overrightarrow{x=N}\rangle\rangle)\overrightarrow{M}R\in\mathcal{R}^{\tau}] & \Rightarrow (IH)\\ \forall R\in\mathcal{R}^{\sigma}\, [((P\langle y=Q\rangle)\langle \overrightarrow{x=N}\rangle)\overrightarrow{M}R\in\mathcal{R}^{\tau}] & \Rightarrow (18)\\ ((P\langle y=Q\rangle)\langle \overrightarrow{x=N}\rangle)\overrightarrow{M}\in\mathcal{R}^{\sigma\to\tau}. & \end{array}$$

$(\sigma\cap\tau)$: For all properties this case is immediate by Def. 18 and induction. ■

We shall prove our strong normalisation result by showing that every typeable term is reducible. For this, we need to prove a stronger property: we will show that if we substitute term variables by reducible terms in a typeable term, then we obtain a reducible term. This gives the soundness of our type interpretation.

Theorem 21 (Soundness). If $\{x_1{:}\mu_1,\ldots,x_n{:}\mu_n\}\vdash M:\sigma$, *and, for* $1\le i\le n$, $N_i\in\mathcal{R}^{\mu_i}$, *such that* $x_i\notin fv(N_j)$, *for all* $1\le i,j\le n$, *then* $M\langle\overrightarrow{x=N}\rangle\in\mathcal{R}^{\sigma}$.

Proof: By induction on the structure of derivations. Let $\Gamma=\{x_1{:}\mu_1,\ldots,x_n{:}\mu_n\}$.

(Ax) : Then $M\equiv x_j$, and $\mu_j=\sigma$, for some $1\le j\le n$. Since $N_j\in\mathcal{R}^{\mu_j}$, $N_j\in\mathcal{R}^{\sigma}$. Then, by Lem. 20:4 & 5, $x_j\langle\overrightarrow{x=N}\rangle\in\mathcal{R}^{\sigma}$.

$(\to I)$: Then $M\equiv\lambda y.M'$, $\sigma=\rho{\rightarrow}\tau$, and $B,y{:}\rho\vdash M':\tau$. Let $N\in\mathcal{R}^{\rho}$, then, by induction, $M'\langle\overrightarrow{x=N}\rangle\langle y=N\rangle\in\mathcal{R}^{\tau}$. So, by Lem. 20:1, $(\lambda y.M'\langle\overrightarrow{x=N}\rangle)N\in\mathcal{R}^{\tau}$, and, by Def. 18, $\lambda y.M'\langle\overrightarrow{x=N}\rangle\in\mathcal{R}^{\rho\to\tau}$. We can assume $y\notin fv(N)$, so, by Lem. 20:3, $(\lambda y.M')\langle\overrightarrow{x=N}\rangle\in\mathcal{R}^{\rho\to\tau}$.

$(\to E)$: Then $M\equiv M_1M_2$ and there exists τ such that $\Gamma\vdash M_1:\tau{\rightarrow}\sigma$ and $\Gamma\vdash M_2:\tau$. By induction, $M_1\langle\overrightarrow{x=N}\rangle\in\mathcal{R}^{\tau\to\sigma}$ and $M_2\langle\overrightarrow{x=N}\rangle\in\mathcal{R}^{\tau}$. But then, by Def. 18, $M_1\langle\overrightarrow{x=N}\rangle M_2\langle\overrightarrow{x=N}\rangle\in\mathcal{R}^{\sigma}$, so, by Lem. 20:2, $(M_1M_2)\langle\overrightarrow{x=N}\rangle\in\mathcal{R}^{\sigma}$.

$(\cap I)$: Then $\sigma\equiv\sigma_1\cap\sigma_2$ and, for $i\in\underline{2}$, $\Gamma\vdash M:\sigma_i$. So, by induction, $M\langle\overrightarrow{x=N}\rangle\in\mathcal{R}^{\sigma_1}$ and $M\langle\overrightarrow{x=N}\rangle\in\mathcal{R}^{\sigma_2}$, so, by Def. 18, $M\langle\overrightarrow{x=N}\rangle\in\mathcal{R}^{\sigma}$.

$(\cap E)$: Then there exists τ such that $\Gamma\vdash M:\sigma\cap\tau$, and, by induction, $M\langle\overrightarrow{x=N}\rangle\in\mathcal{R}^{\sigma\cap\tau}$. Then, by Def. 18, $M\langle\overrightarrow{x=N}\rangle\in\mathcal{R}^{\sigma}$.

$(cut\mathbf{I})$: Then $M\equiv P\langle y=Q\rangle$, and there exists τ such that $\Gamma,y{:}\tau\vdash P:\sigma$ and $\Gamma\vdash Q:\tau$. Then, by induction, $Q\langle\overrightarrow{x=N}\rangle\in\mathcal{R}^{\tau}$, so, again by induction, $P\langle\overrightarrow{x=N}\rangle\langle y=Q\langle\overrightarrow{x=N}\rangle\rangle\in\mathcal{R}^{\sigma}$. So, by Lem. 20:6, $(P\langle y=Q\rangle)\langle\overrightarrow{x=N}\rangle\in\mathcal{R}^{\sigma}$.

$(cut\mathbf{K})$: Then $M\equiv P\langle y=Q\rangle$, $\Gamma\vdash P:\sigma$, $y\notin\Gamma$ and there exist Δ,τ such that $\Delta\vdash Q:\tau$. By induction $P\langle\overrightarrow{x=N}\rangle\in\mathcal{R}^{\sigma}$. Let $\Delta=\{z_1{:}\rho_1,\ldots,z_m{:}\rho_m\}$. By Lem. 19:2, for all $j\in\underline{m}$, $z_j\in\mathcal{R}^{\rho_j}$, and therefore, by induction, $Q\langle\overrightarrow{z=z}\rangle\in\mathcal{R}^{\tau}$. By Lem. 19:1 we get $Q\langle\overrightarrow{z=z}\rangle\in\mathcal{SN}$, which implies $Q\in\mathcal{SN}$. By Lem. 12:4, $y\notin pfv(P)$. So, by Lem. 20:5, we conclude $(P\langle y=Q\rangle)\langle\overrightarrow{x=N}\rangle\in\mathcal{R}^{\sigma}$. ■

Theorem 22. If $\Gamma\vdash M:\sigma$ *for some* Γ,σ *then* $M\in\mathcal{SN}$.

Proof: By Lem. 19:2, all term variables are reducible for any type, so, by Thm. 21, for all M, $M\langle\overrightarrow{x=x}\rangle$ is reducible. Strong normalisation of $M\langle\overrightarrow{x=x}\rangle$ then follows from Lem. 19:1. Since $M\langle\overrightarrow{x=x}\rangle\to M$, also $M\in\mathcal{SN}$. ■

5 Final Remarks

The only difference between the present type assignment system and that one of [18] is the formulation of the (*cut*K)-rule which there becomes:

$$\frac{\Gamma \vdash M:\tau \quad \Delta \vdash N:\sigma}{\Gamma \vdash M\langle x=N\rangle:\tau}\ (x \notin \mathit{pfv}(M))$$

By Lem. 12:4 & 5 the two systems deduce the same types for the same terms. This was first observed by Dougherty and Lescanne (private communication). Deeper relations between the two systems will be object of further investigations.

If we add an universal type Ω with the axiom (Ω) [9]

$$\frac{}{\Gamma \vdash M:\Omega}$$

then in the so obtained system, rule (*cut*K) becomes admissible. But using axiom (Ω) we can type trivially all terms, and we can also type terms which are not strongly normalising with types different from Ω. An example is $y{:}\varphi \vdash (\lambda x.y)((\lambda z.zz)(\lambda z.zz)):\varphi$. So we cannot obtain a characterisation of the set $\mathcal{SN}$.

We like to point out that the fact that the calculus we considered enjoys the preservation of strong normalization does not trivialize our result. In fact there are non-strongly normalizing λ-terms which reduce to strongly normalizing Λx-terms which are not λ-terms, see Example 11.

In the line of [10], we plan to characterize other compositional properties of calculi of explicit substitutions by means of intersection type disciplines, like that of reducing to a closed term or of being a weak head-normalising term.

Finally, we will explore the possibility of building suitable intersection type assignment systems in which it will be possible to show that some calculi of explicit substitutions enjoy the weak or strong normalisation property.

Ackowlegments We are very grateful to Dan Dougherty, Stéphane Lengrand and Pierre Lescanne for many crucial suggestions and in particular to Pierre Lescanne for pointing out a mistake in a previous version of the present paper. We thank also the referees for their useful comments.

References

[1] M. Abadi, L. Cardelli, P.-L. Curien, and J.-J. Lévy. Explicit substitutions. *Journal of Functional Programming*, 1(4):375–416, 1991.

[2] R. M. Amadio and P.-L. Curien. *Domains and lambda-calculi*. Cambridge University Press, Cambridge, 1998.

[3] S. van Bakel. Complete restrictions of the intersection type discipline. *Theoretical Computer Science*, 102(1):135–163, 1992.

[4] S. van Bakel. Intersection Type Assingment Systems. *Theoretical Computer Science*, 151(2):385–435, 1995.

[5] H. Barendregt, M. Coppo, and M. Dezani-Ciancaglini. A filter lambda model and the completeness of type assignment. *The Journal of Symbolic Logic*, 48(4):931–940, 1983.

[6] Z. Benaissa, D. Briaud, P. Lescanne, and J. Rouyer-Degli. λv, a calculus of explicit substitutions which preserves strong normalization. *Journal of Functional Programming*, 6(5):699–722, 1996.

[7] R. Bloo and K. Rose. Preservation of strong normalization in named lambda calculi with explicit substitution and garbage collection. In *Computer Science in the Netherlands*, pages 62–72. Koninklijke Jaarbeurs, 1995.

[8] M. Coppo and M. Dezani-Ciancaglini. An extension of the basic functionality theory for the λ-calculus. *Notre Dame Journal of Formal Logic*, 21(4):685–693, 1980.

[9] M. Coppo, M. Dezani-Ciancaglini, and B. Venneri. Principal type schemes and λ-calculus semantics. In *To H. B. Curry: essays on combinatory logic, lambda calculus and formalism*, pages 535–560. Academic Press, London, 1980.

[10] M. Dezani-Ciancaglini, F. Honsell, and Y. Motohama. Compositional characterization of λ-terms using intersection types. In *Mathematical Foundations of Computer Science 2000*, volume 1893 of *Lecture Notes in Computer Science*, pages 304–313. Springer, 2000.

[11] D. Dougherty and P. Lescanne. Reductions, intersection types and explicit substitutions. In *Typed Lambda Calculi and Applications 2001*, volume 2044 of *Lecture Notes in Computer Science*, pages 121–135. Springer, 2001.

[12] J. Gallier. Typing untyped λ-terms, or reducibility strikes again! *Annals of Pure and Applied Logic*, 91:231–270, 1998.

[13] S. Ghilezan. Strong normalization and typability with intersection types. *Notre Dame Journal of Formal Logic*, 37(1):44–52, 1996.

[14] J.-Y. Girard. Une extension de l'interprétation de Gödel à l'analyse, et son application à l'elimination des coupures dans l'analyse et la théorie des types. In *2nd Scandinavian Logic Symposium*, pages 63–92. North-Holland, 1971.

[15] F. Kamareddine and A. Rios. Extending a λ-calculus with explicit substitutions which preserves strong normalization into a confluent calculus of open terms. *Journal of Functional Programming*, 7(4):395–420, 1997.

[16] J.-L. Krivine. *Lambda-calcul Types et modèles*. Masson, Paris, 1990.

[17] D. Leivant. Typing and computational properties of lambda expressions. *Theoretical Computer Science*, 44(1):51–68, 1986.

[18] S. Lengrand, D. Dougherty, and P. Lescanne, An improved system of intersection types for explicit substitutions, Internal Report, ENS de Lyon, 2001.

[19] P.-A. Melliès. Typed λ-calculi with explicit substitution may not terminate. In *Typed Lambda Calculi and Applications 2001*, volume 902 of *Lecture Notes in Computer Science*, pages 328–334. Springer, 1995.

[20] J. C. Mitchell. Type systems for programming languages. In *Handbook of Theoretical Computer Science*, volume B, pages 415–431. Elsevire, Amsterdam, 1990.

[21] J. C. Mitchell. *Foundation for Programmimg Languages*. MIT Press, 1996.

[22] G. Pottinger. A type assignment for the strongly normalizable λ-terms. In *To H. B. Curry: essays on combinatory logic, lambda calculus and formalism*, pages 561–577. Academic Press, London, 1980.

[23] E. Ritter. Characterizing explicit substitutions which preserve termination. In *Typed Lambda Calculi and Applications 1999*, volume 1581 of *Lecture Notes in Computer Science*, pages 325–339. Springer, 1999.

[24] P. Severi. *Normalisation in lambda calculus and its relation to type inference*. PhD thesis, Eindhoven University of Technology, 1996.

[25] W. W. Tait. Intensional interpretations of functionals of finite type I. *Journal of Symbolic Logic*, 32:198–212, 1967.

[26] W. W. Tait. A realizability interpretation of the theory of species. In *Logic Colloquium*, volume 453 of *Lecture Notes in Mathematics*, pages 240–251. Springer, 1975.

Parameters in Pure Type Systems

Roel Bloo[1]*, Fairouz Kamareddine[2], Twan Laan[3], and Rob Nederpelt[1]

[1] Eindhoven University of Technology, Computing science,
PO-Box 513, 5600 MB Eindhoven, The Netherlands,
c.j.bloo@tue.nl, r.p.nederpelt@tue.nl

[2] Heriot-Watt University, dept. of Computing and Electrical Eng., Edinburgh, Scotland,
fairouz@cee.hw.ac.uk

[3] twan.laan@wxs.nl

Abstract. In this paper we study the addition of parameters to typed λ-calculus with definitions. We show that the resulting systems have nice properties and illustrate that parameters allow for a better fine-tuning of the strength of type systems as well as staying closer to type systems used in practice in theorem provers and programming languages.

1 What Are Parameters?

Parameters occur when functions are only allowed to occur when provided with arguments. As we will show below, both in mathematics and in programming languages the use of parameters is abundant and closely connected to the use of constants and definitions. If we want to be able to use type systems in accordance with practice and yet described in a precise manner, we therefore need parameters, constants, and definitions in type theory as well.

Parameters, constants and definitions in theorem proving It is interesting to note that the first tool for mechanical representation and verification of mathematical proofs, AUTOMATH, already has a combined constant, definition and parameter mechanism and was developed from the viewpoint of mathematicians (see [6]). The representation of a mathematical text in AUTOMATH consists of a finite list of *lines* where every line has the format: $x_1 : A_1, \ldots, x_n : A_n \vdash g(x_1, \ldots, x_n) = t : T$. Here g is a new name; a constant if t is the acronym 'primitive notion', or an abbreviation (definition) for the expression t of type T, and $x_1, \ldots, x_n$ are the parameters of g, with respective types $A_1, \ldots, A_n$. Use of the definition g in the rest of the list of lines is only allowed when g is supplied with a list of arguments $t_1, \ldots, t_n$ (with types conforming to $A_1, \ldots, A_n$, see again [6]) to be substituted for the parameters of g.

We see that parameters and definitions are a very substantial part of AUTOMATH since each line introduces a new constant or definition which is *inherently parameterized* by the variables occurring in the context needed for it. Actual development of ordinary mathematical theory in the AUTOMATH system by e.g. van Benthem Jutting (cf. [2]) revealed that this combined definition and parameter mechanism is vital for keeping proofs manageable and sufficiently readable for humans.

* Contact author.

S. Rajsbaum (Ed.): LATIN 2002, LNCS 2286, pp. 371–385, 2002.

Similar but more recent experience with the Coq proof system [9] suggests the same necessity of parameterized definitions, and indeed, state of the art theorem provers have parameter mechanisms as well.

There is another advantage to the use of parameters. Allowing only parameters of a certain type and not the corresponding abstractions may yield a weaker type system. This can have advantages such as a first-order system instead of a higher-order one, or a simpler typecheck algorithm as has been observed for the type system $\lambda P-$ in [13].

Parameters, constants and definitions in programming languages Most non-assembly level programming languages have parameterized definitions as part of the syntax. Consider the Pascal definition of a function double:

```
function  double(z : Integer) : Integer;
begin
    double := z + z
end;
```

The argument (`z : Integer`) is a parameter in our sense: the function `double` can only be used when given an argument, ergo `double` is a non-abstracted function. In ordinary λ-calculus this function `double` can only be represented by the λ-abstraction $(\lambda x : \texttt{Int}.(x + x))$. This representation is unfaithful since this way `double` is a term on its own, of 'higher-order character' (it can be used without a parameter).

For an example of the use of constants, we consider the programming language ML, which has the basic types `int` and `list`. However, `list` is not just an ordinary constant. It can only be used when given an argument (which is written prefix in ML) as in `int list`. We see that `list` is in fact a parameterized constant.

1.1 Extending Pure Type Systems with Parameters, Constants, and Definitions

There are many other examples of the frequent use of parameters in mathematics and computer science, occurring in combination with both definitions and constants. The general framework used to describe type systems, Pure Type Systems (PTSs, [1]), does not possess constants[1] or parameters nor does it have syntax for definitions. Therefore we set out to extend pure type systems with parameters, constants and definitions in order to better be able to describe type systems used in practice. This work is based on the parameterized type systems of Laan in [11], although there are several subtle differences[2] in the precise definition of the system. We first discuss work that has already been done in this direction.

Two approaches are known for extending type theory with definitions. The first, by Severi and Poll, extends the syntax of λ-terms to include definitions [15]. The second only extends PTSs with global definitions (i.e., definitions in the context of derivations), and treats local definitions as ordinary β-redexes [4].

[1] The role of unparameterized constants is usually imitated by variables, by agreeing not to make any abstraction over such variables. This is done in ordinary PTSs for the sorts $*$ and $\Box$. Another extension of PTSs in which constants play an essential role are Modal PTSs, cf. [5].

[2] Definition 1 is slightly different, the rules (C^p-weak) and (D^p-weak) are now correct and we have a different treatment of topsorts in the rules for definitions.

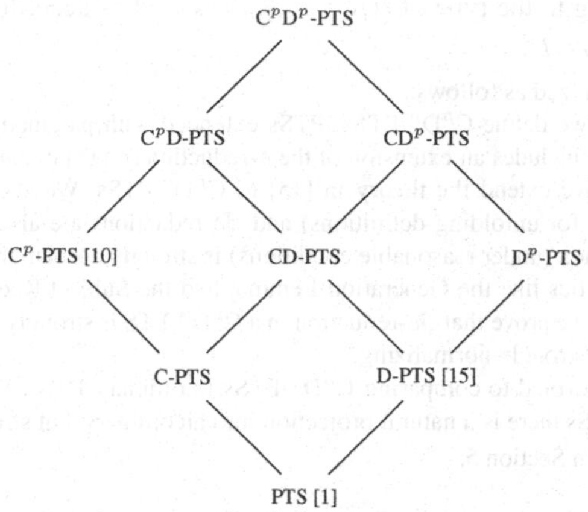

Fig. 1. The hierarchy of parameters, constants and definitions

Both these approaches fail to model the definition of the double function from section 1 above, since they don't have parameterized definitions. The best they can do is adding $\texttt{double} = (\lambda x{:}\texttt{Int}.x + x) : \Pi x{:}\texttt{Int}.\texttt{Int}$ to the context, which clearly isn't in the spirit of Pascal where double on its own is not a valid expression.

We shall call the extension of PTSs with unparameterized definitions D-PTSs. In this paper, we go one step further and introduce D^p-PTSs, *pure type systems extended with parameterized definitions*, so that we can imitate double by adding $\texttt{double}(z : \texttt{Int}) = z + z : \texttt{Int}$ to the context. This will *not* allow the use of double unless it is provided an argument for its parameter z. This is an extension of the work of [15] on unparameterized definitions.

Orthogonally, one can extend PTSs with parameterized constants as has been studied in [10]. We shall call these systems C^p-PTSs. Similar to the extension of PTSs with unparameterized definitions, one might consider PTSs extended with unparameterized constants only (C-PTSs). Although C-PTSs are not very interesting on their own, we include them here for symmetry reasons.

Combining the various extensions, we obtain a hierarchy that can be depicted as in Figure 1.

In this paper we study the top system in Figure 1, that is, PTSs extended with parameterized constants as well as parameterized definitions.

Similar to the restrictions on the formation of abstractions in ordinary PTSs, it is natural to put restrictions on the formation of parameters as well. Although an unrestricted use of parameters may seem elegant from a theoretical point of view, this is not custom in programming languages. For instance, in many Pascal versions, parametric *terms* can only have parameters at *term level*. Therefore, in the C^pD^p-PTSs we study in this paper, in a parameterized term $t(p_1, \ldots, p_n)$ we might want to restrict its formation according to the type of $t(p_1, \ldots, p_n)$ as well as according to the types of the parameters $p_1, \ldots, p_n$.

This paper is organized as follows:

In Section 2, we define C^{p}D^p-PTSs, PTSs extended with parametric constants and definitions. This section includes an extension of the δ-reduction of [15] to parametric definitions.

In Section 3 we extend the theory in [15] to C^{p}D^p-PTSs. We show that our extended δ-reduction (needed for unfolding definitions) and $\beta\delta$-reductions are also confluent, and that the extended δ-reduction (under reasonable conditions) is strongly normalizing. Then we show some elementary properties like the Generation Lemma, and the Subject Reduction property for $\beta\delta$-reduction. Finally we prove that $\beta\delta$-reduction in a C^{p}D^p-PTS is strongly normalizing if a slightly stronger PTS is β-strongly normalizing.

Section 4 is devoted to comparing C^{p}D^p-PTSs to ordinary PTSs. We show that for a large class of C^{p}D^p-PTSs there is a natural projection into an ordinary, but stronger, PTS.

We conclude in Section 5.

2 Extending PTSs with Parametric Constants and Definitions

In this section, we extend Pure Type Systems (PTSs) (cf. [1]) with parameterized constants and parameterized definitions.

Pure Type Systems (PTSs) were introduced by Berardi [3] and Terlouw [16] as a general framework in which many current type systems can be described. Though PTSs were not introduced before 1988, many rules in PTSs are highly influenced by rules of known type systems like Church's Simple Theory of Types [7] and AUTOMATH (see 5.5.4. of [8]). The description of our extension of PTSs with parametric constants and definitions is based on the description of PTSs in [1].

Definition 1 Let $\mathcal{V}$, $\mathcal{C}$ and $\mathcal{S}$ be disjoint sets of respectively variables, constants and sorts.[3] The set $\mathcal{T}_P$ of *parametric terms* is defined together with the set $\mathcal{L}_\mathcal{V}$ of *lists of typed variables* and the set $\mathcal{L}_\mathcal{T}$ of *lists of terms* by:

$$\mathcal{T}_P ::= \mathcal{V} \mid \mathcal{S} \mid \mathcal{C}(\mathcal{L}_\mathcal{T}) \mid (\mathcal{T}_P\mathcal{T}_P) \mid (\lambda\mathcal{V}{:}\mathcal{T}_P.\mathcal{T}_P) \mid (\Pi\mathcal{V}{:}\mathcal{T}_P.\mathcal{T}_P) \mid (\mathcal{C}(\mathcal{L}_\mathcal{V}){=}\mathcal{T}_P{:}\mathcal{T}_P \texttt{ in } \mathcal{T}_P);$$

$$\mathcal{L}_\mathcal{V} ::= \varnothing \mid \langle \mathcal{L}_\mathcal{V}, \mathcal{V}{:}\mathcal{T}_P \rangle; \qquad \mathcal{L}_\mathcal{T} ::= \varnothing \mid \langle \mathcal{L}_\mathcal{T}, \mathcal{T}_P \rangle.$$

Instead of $\langle \cdots \langle \langle \varnothing, x_1{:}A_1 \rangle, x_2{:}A_2 \rangle \cdots x_n{:}A_n \rangle$, we write $\langle x_1{:}A_1, \ldots, x_n{:}A_n \rangle$ or $x_1{:}A_1, \ldots, x_n{:}A_n$. A similar convention is adopted for lists of terms. In a parametric term of the form $c(b_1, \ldots, b_n)$, the subterms $b_1, \ldots, b_n$ are called the *parameters* of the term.

Terms of the form $\mathcal{C}(\mathcal{L}_\mathcal{V}){=}\mathcal{T}_P{:}\mathcal{T}_P \texttt{ in } \mathcal{T}_P$ represent parametric local definitions. An example of such a term is `double(x:Int)=(x+x):Int in` A which indicates that a subterm of A of the form `double`(P) is to be interpreted as $P + P$, and has type `Int`. The definition is local, that is: the scope of the definition is the term A. Local definitions contrast with global definitions which are given in a context Γ, and refer to any term that is considered within Γ (see Definition 5). The definition system in AUTOMATH is similar to the system of global definitions in this paper. However, there are no local definitions in AUTOMATH.

[3] Note that, in contrast to PTSs, we require the set of sorts to be disjoint from the set of (parametric) constants.

Definition 2 Let $\overrightarrow{x}{:}\overrightarrow{A}$ denote the list $x_1{:}A_1, \ldots, x_n{:}A_n$. $FV(A)$, the set of *free variables* of a parametric term A is defined as usual with the extra cases for constants and definitions:

$FV(c(a_1, \ldots, a_n)) = \bigcup_{i=1}^{n} FV(a_i)$; and

$FV(c(\overrightarrow{x}{:}\overrightarrow{A}){=}A{:}B \texttt{ in } C) =$

$\bigcup_{i=1}^{n}(FV(A_i) \setminus \{x_1, \ldots, x_{i-1}\}) \cup ((FV(A) \cup FV(B)) \setminus \{x_1, \ldots, x_n\}) \cup FV(C)$.

We similarly define $Cons(A)$, the set of *constants and global definitions* of A as follows:

$Cons(s) = Cons(x) = \emptyset$;

$Cons(c(a_1, \ldots, a_n)) = \{c\} \cup \bigcup_{i=1}^{n} Cons(a_i)$;

$Cons(AB) = Cons(\lambda x{:}A.B) = Cons(\Pi x{:}A.B) = Cons(A) \cup Cons(B)$;

$Cons(c(\overrightarrow{x}{:}\overrightarrow{A}){=}A{:}B \texttt{ in } C) =$

$\bigcup_{i=1}^{n} Cons(A_i) \cup Cons(A) \cup Cons(B) \cup (Cons(C) \setminus \{c\})$.

$FV(A) \cup Cons(A)$ forms the *domain* $Dom(A)$ of A.

We omit parentheses in parametric terms when possible. As usual in PTSs (cf. [1]), we do not distinguish terms that are equal up to renaming of bound variables. Moreover, we assume the Barendregt variable convention so that names of bound variables and constants will always be chosen such that they differ from the free ones in a term.

Definition 3 We extend the usual definition of substitution of a term a for a variable x in a term b, $b[x{:=}a]$, to parametric terms, assuming that x is not a bound variable of either b or a:

$$c(b_1, \ldots, b_n)[x{:=}a] \equiv c(b_1[x{:=}a], \ldots, b_n[x{:=}a]);$$

$$(c(\overrightarrow{x}{:}\overrightarrow{A}) = A{:}B \texttt{ in } C)[x{:=}a] \equiv$$

$$c(x_1{:}A_1[x{:=}a], \ldots, x_n{:}A_n[x{:=}a]) = A[x{:=}a]{:}B[x{:=}a] \texttt{ in } C[x{:=}a].$$

Definition 4 The set $\mathcal{C}_P$ of *contexts*, which we denote by $\Gamma, \Gamma', \ldots$, is given by:

$$\mathcal{C}_P ::= \varnothing \mid \langle \mathcal{C}_P, \mathcal{V}{:}\mathcal{T}_P \rangle \mid \langle \mathcal{C}_P, \mathcal{C}(\mathcal{L}_V){=}\mathcal{T}_P{:}\mathcal{T}_P \rangle \mid \langle \mathcal{C}_P, \mathcal{C}(\mathcal{L}_V){:}\mathcal{T}_P \rangle.$$

Notice that $\mathcal{L}_V \subseteq \mathcal{C}_P$: all lists of variable declarations are contexts as well.

Definition 5 Let Γ be a context. *Declarations* are elements of Γ as follows:

- $x{:}A$ is a *variable declaration* with *subject* x and *type* A;
- $c(x_1{:}B_1, \ldots, x_n{:}B_n){:}A$ is a *constant declaration* with *subject* c (also called *primitive constant), parameters* $x_1, \ldots, x_n$ and *type* A;
- $c(x_1{:}B_1, \ldots, x_n{:}B_n){=}a{:}A$ is a *global definition (declaration)* with *subject* c (also called *globally defined constant), parameters* $x_1, \ldots, x_n$, *definiens* a and *type* A.

Notation In the rest of this paper, Δ denotes a context $x_1{:}B_1, \ldots, x_n{:}B_n$ consisting of variable declarations only. Such a context is typically used as a list of parameters in a definition $c(\Delta){=}a{:}A$. We write $\Delta_i \equiv x_1{:}B_1, \ldots, x_{i-1}{:}B_{i-1}$ for $i \leq n$. We extend the definition of substitution to contexts in the usual way.

Definition 6 For a context Γ we define $FV(\Gamma)$ to be the set of subjects of variable declarations in Γ and $Cons(\Gamma)$ the set of subjects of constant declarations and global definitions in Γ. The *domain* of Γ, $Dom(\Gamma)$, is defined as $FV(\Gamma) \cup Cons(\Gamma)$.

In ordinary PTSs we have that, for a legal term A in a legal context Γ, $FV(A) \subseteq FV(\Gamma)$. In our C^pD^p-PTSs we will have: $FV(A) \subseteq FV(\Gamma)$ and $Cons(A) \subseteq Cons(\Gamma)$.

A natural condition on contexts is that all variables, primitive constants and defined constants are declared only once. Furthermore it is also natural to require that variables and constants are declared *before* they are being used. For this we introduce the notion of *sound* context:

Definition 7 $\Gamma \in \mathcal{C}_P$ is *sound* if variables, primitive constants and defined constants are declared only once and if $\Gamma \equiv \Gamma_1, c(\Delta){=}a{:}A, \Gamma_2$ then $Dom(a) \cup Dom(A) \subseteq Dom(\Gamma_1) \cup Dom(\Delta)$ and for $i = 1, \dots, n$: $Dom(B_i) \subseteq Dom(\Gamma_1, \Delta_i)$.

The contexts occurring in the type systems proposed in this paper are all sound (see Lemma 15). This fact will be useful when proving properties of these systems.

We now start a more detailed description of the top system in Figure 1, the system with both parameterized defined constants and parameterized primitive constants. We define two reduction relations, namely the δ- and β-reduction. β-reduction is defined as usual, and we use $\rightarrow_\beta$, $\twoheadrightarrow_\beta$, $\stackrel{+}{\rightarrow}_\beta$, and $=_\beta$ as usual. As far as global definitions are concerned, δ-reduction is comparable to δ-reduction in AUTOMATH. This is reflected in rule ($\delta 1$) in Figure 2 and Definition 8 below. But now, a δ-reduction step can also unfold *local* definitions. Therefore, two new reduction steps are introduced. Rule ($\delta 2$) removes the declaration of a local definition if there is no position within its scope where it can be unfolded ('removal of void local definitions'). Rule ($\delta 3$) shows how one can treat a local definition as a global definition, and thus how the problem of unfolding local definitions can be reduced to unfolding global definitions ('localization of global definitions'). Remember that $\Delta \equiv x_1{:}B_1, \dots, x_n{:}B_n$.

Definition 8 (δ-reduction) δ-reduction is defined as the smallest relation $\rightarrow_\delta$ on $\mathcal{C}_P \times \mathcal{T}_P \times \mathcal{T}_P$ closed under the rules ($\delta 1$), ($\delta 2$), ($\delta 3$) and the compatibility rules of Figure 2.

$\Gamma \vdash \cdot =_\delta \cdot$ denotes the reflexive, symmetric and transitive closure of $\Gamma \vdash \cdot \rightarrow_\delta \cdot$. When Γ is the empty context, we write $a \rightarrow_\delta a'$ instead of $\Gamma \vdash a \rightarrow_\delta a'$.

Furthermore δ-reduction between contexts is the smallest relation $\rightarrow_\delta$ on $\mathcal{C}_P \times \mathcal{C}_P$ closed under the rules in Figure 3.

Before describing the typing rules for C^pD^p-PTSs, we introduce the concepts of specification (taken from [1]) and parametric specification.

Definition 9 (Specification) A *specification* is a triple $(\boldsymbol{S}, \boldsymbol{A}, \boldsymbol{R})$, such that $\boldsymbol{S} \subseteq \mathcal{C}$, $\boldsymbol{A} \subseteq \boldsymbol{S} \times \boldsymbol{S}$ and $\boldsymbol{R} \subseteq \boldsymbol{S} \times \boldsymbol{S} \times \boldsymbol{S}$. The specification is called *singly sorted* if $\boldsymbol{A}$ is a (partial) function $\boldsymbol{S} \rightarrow \boldsymbol{S}$, and $\boldsymbol{R}$ is a (partial) function $\boldsymbol{S} \times \boldsymbol{S} \rightarrow \boldsymbol{S}$. $\boldsymbol{S}$ is called the set of *sorts*, $\boldsymbol{A}$ is the set of *axioms*, and $\boldsymbol{R}$ is the set of (Π-formation) *rules* of the specification.

A *parametric specification* is a quadruple $(\boldsymbol{S}, \boldsymbol{A}, \boldsymbol{R}, \boldsymbol{P})$ such that $(\boldsymbol{S}, \boldsymbol{A}, \boldsymbol{R})$ is a specification, and the set of *parametric rules* $\boldsymbol{P} \subseteq \boldsymbol{S} \times \boldsymbol{S}$. The parametric specification is called *singly sorted* if the specification $(\boldsymbol{S}, \boldsymbol{A}, \boldsymbol{R})$ is singly sorted.

We first give the typing rules for ordinary terms of PTSs. These can also be found in [1].

$$(\delta 1):\ \Gamma_1, c(\Delta){=}a{:}A, \Gamma_2 \vdash c(b_1,\ldots,b_n) \rightarrow_\delta a[x_i{:=}b_i]_{i=1}^n$$

$$(\delta 2):\ \frac{c \notin Cons(b)}{\Gamma \vdash c(\Delta){=}a{:}A \;\mathtt{in}\; b \rightarrow_\delta b} \qquad (\delta 3):\ \frac{\Gamma, c(\Delta){=}a{:}A \vdash b \rightarrow_\delta b'}{\Gamma \vdash c(\Delta){=}a{:}A \;\mathtt{in}\; b \rightarrow_\delta c(\Delta){=}a{:}A \;\mathtt{in}\; b'}$$

$$\frac{\Gamma, \Delta \vdash a \rightarrow_\delta a'}{\Gamma \vdash c(\Delta){=}a{:}A \;\mathtt{in}\; b \rightarrow_\delta c(\Delta){=}a'{:}A \;\mathtt{in}\; b} \qquad \frac{\Gamma, \Delta \vdash A \rightarrow_\delta A'}{\Gamma \vdash c(\Delta){=}a{:}A \;\mathtt{in}\; b \rightarrow_\delta c(\Delta){=}a{:}A' \;\mathtt{in}\; b}$$

$$\frac{\Gamma, \Delta_i \vdash B_i \rightarrow_\delta B_i'}{\Gamma \vdash c(\Delta){=}a{:}A \;\mathtt{in}\; b \rightarrow_\delta c(x_1{:}B_1,\ldots,x_i{:}B_i',\ldots,x_n{:}B_n){=}a{:}A \;\mathtt{in}\; b}$$

$$\frac{\Gamma \vdash a \rightarrow_\delta a'}{\Gamma \vdash ab \rightarrow_\delta a'b} \qquad \frac{\Gamma \vdash b \rightarrow_\delta b'}{\Gamma \vdash ab \rightarrow_\delta ab'}$$

$$\frac{\Gamma, x{:}A \vdash a \rightarrow_\delta a'}{\Gamma \vdash \lambda x{:}A.a \rightarrow_\delta \lambda x{:}A.a'} \qquad \frac{\Gamma \vdash A \rightarrow_\delta A'}{\Gamma \vdash \lambda x{:}A.a \rightarrow_\delta \lambda x{:}A'.a}$$

$$\frac{\Gamma, x{:}A \vdash a \rightarrow_\delta a'}{\Gamma \vdash \Pi x{:}A.a \rightarrow_\delta \Pi x{:}A.a'} \qquad \frac{\Gamma \vdash A \rightarrow_\delta A'}{\Gamma \vdash \Pi x{:}A.a \rightarrow_\delta \Pi x{:}A'.a}$$

$$\frac{\Gamma \vdash a_j \rightarrow_\delta a_j'}{\Gamma \vdash c(a_1,\ldots,a_n) \rightarrow_\delta c(a_1,\ldots,a_j',\ldots,a_n)}$$

Fig. 2. Reduction rules and compatibility rules for $\rightarrow_\delta$

$$\frac{\Gamma_1 \vdash A \rightarrow_\delta A'}{\Gamma_1, x{:}A, \Gamma_2 \rightarrow_\delta \Gamma_1, x{:}A', \Gamma_2} \qquad \frac{\Gamma_1, \Delta \rightarrow_\delta \Gamma_1, \Delta'}{\Gamma_1, c(\Delta){:}A, \Gamma_2 \rightarrow_\delta \Gamma_1, c(\Delta'){:}A, \Gamma_2}$$

$$\frac{\Gamma_1, \Delta \vdash A \rightarrow_\delta A'}{\Gamma_1, c(\Delta){:}A, \Gamma_2 \rightarrow_\delta \Gamma_1, c(\Delta){:}A', \Gamma_2} \qquad \frac{\Gamma_1, \Delta \vdash a \rightarrow_\delta a'}{\Gamma_1, c(\Delta){=}a{:}A, \Gamma_2 \rightarrow_\delta \Gamma_1, c(\Delta){=}a'{:}A, \Gamma_2}$$

$$\frac{\Gamma_1, \Delta \rightarrow_\delta \Gamma_1, \Delta'}{\Gamma_1, c(\Delta){=}a{:}A, \Gamma_2 \rightarrow_\delta \Gamma_1, c(\Delta'){=}a{:}A, \Gamma_2} \qquad \frac{\Gamma_1, \Delta \vdash A \rightarrow_\delta A'}{\Gamma_1, c(\Delta){=}a{:}A, \Gamma_2 \rightarrow_\delta \Gamma_1, c(\Delta){=}a{:}A', \Gamma_2}$$

Fig. 3. Rules for δ-reduction on contexts

Definition 10 (Typing rules for ordinary terms) Let $\mathcal{S} = (\boldsymbol{S}, \boldsymbol{A}, \boldsymbol{R})$ be a specification. The Pure Type System $\lambda\mathcal{S}$ describes in which ways judgements $\Gamma \vdash_\mathcal{S} A : B$ (or $\Gamma \vdash A : B$, if it is clear which $\mathcal{S}$ is used) can be derived. $\Gamma \vdash A : B$ states that A has type B in context Γ. $\lambda\mathcal{S}$ consists of the derivation rules given in Figure 4.

A term a is *legal* (with respect to a certain type system) if there are Γ, b such that either $\Gamma \vdash a : b$ or $\Gamma \vdash b : a$ is derivable (in that type system). Similarly, a context Γ is *legal* if there are a, b such that $\Gamma \vdash a : b$. A sort $s \in \boldsymbol{S}$ is called *topsort* if there is no context Γ and $s' \in \boldsymbol{S}$ such that $\Gamma \vdash s : s'$.

An important class of examples of PTSs is formed by the eight PTSs of the so-called Barendregt Cube. The Barendregt Cube is a three-dimensional presentation of eight well-known PTSs. All systems have sorts $\boldsymbol{S} = \{*, \square\}$, and axioms $\boldsymbol{A} = \{(*, \square)\}$. Moreover, all the systems have rule $(*, *, *)$. System $\lambda\!\rightarrow$ has no extra rules, but the other seven systems all have one or more of the rules $(*, \square, \square)$, $(\square, *, *)$ and $(\square, \square, \square)$.

$$\text{(axiom)} \quad \langle\rangle \vdash s_1 : s_2 \qquad (s_1, s_2) \in \boldsymbol{A}$$

$$\text{(start)} \quad \frac{\Gamma \vdash A : s}{\Gamma, x{:}A \vdash x : A} \qquad x \notin \mathit{Dom}(\Gamma)$$

$$\text{(weak)} \quad \frac{\Gamma \vdash A : B \qquad \Gamma \vdash C : s}{\Gamma, x{:}C \vdash A : B} \qquad x \notin \mathit{Dom}(\Gamma)$$

$$(\Pi) \quad \frac{\Gamma \vdash A : s_1 \qquad \Gamma, x{:}A \vdash B : s_2}{\Gamma \vdash (\Pi x{:}A.B) : s_3} \qquad (s_1, s_2, s_3) \in \boldsymbol{R}$$

$$(\lambda) \quad \frac{\Gamma, x{:}A \vdash b : B \qquad \Gamma \vdash (\Pi x{:}A.B) : s}{\Gamma \vdash (\lambda x{:}A.b) : (\Pi x{:}A.B)}$$

$$\text{(appl)} \quad \frac{\Gamma \vdash F : (\Pi x{:}A.B) \qquad \Gamma \vdash a : A}{\Gamma \vdash Fa : B[x{:=}a]}$$

$$\text{(conv)} \quad \frac{\Gamma \vdash A : B \qquad \Gamma \vdash B' : s \qquad B =_\beta B'}{\Gamma \vdash A : B'}$$

Fig. 4. Typing rules for PTSs

$$(\mathrm{C}^p\text{-weak}) \quad \frac{\Gamma \vdash^{C^p} b{:}B \quad \Gamma, \Delta \vdash^{C^p} A{:}s \quad \Gamma, \Delta_i \vdash^{C^p} B_i{:}s_i \quad (s_i, s) \in \boldsymbol{P} \quad (i = 1, \ldots, n)}{\Gamma, c(\Delta) : A \vdash^{C^p} b : B}$$

$$(\mathrm{C}^p\text{-app}) \quad \frac{\begin{array}{ll} \Gamma_1, c(\Delta){:}A, \Gamma_2 \vdash^{C^p} b_i{:}B_i[x_j{:=}b_j]_{j=1}^{i-1} & (i = 1, \ldots, n) \\ \Gamma_1, c(\Delta){:}A, \Gamma_2 \vdash^{C^p} A : s & (\text{if } n = 0) \end{array}}{\Gamma_1, c(\Delta){:}A, \Gamma_2 \vdash^{C^p} c(b_1, \ldots, b_n) : A[x_j{:=}b_j]_{j=1}^{n}}$$

Fig. 5. Typing rules for parametric constants

Now we add rules for typing parametric constants. In these rules we use the set of parametric rules $\boldsymbol{P}$ to govern which types parameters can have.

Definition 11 (Typing rules for parametric constants) Let $\mathcal{S} = (\boldsymbol{S}, \boldsymbol{A}, \boldsymbol{R}, \boldsymbol{P})$ be a parametric specification. The *typing relation* $\vdash^{C^p}$ is the smallest relation on $\mathcal{C}_P \times \mathcal{T}_P \times \mathcal{T}_P$ closed under the rules in Definition 10 and the rules (C^p-weak) and (C^p-app) in Figure 5. Recall that $\Delta \equiv x_1{:}B_1, \ldots, x_n{:}B_n$ and $s \in \boldsymbol{S}$. The parameterized constant c in the C^p-weakening rule is, due to the Barendregt variable convention, assumed to be Γ-fresh, that is, $c \notin \mathit{Cons}(\Gamma)$.

At first sight one might miss a C^p-introduction rule. Such a rule, however, is not necessary, as c (on its own) is not a term. A parameterized constant c can only be (part of) a term in the form $c(b_1, \ldots, b_n)$, and such terms can be typed by the C^p-application rule. The extra condition $\Gamma_1, c(\Delta){:}A, \Gamma_2 \vdash^{C^p} A : s$ in the C^p-application rule for $n = 0$ is necessary to prevent an empty list of premises. Such an empty list of premises would make it possible to have almost arbitrary contexts in the conclusion. The extra condition is only needed to assure that the context in the conclusion is a legal context.

$$(\mathrm{D}^p\text{-weak})\quad \frac{\Gamma \vdash^{D^p} b{:}B \quad \Gamma, \Delta \vdash^{D^p} a{:}A{:}s \quad \Gamma, \Delta_i \vdash^{D^p} B_i{:}s_i \quad (s_i, s) \in \boldsymbol{P} \quad (i{=}1,\ldots,n)}{\Gamma, c(\Delta){=}a{:}A \vdash^{D^p} b : B}$$

$$(\mathrm{D}^p\text{-app})\quad \frac{\begin{array}{ll}\Gamma_1, c(\Delta){=}a{:}A, \Gamma_2 \vdash^{D^p} b_i : B_i[x_j{:=}b_j]_{j=1}^{i-1} & (i = 1,\ldots,n) \\ \Gamma_1, c(\Delta){=}a{:}A, \Gamma_2 \vdash^{D^p} a : A & (\text{if } n = 0)\end{array}}{\Gamma_1, c(\Delta){=}a{:}A, \Gamma_2 \vdash^{D^p} c(b_1,\ldots,b_n) : A[x_j{:=}b_j]_{j=1}^{n}}$$

$$(\mathrm{D}^p\text{-form})\quad \frac{\Gamma, c(\Delta){=}a{:}A \vdash^{D^p} B : s}{\Gamma \vdash^{D^p} c(\Delta){=}a{:}A \;\mathtt{in}\; B : s}$$

$$(\mathrm{D}^p\text{-intro})\quad \frac{\Gamma, c(\Delta){=}a{:}A \vdash^{D^p} b : B \qquad \Gamma \vdash^{D^p} c(\Delta){=}a{:}A \;\mathtt{in}\; B : s}{\Gamma \vdash^{D^p} c(\Delta){=}a{:}A \;\mathtt{in}\; b : c(\Delta){=}a{:}A \;\mathtt{in}\; B}$$

$$(\mathrm{D}^p\text{-conv})\quad \frac{\Gamma \vdash^{D^p} b : B \qquad \Gamma \vdash^{D^p} B' : s \qquad \Gamma \vdash B =_\delta B'}{\Gamma \vdash^{D^p} b : B'}$$

Fig. 6. Typing rules for parametric definitions

Note that in the (C^p-weak) rule it is not necessary that all the s_i are equal: in one application of rule (C^p-weak) it is possible to rely on more than one element of $\boldsymbol{P}$.

Remark 12 If we have a parametric constant $\mathit{plus}(x{:}\mathtt{Int}, y{:}\mathtt{Int}){:}\mathtt{Int}$ in the context, then it is tempting to think of *plus* as a parametric function. Note however that in PTS-terms it is not a function anymore since the only way to obtain a legal term with it is in its parameterized form $\mathit{plus}(x, y)$ which has type $\mathtt{Int}$; $\mathit{plus}(x{:}\mathtt{Int}, y{:}\mathtt{Int})$ itself is not a legal term. In order to talk about properties of *plus* 'as a function' we are forced to consider $\lambda x{:}\mathtt{Int}.\lambda y{:}\mathtt{Int}.\mathit{plus}(x, y)$.

Adapting the rules from Definition 11 and the rules for definitions of [15] results in rules for *parametric definitions* (by $\Gamma \vdash a : A : s$ we mean $\Gamma \vdash a : A$ and $\Gamma \vdash A : s$):

Definition 13 (Typing rules for parametric definitions) The *typing relation* $\vdash^{D^p}$ is the smallest relation on $\mathcal{C}_P \times \mathcal{T}_P \times \mathcal{T}_P$ closed under the rules in Definition 10 and those in Figure 6, where $s \in \boldsymbol{S}$, and the parameterized definition c that is introduced in the D^p-weakening rule is assumed to be Γ-fresh. Again it is not necessary that all the s_i in the **(D^p-weak)** rule are equal.

Definition 14 ($\mathrm{C}^p\mathrm{D}^p$-PTSs) Let $\mathcal{S} = (\boldsymbol{S}, \boldsymbol{A}, \boldsymbol{R}, \boldsymbol{P})$ be a parametric specification. Then $\vdash^{C^pD^p}$ is the smallest relation on $\mathcal{C}_P \times \mathcal{T}_P \times \mathcal{T}_P$ that is closed under the rules of Definitions 10, 11 and 13. The $\mathrm{C}^p\mathrm{D}^p$-PTS $\lambda^{C^pD^p}\mathcal{S}$ is the system with typing relation $\vdash^{C^pD^p}$.

All contexts occurring in $\mathrm{C}^p\mathrm{D}^p$-PTSs are sound (see Definition 7). As $\mathrm{C}^p\mathrm{D}^p$-PTSs are clearly extensions of PTSs, C^p-PTSs and D^p-PTSs, then all contexts occurring in PTSs, C^p-PTSs and D^p-PTSs are sound.

Lemma 15 *Assume* $\Gamma \vdash^{C^pD^p} b : B$. *Then*

1. $\mathit{Dom}(b), \mathit{Dom}(B) \subseteq \mathit{Dom}(\Gamma)$; *2.* Γ *is sound.*

PROOF: We prove 1 and 2 simultaneously by induction on the derivation of $\Gamma \vdash^{C^p D^p} b : B$. ⊠

In the specific case of the Barendregt Cube, the combination of $\boldsymbol{R}$ and $\boldsymbol{P}$ leads to a refinement of the Cube, thus making it possible to classify more type systems within one and the same framework. This is studied in detail for PTSs extended with parameterized constants (without definitions) in [10]; it is shown that the type systems of AUTOMATH, LF and ML can be described more naturally and accurately than in ordinary PTSs.

Remark 16 Let $\mathcal{S} = (\boldsymbol{S}, \boldsymbol{A}, \boldsymbol{R})$ be a specification, and observe the parametric specification $\mathcal{S}' = (\boldsymbol{S}, \boldsymbol{A}, \boldsymbol{R}, \varnothing)$. The fact that the set of parametric rules is empty does not exclude the existence of definitions: it is still possible to apply the rules (D^p-weak) and (D^p-app) for $n = 0$. In that case, we obtain only definitions without parameters, and the rules of the parametric system reduce to the rules of a D-PTS with specification $\mathcal{S}$ as introduced by [15].[4] There is however one case of the rules in [15] that a $\mathrm{C}^p\mathrm{D}^p$-PTS cannot simulate, since in rule (D^p-weak) we require $\Gamma, \Delta \vdash^{D^p} A : s$ which [15] does not. Therefore, the system of [15] can abbreviate inhabitants of topsorts which is impossible in the $\mathrm{C}^p\mathrm{D}^p$-PTSs of Definition 14. We feel that abbreviating inhabitants of topsorts is not very useful. It is however routine to check that adding an extra rule

$$(\mathrm{D}^p\text{-weak-top}) \quad \frac{\Gamma \vdash^{D^p} b : B \quad \Gamma, \Delta \vdash^{D^p} a : A}{\Gamma, c() = a : A \vdash^{D^p} b : B}$$

does not change the theory of $\mathrm{C}^p\mathrm{D}^p$-PTSs developed in sections 3 and 4, and yields exactly the same power for unparameterized definitions as the D-PTSs of [15].

3 Meta-properties

We first list the properties of terms which are not dependent of their being legal. However, we often demand that the free variables and constants of a term are contained in the domain of a sound context.

First we show that δ-reduction is invariant under enlarging of the context ($\to_\delta$-weakening), then we establish a relation between substitution and $\to_{\beta\delta}$ (substitutivity), then we can establish confluence for $\to_{\beta\delta}$ and termination of $\to_\delta$. Most proofs are similar to those of [17, 15]; more details can be found in the technical report [12].

Lemma 17 **($\to_\delta$-weakening and Substitutivity)**

1. *If* $\langle \Gamma_1, \Gamma_2, \Gamma_3 \rangle \in \mathcal{C}_P$ *is such that* $\Gamma_1, \Gamma_3 \vdash b \to_\delta b'$, *then* $\Gamma_1, \Gamma_2, \Gamma_3 \vdash b \to_\delta b'$;
2. *If* $a \to_\beta a'$ *then* $a[x{:=}b] \to_\beta a'[x{:=}b]$;
3. *If* $\Gamma \vdash a \to_\delta a'$ *then* $\Gamma[x{:=}b] \vdash a[x{:=}b] \to_\delta a'[x{:=}b]$;
4. *If* $\Gamma \vdash b \to_{\beta\delta} b'$ *then* $\Gamma \vdash a[x{:=}b] \twoheadrightarrow_{\beta\delta} a[x{:=}b']$.

[4] The parametric system with specification $\mathcal{S}'$ has a C^p-weakening rule while the system of [15] does not. But the C^p-weakening rule can only be used for $n = 0$, and in that case C^p-weakening can be imitated by the normal weakening rule of PTSs: a parametric constant with zero parameters is in fact a parameter-free constant, and for such a constant one can use a variable as well.

Theorem 18 (Confluence for $\rightarrow_\beta$, $\rightarrow_\delta$ and Strong Normalization for $\rightarrow_\delta$)

1. *$\rightarrow_\beta$ is confluent.*
2. *$\rightarrow_{\beta\delta}$ is confluent when taking contexts into account: if Γ is sound, $\Gamma \vdash a \twoheadrightarrow_{\beta\delta} b_1$ and $\Gamma \vdash a \twoheadrightarrow_{\beta\delta} b_2$ then there exists a term d such that $\Gamma \vdash b_1 \twoheadrightarrow_{\beta\delta} d$ and $\Gamma \vdash b_2 \twoheadrightarrow_{\beta\delta} d$.*
3. *$\rightarrow_\delta$, when restricted to sound contexts Γ and terms a with $Dom(a) \subseteq Dom(\Gamma)$, is strongly normalizing, i.e. there are no infinite δ-reduction paths.*

Without the restriction to sound contexts Γ and terms a with $Dom(a) \subseteq Dom(\Gamma)$, we do not even have weak normalization: consider $\Gamma \equiv \langle \mathtt{c()}{=}\mathtt{d()}{:}\mathtt{A}, \mathtt{d()}{=}\mathtt{c()}{:}\mathtt{A}\rangle$. The term $\mathtt{c()}$ does not have a Γ-normal form.

3.1 Properties of Legal Terms

The properties in this section are proved for all *legal* terms, i.e. for terms a for which there are A, Γ such that $\Gamma \vdash^{C^pD^p} a : A$ or $\Gamma \vdash^{C^pD^p} A : a$. The main property we prove is that strong normalization of a PTS is preserved by certain extensions.

Many of the standard properties of PTSs in [1] hold for C^{p}D^p-PTSs as well. In the same way as in [1], we can prove the following theorem:

Theorem 19 *Let $\mathcal{S}$ be a parametric specification. The type system $\lambda^{C^pD^p}\mathcal{S}$ has the following properties:*[5]

1) Substitution Lemma; 2) Correctness of Types; 3) Subject Reduction (for $\rightarrow_{\beta\delta}$). 4) If $\mathcal{S}$ is singly sorted then $\lambda^{C^pD^p}\mathcal{S}$ has Uniqueness of Types.

The Generation Lemma is extended with two extra cases:

Lemma 20 (Generation Lemma, extension)

1. *If $\Gamma \vdash^{C^pD^p} \mathtt{c}(b_1,\ldots,b_n) : D$ then there exist sort s, $\Delta \equiv x_1{:}B_1,\ldots,x_n{:}B_n$ and term A such that $\Gamma \vdash D =_{\beta\delta} A[x_i{:=}b_i]_{i=1}^n$, and $\Gamma \vdash^{C^pD^p} b_i : B_i[x_j{:=}b_j]_{j=1}^{i-1}$. Besides we have one of these two possibilities: (a) $\Gamma = \langle \Gamma_1, \mathtt{c}(\Delta){:}A, \Gamma_2\rangle$ and $\Gamma_1, \Delta \vdash^{C^pD^p} A : s$; or (b) $\Gamma = \langle \Gamma_1, \mathtt{c}(\Delta){=}a{:}A, \Gamma_2\rangle$ and $\Gamma_1, \Delta \vdash^{C^pD^p} a : A : s$ for some sort s;*
2. *If $\Gamma \vdash^{C^pD^p} \mathtt{c}(\Delta){=}a{:}A \;\mathtt{in}\; b : D$ then either we have (a) or (b) below:*
   (a) $\Gamma, \mathtt{c}(\Delta){=}a{:}A \vdash^{C^pD^p} b : B$, $\Gamma \vdash^{C^pD^p} (\mathtt{c}(\Delta){=}a{:}A \;\mathtt{in}\; B) : s$ and $\Gamma \vdash D =_{\beta\delta} \mathtt{c}(\Delta){=}a{:}A \;\mathtt{in}\; B$;
   (b) $\Gamma, \mathtt{c}(\Delta){=}a{:}A \vdash^{C^pD^p} b : s$ and $\Gamma \vdash D =_{\beta\delta} s$.

Also Correctness of Contexts has some extra cases compared to usual PTSs. Recall that Γ is *legal* if there are b, B such that $\Gamma \vdash^{C^pD^p} b : B$.

---

[5] Substitution Lemma: if $\Gamma, x : A, \Delta \vdash b : B$ and $\Gamma \vdash a : A$ then also $\Gamma, \Delta[x := a] \vdash b[x := a] : B[x := a]$. Correctness of Types: if $\Gamma \vdash a : A$ then $\Gamma \vdash A : s$ for some sort s or A is a topsort. Subject Reduction: if $\Gamma \vdash a : A$ and $a \rightarrow_{\beta\delta} a'$ then also $\Gamma \vdash a' : A$. Uniqueness of Types: if $\Gamma \vdash a : A$ and $\Gamma \vdash a : B$ then $\Gamma \vdash A =_{\beta\delta} B$.

Lemma 21 (Correctness of Contexts)

1. *If $\Gamma, x{:}A, \Gamma'$ is legal then there exists a sort s such that $\Gamma \vdash^{C^pD^p} A : s$;*
2. *If $\Gamma, c(\Delta){:}A, \Gamma'$ is legal then $\Gamma, \Delta \vdash^{C^pD^p} A : s$;*
3. *If $\Gamma, c(\Delta){=}a{:}A, \Gamma'$ is legal then $\Gamma, \Delta \vdash^{C^pD^p} a : A : s$.*

Now we prove that $\lambda^{C^pD^p}\mathcal{S}$ is $\beta\delta$-strongly normalizing if a slightly larger PTS $\lambda\mathcal{S}'$ is β-strongly normalizing. We follow the idea of [15] for PTSs extended with only definitions; the extension is tedious but fairly routine.

For legal terms $a \in \mathcal{T}_P$ in a context Γ, we define a lambda term $\|a\|_\Gamma$ without definitions and without parameters. If a is typable in a C^{p}D^p-PTS $\lambda^{C^pD^p}\mathcal{S}$, then $\|a\|_\Gamma$ will be typable in a PTS $\lambda\mathcal{S}'$, where $\mathcal{S}'$ is a so-called *completion* (see Definition 24) of the specification $\mathcal{S}$. Moreover, we take care that if $a \to_\beta a'$, then $\|a\|_\Gamma \to^+_\beta \|a'\|_\Gamma$ (that is: $\|a\|_\Gamma \twoheadrightarrow_\beta \|a'\|_\Gamma$ and $\|a\|_\Gamma \not\equiv \|a'\|_\Gamma$). Together with strong normalization of δ-reduction (Theorem 18.3), this guarantees that $\lambda^{C^pD^p}\mathcal{S}$ is $\beta\delta$-strongly normalizing whenever $\lambda\mathcal{S}'$ is β-strongly normalizing.

Definition 22 For $a \in \mathcal{T}_P$ and $\Gamma \in \mathcal{C}_P$ we define $\|a\|_\Gamma$ as in [15], but with the extra cases:

$$\|c(b_1,\ldots,b_n)\|_\Gamma \equiv \begin{cases} \|\lambda_{i=1}^n x_i{:}B_i.a\|_{\Gamma_1}\|b_1\|_\Gamma\cdots\|b_n\|_\Gamma \text{ if } \Gamma{=}\langle\Gamma_1, c(\Delta){=}a{:}A, \Gamma_2\rangle; \\ c\,\|b_1\|_\Gamma\cdots\|b_n\|_\Gamma \text{ otherwise;} \end{cases}$$

$$\|c(\Delta){=}a{:}A \texttt{ in } b\|_\Gamma \equiv \left(\lambda c{:}(\|\textstyle\prod_{i=1}^n x_i{:}B_i.A\|_\Gamma).\,\|b\|_{\Gamma,c(\Delta)=a:A}\right)\|\lambda_{i=1}^n x_i{:}B_i.a\|_\Gamma\,.$$

Note: constants in $\mathcal{T}_P$ are translated to similarly named variables in λ-calculus without definitions and parameters.

We now show that $\|\_\|_\_$ translates a δ-reduction into zero or more β-reductions, and that it translates a β-reduction into one or more β-reductions.

Lemma 23 *Let Γ be sound, and assume $Dom(a) \subseteq Dom(\Gamma)$. The following holds: If $\Gamma \vdash a \to_\delta b$ then $\|a\|_\Gamma \twoheadrightarrow_\beta \|b\|_\Gamma$. Similarly: If $a \to_\beta b$ then $\|a\|_\Gamma \to^+_\beta \|b\|_\Gamma$.*

Definition 24 The specification $\mathcal{S} = (\boldsymbol{S}, \boldsymbol{A}, \boldsymbol{R})$ is called *quasi full* if for all $s_1, s_2 \in \boldsymbol{S}$ there exists $s_3 \in \boldsymbol{S}$ such that $(s_1, s_2, s_3) \in \boldsymbol{R}$.
A specification $\mathcal{S}' = (\boldsymbol{S}', \boldsymbol{A}', \boldsymbol{R}')$ is a *completion* of a parametric specification $\mathcal{S} = (\boldsymbol{S}, \boldsymbol{A}, \boldsymbol{R}, \boldsymbol{P})$ if $\boldsymbol{S} \subseteq \boldsymbol{S}'$, $\boldsymbol{A} \subseteq \boldsymbol{A}'$, and $\boldsymbol{R} \subseteq \boldsymbol{R}'$, $\boldsymbol{S}'$ is quasi full, and $\forall s \in \boldsymbol{S} \exists s' \in \boldsymbol{S}'[(s, s') \in \boldsymbol{A}']$.[6]

Theorem 25 *Let $\mathcal{S} = (\boldsymbol{S}, \boldsymbol{A}, \boldsymbol{R}, \boldsymbol{P})$ and $\mathcal{S}' = (\boldsymbol{S}', \boldsymbol{A}', \boldsymbol{R}')$ be such that $\mathcal{S}'$ is a completion of $\mathcal{S}$. If $\Gamma \vdash^{C^pD^p}_{\mathcal{S}} a : A$ then $\|\Gamma\| \vdash_{\mathcal{S}'} \|a\|_\Gamma : \|A\|_\Gamma$.*

Using Theorem 18.3, Lemma 23 and Theorem 25, we can now prove our normalization result for C^{p}D^p-PTSs.

[6] Note that there are no requirements on $\boldsymbol{P}$ in the definition of completion.

Theorem 26 *Let $\mathcal{S} = (\boldsymbol{S}, \boldsymbol{A}, \boldsymbol{R}, \boldsymbol{P})$ and $\mathcal{S}' = (\boldsymbol{S}', \boldsymbol{A}', \boldsymbol{R}')$ be such that $\mathcal{S}'$ is a completion of $\mathcal{S}$. If the PTS $\lambda\mathcal{S}'$ is β-strongly normalizing, then the $\mathrm{C^pD^p}$-PTS $\lambda^{C^pD^p}\mathcal{S}$ is $\beta\delta$-strongly normalizing.*

Since ECC of [14] is β-strongly normalizing and is a completion of all systems of the extended λ-cube, Theorem 26 guarantees that all extended systems of the λ-cube are $\beta\delta$-strongly normalizing. Note that λC itself is not a completion since it has a topsort $\Box$.

4 Comparison of C^pD^p-PTSs with D-PTSs

In this section we show that the parameter mechanism in $\mathrm{C^pD^p}$-PTSs can be seen as a system for abstraction and application that is weaker than the λ-calculus mechanism. We will make this precise by proving (in Theorem 31) that a $\mathrm{C^pD^p}$-PTS with parametric specification $(\boldsymbol{S}, \boldsymbol{A}, \boldsymbol{R}, \varnothing)$ is as powerful as any $\mathrm{C^pD^p}$-PTS with parametric specification $(\boldsymbol{S}, \boldsymbol{A}, \boldsymbol{R}, \boldsymbol{P})$ for which $(s_1, s_2) \in \boldsymbol{P}$ implies $(s_1, s_2, s_2) \in \boldsymbol{R}$. We call such a $\mathrm{C^pD^p}$-PTS *parametrically conservative*; each $\mathrm{C^pD^p}$-PTS with $\boldsymbol{P} \subseteq \boldsymbol{S} \times \boldsymbol{S}$ can be extended to a parametrically conservative one by taking its *parametric closure.*

Definition 27 Let $\mathcal{S} = (\boldsymbol{S}, \boldsymbol{A}, \boldsymbol{R}, \boldsymbol{P})$ be a parametric specification. $\mathcal{S}$ is *parametrically conservative* if for all $s_1, s_2 \in \boldsymbol{S}$, $(s_1, s_2) \in \boldsymbol{P}$ implies $(s_1, s_2, s_2) \in \boldsymbol{R}$.

$\mathrm{CL}(\mathcal{S})$, the *parametric closure* of $\mathcal{S}$, is defined as $(\boldsymbol{S}, \boldsymbol{A}, \boldsymbol{R}', \boldsymbol{P})$, where $\boldsymbol{R}' = \boldsymbol{R} \cup \{(s_1, s_2, s_2) \mid s_1, s_2 \in \boldsymbol{S} \text{ and } (s_1, s_2) \in \boldsymbol{P}\}$.

Lemma 28 *Let $\mathcal{S}$ be a parametric specification. The following holds:*
1. $\mathrm{CL}(\mathcal{S})$ is parametrically conservative; and 2. $\mathrm{CL}(\mathrm{CL}(\mathcal{S})) = \mathrm{CL}(\mathcal{S})$.

Let $\mathcal{S} = (\boldsymbol{S}, \boldsymbol{A}, \boldsymbol{R}, \boldsymbol{P})$ be a parametric specification. If $\mathcal{S}$ is parametrically conservative, then each parametric rule (s_1, s_2) of $\mathcal{S}$ has a corresponding Π-formation rule (s_1, s_2, s_2). We show that this Π-formation rule can indeed take over the role of the parametric rule (s_1, s_2). This means that $\mathcal{S}$ has the same 'power' (see Theorem 31) as $(\boldsymbol{S}, \boldsymbol{A}, \boldsymbol{R}, \varnothing)$. This even means that $\mathcal{S}$ has the same power as the D-PTS with specification $(\boldsymbol{S}, \boldsymbol{A}, \boldsymbol{R})$. In order to compare $\mathcal{S} = (\boldsymbol{S}, \boldsymbol{A}, \boldsymbol{R}, \boldsymbol{P})$ with $\mathcal{S}' = (\boldsymbol{S}, \boldsymbol{A}, \boldsymbol{R}, \varnothing)$, we need to remove the parameters from the syntax of $\lambda^{C^pD^p}\mathcal{S}$. This can be obtained as follows:

- The parametric application in a term $c(b_1, \ldots, b_n)$ is replaced by a function application $cb_1 \cdots b_n$;
- A local parametric definition is translated by a parameter-free local definition, and the parameters are replaced by λ-abstractions;
- A global parametric definition is translated by a parameter-free global definition, and the parameters are replaced by λ-abstractions.

This leads to the following definitions (which can be extended to contexts in the obvious way):

Definition 29 We define the parameter-free translation $\{t\}$ of a term $t \in \mathcal{T}_P$ inductively as follows:
$\{a\} \equiv a$ if $a \equiv x$ or $a \equiv s$; $\{c(b_1, \ldots, b_n)\} \equiv c\,\{b_1\} \cdots \{b_n\}$; $\{ab\} \equiv \{a\}\,\{b\}$;
$\{\mathcal{O}x{:}A.B\} \equiv \mathcal{O}x{:}\,\{A\}\,.\,\{B\}$ if $\mathcal{O}$ is λ or Π;
$\{c(\Delta){=}a{:}A \;\mathtt{in}\; b\} \equiv c{=}\{\lambda\Delta.a\} : \{\prod\Delta.A\} \;\mathtt{in}\; \{b\}$.

The mapping $\{\_\}$ maintains β-reduction. A δ-reduction is translated into a δ-reduction followed by zero or more β-reductions. These β-reductions take over the n substitutions that are needed in a δ-reduction $c(b_1, \ldots, b_n) \rightarrow_\delta a[x_i{:=}b_i]_{i=1}^n$. Note that the restriction on the formation of parameters induced by the parametric rules $\boldsymbol{P}$ prevents the creation of illegal abstractions $\Pi\Delta.A$ with A a topsort.

Lemma 30 *1. For $a, b \in \mathcal{T}_P$: $\{a[x{:=}b]\} \equiv \{a\}\,[x{:=}\,\{b\}]$;*
2. If $a \rightarrow_\beta a'$ then $\{a\} \rightarrow_\beta^+ \{a'\}$;
3. If $\Gamma \vdash a \rightarrow_\delta a'$ then there is a'' such that $\{\Gamma\} \vdash \{a\} \rightarrow_\delta^+ a'' \twoheadrightarrow_\beta \{a'\}$;
4. If $\Gamma \vdash a \twoheadrightarrow_{\beta\delta} a'$ then $\{\Gamma\} \vdash \{a\} \twoheadrightarrow_{\beta\delta} \{a'\}$.

Now we show that the parameter-free translation $\{\_\}$ embeds the C^{p}D^p-PTS with parametric specification $\mathcal{S} = (\boldsymbol{S}, \boldsymbol{A}, \boldsymbol{R}, \boldsymbol{P})$ in the C^{p}D^p-PTS with parametric specification $\mathcal{S}' = (\boldsymbol{S}, \boldsymbol{A}, \boldsymbol{R}, \varnothing)$, provided that $\mathcal{S}$ is parametrically conservative.

Thus we can conclude that the (restrictive) use of parameters does not yield a stronger type system than using abstraction, application and unparameterized constants and definitions only.

Theorem 31 *Let $\mathcal{S} = (\boldsymbol{S}, \boldsymbol{A}, \boldsymbol{R}, \boldsymbol{P})$ be a parametric specification. Assume that $\mathcal{S}$ is parametrically conservative. Let $\mathcal{S}' = (\boldsymbol{S}, \boldsymbol{A}, \boldsymbol{R}, \varnothing)$. Then $\Gamma \vdash_{\mathcal{S}}^{C^pD^p} a : A$ implies $\{\Gamma\} \vdash_{\mathcal{S}'}^{CD} \{a\} : \{A\}$.*

PROOF: Induction on the derivation of $\Gamma \vdash_{\mathcal{S}}^{C^pD^p} a : A$. ⊠

Since unparameterized constants can be mimicked by dedicated variables, Theorem 31 can be paraphrased as "if $\mathcal{S}$ and $\mathcal{S}'$ are parametric specifications such that $\mathcal{S}'$ is CL$(\mathcal{S})$ with $\boldsymbol{P}$ replaced by $\varnothing$, then type derivations in $\lambda^{C^pD^p}\mathcal{S}$ can be mapped to type derivations in the D-PTS $\lambda\mathcal{S}'$".

Now [15] shows that type derivations in a D-PTS $\lambda\mathcal{S}'$ can be mapped to type derivations in the PTS which is a completion of the D-PTS. We conclude that parametrically conservative C^{p}D^p-PTSs are conservative extensions of PTSs.

Note however that, in order to mimic type derivations in $\lambda^{C^pD^p}\mathcal{S}$ in a PTS, we need to strengthen the system twice: first by considering the parametric closure and then by considering a completion of the D-PTS.

5 Conclusion

In recent literature, extensions of Pure Type Systems with *unparameterized* definitions and with parameterized constants have been proposed. However, parameterized constants and definitions are required for PTSs in order to be suitable for studying semantics and implementations of theorem provers and programming languages and for reasoning about mathematics, since these all depend heavily on the use of parameterized constants as well as parameterized definitions.

In this paper we studied an extension of PTSs with both parameterized constants and parameterized definitions. Extending the existing theory we showed that our extension has all the desired properties and yields well-behaved type systems.

References

[1] H.P. Barendregt. Lambda calculi with types. In [?], pages 117–309. Oxford University Press, 1992.

[2] L.S. van Benthem Jutting. *Checking Landau's "Grundlagen" in the Automath system*. PhD thesis, Eindhoven University of Technology, 1977. Published as Mathematical Centre Tracts nr. 83 (Amsterdam, Mathematisch Centrum, 1979).

[3] S. Berardi. Towards a mathematical analysis of the Coquand-Huet calculus of constructions and the other systems in Barendregt's cube. Technical report, Dept. of Computer Science, Carnegie-Mellon University and Dipartimento Matematica, Universita di Torino, 1988.

[4] R. Bloo, F. Kamareddine, and R. Nederpelt. The Barendregt Cube with Definitions and Generalised Reduction. *Information and Computation*, 126(2):123–143, 1996.

[5] V.A.J. Borghuis. Modal Pure Type Systems. *Journal of Logic, Language, and Information*, 7:265–296, 1998.

[6] N.G. de Bruijn. Reflections on Automath. Eindhoven University of Technology, 1990. Also in [?], pages 201–228.

[7] A. Church. A formulation of the simple theory of types. *The Journal of Symbolic Logic*, 5:56–68, 1940.

[8] D.T. van Daalen. A description of Automath and some aspects of its language theory. In P. Braffort, editor, *Proceedings of the Symposium APLASM*, volume I, pages 48–77, 1973. Also in [?], pages 101–126.

[9] J.H. Geuvers, F. Wiedijk, J. Zwanenburg, R. Pollack, and H. Barendregt. Personal communication on the "Fundamental Theorem of Algebra" project. FTA web page available at `http : //www.cs.kun.nl/gi/projects/fta/index.html`.

[10] F. Kamareddine, L. Laan, and R.P. Nederpelt. Refining the Barendregt cube using parameters. *Fifth International Symposium on Functional and Logic Programming, FLOPS 2001*, Lecture Notes in Computer Science:375–389, 2001.

[11] T. Laan. *The Evolution of Type Theory in Logic and Mathematics*. PhD thesis, Eindhoven University of Technology, 1997.

[12] T. Laan, R. Bloo, F. Kamareddine, and R. Nederpelt. Parameters in pure type systems. Technical Report 00-18, TUE Computing Science Reports, Eindhoven University of Technology, 2000. Available from `http : //www.win.tue.nl/~bloo/parameter − report.ps.gz`.

[13] Twan Laan and Michael Franssen. Parameters for first order logic. *Logic and Computation*, 2001.

[14] Z. Luo. *An Extended Calculus of Constructions*. PhD thesis, 1990.

[15] P. Severi and E. Poll. Pure type systems with definitions. In A. Nerode and Yu.V. Matiyasevich, editors, *Proceedings of LFCS'94 (LNCS* **813***)*, pages 316–328, New York, 1994. LFCS'94, St. Petersburg, Russia, Springer Verlag.

[16] J. Terlouw. Een nadere bewijstheoretische analyse van GSTT's. Technical report, Department of Computer Science, University of Nijmegen, 1989.

[17] R. de Vrijer. A direct proof of the finite developments theorem. *The Journal of Symbolic Logic*, 50(2):339–343, 1985.

Category, Measure, Inductive Inference: A Triality Theorem and Its Applications

Rūsiņš Freivalds[1]* and Carl H. Smith[2]**

[1] Institute of Math and Computer Science, University of Latvia,
Raiņa bulvāris 29,LV-1459, Riga, Latvia
Rusins.Freivalds@mii.lu.lv

[2] Department of Computer Science, University of Maryland,
A.V.Williams Bldg., College Park, MD 20742, USA
smith@cs.umd.edu

Abstract. The famous Sierpinski-Erdös Duality Theorem [Sie34b, Erd43] states, informally, that any theorem about effective measure 0 and/or first category sets is also true when all occurrences of "effective measure 0" are replaced by "first category" and vice versa. This powerful and nice result shows that "measure" and "category" are equally useful notions neither of which can be preferred to the other one when making formal the intuitive notion "almost all sets." Effective versions of measure and category are used in recursive function theory and related areas, and resource-bounded versions of the same notions are used in Theory of Computation. Again they are dual in the same sense.

We show that in the world of recursive functions there is a third equipotent notion dual to both measure and category. This new notion is related to learnability (also known as inductive inference or identifiability). We use the term "triality" to describe this three-party duality.

1 Introduction

Mathematicians have invented many ways to determine the relative sizes of infinite sets. Cardinality is the most natural technique, but it considers infinite sets as either denumerable or undenumerable. Measure is a generalization of volume and area. Hence, it is the most popular method employed to show that one set of infinite cardinality is larger than another set of the same cardinality. Hence, it is widely used to discuss the relative size of of infinite sets. However there are other ways how to say that one set is smaller than the other one. Category has become a popular notion to discuss smallness of sets [Oxt71]. W. Sierpinski [Sie34b] and P. Erdös [Erd43] proved a fundamental theorem with the following important consequence.

* This project was supported by Latvian Science Council Grant No. 01.0354 and Contract IST-1999-11234 from the European Commission

** This project was supported by NSF Grant CCR-9732692

S. Rajsbaum (Ed.): LATIN 2002, LNCS 2286, pp. 386–399, 2002.

Theorem 1 (Duality Principle). *Let P be an assertion containing only notions of "set of measure 0", "set of the first category" and the notions from set theory. Let P^* be an assertion obtained by interchange of the notions "set of measure 0" and "set of the first category." Then each of the assertions P and P^* is implied by the other one, provided the Continuum Hypothesis holds.*

Hence, by the Duality Principle, "measure zero" and "first category" are dual notions (like disjunction and conjuction, or existence and universality) neither of which can be preferred to the other one.

The set of all the (0-1)-valued total recursive functions is denumerable. Hence, from the viewpoint of the classical mathematics this is a small set, e.g. a set of measure zero and of the first category. However two decades ago counterparts of the notions "measure zero" and "first category" were defined for the set of all the (0-1)-valued total recursive functions (see [Lis81]) and they also were found to be dual in the same sense, without any reference to the Continuum Hypothesis. In recent papers resource-bounded versions of the traditional measure and category notions have been developed [AS95, ASMWZ96, BM95a, BM95b, BL96, Fen91], [Fen95, Kum96, Lut93, Lut96]. For instance, K. Melhhorn [Meh73] and J. Lutz [Lut93] used constructive notions of category and measure to study subrecursive degree structures. The Sierpinski and Erdös Duality Principle is known to hold for the resource-bounded versions of the notions of category and measure as well.

It is a well-known general observation that it is difficult to invent natural notions on a high abstract level but it is much easier to invent such natural notions on a lower level where there is a considerable amount of specific information. Hence. it is no wonder that on the level of classical mathematics there have been no attempts to invent a third notion dual to both measure and category while at the level of recursive functions this is achieved in this paper, opening even more possibilities on the resource-bounded level.

For recursive functions there is a third possible approach to make precise the notion "the set of functions is small," namely "the set of functions is identifiable." Gold[Gol67] considered the reconstruction of computer programs from sample computations. Nowadays this process is called "learning," "identification" or "inductive inference." The values $f(0), f(1), f(2), \dots$ of a function are given to the identifying algorithm, and a valid program which computes this function, is to be constructed. Various types of identification are considered (finite identification, identification in the limit, -see [AS83, Fre91]). It turns out that the class of all recursive functions is not identifiable (according to any of these identification types). Identifiable sets of recursive functions might be considered "small," however not in terms of measure or category.

In [FFG+98] an attempt was made to discuss the relative sizes of learnable sets using effective notions of measure and category. It was shown that, for all known identification types, the complements of all identifiable sets of functions are huge both in measure and in category. This seemed to eliminate any possibility of proving an analogue of the Duality Principle. In [AFS96] another, more general inference type, was introduced with the intent of trying to find an identification type which might be useful to construct the third notion dual simultaneously to measure and category. In this paper we have succeeded in finding such an identification type. The new type is very closely related to one considered in [BFS96].

2 Technical Preliminaries

The natural numbers are denoted by **N**. The complement of a set S, with respect to **C** is denoted by $\bar{S}$. For functions f and g, $f \sqsubseteq g$ means that $f(x) = g(x)$ for all x in the domain of f. This work is concerned primarily with the recursive (computable) functions as formalized in [MY78, Smi94a] and many other places. $\mathcal{R}$ is the set of $\{0,1\}$ valued recursive functions. This paper is also concerned with the ideas of measure and category that are defined over point sets in real interval $[0,1]$. In order to compare the learning of recursive functions with with either measure or category, it is necessary to represent functions as points in the interval $[0,1]$. To do so, it will be convenient to represent a function by a sequence of values from its range. Such a representation is called a *string* representation. So, for example, the sequence $01^2 0^4 1^\infty$ represents the (total) function:

$$f(x) = \begin{cases} 0 & \text{if } x = 0 \text{ or } 3 \leq x \leq 6, \\ 1 & \text{if } 1 \leq x \leq 2 \\ 1 & \text{otherwise.} \end{cases}$$

We tersely give review the notions of effective measure and category, omitting many of the details. The idea of *measure* comes from [Bor05, Bor14], see [Oxt71]. In the recursion-theoretic setting it is more convenient to use the martingale functions [Lut93]. "Small" sets in terms of measure are those with effective measure 0.

Next we review the the notion of category [Kur58, Oxt71]. A set A is *nowhere dense* if every interval has a subinterval contained in the complement of A. A set is of the *first category* if it can be represented as a countable union of nowhere dense sets. "Small" sets in terms of category are the sets of the first category. Sets of the first category can also be characterized in terms of the Banach-Mazur game [Oxt71]. The recursion-theoretic counterpart of the "set of the first category" is "effectively meager set" as defined by K. Mehlhorn [Meh73]. L. Lisagor showed that Mehlhorn's definition was equivalent to a Banach-Mazur game [Lis81].

Definition 1. *(Due to Mehlhorn [Meh73]) A set $C \subseteq \mathcal{R}$ is effectively meager iff there is a sequence $\{h_k\}_{k\in\omega}$ of uniformly recursive functions $h_k : \{0,1\}^* \to \{0,1\}^*$ such that $\sigma \sqsubseteq h_k(\sigma)$ for all k, σ, and for every $f \in C$ there is k such that $h_k(\sigma) \not\sqsubseteq f$ for all σ. (This formalizes that C is contained in an effective union of effectively nowhere dense sets.)*

Inductive inference emerged when the philosophers became interested in modeling the scientific method [Put75]. An *inductive inference machine* (IIM) is an algorithmic device that inputs the graph of a recursive function and, while doing so, outputs a sequence of computer programs. Typically, an IIM M learns a recursive function f, if M, when given the graph of f as input, outputs a sequence a of programs that converges to a program that computes f. Each IIM M learns some set of recursive functions denoted by $EX(M)$. EX is used to denote the collection of all such sets as M varies across IIMs. Several variations of the Gold model of learning in the limit have been discussed in the literature. Each such variation results in what is called an *identification type*. Our results will apply to some, but far from all, of the investigated identification types.

3 The Triality Theorem

The proof of our main result makes extensive use of the following lemma.

Lemma 1. *Let $\mathcal{X}$ be a subset of $\mathcal{R}$. Let G be a family of sets $G \subset 2^{\mathcal{X}}$ such that*

1. *G is an ideal in $2^{\mathcal{X}}$, i.e.*
   (a) $[c \in G$ and $d \in G] \Rightarrow c \cup d \in G$; and
   (b) $[a \in 2^{\mathcal{X}}$ and $c \in G] \Rightarrow a \cap c \in G$;
2. *the union of all the sets in G is $\mathcal{X}$;*
3. *G has a countable subfamily K such that an arbitrary element of G is contained in some element of K; and*
4. *the complement of every set in G contains an infinite set from G.*

Then $\mathcal{X}$ can be divided into an infinite sequence of disjoint infinite sets $\{\mathcal{X}_n\}$ such that an arbitrary subset $E \subset \mathcal{X}$ belongs to G iff E is a subset of a finite union of sets in $\{\mathcal{X}_n\}$.

Proof: Since the subfamily K is countable, let $K_0, K_1, \cdots$ be a (non effective) sequence of the sets in K such that K_0 is infinite. Let H_n denote $K_0 \cup K_1 \cup \cdots \cup K_n$. Define a sequence of sets by:

$$L_n = \begin{cases} H_0 & \text{if } n = 0; \\ \left(H_n - \bigcup_{i=0}^{n-1} L_i\right) & \text{if } n \neq 0 \text{ and the set } (H_n - \bigcup_{i=0}^{n-1} L_i) \text{ is infinite;} \\ \emptyset & \text{otherwise} \end{cases}$$

It follows from (4.) that there are infinitely many sets in $\{K_n\}$, and there are infinitely many infinite sets in $\{L_n\}$. Hence the union $\cup_{n=1}^{\infty} L_n = \mathcal{X}$. By construction, the sets L_n are either empty or infinite. Now, we remove the empty sets from the list $\{L_n\}$ but do not change the relative ordering of the remaining sets. The new list is denoted by $\{\mathcal{X}_n\}$. These sets are disjoint and infinite. Furthermore, their union is $\mathcal{X}$. If $E \subset G_\alpha \in G$, then it follows from (3.) that there is a K_n such that $E \subset K_n$ and *either* $E \subset L_n$ (if L_n is nonempty) *or* $E \subset L_m$ (if $L_n, L_{n+1}, \cdots, L_{m-1}$ are all empty and L_m is nonempty). Hence, E is in the finite union $\cup_{i=0}^{n} \mathcal{X}_i$ or $\cup_{i=0}^{m} \mathcal{X}_i$, depending on the emptyness of L_n. By (1a), any finite union of sets in $\{\mathcal{X}_n\}$ is in G, and any subset of E of such a finite union is in G by (1b). ☒

Lemma 1 has three corollaries, one for each of measure, category and identifiability.

Corollary 1. *Let $\mathcal{X}$ be a subset of $\mathcal{R}$. Let G be the family of sets of $\{0,1\}$ valued functions belonging to $\mathcal{X}$ with effective measure 0. Then $\mathcal{X}$ can be divided into an infinite sequence of disjoint infinite sets $\{\mathcal{X}_n\}$ such that an arbitrary subset E of $\mathcal{X}$ is of effective measure 0 iff E is a subset of a finite union of the sets in $\{\mathcal{X}_n\}$.*

Proof: The family G of the sets with effective measure 0 satisfy (1) through (4) of Lemma 1 with the subfamily K consisting of all the sets of $\{0,1\}$ valued functions f such that a certain partial recursive martingale succeeds on f. The family G is of the cardinality $2^{\aleph_0}$ while the family K is countable. ☒

Corollary 2. *Let $\mathcal{X}$ be a subset of $\mathcal{R}$. Let G be the family of all the effectively meager sets of $\{0,1\}$ valued functions belonging to $\mathcal{X}$. Then $\mathcal{X}$ can be divided into an infinite sequence of disjoint infinite sets $\{\mathcal{X}_n\}$ such that an arbitrary subset E of $\mathcal{X}$ is effectively meager iff E is a subset of a finite union of the sets in $\{\mathcal{X}_n\}$.*

Proof: The family G of all the effectively meager sets satisfy (1) through (4) of Lemma 1 with the subfamily K consisting of all the sets of $\{0,1\}$ valued functions defined by a partial recursive strategy of one player in the Banach-Mazur game (against all possible partial recursive strategies of the other player). ☒

To tie these notions in with inductive inference, we must first give names to the collection of identification types that our results apply to.

Definition 2. *An identification type I is* smooth *if:*

1. *All the finite classes of recursive $\{0,1\}$ valued functions are I-identifiable,*
2. *$\mathcal{R}$ is not I-identifiable,*
3. *any subclass of any I-identifiable class is I-identifiable, and*
4. *the union of any two I-identifiable classes is I-identifiable.*

Corollary 3. *Let $\mathcal{X}$ be a subset of $\mathcal{R}$. Let G be the family of the I-identifiable sets of $\{0,1\}$ valued functions belonging to $\mathcal{X}$. Then $\mathcal{X}$ can be divided into an infinite sequence of disjoint infinite sets $\{\mathcal{X}_n\}$ such that an arbitrary subset E of $\mathcal{X}$ is I-identifiable iff E is a subset of a finite union of the sets in $\{\mathcal{X}_n\}$.*

Definition 3. *An identification type I is* powerful *if there is a set U of recursive $\{0,1\}$ valued functions such that*

1. *U is I-identifiable,*
2. *U is not a set of effective measure 0, and*
3. *U is not an effectively meager set.*

Now we come to the triality theorem. The proof of this result makes use of the following theorem due to Lisagor:

Theorem 2. [Lis81] *$\mathcal{R}$ can be divided into two disjoint sets A' and B' such that*

1. A' is a set of effective measure 0, and
2. B' is an effectively meager set.

Theorem 3. *Let I be a smooth and powerful identification type. Then there are three one-to-one mappings ϕ, ψ and η taking $\mathcal{R}$ to $\mathcal{R}$ such that for an arbitrary $f \in \mathcal{R}$ and and arbitrary $E \subset \mathcal{R}$:*

1. $\eta(\psi(\phi(f))) = \psi(\phi(\eta(f))) = \phi(\eta(\psi(f))) = f$,
2. E is I-identifiable iff $\phi(E)$ is effectively meager,
3. E is effectively meager iff $\psi(E)$ is of effective measure 0, and
4. E is of effective measure 0 iff $\eta(E)$ is I-identifiable.

Proof: Since I is a powerful identification type, there is a set $U \subset \mathcal{R}$ such that U is I-identifiable but U is not of partial recursive effective measure 0 and U is not effectively meager. Choose A' and B' from Theorem 2. Let A denote $A' - U$ and B denote $B' - U$. As subsets, A is a set of effective measure 0 and B is a set of the first category. Hence, we have divided $\mathcal{R}$ into three disjoint subsets A, U and B situated as in the following table.

	A	U	B
measure	0	1	
I-indentifiability	no	yes	no
category	second		first

Now we define the (non effective) mappings ϕ, ψ and η such that

1. if $f \in A$, then $\phi(f) \in U$, $\psi(f) \in U$ and $\eta(f) \in U$, and
2. if $f \in U$, then $\phi(f) \in B$, $\psi(f) \in B$ and $\eta(f) \in B$, and
3. if $f \in B$, then $\phi(f) \in A$, $\psi(f) \in A$ and $\eta(f) \in A$.

To do this, we first define, in parallel, $\phi(f)$ for $f \in U$, $\psi(f)$ for $f \in B$ and $\eta(f)$ for $f \in A$. This is the easiest of what will be three cases. Merely list (in a non effective way) the three infinite countable sets of functions $U = \{f_0, f_1, \cdots\}$, $B = \{g_0, g_1, \cdots\}$, $A = \{h_0, h_1, \cdots\}$ and define $\phi(f_i) = g_i$, $\psi(g_i) = h_i$ and $\eta(h_i) = f_i$.

Next, we define in parallel $\phi(f)$ for $f \in B$, $\psi(f)$ for $f \in A$ and $\eta(f)$ for $f \in U$. It follows from Corollary 3 (taking $\mathcal{X} = B$) that B can be divided into an infinite sequence of disjoint infinite sets $\{B_n\}$ such that an arbitrary subset E of B is I-identifiable iff E is a subset of a finite union of the sets $\{B_n\}$. It follows from Corollary 2 (taking $\mathcal{X} = A$) that A can be divided into an infinite sequence of disjoint infinite sets $\{A_n\}$ such that an arbitrary subset E of A is effectively meager iff E is a subset of a finite union of the sets $\{A_n\}$. It follows from Corollary 1 (taking $\mathcal{X} = U$) that U can be divided into an infinite sequence of disjoint infinite sets $\{U_n\}$ such that an arbitrary subset E of U is of effective measure 0 iff E is a subset of a finite union of the sets $\{U_n\}$.

In order to define the needed values of $\phi(f)$, $\psi(f)$ and $\eta(f)$, we list (non effectively), for an arbitrary i, $B_i = \{g^i_0, g^i_1, g^i_2, \cdots\}$, $A_i = \{h^i_0, h^i_1, h^i_2, \cdots\}$, $U_i = \{f^i_0, f^i_1, f^i_2, \cdots\}$, and define $\phi(g^i_j) = h^i_j$, $\psi(h^i_j) = f^i_j$ and $\eta(f^i_j) = g^i_j$.

Finally, we define, in parallel, $\phi(f)$ for $f \in A$, $\psi(f)$ for $f \in U$ and $\eta(f)$ for $f \in B$. It follows from Corollary 3 (taking $\mathcal{X} = A$) that A can be divided into an infinite sequence of disjoint infinite sets $\{A'_n\}$ such that an arbitrary subset E of A is I-identifiable iff E is a subset of a finite union of the sets $\{A'_n\}$. It follows from Corollary 2 (taking $\mathcal{X} = U$) that U can be divided into an infinite sequence of disjoint infinite sets $\{U'_n\}$ such that an arbitrary subset E of A is effectively meager iff E is a subset of a finite union of the sets $\{U'_n\}$. It follows from Corollary 1 (taking $\mathcal{X} = B$) that B can be divided into an infinite sequence of disjoint infinite sets $\{B'_n\}$ such that an arbitrary subset E of U is of effective measure 0 iff E is a subset of a finite union of the sets $\{B'_n\}$.

In order to define the needed values of $\phi(f)$, $\psi(f)$ and $\eta(f)$, we list (non effectively), for an arbitrary i, $A_i = \{\bar{h}^i_0, \bar{h}^i_1, \bar{h}^i_2, \cdots\}$, $U_i = \{\bar{f}^i_0, \bar{f}^i_1, \bar{f}^i_2, \cdots\}$, $B_i = \{\bar{g}^i_0, \bar{g}^i_1, \bar{g}^i_2, \cdots\}$, and define $\phi(\bar{h}^i_j) = \bar{f}^i_k$, $\psi(\bar{f}^i_j) = \bar{g}^i_j$, and $\eta(\bar{g}^i_j) = \bar{h}^i_j$.

As a direct consequence of the construction, assertion 1.) of the theorem holds. Now we prove assertion 2.) of the theorem. Suppose E is I-identifiable. Represent $E = E_A \cup E_U \cup E_B$ where $E_A = E \cap A$, $E_U = E \cap U$ and $E_B = E \cap B$. Since $E_U \subseteq U$, $\phi(E_U) \subseteq B$, and B is effectively meager. Hence, $\phi(E_U)$ is effectively meager as well.

E_A is I-identifiable as it is a subset of the I-identifiable set E. Hence, it follows from Corollary 3 (taking $\mathcal{X} = A$) that E_A is contained in a finite union of the sets A'_n. Choose m such that $E_A \subseteq \bigcup_{n=0}^{m} A'_n$. By the definition of ϕ, $\phi(E_A) \subseteq \bigcup_{n=0}^{m} U'_n$, and $\phi(E_A)$ is effectively meager by Corollary 2 (with $\mathcal{X} = U$).

E_B is also I-identifiable as it is a subset of the I-identifiable set E. Hence, it follows from Corollary 3 (taking $\mathcal{X} = B$) that E_B is contained in a finite union of the sets B_n. Choose s such that $E_B \subseteq \bigcup_{n=0}^{s} B_n$. Then, by the definition of ϕ, $\phi(E_B) \subseteq \bigcup_{n=0}^{s} A_s$ and $\phi(E_B)$ is effectively meager by Corollary 2 (with $\mathcal{X} = A$).

It follows from the fact that $E = E_A \cup E_U \cup E_B$ that $\phi(E) = \phi(E_A) \cup \phi(E_U) \cup \phi(E_B)$, i.e. $\phi(E)$ is effectively meager as the union of three effectively meager sets.

Assertions 3.) and 4.) are proved similarly. ☒

As a consequence of Theorem 3 we have the following "Triality Principle."

Theorem 4. *Let P be an arbitrary assertion using only the notions of pure set theory and the notion "set of effective measure 0." Let $P^\star$ be the assertion obtained by substituting "I-identifiable set" for 'set of effective measure 0" everywhere in P. Let $P^{\star\star}$ be the assertion obtained by substituting "effectively meager set" for 'set of effective measure 0" everywhere in P. Then the assertions P, $P^\star$ and $P^{\star\star}$ are either all true or all false.*

Proof: (sketch) Theorem 4 is easily implied by Theorem 3 in spite of the algorithmic nature of the notions of measure, category and identifiability on the one hand and the non effectiveness of the mappings ϕ, ψ and η on the other hand. The crucial point in these proofs is the presentation of all the considered sets as finite unions from the "small" sets $\{A_N\}$, $\{B_n\}$, $\{U_n\}$, $\{A'_N\}$, $\{B'_n\}$ and $\{U'_n\}$. ⊠

4 Suitable Identification Types

A large variety of identification types have been studied in inductive inference [OSW86]. However, only a few of them are smooth. The earliest defined identification type that turned out to be smooth is *reliable* inference defined in [BB75] and investigated in [Min76]. Unfortunately, this type is not powerful. Hence, Theorem 3 does not apply.

Theorem 5. *Let $U \subset \mathcal{R}$ such that U can be reliably identified. Then U is an effectively meager set with partial recursive effective measure 0.*

An identification type that is suitable is a slight modification to the recently introduced *learning with confidence* [BFS96]. A *Confidence inductive inference machine* (CIIM) works like an IIM, except that it outputs a sequence of hypothesis with belief levels (rational numbers between 0 and 1): $s = p_0/c_0, p_1/c_1, p_2/c_2, \cdots$. We say that a CIIM M learns a recursive function f if on input from the graph of f it outputs a sequence $p_0/c_0, p_1/c_1, p_2/c_2, \cdots$ such that for any i, the belief levels associated with p_i are monotone nondecreasing and there exists a unique p with the belief level for p approaching 1 and $\varphi_p = f$. In this way, each CIIM learns a set of recursive functions. The collection of all such sets is denoted by CEX.

The type CEX has most of the needed features, but lacks something essential, namely, $\mathcal{R}$ is CEX-identifiable[BFS96]. However one cannot CEX-identify the minimal programs for all the set $\mathcal{R}$.

Theorem 6. [BFS96] *Suppose $\varphi_0, \varphi_1, \cdots$ is an acceptable programming system. $\mathcal{R}$ is not CEX-identifiable if we demand that minimal indices in $\varphi_0, \varphi_1, \cdots$ are to be found.*

Since we need smoothness to hold for an identification type, we consider the identification type $RMCEX_\varphi$: "Learning with confidence of the minimal indices in the acceptable programming system $\varphi_0, \varphi_1, \cdots$ with that additional restriction that if the target function is not from the identifiable set, then no hypothesis gets assigned a confidence level growing to 1." This identification type incorporates all the features of reliable identification and much more. It should be noted that this type depends on the particular acceptable programming system used.

Theorem 7. *For an arbitrary acceptable programming system $\varphi_0, \varphi_1, \cdots$, the identification type $RMCEX_\varphi$ is smooth and powerful.*

Proof: Use the set from Theorem 4 of [FFG+95]. Details omitted.

Finally, we mention that $RMCEX_\varphi$ is not a *reasonable* identification type in terms of the definition in [FFG+95]. Theorem 17 of [FFG+95] says that if U is I-identifiable with I being a reasonable identification type, then $\mathcal{R}-U$ is both not effectively meager and not of effective measure 0. If one could construct a $RMCEX_\varphi$-identifiable set of functions whose complement would not be huge, it might lead to a strengthening of our Triality Principle for this identification type.

There are four other smooth and powerful identification types. Notice that they all involve taking the union over some parameter that typically gives rise to a hierarchy.

1. EX-identification by teams of an arbitrary finite size [Smi82, Smi94b],
2. Probabilistic EX-identification with and arbitrary probability [Pit89],
3. EX-identification with procrastination in the number of mindchanges [FS93],
4. EX-identification with procrastination in the number of anomalies [FS93].

5 Applications

In this section we consider several applications of Theorem 4. N. Lusin proved the following theorem about classical, continuous, set theory in 1914 (see [Sie34a]).

Theorem 8. [N. Lusin, 1914] *Assuming the Continuum Hypothesis, an arbitrary set of reals E of the second category has a subset M of cardinality* $2^{\aleph_0}$ *such that every non countable subset of M is a set of the second category.*

A counterpart of this result was subsequently shown for Lebesgue measure in 1934.

Theorem 9. [W. Sierpinski [Sie34a]] *Assuming the Continuum Hypothesis, an arbitrary set of reals E which is not a set of measure 0 has a subset M of cardinality* $2^{\aleph_0}$ *such that every non countable subset of M is a set of measure 0.*

We formulate and prove three theorems that are the counterparts of Theorems 8 and 9 using the set of $\mathcal{R}$ of recursive $\{0,1\}$ valued functions instead of the real numbers. These results appear to be new, despite the recent interest effective category and measure theory. Our triality principle (Theorem 4) allows us to essentially prove all three at the same time. This reveals that the following three theorems are the same from some higher viewpoint.

Theorem 10. *Let I be a smooth identification type. An arbitrary* $E \subset \mathcal{R}$ *which is not I-identifiable has an infinite subset M such that every infinite subset of M is not I-identifiable.*

Theorem 11. *An arbitrary* $E \subset \mathcal{R}$ *which is not a set of effective measure 0 has in infinite subset M such that every infinite subset of M is not a set of effective measure 0.*

Theorem 12. *An arbitrary $E \subset \mathcal{R}$ which is not effectively meager has an infinite subset M such that every infinite subset of M is not effectively meager.*

We end this section with a final trio of results obtained using our powerful new technique.

Theorem 13. *Let I be a smooth identification type. There is a one-to-one mapping F from $\mathcal{R}$ to a subset of $\mathcal{R}$ such that $F(E)$ is I-identifiable iff E is finite.*

Theorem 14. *There is a one-to-one mapping F from $\mathcal{R}$ to a subset of $\mathcal{R}$ such that $F(E)$ is a set of effective measure 0 iff E is finite.*

Theorem 15. *There is a one-to-one mapping F from $\mathcal{R}$ to a subset of $\mathcal{R}$ such that $F(E)$ is effectively meager iff E is finite.*

6 Extended Triality Principle

In the proof of theorem 3 the mappings ϕ, ψ, η are very similar. They map the sets A, B and U into the same A, B and U. This suggests that stronger forms of Theorems 3 and 4 might be true. Consider the following:

Assertion 1 *Let I be a smooth and powerful identification type. Then there is a one-to-one mapping $F : \mathcal{R} \to \mathcal{R}$ such that for an arbitrary $f \in \mathcal{R}$ and for an arbitrary $E \subset \mathcal{R}$:*

1. *$F(F(F(f))) = f$,*
2. *E is I-identifiable $\Leftrightarrow$ $F(E)$ is effectively meager,*
3. *E is effectively meager $\Leftrightarrow$ $F(E)$ is of effective measure 0, and*
4. *E is of effective measure 0 $\Leftrightarrow$ $F(E)$ is I-identifiable.*

Assertion 1 would imply the following stronger form of Theorem 4:

Assertion 2 *Let P be an arbitrary statement using only the notions of pure set theory and the notions "set of effective measure 0," "effectively meager set," and "I-identifiable set." Let $P^\star$ be the statement obtained from P by simultaneously substituting "effectively meager set" for "set of effective measure 0," "I-identifiable set" for "effectively meager set," and "set of effective measure 0" for "I-identifiable set" everywhere in P. Let $P^{\star\star}$ be the statement obtained from P by simultaneously substituting "I-identifiable set" for "set of effective measure 0," "effectively meager set" for "I-identifiable set," and "set of effective measure 0" for "effectively meager set" everywhere in P. Then P, $P^\star$ and $P^{\star\star}$ are either all true or all false.*

Unfortunately, Assertion 2 (and hence Assertion 1 as well) is false for at least some smooth and powerful identification types I. Indeed, the identification type "EX-identification by teams of an arbitrary finite size" is smooth and powerful [Smi82]. However, for any set $U \subset \mathcal{R}$ that is identifiable by this inference type, the set $(\mathcal{R}-U)$ cannot be effectively meager and it cannot be of effective measure 0. Applying Assertion 2 to Theorem 2 yields the following false statement:

$\mathcal{R}$ can be divided into two disjoint sets A' and B' such that

1. A' is an effectively meager set, and
2. B' is learnable by a finite EX team.

Hence, Assertions 1 and 2 are false at least for this one identification type. However, the following extension does hold.

Theorem 16. *Let P be an arbitrary statement using only the notions of pure set theory and the notions "sets of effective measure 0," "effectively meager sets," and "I-identifiable sets." Let $P^\star$ be the statement obtained from P by interchanging the notions "sets of effective measure 0" and "effectively meager sets." Then the statements P and $P^\star$ are either both true of both false.*

Proof: (sketch) The proof is more complicated than merely referring to Theorem 2. Again a mapping $F : \mathcal{R} \to \mathcal{R}$ is constructed such that:

1. $F(F(f)) = f$,
2. E is effectively meager $\Leftrightarrow F(E)$ is of effective measure 0, and
3. E is of effective measure 0 $\Leftrightarrow F(E)$ is effectively meager.
   However, this time we also demand:
4. E is I-identifiable $\Leftrightarrow F(E)$ is I-identifiable.

☒

Theorem 16 and the failure of Assertion 2 imply that not all combinations of the notions measure, category and identifiability are equivalent in spite of the fact that these notions are pairwise dual in the sense of Theorem 4. Consider EX identification by a team of arbitrary finite size. Then, Theorem 4 applies, but the following assertion fails.

Assertion 3 *(False) Let P be an arbitrary statement using only the notions of pure set theory and the notions "set of effective measure 0," "effectively meager set," and "I-identifiable set." Let $P^\star$ be the statement obtained from P by interchanging the notions "I-identifiable set" and "effectively meager set." Then the statements P and $P^\star$ are either both true or both false.*

The failure of Assertions 2 and 3 does not exclude a possibility that for some identification types Assertion 2 may hold. For instance, we conjecture that Assertion 2 may hold for $RMCEX_\varphi$-identification.

7 Directions for Further Research

Because of the failure of Assertion 2, there is no hope to have complete interchangeability of the notions of measure, category and identifiability. However, considering such interchanges is a good source of open problems. Indeed, several prior studies could be viewed as following such a program For example, Bennett and Gill [BG81] proved that $P^A \neq NP^A \neq coNP^A$ for nearly all recursive oracles where "nearly all" was understood as "all but a set of effective measure 0." However, Blum and Impagliazzo [BI87] showed that $P^A = NP^A = coNP^A$ for nearly all oracles where "nearly all" was understood as "all but a set of the first category." It was conjectured that "what is true for nearly all oracles, is true for the empty oracle as well." This conjecture has been disproved for both "all oracles but a set with effective measure 0" and "all oracles except a set of the first category" [Kur83, LFKN92, Fos93, CCG+94]. This conjecture might still be true for a more appropriate notion of "nearly all sets".

Since then, the notions of measure and category have appeared frequently, not only in their effective versions, but also in resource-bounded versions as well. Our Theorem 4 offers a third notion dual to measure and category. It is quite possible that a third approach to the intuitive notion of "nearly all sets" can be useful. In Lebesgue measure, typical reals are called "random," in Baire category, they are called "generic." Generic sets, like random sets, have proved useful in providing a coherent picture of relativized computation. They embody the method of diagonalization in oracle constructions, i.e. requirements which can always be satisfied by a finite extension of the oracle.

Lutz [Lut90] has introduced resource bounded versions of both measure and category. Fenner [Fen91] has studied resource bounded generics. It seems promising to consider the identifiability counterparts of randomness and generics.

8 Acknowledgements

Conversations with Sebastiaan Terwijn were most helpful in sorting out the various notions of effective measure.

References

[AFS96] A. Ambainis, R. Freivalds, and C. Smith. General inductive inference types based on linearly ordered sets. In C. Puech and R. Reischuk, editors, *Proceedings of the* 13th *Symposium on the Theoretical Aspects of Computer Science*, volume LNCS1046, pages 243–253. Springer, 1996.

[AS83] D. Angluin and C. H. Smith. Inductive inference: Theory and methods. *Computing Surveys*, 15:237–269, 1983.

[AS95] K. Ambos-Spies. Resource-bounded genericity. In *Proceedings of the* 10th *Conference on Structure in Complexity Theory,* pages 162–181. IEEE Computer Society, 1995.

[ASMWZ96] K. Ambos-Spies, E. Mayordomo, Y. Wang, and X. Zheng. Resource-bounded genericity, stochasticity and weak randomness. In C. Puech and R. Reischuk, editors, *Proceedings of the* 13th *Symposium on the Theoretical Aspects of Computer Science*, volume LNCS1046, pages 63–74. Springer, 1996.

[BB75] L. Blum and M. Blum. Toward a mathematical theory of inductive inference. *Information and Control*, 28:125–155, 1975.

[BFS96] J. Bārzdiņš, R. Freivalds, and C. Smith. Learning with confidence. In *Proceedings of STACS'96*, 1996.

[BG81] C.G. Bennet and J. Gill. Relative to a random oracle A, $P^a \neq NP^a \neq co\text{-}NP^a$ with probability 1. *SIAM Journal of Computer Science*, 10:96–113, 1981.

[BI87] M. Blum and R. Impagliazzo. Generic oracles and oracle classes. In *Proceedings of the* 28th *Symposium on the Foundations of Computer Science*, pages 118–126. IEEE Computer Soceity, 1987.

[BL96] H. Buhrman and L. Longpré. Compressibility and resource bounded measure. In C. Puech and R. Reischuk, editors, *Proceedings of the* 13th *Symposium on the Theoretical Aspects of Computer Science*, volume LNCS1046, pages 13–24. Springer, 1996.

[BM95a] J. Balcázar and E. Mayordomo. A note on genericity and bi-immunity. In *Proceedings of the* 10th *Conference on Structure in Complexity Theory*, pages 193–196. IEEE Computer Society, 1995.

[BM95b] H. Buhrman and E. Mayordomo. An excursion to the Kolmogorov random strings. In *Proceedings of the* 10th *Conference on Structure in Complexity Theory*, pages 197–205. IEEE Computer Society, 1995.

[Bor05] E. Borel. *Leçons sur les fonctions de variables réeles.* Gauthier-Villars, Paris, 1905.

[Bor14] E. Borel. *Leçons sur la théorie des fonctions.* Gauthier-Villars, Paris, 1914.

[CCG+94] Richard Chang, Benny Chor, Oded Goldreich, Juris Hartmanis, Johan Håstad, Desh Ranjan, and Pankaj Rohatgi. The random oracle hypothesis is false. *Journal of Computer and System Sciences*, 49(1):24–39, August 1994.

[Erd43] P. Erdös. Some remarks on set theory. *Ann. Math*, 44(2):643–646, 1943.

[Fen91] S. Fenner. Notions of resource-bounded category and genericity. In *Proceedings of the* 6th *Conference on Structure in Complexity Theory*, pages 196–212. IEEE Computer Soceity, 1991.

[Fen95] S. Fenner. Resource-bounded Baire category: A stronger approach. In *Proceedings of the* 10th *Conference on Structure in Complexity Theory*, pages 182–192. IEEE Computer Society, 1995.

[FFG+98] L. Fortnow, R. Freivalds, W. Gasarch, M. Kummer, S. Kurtz, C. Smith, and F. Stephan. On the relative sizes of learnable sets. *Theoretical Computer Science*, 197(1-2):139–156, 1998.

[FFG+95] L. Fortnow, R. Freivalds, W. Gasarch, M. Kummer, S. Kurtz, C. Smith, and F. Stephan. Measure, category and learning theory. In *Proceedings of ICALP'95, LNCS Vol. 944*, pages 558–569, 1995.

[Fos93] J. Foster. The generic oracle hypothesis is false. *Information Processing Letters*, 45(2):59–62, 1993.

[Fre91] R. Freivalds. Inductive inference of recursive functions: Qualitative theory. In J. Bārzdiņš and D. Bjørner, editors, *Baltic Computer Science*, pages 77–110. Springer Verlag, 1991. Lecture Notes in Computer Science, Vol. 502.

[FS93] R. Freivalds and C. Smith. On the power of procrastination for machine learning. *Information and Computation*, 107:237–271, 1993.

[Gol67] E. M. Gold. Language identification in the limit. *Information and Control*, 10:447–474, 1967.

[Kum96] M. Kummer. On the complexity of random strings. In C. Puech and R. Reischuk, editors, *Proceedings of the* 13th *Symposium on the Theoretical Aspects of Computer Science*, volume LNCS1046, pages 25–38. Springer, 1996.

[Kur58] C. Kuratowski. *Topologie,* 4th *edition*, volume 20-21 of *Monografie Matematyczne*. Panstwowe Wydawnictwo Naukowe, 1958.

[Kur83] Stuart A. Kurtz. On the random oracle hypothesis. *Information and Control*, 57(1):40–47, 1983.

[LFKN92] C. Lund, L. Fortnow, H. Karloff, and N. Nisan. Algebraic methods for interactive proof systems. *Journal of the ACM*, 39(4), 1992.

[Lis81] L. Lisagor. The Banach-Mazur game. Translated Version of *Matematicheskij Sbornik*, 38:201–206, 1981.

[Lut90] J. Lutz. Category and measure in complexity classes. *SIAM Journal of Computing*, 19:1100–1131, 1990.

[Lut93] J. Lutz. The quantitative structure of exponential time. In *Proceedings of the* 8th *Structure in Complexity Theory Conference*, pages 158–175. IEEE Computer Society, 1993.

[Lut96] J. Lutz. Observations on measures and lowness for Δ_2^P. In C. Puech and R. Reischuk, editors, *Proceedings of the* 13th *Symposium on the Theoretical Aspects of Computer Science*, volume LNCS1046, pages 87–98. Springer, 1996.

[Meh73] K. Mehlhorn. On the size of sets of computable functions. In *Proceedings of the* 14th *Symposium on Switching and Automata Theory*, pages 190–196. IEEE Computer Soceity, 1973.

[Min76] E. Minicozzi. Some natural properties of strong-identification in inductive inference. *Theoretical Computer Science*, 2:345–360, 1976.

[MY78] M. Machtey and P. Young. *An Introduction to the General Theory of Algorithms*. North-Holland, New York, 1978.

[OSW86] D. Osherson, M. Stob, and S. Weinstein. *Systems that Learn*. MIT Press, Cambridge, Mass., 1986.

[Oxt71] J. Oxtoby. *Measure and Category*. Springer-Verlag, 1971.

[Pit89] L. Pitt. Probabilistic inductive inference. *Journal of the ACM*, 36(2):383–433, 1989.

[Put75] H. Putnam. Probability and confirmation. In *Mathematics, Matter and Method*, volume 1. Cambridge University Press, 1975.

[Sie34a] W. Sierpinski. *Hypothèse du continu*, volume 4. Monografie Matematyczne, 1934.

[Sie34b] W. Sierpinski. Sur la dualité entre la première catégore et la mesure nulle. *Fund. Math.*, 28:276–280, 1934.

[Smi82] C. H. Smith. The power of pluralism for automatic program synthesis. *Journal of the ACM*, 29(4):1144–1165, 1982.

[Smi94a] C. Smith. *A Recursive Introduction to the Theory of Computation*. Springer-Verlag, 1994.

[Smi94b] C. Smith. Three decades of team learning. In *Proceedings of AII/ALT'94, Lecture Notes in Artificial Intelligence*. Springer Verlag, 1994.

Verification of Embedded Reactive Fiffo Systems

Frédéric Herbreteau[1,2], Franck Cassez[1], Alain Finkel[3], Olivier Roux[1], and Grégoire Sutre[2,4]

[1] IRCCyN (CNRS UMR 6597), 1, rue de la Noe, 44321 Nantes cedex 3, France
{Franck.Cassez, Olivier.Roux}@irccyn.ec-nantes.fr

[2] LaBRI (CNRS UMR 5800), 351, cours de la Libération, 33405 Talence Cedex, France
{Frederic.Herbreteau, Gregoire.Sutre}@labri.fr

[3] LSV (CNRS UMR 8643), 61, avenue du Président. Wilson, 94235 Cachan cedex, France
finkel@lsv.ens-cachan.fr

[4] ERL, 253 Cory Hall, University of California, Berkeley, CA 94720, USA

Abstract. Reactive Fiffo Systems (RFS) are used to model reactive systems which are able to memorize the events that cannot be processed when they occur. In this paper we investigate the decidability of verification problems for Embedded RFS which are RFS running under some environmental constraints. We show that almost all the usual verification problems are undecidable for the class of Periodically Embedded RFS with two memorizing events, whereas they become decidable for Regularly Embedded RFS with a single memorizing event. We then focus on Embedded Lossy RFS and we show in particular that for Regularly Embedded Lossy RFS the set of predecessors $Pred^*$ is upward closed and effectively computable.

Keywords: Reactive Fiffo Systems, Embedded Systems, Verification, Real-Time Systems, Infinite-State Systems, Decidability.

1 Introduction

Context. Model-checking has become a very popular verification method, since it is fully automatic for finite state systems and it has been applied successfully in particular for VLSI circuits [7]. However, since most systems are intrinsically infinite state, there is a need to design an efficient verification methodology for infinite state systems.

We focus in this paper on *Embedded Reactive Fiffo Systems (Embedded RFS)*, a model for embedded (asynchronous) reactive systems with event memorization, close to SDL [9]. Any Embedded RFS is actually a *closed system*, obtained by synchronization of an environment defining the sequences of input events with a *Reactive Fiffo System* modeling the reactions to these events.

Reactive Fiffo Systems (RFS) were introduced in [8, 22] (and called *Reactive Fiffo Automata* there) to model (asynchronous) reactive systems with event

S. Rajsbaum (Ed.): LATIN 2002, LNCS 2286, pp. 400–414, 2002.

memorizations that are specified in the Electre [8] reactive language. A major feature of this language is that it is possible to store occurrences of events in order to process them later. The way the events are processed is called the *Fiffo* order for *First In First Fireable Out* i.e. we take into account the stored events as soon as possible and in case of conflict the oldest stored occurrence is processed. The number of stored occurrences in the Fiffo queue is theoretically unbounded. Consequently the behavioral model of an Electre program is a RFS which has an infinite number of states [22].

Related work. In a previous work [22], we analyzed RFS *with an implicit environment* of the form Σ^*, that is an environment generating every sequence of events. We proved the following results for RFS: (1) the set *Post** of the reachable states of a RFS is recognizable and effectively computable; (2) the linear temporal logic LTL without the next operator is decidable for RFS.

These results were surprising since RFS are close to *Communicating Finite State Machines (CFSMs)* or *Fifo Automata*, and it is well known that this class has the power of Turing Machines [6, 13]. This difference is mainly due to the fact that RFS (with an implicit environment) are open systems (whereas CFSMs are closed) on the one hand, and from the looping form of the memorizing transitions (which can thus be executed any number of times) which entails that the state space of RFS are downward closed and thus recognizable.

Our Contribution. The purpose of this paper is to analyze RFS *with an explicit environment*, the so-called *Embedded RFS*, which are RFS that run under some *environmental constraints*. Indeed, without an explicit environment, we are able to compute the set of reachable states *Post** [22], but many of these states would not be reachable in a "realistic" environment that would constrain the system.

Embedded RFS are naturally defined as the synchronization between a RFS and an environment given as a labeled transition system. We in particular consider *regular* and *periodic* environments, where the languages of event sequences are regular or periodic respectively. These classes of environments are particularly relevant in the area of real-time systems as for instance, many applications are strongly periodic (e.g. *scheduling*) or mainly composed of *regular* cyclic processes (control of nuclear plants, avionic systems, ...).

The main results of the paper are two fold; we first show that:

(1) Periodically Embedded RFS have the power of Turing machines: most verification problems are undecidable for this class.

Since Embedded RFS are too powerful, even in the "simple" case of periodic environments, we then try to lower their expressiveness in order to obtain decidability results. We show that:

(2) When there is at most one memorizing event, most verification problems are decidable for Regularly Embedded RFS.

(3) Embedded Lossy RFS are well-structured transition systems (WSTS) with effective pred-basis (as defined in [14]), provided that the environment is a WSTS with effective pred-basis itself. Hence the covering problem is decidable, and we can model-check safety properties on these systems.
(4) Moreover, we show that *Pred**, the set of predecessors, is effectively computable for Regularly Embedded Lossy RFS. Consequently model-checking the existential fragment of CTL is decidable for this class.

As emphasized in the conclusion these results give a foundation for the verification of Embedded RFS.

Outline of the paper. The next section recalls some basics and introduces the model of *Embedded RFS*. In Section 3 we prove one of our main results: Periodically Embedded RFS have the power of Turing machines. In Section 4, we focus on a *lossy* version of RFS and we prove our second main result: Embedded Lossy RFS are well structured transition systems. Finally in Section 5 we draw some conclusions and give some hints for future work.

2 Embedded Reactive Fiffo Systems

2.1 Preliminaries

Ordering and quasi-ordering. For a given set X, a binary relation $\leq$ is a *quasi-ordering* if it is transitive and reflexive. The *upward-closure* of $Y \subseteq X$ is $\uparrow Y = \{x' \in X \mid \exists x \in Y,\, x \leq x'\}$. A finite set X is a *finite basis* for an upward-closed set X' if $\uparrow X = X'$. A *well quasi-ordering (wqo)* $(X, \leq)$ is a quasi-ordering $\leq$ such that in every infinite sequence: $x_1, x_2, \ldots, x_i, \ldots$ of elements from X, there exist two integers $i < j$ such that $x_i \leq x_j$. Every pair $(X, \leq)$ such that X is finite is *de facto* a wqo. Furthermore, by definition, a well quasi-ordering $(X, \leq)$ has a finite number of minimal elements (it is well-founded).

Considering a finite alphabet $\Sigma = \{a_1, \ldots, a_n\}$, a finite (resp. infinite) word $u \in \Sigma^*$ is a finite (resp. infinite) sequence of symbols from Σ; u_i denotes the ith letter of u and ε is the empty word. The *shuffle* of two finite words u and v is the set defined by: $u \sqcup\!\sqcup\, v = \{u_1 v_1 \ldots u_n v_n \mid u_1, \ldots, v_n \in \Sigma^* \text{ and } u = u_1 \ldots u_n,\ v = v_1 \ldots v_n\}$. The *sub-word ordering* $\preccurlyeq$ is defined by: $\forall u, v \in \Sigma^*$, $u \preccurlyeq v$ iff u can be obtained from v by removing some letters. The *Parikh mapping* [21] Ψ defines the vector $\Psi(u) = (k_1 \;\ldots\; k_n)$ where k_i is the number of occurrences of a_i in the finite word u.

Labeled Transition Systems. A *labeled transition system* is a 4-tuple $S = (Q, q_0, L, \rightarrow)$ where Q is the set of states, $q_0 \in Q$ is the initial state, L is the set of labels and $\rightarrow \subseteq Q \times L \times Q$ is the transition relation. A *run* of S is a finite or infinite sequence of transitions $q_0 \xrightarrow{l_0} q_1 \xrightarrow{l_1} \cdots \xrightarrow{l_n} q_{n+1} \cdots$. We denote $\rightarrow^*$ the reflexive and transitive closure of $\rightarrow$. A state $q' \in Q$ is *reachable* in S from state $q \in Q$ iff $q \rightarrow^* q'$. *Post**(S) is the set of reachable states in S from q_0. For a given state $q \in Q$ and a given label $l \in L$, $Pred_l(q) = \{q' \in Q \mid q' \xrightarrow{l} q\}$ is the set of the predecessors of q by l. Let $in(q)$ (resp. $out(q)$) denotes the sets of *incoming labels* (resp. *outgoing labels*) from $q \in Q$. Finally, the sets of the *predecessors* of q is $Pred = \bigcup_{l \in in(q)} Pred_l(q)$ and $Pred^*$ denotes its transitive and reflexive closure.

Problems of interest. For a given labeled transition system $S = (Q, q_0, L, \rightarrow)$ and a wqo $(Q, \leq)$, we can define the following problems :

- The *boundedness problem*: Is $Post^*(S)$ finite ?
- The *covering problem*: For any states $q \in Post^*(S)$ and $q' \in Q$, is there any $q'' \in Q$ such that $q' \leq q''$ and q'' is reachable from q ?
- The *termination problem*: Is every run of S finite ?
- The *(resp. effective) recognizability problem*: Is there a (resp. computable) finite representation for $Post^*(S)$?
- The *recurrent locality problem*: Is there a run of S that visits a given $q \in Q$ infinitely often ?
- The *LTL (resp. CTL, EG, EF) model-checking problem*: For a given LTL (resp. CTL, EG, EF) formula[1] ϕ, does every run of S satisfy ϕ ?

2.2 Reactive Fiffo Automata

Reactive systems are usually composed of a set of tasks which are activated, preempted or ended when some events occur. Reactive Fiffo Automata (RFA) [8] aim at modeling such *asynchronous* systems, thus they follow two main guidelines: *memorization* and *instantaneous reaction.* Indeed, the asynchronous assumption implies that the tasks have a non null duration, thus they may prevent the processing of events, which may thus need to be memorized. For reactive systems, we have chosen to batch process memorized events *as soon as possible* and in case two memorized events can be processed, to give the priority to the older one. This guarantees fairness among the memorized occurrences. Moreover, batch processing has priority against processing of new incoming occurrences. This type of processing is called *Fiffo*[2]. Hence, RFA can serve as a semantic model for Electre programs [8], as well as for SDL [9] specifications, and for any formalism for reactive systems with memorization.

Figure 1(a) depicts a RFA modeling a readers/writers mutual exclusion protocol with two readers R_1 and R_2 and one writer W. In each location the running tasks are given ($\varnothing$ means no task is running, R_1 that reader 1 is reading and so on.) The events $\overline{W}, \overline{R_1}, \overline{R_2}$ correspond to the end of a writing or reading task. A transition of the form $\varnothing \xrightarrow{r_1} R_1$ means that event r_1 can be processed in location $\varnothing$ and leads to a state where R_1 is running; $\varnothing \xrightarrow{?r_1} R_1$ means that processing a memorized occurrence of event r_1 leads to R_1 and finally $R_1 \xrightarrow{!w} R_1$ means that if w occurs when the system is in location R_1 it has to be stored in the queue. (Notice that two states correspond to task R_1 running but they are not bisimilar.)

---

[1] Recall that the EF (resp. EG) fragment of CTL uses predicates, boolean operators, the one-step next and the EF (resp. EG) operators.

[2] First In First Fireable Out.

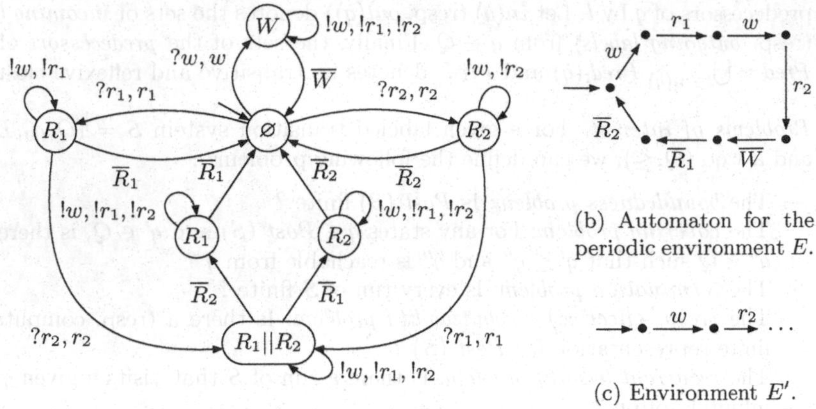

(a) A RFA for a 2 readers/1 writer problem.

(b) Automaton for the periodic environment E.

(c) Environment E'.

Fig. 1. Some examples of RFA and environments.

Definition 1 (Reactive Fiffo Automaton (RFA)). *A* Reactive Fiffo Automaton (RFA) *is a 4-tuple* $R = (Q_R, q_R^0, A_R, \rightarrow_R)$ *where:* Q_R *is a finite set of* locations *with the distinguished* initial location q_R^0, $A_R = \Sigma \cup (\{!, ?\} \times \Sigma_M)$ *is the finite set of* actions *(labels) where* Σ *is the finite set of* events, $\Sigma_M \subseteq \Sigma$ *is the finite set of* memorized events *and* $\Sigma_F = \Sigma \setminus \Sigma_M$ *is the set of* fleeting events *and finally,* $\rightarrow_R \subseteq Q_R \times A_R \times Q_R$ *is the deterministic transition relation such that* $\forall q \in Q_R, \forall e \in \Sigma_M$*:*

1. *either* $q \xrightarrow{!e}_R q$,
2. *or* $\exists q' \in Q_R$ *such that* $q \xrightarrow{e}_R q'$ *and* $q \xrightarrow{?e}_R q'$. □

The memorizing transitions are loops: the processing of the event is delayed and the system does not allow any state change. Notice that the memorizing capability does not apply to all the events of the RFA. Indeed, the events' set Σ is divided in two parts: the memorized events Σ_M and the fleeting events Σ_F, with $\Sigma_M \cap \Sigma_F = \emptyset$. One could also consider static priority between events, like "if e_1 is in the Fiffo list, batch process e_1 even if some other e_2 is before e_1 in the list and can be processed." This can already be specified by RFA, and thus our work includes this case.

Following points (1) and (2) the RFA never *looses* a memorized event occurrence, and as entailed by (2), the processing (e) and the batch processing ($?e$) of an event have the same effect w.r.t. to the state's change they bring about.

Definition 2 (Reactive Fiffo System (RFS)). *A* Reactive Fiffo System (RFS)*, giving the semantics of the RFA* $R = (Q_R, q_R^0, A_R, \rightarrow_R)$*, is the (infinite) labeled transition system* $S = (Q_S, q_S^0, A_S, \rightarrow_S)$ *defined by:* $Q_S = Q_R \times \Sigma_M^*$ *is the set of states where one distinguishes the initial state* $q_S^0 = (q_R^0, \varepsilon)$, $A_S = A_R$ *is the finite set of actions (labels) and the transition relation* $\rightarrow_S$ *is the smallest subset of* $Q_S \times A_S \times Q_S$ *such that:*

1. $(q,w) \xrightarrow{!e}_S (q,w.e)$ *if* $q \xrightarrow{!e}_R q$ *and* $w \in \left(\Sigma^!_q\right)^*$ *(memorizing)*
2. $(q,w) \xrightarrow{e}_S (q',w)$ *if* $q \xrightarrow{e}_R q'$ *and* $w \in \left(\Sigma^!_q\right)^*$ *(processing)*
3. $(q,w_1.e.w_2) \xrightarrow{?e}_S (q',w_1.w_2)$ *if* $q \xrightarrow{?e}_R q'$ *and* $w_1 \in \left(\Sigma^!_q\right)^*$ *(batch processing)*
4. $(q,w) \xrightarrow{e}_S (q,w)$ *if* $e \notin (out(q) \cup \Sigma_M)$ *and* $w \in \left(\Sigma^!_q\right)^*$

with $\Sigma^!_q = \left\{e \in \Sigma_M \mid q \xrightarrow{!e}_R q\right\}$. □

A state (q,w) is *stable* if $w \in \left(\Sigma^!_q\right)^*$ otherwise it is *unstable* (in the latter case an event is to be batch processed). Priority is given to the batch processings since the other transitions are only enabled in stable states. Moreover the batch processed occurrence (if any) is the oldest one as condition $w_1 \in \left(\Sigma^!_q\right)^*$ of (3) above implies. Finally, by point (4), the RFS is complete and thus reacts to every event (whereas the RFA may not be complete w.r.t fleeting events).

Example 1. The sequence below depicts a possible run of the readers/writers RFA in Figure 1(a).

$$(\varnothing,\varepsilon) \xrightarrow{w}_S (W,\varepsilon) \xrightarrow{!r1}_S (W,r_1) \xrightarrow{!w}_S (W,r_1w) \xrightarrow{!r_2}_S (W,r_1wr_2) \xrightarrow{\overline{W}}_S$$
$$(\varnothing,r_1wr_2) \xrightarrow{?r_1}_S (R_1,wr_2) \xrightarrow{?r_2}_S (R_1\|R_2,w) \xrightarrow{\overline{R_1}}_S (R_2,w) \xrightarrow{\overline{R_2}}_S (\varnothing,w)\ldots$$

The transition $(R_1,wr_2) \xrightarrow{?r_2}_S (R_1\|R_2,w)$ clearly depicts the Fiffo (First In First Fireable Out) policy. Indeed, r_2 is batch processed because (i) the configuration (R_1,wr_2) is unstable, and (ii) the events are batch processed as soon as possible and with priority to the oldest one.

2.3 Embedded Reactive Fiffo Systems

An *environment* $E = (Q_E, q^0_E, A_E, \rightarrow_E)$ for a RFS $S = (Q_S, q^0_S, A_S, \rightarrow_S)$ is a labeled transition system, with $A_E = \Sigma$, defining the input words of the system (i.e. sequences of events in Σ^*). The *Embedded Reactive Fiffo System* $S \parallel E$ is obtained by synchronization of S and E:

Definition 3 (Embedded Reactive Fiffo Systems). *An* Embedded Reactive Fiffo System (Embedded RFS) *is a (infinite) labeled transition system* $S \parallel E = (Q_{S\|E}, q^0_{S\|E}, A_{S\|E}, \rightarrow_{S\|E})$ *defined by:* $Q_{S\|E} = Q_R \times (\Sigma_M)^* \times Q_E$ *is the set of configurations and* $q^0_{S\|E} = \langle q^0_R, \varepsilon, q^0_E\rangle$ *is the initial configuration,* $A_{S\|E} = A_S$ *is the set of actions, and finally the transition relation* $\rightarrow_{S\|E}$ *is the smallest subset of* $Q_{S\|E} \times A_{S\|E} \times Q_{S\|E}$ *such that:*

1. $\langle q_R, w, q_E\rangle \xrightarrow{!e}_{S\|E} \langle q_R, we, q'_E\rangle$ *if* $(q_R,w) \xrightarrow{!e}_S (q_R, we)$ *and* $q_E \xrightarrow{e}_E q'_E$,
2. $\langle q_R, w, q_E\rangle \xrightarrow{e}_{S\|E} \langle q'_R, w, q'_E\rangle$ *if* $(q_R,w) \xrightarrow{e}_S (q'_R, w)$ *and* $q_E \xrightarrow{e}_E q'_E$,
3. $\langle q_R, w_1ew_2, q_E\rangle \xrightarrow{?e}_{S\|E} \langle q'_R, w_1w_2, q_E\rangle$ *if* $(q_R, w_1ew_2) \xrightarrow{?e}_S (q'_R, w_1w_2)$.

Notice that by definition 2, rule 3 has priority over rules 1 and 2. □

The runs of $S \parallel E$ are called the *constrained runs* of the system. The notions of stability and unstability naturally extend to Embedded RFS. In the sequel, Embedded RFS(n) denotes the class of Embedded RFS with n memorized events ($|\Sigma_M| = n$).

Considering the environment E in Figure 1(b), the execution of Example 1 is the prefix of a constrained run of the RFA in Figure 1(a). But this execution is discarded by E' in Figure 1(c) since the transition $(W, \varepsilon) \xrightarrow{!r_1}_S (W, r_1)$ is disabled by r_2 in E'.

Observe that, provided the set of events Σ is non empty, a RFS has no deadlock state, since in every state it is able to react to an event occurrence, either by memorizing it, by losing it (if this event is fleeting, by (4) in Definition 2), or by processing it. Hence any Embedded RFS with a non-terminating environment does not terminate. In the rest of the paper, we won't discuss the termination problem anymore, and by "problems of interest" we mean all the problems defined in Section 2.1 except the termination problem.

3 The Power of Embedded RFS

In this section, we first prove that RFS having two memorized events and with a periodic environment are deterministic counter automata. It follows that all the problems of interest are undecidable. Then, we focus on RFS with at most one memorized event constrained by regular environments and prove that their reachability set is effectively recognizable.

3.1 Undecidability of Periodically Embedded RFS(2)

A *Periodically Embedded RFS* is an Embedded RFS the environment of which recognizes a periodic sequence of events u^* where u is a finite word.

Definition 4 (n-counter automata [20]). *A* n-counter automaton C *is defined by the tuple* $(Q_C, q_C^0, \{c_1, \ldots, c_n\}, \rightarrow_C)$ *where:* Q_C *is a finite set of states with initial state* q_C^0*;* $\{c_1, \ldots, c_n\}$ *is a finite set of* counters *with values in* $\mathbb{N}$ *and* $\rightarrow_C \subseteq Q_C \times (\{c_1, \ldots, c_n\} \times \{{+}{+}, {-}{-}, {=}0?\}) \times Q_C$ *is the transition relation where* c_{k++} *denotes the increasing of* c_k *whereas* c_{k--} *is its decreasing and* $c_{k=0?}$ *represents the zero testing operation. A n-counter automaton is* deterministic *provided for each state* q *there is:*

- *either an increasing transition:* $q \xrightarrow{c_{k++}}_C q'$,
- *or a decreasing and a zero testing transition dealing with the same counter:* $q \xrightarrow{c_{k--}}_C q''$ *and* $q \xrightarrow{c_{k=0?}}_C q'$. □

The semantics of a n-counter automaton C is the labeled transition system $S = (Q \times \mathbb{N}^2, \langle q_0, 0, \ldots, 0\rangle, \rightarrow)$ where $\rightarrow$ is the smallest subset of $(Q \times \mathbb{N}^2) \times (Q \times \mathbb{N}^2)$ given by ($m_1, \ldots, m_n \in \mathbb{N}$):

1. $\langle q, m_1, \ldots, m_n\rangle \rightarrow \langle q', m_1 + 1, \ldots, m_n\rangle$ if $q \xrightarrow{c_{1++}}_C q'$,
2. $\langle q, 0, m_2, \ldots, m_n\rangle \rightarrow \langle q', 0, m_2, \ldots, m_n\rangle$ if $q \xrightarrow{c_{1=0?}}_C q'$,
3. $\langle q, m_1, \ldots, m_2\rangle \rightarrow \langle q', m_1 - 1, \ldots, m_2\rangle$ if $q \xrightarrow{c_{1--}}_C q'$ and $m_1 > 0$.

and the symmetric transitions for $c_2, \ldots, c_n$. A *run* of a n-counter automaton C is a run in S.

In the sequel we say that a transition system S *simulates* a transition system S' iff $Post^*(S') = f(Post^*(S))$ with f a simple mapping (e.g. projection on a subset of the set of states in Q so that $Post^*(S')$ can be computed easily from $Post^*(S)$).

Theorem 1 ([20]). *For any Turing machine T, there is a deterministic 2-counter automaton that simulates T.* □

We now prove how to simulate a Deterministic 2-counter automaton by a Periodically Embedded RFS(2). Assume C is a deterministic 2-counter automaton with counters c_1 and c_2. Let $S \parallel E$ be the Periodically Embedded RFS built in the following way. The two counters c_1 and c_2 are respectively defined by the two memorized events e_1 and e_2 so that the number of memorized occurrences of e_1 (resp. e_2) is the value of c_1 (resp. c_2).

Every transition in C is translated in a RFA structure (a widget). Our construction runs as follows: to simulate one step of the 2-counter automaton, we constrain the RFA widgets with a word u. This way, by processing u^* we can simulate an execution of the counter machine. The key point here is to find a word u to be processed by a RFA widget in order to simulate a step. The environment we take here recognizes $(\tau e_1 e_2 \tau)^*$ where τ is a fleeting event and e_1 and e_2 are memorized events.

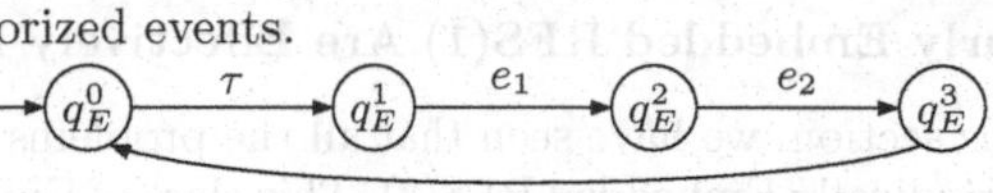

The widgets for the two types of transitions using counter c_1 in a 2-counter automaton are given by[3]:

1. widget for $q \xrightarrow{c_1++}_C q'$:

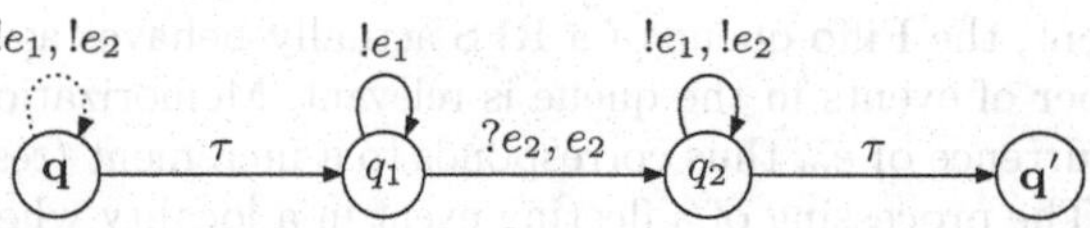

2. widget for $q \xrightarrow{c_1=0?}_C q'$ and $q \xrightarrow{c_1--}_C q''$:

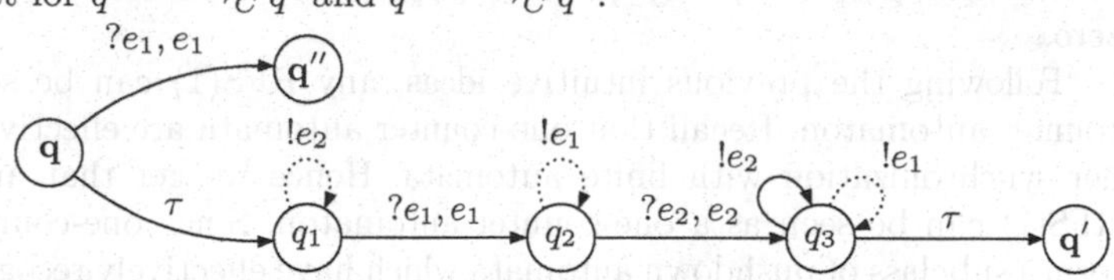

The fleeting event τ is introduced to allow zero testing (see widget (2)) and the two τ transitions: $q_2 \xrightarrow{\tau} q'$ (on widget (1)) and $q_3 \xrightarrow{\tau} q'$ (on widget (2)) are needed to put together the widgets translating the transitions of C.

[3] All the dotted loops are necessary to match Definition 1 point (1), but they are never executed during the simulation of C.

Since increasing c_k is modeled by memorizing e_k, the environment must produce e_1 and e_2 sequentially since (i) it is periodic, and (ii) widget (1) needs to apply to both counters.

Thus, on widget (1), from q_1, either there is one memorized e_2: one of them is batch processed (leading from q_1 to q_2) and e_1 then e_2 are memorized in q_2, or e_1 is memorized in q_1 and e_2 is processed, leading to q_2. In both cases, the number of memorized e_1 has been increased by one, whereas the number of stored e_2 is invariant from q to q'. In widget (2), e_1 is immediately processed from q_1 to q_2 since q_1 is reachable iff there is no memorized e_1. Then, from q_2, e_2 is applied the same treatment as in widget (1).

It follows that the 2-counter automaton C can move from $\langle q, m_1, m_2\rangle$ to $\langle q', m_1+1, m_2\rangle$ iff widget (1) can move from $\langle q, w, q_E^0\rangle$, with $\Psi(w) = (m_1 \quad m_2)$, to $\langle q', w', q_E^0\rangle$, with $\Psi(w') = (m_1+1 \quad m_2)$ i.e. by processing exactly the word $\tau e_1 e_2 \tau$. The same remark applies for the other widget and the decreasing transition.

Theorem 2. *For every deterministic 2-counter automaton, there exists a Periodically Embedded RFS(2) that simulates C.* □

As a consequence, all the problems of interest are undecidable for Embedded RFS with at least two memorizing events and a periodic environment.

3.2 Regularly Embedded RFS(1) Are Effectively Recognizable

In the previous section, we have seen that all the problems of interest are undecidable for Periodically Embedded RFS(2). This class of Embedded RFS includes the huge majority of practical systems. We now focus on Embedded RFS with only one memorizing event e_m in order to complete our study. Particularly, we show that this restriction heavily decreases the power of Embedded RFS as this is the case for Fifo automata. Indeed, note that when there is a unique memorizing event, the Fiffo queue of a RFS actually behaves as a counter, since only the number of events in the queue is relevant. Memorization (batch processing) of an occurrence of e_m thus corresponds to a increment (resp. decrement) of the counter. The processing of a fleeting event in a locality where e_m can be (batch) processed is possible iff the queue is empty, that is iff the counter is equal to zero.

Following the previous intuitive ideas, any RFS(1) can be seen as a one-counter automaton. Recall that one-counter automata are effectively closed under synchronization with finite automata. Hence we get that any Embedded RFS(1) can be seen as a one-counter automaton. Since one-counter automata form a subclass of pushdown automata which have effectively recognizable reachability sets and also decidable LTL and CTL model-checking [4, 15], we obtain the following result:

Theorem 3. *Regularly Embedded RFS with at most one memorizing event have effectively recognizable reachability sets. Moreover LTL and CTL model-checking are decidable for this class.* □

As a consequence, all the problems of interest are decidable for Embedded RFS with at most one memorizing event and a regular environment.

4 Embedded Lossy Reactive Fiffo Systems

In this section, we consider unreliable Embedded RFS that may loose some memorized events non deterministically. Embedded Reactive Fiffo Systems are equipped with the lossy capability by extending definition 3 with the following rule :

4. $\langle q_R, w, q_E\rangle \xrightarrow{l}_{S\|E} \langle q_R, w', q_E\rangle$ if $w' \preccurlyeq w$.

The lossy action l must also be added to $A_{S\|E}$ to obtain an *Embedded Lossy Reactive Fiffo System (Embedded Lossy RFS)*. Notice that the losing action only acts on the memorizing queue, while R and E are not affected by it. We prove that Lossy RFS with well-structured environments are well-structured themselves.

4.1 Well-Structured Transition Systems

Well-Structured Transition Systems (WSTS) [14, 12, 11, 18, 3] provide a general framework for the analysis of infinite state systems. Several definitions of WSTS exist, from the first ones of Finkel [11, 12] and Abdulla *et al.* [3] to the unifying framework of Finkel and Schnoebelen [14].

WSTS are transition systems equipped with a wqo $\leq$ on the states which is compatible with their transition relation $\rightarrow$: "for all $q_1 \leq q_1'$ and transition $q_1 \rightarrow q_2$ there exists a sequence $q_1' \xrightarrow{*} q_2'$ such that $q_2 \leq q_2'$."

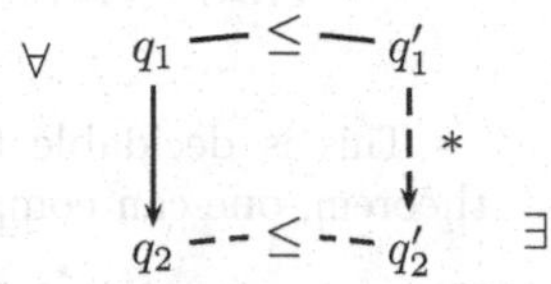

Depending on the sequence $\sigma = q_1' \xrightarrow{*} q_2'$ and following [14], the compatibility is *strong* if σ is of length 1. If the sequence $\sigma = q_1' \rightarrow q_{11}' \rightarrow \cdots \rightarrow q_{1n}' \rightarrow q_2'$ contains at least one transition and moreover $q_1 \leq q_{1i}'$, $\forall i \in \{1, \ldots, n\}$ then it is a *stuttering* compatibility.

Taking into account the labels of the transition system, we obtain a (strongly compatible) *Well-Structured Labeled Transition System (WSLTS)* [14]

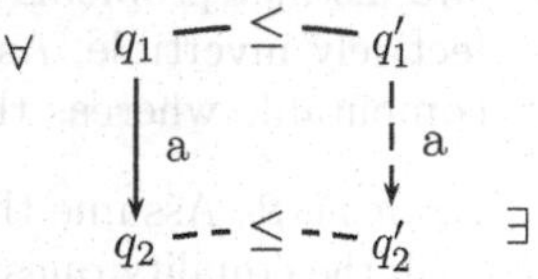

A WS(L)TS has *effective pred-basis* [14] if there exists an algorithm computing a finite-basis of $\uparrow$*Pred*$= (\uparrow q)$ for a given state q (denoted $pb(q)$). Then, for a WS(L)TS with decidable $\leq$ and effective pred-basis, it is possible to compute a finite basis of *Pred**$(\uparrow q)$, and it follows that the covering problem is decidable [14]. Furthermore, if the WS(L)TS has stuttering compatibility the model-checking of the fragment EG of CTL is decidable [18, 14].

4.2 WS–Embedded Lossy RFS Are Well-Structured

Definition 5 (WS–Embedded Lossy RFS). *A* WS–Embedded Lossy RFS *is an Embedded Lossy RFS where the environment is a WSLTS.* □

Let $S \parallel E$ be an WS–Embedded Lossy RFS with a well-structured environment $(E, \sqsubseteq)$. We denote $\leq$ the binary relation on the configurations in $Q_{S\parallel E}$ defined by: $\langle q_R, w, q_E\rangle \leq \langle q'_R, w', q'_E\rangle \Leftrightarrow q_R = q'_R$ and $w \preccurlyeq w'$ and $q_E \sqsubseteq q'_E$.

Notice that $(Q_{S\parallel E}, \leq)$ is a wqo since $=$ is a wqo on finite sets and Q_R is finite, $(\Sigma^*, \preccurlyeq)$ is a wqo by Higman's lemma [16] and $\sqsubseteq$ is a wqo as we assume that the environment is well-structured.

Theorem 4. *Every WS–Embedded Lossy RFS $(S \parallel E, \leq)$ is a WSTS with stuttering compatibility.* □

Now, deciding if a given configuration $\langle q_R, w, q_E\rangle$ is covered from $\langle q'_R, w', q'_E\rangle$ consists in checking whether $\langle q'_R, w', q'_E\rangle \in \mathit{Pred}^*(\uparrow\langle q_R, w, q_E\rangle)$, where *Pred* is given by:

$$\begin{aligned}
\mathit{Pred}(\langle q_R, w, q_E\rangle) = &\left\{\langle q'_R, w, q'_E\rangle \mid q'_R \xrightarrow{e}_R q_R \ \wedge\ q'_E \xrightarrow{e}_E q_E \ \wedge\ w \in (\Sigma^!_{q'_R})^*\right\}\\
&\cup \left\{\langle q_R, w', q'_E\rangle \mid q_R \xrightarrow{\text{!e}}_R q_R \ \wedge\ q'_E \xrightarrow{e}_E q_E \ \wedge\ w = w'e \ \wedge\ w' \in (\Sigma^!_{q'_R})^*\right\}\\
&\cup \left\{\langle q'_R, w', q_E\rangle \mid q'_R \xrightarrow{?e}_R q_R \ \wedge\ \forall w'_1 \in (\Sigma^!_{q'_R})^*,\ w'_2 \in \Sigma^*_M \text{ s.t. } w = w'_1 w'_2,\right.\\
&\qquad \left. w' = w'_1 e w'_2\right\}\\
&\cup \{\langle q_R, w', q_E\rangle \mid w' \in (\Sigma^*_M \sqcup\!\sqcup\, w)\}
\end{aligned}$$

This is decidable for WS–Embedded Lossy RFS since from the following theorem, one can compute a finite basis for $\mathit{Pred}^*(\uparrow\langle q_R, w, q_E\rangle)$.

Theorem 5. *A WS–Embedded Lossy RFS has effective pred-basis if its environment has effective pred-basis.* □

Notice that computing a finite representation for *Post** (*S*) (the effective recognizability problem) and computing a finite representation for *Pred** (*S*) are two distinct problems since RFS (and thus (Lossy) Embedded RFS) are not effectively invertible. As a result, the *Post** sets of RFS are recognizable but not computable whereas the *Pred** sets are effectively recognizable.

Example 2. Assume that the environment E is finite-state. Clearly, E equipped with the equality quasi-ordering is a WSLTS with decidable wqo and with effective pred-basis.

Recalling that the loosing action l is not visible by the environment, we obtain that $\uparrow\langle q_R, w, q_E\rangle \subseteq \mathit{Pred}^*(\langle q_R, w, q_E\rangle)$ for any configuration $\langle q_R, w, q_E\rangle$ of $S \parallel E$. Therefore, we can compute a finite basis for $\mathit{Pred}^*(\langle q_R, w, q_E\rangle)$ by applying the principle of [1] for Lossy communicating machines.

Theorem 6. *For any configuration $\langle q_R, w, q_E \rangle$ of a WS–Embedded Lossy RFS, the upward closed set $Pred^*(\langle q_R, w, q_E \rangle)$ is effectively computable.* □

Thus it follows that the coverability problem is decidable for WS–Embedded Lossy RFS and as a consequence the reachability problem and the EF model-checking problem are decidable. The EG model-checking problem is also decidable since WS–Embedded Lossy RFS have stuttering compatibility.

4.3 Undecidability Results

The strong connection between deterministic two counters automata and Periodically Embedded RFS(2) (see Section 3.1) leads us to focus on the link between *Lossy Counters Automata* [5] and Periodically Embedded Lossy RFS.

Lossy Counters Automata (Lossy CA) are defined from counter automata (see Definition 4) by extending their semantics with the loosing rule:

4. $(q, m_1, \ldots, m_n) \xrightarrow{l} (q, m'_1, \ldots, m'_n)$ if $(\forall i,\ m'_i \leq m_i)$

We first prove the following result, which is easily derived from [10].

Theorem 7. *The boundedness problem and the recurrent locality problem are undecidable for Lossy 3–Counter Automata.* □

Now, the construction given in Section 3.1 to simulate deterministic 2-counter automata with Periodically Embedded RFS(2) can be adapted for deterministic 3-counter automata and Periodically Embedded RFS(3). Observe that a loss in a Periodically Embedded Lossy RFS corresponds (w.r.t. Parikh's mapping) to a loss in the counters automaton. Moreover, correctness of the construction is preserved by losiness.

The undecidability of the recurrent locality problem stated in the following theorem is proved in [19], using the idea in [2] for communicating finite state machines. The proof uses a machine that guesses its initial configuration in order to have an infinite execution. The existence of such a configuration is itself undecidable since the termination problem is. One needs a non-deterministic counter automaton for that "initial guess", but periodic environments constrain RFS in a strong deterministic way : our construction in Section 3.1 would not apply. The solution consists in considering *ultimately periodic environments* which first allow this "initial guess", and then constrain the considered RFS in the same periodic way that the environment we used in Section 3.1, thus preserving our construction.

An *Ultimately Periodic Embedded Lossy RFS* is an Embedded RFS with an ultimately periodic environment : it recognizes a language $L.(u)^\omega$ where L is a regular language and u a finite word. We get the following theorem.

Theorem 8. *The recurrent locality problem, the LTL model-checking problem and the boundedness problem are undecidable for Ultimately Periodic Embedded Lossy RFS(3). The CTL model-checking problem is undecidable for Ultimately Periodic Embedded Lossy RFS(4).* □

The difference between the LTL and the CTL model-checking problems lies in the impossibility to model the recurrent locality problem in CTL. As far as we know, the (un)decidability of CTL model-checking problem for Lossy 3–Counter Automata is still an open problem.

Notice that ultimately periodic environments are well-structured with effective pred-basis, since they can be modeled by finite state transition systems, and = is a wqo on finite states. Thus, decidability results for WS–Embedded Lossy RFS apply to Ultimately Periodic Embedded Lossy RFS.

Theorem 9. *The covering problem and the model-checking problems for the fragments EG and EF of CTL are decidable for Ultimately Periodic Embedded Lossy RFS.* □

And the undecidability results for Ultimately Periodic Embedded Lossy RFS extend to WS–Embedded Lossy RFS.

Theorem 10. *The boundedness problem, the recurrent locality problem and the LTL and CTL model-checking problems are undecidable for WS–Embedded Lossy RFS.* □

5 Conclusion

Our results are summarized in table 1. In this table, *U* stands for "undecidable", *D* for "decidable", *Y* and *N* indicates if the state-space is recognizable, and its effectiveness between parentheses.

	Regularly Embedded RFS(1)	WS–Embedded Lossy RFS	Periodically Embedded RFS(2)
Boundedness	D	U*	U
Covering	D	D	U
Recognizability(effective)	Y(Y)	Y(N)	N
Recurrent locality	D	U*	U
LTL Model-Checking	D	U*	U
CTL/EG/EF Model-checking	D/D/D	U*/D/D	U/U/U

Table 1. Summary of the results.

For Embedded Lossy RFS, the star * denotes that these problems are still open when the RFS has two memorized events (and also for 3 memorized events in the case of the CTL model-checking problem).

Notice that the termination problem for Embedded RFS reduces to the termination problem for its environment. The decidability of the covering problem for WS–Embedded Lossy RFS allows us to verify safety properties on this class.

Since the lossy semantics yields an upper approximation of an Embedded RFS, we get an (of course incomplete) method to verify safety properties on (non-lossy) Embedded RFS.

Our future work is concerned with the boundedness problem. Since it is undecidable for Embedded RFS and WS–Embedded Lossy RFS, we intend to design a semi-algorithm based on a test close to the one defined for Fifo Automata [17] in order to detect unboundedness of Embedded RFS.

References

[1] P. Abdulla and B. Jonsson. Verifying programs with unreliable channels. In *Proceedings, Eighth Annual IEEE Symposium on Logic in Computer Science*, pages 160–170. IEEE Computer Society Press, 1993.

[2] P. Abdulla and B. Jonsson. Undecidable verification problems for programs with unreliable channels. In S. Abiteboul and E. Shamir, editors, *Automata, Languages and Programming, 21st International Colloquium*, volume 820 of *Lecture Notes in Computer Science*, pages 316–327, Jerusalem, Israel, 11–14 July 1994. Springer-Verlag.

[3] P. A. Abdulla, K. Čerāns, B. Jonsson, and Y-K. Tsay. Algorithmic analysis of programs with well quasi-ordered domains. *INFCTRL: Information and Computation (formerly Information and Control)*, 160, 2000.

[4] A. Bouajjani, J. Esparza, and O. Maler. Reachability analysis of pushdown automata: Application to model-checking. In *Proc. 8th Int. Conf. Concurrency Theory (CONCUR'97), Warsaw, Poland, Jul. 1997*, volume 1243 of *Lecture Notes in Computer Science*, pages 135–150. Springer, 1997.

[5] A. Bouajjani and R. Mayr. Model checking lossy vector addition systems. In *Proc. 16th Ann. Symp. Theoretical Aspects of Computer Science (STACS'99), Trier, Germany, Mar. 1999*, volume 1563 of *Lecture Notes in Computer Science*, pages 323–333. Springer, 1999.

[6] D. Brand and P. Zafiropulo. On communicating finite-state machines. *Journal of the ACM*, 30(2):323–342, April 1983.

[7] J. R. Burch, E. M. Clarke, K. L. McMillan, D. L. Dill, and J. Hwang. Symbolic model checking: 10^{20} states and beyond. *Information and Computation*, 98(2):142–170, 1992.

[8] F. Cassez and O. Roux. Compilation of the ELECTRE reactive language into finite transition systems. *Theoretical Computer Science*, 146(1–2):109–143, 24 July 1995.

[9] CCITT. *Recommendation Z.100: Specification and Description Language SDL*, blue book, volume x.1 edition, 1988.

[10] C. Dufourd, P. Jančar, and Ph. Schnoebelen. Boundedness of Reset P/T nets. In *Proc. 26th Int. Coll. Automata, Languages, and Programming (ICALP'99), Prague, Czech Republic, July 1999*, volume 1644 of *Lecture Notes in Computer Science*, pages 301–310. Springer, 1999.

[11] A. Finkel. A generalization of the procedure of Karp and Miller to well structured transition systems. In Thomas Ottmann, editor, *Proceedings of the 14th International Colloquium on Automata, Languages, and Programming*, volume 267 of *LNCS*, pages 499–508, Karlsruhe, FRG, July 1987. Berlin: Springer.

[12] A. Finkel. Reduction and covering of infinite reachability trees. *Information and Computation*, 89(2):144–179, December 1990.

[13] A. Finkel and P. McKenzie. Verifying identical communicating processes is undecidable. *Theoretical Computer Science*, 174(1–2):217–230, 15 March 1997.

[14] A. Finkel and Ph. Schnoebelen. Well structured transition systems everywhere! *Theoretical Computer Science*, 256(1–2):63–92, 2001.

[15] A. Finkel, B. Willems, and P. Wolper. A direct symbolic approach to model checking pushdown systems. In *Proc. 2nd Int. Workshop on Verification of Infinite State Systems (INFINITY'97), Bologna, Italy, July 1997*, volume 9 of *Electronic Notes in Theor. Comp. Sci.*, pages 30–40. Elsevier Science, 1997.

[16] G. Higman. Ordering by divisibility in abstract algebras. *Proceedings of the London Mathematical Society (3)*, 2(7):326–336, September 1952.

[17] T. Jéron and C. Jard. Testing for unboundedness of fifo channels. *Theoretical Computer Science*, 113(1):93–117, 1993.

[18] O. Kushnarenko and Ph. Schnoebelen. A formal framework for the analysis of recursive-parallel programs. *Lecture Notes in Computer Science*, 1277:45–??, 1997.

[19] R. Mayr. Undecidable problems in unreliable computations. In *International Symposium on Latin American Theoretical Informatics (LATIN'2000)*, volume 1776 of *Lecture Notes in Computer Science*, Punta del Este, Uruguay, 2000. Springer-Verlag.

[20] M. L. Minsky. *Computation: Finite and Infinite Machines.* Prentice Hall, London, 1 edition, 1967.

[21] R. J. Parikh. On context-free languages. *Journal of the ACM*, 13(4):570–581, October 1966.

[22] G. Sutre, A. Finkel, O. Roux, and F. Cassez. Effective recognizability and model checking of reactive fiffo automata. In *Proc. 7th Int. Conf. Algebraic Methodology and Software Technology (AMAST'98), Amazonia, Brazil, Jan. 1999*, volume 1548 of *Lecture Notes in Computer Science*, pages 106–123. Springer, 1999.

Electronic Jury Voting Protocols

Alejandro Hevia[1]* and Marcos Kiwi[2]**

[1] Dept. of Computer Science & Engineering, U. of California, San Diego, CA, and Dept. Cs. de la Computación, U. Chile,
aheviaOcs.ucsd.edu

[2] Dept. Ing. Matemática, U. Chile & Ctr. de Modelamiento Matemático, UMR 2071 U. Chile–CNRS, Santiago 170-3, Chile,
mkiwi@dim.uchile.cl

Abstract. This work elicits the fact that all current proposals for electronic voting schemes disclose the final tally of the votes. In certain situations, like jury voting, this may be undesirable. We present a robust and universally verifiable Membership Testing Scheme (MTS) that allows, among other things, a collection of voters to cast votes and determine whether their tally belongs to some pre–specified small set (e.g., exceeds a given threshold) — our scheme discloses no additional information than that implied from the knowledge of such membership. We discuss several extensions of our basic MTS. All the constructions presented combine features of two parallel lines of research concerning electronic voting schemes, those based on MIX–networks and in homomorphic encryption.

1 Introduction

In a typical trial by jury in the United States, twelve jurors deliberate in private. A foreman appointed by the judge among the jurors presides the deliberations. Jurors might be called upon to decide on several different counts according to a policy which may be complicated. Nevertheless, the simplest and most important jury verdicts are of the binary type, e.g., innocent/guilty. In criminal cases unanimity is required in order to reach a verdict. In civil cases there are different standards, nine out of twelve votes are representative numbers. Jury deliberations proceed in discussion rounds followed by voting rounds. Voting is performed by raising hands. Hence, a typical requirement of an election protocol, privacy of the votes, is not achieved. This opens the possibility of biases on decisions due to jurors fear of rejection, a posteriori reprisals by interested parties, and/or follow-the-leader kind of behavior. In fact, just knowledge of tallies can cause undesirable follow-the-pack type conducts among jurors.

A ballot box system could be implemented in order to guarantee privacy. A subset of the jury might be held responsible for tallying the votes and communicating to the others whether a verdict has been reached. Still, this discloses the

* Partially supported by Conicyt via Fondap in Applied Mathematics 1999–2000, and Fondecyt No. 1981182, and by NSF CCR-0093029.

** Gratefully acknowledges the support of Conicyt via Fondecyt No. 1981182 and Fondap in Applied Mathematics 1999–2000.

S. Rajsbaum (Ed.): LATIN 2002, LNCS 2286, pp. 415–429, 2002.

final tally to a subset of the jury and allows them to manipulate the deliberation process. An outside third party (e.g., a judge, government employee, etc.) could be responsible for tallying the votes, but this would cast doubts on the whole process since it allows for outside jury manipulation, could cause undesirable leaks on how the jury is leaning, etc.

We provide an electronic drop in procedure for jury voting in the presence of a curious media, interested parties, dishonest court employees, and conflictive jury members, that reveals nothing besides whether the final tally exceeds or not a given threshold value. We stress that we do not question the adequacy of the way in which juries deliberate. There are good reasons to encourage jurors to express clearly and openly their opinions. The point is that the way in which juries deliberate is just one familiar example, among many, where it is clear that the voting procedure itself has an effect on the final outcome. In particular, our work is motivated by the observation that voting procedures that disclose final tallies may be undesirable. This situation occurs whenever small groups wish to make a yes/no type decision by majority vote, e.g., whether to accept or reject a paper submitted to a cryptology conference — the cryptographers program committee problem, to confirm or not someone as president of a committee or chair of a department, whether or not to send an invitation to a speaker, to decide whether to go forth with a given investment.

Our main procedure also provides a novel solution for the problem of computing partial information from private data, which includes among others, the 'scoring' problem. In the latter, a person is willing to answer some very sensitive questions to a group of evaluators (say for a job interview or insurance application). Answers are coded as integer values and might be weighted differently depending on the question. Evaluators would like to learn whether the weighted score of the answers T exceeds a given threshold or belongs to a set S of "satisfactory" values. The respondent wishes to keep private the answers to each individual question. A solution satisfying both requirements can be obtained by using a threshold voting scheme. Here, answers to different questions are seen as votes coming from different individuals. The (weighted) sum T of these "votes" is tested for membership in the set S of "satisfactory" values. This work's main scheme provides a solution to these problem and guarantees that only one bit of information is released: whether the "tally" T belongs or not to the given set S.

The first electronic voting scheme proposals focused on breaking the correspondence between the voters and the vote casted. Afterward, several other desirable properties of electronic voting schemes (besides correctness, privacy, and efficiency) were identified, e.g., robustness, availability, non–duplication, universal verifiability, non–coercibility. Electronic voting protocols satisfying different subsets of the latter properties were designed. Nevertheless, all of them reveal the final vote tally. In this work we propose a cryptographic procedure for addressing this problem and stress its relevance by describing other applications.

1.1 Related Work

Voting schemes where one wants only one bit of information regarding the outcome, like the ones discussed in the previous section, can be cast in the framework of secure multi–party computation. Thus, plausibility results, asserting that such voting schemes can be in principle built, can be obtained. Indeed, the application of general techniques like the ones proposed in [GMW86, CCD87, BGW88] yield such constructions. Unfortunately, the solutions thus obtained do not exhibit some of the properties one desires of an electronic voting scheme (e.g., non–interaction among voters). On the contrary, homomorphic voting protocols, MIX–network based protocols, and verifiable secret sharing protocols are, in general, more efficient and require less communication than general purpose secure multi-party computation protocols.

Electronic voting schemes are one of the prime examples of secure multi–party computation. This partly explains why they have been intensively studied. The first electronic election scheme in the literature was proposed by Chaum [Cha81]. His work is based on a realization of a computational secure anonymous channel called the MIX–network. Anonymous channels and election schemes are closely related. Indeed, an anonymous channel hides the correspondence between senders and receivers. An election scheme requires hiding the correspondence between the voters and their votes. Since Chaum's work, several other electronic election schemes based on untraceability networks have been proposed. Among the earlier ones are [Cha88, Boy89, PIK93, SK95]. More recent proposals of these type are those of [FS01, N01, OA00, Abe98, Abe99, DK00, JJ99]. (For actual implementations of MIX–networks see [SGR97] and the references therein.)

In contrast to the above mentioned schemes [CF85, BY86, Ben87] introduced ones that do not rely on the use of anonymous channels. In these schemes, ballots are distributed over a number of tallying authorities through a special type of broadcast channel. Rather than hiding the correspondence between the voter and his ballot, the value of the vote is hidden. Among these latter type of schemes are [SK94, BT94]. More recent proposals are [CFSY96, CGS97, Sch99].

1.2 Our Contributions

This work's first contribution is that it elicits the fact that all current proposals for electronic voting schemes disclose the final tally of the votes. As discussed above this may be undesirable in some situations. Our main technical contribution is a cryptographic protocol to which we refer as *Membership Testing Scheme (MTS)*. Given a fixed sequence of integers $c_1, \ldots, c_n$ and sets $S_1, \ldots, S_n$, it allows a collection of parties $P_1, \ldots, P_n$ to cast values $v_1, \ldots, v_n$, where $v_i \in S_i$, and determine whether $\sum_i c_i v_i$ belongs to some pre–specified small set S.

Based on our MTS we obtain a drop in replacement electronic procedure for a civil case jury voting protocol by letting $n = 12$, $c_1 = \ldots = c_n = 1$, $S_1 = \ldots = S_n = \{0, 1\}$, and $S = \{9, 10, 11, 12\}$ (simpler schemes can be devised for criminal type trials, so we will focus on the more challenging civil type case). For the sake of simplicity of exposition, we discuss our results in the terminology of jury systems. Thus, for notational and mnemonic purposes we refer to parties $P_1, \ldots, P_n$ as voters and denote them by $V_1, \ldots, V_n$, to the values $v_1, \ldots, v_n$ as votes, and to $\sum_i c_i v_i$ as the tally. Our main MTS and variants satisfy some subset of the following properties:

- Eligibility: Only authorized voters can vote and none more than once.
- Correctness: If all participants are honest, the correct output is generated.
- Robustness: The system can recover from the faulty or malicious behavior of any (reasonably sized) coalition of participants.
- Computational Privacy: A voter ballot's content will be kept secret from any (reasonably sized) coalition of parties that does not include the voter.
- Universal Verifiability: Ensures that any party, even a passive observer, can check that ballots are correctly cast, only invalid ballots are discarded, and the published final tally is consistent with the correctly cast ballots.
- No–duplication: No one can duplicate anyone else's vote.

In our scheme the voters send in a ballot identical to those proposed in [CGS97], i.e., an ElGamal ciphertext representing his/her vote plus a proof that the ciphertext is indeed a valid ballot. Hence, as in [CGS97], both the computational and communication complexity of the voter's protocol is linear in the security parameter k — thus optimal.[1] Moreover, for any reasonable security parameter, the voters' protocol remains the same even if the number of voters varies. Assuming m authorities, the work performed by each authority is $O(((m+k)|S|+n)k)$. Moreover, the computational complexity of verifying each authority's work is proportional to the work performed by each authority. As in [CGS97] the work needed to verify that a voter sent in a well formed ballot is $O(k)$ per voter.

Our MTS proposal combines features of two parallel lines of research concerning electronic voting schemes, those based on MIX–networks (a la [Cha81]) and in homomorphic encryption schemes (a la [CF85, BY86, Ben87]). We use homomorphic (ElGamal) encryption in order to hide the vote tallies. We rely on special properties of the ElGamal cryptosystem in order to perform an equality test between the tally and members of S. We use MIX–networks (ElGamal based) in order to hide the value of the member of S involved in each equality test. To the best of our knowledge, the only other cryptographic protocols which rely both on homomorphic encryption schemes and MIX–networks are the independent recent proposals of Hirt and Sako [HS00] and Jakobsson and Juels [JJ00]. But, our MTS combines both theses schemes in a novel way. Indeed, Hirt and Sako's proposal uses a MIX–network in order to randomly permute, for each voter, potential ballots, while Jakobsson and Juels' scheme permutes truth tables rows to compute the output of each Boolean gate of a circuit. In contrast, our MTS relies on MIX–networks in order to randomly permute the elements of the pre–specified set S on which one desires to test membership.

The applications we provide for our MTS constitute novel uses of MIX–networks. A feature of these applications is that they rely on the capacity, that

[1] Throughout, a modular multiplication of $O(k)$ bit sized numbers will be our unit with respect to which we measure computational costs.

the overwhelming majority of MIX–network proposals exhibit, to randomly permute and encrypt a list of ElGamal ciphertexts. On the contrary, they do not use the decryption capabilities that accompany most MIX–network proposals.[2] By combining MIX–networks with efficient and available cryptographic protocols (namely, verifiable secret sharing and homomorphic voting), this paper gives a first (practical) solution to the mentioned jury voting problem that does not rely on general secure multi-party computation techniques.

We propose several implementations of a MTS. Our first proposal relies on the homomorphic encryption based electronic election scheme of Cramer, Gennaro and Schoenmakers [CGS97] and the MIX–network of Abe [Abe98]. We also discuss alternative implementations of our MTS based on the work of Jakobsson and Desmedt and Kurosawa [Jak98, DK00] and Furukawa and Sako [FS01] as opposed to that of Abe [Abe98]. Our different MTS implementations exhibit different properties depending on the previous work we use to build them.

Organization: In Sect. 2, we informally outline the protocol and discuss the building blocks on which our basic MTS proposal relies. In Sect. 3, we describe and analyze our MTS and use it for building an electronic drop in replacement for a jury voting protocol that reveals nothing besides whether the final tally exceeds or not a given threshold. In Sect. 5 and Sect. 6 we discuss variants and other applications of our basic scheme. We conclude in Sect. 7 discussing a feature of all of the MTSs that we propose and some desirable future developments.

2 Preliminaries

We work in the model introduced by Benaloh et al. (see [CF85, BY86, Ben87] and [CGS97]), where participants are divided into n voters $V_1, \ldots, V_n$ and m authorities $A_1, \ldots, A_m$ called active parties. Al parties are limited to have polynomially–bounded computational resources and have access to a so called bulletin board whose characteristics we describe below.

In the sequel we assume that a designated subset of active participants on input 1^k, where k is a security parameter, jointly generate the following system values: a k bit long prime p, a large prime q such that q divides $p-1$, and generators g and h of an order q multiplicative subgroup G_q of $\mathbb{Z}_p^*$. One way for participants to collectively generate these system values is to run the same probabilistic algorithm over jointly generated uniformly and independent coinflips.

Conventions: Henceforth, unless otherwise specified, all arithmetic is performed modulo p except for arithmetic involving exponents which is performed modulo q. Throughout this paper, $x \in_R \Omega$ means that x is chosen uniformly at random from Ω. Furthermore, negligible and overwhelming probability correspond to probabilities that are at most $\nu(k)$ and at least $1 - \nu(k)$ respectively, where $\nu(k)$

[2] Recently, Neff [N01] suggested another application, namely using exponentiating MIX–networks (as opposed to re-encrypting MIX–networks as we do) to implement universally-verifiable voting schemes.

is a function vanishing faster than the inverse of any polynomial in the security parameter k. A non–negligible probability is said to be significant.

2.1 Protocol Overview

The protocol consists of five main stages. (Setup, MIX, Verification, Voting and Output). In the setup phase, shared parameters subsequently used in the protocol are selected. In the mix phase, the list of encryptions of the elements in a fixed set S is shuffled by a MIX–network. To shuffle the list means permuting it while re–randomizing each of its entries. Hence, the MIX-network's output is a randomly permuted list of re–encryptions of elements in S. Next, in the voting stage, each voter posts an encryption of his vote (using the authorities' jointly generated public–key) and a publicly verifiable proof that the encryption corresponds to a valid vote (a la [CGS97]). Using the homomorphic property of the underlying encryption scheme the authorities proceed to compute the encryption of the tally. Finally, in the output stage, each element of the MIX–network's output list is compared with the encryption of the tally to test whether they encrypt the same plaintext. This stage is performed in such a way that the authorities do not actually decrypt the tally nor the encryption of any element in the shuffled list. Moreover, no information concerning any of the plaintexts involved is revealed. Instead a "blinded" copy of the difference between the tally and each element in the shuffled list is implicitly decrypted. The protocol relies on the fact that discrete exponentiation is injective to unequivocally identify an encryption of 0, and therefore, when two encrypted values are the same. The verification stage checks whether all previous phases were correctly performed.

2.2 Building Blocks

Bulletin board: The communication model used in our MTS consists of a public broadcast channel with memory, usually referred too in the literature as bulletin board. Messages that pass through this communication channel can be observed by any party including passive observers. Nobody can erase, alter, nor destroy any information. Every active participant can post messages in his own designated section of a bulletin board. This requires the use of digital signatures to control access to distinct sections of the bulletin board. Here we assume a public–key infrastructure is already in place. This suffices for computational security. Note that it is implicitly assumed that denial–of–service attacks are excluded from consideration (see [CGS97] for a discussion of how to implement a bulletin board in order to achieve this).

Distributed Key Generation Protocol (DKG): A DKG protocol allows parties $A_1, \ldots, A_m$ to respectively generate private outputs $s_1, \ldots, s_m$, called shares, and a public output $y = g^s$ such that the following requirements hold:

- **Correctness:** There is an efficient procedure that on at least $t+1$ shares submitted by honest parties and the public values produced by the DKG

protocol, outputs the unique secret value s, even if up to t shares come from faulty parties. Honest parties coincide on the public key $y = g^s$ and $s \in_R \mathbb{Z}_q$.
- **Secrecy:** No information on s can be learned by the adversary except what is implied by the value $y = g^s$ (for a more formal definition in terms of simulatability see [GJKR99]).

The first DKG protocol was proposed by Pedersen [Ped91]. Henceforth in this work, DKG refers to the protocol presented in [GJKR99] and shown to be secure in the presence of an active adversary that can corrupt up to $t < n/2$ parties.

ElGamal Encryption and Robust (threshold) Proof of Equality of Encryptions: Our MTS relies on a robust threshold version of the ElGamal cryptosystem [ElG85] proposed in [CGS97]. Recall that in ElGamal's cryptosystem $x \in G_q$ is encrypted as $(\alpha, \beta) = (g^r, y^r x)$ for $r \in_R \mathbb{Z}_q$, where $y = g^s$ is the public key and s is the secret key. In a robust threshold version of the ElGamal cryptosystem, the secret key and public key are jointly generated by the intended ciphertext recipients by means of a DKG protocol like the one described above.

A robust threshold ElGamal cryptosystem has a feature on which all our MTS proposals rely. This property allows checking whether a ciphertext encodes the plaintext 1 without either decrypting the message nor reconstructing the secret s. Indeed, assume (α, β) is an ElGamal encryption of message x, that is $(\alpha, \beta) = (g^r, y^r x)$. Verifying whether it is an encryptions of $x = 1$ boils down to checking if $(\alpha^{s'})^s = \beta^{s'}$, where s' is a randomly distributed shared secret that effectively "blinds" the decryption of (α, β). (This technique for checking equality of plaintexts has also been used in [CG99, JJ00, BST01]. But, it at least dates back to [Gen95a] where the case of ElGamal encryptions is considered.) The aforementioned equality can be verified by m parties each holding distinct shares $s_1, \ldots, s_m$ and $s'_1, \ldots, s'_m$ of the secrets s and s' without reconstructing either secret. To achieve this, participant j commits to her secret shares s_j and s'_j by posting $y_j = g^{s_j}$ and $y'_j = g^{s'_j}$ in her designated area of the bulletin board. Then, three Distributed Exponentiation (DEx) protocols are executed (two for computing $\alpha' = \alpha^{s'}$ and $\beta' = \beta^{s'}$ and the last one to check that $(\alpha')^s = \beta'$). Such protocol on input α outputs α^s by means of the following steps:

1. Participant j posts $\omega_j = \alpha^{s_j}$ and proves in zero knowledge that $\log_g y_j = \log_\alpha \omega_j$ using the protocol of [CP92] for proving equality of discrete logs. The protocol is honest–verifier zero–knowledge [CGS97]. This suffices for our application. In order to make the protocol non–interactive the Fiat–Shamir heuristic is used. This requires a cryptographically strong hash function. We henceforth refer to this non–interactive proof as **Proof–Log**$(g, y_j; \alpha, \omega_j)$.
2. Let Λ denote any subset of t participants who successfully passed the zero knowledge proof and let $\lambda_{j,\Lambda}$ denote the appropriate Lagrange interpolation coefficients. The desired value can be obtained from the following identities:

$$\alpha^s = \prod_{j \in \Lambda} \omega_j^{\lambda_{j,\Lambda}}, \qquad \lambda_{j,\Lambda} = \prod_{l \in \Lambda \setminus \{j\}} \frac{l}{l-j}.$$

ELGAMAL BALLOTS AND EFFICIENT PROOFS OF VALIDITY: In our MTS each voter will post on the bulletin board an ElGamal encryption. The encryption is accompanied by a proof of validity that shows that the ballot is indeed of the correct form. To implement this, consider a prover who knows $x \in \{x_0, x_1\}$ and wants to show that an ElGamal encryption of x, say $(\alpha, \beta) = (g^r, y^r x)$, is indeed of this form without revealing the value of x. The prover's task amounts to showing that the following relation holds:

$$\log_g \alpha \in \{\log_y(\beta/x_0), \log_y(\beta/x_1)\}.$$

Building on [CDS94], an efficient witness indistinguishable (honest–verifier zero–knowledge) proof of knowledge for the above relation was proposed in [CGS97]. Henceforth, **Proof–Ballot**$_{\{x_0,x_1\}}(\alpha, \beta)$ denotes this (non–interactive) proof.

UNIVERSALLY VERIFIABLE MIX–NETWORK: A MIX–network for ElGamal ciphertexts consists of a bulletin board and a collection of authorities called the MIX–servers. It takes a list of ElGamal ciphertexts, permutes them according to some (secret) permutation and outputs an ElGamal re–encryption of the original list (without ever decrypting the original list of ciphertexts).

We now describe a MIX–network proposal due to Abe [Abe98] which in addition to the aforementioned properties also satisfies: correctness, robustness, privacy, and universal verifiability. MIX–servers first run the DKG protocol and jointly generate a secret s and a public y. Initially, the bulletin board contains a list of ElGamal ciphertexts $((G_{0,l}, M_{0,l}))_l$ where $M_{0,l} = m_l y^{t_{0,l}}$ and $G_{0,l} = g^{t_{0,l}}$ for $m_l \in G_q$ and $t_{0,l} \in_R \mathbb{Z}_q$. (To avoid the attack shown in [Pfi95] a proof of knowledge of $t_{0,l}$ must accompany $(G_{0,l}, M_{0,l})$.) The list of ElGamal ciphertexts is re-randomized and permuted by the cascade of MIX–servers. Server j chooses a random permutation π_j of S, picks $t_{j,l} \in_R \mathbb{Z}_q$ for each l, reads $((G_{j-1,l}, M_{j-1,l}))_l$ from the bulletin board, and posts in the bulletin board $((G_{j,l}, M_{j,l}))_l$ where

$$G_{j,l} = G_{j-1,\pi_j(l)} g^{t_{j,l}}, \quad \text{and} \quad M_{j,l} = M_{j-1,\pi_j(l)} y^{t_{j,l}}.$$

Processing proceeds sequentially through all servers.

Lemma 1. ([Abe98]) *Under the intractability of the Decision Diffie–Hellman problem, given correctly formed* $((G_{j-1,l}, M_{j-1,l}))_l$ *and* $((G_{j,l}, M_{j,l}))_l$*, no adversary can determine* $\pi_j(l)$ *for any* l *with probability significantly better than* $1/|S|$.

An additional protocol, referred to as **Protocol–Π**, is executed in order to prove the correctness of randomization and permutation to external verifiers as well as convince honest servers that they have contributed to the output, i.e., no one has canceled the randomization and permutation performed by the honest servers (with success probability significantly better than a random guess). A non–interactive version of **Protocol–Π** can be derived through standard techniques. We henceforth denote this (non–interactive) version by **Proof–Π**. (See details in [Abe98].)

Our MTS can be based on any re-encrypting MIX–network with the mentioned characteristics. Other alternatives will be discussed later on.

3 Membership Testing Scheme (MTS)

In what follows, N denotes the cardinality of the set S for which one seeks to verify whether it contains the vote tally. Also, henceforth, i runs over $\{1,\ldots,n\}$, j runs over $\{1,\ldots,m\}$, and l runs over S. We work in the model described in the previous section where the active set of participants is $V_1,\ldots,V_n$ (the voters) and $A_1,\ldots,A_m$ (the authorities). Voters and authorities might overlap.

BASIC MTS PROTOCOL

Input

1. Public Input: System parameters, i.e., a k bit long prime p, a prime $q > n$ that divides $p-1$ and generators g and h of an order q multiplicative subgroup G_q of $\mathbb{Z}_p^*$ (elements g and h are independently generated). A set $S \subset \{1,\ldots,n\}$.
2. Private Input for voter V_i: A vote $v_i \in \{0,1\}$.

Goal

To determine whether $\sum_i v_i$ belongs to S without revealing anything else besides this bit of information.

Setup Phase

1. Using the DKG protocol $A_1,\ldots,A_m$ jointly generate the public value $y = g^s$ where $s \in_R G_q$ and the private shares $s_1,\ldots,s_m$. Authorities commit to their share s_j of s by posting $y_j = g^{s_j}$ in their designated bulletin board area.
2. Using the DKG protocol, authorities jointly generate, for each $l \in S$, the public value $y_l = g^{s'_l}$ where $s'_l \in_R \mathbb{Z}_q$ and the private shares $s'_{l,1},\ldots,s'_{l,m}$. Authorities commit to their share $s'_{l,j}$ of s'_l by posting $y'_{l,j} = g^{s'_{l,j}}$ in their designated area of the bulletin board.

MIX Phase

1. Let $((G_{0,l}, M_{0,l}))_l$ be a list such that $G_{0,l} = 1$ and $M_{0,l} = h^{-l}$ for each $l \in S$.
2. Authority A_j chooses at random a permutation π_j of $\{1,\ldots,N\}$, for each l picks $t_{j,l} \in_R \mathbb{Z}_q$, and posts the list $((G_{j,l}, M_{j,l}))_l$ such that for each $l \in S$,

$$G_{j,l} = G_{j-1,\pi_j(l)} g^{t_{j,l}} \quad \text{and} \quad M_{j,l} = M_{j-1,\pi_j(l)} y^{t_{j,l}} .$$

Verification Phase

Authorities cooperate to issue **Proof–Π** — a honest–verifier zero–knowledge (non–interactive) proof that shows that they know random factors and permutations that relate $((G_{0,l}, M_{0,l}))_l$ with $((G_{m,l}, M_{m,l}))_l$. Each authority signs **Proof–Π** in order to insure verifiers of the presence of an authority they can trust. Each authority checks the proof. If the check succeeds the result is declared VALID. If it fails, dishonest authorities are identified (and removed) by means of the tracing capabilities that **Proof–Π** provides. The remaining authorities restart from the beginning of the MIX Phase.

Voting Phase

Voter V_i chooses $r_i \in_R \mathbb{Z}_q$ and posts both an ElGamal encryption representing his vote v_i, say $(\alpha_i, \beta_i) = (g^{r_i}, y^{r_i} h^{v_i})$, and **Proof–Ballot**$_{\{h^0,h^1\}}(\alpha_i, \beta_i)$.

Output Phase

1. Each authority computes $\alpha = \prod_i \alpha_i$ and $\beta = \prod_i \beta_i$.
2. Using the DEx protocol, for each $l \in S$, authorities compute

$$G'_l = (G_{m,l}\,\alpha)^{s'_l} \quad \text{and} \quad M'_l = (M_{m,l}\,\beta)^{s'_l} .$$

Then, using the DEx protocol again, authorities verify whether $(G'_l)^s = M'_l$ for some l in S. In the affirmative case they output MEMBER, otherwise NON–MEMBER.

Remark 1. Note that both the MIX Phase and the Verification Phase may be pre-computed before the voting begins. In fact, if the Verification Phase is not declared VALID, there is no need to perform the Voting Phase.

ELECTRONIC JURY VOTING PROTOCOL: We conclude this section with a simple observation; an electronic analog of a 12–juror civil case voting protocol where 9 votes suffice to reach a verdict can be derived from our Basic MTS by letting $n = 12$ and $S = \{9, 10, 11, 12\}$.

4 Analysis

ELIGIBILITY: The non–anonymity of ballot casting insures that only authorized voters cast ballots. Indeed, recall that voters must identify themselves through digital signatures in order to post their vote onto their designated area of the bulletin board. This also insures that no voter can cast more than one ballot.

NO–DUPLICATION: Follows from requiring each voter to compute the challenge in the (non–interactive) proof of validity of ballots as a hash of, among others, a unique public key identifying the voter.

CORRECTNESS: Clearly, an honest voter can construct a ballot and its accompanying proof of validity. Moreover, the following holds (proof can be seen in full version [HK00]):

Theorem 1. *If all participating authorities are honest, then they will output MEMBER if and only if the tally of the validly cast votes belongs to the set S.*

ROBUSTNESS: First we observe that robustness with respect to malicious voters is achieved.

Lemma 2. ([CGS97]) *An incorrectly formed ballot will be detected with overwhelming probability.*

Still, we need to show that the protocol cannot be disrupted by dishonest authorities. We will need the following:

Lemma 3. ([Abe98]) **Protocol–Π** *is a honest verifier zero-knowledge proof of knowledge for* π *and* $\tau_{m,l}$*'s. The protocol is also honest verifier zero-knowledge proof of knowledge for* π_j*'s and* $t_{j,l}$*'s held by honest provers.*

Robustness with respect to malicious authorities is now guaranteed by the following result (proof can be seen in full version [HK00]):

Theorem 2. *Assume there are at most* $m - t - 1$ *participating authorities controlled by an adversary. The goal of the adversary is to force the output of the scheme to be incorrect (i.e., to be MEMBER when it should be NON-MEMBER and vice versa). The adversary cannot succeed with non–negligible probability and the identity of the authorities controlled by the adversary will be learned with overwhelming probability.*

Privacy: We now show that under a standard computational assumption our Basic MTS does not disclose any information pertaining the honest voter's ballots besides that implied by the output of the scheme. Specifically, the following holds (proof can be seen in full version [HK00]):

Theorem 3. *Assume there are less than* t *dishonest authorities and* n' *dishonest voters controlled by an adversary. Let* T_h *and* T_d *be the tally of the correctly emitted ballots among the honest and dishonest voters, respectively. The goal of the adversary is to learn any additional information concerning the votes cast by honest voters, besides that implied by whether or not* T_h *belongs to* $(S - T_d) \cap \{0, \ldots, n - n'\}$.[3] *Under the Diffie–Hellman assumption, the adversary has a negligible probability of success.*

Universal Verifiability: Follows from the public verifiability of the proofs of ballot validity (**Proof–Ballot**), the proof of randomization and permutation (**Proof–Π**), the proof of knowledge of equality of discrete logarithms (**Proof–Log**) and the correctness proof associated to the DKG protocol. Note that even in the case that there are more than t dishonest authorities, although privacy might be compromised, passive observers will still be able to ascertain whether the protocol was correctly performed.

Efficiency: We make the (realistic) assumption that $n \geq N \geq t$. Recall that a modular multiplication of $O(k)$ bit sized numbers is our unit measure of computational costs.

The voter's ballot consists of an ElGamal ciphertext and a (non–interactive) proof that it is indeed a valid ballot. The size of both components is linear in the size of an element of $\mathbb{Z}_p^*$, i.e., $O(k)$. The work involved in the computation of both ballot components is dominated by the modular exponentiations, of which there are a constant number, each one requiring $O(k)$ work. Hence, the

[3] For a set of integers Ω and an integer x the set $\{\, \omega - x \,:\, \omega \in \Omega \,\}$ is denoted by $\Omega - x$.

computational and communication complexity of the voter's protocol is linear in the security parameter k — thus optimal. Moreover, for any reasonable security parameter, a voter's protocol remains the same even if the number of voters varies. The work needed to verify that a voter sent in a well formed ballot equals the computational cost of making the ballot, i.e., $O(k)$ per voter. We stress that all the above characteristics of a voter's protocol are inherited from the electronic election scheme proposed in [CGS97].

The work performed by the j–th authority during the MIX Phase is dominated by the cost of computing $((G_{j,l}, M_{j,l}))_l$. Since

$$G_{j,l} = G_{j-1,\pi_j(l)} g^{t_{j,l}} \quad \text{and} \quad M_{j,l} = M_{j-1,\pi_j(l)} y^{t_{j,l}} ,$$

the work performed by each authority during this phase is $O(Nk)$. Analogously, the work performed by each authority during the Verification Phase is $O(Nk^2)$. Finally, since each run of the DEx protocol costs $O(mk)$ per authority, the work performed by each authority during the Output Phase is $O((mN + n)k)$ (the $O(nk)$ term is due to the work performed in order to compute $\alpha = \prod_i \alpha_i$, and $\beta = \prod_i \beta_i$). The other tasks performed by the authorities are not relevant in terms of computational costs. Thus, the work performed by each authority is $O((mN + n)k)$ provided they spend $O(Nk^2)$ work during pre-computation. The communication complexity (in bits) incurred by each authority exceeds the computational complexity by a factor of k.

The computational complexity of verifying the authorities work is proportional to the computational work performed by each authority during the corresponding phase.

5 Variants

More Efficient and Alternative MTSs: If one is willing to forgo universal verifiability, more efficient MIX–networks like the one proposed by Desmedt and Kurosawa [DK00] might be used instead of Abe's MIX–network in the MTS of Sect. 3. In this case, the work done by each authority during the pre-computation stage is reduced to $O(kN)$. In fact, the only essential characteristic our MTS scheme requires from the underlying MIX–network is that it performs a random secret permutation and ElGamal re–encryption of an input list of ElGamal ciphertexts. (The threshold decryption capabilities utilized in the DEx protocol is a feature from the underlying encryption scheme, not of the MIX–network). Thus, other more efficient recent MIX–network proposals like those of Furukawa and Sako [FS01], Abe [Abe99], and Jakobsson and Juels [JJ99] are good candidates for drop in replacements in the MIX module of the MTS of Sect. 3.

Unanimity Voting: In case of unanimity voting S is a singleton. Therefore, there is no need to use a MIX–network. Thus, the computational cost of the scheme is reduced by skipping the pre-computation phase.

6 Applications

TESTING MEMBERSHIP OF LINEAR FUNCTIONS: We can modify our Basic MTS to allow parties $P_1, \ldots, P_n$ to determine whether their private inputs $v_i \in S_i$, for $i \in \{1, \ldots, n\}$, are such that $\sum_i c_i v_i \in S$ without revealing $\sum_i c_i v_i$. Here, $S_1, \ldots, S_n$ and S are publicly known subsets of $\mathbb{Z}_q$, and $c_1, \ldots, c_n$ is a publicly available fixed sequence of integers. This modification of our Basic MTS allows to implement a weighted majority voting electronic election scheme.

SCORING: Consider a person/entity which is willing to answer n very sensitive questions to a group of m evaluators. Assume the i–th question accepts as answer any element of S_i. Each evaluator would like to learn whether the weighted score of the answers $\sum_i c_i a_i$ exceeds a threshold (here again $c_1, \ldots, c_n$ is a publicly available fixed sequence of integers). But, the respondent wishes to keep private the answers to each individual question. This problem clearly reduces to the one discussed in the previous paragraph. Thus, it follows that our Basic MTS can be used to solve it.

PRIVATE INFORMATION RETRIEVAL: When restricted to a single voter V with a vote $v \in S' \supset S$, the proposed scheme yields a method of searching for v on the "database" S without revealing neither the target v nor the contents of the database. The corresponding proof of validity can be designed using 1-out-of-n proofs [CDS94] as suggested in [CFSY96]. This problem is a special case of what is known as private information retrieval. It encompasses situations where users are likely to be highly motivated to hide what information they query from a database that contains particularly sensitive data, e.g., stock quotes, patents or medical data.

7 Final Comments

An interesting feature of our electronic jury voting scheme is that it combines parallel lines of research concerning electronic voting, one based on MIX–networks [Cha81] and another on homomorphic encryptions [CF85, BY86, Ben87]. We need homomorphic encryption in order to hide the ballots content and compute the tally while keeping it secret. We need ElGamal based MIX–networks in order to hide the value of the elements of S to which the ElGamal encryption of the vote tally is compared. It is an interesting challenge to design an electronic jury voting scheme in the model introduced in [CF85, BY86, Ben87] which does not rely on MIX–networks.

Acknowledgments

We would like to thank Daniele Micciancio, Rosario Gennaro, and Mihir Bellare for interesting and helpful discussions, Martin Loebl for suggesting the application to the scoring problem, an anonymous referee for bringing to our attention [HS00] and suggesting the PIR application, and all the reviewers for comprehensive comments that helped us improve the presentation of this work.

References

[Abe98] M. Abe. Universally verifiable MIX-net with verification work independent of the number of MIX-servers. *Proc. of EuroCrypt'98*, 437–447. Springer-Verlag. LNCS Vol. 1403.

[Abe99] M. Abe. Mix–Networks on Permutation Networks. *Proc. of Asiacrypt'99*, 258–273. Springer-Verlag. LNCS Vol. 1716.

[Ben87] J. Benaloh. *Verifiable Secret–Ballot Elections*. PhD thesis, Yale University, Dept. of Computer Science, Sep. 1987.

[BGW88] M. Ben-Or, S. Goldwasser and A. Wigderson. Completeness theorems for non-cryptographic fault-tolerant distributed computation. *Proc. of STOC'88*, 1–10, ACM.

[Boy89] C. Boyd. A new multiple key cipher and an improved voting scheme. *Proc. of EuroCrypt'89*, 617–625. Springer-Verlag. LNCS Vol. 434.

[BST01] F. Boudot, B. Schoenmakers and J. Traore. A fair and efficient solution to the socialist millionaires' problem. *Discrete Applied Mathematics*, 111(1-2):23–36, 2001.

[BT94] J. Benaloh and D. Tuinstra. Receipt-free secret-ballot elections. *Proc. of STOC'94*, 544–553, ACM.

[BY86] J. Benaloh and M. Yung. Distributing the power of a government to enhance the privacy of voters. *Proc. of PODC'86*, 52–62. ACM.

[CCD87] D. Chaum, C. Crépeau and I. B. Damgård. Multiparty unconditionally secure protocols. *Proc. of Crypto'87*, 462–462. Springer-Verlag. LNCS Vol. 293.

[CDS94] R. Cramer, I. Damgård and B. Schoenmakers. Proofs of partial knowledge simplified design of witness. *Proc. of Crypto'94*, 174–187. Springer-Verlag. LNCS Vol. 839.

[CF85] J. D. Cohen and M. J. Fischer. A robust and verifiable cryptographically secure election scheme. *Proc. of FOCS'85*, 372–382, IEEE.

[CFSY96] R. Cramer, M. K. Franklin, B. Schoenmakers and M. Yung. Multi-authority secret-ballot elections with linear work. *Proc. of EuroCrypt'96*, 72–83. Springer-Verlag. LNCS Vol. 1070.

[CGS97] R. Cramer, R. Gennaro and B. Schoenmakers. A secure and optimally efficient multi-authority election scheme. *Proc. of EuroCrypt'97*, 103–118. Springer-Verlag. LNCS Vol. 1233.

[CG99] R. Canetti and S. Goldwasser. An efficient threshold public key cryptosystem secure against adaptive chosen ciphertext attack. *Proc. of EuroCrypt'99*, 90–106. Springer-Verlag. LNCS Vol. 1592.

[Cha81] D. L. Chaum. Untraceable electronic mail, return addresses, and digital pseudonyms. *Communications of the ACM*, 24(2):84–88, 1981. ACM.

[Cha88] D. Chaum. Elections with unconditionally-secret ballots and disruption equivalent to breaking RSA. *Proc. of EuroCrypt'88*, 177–182. Springer-Verlag. LNCS Vol. 330.

[CP92] D. Chaum and T. P. Pedersen. Wallet databases with observers. *Proc. of Crypto'92*, 89–105. Springer-Verlag. LNCS Vol. 740.

[DK00] Y. Desmedt and K. Kurosawa. How to break a practical MIX and design a new one. *Proc. of EuroCrypt'00*, 557–572. Springer-Verlag. LNCS Vol. 1807.

[ElG85] T. ElGamal. A public key cryptosystem and a signature scheme based on discrete logarithms. *IEEE Transactions of Information Theory*, IT-31(4):469–472, 1985.

[FS01] J. Furukawa and K. Sako. An efficient scheme for proving a shuffle. *Proc. of Crypto'01*, 368–387. Springer-Verlag. LNCS Vol. 2139.

[Gen95a] R. Gennaro. Manuscript, 1995.

[Gen95b] R. Gennaro. Achieving independence efficiently and securely. *Proc. of PODC'95*, 130–136, ACM.

[GJKR99] R. Gennaro, S. Jarecki, H. Krawczyk and T. Rabin. Secure distributed key generation for discrete-log based cryptosystems. *Proc. of EuroCrypt'99*, 295–310. Springer-Verlag. LNCS Vol. 1592.

[GMW86] O. Goldreich, S. Micali and A. Wigderson. Proofs that yield nothing but their validity and a methodology of cryptographic protocol design. *Proc. of FOCS'86* 174–187, IEEE.

[HK00] A. Hevia and M. Kiwi. Electronic Jury Voting Protocols. Cryptology ePrint Archive, Report 2000/035, Jul. 2000.

[HS00] M. Hirt and K. Sako. Efficient receipt–free voting based on homomorphic encryption. *Proc. of EuroCrypt'00*, 539–556. Springer-Verlag. LNCS Vol. 1807.

[Jak98] M. Jakobsson. A practical mix. *Proc. of EuroCrypt'98*, 448–461. Springer-Verlag. LNCS Vol. 1403.

[JJ99] M. Jakobsson and A. Juels. Millimix: Mixing in small batches. Technical Report 99–33, DIMACS, 1999.

[JJ00] M. Jakobsson and A. Juels. Mix and match: Secure function evaluation via ciphertexts. *Proc. of Asiacrypt'00*, 162–178. Springer-Verlag. LNCS Vol. 1976.

[MK00] M. Mitomo and K. Kurosawa. Attack for flash MIX. *Proc. of Asiacrypt'00*, 192–204. Springer-Verlag. LNCS Vol. 1976.

[N01] C. A. Neff. A verifiable secret shuffle and its application to e-voting. *Proc. of CCS'01*. ACM.

[OA00] M. Ohkubo and M. Abe. A length-invariant hybrid mix. *Proc. of Asiacrypt'00*, 178–191. Springer-Verlag. LNCS Vol. 1976.

[Ped91] T. P. Pedersen. Distributed provers with applications to undeniable signatures. In *EuroCrypt'91*, 221–242. Springer-Verlag. LNCS Vol. 547.

[Pfi95] B. Pfitzmann. Breaking an efficient anonymous channel. *Proc. of EuroCrypt'94*, 332–340. Springer-Verlag. LNCS Vol. 950.

[PIK93] C. Park, K. Itoh and K. Kurosawa. Efficient anonymous channel and all/nothing election scheme. *Proc. of EuroCrypt'93*, 248–259. Springer-Verlag. LNCS Vol. 765.

[Sch99] B. Schoenmakers. A simple publicly verifiable secret sharing scheme and its application to electronic voting. *Proc. of Crypto'99*, 148–164. Springer-Verlag. LNCS Vol. 1666.

[SGR97] P. Syverson, D. Goldschlag and M. Reed. Anonymous connections and onion routing. *Proc. of IEEE Symposium on Security and Privacy*, 44–54, IEEE, 1997.

[SK94] K. Sako and J. Kilian. Secure voting using partially compatible homomorphisms. *Proc. of Crypto'94*, 411–424. Springer-Verlag. LNCS Vol. 839.

[SK95] K. Sako and J. Kilian. Receipt-free mix-type voting scheme — A practical solution to the implementation of a voting booth. *Proc. of EuroCrypt'95*, 393–403. Springer-Verlag. LNCS Vol. 921.

Square Roots Modulo p

Gonzalo Tornaría*

Department of Mathematics,
University of Texas at Austin,
Austin, Texas 78712, USA,
tornaria@math.utexas.edu

Abstract. The algorithm of Tonelli and Shanks for computing square roots modulo a prime number is the most used, and probably the fastest among the known algorithms when averaged over all prime numbers. However, for some particular prime numbers, there are other algorithms which are considerably faster.

In this paper we compare the algorithm of Tonelli and Shanks with an algorithm based in quadratic field extensions due to Cipolla, and give an explicit condition on a prime number to decide which algorithm is faster.

Finally, we show that there exists an infinite sequence of prime numbers for which the algorithm of Tonelli and Shanks is asymptotically worse.

1 Introduction

When p is an odd prime, we denote by $\mathbb{F}_p$ the finite field of p elements, and by $\mathbb{F}_p^\times$ its multiplicative group. We will consider two algorithms for solving the following:

Problem. Let p be an odd prime, and $a \in \mathbb{F}_p$ a quadratic residue. Find $x \in \mathbb{F}_p$ such that $x^2 = a$.

From now on p will be a fixed prime number. Write $p = 2^e q + 1$ with q odd; this determines e and q. Let n be the number of binary digits of p, and let k be the number of ones in the binary representation of p. We denote by $\mathbf{G}_p$ the Sylow 2-subgroup of $\mathbb{F}_p^\times$, which is cyclic of order 2^e.

2 The Algorithm of Tonelli and Shanks

The algorithm of Tonelli [9], is based in this observation: it's easy to reduce the problem to the case $a \in \mathbf{G}_p$, because $[\mathbb{F}_p^\times : \mathbf{G}_p]$ is odd. Then one can use the Legendre symbol to find a generator of $\mathbf{G}_p$, and compute the square root of a by means of the discrete logarithm of a with respect to that generator, which is fast because most of the time the group $\mathbf{G}_p$ is small. Moreover, there is a binary algorithm for computing discrete logarithms in a cyclic group of order 2^e; namely, compute the discrete logarithm bit by bit.

* This work was partially supported by a scholarship of PEDECIBA Matemática

S. Rajsbaum (Ed.): LATIN 2002, LNCS 2286, pp. 430–434, 2002.

The original method of Tonelli required about $e^2/2$ operations for computing the discrete logarithm. It was improved by Shanks [8], who rearranged the algorithm in a clever way such that the operations done for computing a 0 bit include the operations needed for computing the next bit. Thus, while the number of operations in the worst case is the same, the number is roughly halved in average.

Indeed one has

Proposition 2.1 (Lindhurst [5, Lemma 1]). *Averaged over all quadratic residue and non-residue inputs, and ignoring the initialization stage, the algorithm of Tonelli and Shanks requires* $\frac{1}{4}(e^2+7e-12)+\frac{1}{2^{e-1}}$ *modular multiplications.* □

Remark. In this result the average is taken over a uniformly distributed input $a \in \mathbb{F}_p^\times$; if the distribution of a is a concern, one can compute instead $\frac{\sqrt{ab^2}}{b}$, where b is a uniformly distributed random element of $\mathbb{F}_p^\times$.

Of course, the density of the prime numbers $p = 2^e q + 1$ for a fixed e is, by Dirichlet's theorem, 2^{-e}, and one can conclude that:

Corollary 2.2 (Lindhurst [5, Theorem 2]). *Averaged over all prime numbers, quadratic residues and non-residues, the algorithm of Tonelli and Shanks requires* $8/3$ *multiplications after the initialization stage.* □

Remark. The average over all prime numbers is, as usual, the limit for $N \to \infty$ of the (uniform) average over all primes $\leq N$.

The initialization stage has two steps:

- Find a generator of $\mathbf{G}_p$. For this we take $t \in \mathbb{F}_p^\times$ at random and compute the Legendre symbol of t until we get a quadratic non-residue. Then $z = t^q$ is a generator of $\mathbf{G}_p$. The probability of getting a quadratic non-residue is $1/2$ for each try; the expected number of Legendre symbol computations is therefore 2. Also, q has $n - e$ binary digits, of which $k - 1$ are ones, so computing t^q will require $n + k - e - 3$ multiplications.
- Compute $a^{(q+1)/2}$ and a^q. The binary expression of $(q-1)/2$ has $n - e - 1$ digits, of which $k - 2$ are ones. Using a binary powering algorithm, one can compute $a^{(q-1)/2}$ with $n + k - e - 5$ multiplications, and we need one more for computing $a^{(q+1)/2}$ and another one to compute $a^q = a^{(q-1)/2}a^{(q+1)/2}$.

Corollary 2.3. *Averaged over all quadratic residue and non-residue inputs, the number of operations required by the algorithm of Tonelli and Shanks is*

$$2n + 2k + \frac{e(e-1)}{4} + \frac{1}{2^{e-1}} - 9 \tag{1}$$

multiplications, and 2 *expected (with respect to choosing* t*, which is independent of the input) computations of the Legendre symbol.* □

These results support our assertion that the algorithm of Tonelli and Shanks is very good when the modulus is fairly random, but they also show that there could be room for improvements when e is large with respect to n, provided that such prime numbers exist.

3 The Algorithm of Cipolla

An alternative to using discrete logarithms is the algorithm of Cipolla [3]. Let $a \in \mathbb{F}_p^\times$, and assume that we know $t \in \mathbb{F}_p$ such that $t^2 - a$ is a quadratic non-residue. Then $X^2 - (t^2 - a)$ is irreducible over $\mathbb{F}_p$, and $\mathbb{F}_p[\alpha]$, with $\alpha^2 = t^2 - a$, is a finite field of p^2 elements, a quadratic extension of $\mathbb{F}_p$.

It's enough to compute the square root of a in $\mathbb{F}_p[\alpha]$. If a is a quadratic residue, its two square roots will be in $\mathbb{F}_p$; otherwise we will get something not in $\mathbb{F}_p$, and we will be able to conclude that a is a quadratic non-residue.

In $\mathbb{F}_p[\alpha]$ we have

$$(t+\alpha)^{p+1} = (t+\alpha)(t+\alpha)^p = (t+\alpha)(t-\alpha) = t^2 - \alpha^2 = a \ , \tag{2}$$

where the second equality follows because the Frobenius automorphism carries $t + \alpha$ to its conjugate, $t - \alpha$. Therefore, if we compute $x = (t + \alpha)^{(p+1)/2}$, we have $x^2 = a$.

To find t, we just take $t \in \mathbb{F}_p^\times$ at random and compute the Legendre symbol of $t^2 - a$ until we get one such that $t^2 - a$ is a quadratic non-residue or 0 (if we are so lucky, t itself is a square root).

Lemma 3.1. *Let $a \in \mathbb{F}_p^\times$. The number of $t \in \mathbb{F}_p$ such that $t^2 - a$ is a quadratic non-residue is $(p-1)/2$ if a is a quadratic residue and $(p+1)/2$ if a is a quadratic non-residue.*

Proof. Notice that the set of t such that $t^2 - a$ is a quadratic residue is exactly the same as the set of different t which appear among the pairs (s, t) such that $s^2 = t^2 - a$. This equation is the same as $(t - s)(t + s) = a$, and so it clearly has $p - 1$ solutions.

Now, for each solution (s, t) we get a different solution $(-s, t)$ with the same t, unless $s = 0$. We have two cases to consider:

- If a is a quadratic residue, then there are two different solutions with $s = 0$, and so the number of different t which appear in the set of solutions is $(p - 3)/2 + 2 = (p + 1)/2$.
- If a is a quadratic non-residue, then there are no solutions with $s = 0$, and the number of different t is $(p - 1)/2$. □

This lemma shows that in practice is very easy to find such a t, the probability for each try being slightly more than $1/2$; the expected number of tries is therefore less than 2, and we only need one multiplication, one sum, and one computation of the Legendre symbol for each try.

After finding t, we only have to compute a $(p+1)/2$ power of $(t+\alpha)$ in $\mathbb{F}_p[\alpha]$. For this we can use a binary powering algorithm, using the following formulas for multiplication in $\mathbb{F}_p[\alpha]$:

- $(u + v\alpha)^2 = (u^2 + v^2 r) + ((u + v)^2 - u^2 - v^2)\alpha$, where $r = t^2 - a$ is known in advance, which needs 4 multiplications and 4 sums;

– $(u + v\alpha)^2(t + \alpha) = (td^2 - b(u + d)) + (d^2 - bv)\alpha$, where $d = (u + vt)$ and $b = av$, which needs 6 multiplications and 4 sums.

We can assume that $p \equiv 1 \pmod 4$, as Shanks' algorithm is obviously better otherwise. In that case, the binary expression of $(p + 1)/2$ has $n - 1$ digits, of which k are ones. For computing a $(p + 1)/2$ power of $(t + \alpha)$ we therefore need to use $n - k - 1$ times the first formula, and $k - 1$ times the second formula.

Proposition 3.2. *For any quadratic residue or non-residue input, the expected number of operations required for the algorithm of Cipolla is $4n + 2k - 4$ multiplications, $4n - 2$ sums, and 2 computations of the Legendre symbol.*

Combining Corollary 2.3 and Proposition 3.2 we obtain

Theorem 3.3. *Given a prime number p, let n be the number of binary digits of p, and let 2^e be the maximum power of 2 which divides $p-1$. With respect to the expected number of operations, and averaged over all quadratic residue and non-residue inputs, the algorithm of Cipolla (neglecting the sums) is better than the algorithm of Tonelli and Shanks if and only if $e(e - 1) > 8n + 20$.* □

4 The Existence of Primes

For each $i = 1, 2, \ldots$, we define p_i to be the least prime number such that $p_i \equiv 2^i + 1 \pmod{2^{i+1}}$. Let n_i be the number of binary digits of p_i, and let 2^{e_i} be the maximum power of 2 which divides $p_i - 1$. From the definition is clear that $e_i = i$, and $n_i > i$. We now give an upper bound for n_i.

Lemma 4.1. *There exists absolute constants L, C such that $e_i L + C > n_i$.*

Proof. A theorem of Linnik [6,7] states that if $(a, m) = 1$ then the least prime number congruent to a modulo m is less than $C_0 m^L$ for some absolute constants C_0, L. Applying this to p_i we get

$$p_i < C_0 2^{(e_i+1)L} \ . \tag{3}$$

Taking base 2 logarithms, we conclude that $n_i < e_i L + C$. □

Remark. The best known unconditional value for L is $11/2$, due to Heath-Brown [4]. Assuming the Generalized Riemann Hypothesis, one can use $L = 2+\epsilon$ for arbitrary $\epsilon > 0$ [1,4]. In the case in hand, where the modulus are all powers of two, there may be even stronger results. For example, in [1] the authors prove that one can use $L = 8/3 + \epsilon$ provided the modulus are restricted to powers of a fixed *odd* prime.

In the following theorem, $\mathsf{TS}(p)$ and $\mathsf{Cip}(p)$ are the expected number of operations required for the prime p by the algorithms of Tonelli and Shanks and by the algorithm of Cipolla respectively, averaged over all quadratic residue and non-residue inputs.

Theorem 4.2.

$$\limsup_{p\ prime} \frac{\mathsf{TS}(p)}{\mathsf{Cip}(p)} = \infty \ . \tag{4}$$

Proof. From Corollary 2.3 and Proposition 3.2 we know that

$$\mathsf{TS}(p_i) > \frac{e_i(e_i-1)}{4} > \frac{(n_i-C)(n_i-C-L)}{4L^2} = \Omega(n_i^2) \tag{5}$$

and that $\mathsf{Cip}(p_i) = O(n_i)$ (even counting the sums and the Legendre symbol computations). Therefore $\frac{\mathsf{TS}(p_i)}{\mathsf{Cip}(p_i)} = \Omega(n_i)$, and the theorem follows. □

5 Last Remarks

I thank the referee for pointing me to the work of Bernstein [2]. In this work, Bernstein improves the algorithm of Tonelli and Shanks. He computes discrete logarithms several bits at a time by means of some auxiliar precomputations. This is especially appealing if one needs to compute several square roots modulo the same prime number, but the improvement is still good for the casual use, if the number of bits computed at a time, and with it the amount of precomputation, is choosen apropriately.

Taking into account the precomputations for this new algorithm, Theorem 4.2 is still valid, but Theorem 3.3 would have to be changed; according to [2] the new algorithm is better than the algorithm of Cipolla when $e^2 = O(n(\lg n)^2)$.

References

1. Barban, M.B., Linnik, Y.V., Tshudakov, N.G.: On prime numbers in an arithmetic progression with a prime-power difference. Acta Arith. **9** (1964) 375–390
2. Bernstein, D.J.: Faster square roots in annoying finite fields, draft. Available from `http://cr.yp.to/papers.html` (2001)
3. Cipolla, M.: Un metodo per la risoluzione della congruenza di secondo grado. Rend. Accad. Sci. Fis. Mat. Napoli **9** (1903) 154–163
4. Heath-Brown, D.R.: Zero-free regions for Dirichlet L-functions, and the least prime in an arithmetic progression. Proc. London Math. Soc. (3) **64** (1992) 265–338
5. Lindhurst, S.: An analysis of Shanks's algorithm for computing square roots in finite fields. In: Number theory (Ottawa, ON, 1996). Amer. Math. Soc., Providence, RI (1999) 231–242
6. Linnik, U.V.: On the least prime in an arithmetic progression. I. The basic theorem. Rec. Math. [Mat. Sbornik] N.S. **15(57)** (1944) 139–178
7. Linnik, U.V.: On the least prime in an arithmetic progression. II. The Deuring-Heilbronn phenomenon. Rec. Math. [Mat. Sbornik] N.S. **15(57)** (1944) 347–368
8. Shanks, D.: Five number-theoretic algorithms. In: Proceedings of the Second Manitoba Conference on Numerical Mathematics (Univ. Manitoba, Winnipeg, Man., 1972). Utilitas Math., Winnipeg, Man. (1973) 51–70. Congressus Numerantium, No. VII
9. Tonelli, A.: Bemerkung über die Auflösung quadratischer Congruenzen. Göttinger Nachrichten (1891) 344–346

Finding Most Sustainable Paths in Networks with Time-Dependent Edge Reliabilities

Goran Konjevod, Soohyun Oh*, and Andréa W. Richa*

Department of Computer Science and Engineering
Arizona State University
Tempe, AZ 85287-5406, USA
{goran, soohyun, aricha}@asu.edu

Abstract. In this paper, we formalize the problem of finding a routing path for the streaming of continuous media (e.g., video or audio files) that maximizes the probability that the streaming is successful, over a network with nonuniform edge delays and capacities, and arbitrary time-dependent edge reliabilities. We call such a problem the *most sustainable path (MSP)* problem. We address the MSP problem in two network routing models: the wormhole and the circuit-switching routing models. We present fully-distributed polynomial-time algorithms for the streaming of constant-size data in the wormhole model, and for arbitrary-size data in the circuit-switching model. Our algorithms are simple and assume only local knowledge of the network topology at each node. The algorithms evolved from a variation of the classical Bellman-Ford shortest-path algorithm. One of the main contributions of this paper was to show how to extend the ideas in the Bellman-Ford algorithm to account for arbitrary time-dependent edge reliabilities.

1 Introduction

Continuous media applications — such as audio and video applications — became standards in today's wide-area internetworks. It is expected that by 2003, continuous media will account for more than 50% of data available on the web servers [6], [8]. When streaming such an application through the network, it is desirable not to make flow stops to correct errors nor to resend lost packets. Since current networks are not fully reliable, we would like to ensure QoS guarantees by finding a path in the network with the lowest probability of failure for the entire streaming procedure. Conventional routing algorithms for connection-oriented traffic tend to use shortest paths, without taking into consideration how reliable the paths are. A problem in finding a path to be used in a long duration streaming application lies on how to properly capture and handle the relevant dynamic changes of the network conditions.

The reliability of a link (or edge) in the network is a function of many parameters — such as network traffic pattern, link bandwidth, routing paths' latency

* Supported in part by NSF CAREER Award CCR–9985284.

S. Rajsbaum (Ed.): LATIN 2002, LNCS 2286, pp. 435–450, 2002.

and congestion — which vary with time. Thus, when estimating how reliable a path is for a streaming application, one needs to account for the reliabilities of all edges involved in the streaming procedure, over all time steps. Moreover, estimating the reliability of a path for a streaming application must account not only for the time-dependent reliabilities of the edges used, but also for the congestion of the path and the size of the data: The size of the message over the capacity of the path chosen indicates the number of time steps each edge in the path will be used during the transmission.

More specifically, we consider the problem of finding a path for the continuous transmission (streaming) of a message of size σ (i.e., consisting of σ units of data) from a given source node s to a given destination node d, in a network with arbitrary polynomial (on the number fo network nodes) edge capacities and delays, and with time-dependent edge reliabilities. Since a message consists of multiple units of data, the duration of the connection between s and d for transmitting the message may be much longer than a short time interval. Thus we define the *sustainability* of a path with respect to transmitting a message to be the probability that all the edges used in the transmission at a given time step will be alive during all time steps used for the transmission. Note that the only parameter of the data relevant to computing the sustainability of a path is the size of the message. Hence we will, in the remainder of this paper, talk about most sustainable paths in a network for a given message size.

We consider two different network routing models [13]: wormhole routing and circuit-switching.

In **wormhole routing**, a message is a sequence of fixed-size packets called *flits*. The first flit is called the head and the rest the body of the message. The flits are sent in a pipelined fashion. Only the head of a message has the routing information, and the body follows the head contiguously. Since the message moves flit by flit, no buffering is required. This model works on top of a packet-forwarding network model, thus only the edges that are actively transmitting packets at any given time step need to be alive during that time step.

In the **circuit-switching model**, we reserve a connection path between the source and destination before the transmission begins. Once the transmission is initiated, the message is never blocked. Since the edges in the path are reserved, no buffering is necessary. In this model, which is connection-oriented, all the edges of the reserved path must be alive throughout the entire transmission of the message, otherwise the transmission will fail. The popular Asynchronous Transfer Mode (ATM) networks [7], [10] provide circuit-switching network services on a packet-switching environment via virtual circuits.

One problem that can occur in the wormhole routing model is called the *head blocking* deadlock: If the head of a message cannot advance because it competes for the traversal of an edge with some other flits in the body of the message, then all the flits must stall, and the transmission of the message is halted. This scenario occurs when there exists a cycle in the selected routing path which is not long enough to accommodate the entire message: i.e., the length of the cycle is less than the size of the message over the capacity of the path. This problem is avoided in the circuit-switching model, because a most sustainable path in this model cannot contain cycles.

In any continuous media streaming application, it is important that the data to be transmitted arrives at the destination node in the order it left the source node (we call this property the *nonpassing property*). Both the wormhole and the circuit-switching model guarantee that this property is satisfied.

In Section 3, we propose a polynomial time algorithm to compute a most sustainable path for the transmission of constant-size messages in the wormhole routing model. We call such a path a *σ-most sustainable path* from s to d, where σ is the size of the message to be transmitted, or *σ-MSP* for short. In our wormhole routing scenario, no waiting or buffering at a node is allowed once the transmission starts. In Section 4, we propose a polynomial time algorithm to compute a most sustainable path for the transmission of a message of an arbitrary size σ in the circuit-switching model. We call such a path a *σ-most sustainable reserved path* from s to d, or a *σ-MSRP* for short.

To the best of our knowledge, this is the first work that handles time-dependent edge reliabilities when computing the best path for data streaming. All the algorithms we present are simple and fully distributed. They can be implemented using only local knowledge of the network at the nodes. (Each node only needs information about its local topology: the edges adjacent to it and their capacities, delays, and reliabilities.)

We assume that each node has full knowledge of its edge reliability function. We understand that this may not be strictly true in a real setting. In a more realistic scenario, these functions would be estimated using prior traffic data. Also, in a realistic setting, we can assume the reliability of each edge to be at least $\frac{1}{f(n)}$ and no more than $1 - \frac{1}{f(n)}$, for some (positive) polynomially increasing function f on the number of nodes in the network n. Without this assumption, we could have most sustainable paths of exponential total delay, even for short messages, which would result in an unreasonable over-utilization of network resources.

The algorithms evolved from the distributed Bellman-Ford shortest path algorithm, (which can also be used directly to compute most sustainable paths for unit-size messages). Our main contribution is to show how the Bellman-Ford framework can be extended in order to account for time-dependent edge reliabilities. Also, we introduce pruning techniques for incremental construction of most sustainable subpaths.

The remainder of the paper is organized as follows. In the next section, we give a brief literature review of related network routing problems. The description of the MSP problem is presented in Section 3, as well as a polynomial-time algorithm for solving this problem for constant size data. In Section 4, we define the MSRP problem and propose an algorithm for solving this problem in polynomial time. In Section 5 we conclude and discuss future work.

2 Related Work

Over the last few decades, routing algorithms have been the subject of extensive research. The shortest-path routing problem has certainly been the most widely studied routing problem. Various shortest-path algorithms for networks with nonnegative edge delays have been proposed and implemented, the best-known being Dijkstra's algorithm and the Bellman-Ford algorithm [3]. Both these algorithms satisfy the optimal subpath property, which says that any subpath of an optimal path is also optimal.

One variation of the shortest path problem is finding a path to send a message of size σ from a source to a destination with minimal transmission time in a network with nonuniform edge delays and capacities [4]. The most important observation here is that due to the variance in edge capacities, the transmission time is closely related to the size of the message. Another variation, closer to the problems we study, is finding a most reliable path for a single packet [1].

Tragoudas [15] was the first to formalize the problem of finding a most sustainable path for an arbitrary size message in a network with nonuniform edge delays, capacities, and reliabilities. He presented an *off-line, centralized* algorithm for this problem. However, his result assumes that all the network parameters, including edge reliabilities, are *static*, that is, do not vary over time.

The shortest-path problem in networks in which the delays of the edges change with time was studied by Orda and Rom [11], [12]. In [11], they present algorithms and hardness results for the problem of finding shortest-paths under various *waiting constraints*. In their network model, the *nonpassing property* is violated. (The nonpassing, or FIFO, property says that two packets that traverse an edge (u, v) will arrive at node v in the same order as they left node u.) In network streaming applications, the nonpassing property should be preserved. Therefore, the results of Orda and Rom, in spite of being closely related to our problem, cannot be extended to solve the MSP problem.

A new model, named the *flow speed model*, for time-dependent networks where the flow speed of each link depends on the time interval is suggested by Sung, Bell, Seong, and Park [14]. They use a variation of Dijkstra's algorithm, incorporating a procedure for calculating the link travel time from the flow speed so that their model satisfies the nonpassing property.

Cooke and Halsey [5] modified the Bellman-Ford algorithm to solve a shortest-path problem in networks with time-dependent internodal transit times.

Recently, Miller-Hooks and Patterson [9] proposed a pseudopolynomial algorithm for solving the integral time-dependent quickest flow problem. In their network model, a packet may achieve an earlier arrival time at the destination by waiting at a node while en route, which may violate the nonpassing property.

In [2], various routing problems in time-dependent labeled networks with applications to transportation science are described. A realistic model is given in which Dijkstra's algorithm can be extended to time-dependent edge-delays. (The nonpassing property is essential here.)

Another recently studied routing problem is finding an acceptable path under multiple constraints to meet quality of service (QoS) requirements for the correct processing and transmission of continuous media on integrated communication networks (for example, ATM networks) [16], [17]. In the current routing approaches for connection-oriented network services, routing decisions are made using only information on the network conditions at connection establishment. Since continuous multimedia streams are comparatively long-lived, each routing decision should consider the network conditions over a longer time period.

3 The MSP Problem

In this section, we study the problem of finding most sustainable paths in the wormhole routing model. We first introduce the model and related definitions.

Consider a network $G(V, E)$, where V is the set of nodes and $E \subseteq V \times V$ the set of edges. Let $n = |V|$. Associated with each edge (i, j), we have a delay ℓ_{ij} and a capacity c_{ij}, as well as a time-dependent reliability function $r_{ij}(t)$. (whose value at time t is the probability that the edge (i, j) operates correctly at that time step.) As mentioned in the introduction, we assume that $\frac{1}{f(n)} \leq r_{ij}(t) \leq 1 - \frac{1}{f(n)}$, for some polynomially increasing f. The edge delay ℓ_{ij} denotes the number of time steps required for a unit of data (or *packet*) to be transmitted from i to j through edge (i, j), and the capacity c_{ij} gives the number of units of data that can traverse (i, j) in a single time step. We assume that edge delays and capacities are positive polynomials in n. Without loss of generality, we assume that all capacities and delays are integral and at least 1.

In the wormhole routing model we need only to account for the reliability of an edge at time t if a packet is traversing the edge at that time. The sustainability of a path is thus defined as follows. Let $c(P)$ denote the *capacity of path* P, the smallest capacity of an edge in P. Let P^{ℓ}_{vw} be a path from node v to node w of length ℓ, where the length of a path is equal to the sum of the delays of the edges in the path. Let $P^{\ell}_{v_0 v_k} = (v_0, v_1, \ldots, v_k)$. Let ℓ_j, $j = 1, \ldots, k$, denote the length of the subpath of $P^{\ell}_{v_0 v_k}$ from v_0 to v_j, and let $\ell_0 = 0$. For convenience, we assume that v_0 begins transmission at time zero. If $c = \lceil \sigma / c(P^{\ell}_{v_0 v_k}) \rceil$, the sustainability of the path $P^{\ell}_{v_0 v_k}$ for a message of size σ is given by

$$Re^{\sigma}(P^{\ell}_{v_0 v_k}) = \prod_{i=1}^{k} \prod_{t=\ell_{i-1}}^{c+\ell_i-1} r_{v_{i-1} v_i}(t).$$

The sustainability of an edge (v_i, v_{i+1}) for this transmission is

$$r^{\sigma}_{v_{i-1} v_i}(\ell_{i-1}) = \prod_{t=\ell_{i-1}}^{c+\ell_i-1} r_{v_{i-1} v_i}(t). \qquad (1)$$

Now the path sustainability can be rewritten as

$$Re^{\sigma}(P^{\ell}_{v_0 v_k}) = \prod_{i=1}^{k} r^{\sigma}_{v_{i-1} v_i}(\ell_{i-1}). \qquad (2)$$

We use $Re(P^{\ell}_{v_0 v_k})$ to denote the case when $\sigma = 1$.

For a fixed length ℓ, a path P^{ℓ}_{vw} from v to w which maximizes $Re^{\sigma}(P^{\ell}_{vw})$ is called a *most sustainable path of length* ℓ from v to w for a message of size σ, or for short, a (σ, ℓ)–MSP from v to w. A σ–MSP from v to w is the path that has maximum sustainability among all the (σ, ℓ)-MSPs from v to w for all possible lengths ℓ.

In the wormhole routing, there are no buffers at the nodes. Thus packets must travel contiguously. In case of only one packet (i.e., $\sigma = 1$), our problem is similar to that of finding a shortest path with time-dependent edge delays, where no waiting is allowed [11]. Orda and Rom show that the optimal subpath property is not satisfied in their model and claim NP-hardness of this problem. However, in our problem there exists an optimal subpath property (to be defined in Section 3.1), which allows dynamic programming to find an optimal solution.

3.1 The MRP Algorithm: Unit-Size Messages

We first consider unit-size messages. The ideas presented for this simple case will be generalized in the next section. The algorithm is similar to Bellman-Ford [3]. A most sustainable path for a single unit of message is also called a *most reliable path* (MRP). For a fixed ℓ, a path P^{ℓ}_{sv} from s to v which maximizes $Re(P^{\ell}_{sv})$ is called a *most reliable path of length* ℓ, or for short ℓ-MRP, from s to v. In the remainder of this paper, we will use $\overline{P}^{\ell}_{sv}$ to denote an ℓ-MSP path from s to v.

The source node s initiates the algorithm by informing each of its neighbors v of the reliability of the ℓ_{sv}-MRP which contains edge (s, v) only.

Upon the receipt of the sustainabilities of paths of length ℓ from s to v from its neighbors, node v chooses the one $(\overline{P}^{\ell}_{sv})$ with maximum sustainability. Then v computes the sustainabilities of paths composed of $\overline{P}^{\ell}_{sv}$ followed by each of the edges adjacent to v. The algorithm terminates once the destination node d has computed all the values $Re(\overline{P}^{\ell}_{sd})$ for all possible MRP lengths ℓ. The path with the largest sustainability is then an MRP from s to d. The algorithm keeps track of all possible MRP lengths from s to each node v in G. If there exist two or more paths of the same length, the algorithm keeps the most sustainable one. Any ties are broken arbitrarily.

Each node v maintains a set of tuples of the form (ℓ, R, u) where for some path P from s to v, ℓ is the length and R the sustainability of P, and u is the node preceeding v in P. At each time step ℓ, each node v updates a set $L(v)$ whose elements are tuples corresponding to the k-MRP from s to v, for all possible values of k where $k \leq \ell$ (if such a path exist). Node v chooses the ℓ-MRP based upon learning of a tuple $(\ell, R, \cdot)$ from its neighbors and it ignores the others. Let the chosen tuple be (ℓ, R, u). Then v adds the tuple (ℓ, R, u) to its set $L(v)$. Node v calculates $Re(P^{\ell+\ell_{vw}}_{sw})$, the sustainability of the path of length $\ell + \ell_{vw}$ composed of the path from s to v associated with the tuple (ℓ, R, u) followed by edge (v, w). The sustainability of path $P^{\ell+\ell_{vw}}_{sw}$ is given by

$$Re(P^{\ell+\ell_{vw}}_{sw}) = R \times \prod_{t=\ell}^{\ell+\ell_{vw}-1} r_{vw}(t).$$

An upper bound M of the number of possible MRP lengths in the network can be computed as follows. It is easy to see that there cannot be any MRP of length greater than $M = n + x$ where x is maximum y such that $\frac{1}{f(n)} \leq (1 - \frac{1}{f(n)})^y$, (where $f(n)$ is the polynomial function that bounds the edge reliabilities). Since $\frac{1}{f(n)} > 0$, $\frac{1}{f(n)} \leq e^{-\frac{y}{f(n)}}$. By solving for y, we get

$$y \leq f(n) \ln f(n).$$

Hence M is bounded by a polynomial on n.

To analyze the running time, note that each tuple calculation and comparison takes constant time. Since in the worst case all nodes are adjacent (and all edge lengths are the same), at each time step, every node receives at most n tuples. Hence the new tuple computations and comparisons take $O(n)$ time. Since $L(d)$ will have at most M tuples (one per time step), choosing the tuple with maximum sustainability at time step $M + 1$ takes $O(M)$ time and the overall time bound is $O(nM)$.

In our MRP algorithm, each node v compares the tuples of the form $(t, \cdot, \cdot)$, at time step t only, in order to choose a t-MRP from s to v. Hence if a tuple with associated length t arrives at node v at time t' other than t, the algorithm cannot guarantee that it finds a t-MRP. Therefore it is important that all the tuples with associated length t arrive at node v at time t. We omit the proof here, since it is similar to that of Lemma 1 in Section 4.

In order to retrieve the ℓ-MRP, $\overline{P}^{\ell}_{sd}$, that corresponds to a tuple (ℓ, R, v), follow the node v that corresponds to the third component of the tuple and choose the tuple $(\ell - \ell_{vd}, \cdot, \cdot)$ from $L(v)$. Since $L(v)$ keeps only the tuples with the highest sustainabilities for all possible lengths of paths from s to v, there exists only one tuple with associated length $\ell - \ell_{vd}$ in $L(v)$. We can retrieve the entire path by doing this process recursively until we get a null value at the third component of the tuple.

We now show an optimal subpath property for the MRP problem, which allows the use of dynamic programming. Let $P = (v_0, v_1, \ldots, v_k)$ be the ℓ-MRP from v_0 to v_k. Then the subpath $Q = (v_0, \ldots, v_i)$ of P is the ℓ_i-MRP from v_0 to v_i, where ℓ_i is the length of Q. Suppose for a contradiction, that there exists another path $Q' = (v_0, u_1, \ldots, u_j, v_i)$ with length ℓ_i and $Re(Q') > Re(Q)$. Then at time step ℓ_i, node v_i gets two tuples $q = (\ell_i, R, v_{i-1})$ and $q' = (\ell_i, R', u_j)$ corresponding to the paths Q and Q', respectively. Let the sustainability of path $(v_{i+1}, \ldots, v_k)$ with starting time ℓ_i be $\bar{R}$. Since we assumed that $Re(Q') > Re(Q)$, the path $P' = Q(v_{i+1}, \ldots, v_k)$ has sustainability $Re(Q') \cdot \bar{R} > Re(Q) \cdot \bar{R}$. This contradicts the assumption that P is an ℓ-MRP from v_0 to v_k.

The following theorem summarizes the properties of the MRP algorithm. We omit the full proof of this theorem due to space limitations.

Theorem 1. *The MRP algorithm finds the most reliable path for a unit-size message in $O(nM)$ time, where M is the upper bound of the number of time steps chosen.*

3.2 The MSP Algorithm: Constant-Size Messages

We generalize the algorithm of Section 3.1 to the case where the data size σ may be greater than one. In this case, the edge capacities play a concrete role in computing the path sustainabilities. There are two main differences between the algorithm in this section and the MRP algorithm. One is that when computing σ-MSPs at node v from a σ-MSP from s to a neighbor u of v, due to the nonuniform edge capacities, we may need to recompute the sustainability of the σ-MSP from s to u. The other difference is that the algorithm should be able to check for a cycle that causes a head blocking problem. This can be done by including the last σ nodes of a path in the tuples sent to neighboring nodes.

Before presenting the algorithm, we define the concept of *possible capacities*. Suppose there exists a path Q from node s to node v of capacity $c(Q)$. If $c \leq c(Q)$, then c is said to be a *possible capacity* of path Q. Furthermore, if there exists a path Q from s to v with length ℓ, we say that (c, ℓ) is a *possible capacity-length pair* from s to v, for any $c \leq c(Q)$.

We redefine the sustainability of an edge and the sustainability of a path in order to explicitly relate them to a specific possible capacity. The sustainability of edge (i, j) for the transmission of σ units of data for a possible capacity c of the transmission path is given by

$$r_{ij}^{\sigma,c}(t_i) = \prod_{t=t_i}^{\lceil \sigma/c \rceil + \ell - 1} r_{ij}(t), \tag{3}$$

where t_i denotes the starting time when edge (i, j) is used in the transmission. Equation(3) will be used instead of Equation(1) to account for possible capacities. A (σ, ℓ, c)-MSP from s to v is a (σ, ℓ)-MSP from s to v with possible capacity c. We change the notation P_{sv}^{ℓ} to $P_{sv}^{\ell,c}$ to allow for the possible capacity of a path. Equation(2), which gives the sustainability of a path $P_{v_0 v_k}^{\ell,c} = (v_0, v_1, \ldots, v_k)$ can be rewritten as

$$Re^{\sigma,c}(P_{v_0 v_k}^{\ell,c}) = \prod_{i=0}^{k-1} r_{v_i v_{i+1}}^{\sigma,c}(t_{v_i}),$$

where $\ell = \sum_{i=0}^{k-1} \ell_{v_i v_{i+1}}$ and c is a possible capacity of the path. We may omit the superscript c from $Re^{\sigma,c}$, whenever clear from the context.

Our algorithm keeps track of all possible capacity-length pairs (c, ℓ) in addition to all possible sequences of the last σ nodes on any path to the neighbors of a node v. We call these sequences the *σ-nearest subpaths* of v. If there exist two or more different paths with the same possible capacity-length pair and the same σ-nearest subpath, then our algorithm keeps the path with the highest sustainability. Each node v in the network G maintains a set of tuples of the form (c, ℓ, R, Q) where for some path P from s to v, c is a possible capacity of P, ℓ is the length of P, R is the corresponding sustainability of P, and Q is the σ-nearest subpath of v in P.

The MSP algorithm works just like the MRP algorithm, except for checking and detecting head blocking cycles. Note that the path found by our MSP algorithm may not be simple (it may contain cycles of length greater than σ).

Algorithm MSP(G, s, d)

1. *Initialization: during time step 0*:
   1.1 The source node s initializes $R = Re(P_{ss}^{0,c}) = 1$ for all $c \leq \sigma$ and $L(s) = \{(c, 0, 1, null) | \forall c \leq \sigma\}$
   1.2 Node s sends only the tuples $(c, \ell_{sv}, R \times r_{sv}^{\sigma,c}(t_s), (null, \ldots, null, s))$ with $c \leq \min(\sigma, c_{sv})$ to its neighbor, for all (s, v) in E.
2. *During time step* $t \leq M$:
   2.1 Each node v chooses the tuple $(c, t, R, (u_1, \ldots, u_\sigma))$ that has the maximum sustainability among all tuples of the form $(c, t, \cdot, (u_1, \ldots, u_\sigma))$ that v received during time step t.
   2.2 If $(u_i, u_{i+1}) \neq (u_\sigma, v)$, for all $i = 1, \cdots, \sigma - 1$, insert $(c, t, R, (u_1, \ldots, u_\sigma))$, for all c and for all $(u_1, \ldots, u_\sigma)$, into $L(v)$.
   2.3 If node v is not the destination, then for each tuple $(c, t, R, (u_1, \ldots, u_\sigma)) \in L(v)$, v sends to neighbor w tuples $(c, t + \ell_{vw}, R \times r_{vw}^{\sigma,c}(t_v), (u_2, \ldots, u_\sigma, v))$ with $c \leq c_{vw}$, for all (v, w) in E.
3. *Termination: time step* $M + 1$:
   3.1 Node d computes a MSP of length ℓ from s to d by choosing the tuple $q = (c, \ell, R, Q)$ from $L(d)$ that maximizes the sustainability R.

In the MSP algorithm, Step 2.2 ensures paths without cycles of length less than or equal to σ are considered. This is the basis for the proof of correctness of the MSP algorithm.

Theorem 2. *The MSP algorithm finds the most sustainable path for a message of size* σ *in* $O(M\sigma\alpha n^{\sigma+1})$ *time, where* $\alpha = \min\{\sigma, \max_{(i,j)\in E}(c_{ij})\}$. *In particular, if* σ *is a constant, the MSP algorithm runs in polynomial time on* n.

Proof. By the nature of the MSP algorithm, $L(d)$ will have exactly one tuple with associated length ℓ, for all possible capacities and for all possible σ-nearest subpaths of node d, at the end of time step M. Therefore the tuple computed in Step 3.1 is the σ-MSP from s to d. For the time bound of the algorithm, we observe that it takes constant time to calculate a tuple to be sent to the neighbors and to compare any two tuples. In Step 2.1, each node receives at most αn^σ tuples from an adjacent node. Thus it takes $O(\alpha n^{\sigma+1})$ time to choose the tuple with the maximum sustainability. In Step 2.2, for each selected tuple, only $\sigma - 1$ comparisons are necessary. Hence, Step 2.2 takes $O(\sigma\alpha n^{\sigma+1})$ time, and Step 2.3 takes $O(\alpha n^{\sigma+1})$ time. Therefore, the overall time complexity is $O(M\sigma\alpha n^{\sigma+1})$. If σ is a constant, this running time is polynomial on n (since M is also polynomial on n).

Note that the algorithm works correctly for any message size. However, if the size of the message is not a constant, we can no longer guarantee that it will run in time polynomial in n.

4 The MSRP Problem

In this section we present a polynomial time algorithm for finding most sustainable paths in circuit-switching networks, for arbitrary size messages.

In the circuit-switching routing model, once a path is established at the connection setup phase, it is reserved for the entire duration of the message transmission. Thus, the sustainability of a path is the product of the edge reliabilities of all edges in the path over all the time steps needed to transmit the message. We call the problem of finding a most sustainable path from s to d in circuit-switching networks *the most sustainable reserved path* (MSRP) problem. In particular, if the size of the message to be transmitted is σ, we call it the σ-MSRP problem.

The only difference from the network model of the previous section is the way the sustainability of an edge is defined. We keep the same notation as in the previous section unless specified. Before formalizing edge sustainability in this context, we define *the actual transmission duration* D of a path as the time between transmission of the first packet by the source and the receipt of the last packet by the destination. For each capacity-length pair (c, ℓ) at node v, the actual transmission duration D of a path from s to v of length ℓ and capacity c is $D = \lceil \sigma/c \rceil + \ell - 1$.

The sustainability of edge (i, j) for the entire duration of the transmission of a message of size σ, assuming that c is a possible capacity of a transmission path of length ℓ, is given by

$$R_{ij}^{\sigma,c} = \prod_{t=0}^{D} r_{ij}(t), \tag{4}$$

where D is the actual transmission duration of the path (which depends on ℓ). The sustainability of the path $P = (v_0, \ldots, v_k)$ is given by

$$Re^{\sigma,c}(P) = \prod_{i=0,\ldots,k-1} R_{v_i v_{i+1}}^{\sigma,c}.$$

Let P be a path whose actual transmission time is at most T. A (T, σ)-*most sustainable reserved path*, or for short (T, σ)-MSRP, for a message of size σ from s to d is a path P which maximizes $Re^{\sigma,c}(P)$. In this scenario, head blocking is no longer a concern since an MSRP is always simple, by the way the edge sustainabilities are defined in this model.

We now propose a polynomial time algorithm for the σ-MSRP problem. This algorithm uses the T-MSRP algorithm, which finds a (T, σ)-MSRP from s to d, as a subroutine. If path P is a (T, σ)-MSRP, where T is equal to the actual transmission duration of P, then P is a candidate for being a σ-MSRP. In other words, a σ-MSRP is the one with the highest path sustainability among all (T, σ)-MSRPs whose actual transmission durations are equal to T.

Our T-MSRP algorithm is similar in essence to the MSP algorithm. However, we introduce some pruning rules which help us reduce the number of subpaths we need to keep (these prunning rules do not apply to the MSP algorithm). At the beginning of the algorithm we do not know the actual transmission duration of the paths from s to t. Hence, we will choose different maximum transmission durations T arbitrarily, and then check whether there is a candidate path for being an MSRP, whose actual transmission duration is in fact equal to T.

The algorithm works as follows. Each node v will maintain a set of tuples $L(v)$ of the form (c, ℓ, R, T, u). Node v will update $L(v)$ upon learning the tuple (c, ℓ, R, T, u) received from one of its neighbors u, where $c \leq \min\{\sigma, c_{uv}\}$. Node v will select the tuple which maximizes R among the tuples $(c, \ell, R, T, \cdot)$ received from its neighbors and will add it to $L(v)$. Node v will calculate $Re^{\sigma,c}(P)$, the sustainability for possible capacity c of the path P of length $\ell + \ell_{vw}$ composed of the path from s to v associated with the tuple $(c, \ell, R, T, \cdot)$ followed by edge (v, w), for all $(v, w) \in E$. Then, if $\lceil \sigma/c \rceil + \ell + \ell_{vw} - 1 \leq T$ for a possible capacity c, the sustainability of path P *as a subpath of a path from s to d* is given by

$$Re^{\sigma,c}(P) = R \cdot R_{vw}^{\sigma,c}(T), \tag{5}$$

where $R_{vw}^{\sigma,c}(T) = \prod_{t=0}^{T} r_{vw}(t)$. We use $R_{ij}^{\sigma,c}(T)$ to denote the sustainability of edge (i, j) for maximum transmission duration T, in order to distinguish it from Equation (4); we will also use the notation $R_P^{\sigma,c}(T)$ to denote the sustainability of path P for maximum transmission duration T. If node v does not have another tuple of the form $(c, \ell + \ell_{vw}, \cdot, T, \cdot)$ in $L(v)$, then v will add the tuple $(c, \ell + \ell_{vw}, Re^{\sigma,c}(P), T, v)$ to $L(v)$. Otherwise, v will keep the tuple with higher path sustainability and ignore the other tuples of the form $(c, \ell + \ell_{vw}, \cdot, T, \cdot)$.

In some cases, we may be able to prune a path from s to node v, if at some point we can guarantee that this path will not lead to a σ-MSRP from s to d. In the following paragraphs, we present three pruning rules which we use in our algorithm. Let $(c^*, t^*, R^*, T^*, \cdot)$ where $T^* = \lceil \sigma/c^* \rceil + t^* - 1$ is the tuple with the best sustainability found by our algorithm so far at node d. Suppose there exists a tuple $q = (c, t, R, T, u)$ at node v — associated with path Q from s to v — such that the actual transmission duration $D = \lceil \sigma/c \rceil + t - 1$ of Q is greater than T^* and the sustainability R is less than R^*. Since any path P from s to d which extends the path Q has longer transmission duration than that of Q, the final sustainability of path P will be less than R which is less than R^*. Thus the first pruning rule says that we can prune the path Q at node v — that means node v does not propagate the tuple calculated from q to its neighbors.

We can also prune path Q if $D > T$, since in this case, we cannot transmit the whole message in time T through this path, and so the information about the sustainability R is not well-defined. Thus the second pruning rule says that node v does not add the tuple q to $L(v)$ if $D > T$.

The third pruning rule compares paths from s to node $v \neq d$, with respect to the same total transmission duration. Suppose there are two paths $P = (s = u_1, u_2, \ldots, u_p = v)$ and $P' = (s = w_1, w_2, \ldots, w_{p'} = v)$ from s to node v associated with the tuples $q = (c, \ell, R, T, \cdot)$ and $q' = (c', \ell', R', T, \cdot)$, respectively. Assume that $R > R'$ and $c \geq c'$. Consider a path $Q = (v = v_1, v_2, \ldots, v_k = d)$ from v to d whose capacity is at least c. Then

$$R^{\sigma,c}_{P\cup Q}(T) = \prod_{i=1}^{p-1} R^{\sigma,c}_{u_i u_{i+1}}(T) \times \prod_{j=1}^{k-1} R^{\sigma,c}_{v_j v_{j+1}}(T)$$

$$> \prod_{i=1}^{p'-1} R^{\sigma,c'}_{w_i w_{i+1}}(T) \times \prod_{j=1}^{k-1} R^{\sigma,c'}_{v_j v_{j+1}}(T) = R^{\sigma,c'}_{P'\cup Q}(T).$$

Since $Re^{\sigma,c}_{P\cup Q}(T)$ is strictly greater than $Re^{\sigma,c'}_{P'\cup Q}(T)$ for any choice of Q, intuitively, we might want to prune the path P' at node v. However if $\ell > \ell'$, it is still possible that P' will lead to a T-most sustainable path, whereas P will not. path from s to d. It may happen that the actual transmission duration of the path obtained by extending path P becomes greater than T, whereas that of the path obtained by extending path P' is still less than T. A similar scenario can happen when $R > R'$, $\ell \leq \ell'$ and $c < c'$. Hence we can safely prune the path P' only if $\ell \leq \ell'$, $c \geq c'$, and $R > R'$.

Before presenting the T-MSRP algorithm, we briefly describe our MSRP algorithm for a message of size σ. The algorithm will call the T-MSRP algorithm for all $1 \leq T \leq M$, and keep the tuple $(c, \ell, R, T, \cdot)$ returned by T-MSRP algorithm if the actual transmission duration $\lceil \sigma/c \rceil + \ell - 1$ is equal to T. At the end of the algorithm, we will have all the tuples that correspond to (T, σ)-MSRPs from s to d, for all T. Hence the σ-MSRP from s to d will be obtained by choosing the tuple with maximum sustainability.

Now we present our algorithm T-MSRP, which computes a T-most sustainable path from s to d for a message of size σ. The parameters c^*, R^*, and T^* are obtained from the tuple $(c^*, \ell, R^*, T^*, \cdot)$ with maximum path sustainability where $\lceil \sigma/c^* \rceil + \ell - 1 = T^*$ that the MSRP algorithm found so far at node d.

Algorithm T-MSRP(G, s, d, σ,T, c^*, T^*, R^*)

1. *During time step $t = 0$:*
   1.1 The source node s initializes $R = Re^{\sigma,c}(P^0_{ss}) = 1$, for all $c \leq \sigma$, and $L(s) = \{(c, 0, R, T, null) \mid \forall c \leq \sigma\}$.
   1.2 For each neighbor v, node s sends to v only the tuples $(c, \ell_{sv}, R \times R^{\sigma,c}_{sv}(T), T, s)$, such that either $\lceil \sigma/c \rceil + \ell_{sv} - 1 \leq T^*$ or $R \geq R^*$, and $\lceil \sigma/c \rceil + \ell_{sv} - 1 \leq T$, where $c \leq \min\{\sigma, c_{sv}\}$. (by the first and second pruning rules)
2. *During time step $t \leq T$:*
   2.1 Each node v compares all the tuples of the form $(c, t, \cdot, T, \cdot)$ received from its neighbors at time step t.
   2.2 For each possible capacity c at node v,
      2.2.1 Choose the tuple $(c, t, R, T, \cdot)$ that has maximum sustainability among all tuples of the form $(c, t, \cdot, T, \cdot)$ v received.
      2.2.2 Node v compares $(c, t, R, T, \cdot)$ to each tuple $(c', t', R', T, \cdot)$ in $L(v)$.
         2.2.2.1 If there is a tuple $(c', t', R', T, \cdot)$ with $c \geq c'$, $t < t'$, and $R > R'$ then remove the tuple from $L(v)$. (by the third pruning rule)
         2.2.2.2 If there does not exist any tuple $(c', t', R', T, \cdot)$ with $c \leq c'$, $t > t'$, and $R < R'$ then insert $(c, t, R, T, \cdot)$ into $L(v)$. Otherwise, discard the tuple $(c', t', R', T, \cdot)$. (by the third pruning rule)

2.2.3 If node v is not the destination and $(c, t, R, T, \cdot)$ is not discarded, then v computes $\bar{t} = t + \ell_{vw}$ and $\bar{R} = R \times R_{vw}^{\sigma,c}(T)$ and update the previous node to v, for all (v, w) in E.

2.2.4 Node v sends to neighbor w tuples $(c, \bar{t}, \bar{R}, T, \cdot)$, such that either $\lceil \sigma/c \rceil + \bar{t} - 1 \leq T^*$ or $R \geq R^*$, and $\lceil \sigma/c \rceil + \bar{t} - 1 \leq T$, where $c \leq \min\{\sigma, c_{vw}\}$. (by the first and second pruning rules)

3. *During time step* $t = T + 1$:

3.1 Node d computes a (T, σ)-MSRP from s to d by choosing the tuple $q = (c, t, R, T, v)$ that maximizes the sustainability R.

3.2 Return the tuple q.

We now present three lemmas that form the foundation for the proof of correctness of our MSRP algorithm.

Lemma 1. *At time step t, a node v only receives tuples with associated length t from its neighbors.*

Proof. We prove the lemma by induction on time step t. When $t = 0$, no node in the network receives any tuple. Assume the above lemma is true at any time $t' < t$. We prove the lemma also holds for time step t.

At time step $t - \ell_{vw}$, node v received all tuples of the form $(c, t - \ell_{vw}, \cdot, T, \cdot)$, by the induction hypothesis. After selecting the tuple $(c, t - \ell_{vw}, R, T, \cdot)$ which has the highest sustainability among tuples of the form $(c, t - \ell_{vw}, \cdot, T, \cdot)$, v sends $q = (c, t - \ell_{vw} + \ell_{vw}, \cdot, T, v)$ to its neighbor w still at time step $t - \ell_{vw}$. Since v does not receive any other tuples of the form $(c, t - \ell_{vw}, \cdot, T, \cdot)$ at any time step other than $t - \ell_{vw}$, q is the only tuple sent by node v to node w with associated capacity-length pair (c, t). Since it takes ℓ_{vw} time steps for all tuples bound from v to w to traverse edge (v, w) (there are at most as many possible capacities — and thus tuples — as the capacity of edge (v, w); therefore we assume without loss of generality that all tuples in consideration can be sent from v to w at time $t - l_{vw}$) all the tuples that left v to w at time step $t - \ell_{vw}$ must reach node w at time t. In particular, q reaches node w at time t. Since this is true for any pair of adjacent nodes v and w and any possible capacity, the lemma holds.

Lemma 2. *A T-most sustainable reserved path obtained by the T-MSRP algorithm does not contain a cycle.*

Proof. Suppose a path P from s to d contains a cycle C. Then we can construct a path P' from s to d by removing the edges in C from P, leaving s and d still connected by the edges in P'. Thus the number of edges in P' is less than the number of edges in P and P' is a subpath of P. Then $R_P^{\sigma,c}(T) = R_{P'}^{\sigma,c}(T) \times R_C^{\sigma,c}(T)$ for any possible capacity c. Since $0 < r_{ij}(t) < 1$, for all (i, j) in E, $0 < R_C^{\sigma,c}(T) < 1$. Hence $R_P^{\sigma,c}(T) < R_{P'}^{\sigma,c}(T)$

Lemma 3. *Algorithm T-MSRP finds a T-most sustainable reserved path for a message of size σ from s to d.*

Proof. The tuples that a node v receives from a neighbor u at time step t have length t. For each possible capacity c, node v chooses the tuple $q = (c, t, R, T, \cdot)$ that has the maximum sustainability R among all the tuples of the form $(c, t, R, T, \cdot)$ that v received. Then v computes a new tuple $\bar{q} = (c, t + \ell_{vw}, R \times R_{vw}^{\sigma,c}(T), T, \cdot)$, for each $(v, w) \in E$ based on the tuple q, and sends $\bar{q}$ to its neighbor w if the tuple does not satisfy any one of the pruning conditions. By Lemma 1, q is the only tuple kept in $L(v)$ at time step t with associated capacity-length pair $(c, t + \ell_{vw})$. Thus, at time step $T + 1$, node d will have only one tuple in $L(d)$ for each capacity-length pair (c, ℓ) if there exists a path from s to d with length ℓ and a possible capacity c. The tuple with the maximum sustainability among the tuples in $L(d)$ represents the T-most sustainable path from s to d for a message of size σ and transmission duration T.

Since an MSRP has to be simple, it suffices to let the upper bound on the maximum path length, M, be equal to the number of edges in a longest path multiplied by the largest edge delay in the network. Since the delay of an edge is bounded by a polynomial on n, M is also bounded by a polynomial on n. For most multimedia applications, in particular for real-time ones, the end-to-end delay is one of the most important QoS requirements. Thus one may want to choose M to be the deadline that guarantees the desired timely delivery of messages a the upper bound on the maximum allowed path length.

Finally, we present an alorithm for computing a σ-MSRP of G.

Algorithm MSRP(G, s, d, σ)

1 Initialize $c^* = 1$, $R^* = 0$, and $T^* = 0$

2 For each possible transmission duration T $(1 \leq T \leq M)$,

  2.1 Call T-MSRP$(G, s, d, \sigma, T, c^*, T^*, R^*)$ and let $q = (c, \ell, R, T, u)$ be the tuple returned by T-MSRP.

  2.2 If $\lceil \sigma / c \rceil + \ell - 1 = T$, then keep the tuple q.

  2.3 If $R^* \leq R$, then let $c^* = c$, $R^* = R$, and $T^* = T$.

3 The σ-MSRP is the path that corresponds to the tuple $(c^*, T^*, R^*, T^*, \cdot)$.

We can retrieve the entire path which is selected by MSRP algorithm as we did in Section 3.1.

Theorem 3. *The MSRP algorithm finds the most sustainable reserved path for a message of size σ in $O(\alpha^2 M^3)$ time, where $\alpha = \min\{\sigma, \max_{(i,j)\in E}(c_{ij})\}$.*

Proof. The correctness of the algorithm follows directly from the previous lemmas. We then show the bound on the running time. In the T-MSRP algorithm, Step 2.2 will be executed at most α times. Since at time step t, node v has at most αM tuples in $L(v)$, the comparisons in Step 2.2.2 takes $O(\alpha M)$ time. The computations in Step 2.2.3 takes $O(n)$ time in the worst case. Hence the T-MSRP algorithm takes $O(\alpha M(\alpha M + n)) = O(\alpha^2 M^2)$ time (we can assume that $M \geq n$, without loss of generality). Therefore, the time complexity of the MSRP algorithm is $O(\alpha^2 M^3)$.

5 Conclusion

In this paper, we formalized the problem of finding most sustainable paths in networks with time-dependent edge reliabilities. We considered two network models: the wormhole and circuit-switching models. For the former model, we presented a fully-distributed polynomial time algorithm for selecting a MSP for the transmission of constant-size messages. Even though it works correctly for any message size, due to the size of tuples that are exchanged between nodes during the execution of the algorithm, we can no longer guarantee that it will run in polynomial time on n if the sizes of messages are not constant. For the latter model, we presented the MSRP algorithm, which finds a σ-MSRP from the source to the destination in polynomial time, where σ is an arbitrary polynomial on n.

One natural extension of the work presented in this paper would be to seek for polynomial time algorithms that compute a σ-MSP in the wormhole routing model for an arbitrary message size σ, with time-dependent edge reliabilities. This problem may prove to be very hard to solve. Thus, approximation algorithms for this problem would also be of great interest.

References

1. Ahuja, R. K.: Minimum cost reliability problem. Computers and Operations Research 16 (1988) 83-89
2. Barrett, C., Bisset, K., Konjevod, G., Jacob, R., Marathe, M.: Routing problems in time-dependent and labeled networks, submitted to SODA 2002
3. Bertsekas, D., Gallager, R.: Data Networks. Prentice Hall, (1992)
4. Chen, Y. L., Chin, Y. H.: The quickest path problem. Computers and Operations Research 17, no. 2 (1990) 153-161
5. Cooke, K., Halsey, E.: The shortest route through a network with time dependent internodal transit times. Journal of mathematical analysis and applications 14 (1966) 493-498
6. Gibson, G. A., Vitter, J. S., Wilkes, J.: Strategic Direction in Storage I/O issues in large-scale computing. In Proceedings of the ACM Workshop on Strategic Directions in Computing Research, ACM Computing Surveys 28, no. 4 (1996) 779-793
7. Handel, R., Huber, M. N.: Integrated Broadband Networks: An Introduction to ATM-Based Networks. Addison-Wesley (1991)
8. Inktomi Inc. Streaming media caching white paper. Inktomi Corporation, Technical Report (1999)
9. Miller-Hooks, E., Patterson S. S.: The time-dependent quickest flow problem. Submitted to Operations Research 2000
10. McDysan, D. E., Spohn, D. L.: ATM: Theory and Application. McGraw-Hill (1995)
11. Orda, A., Rom, R.: Shortest-path and minimum-delay algorithms in networks with time-dependent edge-length. Journal of the Association for Computing 37, no. 3 (1990) 607-625
12. Orda, A., Rom, R.: Distributed shortest-path protocols for time-dependent networks. Distributed Computing 10 (1996) 49-62
13. Scheideler, C.: Universal routing strategies for interconnection networks. Springer (1998)

14. Sung, K., Bell, M., Seong, M., Park, S.: Shortest paths in a network with time-dependent flow speeds. European Journal of Operational Research 121 (2000) 32-39
15. Tragoudas, S.: The most reliable path transmission. Performance, Computing, and Communications Conference, 1999 IEEE International (1999) 15-19
16. Vogel, R., Herrtwich, R.G., Kalfa, W., Wittig, H., Wolf, L.C.: QoS-based routing of multimedia streams in computer networks. IEEE Journal on Selected Areas in Communications 14, no. 7 (1996) 1235-1244
17. Wang, Z., Crowcroft, J.: Quality-of-service routing for supporting multimedia applications. IEEE Jornal on Selected Areas in Communications 14, no. 7 (1996) 1228-1234

Signals for Cellular Automata in Dimension 2 or Higher

Jean-Christophe Dubacq[1] and Véronique Terrier[2]

[1] Université de Paris-Sud, LRI, Bâtiment 490,
F-91405 Orsay Cedex, France
[2] GREYC, Campus II, Université de Caen,
F-14032 Caen Cedex, France

Abstract. We investigate how increasing the dimension of the array can help to draw signals on cellular automata. We show the existence of a gap of constructible signals in any dimension. We exhibit two cellular automata in dimension 2 to show that increasing the dimension allows to reduce the number of states required for some constructions.

1 Introduction

Cellular automata (CA) are simple mechanisms that appear in many fields. They are best described as simple cells regularly arranged in an array of dimension k. All these cells have a finite number of states, and change all at the same time (synchronously) of state according to the same rules, looking at their neighbors. Physical systems containing many discrete elements with local interactions are conveniently modeled as cellular automata, such as dendritic crystals growth, evolution of biological populations...

Introduced by von Neumann in [vN66] to study self-reproduction, cellular automata emerge as a key model of massively parallel computation. Exact mathematical computations are possible, since one can simulate a Turing machine, but cellular automata have a very different way to represent data. The geometrical aspect of cellular automata induces specific questions that do not appear in sequential models.

Whereas the work of a CA is based on local exchange in the nearest neighborhood, at global scale the collective behavior of the CA often emerges as signals, i.e. continuous lines in the space-time diagram, which capture the organization and the sending of information through the network. Cellular automata as computational systems can be seen from two main points of view: either a CA is designed to fill a specific task, or a given CA is analyzed in terms of general properties and dynamics. In both cases, the notion of signal appears. To build a CA, signals are a tool that makes the transition from the local to the global behavior, to geometrically describe the organization and the motion of information between cells (see e.g. [Fis65, Maz87]). When analyzing a CA, the behavior of many CA shows "particles in motions", whose trajectories can be interpreted as signals (see [Mar00], or even the gliders in the game of Life [BCG82]).

S. Rajsbaum (Ed.): LATIN 2002, LNCS 2286, pp. 451–464, 2002.

Intuitively, signals are some paths through the space-time diagram which encode and combine the information, but an all-encompassing formalization is lacking. Nevertheless, some attempt has been done (see [MT99]). We propose an alternative definition for CA that generate signals.

In dimension 1, it has been shown that some signals around the diagonal axis can not be set up by any CA: a signal set up by any CA either becomes parallel to the diagonal axis or takes at least a logarithmic slow-down. Surprisingly, in higher dimensions, although more cells are involved around the diagonal axis, we will show that the same gap occurs. So, increasing the dimension does not help to construct such signals around the diagonal axis. This partially answers the problem #51 of the list of open problems on CA (see [DFM00]).

However, we have a gain in terms of number of states. In dimension 1, performing along the diagonal axis a logarithmic slow-down requires at least 4 states (it is not difficult to review the few CA with 3 states). But in dimension 2, we exhibit a CA with 3 states (including the quiescent state) which performs a logarithmic slow-down along the diagonal axis. Furthermore we show that this CA is optimal in terms of number of states.

To complete the analysis of the gain of working in higher dimension, we describe a CA that supports other logarithmic slow-downs with less states in dimension 2 than in dimension 1.

2 Definition of a Signal

A k-dimensional cellular automata is a k-dimensional array of finite automata (cells) indexed by $\mathbf{Z}^k$. All cells evolve synchronously at discrete time steps. At each step, each cell enters a new state according to a transition function involving only its local neighborhood.

We use the notation $\mathbf{u} = (u_1, \ldots, u_k)$ to designate a k-vector. $\mathbf{0}$ is the null vector $(0, \ldots, 0)$. $\mathbf{1}$ is the unary vector $(1, \ldots, 1)$ and $t \cdot \mathbf{u}$ is the product of $\mathbf{u}$ by a scalar t.

Formally a k-CA is defined by $(\mathcal{S}, V, f, \lambda)$ where: $\mathcal{S}$ is the set of states, $V = \{\mathbf{x}^1, \ldots, \mathbf{x}^v\} \subset \mathbf{Z}^k$ is the neighborhood, f from $\mathcal{S}^v$ into $\mathcal{S}$ is the transition function, $\lambda \in \mathcal{S}$ is the quiescent state which verifies $f(\lambda, \ldots, \lambda) = \lambda$.

A site $(\mathbf{u}, t)$ refers to the cell $\mathbf{u}$ at time t and $\langle \mathbf{u}, t \rangle$ denotes its state at time t. We refer to the whole mapping $(\mathbf{u}, t) \mapsto \langle \mathbf{u}, t \rangle$ as the space-time diagram of the CA.

For time $t \geq 0$ we have

$$\langle \mathbf{u}, t+1 \rangle = f\left(\langle \mathbf{u} + \mathbf{x}^1, t \rangle, \ldots, \langle \mathbf{u} + \mathbf{x}^v, t \rangle\right)$$

We will consider three different neighborhoods: the Von Neumann neighborhood, the Moore neighborhood and the trellis neighborhood.

$$V_{\text{Von Neumann}} = \left\{\mathbf{x} \in \mathbf{Z}^k : \sum |x_i| \leq 1\right\},$$

$$V_{\text{Moore}} = \left\{\mathbf{x} \in \mathbf{Z}^k : |x_i| \leq 1\right\},$$

$$V_{\text{trellis}} = \left\{\mathbf{x} \in \mathbf{Z}^k : |x_i| = 1\right\}.$$

Note that, with the trellis neighborhood, the states $\langle \mathbf{u}, t \rangle$ and $\langle \mathbf{u}', t' \rangle$ do not interfere if for some i the sums $u_i + t$ and $u'_i + t'$ are not of same parity. So at time t we will deal only with cells $\mathbf{u} = (u_1, \cdots, u_k)$ such that $u_1, \cdots, u_k, t$ are of same parity, the other sites are considered as quiescent or non-existent.

Observe that the graph of dependencies of a k-dimensional cellular automata with Moore neighborhood contains the graph of dependencies of a k-dimensional cellular automata with Von Neumann neighborhood; so the simulation of a Von Neumann CA can be done in real time by a Moore CA. The graph of dependencies of a k-CA with Moore neighborhood also contains the graph of dependencies of a k-dimensional trellis. And as shown in dimension 1 (see [CČ84, IKM85]), provided the cells $\mathbf{u}$ of a CA with trellis neighborhood correspond to the set of cells $\{\mathbf{u} + \mathbf{x} : \mathbf{x} \in \{0, 1\}^k\}$ of a CA with Moore neighborhood, the trellis CA and the Moore CA are time-wise equivalent. Hence a trellis CA which performs the same task than a Moore CA, might have more states but always with less interconnections.

We recall the definition of impulse CA's and signals (see [MT99]):

Definition 1 (Impulse CA) *An impulse CA is a* 5*-tuple* $(\mathcal{S}, V, f, G, \lambda)$ *where* $(\mathcal{S}, V, f, \lambda)$ *is a CA and* G *a distinguished state of* $\mathcal{S}$ *such that at initial time* $t = 0$ *all cells are in the quiescent state* λ *but the cell* $\mathbf{0}$ *which is in state* G*:*

$$\begin{cases} \langle \mathbf{x}, 0 \rangle = \lambda & \text{if } \mathbf{x} \neq \mathbf{0}, \\ \langle \mathbf{0}, 0 \rangle = G. \end{cases}$$

Definition 2 (Signal) *For a given neighborhood* V*, a* V*-signal* Γ *is a sequence of sites* $\{(\mathbf{u}(t), t)\}_{t \geq 0}$ *such that*

- $\mathbf{u}(0) = \mathbf{0}$.
- *For all* $t \geq 0$*:* $\mathbf{u}(t+1) - \mathbf{u}(t) \in V$.

Fundamentally, a signal is a continuous path in the graph of dependencies of the CA.

To emphasize the elementary moves of the V-signal Γ, we denote by $\Gamma_\mathbf{x}$ where $\mathbf{x} \in V$, the set of sites of Γ which reach the next one by a $-\mathbf{x}$ move: $\Gamma_\mathbf{x} = \{(\mathbf{u}(t), t) \in \Gamma : (\mathbf{u}(t) - \mathbf{x}, t+1) \in \Gamma\}$. Note that $\{\Gamma_\mathbf{x}\}_{\mathbf{x} \in V}$ defines a partition of Γ.

We recall the definition of impulse CA which draw explicitly a signal.

Definition 3 (Construction of a signal) *An impulse CA* $A = (\mathcal{S}, V, f, G, \lambda)$ *constructs a* V*-signal* Γ *if there exists a subset* $\mathcal{S}_0$ *of* $\mathcal{S}$ *such that* $(\mathbf{u}, t) \in \Gamma$ *if and only if* $\langle \mathbf{u}, t \rangle \in \mathcal{S}_0$.

We propose also two alternative definitions of impulse CA which draw implicitly signals.

Definition 4 (Detection of a signal) *An impulse CA* $A = (\mathcal{S}, V, f, G, \lambda)$ *detects a* V*-signal* Γ *if there exists a partition* $\{\mathcal{S}_\mathbf{x}\}_{\mathbf{x} \in V}$ *of the set of states* $\mathcal{S}$ *such that if* $(\mathbf{u}, t) \in \Gamma_\mathbf{x}$ *then* $\langle \mathbf{u}, t \rangle \in \mathcal{S}_\mathbf{x}$.

Definition 5 (Supporting a signal) *An impulse CA* $A = (\mathcal{S}, V, f, G, \lambda)$ *supports a V-signal* Γ *if there exists a finite automaton* $F = (\mathcal{S}, Q, \delta, q_0)$ *with* $\mathcal{S}$ *the input alphabet,* Q *the set of states,* δ *from* $Q \times \mathcal{S}$ *into* $Q \times V$ *the transition function and* q_0 *the initial state and a sequence of states* $\{q(t)\}_{t\geq 0}$ *such that* $q(0) = q_0$ *and for all* $t \geq 0$: $\delta(q(t), \langle \mathbf{u}(\mathbf{t}), t\rangle) = (q(t+1), \mathbf{u}(t+1) - \mathbf{u}(t))$.

The construction of a signal is a characterization by marking all the sites of the signal with a special set of states, whereas supporting a signal is a more dynamic tool, enabling the use of a finite automaton to retrieve the signal from the space-time diagram. Detection is a special case of support.

Actually the three notions are equivalent. If an impulse CA A constructs a V-signal Γ then it detects it and if an impulse CA A detects a V-signal Γ then it supports it. Furthermore, we get:

Proposition 1 *If an impulse CA A supports a V-signal* Γ *then there exists an impulse CA* A' *which constructs it.*

Proof. Suppose that Γ is supported by the impulse CA $A = (\mathcal{S}, V, f, G, \lambda)$ with the finite automata $F = (\mathcal{S}, Q, \delta, q_0)$. Consider the new impulse CA $A' = (\mathcal{S} \times (\{0\} \cup Q), V, f', (G, q_0), (\lambda, 0))$ with

$$f'(\underbrace{(s_{\mathbf{x}}, m_{\mathbf{x}}), \ldots}_{\mathbf{x}\in V}) = (f(\underbrace{s_{\mathbf{x}}, \ldots}_{\mathbf{x}\in V}), m))$$

where $m \in Q$ if and only if there exists $\mathbf{a} \in V$ such that $m_{\mathbf{a}} \in Q$ and $\delta(m_{\mathbf{a}}, s_{\mathbf{a}}) = (m, \mathbf{a})$. Then the subset $\mathcal{S}_0 = \mathcal{S} \times Q$ marks exactly the sites of Γ.

Definition 6 (Basic signals) *A V-signal is basic if the sequence of its elementary moves (whose values are in V)* $\{\mathbf{u}(t+1) - \mathbf{u}(t)\}_{t\geq 0}$ *is ultimately periodic.*

Actually the basic signals do not use the parallelism of the CA:

Claim 2 *The impulse CA* $\mathcal{A} = (\{\lambda\}, V, f, \lambda, \lambda)$ *supports exactly the basic V-signals.*

Proof. Any impulse CA $(\mathcal{S}, V, f, \lambda)$, in particular the CA $\mathcal{A} = (\{\lambda\}, V, f, \lambda, \lambda)$, supports any basic V-signal. Conversely, a quiescent background can only support ultimately periodic moves.

3 A Gap on Constructible Signals

In dimension 1, it has been shown that the signals $\{(t - u(t), t)\}_{t\geq 0}$ such that $u(t) = o(\log(t))$ and $u(t) \neq \Theta(1)$ are not constructible (see [MT99]). Here we will show for Moore neighborhood (and therefore trellis neighborhood) that even in higher dimension, the signal of maximal speed $\{(t \cdot \mathbf{1}, t)\}_{t\geq 0}$ can not be slowed down below the logarithm.

First we define, for $\mathbf{i} \in \mathbf{Z}^k$ and $t \in \mathbf{N}$, $\mathcal{D}_\mathbf{i}^t$ to be the state $\langle t \cdot \mathbf{1} - \mathbf{i}, t\rangle$. The states of the neighbor cells of $\mathcal{D}_\mathbf{i}^{t+1}$ with relative coordinates $\mathbf{x} \in V_{\text{Moore}}$ are $\mathcal{D}_{\mathbf{i}-\mathbf{x}-\mathbf{1}}^t$. Thus, $\mathcal{D}_\mathbf{i}^{t+1} = f(\underbrace{\mathcal{D}_{\mathbf{i}-\mathbf{x}-\mathbf{1}}^t, \ldots}_{\mathbf{x}\in V_{\text{Moore}}})$. And at initial time, only the cell $\mathbf{0}$ is in a non-quiescent state G:

$$\mathcal{D}_\mathbf{i}^0 = \begin{cases} G & \text{if } \mathbf{i} = \mathbf{0} \\ \lambda & \text{else.} \end{cases}$$

Claim 3 *$\mathcal{D}_\mathbf{i}^t = \lambda$ if $\mathbf{i} \in \mathbf{Z}^k \setminus \mathbf{N}^k$ or $2t < \max(i_1, \ldots, i_k)$.*

Proof. As $\mathbf{0}$ is the only active cell at time 0, at time $t \geq 0$ (with Moore neighborhood) a cell $\mathbf{c}$ is in a quiescent state if any of its coordinates c_a is such that $|c_a| > t$. In particular, with $\mathbf{c} = t \cdot \mathbf{1} - \mathbf{i}$, if any i_a is such that $|t - i_a| > t$, i.e. $i_a < 0$ or $i_a > 2t$, we have $\mathcal{D}_\mathbf{i}^t = \lambda$.

We consider the words $\mathcal{D}_\mathbf{i} \in \mathcal{S}^\infty$ corresponding to the significant part of the diagonals:

$$\mathcal{D}_\mathbf{i} = \begin{cases} (\mathcal{D}_\mathbf{i}^t)_{t \geq \lceil \max(i_1,\ldots,i_k)/2 \rceil} & \text{if } \mathbf{i} \in \mathbf{N}^k \\ \lambda^\infty & \text{else.} \end{cases}$$

The next proposition states the periodic behavior of $\mathcal{D}_\mathbf{i}$. As $\mathcal{D}_\mathbf{i}^{t+1}$ is defined by $\mathcal{D}_{\mathbf{i}-\mathbf{x}-\mathbf{1}}^t$ where $\mathbf{x} \in V_{\text{Moore}}$, the periodic behavior of $\mathcal{D}_\mathbf{i}$ can be characterized by the periodic behavior of the lower diagonals $\mathcal{D}_{\mathbf{i}-\mathbf{x}-\mathbf{1}}$ where $\mathbf{x} \in V_{\text{Moore}} \setminus \{\mathbf{1}\}$.

Proposition 4 *For all $\mathbf{i} \in \mathbf{Z}^k$, there exists $\alpha_\mathbf{i} \in \mathcal{S}^\star$, $\beta_\mathbf{i} \in \mathcal{S}^\star$, $u_\mathbf{i} \in \mathbf{N}$ and $v_\mathbf{i} \in \mathbf{N}$ such that:*

- $\mathcal{D}_\mathbf{i} = \alpha_\mathbf{i} (\beta_\mathbf{i})^\infty$.
- $u_\mathbf{i} + v_\mathbf{i} \leq |\mathcal{S}|$ *and* $1 \leq v_\mathbf{i} \leq |\mathcal{S}|$.
- $|\alpha_\mathbf{i}| \leq M_\mathbf{i} + u_\mathbf{i} P_\mathbf{i}$, *where $M_\mathbf{i}$ is* $\max_{\mathbf{x}\in V_{Moore}\setminus\{\mathbf{1}\}} (|\alpha_{\mathbf{i}-\mathbf{x}-\mathbf{1}}|)$.
- $|\beta_\mathbf{i}|$ *divides $v_\mathbf{i} P_\mathbf{i}$, where $P_\mathbf{i}$ is* $\operatorname{lcm}_{\mathbf{x}\in V_{Moore}\setminus\{\mathbf{1}\}} (|\beta_{\mathbf{i}-\mathbf{x}-\mathbf{1}}|)$.

Proof. We do an induction on $r = i_1 + \cdots + i_k$. Remark that the sum of all coordinates of $\mathbf{i} - \mathbf{x} - \mathbf{1}$ is always smaller than the sum of all coordinates of $\mathbf{i}$, for $\mathbf{x} \in V_{\text{Moore}} \setminus \{\mathbf{1}\}$. The proposition is true for $r < 0$. Indeed in this case $\mathbf{i} \in \mathbf{Z}^k \setminus \mathbf{N}^k$. So $\mathcal{D}_\mathbf{i}$ is λ^∞ and we can set $\alpha_\mathbf{i}$ to be the empty word, $\beta_\mathbf{i} = \lambda$, $u_\mathbf{i} = 0$ and $v_\mathbf{i} = 1$. We suppose the proposition true up to r and we will prove it for $r + 1 = i_1 + \cdots + i_k$. Let τ stand for $\lceil \max(i_1, \ldots, i_k)/2 \rceil$. By hypothesis of recurrence, we have $\mathcal{D}_{\mathbf{i}-\mathbf{x}-\mathbf{1}}^t = \mathcal{D}_{\mathbf{i}-\mathbf{x}-\mathbf{1}}^{t'}$ for all $\mathbf{x} \in V_{\text{Moore}} \setminus \{\mathbf{1}\}$ and all t, t' such that $t > t' \geq \tau + M_\mathbf{i}$ and $P_\mathbf{i}$ divides $(t - t')$ ($P_\mathbf{i}$ is the least common multiple of all the periods for $\mathbf{x} \in V_{\text{Moore}} \setminus \{\mathbf{1}\}$, hence the periodicity).

Among the $|\mathcal{S}| + 1$ states $\left\{\mathcal{D}_\mathbf{i}^{\tau+M_\mathbf{i}}, \mathcal{D}_\mathbf{i}^{\tau+M_\mathbf{i}+P_\mathbf{i}}, \ldots, \mathcal{D}_\mathbf{i}^{\tau+M_\mathbf{i}+|\mathcal{S}|P_\mathbf{i}}\right\}$ at least two are equal: for some a and b with $0 \leq a < b \leq |\mathcal{S}|$, $\mathcal{D}_\mathbf{i}^{\tau+M_\mathbf{i}+aP_\mathbf{i}} = \mathcal{D}_\mathbf{i}^{\tau+M_\mathbf{i}+bP_\mathbf{i}}$. Thus, by induction, it follows that for all u:

$$\mathcal{D}_{\mathbf{i}}^{\tau+M_{\mathbf{i}}+aP_{\mathbf{i}}+u+1} = f(\underbrace{\mathcal{D}_{\mathbf{i-x-1}}^{\tau+M_{\mathbf{i}}+aP_{\mathbf{i}}+u},\dots)}_{\mathbf{x}\in V_{\text{Moore}}} = f(\underbrace{\mathcal{D}_{\mathbf{i-x-1}}^{\tau+M_{\mathbf{i}}+bP_{\mathbf{i}}+u},\dots)}_{\mathbf{x}\in V_{\text{Moore}}}$$
$$= \mathcal{D}_{\mathbf{i}}^{\tau+M_{\mathbf{i}}+bP_{\mathbf{i}}+u+1}.$$

We can choose $\alpha_{\mathbf{i}} = \mathcal{D}_{\mathbf{i}}^{\tau} \cdots \mathcal{D}_{\mathbf{i}}^{\tau+M_{\mathbf{i}}+aP_{\mathbf{i}}-1}$, $\beta_{\mathbf{i}} = \mathcal{D}_{\mathbf{i}}^{\tau+M_{\mathbf{i}}+aP_{\mathbf{i}}} \cdots \mathcal{D}_{\mathbf{i}}^{\tau+M_{\mathbf{i}}+bP_{\mathbf{i}}-1}$, $u_{\mathbf{i}} = a$ and $v_{\mathbf{i}} = b - a$ which verify the desired properties.

The following corollary specifies the length of the periodic and non-periodic parts of $\mathcal{D}_{\mathbf{i}}$.

Corollary 5 *For all* $\mathbf{i} \in \mathbf{N}^k$, *there exists* $\alpha_{\mathbf{i}} \in \mathcal{S}^\star$, $\beta_{\mathbf{i}} \in \mathcal{S}^\star$ *such that*

- $\mathcal{D}_{\mathbf{i}} = \alpha_{\mathbf{i}}\,(\beta_{\mathbf{i}})^\infty$.
- $|\alpha_{\mathbf{i}}| < |\mathcal{S}|\,\mathrm{lcm}(1,\dots,|\mathcal{S}|)^{i_1+\dots+i_k}$.
- $|\beta_{\mathbf{i}}|$ *divides* $\mathrm{lcm}(1,\dots,|\mathcal{S}|)^{i_1+\dots+i_k+1}$.

Proof. We do a recurrence on $r = i_1 + \dots + i_k$. For $r = 0$, according the proposition we have $|\alpha_0| \le u_0 < |\mathcal{S}|$, $|\beta_0|$ divides v_0 which divides $\mathrm{lcm}(1,\dots,|\mathcal{S}|)$.

Now we suppose the corollary true up to r. Then for $r+1 = i_1 + \dots + i_k$ we have $M_{\mathbf{i}} < |\mathcal{S}|\,\mathrm{lcm}(1,\dots,|\mathcal{S}|)^{i_1+\dots+i_k-1} \le \mathrm{lcm}(1,\dots,|\mathcal{S}|)^{i_1+\dots+i_k}$; $P_{\mathbf{i}}$ divides $\mathrm{lcm}(1,\dots,|\mathcal{S}|)^{i_1+\dots+i_k}$. So

$$\begin{aligned}|\alpha_{\mathbf{i}}| &\le M_{\mathbf{i}} + u_{\mathbf{i}} P_{\mathbf{i}}\\ &< \mathrm{lcm}(1,\dots,|\mathcal{S}|)^{i_1+\dots+i_k} + (|\mathcal{S}|-1)\,\mathrm{lcm}(1,\dots,|\mathcal{S}|)^{i_1+\dots+i_k}\\ &< |\mathcal{S}|\,\mathrm{lcm}(1,\dots,|\mathcal{S}|)^{i_1+\dots+i_k}.\end{aligned}$$

And $|\beta_{\mathbf{i}}|$ divides $v_{\mathbf{i}} P_{\mathbf{i}}$ which divides $\mathrm{lcm}(1,\dots,|\mathcal{S}|)^{i_1+\dots+i_k+1}$.

The following claim emphasizes a first constraint on constructible signals implied by proposition 4.

Claim 6 *If a V-signal* $\Gamma = \{(\mathbf{u}(t), t)\}_{t\ge 0}$, *constructed by an impulse CA (with Moore neighborhood) enters the periodic part of the CA at some step* t_0 *then the V-signal becomes constant: for all* $t \ge t_0$, $\mathbf{u}(t) = \mathbf{u}(t_0) + (t - t_0)\cdot\mathbf{1}$.

Proof. The value of $(\mathbf{u}(t_0), t_0)$ belongs to the subset $\mathcal{S}_0$ which marks the V-signal Γ. Moreover, it belongs to the diagonal $\mathcal{D}_{t_0\cdot\mathbf{1}-\mathbf{u}(t_0)}$. Suppose this site belongs to the periodic part $\beta_{t_0\cdot\mathbf{1}-\mathbf{u}(t_0)}$. Then from $(\mathbf{u}(t_0), t_0)$ onward, there is an infinite number of sites of $\mathcal{D}_{t_0\cdot\mathbf{1}-\mathbf{u}(t_0)}$ which belong to $\mathcal{S}_0$. As Γ must go through all sites whose states belong to $\mathcal{S}_0$, the signal Γ always remains on the diagonal $\mathcal{D}_{t_0\cdot\mathbf{1}-\mathbf{u}(t_0)}$.

Finally we exhibit the gap relating to constructible signals.

Proposition 7 *Let a V-signal* $\{(t - \mathbf{u}(t), t)\}_{t\ge 0}$ *and* $m(t) = \max(u_1(t),\dots,u_k(t))$ *be such that:*

- $m(t)$ *is not constant:* $m(t) \neq \Theta(1)$;
- $m(t)$ *is below the logarithm:* $m(t) = o(\log(t))$.

Then there exists no impulse CA with Moore neighborhood which supports such V-signal.

Proof. According to claim 6, a V-signal $\{(t - \mathbf{u}(t), t)\}_{t\geq 0}$ constructible by an impulse CA with Moore neighborhood , providing $m(t) \neq \Theta(1)$, never enters the periodic part of the CA. Moreover, observe that $\langle t - \mathbf{u}(t), t\rangle$ belongs to $\mathcal{D}_{\mathbf{u}(t)}$.

The non-quiescent part of $\mathcal{D}_{\mathbf{u}(t)}$ begins on the site $(\lceil m(t)/2\rceil \cdot \mathbf{1} - \mathbf{u}(t), \lceil m(t)/2\rceil)$ and so the periodic part of $\mathcal{D}_{\mathbf{u}(t)}$ begins on the site $\left(\left(\lceil m(t)/2\rceil + |\alpha_{\mathbf{u}(t)}|\right) \cdot \mathbf{1} - \mathbf{u}(t), \lceil m(t)/2\rceil + |\alpha_{\mathbf{u}(t)}|\right)$. Hence if the signal is constructible, we get for all t, $t < \lceil m(t)/2\rceil + |\alpha_{\mathbf{u}(t)}|$; and according to corollary 5, $t < \lceil m(t)/2\rceil + |\mathcal{S}|^{1+k\cdot m(t)}$. So for some constant C, we have for all t: $t < C^{m(t)}$. In other words $m(t) = \Omega(\log(t))$.

Remark 1. Remark that the periodic phenomenon we just have examined along the signal of maximal speed $\{(t \cdot \mathbf{1}, t)\}_{t\geq 0}$, by an adequate rotation, occurs along all signals $\{(t \cdot \mathbf{x}, t)\}_{t\geq 0}$ with $\mathbf{x} \in V_{\text{trellis}}$. The proposition 7 remains true for any V-signal $\{(\mathbf{c}(t), t)\}_{t\geq 0}$ and $m(t) = \max(|c_1(t)| - t, \ldots, |c_k(t)| - t)$ with $m(t) \neq \Theta(1)$ and $m(t) = o(\log(t))$.

Remark 2. Due to the equivalence of Moore CA and trellis CA, the same limitation operates for the construction of V_{trellis}-signals on CA with trellis neighborhood.

Remark 3. With von Neumann neighborhood, the signals $\{((t - m(t)) \cdot \mathbf{x}, t)\}_{t\geq 0}$ with $\mathbf{x} \in V_{\text{Von Neumann}}$, $m(t) \neq \Theta(1)$ and $m(t) = o(\log(t))$ are not constructible by any impulse CA with Von Neumann neighborhood; otherwise using an adequate rotation, signals such $\{((t - m(t)) \cdot \mathbf{1}, t)\}_{t\geq 0}$ would be constructible by an impulse CA with Moore neighborhood, contradicting proposition 7.

4 Construction of the Logarithm in Dimension 2

Let $\mathcal{A}$ be the following impulse 2-CA with the neighborhood $V_{\text{trellis}} = \left(\binom{1}{1}, \binom{-1}{-1}, \binom{1}{-1}, \binom{-1}{1}\right)$. The set of states $\mathcal{S}$ is $\{\lambda, 0, 1\}$, the initial distinguished state is 1, the quiescent state is λ and the transition function f is depicted in figure 1.

Proposition 8 *Let ℓ be the function $t \mapsto \lfloor \log_2(t+1)\rfloor$. $\mathcal{A}$ detects the signal* $\left\{\begin{pmatrix} t-\ell(t) \\ t-\ell(t) \\ t+\ell(t)\end{pmatrix}\right\}_{t\geq 0}$, *with the following partition:* $\mathcal{S}_{\binom{1}{1}} = \{0\}$ *and* $\mathcal{S}_{\binom{-1}{-1}} = \{1\}$.

Proof. **Claim 9** *All cells with state* 1 *or* 0 *(called "active cells") have coordinates* $\begin{pmatrix} x \\ y \\ z\end{pmatrix}$ *such that* $-z \leq y \leq x \leq z$, $z \geq 0$ *and* $x + z$ *and* $y + z$ *are both even ("even" cells).*

The property of $x + z$ and $y + z$ is always true in such a trellis (see the definition). The condition $z \geq 0$ is quite obvious, as this is mandated by the definition of a space-time diagram. Now, let us look at the relation $-z \leq x \leq z$ ($-z \leq y \leq z$ can be proved by the same arguments). As the only cell at time

a b c d	$f(a,b,c,d)$	Rule number
$\lambda\ \lambda\ \lambda\ \lambda$	λ	#0
$1\ \lambda\ \lambda\ \lambda$	0	#1
$0\ \lambda\ \lambda\ \lambda$	1	#2
$\lambda\ \lambda\ 0\ 1$	1	#3
$1\ \lambda\ 0\ 1$	0	#4
$0\ \lambda\ 0\ 1$	1	#5
$1\ \lambda\ 1\ 0$	1	#6
$1\ \lambda\ 0\ 0$	1	#7
$0\ \lambda\ 1\ 0$	0	#8
$0\ \lambda\ 0\ 0$	0	#9
$\star\ 1\ \lambda\ \star$	1	#10
$\star\ 1\ 1\ \star$	1	#11
$\star\ 1\ 0\ \star$	0	#12
$\star\ 0\ \star\ \star$	0	#13
$\star\ \star\ \star\ \star$	λ	#14

a, b, c and d are the cells with the following relative coordinates in the space-time diagram:

- a is $\begin{pmatrix} -1 \\ -1 \\ -1 \end{pmatrix}$,
- b is $\begin{pmatrix} -1 \\ 1 \\ -1 \end{pmatrix}$,
- c is $\begin{pmatrix} 1 \\ 1 \\ -1 \end{pmatrix}$
- d is $\begin{pmatrix} 1 \\ -1 \\ -1 \end{pmatrix}$.

Rules are sorted by order of precedence.

Fig. 1. Transition function for $\mathcal{A}$.

$z = 0$ that has a non-quiescent state has the property that $x = 0$, and that the neighborhood's x-range is $\{1, -1\}$, no cell can enter a non-quiescent state unless it includes the cell $\begin{pmatrix} 0 \\ 0 \\ 0 \end{pmatrix}$ in its dependencies, i.e. if $-z \leq x \leq z$.

The remaining condition $y \leq x$ is due to the transition function. Let us suppose that the condition is true up to some value z, and let us check that the condition at $z+1$ still holds true, i.e. that no rule numbered from #1 to #13 is applied to any cell such that $x > y$. If the rule holds true up to z, then all neighbors except the one with relative coordinates $\begin{pmatrix} 1 \\ -1 \\ -1 \end{pmatrix}$ are such that $x > y$. So, their states are the quiescent state λ. Thus, the only possibility for cells such that $x > y$ is to meet the quadruplet $(\lambda, \lambda, \lambda, 1)$ or $(\lambda, \lambda, \lambda, 2)$ or the quiescent rule (#0). As both these quadruplets fall in the catch-all rule (#14), the induction is proved (it is obviously true for $z = 0$).

Let us now define $\overline{k}$ by $k = \sum_i \overline{k}^{(i)} 2^i$) (i.e. $\overline{k}$ is the binary writing of k). We also define $\tilde{k}$ to be the number of consecutive 1's at the beginning of $\overline{k}$. Let us define a spatial transformation for the space-time diagram with:

$$\mathcal{W}_{k,l}^{(i)} = \mathcal{V}\begin{pmatrix} k-i+l \\ k-i-l \\ k+i+l \end{pmatrix}.$$

Let us call $\mathcal{A}_{k,l}^{(i)}$, $\mathcal{B}_{k,l}^{(i)}$, $\mathcal{C}_{k,l}^{(i)}$ and $\mathcal{D}_{k,l}^{(i)}$ the four neighbors of cell $\mathcal{W}_{k,l}^{(i)}$. We have the following equalities:

$$\mathcal{A}_{k,l}^{(i)} = \mathcal{W}_{k-1,l}^{(i)} \quad \mathcal{B}_{k,l}^{(i)} = \mathcal{W}_{k,l-1}^{(i)} \qquad \mathcal{C}_{k,l}^{(i)} = \mathcal{W}_{k,l}^{(i-1)} \quad \mathcal{D}_{k,l}^{(i)} = \mathcal{W}_{k-1,l+1}^{(i-1)}$$

As proven by claim 1, $\mathcal{W}_{k,l}^{(i)}$ is λ if any of the indices is -1. We consider the words $\mathcal{W}_{k,l}$ that we get by the concatenation of all $\mathcal{W}_{k,l}^{(i)}$, for $i \geq 0$:

Claim 10 *$\mathcal{W}_{k,0}$ is $\overline{k+1}\lambda^\infty$, and for all $l > 0$, $\mathcal{W}_{k,l}$ is $1^{\widetilde{k+1}}0^{|\overline{k+1}|-\widetilde{k+1}}\lambda^\infty$.*

The execution can be seen on figure 2. In fact, we can read the binary writing of $k+1$ according to some spatial transformation (the one performed by $\mathcal{W}_{k,l}^{(i)}$).

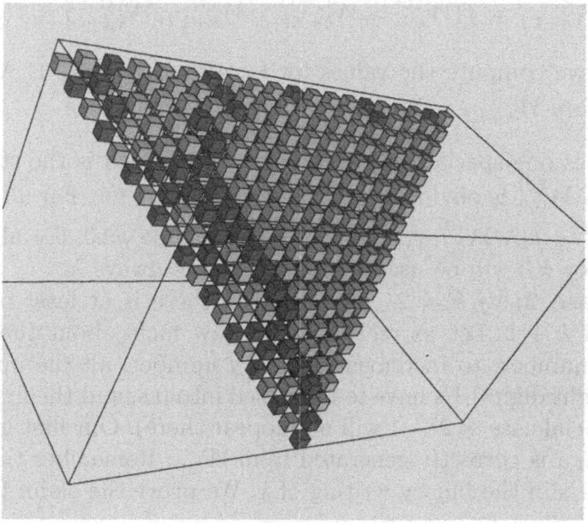

Fig. 2. Sample execution of $\mathcal{A}$. White cubes are sites instate 0, dark cubes are sites in state 1. The binary writingcan be seen horizontally on the picture. Depth is difficult to perceive.

We shall prove the claim with an induction on k, and dividing the proof in two subcases: whether $k = 2^j - 2$ for some j or not.

Subcase 1: $k = 2^j - 2$. We suppose the claim true up to k, and we shall prove the claim for $k+1$. The claim can be transformed this way for $k = 2^j - 2$: $\mathcal{W}_{k,l}^{(i)} = 1$ for any l (even $l = 0$) and $0 \leq i < j$, else $\mathcal{W}_{k,l}^{(i)} = \lambda$. We get the following items:

- $\mathcal{B}_{k+1,0}^{(i)} = \lambda$ for all values of i (see claim 1).
- $\mathcal{W}_{k+1,0}^{(0)} = 0$ (according to rule #1).
- With a quick induction (for $\mathcal{C}_{k+1,0}^{(m)}$), $\mathcal{W}_{k+1,0}^{(m)} = f(1, \lambda, 0, 1) = 0$ for all $m < j$ (rule #4).
- $\mathcal{W}_{k+1,0}^{(j)} = f(\mathcal{W}_{k,0}^{(j)}, \lambda, \mathcal{W}_{k+1,0}^{(j-1)}, \mathcal{W}_{k,1}^{(j-1)}) = f(\lambda, \lambda, 0, 1) = 1$ (rule #3).
- $\mathcal{W}_{k+1,0}^{(j+1)} = f(\mathcal{W}_{k,0}^{(j+1)}, \lambda, \mathcal{W}_{k+1,0}^{(j)}, \mathcal{W}_{k,1}^{(j)}) = f(\lambda, \lambda, 1, \lambda) = \lambda$ (rule #14).

- With a quick induction (for $\mathcal{C}_{k+1,0}^{(m)}$), $\mathcal{W}_{k+1,0}^{(m)} = \lambda$ for all $m > j+1$ (rule #0).
- Now, we use an induction on l, since we proved the claim for $l = 0$. $\mathcal{B}_{k+1,l}^{(i)} = 0$ for all values of i such that $0 \leq i \leq j$, except if $l = 1$ (where $\mathcal{B}_{k+1,1}^{(j)} = 1$). So, rule #13 applies for all i such that $0 \leq i \leq j$ and $\mathcal{W}_{k+1,l}^{(i)} = 0$. The only exception is $\mathcal{W}_{k+1,1}^{(j)} = f(\lambda, 1, 0, 1) = 0$ according to rule #12.
- We continue the induction on l to use rule #14 for $\mathcal{W}_{k+1,l}^{(j+1)}$.

$$\mathcal{W}_{k+1,l}^{(j+1)} = f(\mathcal{W}_{k,l}^{(j+1)}, \mathcal{W}_{k+1,l-1}^{(j+1)}, \mathcal{W}_{k+1,l}^{(j)}, \mathcal{W}_{k,l+1}^{(j)}) = f(\lambda, \lambda, 0, \lambda) = \lambda.$$

- Last, we compute the values for $l > 0$ and $i > j+1$. All neighbors are λ, therefore $\mathcal{W}_{k+1,l}^{(i)} = \lambda$.

There is one special case, the case $k = 0$, which is the starting point of the induction. $\mathcal{W}_{0,0}^{(0)}$ is obviously what we are looking for. For all values of $l > 0$, we have $\mathcal{W}_{0,l}^{(0)} = f(\lambda, \mathcal{W}_{0,l-1}^{(0)}, \lambda, \lambda) = 1$, by using rule #10. For all other $\mathcal{W}_{0,l}^{(i)}$, either rule #14 or #0 will be used, the result being always λ.

Subcase 2: $\forall j, k \neq 2^j - 2$. As such, there is at least one 0 in the binary writing of $k+1$. Let us recapitulate a few facts about the incrementation of a binary number: to increment a binary number, all the initial (starting from lower-weight digits) 1's have to be turned into 0's, and the first 0 is turned into a 1 (the special case of $2^j - 1$ will not appear there). Our first goal will be to check that $\mathcal{W}_{k+1,0}$ is correctly generated from $\mathcal{W}_{k,0}$. Remember that $\tilde{k}$ is the number of initial 1's in the binary writing of k. We prove the claim by induction:

- $\mathcal{B}_{k+1,0}^{(i)} = \lambda$ for all values of i (see claim 1). If k is even then $\mathcal{W}_{k+1,0}^{(0)} = 0$ (according to rule #1). If k is odd, then $\mathcal{W}_{k+1,0}^{(0)} = 1$ (rule #2).
- For $0 < i \leq \widetilde{k+1}$, the value of $\mathcal{D}_{k+1,0}^{(i)} = 1$. With a quick induction, we can prove that $\mathcal{W}_{k+1,0}^{(i)} = 0$ (rule #4) as all $\mathcal{A}_{k+1,0}^{(i)} = 1$. This condition is not met if k is odd.
- For $i = \widetilde{k+1}$, rule #5 is triggered ($\mathcal{W}_{k+1,0}^{(i)} = 1$). This is not done if k was odd (in fact the incrementation is already over if k is odd).
- For $\widetilde{k+1} < i < |\overline{k+1}|$ (and there is at least one value of i for which this is true), the value of $\mathcal{D}_{k+1,0}^{(i)} = 0$. As $\mathcal{B}_{k+1,0}^{(i)} = \lambda$, any of the rules #6, #7, #8 or #9 will be used. All those rules state that $\mathcal{W}_{k+1,0}^{(i)} = \mathcal{A}_{k+1,0}^{(i)} = \mathcal{W}_{k,0}^{(i)}$.
- For $i = |\overline{k+1}|$, we have $\mathcal{A}_{k+1,0}^{(i)} = \mathcal{B}_{k+1,0}^{(i)} = \lambda$, $\mathcal{C}_{k+1,0}^{(i)} = 1$ (since all binary writings end with a 1). So, rule #14 applies, and $\mathcal{W}_{k+1,0}^{(i)} = \lambda$.
- For $i > |\overline{k+1}|$, a quick induction shows that $\mathcal{C}_{k+1,0}^{(i)}$ is λ, thus making $\mathcal{W}_{k+1,0}^{(i)}$ be λ.
- We proved the claim for $l = 0$ (see the preliminary explanation on binary incrementation). As $\mathcal{C}_{k+1,l}^{(0)} = \lambda$, it's easy to prove that for all $l > 0$, $\mathcal{B}_{k+1,l}^{(0)}$ is not λ, so $\mathcal{W}_{k+1,l}^{(0)} = \mathcal{W}_{k+1,0}^{(0)}$ (using either rule #10 or rule #13).

- Let us consider now the case $0 < i < \widetilde{k+2}$ (this case may not happen if k is even). We prove by induction on $X = l + i$ that in this case, the value is always 1. $\mathcal{W}^{(i)}_{k+1,l} = f(\mathcal{A}^{(i)}_{k+1,l}, 1, 1, \mathcal{D}^{(i)}_{k+2,l})$, since $l - 1$ and $i - 1$ will both have a smaller sum than X. Thus, this sub-proof is done (using rule #11, the value is always 1). The proof for the case $l = 1$ or $i = 1$ is very easy (using the previous item).
- Now, let us consider the case where $i = \widetilde{k+2}$, with $i \neq 0$ and $i \neq |\overline{k+2}|$ (i.e. k is odd and there is at least a 0 in the binary writing of $k+2$). With a quick induction on l, as $\mathcal{W}^{(i)}_{k+1,0}$ is 0, we have $\mathcal{W}^{(i)}_{k+1,l} = 0$ for all l (using rule #13) (this is the first 0 in the binary writing of $k + 2$).
- Now, we study the case $\widetilde{k+2} < i < |\overline{k+2}|$. We still do an induction of $l + i$ and prove that the value is always 0. $\mathcal{W}^{(i)}_{k+1,l} = f(\mathcal{A}^{(i)}_{k+1,l}, \mathcal{B}^{(i)}_{k+1,l}, 0, \mathcal{D}^{(i)}_{k+2,l})$, with $\mathcal{B}^{(i)}_{k+1,l}$ being 0 or 1 (if $l = 1$). Either way, rule #12 or #13 is used, and the value still ends up being 0.
- We consider $i = |\overline{k+2}|$, and increasing values of l. $\mathcal{A}^{(i)}_{k+1,l} = \lambda$. $\mathcal{B}^{(i)}_{k+1,1}$ is also λ. We will prove with an induction on l that $\mathcal{B}^{(i)}_{k+1,l} = \lambda$ for any $l > 0$. Let us presume it's true. $\mathcal{C}^{(i)}_{k+1,l}$ may be 0, but $\mathcal{D}^{(i)}_{k+1,l} = \mathcal{W}^{(i-1)}_{k,l+1}$ is always 0. If it was not 0, then $k+1$ would have no 0 in its writing, and this is excluded in this subcase. So, rule #3 does not apply, hence the result (rule #14 is used).
- For larger values of i, the claim is straightforward, since only rule #0 will be used.

Proposition 11 (Optimality) *The result of proposition 8 is optimal for dimension 2, that means it is not possible to detect the signal* $\left\{\begin{pmatrix} t-\ell(t) \\ t-\ell(t) \\ t+\ell(t) \end{pmatrix}\right\}_{t\geq 0}$ *with only two states.*

Proof. Let us suppose that there exists an impulse CA with only two states 0 and 1 contradicting the proposition. Referring to claim 2, the general state of the impulse CA must not be the quiescent state. Let us decide that 0 is the quiescent state. As we want to detect the aforementioned signal, then we can only choose one partition (because the state 1 has to be in $\Gamma_{\binom{-1}{-1}}$). Thus, we have $\Gamma_{\binom{1}{1}} = \{0\}$ and $\Gamma_{\binom{-1}{-1}} = \{1\}$.

Let us now consider a few sites of the signal. We must obtain:

$$\mathcal{V}\begin{pmatrix} 0 \\ 0 \\ 0 \end{pmatrix} = 1 \quad \mathcal{V}\begin{pmatrix} 1 \\ 1 \\ 1 \end{pmatrix} = 0 \quad \mathcal{V}\begin{pmatrix} 0 \\ 0 \\ 2 \end{pmatrix} = 1 \quad \mathcal{V}\begin{pmatrix} 1 \\ 1 \\ 3 \end{pmatrix} = 1.$$

Recall that $\mathcal{V}\begin{pmatrix} x \\ y \\ z \end{pmatrix}$ is 0 if $|x| > z$ or $|y| > z$. We can rewrite two of these values the way they are computed from f:

$$\mathcal{V}\begin{pmatrix} 1 \\ 1 \\ 1 \end{pmatrix} = f(\mathcal{V}\begin{pmatrix} 0 \\ 0 \\ 0 \end{pmatrix}, \mathcal{V}\begin{pmatrix} 2 \\ 0 \\ 0 \end{pmatrix}, \mathcal{V}\begin{pmatrix} 2 \\ 2 \\ 0 \end{pmatrix}, \mathcal{V}\begin{pmatrix} 0 \\ 2 \\ 0 \end{pmatrix}) = f(1, 0, 0, 0) = 0$$

Thus, we get $f(1,0,0,0) = f(0,0,0,0) = 0$. Now we write:

$$\mathcal{V}\begin{pmatrix}1\\1\\3\end{pmatrix} = f(\mathcal{V}\begin{pmatrix}0\\0\\2\end{pmatrix}, \mathcal{V}\begin{pmatrix}2\\0\\2\end{pmatrix}, \mathcal{V}\begin{pmatrix}2\\2\\2\end{pmatrix}, \mathcal{V}\begin{pmatrix}0\\2\\2\end{pmatrix}) = f(1,a,b,c) = 1$$

with the following values for a, b and c:

$$a = \mathcal{V}\begin{pmatrix}2\\0\\2\end{pmatrix} = f(\mathcal{V}\begin{pmatrix}1\\-1\\1\end{pmatrix}, \mathcal{V}\begin{pmatrix}3\\-1\\1\end{pmatrix}, \mathcal{V}\begin{pmatrix}3\\1\\1\end{pmatrix}, \mathcal{V}\begin{pmatrix}1\\1\\1\end{pmatrix}) = f(\mathcal{V}\begin{pmatrix}1\\-1\\1\end{pmatrix},0,0,0) = 0$$

$$b = \mathcal{V}\begin{pmatrix}2\\2\\2\end{pmatrix} = f(\mathcal{V}\begin{pmatrix}1\\1\\1\end{pmatrix}, \mathcal{V}\begin{pmatrix}3\\1\\1\end{pmatrix}, \mathcal{V}\begin{pmatrix}3\\3\\1\end{pmatrix}, \mathcal{V}\begin{pmatrix}1\\3\\1\end{pmatrix}) = f(0,0,0,0) = 0$$

$$c = \mathcal{V}\begin{pmatrix}0\\2\\2\end{pmatrix} = f(\mathcal{V}\begin{pmatrix}-1\\1\\1\end{pmatrix}, \mathcal{V}\begin{pmatrix}-1\\3\\1\end{pmatrix}, \mathcal{V}\begin{pmatrix}1\\3\\1\end{pmatrix}, \mathcal{V}\begin{pmatrix}1\\1\\1\end{pmatrix}) = f(\mathcal{V}\begin{pmatrix}-1\\1\\1\end{pmatrix},0,0,0) = 0$$

Thus, $a = b = c = 0$, and $\mathcal{V}\begin{pmatrix}1\\1\\3\end{pmatrix} = f(1,0,0,0) = 0$. This is in contradiction with the fact that $\mathcal{V}\begin{pmatrix}1\\1\\3\end{pmatrix}$ must be 1.

5 Building Non-primal Logarithmic Signals

Recall that, in dimension 1, to build a logarithmic slow-down (in base b) requires at least b states. If b is not primal, then it can be written as the product of two numbers x and y whose gcd is 1. Here, we exhibit a 2-dimensional CA that supports such a logarithmic slow-down with only $x+y+2$ states instead of xy.

Let x and y be two integers such that $\gcd(x,y) = 1$. Let $\mathcal{A}$ be the following impulse 2-CA with the neighborhood $V_{\text{trellis}} = (\left(\begin{smallmatrix}1\\1\end{smallmatrix}\right), \left(\begin{smallmatrix}-1\\1\end{smallmatrix}\right), \left(\begin{smallmatrix}1\\-1\end{smallmatrix}\right), \left(\begin{smallmatrix}-1\\-1\end{smallmatrix}\right))$. The set of states $\mathcal{S}$ is $\{\lambda, \pi_0, \pi_1, \ldots, \pi_x, \kappa_0, \kappa_1, \ldots, \kappa_y\}$, the initial distinguished state is π_1, the quiescent state is λ and the transition function f is as in figure 3.

Proposition 12 *Let ℓ be the function $t \mapsto \lfloor \log_{xy}(t+1) \rfloor$. Then $\mathcal{A}$ supports the signal $\left\{\begin{pmatrix}t-\ell(t)\\t-\ell(t)\\t+\ell(t)\end{pmatrix}\right\}_{t\geq 0}$, using the following finite automaton:*

$$F = (\mathcal{S}, \{a_1, \cdots, a_y\}, \delta, a_1) \quad \textit{with}$$

$$\begin{cases}\delta(a_y, \pi_x) = (a_1, \left(\begin{smallmatrix}1\\1\end{smallmatrix}\right))\\ \forall j \neq y, \delta(a_j, \pi_x) = (a_{j+1}, \left(\begin{smallmatrix}-1\\-1\end{smallmatrix}\right))\\ \forall i \neq x, \forall j, \delta(a_j, \pi_i) = (a_j, \left(\begin{smallmatrix}-1\\-1\end{smallmatrix}\right))\end{cases}$$

Proof. Let us use the same spatial transformation as in the preceding section.

$$\mathcal{W}_{k,l}^{(i)} = \mathcal{V}\begin{pmatrix}k-i+l\\k-i-l\\k+i+l\end{pmatrix}.$$

Let us call $\mathcal{A}_{k,l}^{(i)}$, $\mathcal{B}_{k,l}^{(i)}$, $\mathcal{C}_{k,l}^{(i)}$ and $\mathcal{D}_{k,l}^{(i)}$ the four neighbors of cell $\mathcal{W}_{k,l}^{(i)}$. We have the following equalities:

$$\mathcal{A}_{k,l}^{(i)} = \mathcal{W}_{k-1,l}^{(i)} \quad \mathcal{B}_{k,l}^{(i)} = \mathcal{W}_{k,l-1}^{(i)} \quad \mathcal{C}_{k,l}^{(i)} = \mathcal{W}_{k,l}^{(i-1)} \quad \mathcal{D}_{k,l}^{(i)} = \mathcal{W}_{k-1,l+1}^{(i-1)}$$

A	B	C	D	$f(A,B,C,D)$	Rule number
λ	λ	λ	λ	λ	#0
Rules for $l = 0$					
π_j	λ	λ	λ	π_{j+1} (or π_1 if $j = k$)	#1
π_x	λ	π_k	$\kappa_\star$	π_0 $(k \neq x)$	#2
π_j	λ	π_k	$\kappa_\star$	π_j $(j, k \neq x)$	#3
π_x	λ	π_x	κ_k	π_0 $(k \neq y-1)$	#4
π_j	λ	π_x	κ_k	π_j $(j \neq x, k \neq y-1)$	#5
π_j	λ	π_x	κ_{y-1}	π_{j+1} (or π_1 if $j = k$)	#6
λ	λ	π_x	κ_{y-1}	π_1	#7
Rules for $l = 1$					
κ_{y-1}	π_x	λ, κ_y	λ	κ_y	#8
κ_{y-1}	π_k	λ, κ_y	λ	κ_0 $(k \neq x)$	#9
κ_y	$\pi_\star$	λ, κ_y	λ	κ_1	#10
κ_j	$\pi_\star$	λ, κ_y	λ	κ_{j+1} $(j \neq y-1, y)$	#11
κ_y	$\pi_\star$	κ_k	λ	κ_0 $(k \neq y)$	#12
κ_j	$\pi_\star$	κ_k	λ	κ_j $(j \neq y, k \neq y)$	#13
λ	π_1	λ, κ_y	λ	κ_1	#14
$\star$	$\star$	$\star$	$\star$	λ	#15

a, b, c and d are the cells with the following relative coordinates in the space-time diagram:

- a is $\begin{pmatrix}-1\\-1\\-1\end{pmatrix}$,
- b is $\begin{pmatrix}-1\\1\\-1\end{pmatrix}$,
- c is $\begin{pmatrix}1\\1\\-1\end{pmatrix}$
- d is $\begin{pmatrix}1\\-1\\-1\end{pmatrix}$.

Please note that $\pi_\star$ is any state π_j, $\kappa_\star$ is any state κ_j, and $\star$ is any state. Rules are sorted by order of precedence.

Fig. 3. Transition function for $\mathcal{A}$.

We have the following fact: only cells with $l = 1$ or $l = 0$ will be non-quiescent. In fact, cells with $l = 0$ will only have states in $\pi_0, \ldots, \pi_x, \lambda$ and cells with $l = 1$ will only have states in $\kappa_0, \ldots, \kappa_y, \lambda$. This is quite easy to check: rules #1 to #7 always require the neighbor b to be λ, and rules #8 to #14 always require d to be λ.

The proof is on the same lines as the preceding proof: we can read the writing of $k+1$ with the values of $\mathcal{W}_{k,l}$. However, the writing is a bit more complicated. In fact, x and y define a mapping of $\{0, \ldots, x-1\} \times \{0, \ldots, y-1\}$ into $\{0, \ldots, xy-1\}$ through the Chinese remainder lemma. Let us call this mapping μ, with the added fact that π_0 and π_x are made equivalent (idem for κ_0 and κ_y). Then, we can read the writing in base xy of $k+1$ by considering the i-th bit to be $\mu(\mathcal{W}_{k,0}^{(i)}, \mathcal{W}_{k,1}^{(i)})$.

The proof is quite cumbersome, and is reminiscent of the proof of proposition 8. The main point is that when $\mathcal{W}_{k,1}^{(i)} = \kappa_y$ is used instead of κ_0, it conveys the information that the neighbor $\mathcal{W}_{k,1}^{(i)}$ is exactly π_x, and thus that the digit $\mu(\mathcal{W}_{k,0}^{(i+1)}, \mathcal{W}_{k,1}^{(i+1)})$ has to be increased. The fact that $\mathcal{W}_{k,0}^{(i)}$ is π_x instead of being π_0 carries the fact that the corresponding $\mathcal{W}_{k,1}^{(i)}$ has to be increased by 1.

Possible enhancement: It is possible to use the same set of states for π and κ. That is, the CA has just a set of states $\{\pi_0, \ldots, \pi_{\max(x,y)}, \lambda\}$. It is necessary that x is the smallest of the two numbers x and y. The transformation of the transition function is as follows: each $\acute{\kappa}_i$ becomes π_i, and each λ in the column D of figure 3 becomes a $\star$ (any state).

6 Prospectives

Note that limitations in the construction of signals are likely correlated to limitations in terms of language recognition. In particular, the hierarchy between time n and $n+\log n$ set up for one-way cellular automata in dimension 1 (see [KK01]) might be generalized to higher dimensions.

It should be possible to extend the last proposition in dimension k as follows: one can get a $\log_a$ slow-down with the sum of k factors whose gcd is 1 and whose product is a, plus 2 (the distinguished π_x state and λ). Thus, the number of states needed to get a $\log_a$ slow-down in dimension k looks strongly related to the decomposition of a in prime numbers and the number of its factors.

References

[BCG82] E. R. Berlekamp, John H. Conway, and R. K. Guy. *Winning Ways for Your Mathematical Plays*, volume 2, chapter 25. Academic Press, 1982.

[CČ84] Christian Choffrut and Karel Čulik, II. On real-time cellular automata and trellis automata. *Acta Informatica*, 21:393–407, 1984.

[DFM00] Marianne Delorme, Enrico Formenti, and Jacques Mazoyer. Open problems on cellular automata. Research report 2000-25, École normale supérieure de Lyon, July 2000.

[Fis65] Patrick C. Fischer. Generation of primes by a one-dimensional real-time iterative array. *JACM*, 12:388–394, 1965.

[IKM85] Oscar H. Ibarra, Sam M. Kim, and Shlomo Moran. Sequential machine characterizations of trellis and cellular automata and applications. *SIAM J. Comput.*, 14(2):426–447, 1985.

[KK01] Andreas Klein and Martin Kutrib. A time hierarchy for bounded one-way cellular automata. In J. Sgall, A. Pultr, and P. Kolman, editors, *MFCS 2001*, volume 2136 of *LNCS*. Springer, 2001. To appear.

[Mar00] Bruno Martin. Apparent entropy of cellular automata. *Complex Systems*, 12(2), 2000.

[Maz87] Jacques Mazoyer. A six-states minmal solution to the firing squad synchronization problem. *TCS*, 50:183–238, 1987.

[MT99] Jacques Mazoyer and Véronique Terrier. Signals in one-dimensional cellular automata. *TCS*, 217:53–80, 1999.

[vN66] John von Neumann. *Theory of Self-Reproducing Automata*. University of Illinois, Urbana, 1966.

Holographic Trees

Paolo Boldi and Sebastiano Vigna*

Dipartimento di Scienze dell'Informazione, Università degli Studi di Milano, Via Comelico 39/41, I-20135 Milano MI, Italy.
{vigna,boldi}@dsi.unimi.it

Abstract It is known that computations of anonymous networks can be reduced to the construction of a certain graph, the *minimum base* of the network. The crucial step of this construction is the inference of the minimum base from a finite tree that each processor can build (its *truncated view*). We isolate those trees that make this inference possible, and call them *holographic*. Intuitively, a tree is holographic if it is enough self-similar to be uniquely extendible to an infinite tree. This possibility depends on a *size function* for the class of graphs under examination, which we call a *holographic bound* for the class. Holographic bounds give immediately, for instance, bounds for the quiescence time of self-stabilizing protocols. In this paper we give weakly tight holographic bounds for some classes of graphs.

1 Introduction

This paper investigates combinatorial properties of trees and graphs whose very definition has been inspired by some problems in the study of distributed anonymous and self-stabilizing (synchronous) computations. In particular, we shall define and study *holographic trees* and *holographic bounds*, which turn out to play a major rôle in the construction of distributed anonymous algorithms and self-stabilizing protocols.

The reader might be easily bewildered by the amount of notions that must be absorbed to grasp the concepts above, and by the appearent opaqueness of Definition 2 and 3. Indeed, some knowledge of anonymous computations and theory of graph fibrations is necessary to understand them completely. Thus, the rest of this introduction is devoted to presenting a "historical" reconstruction of the notion of holographic tree and holographic bound, so to place these concepts in a computational perspective.

Consider a *network* of processors, represented by a strongly connected graph[1]. We assume that the network is *anonymous* [1, 13, 2], that is, all processors start from the same initial state and apply the same algorithm. The network is *synchronous*, in the sense that all processors take a step at the same time, and

* The authors have been partially supported by the Italian MURST (Finanziamento di iniziative di ricerca "diffusa" condotte da parte di giovani ricercatori).

[1] Our graphs are directed, and may possess multiple arcs and loops—see Sect. 2.

S. Rajsbaum (Ed.): LATIN 2002, LNCS 2286, pp. 465–478, 2002.

the new state of a processor depends on its own state and on the states of its in-neighbours.

One of the main concerns in the theory of anonymous computation is to establish which problems can be solved on a network[2]. This apparently gigantic task was enormously simplified by the discovery that the state of a processor after the k-th step of *any* anonymous computation depends only on a finite tree, which is the truncation at depth k of an infinite tree, the *view* of the processor. In the bidirectional case, Yamashita and Kameda [13], inspired by the seminal work of Angluin[1], showed that views correspond to a standard graph-theoretic construction, the *universal cover*. Subsequently, it was shown [4] that in the general case views are the *universal total graphs* (in the sense of the theory of graph fibrations [3]) of the processors.

Let G be a network and i a processor of G. The view of G at i, denoted by $\widetilde{G}^i$, has the (finite) paths of G ending into i as nodes (the root of $\widetilde{G}^i$ being the empty path), with an arc from node π to node π' if π is obtained by adding an arc at the beginning of π'. (The reader may want to consult Fig. 2 to see an example of a view.)

It is not difficult to become convinced that each processor can anonymously build its own view truncated at any desired depth: at step k each processor gathers the view truncated at depth k from each of its in-neighbours, and it is thus able to build its own view truncated at depth $k+1$. An example of the first steps of this algorithm is given (for the network of Fig. 2) in Table 1 (for the time being, just ignore the last column).

The goal of each processor would be (in a perfect world) the computation of G and of the node of G that corresponds to the processor itself. However, this is in general impossible, since many processors can possess the same view, and thus cannot reach different states, no matter which algorithm they use. A more feasible goal, which can be indeed achieved, is the computation of the *minimum base* of the network, which is essentially[3] obtained by collapsing those processors that have the same view. (Again, an example can be found in Fig. 2.) Indeed, *every* anonymous computation can be factored into the computation of the minimum base followed by a local computation[4] [7].

The fundamental question we have to answer to is now: How deep must a truncation of the view be for the correct computation of the minimum base? In other words: How long must the algorithm sketched above run to compute a correct result?

The first answers were given originally by Yamashita and Kameda: if a bound N on the number of processors of the network is known, N^2 steps suffice. Sub-

[2] Or on a class of networks, that is, using an algorithm that will work for *every* network out of the class. The class is used to represent the *knowledge* processors possess about the network.

[3] A more precise definition, which requires the introduction of graph fibrations, is given in Sect. 3.

[4] As long as a bound on the network size is known; if no bound on the size is known, this is not possible, and a completely different approach is necessary (see, e.g., [6]).

sequently, using a result of Norris [12], the bound was improved to $2N - 1$, and eventually to $N + D$ (where D is an upper bound on the diameter) in [4].

Once the minimum base B has been computed on the basis of a certain truncated view T, one notes that T is a prefix[5] of a unique view of B. In fact, this infinite tree is the view of one or more processors of the network (among which the processor that computed T).

In other words, we can produce an infinite extension of T, in a way that respects the internal similarities of T, and this infinite extension is entirely described by a graph (the minimum base) with a selected node. With a bit of imagination, we shall call a tree for which this operation is possible (preserving some uniqueness properties) *holographic*, since, as in a hologram, it is a small piece that contains all the information of a much larger (indeed, infinite) picture. For instance, it is reasonable to think that sufficiently deep truncated views will always be holographic. And this is indeed the reason why anonymous computations work: the bounds quoted above quantify (rather roughly, as we shall see) the depth after which a tree becomes holographic—in our terminology, they are *holographic bounds.*

Of course, holographic bounds depend on the class of networks under examination. If the class is small, very few levels (even just one, as shown in an example) will suffice to identify the minimum base. Large classes, instead, will require more information, and thus more levels (see, e.g., the lower bounds given in Sect. 4).

We now must admit that we partially lied to our reader: even if holographic bounds do provide upper bounds on the runs of anonymous computation, their real application lies in the domain of *self-stabilization.* A system is self-stabilizing if, for every initial state, after a finite number of steps it cannot deviate from a specified behaviour. Self-stabilization of distributed systems was introduced by Dijkstra in his celebrated paper [11], and has since become an important framework for the study of fault-tolerant computations.

In [5, 8] we showed the existence of a self-stabilizing protocol that computes the truncated view; essentially, after $D + 1$ steps all processors possess a truncation of their view at least $D + 1$ levels deep, no matter which the initial state was. Similarly to the anonymous case, this allows self-stabilization to arbitrary behaviours (for which a protocol exists) which depend only on the minimum base; thus, the quiescence time (i.e., the number of steps after which the desired behaviour starts) will depend on the number of levels required to compute the minimum base correctly—again, on a holographic bound for the class of networks under examination.

In this paper we show that the function assigning to a graph G the size $n_{\hat{G}} + D_{\hat{G}}$, where $\hat{G}$ is the minimum base of G, is a holographic bound for the class of all strongly connected graphs, and that this bound is weakly tight (there are infinite graphs for which the size must be at least $n_{\hat{G}} + D_{\hat{G}}$). We also show that the bound drops to $D_G + 1$ if the nodes are labelled injectively (i.e., in the

[5] For a formal definition of tree prefix, see Sect. 2.

network interpretation, if processors have unique identifiers). Also in this case, we provide matching lower bounds.

2 Graph-theoretical Definitions

A *(directed) (multi)graph* G is given by a nonempty set $N_G = \{1, 2, \dots, n_G\}$ of nodes and a set A_G of arcs, and by two functions $s_G, t_G : A_G \to N_G$ that specify the source and the target of each arc. A *(arc- and node-)coloured graph* (with set of colours C) is a graph endowed with a colouring function $\gamma : N_G + A_G \to C$ (the symbol $+$ denotes the disjoint union). We write $i \xrightarrow{a} j$ when the arc a has source i and target j, and $i \to j$ when $i \xrightarrow{a} j$ for some $a \in A_G$. We denote with D_G the diameter of G. Subscripts will be dropped whenever no confusion is possible.

A (in-directed) *tree* is a graph[6] with a selected node, the root, such that every node has exactly one directed path to the root. If T is a tree, we write $h(T)$ for its *height* (the length of the longest directed path). In every tree we consider in this paper, all maximal paths have length equal to the height. We write $T \upharpoonright k$ for the tree T truncated at height k, that is, we eliminate all nodes at distance greater than k from the root.

Trees are partially ordered[7] by *prefix*, that is, $T \leq U$ iff $T \cong U \upharpoonright h(T)$; this partial order is augmented with a bottom element $\bot$, with $h(\bot) = -1$ by definition (so h is strictly monotonic). The infimum in this partial order, denoted by $\wedge$, is the tallest common prefix (or $\bot$ if no common prefix exists). The supremum between T and U exists iff T and U are comparable.

3 Graph Fibrations

In this paper we exploit the notion of *graph fibration* [3]. A fibration, which is essentially a local in-isomorphism, formalizes the idea that processors that are connected by arcs with the same colours to processors behaving in the same way (with respect to the colours) will behave alike. In this section we gather (without proof) a number of definitions and results about graph fibrations; although some of the statements are true for all graphs, for sake of simplicity we shall assume that all graphs (except for trees) are strongly connected.

Recall that a *graph morphism* $f : G \to H$ is given by a pair of functions $f_N : N_G \to N_H$ and $f_A : A_G \to A_H$ that commute with the source and target functions, that is, $s_H \circ f_A = f_N \circ s_G$ and $t_H \circ f_A = f_N \circ t_G$. (The subscripts will usually be dropped.) In other words, a morphism maps nodes to nodes and arcs to arcs in such a way to preserve the incidence relation. Colours on nodes and arcs must be preserved too.

---

[6] Since we need to manage infinite trees too, we allow the node set of a tree to be **N**.

[7] We are in fact considering trees up to isomorphism (technically $\leq$ is just a preorder).

Definition 1. *A* fibration[8] *between (coloured) graphs G and B is a morphism $\varphi : G \to B$ such that for each arc $a \in A_B$ and for each node $i \in N_G$ satisfying $\varphi(i) = t(a)$ there is a unique arc $\widetilde{a}^i \in A_G$ (called the* lifting of a at i*) such that $\varphi(\widetilde{a}^i) = a$ and $t(\widetilde{a}^i) = i$.*

We recall some topological terminology. If $\varphi : G \to B$ is a fibration, G is called the *total graph* and B the *base* of φ. We shall also say that G is *fibred (over B)*. The *fibre* over a node $i \in N_B$ is the set of nodes of G that are mapped to i, and will be denoted by $\varphi^{-1}(i)$.

In Fig. 1 we sketched a fibration between two graphs. Note that, because of the lifting property described in Definition 1, all black nodes have exactly two incoming arcs, one (the dotted arc) going out of a white node, and one (the continuous arc) going out of a grey node. In other words, the in-neighbour structure of all black nodes *is the same.*

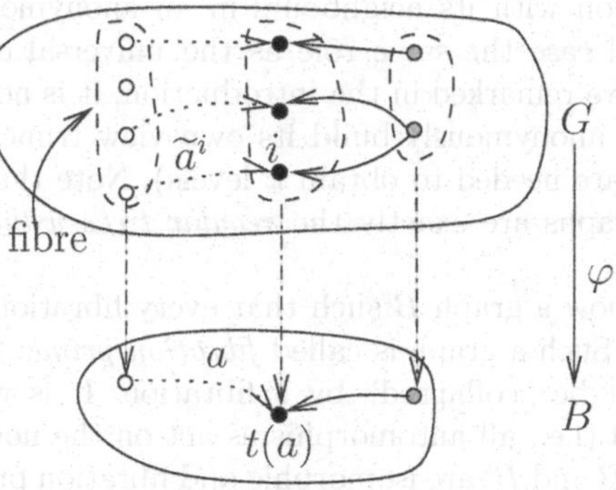

Fig. 1. A fibration.

There is a very intuitive characterization of fibrations based on the concept of local isomorphism. A fibration $\varphi : G \to B$ induces an equivalence relation between the nodes of G, whose classes are precisely the fibres of φ. When two nodes i and j are equivalent (i.e., they are in the same fibre), there is a bijective correspondence between arcs coming into i and arcs coming into j that preserves colours, and such that the sources of any two corresponding arcs are equivalent.

Let now G be a graph and i a node of G. We define an in-directed rooted coloured tree $\widetilde{G}^i$ as follows:

[8] The name "fibration" comes from the categorical and homotopical tradition; indeed, our elementary definition is simply a restatement of the condition that $\varphi : G \to B$ induces a functor that is a fibration [9] between the free categories generated by G and B.

- the nodes of $\widetilde{G}^i$ are the (finite) paths of G ending in i, the root of $\widetilde{G}^i$ being the empty path; each node is given the same colour as the starting node of the path;
- there is an arc from the node π to the node π' if π is obtained by adding an arc a at the beginning of π' (the arc will have the same colour as a).

The tree $\widetilde{G}^i$ (which is always infinite if G is strongly connected and with at least one arc) is called the *universal total graph of G at i*, or, following Yamashita and Kameda, the *view of i*. We can define a graph morphism v^i_G from $\widetilde{G}^i$ to G, by mapping each node π of $\widetilde{G}^i$ (i.e., each path of G ending in i) to its starting node, and each arc of $\widetilde{G}^i$ to the corresponding arc of G. The following important property holds:

Lemma 1. *For every node i of a graph G, the morphism $v^i_G : \widetilde{G}^i \to G$ is a fibration, called the* universal fibration of G at i.

The view at i is a tree representing intuitively "everything processor i can learn from interaction with its neighbours in an anonymous computation"; it plays in the general case the same rôle as the universal covering in the undirected case [13]. As we remarked in the introduction, it is not difficult to see that each processor can anonymously build its own view truncated at any desired depth (and k steps are needed to obtain k levels). Note that views of finite (strongly connected) graphs are exactly the *regular trees without leaves*, in the sense of Courcelle [10].

Consider now a graph B such that every fibration with total graph B is an isomorphism. Such a graph is called *fibration prime*: intuitively, fibration prime graphs cannot be "collapsed" by a fibration. It is worth observing that they are node rigid (i.e., all automorphisms act on the nodes as the identity), so, in particular, if B and B' are isomorphic and fibration prime then all isomorphisms $B \to B'$ coincide on the nodes. We have the following

Lemma 2. *Let $\varphi : G \to B$ and $\varphi' : G \to B'$ be fibrations, with B and B' fibration prime. Then $B \cong B'$.*

In other words, to each graph G we can associate a fibration prime graph $\hat{G}$, the *minimum base* of G, and a *minimal fibration* $\mu_G : G \to \hat{G}$ (in fact, there are several candidates for μ_G, but they are all defined in the same way on the nodes, and in this paper we shall use only the node component of minimal fibrations). In Fig. 2, we illustrate these notions by showing a graph G, its minimum base $\hat{G}$ and one of its views, $\widetilde{G}^1$. (The numbers shown on G are actual node names, and not colours; the numberings on $\hat{G}$ and $\widetilde{G}^1$ illustrate μ_G and v^1_G.) There are three important comments to be made about $\hat{G}$:

- fibration prime graphs (in particular, $\hat{G}$) have distinct views (i.e., $\widetilde{B}^i \cong \widetilde{B}^j$ iff $i = j$);
- $\hat{G}$ can be constructed by identifying isomorphic subtrees of $\widetilde{G}^i$ (the choice of $i \in N_G$ is irrelevant), and μ_G maps node i to the equivalence class containing $\widetilde{G}^i$;

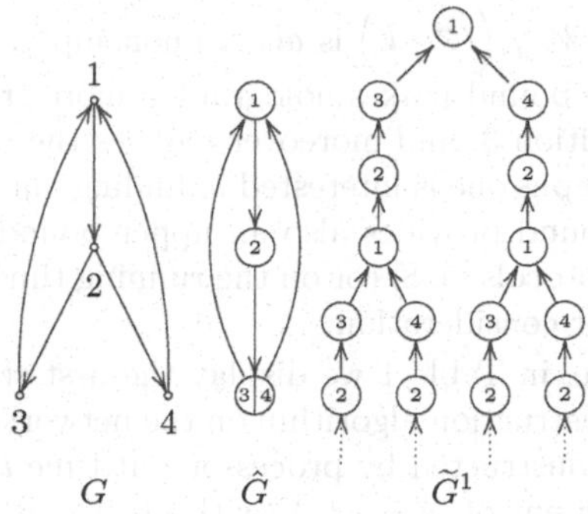

Fig. 2. A graph, its minimum base and a view.

– $\widetilde{G}^i \cong \widetilde{\hat{G}}^{\mu_G(i)}$, so we can compute $\mu_G(i)$ by searching for the node of $\hat{G}$ having $\widetilde{G}^i$ as view.

The fundamental fact we shall use intensively in all proofs is that the above considerations, which involve infinite objects, can be described by means of finite entities using the theorems of Sect. 5.

4 Holographic Bounds

Armed with our basic definitions, we now introduce the main concept we shall deal with. Our interest is in isolating those trees that are enough coherent, and contain enough information, to "replicate" themselves *ad infinitum.*

Definition 2. *Let* $\mathscr{C}$ *be a class of graphs. A* size function *for* $\mathscr{C}$ *is a map* $\nu : \mathscr{C} \to \mathbf{N}$. *Given a class* $\mathscr{C}$*, a size function* ν *and a finite tree* T *we define*

$$\mathscr{U}_{\mathscr{C},\nu}(T) = \{ \langle \hat{H}, \mu_H(j) \rangle \mid H \in \mathscr{C},\ j \in N_H,\ \nu(H) \leq h(T) \text{ and } T \leq \widetilde{H}^j \}.$$

We say that T *is* ($\langle \mathscr{C}, \nu \rangle$-)holographic *iff* $\mathscr{U}_{\mathscr{C},\nu}(T)$ *is nonempty, and for every* $\langle B, i \rangle, \langle B', i' \rangle \in \mathscr{U}_{\mathscr{C},\nu}(T)$ *there is an isomorphism* $\alpha : B \to B'$ *such that* $\alpha(i) = i'$ *(we shall often state this condition by saying that* $\mathscr{U}_{\mathscr{C},\nu}(T)$ *contains* essentially *one element).*

The idea behind the definition above is that the set $\mathscr{U}_{\mathscr{C},\nu}(T)$ contains all the possible candidates for the pointed minimum bases of the graph (network) that generated T. The height of T is also used to confine the search to those graphs whose size is not too large. A holographic tree is a tree that is sufficiently self-similar to identify a single candidate (up to isomorphism).

Definition 3. *We say that* ν *is a* holographic bound *for* $\mathscr{C}$ *if for all* $G \in \mathscr{C}$*, all* $i \in N_G$ *and all* $k \geq \nu(G)$ *we have that* $\widetilde{G}^i \upharpoonright k$ *is* $\langle \mathscr{C}, \nu \rangle$*-holographic.*

Note that the set $\mathscr{U}_{\mathscr{C},\nu}\left(\widetilde{G}^i \restriction k\right)$ is *always* nonempty, as it contains $\langle \hat{G}, \mu_G(i)\rangle$. A small holographic bound makes holographic more trees, as it lowers the height required by Definition 3, and moreover reduces the number of candidates; thus, given a class of graphs one is interested in finding out a small holographic bound. Indeed, such a bound provides also an upper bound on the quiescence time of self-stabilizing protocols [5, 8] (or on the running time of anonymous algorithms) for the class under consideration.

As an example, in Table 1 we display the first steps of the execution of the standard view construction algorithm on the network of Fig. 2, where T_i is $\widetilde{G}^i \restriction t$, that is, the tree constructed by processor i at time t. The last column gives, at each step, the content of $\mathscr{U}_{\mathscr{G},\nu}(T_2)$ (with $\nu(G) = n_{\hat{G}} + D_{\hat{G}}$, and $\mathscr{G}$ the class of all strongly connected graphs). The reader may notice that the second processor "changes his mind" a few times about $\hat{G}$, but ultimately its guess is correct.

t	T_1	T_2	T_3	T_4	$\mathscr{U}_{\mathscr{G},\nu}(T_2)$
0					$\varnothing$
1					
2	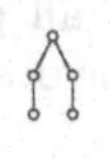				$\varnothing$
3	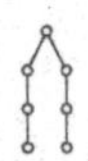	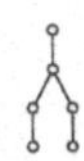			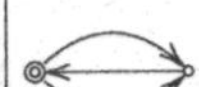
4	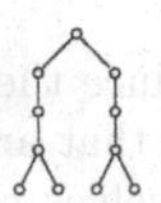	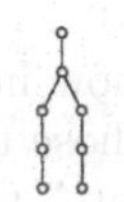	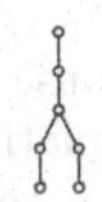	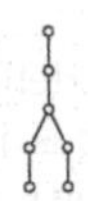	$\varnothing$
5	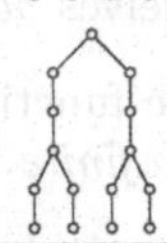	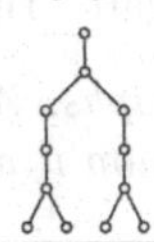	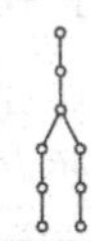	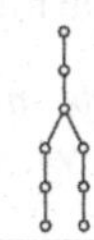	

Table 1. The first steps of the view construction algorithm on the network G of Fig. 2.

We now prove a very general and intuitive property of holographic bounds: a holographic bound for $\mathscr{C}$ works for every subclass of $\mathscr{C}$, and moreover size functions pointwise larger than a holographic bound are still holographic bounds. Formally,

Theorem 1. *Let ν be a holographic bound for a class $\mathscr{C}$. Then every size function $\nu' \geq \nu$ is a holographic bound for every class $\mathscr{C}' \subseteq \mathscr{C}$.*

Proof. Let G be a graph of $\mathscr{C}$, i a node of G and $k \geq \nu'(G)$. We have to prove that $\mathscr{U}_{\mathscr{C}',\nu'}\left(\widetilde{G}^i \restriction k\right)$ contains essentially one element. But since, as it is immediate to check from Definition 2,

$$\mathscr{U}_{\mathscr{C}',\nu'}(T) \subseteq \mathscr{U}_{\mathscr{C},\nu}(T),$$

and the right-hand set contains essentially one element when $T = \widetilde{G}^i \restriction k$, we have the thesis. □

5 A Holographic Bound for All Graphs

In this section we provide a holographic bound for the class of all (strongly connected) graphs (note that by assuming $|C| = 1$, the bound applies also to noncoloured graphs). The bound is based on a number of graph-theoretical results, which show that sufficiently deep truncations of views characterize the (minimum bases of the) graph that generated them. First of all, we recall a result of Norris [12]:

Theorem 2. $\widetilde{G}^i \cong \widetilde{G}^j$ *iff* $h(\widetilde{G}^i \wedge \widetilde{G}^j) \geq n - 1$.

Using the previous theorem, we prove the following

Theorem 3. *Let G be a strongly connected graph and B a fibration prime graph with minimum number of nodes satisfying $h(\widetilde{G}^i \wedge \widetilde{B}^j) \geq n_G + D_G$ for some $i \in N_G$ and $j \in N_B$: then there is a (minimal) fibration $\varphi : G \to B$ such that $\varphi(i) = j$; in particular, $B \cong \hat{G}$.*

Proof. Note that B has at most n nodes, because the minimum base of G satisfies the hypotheses. We shall build a morphism $\varphi : G \to B$ by sending a node l of G to the unique node $\varphi(l)$ of B satisfying $\widetilde{G}^l \restriction (n-1) \cong \widetilde{B}^{\varphi(l)} \restriction (n-1)$. This node can be found as follows: there is certainly a node $l' \in (v_G^i)^{-1}(l)$ which is at depth D at most. Thus, the subtree under l' in $\widetilde{G}^i \restriction n + D$ has height at least $n-1$. Let $\psi : \widetilde{G}^i \restriction (n+D) \to \widetilde{B}^j \restriction (n+D)$ be the isomorphism above. Then $\varphi(l) = (v_B^j \circ \psi)(l')$. Note that the choice of l' is irrelevant, by the considerations about the views of fibration prime graphs made in Sect. 3.

We now define analogously φ on the arcs, by using the lifting property. Let a be an arc of G. We choose, as before, $l \in (v_G^i)^{-1}(t(a))$ which is at depth D at most, and consider the lifting $\widetilde{a}^l$. Then we set $\varphi(a) = (v_B^j \circ \psi)(\widetilde{a}^l)$. Note that this is compatible with our definition on the nodes, because $s(\widetilde{a}^l)$ is at depth $D+1$ at most, and thus its image through $v_B^j \circ \psi$ must be $\varphi(s(a))$, by Theorem 2. It is then easy to check that since φ has been defined by a lifting and composition with isomorphisms and fibrations, it is itself a fibration. Moreover, by its very definition it maps i to j. □

Corollary 1. *Let B_1 and B_2 be fibration prime, i_1 a node of B_1 and i_2 a node of B_2. If*

$$h(\widetilde{B_1}^{i_1} \wedge \widetilde{B_2}^{i_2}) \geq \max\{n_{B_1} + D_{B_1}, n_{B_2} + D_{B_2}\}$$

then there is an isomorphism $\alpha : B_1 \to B_2$ such that $\alpha(i_1) = i_2$.

The previous result shows that fibration prime graphs sharing enough levels of one of their views are isomorphic (and the nodes to which the view are associated are in correspondence). This is all we need to prove our first holographic bound:

Theorem 4. *The function mapping a graph G to $n_{\hat{G}} + D_{\hat{G}}$ is a holographic bound for the class of all graphs (hence for every class).*

Proof. By definition, we have to show that for every graph G, every node i of G and every $k \geq n_{\hat{G}} + D_{\hat{G}}$ the class

$$\{\langle \hat{H}, \mu_H(j) \rangle \mid j \in N_H,\, n_{\hat{H}} + D_{\hat{H}} \leq k \text{ and } \widetilde{G}^i \restriction k \leq \widetilde{H}^j\}$$

contains essentially one element. To this purpose, we take an arbitrary pair $\langle B, l \rangle$ from the set and show how to build an isomorphism between $\hat{G}$ and B that maps $\mu_G(i)$ to l. But since $\widetilde{G}^i \restriction k \leq \widetilde{B}^l$ and $k \geq \max\{n_B + D_B, n_{\hat{G}} + D_{\hat{G}}\}$, we can apply Corollary 1 to $\hat{G}$ and B, with chosen nodes $\mu_G(i)$ and l (recall that $\widetilde{G}^i \cong \widetilde{\hat{G}}^{\mu_G(i)}$). □

An example of trees that are holographic using the bound above is given in Table 2, where we show all the (uncoloured) trees of height at most four and indegree at most two that are holographic, together with the base and the node that generated them. One should compare this list with some of the nonholographic trees appearing in the execution of the view construction algorithm in Table 1. The reader might now be curious to know whether $n_{\hat{G}} + D_{\hat{G}}$ is the best possible holographic bound. We do not know the answer to this question, but we have some partial results. Let $\mathscr{C}$ be any class of graph including the fibration prime graphs shown in Fig. 3. Note that both $G_{n,D}$ and $H_{n,D}$ have n nodes and diameter D (the difference between the two families is given by the positioning of the dotted arc). It is tedious but easy to check that for all D and n we have that $\widetilde{G_{n,D}}^1$ and $\widetilde{H_{n,D}}^1$ are isomorphic up to level $n + D - 1$, but not up to level $n + D$, and this property is the key to the proof of the following lower bounds:

Theorem 5. *Every holographic bound for a class $\mathscr{C}$ as above is at least $n_{\hat{G}} + D_{\hat{G}}$ for an infinite number of graphs.*

Proof. By contradiction, suppose that $\nu(G) \geq n_{\hat{G}} + D_{\hat{G}}$ holds only for finitely many $G \in \mathscr{C}$. Thus, there are n and D such that $\nu(G_{n,D}), \nu(H_{n,D}) < n + D$. But then $\mathscr{U}_{\mathscr{C},\nu}\left(\widetilde{G_{n,D}}^1 \restriction n + D - 1\right)$ contains both $\langle G_{n,D}, 1 \rangle$ and $\langle H_{n,D}, 1 \rangle$, a contradiction. □

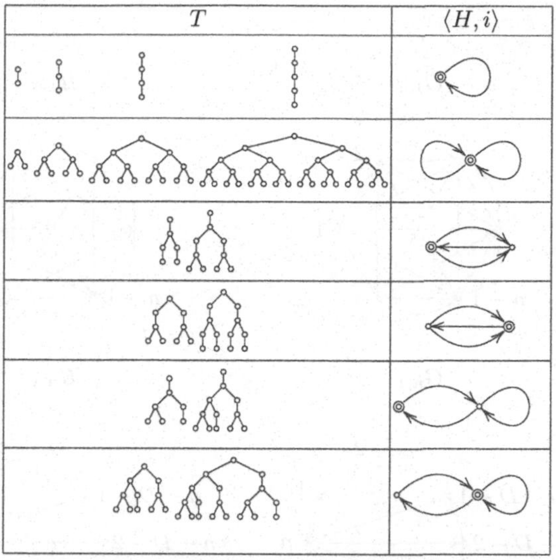

Table 2. Small holographic trees.

Theorem 6. *A class $\mathscr{C}$ as above has no holographic bound depending only on the number of nodes and on the diameter (or on the number of nodes and on the diameter of the minimum base) that is smaller than $n_{\hat G} + D_{\hat G}$ for some $G \in \mathscr{C}$.*

Proof. Let $L \in \mathscr{C}$ be a graph with n nodes and diameter D such that $\nu(L) < n_{\hat L} + D_{\hat L}$; in the first case, $\nu(G_{n,D}) = \nu(H_{n,D}) = \nu(L) < n_{\hat L} + D_{\hat L} \le n + D$, and we proceed as in the proof of Theorem 5. In the second case, we have $\nu(G_{n_{\hat L},D_{\hat L}}) = \nu(H_{n_{\hat L},D_{\hat L}}) = \nu(L) < n_{\hat L} + D_{\hat L}$, and the thesis follows again. □

In the rest of the section, we highlight two worked out examples specializing Theorem 4.

Inregular graphs. Since the inregular graphs are exactly the total graphs over bouquets (i.e., graphs with exactly one node), we have that the constant function 1 is the (obviously minimum) holographic bound for that class, and the holographic trees are exactly the inregular trees with at least one arc.

Complete multipartite graphs. A graph is *complete multipartite* iff its node set can be partitioned into independent[9] sets, and there is exactly one arc from node i to node j when i and j do not belong to the same part. The minimum base of a complete multipartite graph G can be constructed as follows: let $k_1, k_2, \dots, k_{l_G}$ be a list (without repetitions) of the cardinalities of the parts of G, and $m_1, m_2, \dots, m_{l_G}$ be the respective multiplicities (i.e., m_i is the number of parts of

[9] A set of nodes is *independent* iff there are no arcs with source and target in the set.

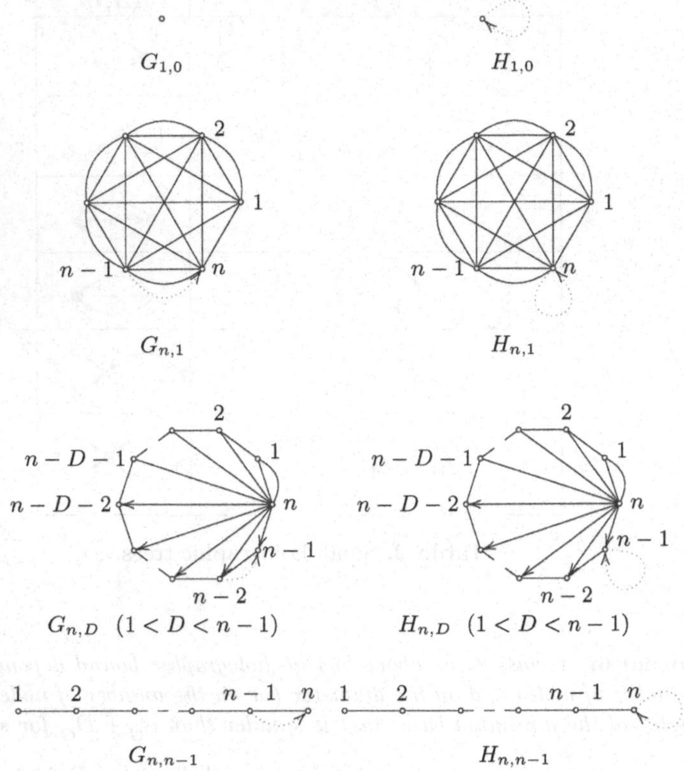

Fig. 3. Graphs with similar views.

cardinality k_i). The graph $\hat{G}$ has l_G nodes, and the number of arcs from node i to node j is $m_i k_i$ if $i \neq j$, $(m_i - 1)k_i$ otherwise. As a consequence, a holographic bound for this class is given by $\nu(G) = l_G + 1$.

6 A Holographic Bound for All Labelled Graphs

In this section we prove that $D_G + 1$ is a holographic bound for the class $\mathscr{L}$ of labelled graphs, that is, graphs whose nodes are coloured injectively. All such graphs are obviously fibration prime. Note that the theory described in Sect. 3 specializes, and that also views carry on the colouring (which however is no longer injective). First, we show that Theorem 3 can be restated as follows:

Theorem 7. *Let G and B be labelled graphs, and suppose $h\bigl(\widetilde{G}^i \wedge \widetilde{B}^j\bigr) \geq D_G + 1$ for some $i \in N_G$ and $j \in N_B$: then there is an isomorphism $\alpha : G \cong B$, and $\alpha(i) = j$.*

Proof. We define α by sending a node of G to the unique node of B having the same colour, and on the arcs by lifting. More in detail, let a be an arc of G. We choose a $l \in (\upsilon^i_G)^{-1}(t(a))$ which is at depth D at most, and consider the lifting $\widetilde{a}^l$. Then we set $\alpha(a) = (\upsilon^j_B \circ \psi)(\widetilde{a}^l)$, where ψ is the isomorphism from $\widetilde{G^i} \upharpoonright D_G + 1$ to $\widetilde{B^j} \upharpoonright D_G + 1$. Note that this is compatible with our definition on the nodes, because $s(\widetilde{a}^l)$ is at depth $D+1$ at most, and thus its image through $\upsilon^j_B \circ \psi$ must have the same colour as $s(a)$. It is then easy to check that since α has been defined by a lifting and composition with isomorphisms and fibrations, it is itself a fibration, and thus an isomorphism by primality of G and B. Moreover, by its very definition it maps i to j. □

Theorem 8. *The function mapping a graph G to D_G+1 is a holographic bound for the class of all labelled graphs (hence for every class of labelled graphs).*

Proof. By definition, we have to show that for every labelled graph G, every node i of G and every $k \geq D_G + 1$ the class

$$\left\{ \langle H, j \rangle \mid H \text{ is labelled}, j \in N_H, D_H + 1 \leq k \text{ and } \widetilde{G^i} \upharpoonright k \leq \widetilde{H^j} \right\}$$

contains essentially one element (we omitted the hat symbols as all labelled graphs are fibration prime). But if $\langle H, j \rangle$ is an arbitrary element from the set, we have $h\left(\widetilde{G^i} \wedge \widetilde{H^j}\right) \geq k \geq D_G + 1$, and by Theorem 7 we obtain the thesis. □

Finally, as in the previous section, we prove some lower bounds:

Theorem 9. *Every holographic bound ν for $\mathscr{L}$ is at least $D+1$ for an infinite number of graphs.*

Proof. Suppose $\nu(G) < D_G + 1$, and consider nodes i and j of G which maximize the distance from j to i. Let H be the graph obtained from G by adding an additional loop at j. Clearly $\mathscr{U}_{\mathscr{L},\nu}\left(\widetilde{G^i} \upharpoonright D_G\right)$ would contain $\langle H, i \rangle$, unless $\nu(H) \geq D_G + 1 = D_H + 1$. This gives immediately the result. □

An absolutely analogous proof shows also that

Theorem 10. *The class $\mathscr{L}$ has no holographic bound depending only on the diameter that is smaller than $D_G + 1$ for some $G \in \mathscr{L}$.*

References

[1] Dana Angluin. Local and global properties in networks of processors. In *Proc. 12th Symposium on the Theory of Computing*, pages 82–93, 1980.

[2] Paolo Boldi, Bruno Codenotti, Peter Gemmell, Shella Shammah, Janos Simon, and Sebastiano Vigna. Symmetry breaking in anonymous networks: Characterizations. In *Proc. 4th Israeli Symposium on Theory of Computing and Systems*, pages 16–26. IEEE Press, 1996.

[3] Paolo Boldi and Sebastiano Vigna. Fibrations of graphs. *Discrete Math.* To appear.

[4] Paolo Boldi and Sebastiano Vigna. Computing vector functions on anonymous networks. In Danny Krizanc and Peter Widmayer, editors, *SIROCCO '97. Proc. 4th International Colloquium on Structural Information and Communication Complexity*, volume 1 of *Proceedings in Informatics*, pages 201–214. Carleton Scientific, 1997. An extended abstract appeared also as a Brief Announcement in *Proc. PODC '97*, ACM Press.

[5] Paolo Boldi and Sebastiano Vigna. Self-stabilizing universal algorithms. In Sukumar Ghosh and Ted Herman, editors, *Self-Stabilizing Systems (Proc. of the 3rd Workshop on Self-Stabilizing Systems, Santa Barbara, California, 1997)*, volume 7 of *International Informatics Series*, pages 141–156. Carleton University Press, 1997.

[6] Paolo Boldi and Sebastiano Vigna. Computing anonymously with arbitrary knowledge. In *Proc. 18th ACM Symposium on Principles of Distributed Computing*, pages 181–188. ACM Press, 1999.

[7] Paolo Boldi and Sebastiano Vigna. An effective characterization of computability in anonymous networks. In Jennifer L. Welch, editor, *Distributed Computing. 15th International Conference, DISC 2001*, number 2180 in Lecture Notes in Computer Science, pages 33–47. Springer–Verlag, 2001.

[8] Paolo Boldi and Sebastiano Vigna. Universal dynamic synchronous self-stabilization. *Distr. Comput.*, 15, 2002.

[9] Francis Borceux. *Handbook of Categorical Algebra 2*, volume 51 of *Encyclopedia of Mathematics and Its Applications.* Cambridge University Press, 1994.

[10] Bruno Courcelle. Fundamental properties of infinite trees. *Theoret. Comput. Sci.*, 25(2):95–169, 1983.

[11] Edsger W. Dijkstra. Self-stabilizing systems in spite of distributed control. *CACM*, 17(11):643–644, 1974.

[12] Nancy Norris. Universal covers of graphs: Isomorphism to depth $n-1$ implies isomorphism to all depths. *Discrete Appl. Math.*, 56:61–74, 1995.

[13] Masafumi Yamashita and Tiko Kameda. Computing on anonymous networks: Part I—characterizing the solvable cases. *IEEE Trans. Parallel and Distributed Systems*, 7(1):69–89, 1996.

On the Spanning Ratio of Gabriel Graphs and β-skeletons

Prosenjit Bose[1]*, Luc Devroye[2]**, William Evans[3]*, and David Kirkpatrick[3]*

[1] School of Computer Science, Carleton University, Ottawa, Ontario, Canada.
jit@cs.carleton.ca
[2] School of Computer Science, McGill University, Montreal, Canada.
luc@cs.mcgill.ca
[3] Department of Computer Science, University of British Columbia, Vancouver, Canada.
{will,kirk}@cs.ubc.ca

Abstract. The spanning ratio of a graph defined on n points in the Euclidean plane is the maximal ratio over all pairs of data points (u, v), of the minimum graph distance between u and v, over the Euclidean distance between u and v. A connected graph is said to be a k-spanner if the spanning ratio does not exceed k. For example, for any k, there exists a point set whose minimum spanning tree is not a k-spanner. At the other end of the spectrum, a Delaunay triangulation is guaranteed to be a 2.42-spanner[11]. For proximity graphs *inbetween* these two extremes, such as Gabriel graphs[8], relative neighborhood graphs[16] and β-skeletons[12] with $\beta \in [0, 2]$ some interesting questions arise. We show that the spanning ratio for Gabriel graphs (which are β-skeletons with $\beta = 1$) is $\Theta(\sqrt{n})$ in the worst case. For all β-skeletons with $\beta \in [0, 1]$, we prove that the spanning ratio is at most $O(n^{\gamma})$ where $\gamma = (1 - \log_2(1 + \sqrt{1 - \beta^2}))/2$. For all β-skeletons with $\beta \in [1, 2)$, we prove that there exist point sets whose spanning ratio is at least $\left(\frac{1}{2} - o(1)\right)\sqrt{n}$. For relative neighborhood graphs[16] (skeletons with $\beta = 2$), we show that there exist point sets where the spanning ratio is $\Omega(n)$. For points drawn independently from the uniform distribution on the unit square, we show that the spanning ratio of the (random) Gabriel graph and all β-skeletons with $\beta \in [1, 2]$ tends to ∞ in probability as $\sqrt{\log n / \log \log n}$.

1 Introduction

Many problems in geometric network design, pattern recognition and classification, geographic variation analysis, geographic information systems, computational geometry, computational morphology, and computer vision use the underlying *structure* (also referred to as the *skeleton* or *internal shape*) of a set of data points revealed by means of a *proximity graph* (see for example [16], [13], [7], [9]). A proximity graph attempts to exhibit the relation between points in

* Research supported by NSERC.
** Research supported by NSERC and by FCAR.

S. Rajsbaum (Ed.): LATIN 2002, LNCS 2286, pp. 479–493, 2002.

a point set. Two points are joined by an edge if they are deemed *close* by some proximity measure. It is the measure that determines the type of graph that results. Many different measures of proximity have been defined, giving rise to many different types of proximity graphs. An extensive survey on the current research in proximity graphs can be found in Jaromczyk and Toussaint [9].

We are concerned with the spanning ratio of proximity graphs. Consider n points in $\mathbb{R}^2$, and define a graph on these points, such as the Gabriel graph [8], or the relative neighborhood graph [16]. For a pair of data points (u, v), the length of the shortest path measured by Euclidean distance is denoted by $L(u, v)$, while the direct Euclidean distance is $D(u, v)$. The *spanning ratio* of the graph is defined by

$$S \stackrel{\text{def}}{=} \max_{(u,v)} \frac{L(u,v)}{D(u,v)} ,$$

where the maximum is over all $\binom{n}{2}$ pairs of data points. Note that if the graph is not connected, the spanning ratio is infinite. In this paper, we will concentrate on connected graphs.

Graphs with small spanning ratios are important in some applications (see [7] for a survey on spanners). The history for the Delaunay triangulation is interesting. First, Chew [2,3] showed that in the worst case, $S \geq \pi/2$. Subsequently, Dobkin et al.[5] showed that the Delaunay triangulation was a $((1+\sqrt{5})/2)\pi \approx 5.08$ spanner. Finally, Keil and Gutwin [10,11] improve this to $2\pi/(3\cos(\pi/6))$ which is about 2.42. It is conjectured that the spanning ratio of the Delaunay triangulation is $\pi/2$. The complete graph has $S = 1$, but is less interesting because the number of edges is not linear but quadratic in n. In this paper, we concentrate on the parameterized family of proximity graphs known as β-skeletons [12] with β in the interval $[0, 2]$. The family of β-graphs contains certain well-known proximity graphs such as the Gabriel graph [8] when $\beta = 1$ and the relative neighborhood graph [16] when $\beta = 2$. As graphs become sparser, their spanning ratios increase. For example, it is trivial to show that there are minimal spanning trees with n vertices for which $S \geq n - 1$, whereas the Delaunay triangulation has a constant spanning ratio.

In this note, we probe the expanse inbetween these two extremes. We show that for any n, in the plane, there exists a point set whose Gabriel graph satisfies $S \geq c\sqrt{n}$, where c is a universal constant. We also show that for any Gabriel graph in the plane, $S \leq c'\sqrt{n}$ for another constant c'. For all β-skeletons with $\beta \in [0, 1]$, we prove that the spanning ratio is at most $O(n^\gamma)$ where $\gamma = (1 - \log_2(1 + \sqrt{1-\beta^2}))/2$. For all β-skeletons with $\beta \in [1, 2)$, we prove that there exist point sets whose spanning ratio is at least $\left(\frac{1}{2} - o(1)\right)\sqrt{n}$. For relative neighborhood graphs, we show that there exist point sets where the spanning ratio is $\Omega(n)$. The second part of the paper deals with point sets drawn independently from the uniform distribution on the unit square. We show that the spanning ratio of the (random) Gabriel graph and all β-skeletons with $\beta \in [1, 2]$ tends to ∞ in probability as $\sqrt{\log n / \log\log n}$.

2 Preliminaries

We begin by defining some of the graph theoretic and geometric terminology used in this paper. For more details see [1] and [15].

A *graph* $G = (V, E)$ consists of a finite non empty set $V(G)$ of *vertices*, and a set $E(G)$ of unordered pairs of vertices known as *edges*. An edge $e \in E(G)$ consisting of vertices u and v is denoted by $e = uv$; u and v are called the *endpoints* of e and are said to be *adjacent* vertices or *neighbors*. The *degree* of a vertex $v \in V(G)$, denoted by $deg_G(v)$ (or just $deg(v)$ when no confusion will result), is the number of edges of $E(G)$ which have v as an endpoint. A *path* in a graph G is a finite non-null sequence $P = v_1 v_2 \dots v_k$ where the vertices $v_1, v_2 \dots v_k$ are distinct and $v_i v_{i+1}$ is an edge for each $i = 1 \dots k-1$. The vertices v_1 and v_k are known as the *endpoints* of the path. A *cycle* is a path whose endpoints are the same. A graph is *connected* if, for each pair of vertices $u, v \in V$, there is a path from u to v.

Intuitively speaking, a *proximity graph* on a finite set $P \subset \mathbb{R}^2$ is obtained by connecting pairs of points of P with line segments if the points are considered to be *close* in some sense. Different definitions of closeness give rise to different proximity graphs. One technique for defining a proximity graph on a set of points is to select a geometric region defined by two points of P—for example the smallest disk containing the two points—and then specifying that a segment is drawn between the two points if and only if this region contains no other points from P. Such a region will be referred to as a *region of influence* of the two points. Four such definitions follow.

Given a set P of points in $\mathbb{R}^2$, the *relative neighborhood graph of P*, denoted by $RNG(P)$, has a segment between points u and v in P if the intersection of the open disks of radius $D(u, v)$ centered at u and v is empty. This region of influence is referred to as the *lune* of u and v. Equivalently, $u, v \in S$ are adjacent if and only if

$$D(u, v) \leq \max[D(u, w), D(v, w)], \quad \text{for all } w \in S, w \neq u, v.$$

The *Gabriel graph of P*, denoted by $GG(P)$, has as its region of influence the closed disk having segment $\overline{uv}$ as diameter. That is, two vertices $u, v \in S$ are adjacent if and only if

$$D^2(u, v) < D^2(u, w) + D^2(v, w), \quad \text{for all } w \in S, w \neq u, v.$$

A *Delaunay triangulation* of a set P of points in the plane, denoted by $DT(P)$, is a triangulation of P such that for each interior face, the triangle which bounds that face has the property that the circle circumscribing the triangle contains no other points of the graph in its interior. A set P may admit more than one Delaunay triangulation, but only if P contains four or more co-circular points. A list of properties of the Delaunay triangulation can be found in [15].

We describe another graph, a *minimum spanning tree*, which is not defined in terms of a region of influence. Given a set P of points in the plane, consider a connected straight-line graph G on P, that is, a graph having as its edge set E a collection of line segments connecting pairs of vertices of P. Define the *weight* of G to be the sum of all of the edge lengths of G. Such a graph is called a *minimum spanning tree of* P, denoted by $MST(P)$, if its weight is no greater than the weight of any other connected straight-line graph on P. (It is easy to see that such a graph must be a tree.) In general, a set P may have many minimum spanning trees (for example, if P consists of the vertices of a regular polygon).

The following relationships among the different proximity graphs hold for any finite set P of points in the plane.

Lemma 1. *[15]* $MST(P) \subseteq RNG(P) \subseteq GG(P) \subseteq DT(P)$

Given a finite set P of distinct points in $\mathbb{R}^2$, we define the β-skeleton of P. β-skeletons are a family of graphs having vertex set P, parameterized by the value of β. For each pair x, y of points in P, we define the region of influence for a given value of β, and denote this region as $R(x, y, \beta)$.

1. *For $\beta = 0$, $R(x, y, \beta)$ is the line segment $\overline{xy}$.*
2. *For $0 < \beta < 1$, $R(x, y, \beta)$ is the intersection of the two disks of radius $D(x, y)/(2\beta)$ passing through both x and y.*
3. *For $1 \leq \beta < \infty$, $R(x, y, \beta)$ is the intersection of the two disks of radius $\beta D(x, y)/2$ and centered at the points $(1 - \beta/2)x + (\beta/2)y$ and $(\beta/2)x + (1 - \beta/2)y$, respectively.*
4. *For $\beta = \infty$, $R(x, y, \beta)$ is the infinite strip perpendicular to the line segment $\overline{xy}$*

Now consider the set of segments $\overline{xy}$ with $x, y \in P$ such that $R(x, y, \beta) \cap P \setminus \{x, y\} = \emptyset$ (i.e. the set of edges $\overline{xy}$ whose region of influence contains no points of $P \setminus \{x, y\}$). This set of distinct points and segments naturally defines a graph called the *β-skeleton* of P [12]. The β-skeleton for a fixed value of $\beta = k$ shall be referred to as the k-skeleton. Notice that different values of the parameter β give rise to different graphs. Note also that different graphs may result for the same value of β if the regions of influence are constructed with open rather than closed disks, however, these boundary effects do not alter our results. When necessary, we will explicitly state whether the region of influence is open or closed. These graphs will be referred to as open β-skeletons and closed β-skeletons, respectively. The closed 1-skeleton is the Gabriel graph and the open 2-skeleton is the relative neighborhood graph.

Also, as the value of β increases, the graphs become sparser since the region of influence increases in size. β-skeletons with $\beta \leq 2$ are connected. Therefore, we will concentrate on the interval $\beta \in [0, 2]$.

Observation 1 *If $k \leq k'$, then the k'-skeleton is a subset of the k-skeleton of a point set.*

3 A Lower Bound on the Spanning Ratio

We begin with a deterministic lower bound on the spanning ratio of β-skeletons. The example developed in this section is essential for the understanding of the results on random Gabriel graphs.

Theorem 1. *For any $n \geq 2$, there exists a set of n points in the plane whose β-skeleton with $\beta \in [1,2]$ has spanning ratio*

$$S \geq \left(\frac{1}{2} - o(1)\right)\sqrt{n}\ .$$

Note: the closed 1-skeleton is the Gabriel graph and that all β-skeletons with $\beta > 1$ are subgraphs of the Gabriel graph. Therefore, it suffices to prove the theorem for the Gabriel graph. Also, the $1/2 - o(1)$ factor can be improved to $2/3$.

Proof. Let $m = \lfloor n/2 \rfloor$. Place points p_i and q_i at locations $(-r_i, y_i)$ and (r_i, y_i) respectively $(1 \leq i \leq m)$ where

$$r_i = 1 - (i-1)/n$$
$$y_i = (i-1)/\sqrt{n}$$

If n is odd place the remaining point at the same location as p_1.

We claim that for each pair p_i, q_i, the circle with diameter p_iq_i contains the points p_{i+1} and q_{i+1} $(1 \leq i \leq m-1)$. Let d be the distance from the center of the circle with diameter p_iq_i to the point p_{i+1}. For p_{i+1} to lie within this circle, d must be at most r_i. By construction,

$$d = \sqrt{(r_i - 1/n)^2 + 1/n}\ .$$

Thus we require $(r_i - 1/n)^2 + 1/n \leq r_i^2$ or, equivalently, $r_i \geq 1/2 + 1/(2n)$, which holds for $1 \leq i \leq m-1$.

It follows that when $i \leq j$, edge p_iq_j does not belong to the Gabriel graph of these points (unless $i = j = m$), since p_{i+1} lies in or on the circle with diameter p_iq_j. Similarly, when $i > j$, edge p_iq_j is precluded by point q_{j+1}.

The Euclidean distance between p_1 and q_1 is two. However, the shortest path from p_1 to q_1 using Gabriel graph edges is at least $2y_m$, which results in a spanning ratio of

$$S = y_m = (\lfloor n/2 \rfloor - 1)/\sqrt{n} = \left(\frac{1}{2} - o(1)\right)\sqrt{n}\ .$$

■

When β is in the interval $(0,1]$, Eppstein[6] presents an elegant fractal construction that provides a non-constant lower bound on the spanning ratio. His result is summarized in the following theorem.

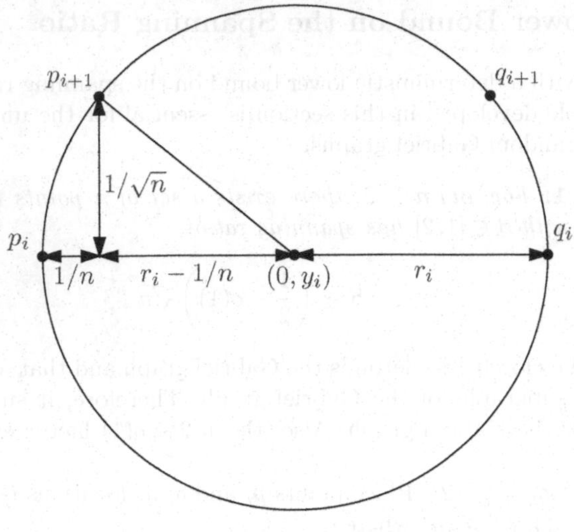

Fig. 1. Illustration for proof of Theorem 1

Theorem 2. *For any $n = 5^k + 1$, there exists a set of n points in the plane whose β-skeleton with $\beta \in (0, 1]$ has spanning ratio $\Omega(n^c)$, where $c = \log_5(5/(3+2sin(\theta)))$ and $\theta < (\pi - sin^{-1}(\beta))/2$.*

As noted before, the spanning ratio leaps to infinity for $\beta > 2$, since past this point, the graph may be disconnected. Therefore, it only makes sense to consider spanning ratios when $\beta \in [0, 2]$. When $\beta = 0$, the β-skeleton of a point set has spanning ratio 1. Note that for Gabriel graphs, the above result implies a ratio of $\Omega(n^c), 0.077 < c < 0.078$, thus for $\beta \geq 1$, Theorem 1 provides a much stronger bound of $\Omega(\sqrt{n})$.

4 Lower Bound for Relative Neighborhood Graphs

In this section, we show that there exist point sets where the spanning ratio for relative neighborhood graphs (open 2-skeletons) is $\Omega(n)$.

Lemma 2. *The spanning ratio for the relative neighborhood graph of a set of n points in the plane can be $\Omega(n)$.*

Proof. Refer to Figure 2. Let $\theta = 60 - \epsilon$ and $\alpha = 60 + 2\epsilon$. We will fix ϵ later. Since $\alpha + 2\theta = \pi$, the points $a_0, a_1, \ldots, a_n$ are colinear. Similarly, the points $b_0, b_1, \ldots, b_n$ are colinear. The point a_{i+1} blocks the edge $\overline{a_i, b_i}$. An edge $\overline{a_i, b_j}$ for $i < j$ is blocked by a_{i+1} and an edge $\overline{a_i, b_j}$ for $i > j$ is blocked by b_{i+1}. Thus, the only edges in the RNG of these points are $\overline{a_i, a_{i+1}}$, $\overline{b_i, b_{i+1}}$ and $\overline{a_n, b_n}$. Let $A_i = \|a_{i+1} - a_i\|$. Let $B_i = \|b_{i+1} - b_i\|$.

Triangle(a_0, a_1, b_0) and Triangle(a_1, b_1, b_0) are similar, therefore, $B_0 = A_0^2/A$. By the same argument, $A_1 = A_0^3/A^2$, and $B_1 = A_0^4/A^3$. In general, $A_i = A_0^{2i+1}/A^{2i}$ and $B_i = A_0^{2i+2}/A^{2i+1}$.

We choose an ϵ so that $A_0/A > (1/2)^{1/2n}$. Let L be the length of the path from a_0 to b_0. $L > \sum_{i=0}^{n-1} A_i + B_i = \sum_{i=0}^{2n-1} A_0(A_0/A)^i$. Since $A_0/A > (1/2)^{1/2n}$, we have that $\sum_{i=0}^{2n-1} A_0(A_0/A)^i > 1/2 \sum_{i=0}^{2n-1} A_0 = A_0 n$. Therefore, $L > A_0 n$. ■

Lemma 3. *For any $\beta \leq 2$, the ratio $L(x,y)/D(x,y) \leq n - 1$.*

Proof. Let G be the β-skeleton of a set of n points P. Note that the minimum spanning tree $MST(P)$ is contained in G. Let x, y be two points in P.

Let $L(x,y)$ be the length of the unique path from x to y in $MST(P)$. This path has at most $n-1$ edges and each edge must have length at most $D(x,y)$, otherwise, $MST(P)$ can be made shorter. Therefore, $L(x,y) \leq (n-1)D(x,y)$. ■

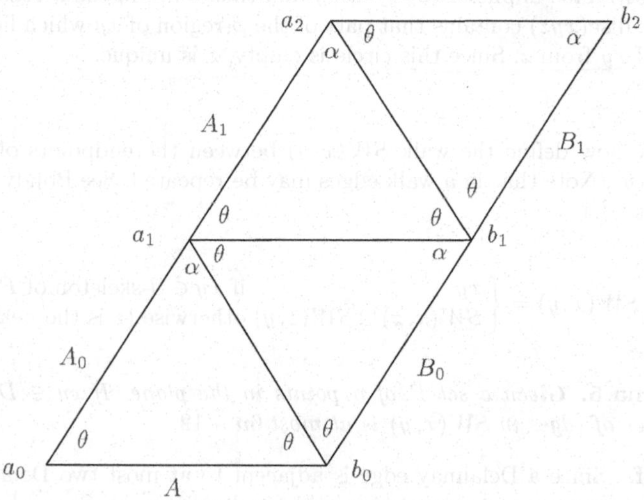

Fig. 2. RNG tower

5 Upper Bound

The upper bound established in this section applies to β-skeletons for $\beta \in [0,1]$. The β-skeleton of a point set P for $\beta \in [0,1]$ is a graph in which points x and y

in P are connected by an edge if and only if there is no other point $v \in P$ such that $\angle xvy > \pi - \arcsin \beta$. Recall that the Gabriel graph is the closed 1-skeleton of P and is a subgraph of the Delaunay triangulation of P.

To upper bound the spanning ratio of β-skeletons, we show that there exits a special walk $SW(x, y)$ in the β-skeleton between the endpoints of any Delaunay edge xy. We upper bound the length $|SW(x, y)|$ of $SW(x, y)$ as a multiple of $D(x, y)$. We then combine this with an upper bound on the spanning ratio of Delaunay triangulations [10,11] to obtain our result.

Let $DT(P)$ be the Delaunay triangulation of a points set P. In order to describe the walk between the endpoints of a Delaunay edge, we define the *peak* of a Delaunay edge.

Lemma 4. *Let xy be an edge of $DT(P)$. Either xy is an edge of the β-skeleton of P or there exists a unique z (called the peak of xy) such that triangle(xyz) is in $DT(P)$ and z lies in the β-region of xy.*

Proof. Suppose $xy \in DT(P)$ is not an edge in the β-skeleton of P. Then there exists a point $v \in P$ such that $\angle xvy > \pi - \arcsin \beta$. Since xy is an edge of $DT(P)$, there exists a unique z on the same side of xy as v such that disc(xyz) is empty. This implies $\angle xzy \geq \angle xvy$ and thus z lies in the β-region of xy. Since $\beta \leq 1$, disc(xyz) contains that part of the β-region of xy which lies on the other side of xy from z. Since this circle is empty, z is unique.

■

We now define the walk $SW(x, y)$ between the endpoints of the Delaunay edge xy. (Note that in a walk edges may be repeated. See Bondy and Murty for details [1].)

$$SW(x,y) = \begin{cases} xy & \text{if } xy \in \beta\text{-skeleton of } P \\ SW(x,z) \cup SW(z,y) & \text{otherwise } (z \text{ is the peak of } xy) \end{cases}$$

Lemma 5. *Given a set P of n points in the plane. If $xy \in DT(P)$ then the number of edges in $SW(x, y)$ is at most $6n - 12$.*

Proof. Since a Delaunay edge is adjacent to at most two Delaunay triangles, an edge can occur at most twice in the walk $SW(x, y)$. Since there are at most $3n - 6$ edges in $DT(S)$ by Euler's Formula, $SW(x, y)$ can consist of at most $6n - 12$ edges.

■

Lemma 6. *Let P be a set of n points in the plane. For all $x, y \in S$, if $xy \in DT(P)$ then*

$$|SW(x,y)| \leq m^{\gamma} D(x,y)$$

where $\gamma = (1 - \log_2(1 + \sqrt{1 - \beta^2}))/2$ and m is the number of edges in $SW(x, y)$.

Proof. The proof is by induction on the number of edges m in $SW(x,y)$. When $m = 1$, i.e. $SW(x,y)$ is simply the line segment from x to y, the lemma clearly holds.

If $m > 1$, then $|SW(x,y)| = |SW(x,z)| + |SW(z,y)|$ for z the peak of xy. Let k be the number of edges in $SW(x,z)$. Thus, $m - k$ is the number of edges in $SW(z,y)$. Let $a = D(x,y)$, $b = D(x,z)$, and $c = D(y,z)$. Since xz and zy are Delaunay edges, by induction, $|SW(x,z)| \leq bk^\gamma$ and $|SW(z,y)| \leq c(m-k)^\gamma$. Thus it suffices to prove that

$$bk^\gamma + c(m-k)^\gamma \leq am^\gamma \ .$$

As a function of k the left-hand side of the equation is maximized when $k = mc^\phi/(b^\phi + c^\phi)$ where $\phi = 1/(\gamma - 1)$. With this substitution for k, after factoring m^γ, it remains to show,

$$b\left(\frac{c^\phi}{b^\phi + c^\phi}\right)^\gamma + c\left(\frac{b^\phi}{b^\phi + c^\phi}\right)^\gamma \leq a \ .$$

By the law of cosines, $a^2 = b^2 + c^2 - 2bc\cos A$ where A is the angle at the peak z. Thus we need only show,

$$b\left(\frac{c^\phi}{b^\phi + c^\phi}\right)^\gamma + c\left(\frac{b^\phi}{b^\phi + c^\phi}\right)^\gamma - \sqrt{b^2 + c^2 - 2bc\cos A} \leq 0 \ .$$

This inequality holds for b,c if and only if it holds for αb,αc for all $\alpha > 0$. Thus we may assume that $b + c = 1$. The left-hand side, as a function of b, is maximized at $b = 1/2$, and the inequality holds as long as $\gamma \geq (1 - \log_2(1 - \cos A))/2$. The angle A is minimized (thus maximizing $(1 - \log_2(1 - \cos A))/2$) when z lies on the boundary of the β-region. For such z, $1 - \cos A = 1 + \sqrt{1 - \beta^2}$. ∎

Theorem 3. *The spanning ratio of the β-skeleton of a set P of n points in the plane is at most*

$$\frac{4\pi(6n-12)^\gamma}{3\sqrt{3}}$$

where $\gamma = (1 - \log_2(1 + \sqrt{1 - \beta^2}))/2$.

Proof. Given two arbitrary points x, y in P, let $M = e_1, e_2, \ldots, e_j$ represent the shortest path between x and y in $DT(P)$. Keil and Gutwin [10,11] have shown that the length of P is at most $2\pi/(3\cos(\pi/6))$ times $D(x,y)$.

For each edge e_i in M, by Lemma 5 and Lemma 6, we know there exists a path in the β-skeleton whose length is at most $(6n - 12)^\gamma$ times the length of e_i. Therefore, the shortest path between x and y in the β-skeleton has length at most $2\pi(6n-12)^\gamma/(3\cos(\pi/6))$ times $D(x,y)$. The theorem follows. ∎

Corollary 1. *The spanning ratio of the Gabriel graph of an n-point set is at most*

$$\frac{4\pi}{3}\sqrt{2n-4}\,.$$

When β lies strictly between 0 and 1, there is a gap between the upper bound and lower bound on the spanning ratio of β-skeletons. As noted in section 3, the spanning ratio is at least $\Omega(n^c)$ where $c = \log_5(5/(3+2\sin(\theta)))$ and $\theta < (\pi - \sin^{-1}(\beta))/2$. We have shown here that the spanning ratio is at most $O(n^\gamma)$ where $\gamma = (1-\log_2(1+\sqrt{1-\beta^2}))/2$. Refer to Figure 3 for a graph of the exponents of the upper and lower bound. The gap is closed for Gabriel graphs ($\beta = 1$). For Gabriel graphs, the lower bound construction given in section 3, together with the upper bound given here, show that the spanning ratio is indeed $\Theta(\sqrt{n})$.

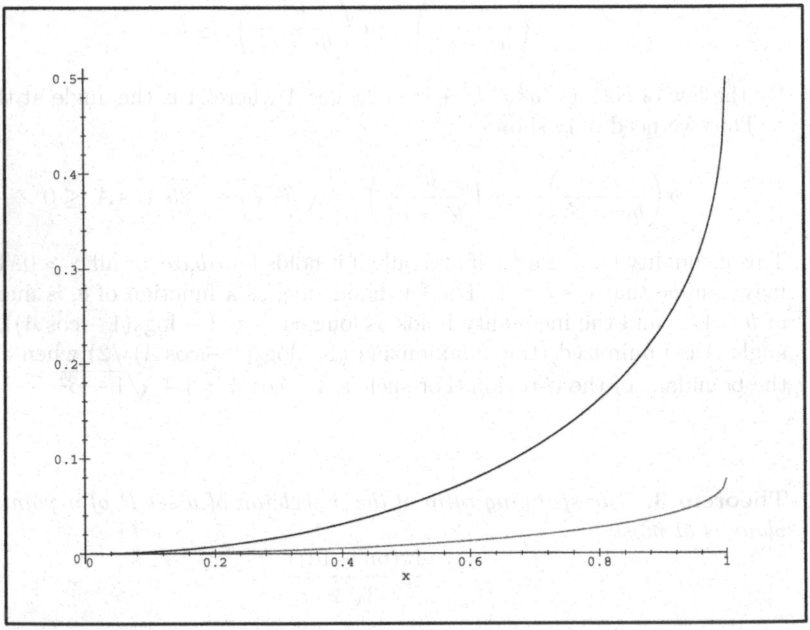

Fig. 3. Gap between upper and lower bound of spanning ratio

6 Random Gabriel Graphs

If n points are drawn uniformly and at random from the unit square $[0,1]^2$, the spanning ratio of the induced Gabriel graph grows unbounded in probability. In particular, we have the following.

Theorem 4. *If n points are drawn uniformly and at random from the unit square $[0,1]^2$, and S is the spanning ratio of the induced Gabriel graph then*

$$\mathbf{P}\left\{S < c\sqrt{\frac{a\log n}{\log\log n}}\right\} \le 2e^{-2n^{1-12a-o(1)}}$$

for constants c and $a < 1/12$. Thus, for $a < 1/12$, with probability tending exponentially quickly to one,

$$S \ge c\sqrt{a\log n/\log\log n}\ .$$

Proof. The main idea is to show that a set of n points randomly distributed in the unit square contains many tower-like structures of size $c\log n/\log\log n$ each of which has spanning ratio approximately the square root of its size. We first define what a tower-like structure is and then show that the expected number of such structures is large.

A tower-like structure resembles the towers of section 3 but the points may be slightly perturbed. For $i = 1, ..., k$, let A_i and B_i be discs both of radius d/k (the constant d will be specified later) located at (r_i, y_i) and $(-r_i, y_i)$ respectively, where the sequences r_i and y_i are given below.

$$r_i = 1 - \frac{i-1}{2k}$$

$$y_i = (i-1)\sqrt{\frac{1/2-(1+\sqrt{2})d}{k}\left(1-\frac{1/2-(1+\sqrt{2})d}{k}\right)}\ .$$

The value of d is chosen so that y_i is positive ($d < 1/(2+2\sqrt{2})$).

Let C be the smallest square enclosing the A_i and B_i within a border of width y_k. (So C extends from $(-3y_k/2-d/k, -y_k-d/k)$ to $(3y_k/2+d/k, 2y_k+d/k)$.)

Assume that each of the A_i and B_i contain exactly one point and C contains no other data point beyond these $2k$ points. We claim that among the points in C, the only edges are those connecting A_1 with A_2, A_2 with A_3, and so forth, up to A_{k-1} and A_k. Then A_k connects with B_k, B_k with B_{k-1} and so forth down to B_1. The proof of this claim is rather technical and can be found in the full version of the paper. Note that the A_i's and B_i's are disjoint.

Let u and v be the points in A_1 and B_1 respectively. We have $D(u,v) \le 2 + 2d/k$. Also, any path from u to v entirely in C must be equal in length to the chain, which is longer than $2y_k$. If the path leaves C, then at least two edges leave C, and those edges have length at least $2y_k$, taken together. Thus, $L(u,v) \ge 2y_k$. Therefore,

$$S \ge \frac{L(u,v)}{D(u,v)} \ge \frac{y_k}{1+d/k} \ge c\sqrt{k}$$

for sufficiently large k where c is a constant that depends on d.

Let bC denote the scaled down set $\{bx : x \in C\}$.

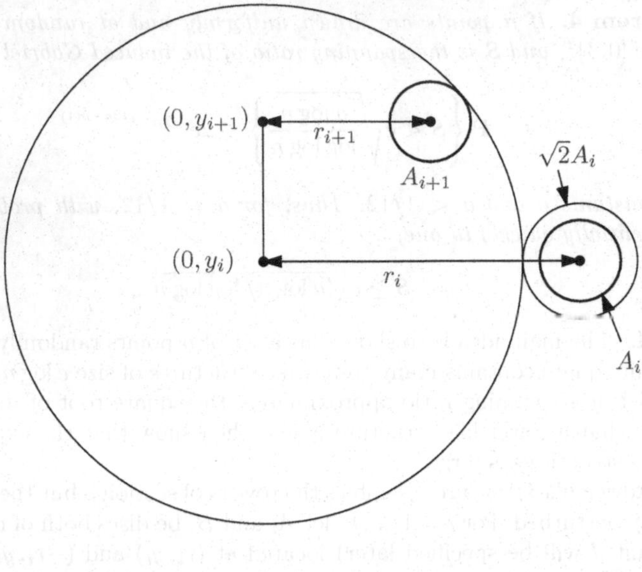

Fig. 4. The construction of A_i and A_{i+1}.

Divide $[0,1]^2$ into n non-overlapping *tiles* of size $1/\sqrt{n} \times 1/\sqrt{n}$. For $b = 1/(4\sqrt{kn})$, bC fits within one of these tiles. Thus we may place n non-overlapping copies of bC within the unit square. For a given data set, we call a tile *tower-like* if it contains exactly $2k$ data points, one each for bA_i and bB_i, $1 \le i \le k$ within it. Let N be the number of tiles that are tower-like.

Clearly, since the distribution is uniform,

$$\mathbf{E}N = n\mathbf{P}\{\text{a tile is tower-like}\}\ .$$

Pick one tile and partition the n data points over the following disjoint sets: the bA_i's, the bB_i's, $bC - \cup bA_i \cup bB_i$, and $[0,1]^2 - bC$. The cardinalities of these sets, taken together, form a multinomial random vector with probabilities given by the areas of the sets involved. For example, area $(bA_i) = b^2\pi d^2/k^2$. According to the formula for the multinomial distribution,

$$\begin{aligned}\mathbf{P}\{\text{a tile is tower-like}\} &= \frac{n!}{(n-2k)!}\left(\frac{b^2\pi d^2}{k^2}\right)^{2k}(1-1/n)^{n-2k}\\ &\ge (n-2k+1)^{2k}\left(\frac{\pi d^2}{16nk^3}\right)^{2k}(1-1/n)^n\\ &\ge \frac{1}{4}\left(\frac{(n-2k+1)\pi d^2}{16nk^3}\right)^{2k}\\ &\ge \frac{1}{4}\left(\frac{\pi d^2}{32k^3}\right)^{2k}\end{aligned}$$

provided that n is sufficiently large and $k < (n+2)/4$. We conclude that

$$\mathbf{E}N \geq \frac{n}{4}\left(\frac{\pi d^2}{32k^3}\right)^{2k} .$$

If $k = a\log n/\log\log n$ for a constant $a < 1/6$, then

$$\mathbf{E}N \geq n^{1-6a-o(1)} \to \infty .$$

For each one of these tower-like squares, there is a pair of data points for which the spanning ratio is at least

$$c\sqrt{k} \geq c\sqrt{\frac{a\log n}{\log\log n}} .$$

Change one of the n data points. That will change the number N by at most one. But then, by McDiarmid's inequality [14], we have

$$\mathbf{P}\{|N - \mathbf{E}N| \geq t\} \leq 2e^{-2t^2/n} .$$

In particular, for fixed $\epsilon > 0$,

$$\mathbf{P}\{|N - \mathbf{E}N| \geq \epsilon\mathbf{E}N\} \leq 2e^{-2\epsilon^2 n^{1-12a-o(1)}} \to 0$$

when $a < 1/12$. This shows that $N/\mathbf{E}N \to 1$ in probability for such a choice of a (and thus k), and thus that for every $\epsilon > 0$,

$$\mathbf{P}\{N < (1-\epsilon)\mathbf{E}N\} \to 0 .$$

As another application, we have

$$\begin{aligned}\mathbf{P}\{S < c\sqrt{a\log n/\log\log n}\} &\leq \mathbf{P}\{N = 0\}\\ &= \mathbf{P}\{N - \mathbf{E}N \leq -\mathbf{E}N\}\\ &\leq 2e^{-2n^{1-12a-o(1)}}\\ &\to 0 .\end{aligned}$$

Note that this probability decreases exponentially quickly with n. ■

Remark 1. We have implicitly shown several other properties of random Gabriel graphs. For example, a Gabriel graph partitions the plane into a finite number of polygonal regions. The outside polygon which extends to ∞ is excluded. Let D_n be the maximal number of vertices in these polygons. Then $D_n \to \infty$ in probability, because D_n is larger than the maximal size of any tower that occurs in the point set, and this was shown to diverge in probability. From what transpired above, this is bounded from below in probability by $\Omega(a\log n/\log\log n)$. □

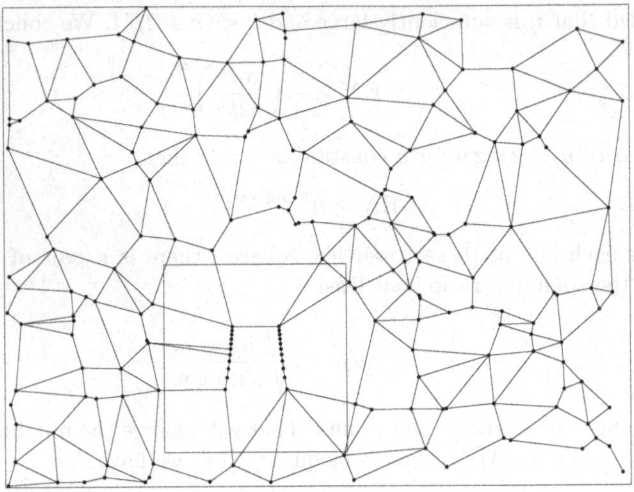

Fig. 5. Gabriel graph with tower-like square.

	$\beta = 0$	$0 < \beta < 1$	$\beta = 1$	$1 < \beta < 2$	$\beta = 2$	$\beta > 2$
Lower Bound	1	$\Omega(n^c)$[6]	$\Omega(\sqrt{n})$	$\Omega(\sqrt{n})$	$\Omega(n)$	∞
Upper Bound	1	$O(n^\gamma)$	$O(\sqrt{n})$	$O(n)$	$O(n)$	∞

$c = \log_5(5/(3 + 2\sin(\theta)))$ and $\theta < (\pi - \sin^{-1}(\beta))/2$.
$\gamma = (1 - \log_2(1 + \sqrt{1 - \beta^2}))/2$.

Table 1. Summary of Results on the Spanning Ratio of β-skeletons

7 Conclusion

We studied the spanning ratio of β-skeletons with β ranging from 0 to 2. This class of proximity graphs includes the Gabriel graph and the relative neighborhood graph. Table 1 summarizes our results. For $\beta > 2$, β-skeletons lose connectivity; thus, their spanning ratio leaps to infinity. For points drawn independently from the uniform distribution on the unit square, we showed that the spanning ratio of the (random) Gabriel graph (and all β-skeletons with $\beta \in [1, 2]$) tends to ∞ in probability as $\sqrt{\log n / \log\log n}$

Several open problems arise from this investigation. It would be interesting to close the gap between upper and lower bounds for β-skeletons in the ranges $0 < \beta < 1$ and $1 < \beta < 2$. Also, for random point sets, it would be interesting to try to find a matching upper bound for the spanning ratio.

References

1. J.A. Bondy and U.S.R. Murty. *Graph theory with applications.* North Holland, 1976.
2. L.P. Chew. There is a planar graph almost as good as the complete graph. In *Proceedings of the 2nd Annual ACM Symposium on Computional Geometry*, pages 169–177, 1986.
3. L.P. Chew. There are planar graphs almost as good as the complete graph. *Journal of Computers and Systems Sciences*, 39:205–219, 1989.
4. L.Devroye. The expected size of some graphs in computational geometry. *Computers and Mathematics with Applications*, 15:53–64, 1988.
5. D.P. Dobkin, S.J. Friedman, and K.J. Supowit. Delaunay graphs are almost as good as complete graphs. In *Proceedings of the 28th Annual Symposium on the Foundations of Computer Science*, pages 20–26, 1987. Also in *Discrete and Computational Geometry*, vol. 5, pp. 399–407, 1990.
6. D. Eppstein. Beta-skeletons have unbounded dilation, Tech. Report 96-15, Dept. of Comp. Sci, University of California, Irvine, 1996.
7. D. Eppstein. Spanning trees and spanners, Handbook of Computational Geometry (J. Sack and J. Urrutia eds.), North Holland, pp. 425–462, 2000.
8. K.R. Gabriel and R.R. Sokal. A new statistical approach to geographic variation analysis. *Systematic Zoology*, 18:259–278, 1969.
9. J. W. Jaromczyk and G. T. Toussaint. Relative neighborhood graphs and their relatives. *Proceedings of the IEEE,* 80(9), pp. 1502-1517, 1992.
10. J.M. Keil and C.A. Gutwin. The Delaunay triangulation closely approximates the complete Euclidean graph. In *Proc. 1st Workshop Algorithms Data Struct.*, volume 382 of *Lecture Notes in Computer Science*, pages 47–56. Springer-Verlag, 1989.
11. J.M. Keil and C.A. Gutwin. Classes of graphs which approximate the complete Euclidean graph. *Discrete and Computational Geometry*, 7:13–28, 1992.
12. D. G. Kirkpatrick and J. D. Radke. A framework for computational morphology. *Computational Geometry, G. T. Toussaint,* Elsevier, Amsterdam, 217-248, 1985.
13. D.W. Matula and R.R. Sokal. Properties of gabriel graphs relevant to geographic variation research and the clustering of points in the plane. *Geographical Analysis*, 12:205–222, 1980.
14. C. McDiarmid. On the method of bounded differences. *Surveys in Combinatorics*, 141:148–188, 1989.
15. F. P. Preparata and M. I. Shamos, *Computational geometry - an introduction.* Springer-Verlag, New York, 1985.
16. G. T. Toussaint. The relative neighborhood graph of a finite planar set. *Pattern Recognition*, 12: 261-268, 1980.

In-Place Planar Convex Hull Algorithms*

Hervé Brönnimann[1], John Iacono[1], Jyrki Katajainen[2], Pat Morin[3], Jason Morrison[4], and Godfried Toussaint[3]

[1] CIS, Polytechnic University, Six Metrotech, Brooklyn, New York, 11201.
{hbr,jiacono}@poly.edu

[2] Department of Computing, University of Copenhagen, Universitetsparken 1, DK-2100 Copenhagen East, Denmark.
jyrki@diku.dk

[3] SOCS, McGill University, 3480 University St., Suite 318, Montréal, Québec, CANADA, H3A 2A7.
{morin,godfried}@cgm.cs.mcgill.ca

[4] School of Computer Science, Carleton University, 1125 Colonel By Dr., Ottawa, Ontario, CANADA, K1S 5B6.
morrison@cs.carleton.ca

Abstract. An in-place algorithm is one in which the output is given in the same location as the input and only a small amount of additional memory is used by the algorithm. In this paper we describe three in-place algorithms for computing the convex hull of a planar point set. All three algorithms are optimal, some more so than others...

1 Introduction

Let $S = \{S[0], \ldots, S[n-1]\}$ be a set of n distinct points in the Euclidean plane. The *convex hull* of S is the minimal convex region that contains every point of S. From this definition, it follows that the convex hull of S is a convex polygon whose vertices are points of S. For convenience, we say that a point p is "on the convex hull of S" if p is a vertex of the convex hull of S.

As early as 1973, Graham [13] gave a convex hull algorithm with $O(n \log n)$ worst-case running time. Shamos [32] later showed that, in any model of computation where sorting has an $\Omega(n \log n)$ lower bound, every convex hull algorithm must require $\Omega(n \log n)$ time for some inputs. Despite these matching upper and lower bounds, and probably because of the many applications of convex hulls, a number of other planar convex hull algorithms have been published since Graham's algorithm [1, 2, 4, 6, 11, 17, 27, 28, 21, 35].

Of particular note is the "Ultimate(?)" algorithm of Kirkpatrick and Seidel [21] that computes the convex hull of a set of n points in the plane in $O(n \log h)$ time, where h is the number of vertices of the convex hull. The same authors show that, on algebraic decision trees of any fixed order, $\Omega(n \log h)$ is a lower

* This research was partly funded by the National Science Foundation, the Natural Sciences and Engineering Research Council of Canada and the Danish Natural Science Research Council under contract 9801749 (project Performance Engineering).

S. Rajsbaum (Ed.): LATIN 2002, LNCS 2286, pp. 494–507, 2002.

bound for computing convex hulls of sets of n points having convex hulls with h vertices.

Because of the importance of planar convex hulls, it is natural to try and improve the running time and storage requirements of planar convex hull algorithms. In this paper, we focus on reducing the intermediate storage used in the computation of planar convex hulls. In particular, we describe in-place and *in situ* algorithms for computing convex hulls. These are algorithms in which the input points are given as an array and the output, namely the vertices of the convex hull sorted in order of appearance on the hull, is returned in the same array. During the execution of the algorithm, additional working storage is kept to a minimum. In the case of in-place algorithms, the extra storage is kept in $O(1)$ while *in situ* algorithms allow an extra memory of size $O(\log n)$. After execution of the algorithm, the array contains exactly the same points, but in a different order.

In-place and *in situ* algorithms have several practical advantages over traditional algorithms. Primarily, because they use only a small additional amount of storage, they allow for the processing of larger data sets. Any algorithm that uses separate input and output arrays will, by necessity, require enough memory to store $2n$ points. In contrast, an *in situ* or in-place algorithm needs only enough memory to store n points plus $O(\log n)$ or $O(1)$ working space, respectively. Related to this is the fact that *in situ* and in-place algorithms usually exhibit greater locality of reference, which makes them very practical for implementation on modern computer architectures with memory hierarchies. A final advantage of *in situ* algorithms, especially in mission critical applications, is that they are less prone to failure since they do not require the allocation of large amounts of memory that may not be available at run time.

We describe three planar convex hull algorithms. The first is in-place, uses Graham's scan in combination with an in-place sorting algorithm, and runs in $O(n \log n)$ time. The second algorithm runs in $O(n \log h)$ time, is *in situ* and is based on an algorithm of Chan *et al.* [4]. The third ("More Ultimate?") algorithm is based on an algorithm of Chan [3], runs in $O(n \log h)$ time and is in-place. The first two algorithms are simple, implementable, and efficient in practice. To justify this claim, we have implemented both algorithms and made the source code freely available [25].

To the best of our knowledge, this paper is the first to study the problem of computing convex hulls using *in situ* and in-place algorithms. This seems surprising, given the close relation between planar convex hulls and sorting, and the large body of literature on *in situ* sorting and merging algorithms [7, 8, 9, 10, 12, 15, 16, 19, 20, 18, 23, 26, 30, 33, 34, 36].

The remainder of the paper is organized as follows: Sections 2, 3 and 4 describe our first, second and third algorithms, respectively. Section 5 summarizes and concludes with open problems.

2 An $O(n \log n)$ Time Algorithm

In this section, we present a simple in-place implementation of Graham's convex hull algorithm [13] or, more precisely, Andrew's modification of Graham's algorithm [1]. The algorithm requires the use of an in-place sorting algorithm. This can be any efficient in-place sorting algorithm (see, for example, [18, 36]), so we refer to this algorithm simply as InPlace-Sort.

Because this algorithm is probably the most practically relevant algorithm given in this paper, we begin by describing the most conceptually simple version of the algorithm, and then describe a slightly more involved version that improves the constants in the running time.

2.1 The Basic Algorithm

The *upper convex hull* of a point set S is the convex hull of $S \cup \{(0, -\infty)\}$. The *lower convex hull* of a point set S is the convex hull of $S \cup \{(0, \infty)\}$. It is well-known that the convex hull of a point set is the intersection of its upper and lower convex hulls [29]. Graham's scan computes the upper (or lower) hull of an x-monotone chain incrementally, storing the partially computed hull on a stack. The addition of each new point involves removing zero or more points from the top of the stack and then pushing the new point onto the top of the stack.

The following pseudo-code uses the InPlace-Sort algorithm and Graham's scan to compute the upper or lower hull of the point set S. The parameter d is used to determine whether the upper or lower hull is being computed. If $d = 1$, then InPlace-Sort sorts the points by increasing order of lexicographic (x, y)-values and the upper hull is computed. If $d = -1$, then InPlace-Sort sorts the points by decreasing order and the lower hull is computed. The value of h corresponds to the number of elements on the stack.

In the following, and in all remaining pseudo-code, $S = S[0], \ldots, S[n-1]$ is an array containing the input points. We use the C pointer notation $S + i$ to denote the array $S[i], \ldots, S[n-1]$.

Graham-InPlace-Scan(S, n, d)

1: Graham-InPlace-Sort(S, n, d)
2: $h \leftarrow 1$
3: **for** $i \leftarrow 1 \ldots n-1$ **do**
4:     **while** $h \geq 2$ **and not** right_turn($S[h-2], S[h-1], S[i]$) **do**
5:         $h \leftarrow h - 1$ { pop top element from the stack }
6:     **end while**
7:     **swap** $S[i] \leftrightarrow S[h]$
8:     $h \leftarrow h + 1$
9: **end for**
10: **return** h

It is not hard to verify that when the algorithm returns in Line 10, the elements of S that appear on the upper (or lower) convex hull are stored in $S[0], \ldots, S[h-1]$. In the case of an upper hull computation ($d = 1$), the hull vertices are sorted left-to-right (clockwise), while in the case of a lower hull computation ($d = -1$), the hull vertices are sorted right-to-left (also clockwise).

To compute the convex hull of the point set S, we proceed as follows (refer to Fig. 1): First we make a call to GRAHAM-INPLACE-SCAN to compute the vertices of the upper hull of S and store them in clockwise order at positions $S[0], \ldots, S[h-1]$. It follows that $S[0]$ is the bottommost-leftmost point of S and that $S[h-1]$ is the topmost-rightmost point of S. We then use $h-1$ swaps to bring $S[0]$ to position $S[h-1]$ while keeping the relative ordering of $S[1], \ldots S[h-1]$ unchanged. Finally, we make a call to GRAHAM-INPLACE-SCAN to compute the lower convex hull of $S[h-2], \ldots, S[n-1]$ (which is also the lower convex hull of S). This stores the vertices of the lower convex hull in $S[h-2], \ldots, S[h+h'-2]$ in clockwise order. The end result is that the convex hull of S is stored in $S[0], \ldots, S[h+h'-2]$ in clockwise order.

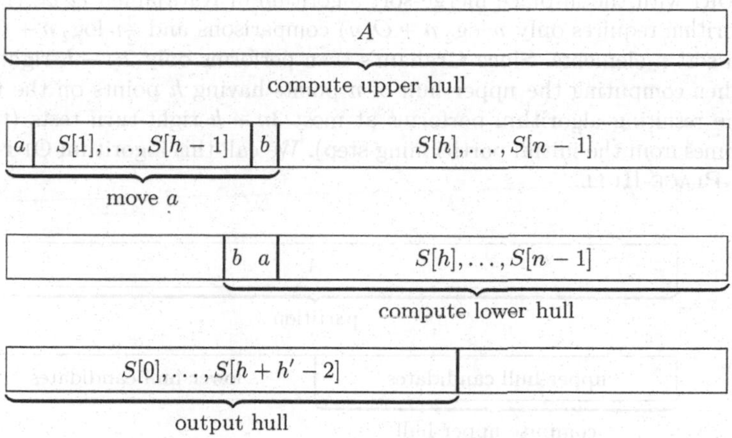

Fig. 1. The execution of the GRAHAM-INPLACE-HULL algorithm.

The following pseudo-code gives a more precise description of the algorithm.

GRAHAM-INPLACE-HULL(S, n)

1: $h \leftarrow$ GRAHAM-INPLACE-SCAN($S, n, 1$)
2: **for** $i \leftarrow 0 \ldots h-2$ **do**
3:    **swap** $S[i] \leftrightarrow S[i+1]$
4: **end for**
5: $h' \leftarrow$ GRAHAM-INPLACE-SCAN($S+h-2, n-h+2, -1$)
6: **return** $h + h' - 2$

Each call to GRAHAM-INPLACE-SCAN executes in $O(n \log n)$ time, and the loop in lines 2–4 takes $O(h)$ time. Therefore, the total running time of the algorithm is $O(n \log n)$. The amount of extra storage used by InPlace-Sort is $O(1)$, as is the storage used by both our procedures.

Theorem 1 *Algorithm* Graham-InPlace-Hull *computes the convex hull of a set of n points in* $O(n \log n)$ *time using* $O(1)$ *additional memory.*

2.2 The Optimized Algorithm

The constants in the running time of Graham-InPlace-Hull can be improved by first finding the extreme points a and b and using these points to partition the array into two parts, one that contains vertices that can only appear on the upper hull and one that contains vertices that can only appear on the lower hull. Fig. 2 gives a graphical description of this. In this way, each point (except a and b) takes part in only one call to Graham-InPlace-Scan.

To further reduce the constants in the algorithm, one can implement InPlace-Sort with the in-place merge-sort algorithm of Katajainen *et al.* [18]. This algorithm requires only $n \log_2 n + O(n)$ comparisons and $\frac{3}{2} n \log_2 n + O(n)$ swaps to sort n elements. Since Graham's scan performs only $2n - h$ right-turn tests when computing the upper hull of n points having h points on the upper hull, the resulting algorithm performs at most $3n - h$ right-turn tests (the extra n comes from the initial partitioning step). We call this algorithm Opt-Graham-InPlace-Hull.

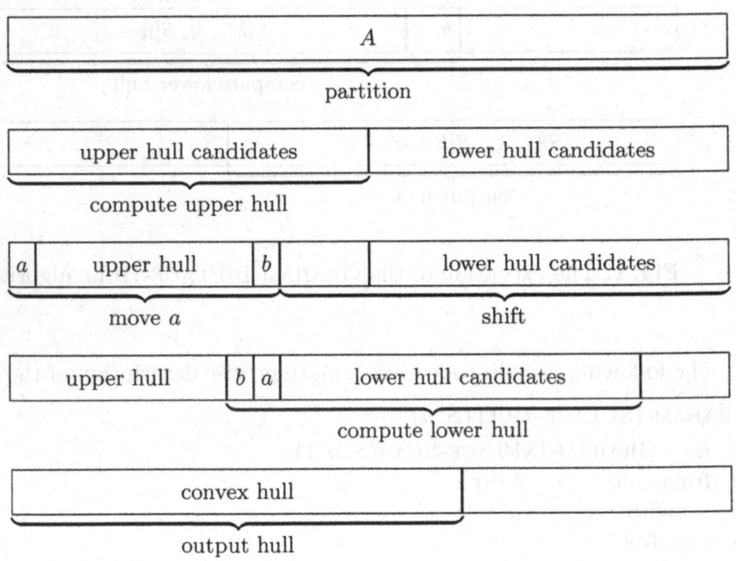

Fig. 2. A faster implementation of Graham-InPlace-Hull.

Theorem 2 OPT-GRAHAM-INPLACE-HULL *computes the convex hull of* n *points in* $O(n \log n)$ *time using at most* $3n - h$ *right turn tests,* $\frac{3}{2} n \log_2 n + O(n)$ *swaps,* $n \log_2 n + O(n)$ *lexicographic comparisons and* $O(1)$ *additional memory, where* h *is the number of vertices of the convex hull.*

Finally, we note that if the array A is already sorted in lexicographic order then no lexicographic comparisons are necessary. One can use an in-place stable partitioning algorithm to partition A into the set of upper hull candidates and the set of lower hull candidates while preserving the sorted order within each set. There exists such an algorithm that runs in $O(n)$ time and perform $O(n)$ comparisons [19]. We call this algorithm SORTED-GRAHAM-INPLACE-HULL

Theorem 3 SORTED-GRAHAM-INPLACE-HULL *computes the convex hull of* n *points given in lexicographic order in* $O(n)$ *time using* $O(n)$ *right turn tests,* $O(n)$ *swaps, no lexicographic comparisons and* $O(1)$ *additional memory.*

3 An $O(n \log h)$ Time Recursive Algorithm

In this section, we show how to compute the upper (and symmetrically, lower) hull of S in $O(n \log h)$ time using an *in situ* algorithm, where h is the number of points of S that on the upper (respectively, lower) hull of S. We begin with a review of the $O(n \log h)$ time algorithm of Chan *et al.* [4].

To compute the upper hull of a point set S, we begin by arbitrarily grouping the elements of S into $\lfloor n/2 \rfloor$ pairs. From these pairs, the pair with median slope s is found using a linear time median finding algorithm.[1] We then find a point $p \in S$ such that the line through p with slope s has all points of S below it.

Let $q.x$ denote the x coordinate of the point q and let $\bar{i}$ denote the index of the element that is paired with $S[i]$. We now use p, and our grouping to partition the elements of S into three groups S^0, S^1, and S^2 as follows (see Fig. 3):

$$S[i] \in \begin{cases} S^0 \text{ if } S[i].x \le p.x \text{ and } (S[\bar{i}], p) \text{ is not above } S[i] \\ S^1 \text{ if } S[i].x > p.x \text{ and } (S[\bar{i}], p) \text{ is not above } S[i] \\ S^2 \text{ otherwise} \end{cases}$$

The algorithm then recursively computes the upper hull of $S^0 \cup \{p\}$ and $S^1 \cup \{p\}$ and outputs the concatenation of the two. For a discussion of correctness and a proof that this algorithm runs in $O(n \log h)$ time, see the original paper [4].

Now we turn to the problem of making this an *in situ* algorithm. The choice of median slope s ensures that $S^0 \le 3n/4$ and $S^1 \le 3n/4$, so the algorithm uses only $O(\log n)$ levels of recursion. Our strategy is to implement each level using $O(1)$ local variables and one call to a median-finding routine that uses $O(\log n)$ additional memory.

[1] Bhattacharya and Sen [2] and Wenger [35] have both noted that median finding can be replaced by choosing a random pair of elements. The expected running time of the resulting algorithm is $O(n \log h)$.

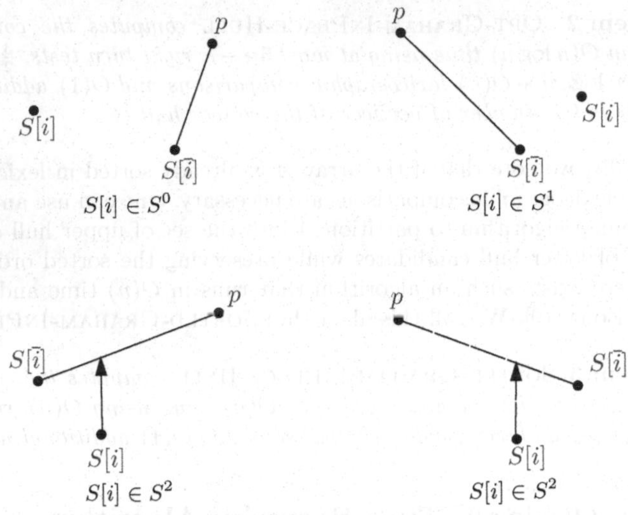

Fig. 3. Partitioning S into S^0, S^1 and S^2.

For simplicity, assume n is even. The case when n is odd is easily handled by processing an extra unpaired element after all the paired elements have been processed. To pair off elements, we pair $S[i]$ with $S[i+1]$ if i is even and with $S[i-1]$ if i is odd. Several *in situ* linear time median finding algorithms exist (see, for example, Horowitz *et al.* [14, Section 3.6] or Lai and Wood [22]), so finding the pair $(S[i], S[i+1])$ with median slope can be done with one of these. The tricky part of the implementation is the partitioning into sets S^0, S^1 and S^2. The difficulty lies in the fact that the elements are grouped into pairs, but the two elements of the same pair may belong to different sets S^i and S^j.

We first note that we can compute the sizes n_0, n_1 and n_2 of these sets in linear time without difficulty by scanning S. Conceptually, we partition S into three *files*, f_0, f_1 and f_2 that contain pairs of points in S. The file f_0 contains the elements $S[0], \ldots, S[2\lfloor n_0/2 \rfloor - 1]$. The file f_1 contains the elements $S[2\lfloor n_0/2 \rfloor], \ldots, S[2\lfloor (n_0+n_1)/2 \rfloor - 1]$. The file f_2 contains the elements $S[2\lfloor (n_0+n_1)/2 \rfloor], \ldots, S[n]$.

It is important to note that these files are only abstractions. Each file f_i is implemented using two integer values r_i and ϕ_i. The value of r_i is initialized to the index of the first record in the file. The value of ϕ_i is initialized to $r_i + k_i$ where k_i is the number of elements in f_i. A READ operation on f_i returns the pair $(S[r_i], S[r_{i+1}])$ and increases the value of r_i by 2. We say that f_i is *empty* if $r_i \geq \phi_i$.

These files are used in conjunction with a stack A that stores pairs of points. The stack and files serve two purposes: (1) when there is no data on the stack we read a pair from one of the files and store it on the stack, and (2) when we are about to overwrite an element from a pair that has not yet been placed on the stack, we read the pair from the file and save it on the stack. In this way no element is ever overwritten without first being saved on the stack, and the initial pairing of elements is preserved.

These ideas are made more concrete by the following pseudo-code, which places the elements of S^0 into array locations $S[0], \ldots, S[n_0 - 1]$, the elements of S^1 into array locations $S[n_0], \ldots, S[n_0 + n_1 - 1]$, and the elements of S^2 into array locations $S[n_0+n_1], \ldots, S[n-1]$. The algorithm repeatedly processes pairs (a, b) of elements by determining which of the three sets a and b belong to and then placing a and b in their correct locations.

```
CSY-PARTITION(S, n, n0, n1)
i0 ← 0
i1 ← n0
i2 ← n0 + n1
m ← 0
while m > 0 or one of f0, f1, f2 is not empty do
  if m = 0 then
    A[m] ← READFROMFILE()
    m ← m + 1
  end if
  m ← m − 1
  P ← A[m] { process this pair }
  for both q ∈ P do
    S^j ← GROUP(q, P)
    PLACE(q, i_j)
    i_j ← i_j + 1
  end for
end while
```

The READFROMFILE function simply reads a pair from one of the non-empty files and returns it. The GROUP(q, P) returns (a pointer to) the group of point q in the pair P. The PLACE(q, k) function places the point q at index k in S, after ensuring that the overwritten element has been read and placed on the stack.

```
PLACE(q, k)
for i ← 0, 1, 2 do
  if k ≥ r_i and k < φ_i then
    { S[k] belongs to f_i and has not yet been read }
    READ(a, b) from f_i
    A[m] ← (a, b)
    m ← m + 1
  end if
end for
S[k] ← q
```

To show that this partitioning step is correct, we make 2 observations. (1) Exactly $n/2$ pairs of elements are read and processed since the file abstraction ensures that no pair is read more than once and the algorithm does not terminate until all files are empty. (2) The code in PLACE ensures that any pair is read and placed on the stack A before an element of the pair is overwritten. Therefore, all of the original $n/2$ pairs of elements are processed and each element is placed into one of S^0, S^1 or S^2.

Since the algorithm uses a stack A that may grow without bound, it is not obvious that the partitioning algorithm's additional memory is of a constant size. To prove that A does not grow without bound note that overwriting k_i elements of f_i causes at most $\lceil k_i/2 \rceil$ read operations. Each iteration of the outer loop places one pair of elements, and each read operation reads one pair of elements. Therefore, the total number of read operations performed after k iterations is at most $k+3$. However, each iteration removes 1 pair of elements from the stack A, so the total number of pairs on the stack after k iterations is not more than 3. Since this holds for any value of k, the stack A never holds more than 3 pairs of elements.

Fig. 4 recaps the algorithm for computing the upper hull of S. First the algorithm partitions S into the sets S^0, S^1 and S^2. It then recurses on the set S^0. After the recursive call, the convex hull of S^0 is stored at the beginning of the array S, and the last element of this hull is the point p that was used for partitioning. The algorithm then shifts S^1 leftward so that it is adjacent to p and recurses on $S^1 \cup \{p\}$. The end result is the upper hull of S being stored consecutively and in clockwise order at the beginning of the array S.

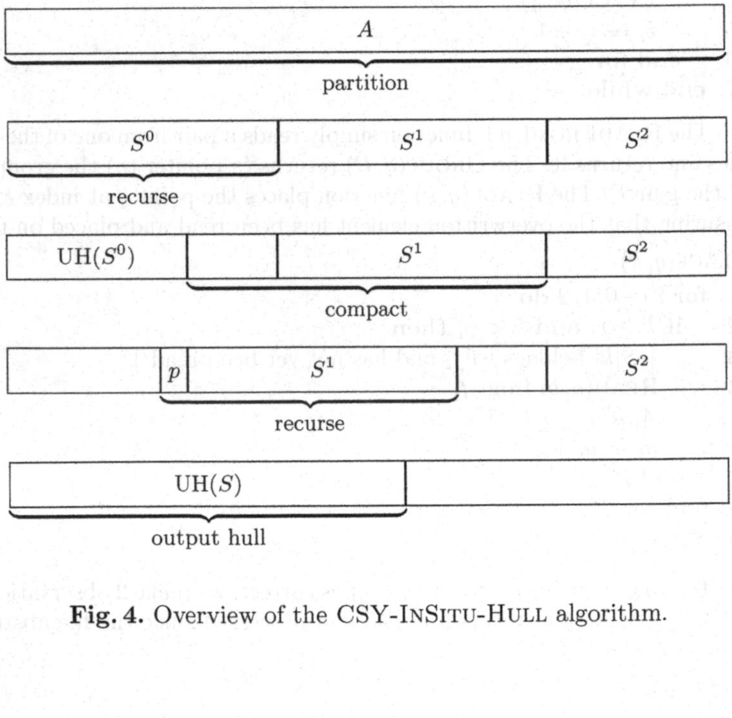

Fig. 4. Overview of the CSY-INSITU-HULL algorithm.

Using the technique from Section 2 (Figures 1 and 2), this upper hull algorithm can be made in to a convex hull algorithm with the same running time and memory requirements.

Theorem 4 *Algorithm* CSY-InSitu-Hull *computes the convex hull of n points in $O(n \log h)$ time using $O(\log n)$ additional storage, where h is the number of vertices of the convex hull.*

4 An $O(n \log h)$ Time Iterative Algorithm

Next, we give an $O(n \log h)$ time in-place planar convex hull algorithm. We begin with a review of Chan's $O(n \log h)$ time algorithm [3], which is essentially a speedup of Jarvis' March [17].

Chan's algorithm runs in rounds. During ith round the algorithm finds the first $g_i = 2^{2^i}$ points on the convex hull. Once $g_i \geq h$ the rounds end as the algorithm detects that it has found all points on the convex hull. During round i, the algorithm partitions the input points into n/g_i groups of size g_i and computes the convex hull of each group. The vertices on the convex hull are output in clockwise order beginning with the leftmost vertex. Each successive vertex is obtained by finding tangents from the previous vertex to each of the n/g_i convex hulls. The next vertex is determined, as in Jarvis' March, by choosing the vertex having largest polar angle with respect to the previously found vertex as origin. In the case where the largest polar angle is not unique, ties are broken by taking the farthest vertex from the previously found vertex.

Finding a tangent to an individual convex hull can be done in $O(\log g_i)$ time if the vertices of the convex hull are stored in an array in clockwise order [5, 29]. There are n/g_i tangent finding operations per iteration and g_i iterations in round i. Therefore, round i takes $O(n \log g_i) = O(n2^i)$ time. Since there are at most $\lceil \log \log h \rceil$ rounds, the total cost of Chan's algorithm is $\sum_{i=1}^{\lceil \log \log h \rceil} O(n2^i) = O(n \log h)$.

Next we show how to implement each round using only $O(1)$ additional storage. Assume for the sake of simplicity that n is a multiple of g_i. For the grouping step, we build n/g_i groups of size g_i by taking groups of consecutive elements in S and computing their convex hulls using Graham-InPlace-Hull. Two questions now arise: (1) Once we start the tangent-finding steps, where do we put the convex hull vertices as we find them? (2) In order to find a tangent from a point to a group in $O(\log g_i)$ time we need to know the size of the convex hull of the group. How can we keep track of all these sizes using only $O(1)$ extra memory?

To answer the first question, we store convex hull vertices at the beginning of the array S in the order that we find them. That is, when we find the kth vertex on the convex hull, we swap it with $S[k-1]$. We say that a group G is *dirty* if one of its members has been found to be on the convex hull of S. A group that is not dirty is *clean*. If a group G is clean, then we can use binary search to find a tangent to G in $O(\log g_i)$ time, otherwise we have to use linear search which takes $O(g_i)$ time. Since after k iterations, the algorithm stores the first

k hull vertices in locations $S[0], \ldots S[k-1]$, the first group consists of elements $S[k], \ldots, S[g_i - 1]$ and is always considered dirty.

To keep track of which other groups are dirty, we mark them by reordering the first two points of the group. In a clean group, the points are stored in lexicographic order. In a dirty group, we store them in reverse lexicographic order. This allows us to test in constant time whether a group is clean or dirty.

To keep track of the size of the convex hull of each clean group without storing the size explicity we use another reordering trick. Let $G[0], \ldots, G[g_i - 1]$ denote the elements of a clean group G and let $<$ denote lexicographic comparison of (x, y) values. We say that the *sign* of $G[j]$ is $+$ if $G[j] < G[j+1]$, and $-$ otherwise. If the convex hull of G contains h vertices, then it follows that the first elements $G[0], \ldots, G[h-2]$ have signs that form a sequence of 1 or more $+$'s followed by 0 or more $-$'s. Furthermore, the elements $G[h], \ldots, G[g_i - 1]$ can be reordered so that the remainder of the signs form an alternating sequence. When we do this, a group element $G[j]$, $0 < j < g_i - 1$, $j \neq h-1$ is on the convex hull of G if and only if $G[j-1]$, $G[j]$, $G[j+1]$ do not have signs that alternate.

As for $G[0]$, $G[g_i - 1]$ and $G[h-1]$ we know that $G[0]$ is always on the convex hull of G. The point $G[g_i - 1]$ is on the convex hull of G if and only if $G[g_i - 2]$ is on the convex hull of G and the triangle $G[g_i - 2], G[g_i - 1], G[0]$ is oriented clockwise.[2] The point $G[h-1]$ is on the convex hull of G if and only if $G[h-2]$ is on the convex hull of G and the triangle $G[h-2], G[h-1], G[0]$ is oriented clockwise. Therefore, for any index $0 \leq j < g_i$, we can test if $G[j]$ is on the convex hull of G in constant time. Using this in conjunction with binary search, we can compute the number of vertices on the convex hull of G in $O(\log g_i)$ time. Thus, we can compute the size of the convex hull of G and find a tangent in $O(\log g_i)$ time, as required.

We have provided all the tools for an in-place implementation of Chan's algorithm. Except for the extra cost of finding tangents in dirty groups, the running time of this implementation is asymptotically the same as that of the original algorithm, so we need only bound this extra cost of finding tangents in dirty groups. During one step of round i, we find tangents of at most g_i dirty groups, at a cost of $O(g_i)$ per group, and there are g_i steps in round i. Therefore, the total cost of searching dirty groups during round i is $O(g_i^3) \subseteq O(n)$ for all $g_i \leq n^{1/3}$. Therefore, the total cost of round i is $O(g_i^3 + n \log g_i) \subseteq O(n \log g_i)$ for any $h < n^{1/3}$. Since we can abort the algorithm when $g_i \geq n^{1/3}$ and use Graham-InPlace-Hull, the overall running time of the algorithm is again $O(n \log h)$.

Theorem 5 *The above algorithm,* Chan-InPlace-Hull, *computes the convex hull of n points in $O(n \log h)$ time using $O(1)$ additional storage, where h is the number of vertices of the convex hull.*

The constants in Chan-InPlace-Hull can be improved using the following trick that is mentioned by Chan [3]. When round i terminates without finding the entire convex hull, the g_i convex hull points that were computed should not

[2] We use the convention that three collinear points are *not* oriented clockwise.

be discarded. Instead, the grouping in round $i+1$ is done on the remaining $n-g_i$ points, thus eliminating the need to recompute the first g_i convex hull vertices. This optimization works perfectly when applied to CHAN-INPLACE-HULL since the first g_i convex hull points are already stored at locations $S[0], \ldots, S[g_i - 1]$.

5 Conclusions

We have given three algorithms for computing the convex hull of a planar point set. The first algorithm is in-place and runs in $O(n \log n)$ time. The second algorithm is *in situ* and runs in $O(n \log h)$ time. The third algorithm is in-place and and runs in $O(n \log h)$ time. The first two algorithms are reasonably simple and implementable. In order to facilitate comparisons with other convex hull implementations, our source code is available for download [25].

This paper came to be when two separate groups of researchers (authors 1–3 and authors 4–6) discovered they were both working on in-place computational geometry algorithms and decided to merge their results. Some of these results have been omitted due to space constraints. These include in-place or *in situ* implementations of Eddy's algorithm (also known as quickhull) [11], Kirkpatrick and Seidel's algorithm [21], Seidel's randomized linear programming algorithm [31] and Megiddo's deterministic linear programming algorithm [24].

The ideas presented in this paper also apply to other problems. The *maximal elements* problem is that of determining all elements $S[i]$ such that $S[j] < S[i]$ for all $0 \leq j < n$. An algorithm almost identical to Graham's scan can be used to solve the maximal elements problems in $O(n \log n)$ time, and this can easily be implemented in-place. Furthermore, an in-place algorithm almost identical to that in Section 4 can be used to solve the maximal elements problem in $O(n \log h)$ time, where h is the number of maximal elements.

The question of *in situ* and in-place algorithms for maximal elements and convex hulls in dimensions $d \geq 3$ is still open. In order for this question to make sense, we ask only that the algorithm identify which input points are maximal or on the convex hull. Testing whether a given point is maximal can be done in $O(dn)$ time using the definition of maximality. Testing whether a single point is on the convex hull is a $d-1$ dimensional linear programming problem that can be solved in-place in $O(d!n)$ expected time using Seidel's algorithm [31]. Thus, the maximal elements problem can be solved in $O(dn^2)$ time and the convex hull problem can be solved in $O(d!n^2)$ time using in-place algorithms. Are there algorithms with reduced dependence on n?

More generally, one might ask what other computational geometry problems admit in-place or *in situ* algorithms. Some problems that immediately come to mind are those of computing k-piercings of sets, finding maximum cliques in intersection graphs, computing largest empty disks, computing smallest enclosing disks, and finding ham-sandwich cuts.

References

[1] A. M. Andrew. Another efficient algorithm for convex hulls in two dimensions. *Information Processing Letters*, 9:216–219, 1979. Corrigendum, *Information Processing Letters*, 10:168, 1980.

[2] B. K. Bhattacharya and S. Sen. On a simple, practical, optimal, output-sensitive randomized planar convex hull algorithm. *Journal of Algorithms*, 25(1):177–193, 1997.

[3] T. Chan. Optimal output-sensitive convex hull algorithms in two and three dimensions. *Discrete & Computational Geometry*, 16:361–368, 1996.

[4] T. Chan, J. Snoeyink, and C. K. Yap. Primal dividing and dual pruning: Output-sensitive construction of four-dimensional polytopes and three-dimensional Voronoi diagrams. *Discrete & Computational Geometry*, 18:433–454, 1997.

[5] B. Chazelle and D. P. Dobkin. Intersection of convex objects in 2 and 3 dimensions. *Journal of the ACM*, 34:1–27, 1987.

[6] K. L. Clarkson and P. W. Shor. Applications of random sampling in computational geometry, II. *Discrete & Computational Geometry*, 4(1):387–421, 1988.

[7] E. W. Dijkstra. Smoothsort, an alternative for sorting in situ. *Science of Computer Programming*, 1(3):223–233, 1982.

[8] E. W. Dijkstra and A. J. M. van Gasteren. An introduction to three algorithms for sorting in situ. *Information Processing Letters*, 15(3):129–134, 1982.

[9] S. Dvorak and B. Durian. Stable linear time sublinear space merging. *The Computer Journal*, 30(4):372–375, 1987.

[10] S. Dvorak and B. Durian. Unstable linear time O(1) space merging. *The Computer Journal*, 31(3):279–282, 1988.

[11] W. Eddy. A new convex hull algorithm for planar sets. *ACM Transactions on Mathematical Software*, 3(4):398–403, 1977.

[12] R. W. Floyd. Algorithm 245, Treesort 3. *Communications of the ACM*, 7:401, 1964.

[13] R. L. Graham. An efficient algorithm for determining the convex hull of a finite planar set. *Information Processing Letters*, 1:132–133, 1972.

[14] E. Horowitz, S. Sahni, and S. Rajasekaran. *Computer Algorithms*. Computer Science Press, 1998.

[15] B.-C. Huang and M. A. Langston. Practical in-place merging. *Communications of the ACM*, 31(3):348–352, 1988.

[16] B.-C. Huang and M. A. Langston. Fast stable merging and sorting in constant extra space. *The Computer Journal*, 35(6):643–650, 1992.

[17] A. Jarvis. On the identification of the convex hull of a finite set of points in the plane. *Information Processing Letters*, 2:18–21, 1973.

[18] J. Katajainen, T. Pasanen, and J. Teuhola. Practical in-place mergesort. *Nordic Journal of Computing*, 3:27–40, 1996.

[19] J. Katajainen and T. A. Pasanen. Stable minimum space partitioning in linear time. *BIT*, 32(4):580–585, 1992.

[20] J. Katajainen and T. A. Pasanen. In-place sorting with fewer moves. *Information Processing Letters*, 70(1):31–37, 1999.

[21] D. G. Kirkpatrick and R. Seidel. The ultimate planar convex hull algorithm? *SIAM Journal on Computing*, 15(1):287–299, 1986.

[22] T. W. Lai and D. Wood. Implicit selection. In *Proceedings of the 1st Scandinavian Workshop on Algorithm Theory*, volume 318 of *Lecture Notes in Computer Science*, pages 14–23. Springer-Verlag, 1988.

[23] H. Mannila and E. Ukkonen. A simple linear-time algorithm for in situ merging. *Information Processing Letters*, 18(4):203–208, 1984.

[24] N. Megiddo. Linear programming in linear time when the dimension is fixed. *Journal of the ACM*, 31(1):114–127, 1984.

[25] P. Morin. insitu.tgz. Available online at `http://cgm.cs.mcgill.ca/~morin/`, 2001.

[26] J. I. Munro, V. Raman, and J. S. Salowe. Stable in situ sorting and minimum data movement. *BIT*, 30(2):220–234, 1990.

[27] F. P. Preparata. An optimal real time algorithm for planar convex hulls. *Communications of the ACM*, 22:402–405, 1979.

[28] F. P. Preparata and S. J. Hong. Convex hulls of finite point sets in two and three dimensions. *Communications of the ACM*, 2(20):87–93, 1977.

[29] F. P. Preparata and M. I. Shamos. *Computational Geometry.* Springer-Verlag, 1985.

[30] J. Salowe and W. Steiger. Simplified stable merging tasks. *Journal of Algorithms*, 8(4):557–571, 1987.

[31] R. Seidel. Small-dimensional linear programming and convex hulls made easy. *Discrete & Computational Geometry*, 6:423–434, 1991.

[32] M. I. Shamos. *Computational Geometry.* PhD thesis, Yale University, 1978.

[33] H. W. Six and L. Wegner. Sorting a random access file in situ. *The Computer Journal*, 27(3):270–275, 1984.

[34] A. Symvonis. Optimal stable merging. *The Computer Journal*, 38(8):681–690, 1995.

[35] R. Wenger. Randomized quick hull. *Algorithmica*, 17:322–329, 1997.

[36] J. W. J. Williams. Algorithm 232, Heapsort. *Communications of the ACM*, 7:347–348, 1964.

The Level Ancestor Problem Simplified

Michael A. Bender[1]* and Martín Farach-Colton[2]**

[1] Department of Computer Science, State University of New York at Stony Brook, Stony Brook, NY 11794-4400, USA.
bender@cs.sunysb.edu

[2] Google Inc, 2400 Bayshore Parkway, Mountain View, California 94043, USA, & Department of Computer Science, Rutgers University.
martin@farach-colton.com

Abstract. We present a very simple algorithm for the *Level Ancestor Problem*. A *Level Ancestor Query* $\mathrm{LA}(v, d)$ requests the depth d ancestor of node v. The Level Ancestor Problem is thus: preprocess a given rooted tree T to answer level ancestor queries. While optimal solutions to this problem already exist, our new optimal solution is simple enough to be taught and implemented.

1 Introduction

A fundamental algorithmic problem on trees is how to find *Level Ancestors* of nodes. A *Level Ancestor Query* $\mathrm{LA}(u, d)$ requests the depth d ancestor of node u. The *Level Ancestor Problem* is thus: preprocess a given n-node rooted tree T to answer level ancestor queries. Thus, one must optimize both the preprocessing time and the query time.

The natural solution of simply climbing up the tree from u is $O(n)$ at query time, and the other solution of precomputing all possible queries has $O(n^2)$ preprocessing.

Solutions with $O(n)$ preprocessing and $O(1)$ query time were given by Dietz [8] and by Berkman and Vishkin [6], though this latter algorithm has a unwieldy constant factor[1], and the former algorithm requires fancy word tricks. A substantially simplified algorithm was given by Alstrup and Holm [1], though their main focus was on dynamic trees, rather than on simplifying LA computations.

We present an algorithm that requires no "heavy" machinery. This algorithm is suitable for teaching data structures to (advanced) undergraduates, unlike previous algorithms. It is perhaps surprising that a problem that heretofore required heavy lifting can be simplified to such an extent, and the algorithm we present is made up of such simple pieces. This last point makes this algorithm particularly suitable for teaching.

---

* Supported in part by HRL Laboratories, Sandia National Laboratories, and NSF ITR grant EIA–0112849.

** Partially supported by NSF CCR 9820879.

[1] In fact, $2^{2^{28}}$.

S. Rajsbaum (Ed.): LATIN 2002, LNCS 2286, pp. 508–515, 2002.

The remainder of the paper is organized as follows. In Section 2, we provide some definitions and initial lemmas. In Section 3, we present an algorithm for Level Ancestors that takes $O(n \log n)$ for preprocessing, and $O(1)$ time for queries. In Section 4, we show how to speed up the preprocessing to an optimal $O(n)$.

2 Definitions

We begin with some basic definitions. The *depth* of a node u in tree T, denoted $\text{depth}(u)$, is the shortest distance from u to the root. Thus, the root has depth 0. The *height* of a node u in tree T, denoted $\text{height}(u)$, is the number of nodes on the path from u to its deepest descendant. Thus, the leaves have height 1.

Let $\text{LA}_T(u, d) = v$ where v is an ancestor of u and $depth(v) = d$, if such a node exists, and `undefined` otherwise. Now we define the *Level Ancestor Problem* formally.

Problem 1. The *Level Ancestor Problem:*

Structure to Preprocess: A rooted tree T having n nodes.
Query: For node u in rooted tree T, query $\text{LevelAncestor}_T(u, d)$ returns $\text{LA}_T(u, d)$, if it exists and `false` otherwise. Thus, $\text{LevelAncestor}_T(u, 0)$ returns the root, and $\text{LevelAncestor}_T(u, \text{depth}(u))$ returns u. (When the context is clear, we drop the subscript T.)

In order to simplify the description of algorithms that have both preprocessing and query complexity, we introduce the following notation. If an algorithm has preprocessing time $f(n)$ and query time $g(n)$, we will say that the algorithm has complexity $\langle f(n), g(n) \rangle$.

One of our notational conventions, which we introduced in [2], is of independent interest.[2] We define the *hyperfloor* of x, denoted $\lfloor\lfloor x \rfloor\rfloor$, to be $2^{\lfloor \log x \rfloor}$, i.e., the largest power of two no greater than x. Thus, $x/2 < \lfloor\lfloor x \rfloor\rfloor \leq x$. Similarly, the *hyperceiling* $\lceil\lceil x \rceil\rceil$ is defined to be $2^{\lceil \log x \rceil}$.

3 An $\langle O(n \log n), O(1) \rangle$ Solution to the Level Ancestor Problem

We now present three simple algorithms for solving the Level Ancestor Problem, which we call the Table Algorithm, the Jump-Pointers Algorithm, and the Ladder Algorithm. At the end of this section we combine the two latter algorithms to obtain a solution with complexity $\langle O(n \log n), O(1) \rangle$. The Table Algorithm will be used in the faster algorithms in the next section.

[2] All logarithms are base 2 if not otherwise specified.

3.1 The Table Algorithm: An $\langle O(n^2), O(1)\rangle$ Solution

We first observe that the Level Ancestor Problem has a solution with complexity $\langle O(n^2), O(1)\rangle$: build a table storing answers to all of the at most n^2 possible queries. Answering a Level-Ancestor query requires just one table lookup.

Lemma 1. *The Table Algorithm solves the Level Ancestor Problem in time* $\langle O(n^2), O(1)\rangle$.

Proof. The lookup table can be filled in $O(n^2)$ by a simple dynamic program.

We make one more note here, which we use in the next section. In the table as described, we store the id of a node as the answer to a query. Instead, we introduce one level of indirection. We assign a depth first search (DFS) number to each node, and store these in the table. Then when we retrieve the DFS number of the answer, we look up the corresponding node in the tree. This extra level of indirection clearly does not increase the asymptotic bounds, but allows us to share preprocessing amongst different subtrees.

3.2 The Jump-Pointers Algorithm: An $\langle O(n \log n), O(\log n)\rangle$ Solution

In the Jump-Pointers Algorithm, we associate $\log n$ pointers with each vertex, which we call *jump pointers*. Jump pointers "jump" up the tree by powers of 2. Thus, there is a pointer from u to u's ℓ-th ancestor, for $\ell = 1, 2, 4, 8, \ldots, \lfloor\lfloor \text{depth}(u) \rfloor\rfloor$. We refer to these pointers as $\text{JUMP}_u[i]$, where $\text{JUMP}_u[i] = \text{LA}(u, \text{depth}(u) - 2^i)$.

We emphasize the following point:

Observation 2 *In a single pointer dereference we can travel at least halfway from u to* $\text{LA}(u, d)$*, for any d. Finding the appropriate pointer takes $O(1)$ time.*

Proof. We let $\delta = \text{depth}(u) - d$. We can travel up by $\lfloor\lfloor \delta \rfloor\rfloor$, which is at least $\delta/2$. The pointer to follow is simply $\text{JUMP}_u[\lfloor \log \delta \rfloor]$.

(Note that since the floor and log operations are word computations, the algorithm is a RAM algorithm.)

As a consequence of Observation 2, we obtain the following lemma:

Lemma 3. *The Jump-Pointers Algorithm solves the Level Ancestor Problem in time* $\langle O(n \log n), O(\log n)\rangle$.

Proof. To achieve $O(n \log n)$ preprocessing, we apply a trivial dynamic program. To answer query $\text{LEVELANCESTOR}_T(u, d)$ in $O(\log n)$ time, we repeatedly follow the pointers that will get us halfway to $\text{LA}_T(u, d)$. Therefore after at most $\log n$ jumps, we locate $\text{LA}_T(u, d)$.

3.3 The Ladder Algorithm: An $\langle O(n), O(\log n)\rangle$ Solution

In the Ladder Algorithm, we decompose the tree T into (nondisjoint) paths, which we call *ladders* because they help us climb up the tree. Our choice of how we do the decomposition may seem peculiar at first, but it is an integral part of the fast algorithms.

To understand why it is advantageous to break the tree into paths, observe that solving the level ancestor problem on a single path of length n is trivial.

Observation 4 *On a path of length n, the Level-Ancestor Problem can trivially be solved with (optimal) complexity $\langle O(n), O(1)\rangle$.*

Proof. We maintain a Ladder array LADDER$[0 \dots n-1]$, where the i-th array position corresponds to the depth-i node on the path. To answer LEVELANCESTOR$_T(u, d)$, we return LADDER$[d]$, which takes $O(1)$ time.

We now describe the *ladder decomposition* of the tree T, which proceeds in two stages: The first stage requires us to find a *long-path decomposition* of the tree T, which greedily decomposes the tree into disjoint paths.

Stage 1: Long-Path Decomposition. We greedily break T into long disjoint paths as follows. We find a longest root-leaf path in T, breaking ties arbitrarily, and remove it from the tree. This removal breaks the remaining tree into subtrees $T_1, T_2, \dots$. We recursively split these subtrees by removing their longest paths. The base case is when the tree is a single path, because the removal yields the empty tree. Note that if a node has height h, it is on a long-path with at least h nodes.

If we now apply the above ladder algorithm to each long-path, we may still have a slow algorithm. In particular, we can only jump up to the top of our long-path. Then we must step to its parent, and jump up its long path, and so forth. The time taken to reach LA(u, d) is the number of long-paths we must traverse. There can be as many as $\Theta(\sqrt{n})$ paths on one leaf-to-root walk.[3]

Stage 2: Extending the Long Paths into Ladders. The problem with the long-path climbing algorithm sketched above is that jumping to the ancestor at the top of a long-path may not help much. Since we have already allocated an array of length h' to a path of length h', we might as well allocate $2h'$. We do this by adding the h' immediate ancestors of the top of the path to the array.

We call these doubled long-paths *ladders*, and note that while ladders overlap, they still have total size at most $2n$. We say that *vertex v's ladder* is the ladder derived from the long path containing v, and note that since long-paths partition the tree, each node v has a unique ladder, but may be listed in many ladders.

Now, if a node has height h, we know that its ladder includes a node of height at least $2h$, or the root, which we can reach in constant time.

This observation will collapse the running time of queries, as the following lemma and corollaries shows:

[3] A heavy path decomposition can reduce this number to $O(\log n)$, but will not ultimately help us for future optimizations.

Lemma 5. *Consider any vertex v of height h. The top of v's ladder is at least distance h above, that is, vertex v has at least h ancestors in its ladder.*

Proof. The top of v's long-path has height $h' \geq h$. Thus, it has h' ancestors in its ladder. Node v has $2h' - h \geq h$ ancestors in its ladder.

Lemma 6. *The Ladder Algorithm solves the Level Ancestor Problem in time $\langle O(n), O(\log n)\rangle$.*

Proof. We find the long-path decomposition of tree T in $O(n)$ time as follows. In linear time, we preprocess the tree to compute the height of every node. Each node picks one of its maximal-height children to be its child on the long-path decomposition. Extending the paths into ladders requires another $O(n)$ time.

We now show how to answer queries. Consider any vertex u of height h. If we travel to the top of u's ladder, we reach a vertex v of height at least $2h$. Since all nodes have height at least 1, after i ladders we reach a node of height at least 2^i, and therefore we find our level ancestor after at most $\log n$ ladders and time.

3.4 Putting It Together: An $\langle O(n \log n), O(1)\rangle$ Solution

The Jump-Pointer Algorithm and the Ladder Algorithm complement each other, since the Jump-Pointer Algorithm makes exponentially decreasing hops up the tree, whereas the Ladder Algorithm makes exponentially increasing hops up the tree.

We combine these approaches into an algorithm that follows a single jump-pointer and climbs only one ladder. Since the jump-pointer transports us halfway there, the ladder climb carries us the rest of the way. Thus, we obtain the following theorem.

Theorem 1. *The Level Ancestor Problem can be solved with complexity $\langle O(n \log n), O(1)\rangle$.*

Proof. We perform the preprocessing of both the Jump-Pointer Algorithm and the Ladder Algorithm in time $O(n \log n)$.

We show that queries can be answered by following a single jump pointer and climbing a single ladder. Consider query $\text{LEVELANCESTOR}_T(u, d)$. Let $\delta = \lfloor\lfloor \text{depth}(u) - d \rfloor\rfloor$. The jump pointer leads to vertex v that has depth $\text{depth}(u) - \delta$ and height at least δ. The distance from v to $\text{LA}_T(u, d)$ is at most δ, so by Lemma 5, v's ladder includes $\text{LA}_T(u, d)$.

4 The Macro-Micro-Tree Algorithm: An $\langle O(n), O(1)\rangle$ Solution

Since ladders only take linear time to precompute, we can afford to use them in the fast solution. The bottleneck is computing jump pointers. Our first step in improving the $\langle O(n \log n), O(1)\rangle$ is to exploit the following observation.

Observation 7 *We need not assign jump pointers to a vertex if a descendant of the vertex has jump pointers. That is, if vertex w is a descendant of vertex v, then $\mathrm{LA}_T(v,d) = \mathrm{LA}_T(w,d)$, for all $d \leq$ depth(v), so w's jump pointers are good enough.*

Since we do not need jump pointers on all vertices, we call vertices having jump pointers assigned to them *jump nodes*. An immediate suggestion based on Observation 7 is to designate only the leaves as jump nodes. Unfortunately, this approach only speeds things up enough in the special case when the tree contains $O(n/\log n)$ leaves.

Our immediate goal is to designate $O(n/\log n)$ jump nodes that "cover" as much of the tree as possible. We define any ancestor of a jump node to be a *macro node* and all others to be *micro nodes*. The macro nodes form a connected subtree of T, which we refer to as the *macrotree*, and we define *microtrees* to be the connected components obtained by deleting all macro nodes.

We can deal with all macro nodes by slightly extending the algorithm from Theorem 1 as noted in Observation 7. We will use a different technique for microtrees.

We pick as jump nodes the maximally deep vertices having at least $\log n/4$ descendants. By maximally deep, we mean that the children of these vertices have *fewer* than $\log n/4$ descendants. The $1/4$ is carefully chosen and will come into play when we take care of microtrees.

Lemma 8. *There are at most $O(n/\log n)$ jump nodes. We can compute all jump node pointers in linear time.*

Proof. In the proof of Lemma 3, we used a simple dynamic program to compute jump pointers at every node. Here, we are only computing jump pointers at a few nodes so do not have all the intermediate values needed for the dynamic program. However, notice that for every jump node we can compute its parent in constant time. The parent has height at least 2, so its ladder will carry us another 2 nodes. We can keep jumping up ladders, and so we can compute the jump pointers for any node in $O(\log n)$ time.

4.1 Dealing with Macro Nodes

Lemma 9. *We can solve the level ancestor problem for all macro nodes in $\langle O(n), O(1)\rangle$.*

Proof. We perform a ladder decomposition and compute the jump pointers of all jump nodes in $O(n)$ time. Then, with one depth first search, we find a jump node descendant JumpDesc(u) for each macro node u. Finally, as noted above, compute LevelAncestor(u, d) by computing LevelAncestor(JumpDesc(u), d) using Theorem 1.

4.2 Dealing with Microtrees

In short, we deal with microtrees by noting that they do not come in too many shapes, $O(\sqrt{n})$ in fact. Therefore we can make an exhaustive list of all microtree shapes and preprocess them via the Table algorithm. We show how to use the preprocessing on these canonical trees to compute level ancestors on micro nodes in T. All that remains are a few details.

Lemma 10. *Microtrees come in at most $\sqrt{n}$ shapes.*

Proof. First, recall that each microtree has fewer than $\log n/4$ vertices. For a DFS, call a *down edge* an edge being traversed from parent to child, and an *up edge* those from child to parent. The shape of a tree is characterized by the pattern of up and down edges. A microtree has fewer than $\log n/4$ edges, each of which is traversed twice. While not every pattern of up and down edges is a valid tree, every valid tree forms some such pattern, and so we have at most $2^{\log n/2} = \sqrt{n}$ possible trees. This bound is not tight, but a tighter bound is not necessary. Also, we now see why we selected $\log n/4$ as the jump node threshold, rather than, e.g., $\log n$.

We conclude with the following.

Theorem 2. *The Level Ancestor problem can be solved in $\langle O(n),\, O(1)\rangle$ time.*

Proof. The only thing that remains is a few details about how to handle microtrees. First, we enumerate all microtree shapes and apply the Table algorithm to these. This takes $O(\sqrt{n}\log^2 n)$ time. Furthermore, we address the tables so produced by the bit pattern of ups and downs of the DFS of the trees. Thus, we do all precomputation for all microtrees in T in $O(n)$ time.

Finally, note that in the Table Algorithm, we added one level of indirection based on DFS numbers. This means that we need only assign DFS numbers to the nodes in each microtree in T. Then, when we lookup a level ancestor in the tables, we can use the local DFS numbering to decode which actual node is the desired ancestor.

This finishes the problem of finding a level ancestor within a microtree. The other case is when a micro node wants an ancestor outside of its microtree. In this case, we can jump to the root of the microtree in constant time, and then to its parent. This will be a macro node, and so we revert to the macro node algorithm.

Summing up, the preprocessing time for micro nodes is $O(n)$, as for macro nodes, and in either case the query time is $O(1)$.

References

[1] S. Alstrup and J. Holm. Improved algorithms for finding level-ancestors in dynamic trees. In *27th International Colloquium on Automata, Languages and Programming (ICALP '00), LNCS. 1853*, pages 73–84, 2000.

[2] M. A. Bender, E. Demaine, and M. Farach-Colton. Cache-oblivious B-trees. In *41st Annual Symposium on Foundations of Computer Science (FOCS)*, pages 399–409, 2000.

[3] M. A. Bender and M. Farach-Colton. The LCA problem revisited. In *LATIN*, pages 88–94, 2000.

[4] O. Berkman and U. Vishkin. Recursive *-tree parallel data-structure. In *Proc. of the 30th IEEE Annual Symp. on Foundation of Computer Science*, pages 196–202, 1989.

[5] O. Berkman and U. Vishkin. Recursive star-tree parallel data structure. *SIAM J. Comput.*, 22(2):221–242, Apr. 1993.

[6] O. Berkman and U. Vishkin. Finding level-ancestors in trees. *J. Comput. Syst. Sci.*, 48(2):214–230, Apr. 1994.

[7] R. Cole and R. Hariharan. Dynamic LCA queries on trees. In *Proc. of the 10th Annual ACM-SIAM Symposium on Discrete Algorithms*, pages 235–244, 1999.

[8] P. F. Dietz. Finding level-ancestors in dynamic trees. In *Workshop on Algorithms and Data Structures*, pages 32–40, 1991.

[9] H. N. Gabow, J. L. Bentley, and R. E. Tarjan. Scaling and related techniques for geometry problems. In *Proc. of the 16th Ann. ACM Symp. on Theory of Computing*, pages 135–143, 1984.

[10] D. Harel and R. E. Tarjan. Fast algorithms for finding nearest common ancestors. *SIAM J. Comput.*, 13(2):338–355, 1984.

[11] B. Schieber and U. Vishkin. On finding lowest common ancestors: Simplification and parallelization. *SIAM J. Comput.*, 17:1253–1262, 1988.

[12] R. E. Tarjan. Applications of path compression on balanced trees. *Journal of the ACM*, 26(4):690–715, Oct. 1979.

[13] B. Wang, J. Tsai, and Y. Chuang. The lowest common ancestor problem on a tree with unfixed root. *Information Sciences*, 119:125–130, 1999.

[14] Z. Wen. New algorithms for the LCA problem and the binary tree reconstruction problem. *Inf. Process. Lett.*, 51(1):11–16, 1994.

Flow Metrics

Claudson F. Bornstein[1]* and Santosh Vempala[2]**

[1] Instituto de Matemática, UFRJ
24445-020, Rio de Janeiro RJ Brazil
cfb@cos.ufrj.br

[2] Department of Mathematics, MIT
Cambridge MA 02139 USA
vempala@math.mit.edu

Abstract. We introduce *flow metrics* as a relaxation of path metrics (i.e. linear orderings). They are defined by polynomial-sized linear programs and have interesting properties including *spreading*. We use them to obtain relaxations for several NP-hard linear ordering problems such as the minimum linear arrangement and minimum pathwidth. Our approach has the advantage of achieving the best-known approximation guarantees for these problems using the same relaxation and essentially the same rounding for all the problems and varying only the objective function from problem to problem. This is in contrast to the current state of the literature where each problem warrants either a new relaxation or a new rounding or both. We also characterize a natural projection of the relaxation.

1 Introduction

Many problems in combinatorial optimization can be expressed in the framework of finding a metric (assignment of pairwise distances) of a specified type which minimizes (or maximizes) a prescribed cost function on the distances. For example, in the *maximum cut* problem [8], the metric is restricted to be a cut metric [4] and the cost function is defined as the sum of the distances between all pairs of vertices that are adjacent in the input graph. Similarly in the *sparsest cut* problem [9] the metric is also a cut metric. Other metrics that model interesting problems include Euclidean metrics (possibly of bounded dimension)[8,2], path metrics [3,6,10] and tree metrics [1].

The problems modeled in this way are often NP-hard (as in the case of the examples above) and one approach to solving them is to consider relaxations of the associated metrics. In [9], cut metrics were relaxed to just metrics, i.e.

* Supported in part by CNPq grant 300083/99-8 and a Protem CNPq/NSF joint Grant.

** Supported in part by NSF Career Award CCR-9875024.

S. Rajsbaum (Ed.): LATIN 2002, LNCS 2286, pp. 516–527, 2002.

distances satisfying the triangle inequality and this led to an $O(\log n)$ approximation for the sparsest cut. In [8] cut metrics were relaxed to Euclidean metrics (of finite but unbounded dimension) and this led to a 0.87856 approximation for the maximum cut. In these relaxations the key is to ensure that the optimization problem over the relaxed metric can be solved in polynomial time and the cost function remains an approximation of the original (i.e. the integrality ratio of the relaxation is small).

In this paper, we consider problems modeled by path metrics, i.e. the metrics induced by line graphs. An example is *minimum linear arrangement* which asks for a linear ordering of the vertices of a given graph that minimizes the sum of differences between endpoints of edges of the graph. In our framework, it asks for a path metric that minimizes the sum of the edge lengths. Other interesting problems that can be modeled by this framework include *minimum interval graph completion* and *minimum pathwidth* and *minimum bandwidth*. These problems are all NP-hard, and a variety of approaches have been used to obtain efficient approximation algorithms for them.

In [6], path metrics were relaxed to *spreading* metrics while in [2,5] they were relaxed to Euclidean spreading metrics. Roughly speaking, a spreading metric satisfies the property that the average distance between vertices in any subset of k vertices is about k, as in the case of a path. This property can be modeled using a set of linear constraints. Although the number of constraints in exponential in the size of the input graph, they admit an efficient separation oracle [7] and so the resulting relaxations can be solved in polynomial time.

In this paper, we present a new relaxation for a path metric, which we call a *flow* metric. The relaxation can be viewed as a polyhedral relaxation of a linear ordering. Further, it is succinct in that it only has a polynomial number of linear inequalities. Flow metrics provide a uniform framework for all the optimization problems associated with path metrics mentioned above. The cost functions for all these problems can be modeled using the variables of the relaxation.

To prove approximation guarantees for the problems, we use (extensions of) the rounding algorithm of [10] to map a fractional solution of the relaxation to a true linear ordering. While we are unable to improve any approximation bounds, our approach has the advantage of achieving the best-known approximation guarantees for these problems using the same relaxation and essentially the same rounding algorithm for all the problems and varying only the objective function from problem to problem. This is in contrast to the current state of the literature where each problem warrants either a new relaxation or a new rounding or both.

The new relaxation has some interesting structural properties as well. We prove that these metrics are (approximately) polyhedral liftings of spreading metrics. We do this by constructing a flow metric that approximates the distances of any given spreading metric. As a result, we obtain a succinct description of the spreading metric polyhedron. In conclusion, we mention various directions for strengthening flow metrics that might lead to improved approximation guarantees for these and other problems.

2 A New Relaxation for Linear Ordering

Consider a path on n vertices numbered $1, 2, \ldots, n$. The distance between u and v in the corresponding path metric is $|u - v|$.

Now imagine sending a flow of one unit from each vertex u to every vertex v to the right of u (i.e $v > u$). These flows satisfy the following properties:

(i) For each pair of vertices u, v, the value of the flow from u to v plus the flow from v to u is one.

(ii) For each triple u, v, w, the flow between u and v that goes through w, the flow between v and w that goes through u and the flow between w and u that goes through v sum to one.

Any metric that satisfies the above properties is a flow metric. To define this formally, let the variables $f_{i,j}^{u,v}$ represent the value of the flow from u to v on the edge (i, j). These variables satisfy flow conservation at every vertex except the source and the sink. Define a set of auxiliary variables

$$g_w^{u,v} = \sum_i f_{i,w}^{u,v}.$$

In other words, $g_w^{u,v}$ represents the flow from u to v that goes through the vertex w.

Definition 1. *A* flow metric *is a solution to the following linear program (FP), where f and g are flow variables as defined above.*

$$g_v^{u,v} + g_u^{v,u} = 1 \qquad \forall u, v \in V \tag{1}$$

$$(g_w^{u,v} + g_w^{v,u}) + (g_u^{v,w} + g_u^{w,v}) + (g_v^{w,u} + g_v^{u,w}) = 1 \qquad \forall u, v, w \in V \tag{2}$$

$$d_f(u, v) = \sum_{i,j} f_{i,j}^{u,v} \quad \forall u, v \in V \tag{3}$$

$$d_f(u, v) + d_f(v, w) \geq d_f(u, w) \quad \forall u, v, w \in V \tag{4}$$

Constraint (3) defines a set of directed distances that form the "metric" and constraint (4) imposes the triangle inequality on these distances.

An important property of a flow metric is that it satisfies the 1-dimensional spreading constraints [6,2].

Theorem 1. *For any subset S of vertices,*

$$\sum_{u,v \in S} d_f(u, v) \geq \binom{|S|}{3}.$$

Proof.

$$\begin{aligned}
\sum_{u,v \in S} d_f(u,v) &= \sum_{u,v \in S} \sum_{i,w} f_{i,w}^{u,v} \\
&\geq \sum_{u,v,w \in S} \sum_{i} f_{i,w}^{u,v} \\
&= \sum_{u,v,w \in S} g_w^{u,v} \\
&= \sum_{\{u,v,w\} \subset S} g_w^{u,v} + g_w^{v,u} + g_u^{v,w} + g_u^{w,v} + g_v^{w,u} + g_v^{u,w} \\
&= \sum_{\{u,v,w\} \subset S} 1 \\
&= \binom{|S|}{3}.
\end{aligned}$$

□

3 Rounding Algorithms

Here we describe two variants of the rounding algorithm of Rao and Richa [10] which we will use either directly or as the main subroutine in our approximation algorithms for various NP-hard problems (described in the next section). The rounding algorithm takes as input the set of pairwise distances d_f defined by a flow metric and produces a linear ordering of the vertices.

Rounding Algorithm A

1. Select a vertex v for which $\sum_{w \in G} d_f(v,w)$ is maximum.
2. Assign edges to *levels*: An edge $e = (i,j) \in E$ is included in level $L(d)$ if $d_f(v,i) < d \leq d_f(v,j)$.
3. Partition levels into *buckets*: A level $L(d)$ is assigned to bucket b if $2^b < |L(d)| \leq 2^{b+1}$.
4. Select the largest bucket. Let $L(d_1), L(d_2), \ldots, L(d_r)$ be the levels in that bucket in the order of increasing distance from v. Then define H_i to be the subgraph induced by the vertices u that satisfy $d_i \leq d_f(v,u) < d_{i+1}$ (where $d_0 = 0$ and $d_{r+1} = n+1$).
5. Recurse on the subgraphs $H_0, H_1, H_2, \ldots H_r$. Return the ordering obtained by concatenating the orderings of the subgraphs.

The next algorithm differs from Algorithm A in that it assigns vertices to levels.

Rounding Algorithm B

Use algorithm A with steps (2) and (4) replaced by the following:

2'. Assign vertices to levels: A vertex u is included in level $L(d)$ if $d_f(v,u) < d$ and there exists a vertex w adjacent to u such that $d \leq d_f(v,w)$.
4'. Select the largest bucket. Let $L(d_1), L(d_2), \ldots, L(d_r)$ be the levels in that bucket in the order of increasing distance from v. Delete the vertices in these levels from the graph. Then define H_i to be the subgraph induced by the remaining vertices u that satisfy $d_i \leq d_f(v,u) < d_{i+1}$.

4 Approximation Algorithms and Guarantees

Here we give algorithms for various linear ordering problems. All the algorithms start by solving the flow metric relaxation with a suitable objective function. We prove that the final solution found is within the best known approximation factor of the optimum for all these problems.

4.1 Minimum Linear Arrangement

Given a graph $G = (V, E)$ the minimum linear arrangement problem asks for an ordering $v_1, v_2, \cdots, v_n$ of the vertices of G such that the sum of the lengths of the edges of G is minimized, where the length of an edge (v_i, v_j) is $|i - j|$.

The problem can be expressed in terms of a flow metric as obtaining an ordering that minimizes the sum of the distances $d_f(i,j)$ for edges $(i,j) \in E$. The next algorithm finds a linear arrangement of cost at most $O(\log n)$ times the minimum by rounding this flow metric, and matches the approximation factor of [10].

Algorithm

1. Minimize $\sum_{(i,j)\in E} d_f(i,j)$ over (FP) to get a set of distances d_f.
2. Apply rounding algorithm A.

Theorem 2. *The algorithm finds an $O(\log n)$ approximation to the minimum linear arrangement.*

Proof. First we bound the cost in the final ordering of the edges that are deleted in the first iteration of the algorithm (i.e. they go across the components created). We do this by separately bounding the contribution of each component H_i. The number of edges that cross H_i is at most the number edges in levels $L(d_i)$ and $L(d_{i+1})$. This is at most $2 \cdot 2^{k+1}$. In the final ordering the contribution of H_i to the lengths of these edges is at most the number of vertices in H_i. Thus the total length of these edges is at most

$$\sum_i 2 \cdot 2^{k+1} |H_i| = 2^{k+2} n.$$

Next we bound the contribution of these edges to the cost of the relaxation. Let (u, w) be any edge in E. It's contribution to the cost of the relaxation is $d_f(u, w) + d_f(w, u)$. If (u,w) belongs to levels $d_i, \ldots, d_j$ then $d_f(v, u) < d_i \leq d_j \leq d_f(v, w)$ and by the triangle inequality

$$d_f(u, w) \geq d_f(v, w) - d_f(v, u) \geq d_j - d_i.$$

By constraint (1) of (FP), $d_f(u, w) + d_f(w, u) \geq 1$ and so the edge (u,w) contributes at least $\max\{1, d_j - d_i\}$ to the objective function. This is at least half the number of levels crossed by the edge (u, w).

Theorem 1 guarantees that there exists at least one vertex v in G with the property that

$$\sum_{u \in V} d_f(v, u) \geq \frac{(n-1)(n-2)}{6}.$$

This implies that there is an edge (v, u) such that $d_f(v, u)$ is at least $(n-2)/6$ and hence there are at least as many levels. On the other hand, for any flow metric the maximum distance is at most $n - 1$ and thus the number of levels is at most $n - 1$. Thus the total number of buckets is at most $\log n$ and the number of levels in the largest bucket is at least $(n-2)/6 \log n$.

Thus, when we add the contributions of all edges "cut" in the first iteration we get:

$$\sum_{\text{edges cut}} \frac{\#\text{levels crossed by edge}}{2} \leq \frac{(\max \#\text{edges in a level})(\#\text{levels in chosen bucket})}{2}$$

$$= \frac{1}{2} \cdot 2^k \cdot \frac{(n-2)}{6 \log n} = \frac{2^k(n-2)}{12 \log n}.$$

These edges are not present in any of the subgraphs. Their removal reduces the total value of the objective function by the above amount. We can upper bound the cost C of the rounded solution in terms of the cost F of the relaxation:

$$C(F) \leq C(F - 2^k \cdot \frac{n-2}{12 \log n}) + 2^{k+2} \cdot n$$

which yields $C(F) = O(F \log n)$. □

4.2 Minimum Cut Linear Arrangement

Given a graph $G = (V, E)$ the minimum cut linear arrangement problem consists of obtaining a linear ordering $v_1, v_2, \ldots, v_n$ for the vertices of G that minimizes the maximum edge cut between the first i and the last $n - i$ vertices in the ordering. In terms of flow metrics, this can be expressed as the maximum over v of the sum of flows through v between all pairs of vertices i and j that are adjancent in G. The algorithm given here finds a solution of cost at most $O(\log^2 n)$ times the minimum, which matches the approximation factor due to [6].

Algorithm

1. Minimize $\max_{v\in V}\sum_{(i,j)\in E}\left(\sum_{u\in V} f_{u,v}^{i,j}\right)$.
2. Apply rounding algorithm A producing an ordering $v_1, v_2, \ldots, v_n$.
3. Select an index i in the ordering such that $n/4 \leq i \leq 3n/4$ and for which the cost of the cut $(\{v_1, v_2, \ldots, v_i\}, \{v_{i+1}, \ldots, v_n\})$ is minimum.
4. Recurse on the subgraphs induced by $v_1, \ldots, v_i$ and $v_{i+1}, \ldots, v_n$.

Theorem 3. *The algorithm finds an $O(\log^2 n)$ approximation to the minimum cut linear arrangement.*

Proof. We note that

$$\max_{v\in V}\sum_{(i,j)\in E}\left(\sum_{u\in V} f_{u,v}^{i,j}\right) \geq \frac{1}{n}\cdot\sum_{v\in V}\sum_{(i,j)\in E}\left(\sum_{u\in V} f_{u,v}^{i,j}\right) = F/n$$

So F/n as defined above is a lower bound on the objective function for the minimum cut linear arrangement. Given an ordering for the vertices of a graph, the sum of all the $n-1$ cuts of the form $(\{v_1, \ldots, v_i\}, \{v_{i+1}, \ldots, v_n\})$ is exactly the cost of the linear arrangement $v_1, \ldots, v_n$. Rounding algorithm A finds an ordering whose linear arrangement cost is at most $O(F \log n)$. Given an ordering for the vertices of a graph, the sum of all the $n-1$ cuts of the form $(\{v_1, \ldots, v_i\}, \{v_{i+1}, \ldots, v_n\})$ is exactly the cost of the linear arrangement $v_1, \ldots, v_n$. Hence the ordering found has average cut size $O(F \log n/n)$. When we restrict our attention to cuts at the middle $n/2$ vertices the size of the minimum cut is still $O(F \log n/n)$.

Each iteration of the algorithm partitions the graph into subgraphs with at most $3n/4$ vertices while incurring a cost of $O(\log n)$ times the optimum for the minimum cut linear arrangement relaxation. The algorithm terminates in $O(\log n)$ iterations with a total cost of $O(\log^2 n)$ times the optimum. □

4.3 Minimum Interval Graph Completion

The minimal interval graph completion problem consists of obtaining an interval graph I from an input graph G by adding the smallest possible number of edges to make G into an interval graph. This is equivalent to obtaining an order $v_1, v_2, \cdots, v_n$ for the vertices of G which minimizes the total number of edges of the graph I that is obtained from G by inserting edges (v_i, v_j) in G whenever there exists an edge $(v_i, v_k) \in E$ such that $i < j < k$. In terms of a flow metric the number of edges of I can be expressed as the sum over all vertices of the maximum of the directed distances from that vertex to its neighbors in G.

Algorithm

1. Minimize $\sum_{i \in V} \max_{j:(i,j) \in E} d_f(i,j)$ over (FP) to get a set of distances d_f.
2. Apply rounding algorithm B.

Theorem 4. *The algorithm finds an $O(\log n)$ approximation to the minimum interval graph completion.*

Proof. Similar to the proof of Theorem 2, we start by bounding the cost in the final ordering of the vertices cut in the first iteration of the algorithm by separately bounding the contribution of each component H_i.

Each subgraph H_i can contribute at most $|H_i|$ to the cost of the vertices that are in levels d_i and d_{i+1} so that the total cost of the vertices in the cuts is at most

$$\sum_i 2^{k+2}|H_i| \leq 2^{k+2}n.$$

Next we bound the contribution of these vertices to the cost of the relaxation. The total number of buckets is at most $\log n$ and the number of levels in the largest bucket is at least $(n-2)/6\log n$ by the same arguments used in the proof of Theorem 2.

Let u be vertex belonging to levels $d_i, \ldots, d_j$. Then there exists a vertex w adjacent to u such that $d(v,u) < d_i \leq d_j \leq d(v,w)$. Applying the triangle inequality and constraint (1) of (FP) we arrive at a total cost associated with all vertices belonging in the cut of at least

$$\frac{1}{2} \cdot 2^k \cdot \frac{(n-2)}{6\log n} = \frac{2^k(n-2)}{12\log n}.$$

We can upper bound the cost C of the rounded solution in terms of the cost F of the relaxation:

$$C(F) \leq \sum_i C(F - 2^k \cdot \frac{n-2}{12\log n}) + 2^{k+2} \cdot n$$

which yields $C(F) = O(F \log n)$. □

4.4 Minimum Pathwidth

Given a graph $G = (V,E)$ the minimum pathwidth problem consists of obtaining an ordering $v_1, v_2, \ldots, v_n$ that minimizes over all v_i the maximum number of vertices in $\{v_1, v_2, \ldots, v_{i-1}\}$ which have neighbors in $\{v_{i+1}, \ldots, v_n\}$.

In terms of a flow metric this can be expressed as the minimum of the maximum over all vertices v of the number of vertices i that have flow through v to one of its neighbors j. The next algorithm matches the approximation factor of the best previously know algorithm for the minimum pathwidth problem due to Bodlaender et al. [3].

Algorithm

1. Minimize $\max_{v \in V} \sum_{i \in V} \max_{j:(i,j) \in E} g_v^{i,j}$.
2. Apply rounding algorithm B to get an ordering $v_1, v_2, \ldots, v_n$
3. Select an index i in the ordering such that $n/4 \leq i \leq 3n/4$ and for which the number of vertices before v_i with neighbors after v_i is minimized.
4. Recurse on the subgraphs induced by $v_1, \ldots, v_{i-1}$ and $v_{i+1}, \ldots, v_n$.

Theorem 5. *The algorithm finds an $O(\log^2 n)$ approximation to the minimum pathwidth.*

Proof. We note that we can relate the cost of this problem with the average width at each vertex:

$$\begin{aligned}\max_{v \in V} \sum_{i \in V} \max_{j:(i,j) \in E} g_v^{i,j} &\geq \frac{1}{n} \cdot \sum_{v \in V} \sum_{i \in V} \max_{j:(i,j) \in E} g_v^{i,j} \\ &\geq \frac{1}{n} \cdot \sum_{i \in V} \max_{j:(i,j) \in E} \sum_{v \in V} g_v^{i,j} \\ &= \frac{1}{n} \cdot \sum_{i \in V} \max_{j:(i,j) \in E} d_f(i,j) = F/n.\end{aligned}$$

With F as defined above, rounding algorithm B can round the flow metric to a linear ordering whose interval graph completion cost is at most $O(F \log n)$. The average width for the ordering produced is $O(F \log n/n)$ so we can find a vertex in the middle $n/2$ vertices whose width is also $O(F \log n/n)$. Each of the $O(\log n)$ steps incurs a cost of $O(\log n)$ times the optimum value of the pathwidth relaxation thus yielding the result. □

5 The Connection with Spreading Metrics

Theorem 1 shows that flow metrics provide a succinct description of (exponentially many) spreading constraints. Here we consider a stronger set of spreading constraints that have turned out to be useful for many applications. A set of distances $d(u, v)$ satisfy the *strong* spreading constraints if for any vertex $u \in V$,

$$\sum_{v \in S} d_f(u, v) \geq |S|^2/K$$

where K is a fixed constant [6,2].

Using one more set of constraints of poynomial size on the flow variables of a flow metric leads to the flow distances satisfying the stronger spreading constraints. In other words, the projection of the flow metric polyhedron with the new constraints defined below is contained in the spreading metric polyhedron. In fact, the converse also holds approximately: for any spreading metric, there is a corresponding flow metric that approximates the distances to within a constant factor.

Consider a solution f to the linear program (FP) with the following additional constraint:

$$\sum_{i,j\in\{v,w,z\}} g_i^{u,j} + g_i^{j,u} \geq 1 \quad \forall u \in V, \quad v, w, z \in V \setminus u. \tag{5}$$

In terms of a path metric this constraint says that given a vertex u, and three other vertices v,w and z then either v, w or z must be in the path from u to another one of the three vertices. This is enough to obtain strong spreading.

Theorem 6. *For any any subset S of vertices, and any vertex $u \in S$,*

$$\sum_{v\in S\setminus u} d_f(u,v) + d_f(v,u) \geq |S|(|S|-1)/6.$$

Proof.

$$\begin{aligned}
\sum_{v\in S\setminus u} d_f(u,v) + d_f(v,u) &= \sum_{v\in S\setminus u}\sum_{w\in V} g_w^{u,v} + g_w^{v,u} \\
&= \frac{1}{|S|-2}\sum_{v\in S\setminus u}\sum_{w\in V}\sum_{z\in S\setminus\{u,v\}} g_w^{u,v} + g_w^{v,u} \\
&\geq \frac{1}{|S|-2}\sum_{v,w,z\in S}\sum_{i,j\in\{v,w,z\}} g_i^{u,j} + g_i^{j,u} \\
&\geq \frac{1}{|S|-2}\binom{|S|}{3} = |S|(|S|-1)/6.
\end{aligned}$$

□

We proceed to show how to construct a flow metric that approximates a spreading metric.

Theorem 7. *Let d be a strong spreading metric on a set V with n vertices, $1 \leq d(u,v) \leq n$ for all $u, v \in V$. Then there exists a flow metric d_f defined on V such that for all $u, v \in V$,*

$$\frac{1}{2}d(u,v) \leq d_f(u,v) + d_f(v,u) \leq K \cdot d(u,v).$$

Proof. We construct $d_f(u,v)$ by first defining a unit flow from an arbitrary vertex u to all v in $V \setminus u$ so as to approximate $d(u,v)$. We then apply the same construction recursively to all $v \in V \setminus u$.

We construct a set of paths as follows. Order the vertices according to their distance from u. Ties are broken arbitrarily. Let $P = v_1, v_2, \ldots, v_n$ be the path with vertices in the order obtained. Let H be $\{v_{n/2}, \ldots, v_n\}$. We start by sending one unit of flow along P. This will modify this flow and reroute it through other paths via an iterative procedure. Each iteration starts by examining the current flow distances from each u to each vertex. We construct a modified path P' from P as follows: vertices $v \notin H$ at a distance of at least $\min\{d(u,v), n/4\}$ stay in the same positions in both P and P'. The remaining vertices are re-arranged. The vertices in H are inserted in P' in the first $n/2$ positions available. Finally the remaining vertices are assigned to the remaining positions in P', in the same order that they appeared in P. This is illustrated in figure 1. If we re-route a fraction p of flow from P to P' only the distances to the vertices that were "re-arranged" are affected. Vertices in H have their distances decreased. Vertices not in H that are re-arranged have their distances increased by $p \cdot n/2$. Reroute exactly as much flow as needed to make exactly one such vertex reach $d_f(u,v) = d(u,v)$. The process is repeated, until all vertices $v \notin H$ satisfy $d_f(u,v) \geq \min\{d(u,v), n/4\}$. At most 1/2 the flow from P will be re-routed before the process finishes.

As for the vertices in H, we now permute them in all paths that were constructed, and uniformly distribute the flow through each path among its permutation paths. This makes the distance from u to all vertices of H exactly the same.

At least half the flow goes through permutations of P, where these vertices are at an average distance of $3n/4$, while at most the other half of the flow flows though paths in which the average distance to these vertices is at least $n/4$, thus adding to a distance of at least $n/2$. Therefore for the final distances, we have

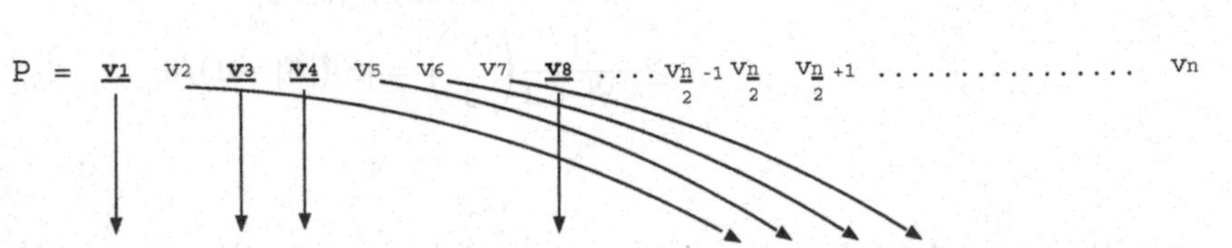

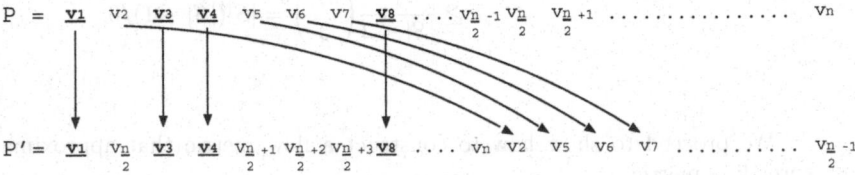

Fig. 1. The path P obtained by ordering the vertices according to their distance to a vertex u is used to route part of the flow. Paths P' are constructed from P by rearranging the vertices of P, so as to not to move vertices v that are already at a distance $\min\{d(u,v), n/4\}$. The underlined vertices depict this set. The remaining vertices are rearranged, with the last $n/2$ vertices being moved to the first $n/2$ positions available.

$$d_f(u,v) \geq \min\{d(u,v), \max\{d(u,v)/2, n/4\}\} \geq d(u,v)/2.$$

The strong spreading constraint implies that the maximum number of vertices at distance x from u is at most $K \cdot x$, so that in fact the distance from any vertex v to u in P is at most $K \cdot d(u,v)$. Since distance to any v is never increased beyond $d(u,v)$ by re-routing flow we have $d_f(u,v) \leq K \cdot d(u,v)$. □

6 Further Research

In our application of flow metrics we have used *independent* flows between pairs of vertices. This could be strengthened by using "capacity" variables $y_{i,j}$ on the edges (i,j) as upper bounds on edge flows and then restricting the sum of the capacities to be $n-1$ (like a path). One could also model trees in a similar manner.

References

1. Y. Bartal, "Probabilistic Approximation of Metric Spaces and its Algorithmic Applications," Proc. of the 37th Ann. IEEE Symp. on Foundations of Computer Science, 184-193, 1996.
2. A. Blum, G. Konjevod, R. Ravi and S. Vempala, "Semi-Definite Relaxation for Minimum Bandwidth and other Vertex-Ordering Problems," *Theoretical Computer Science*, 235 (2000), 25-42. Preliminary version in *Proc. 30th ACM Symposium on the Theory of Computing*, Dallas, 1998.
3. H. L. Bodlaender, J. R. Gilbert, H. Hafsteinsson, T. Klok, "Approximating Treewidth, Pathwidth, and Minimum Elimination Tree Height," Journal of Algorithms 18 (1995) 238-255.
4. M. M. Deza and M. Laurent, *Geometry of cuts and metrics*, Springer-Verlag, 1997.
5. J. Dunagan and S. Vempala, "On Euclidean embeddings and bandwidth minimization," *Proc. of the 5th Intl. Symp. on Randomization and Approximation techniques in Computer Science*, 229-240, 2001.
6. G. Even. J. Naor, S. Rao and B. Schieber, "Divide-and-conquer approximation algorithms via spreading metrics," *Proceedings of the 35th Annual Conference on Foundations of Computer Science*, 62–71, 1995.
7. M. Grötschel, L. Lovász, A. Schrijver, *Geometric Algorithms and Combinatorial Optimization*, Springer, 1988.
8. M. Goemans and D. Williamson, Improved approximation algorithms for maximum cut and satisfiability problems using semidefinite programming, *JACM*, 42: 1115–1145, 1995.
9. T. Leighton and S. Rao. "An approximate max-flow min-cut theorem for uniform multicommodity flow problems with applications to approximation algorithms." In *Proc. of 28th FOCS*, pp256-69, 1988.
10. S. Rao and A. Richa, "New Approximation Techniques for Some Ordering Problems," Proceedings of the Ninth Annual ACM-SIAM Symposium on Discrete Algorithms, 211-218, 1998.

On Logical Descriptions of Regular Languages

Howard Straubing

Computer Science Department
Boston College
Chestnut Hill, Massachusetts USA 02476
straubin@cs.bc.edu

Abstract. There are many examples in the research literature of families of regular languages defined by purely model-theoretic means (that is, in terms of the kinds of formulas of predicate logic used to define them) that can be characterized algebraically (that is, in terms of the syntactic monoids or syntactic morphisms of the languages). In fact the existence of such algebraic characterizations appears to be the rule. The present paper gives an explanation of the phenomenon: A generalization of Eilenberg's variety theorem is proved, and then applied to logic. We find that a very wide assortment of families of regular languages defined in model-theoretic terms form varieties in this new sense,and that consequently membership in the family depends only on the syntactic morphism of the language.

1 Introduction: Why Logic Leads to Algebra

There is by now an extensive literature on what might be called "descriptive automata theory", in which families of regular languages are classified according to the kinds of logical formulas used to define them. This research began with the work of Büchi on the connection between finite automata and weak second-order arithmetic [4], and continued with results of McNaughton and Papert on first-order definable languages [6], Thomas [14] on the Σ_k-hierarchy, Straubing, Thérien and Thomas [12] on modular quantifiers, and others. Straubing [10] presents a large assortment of such results. More recent work has concentrated on descriptions in temporal logic and the expressive power of formulas with a bounded number of variables (see Thérien and Wilke [13] and Straubing and Thérien [11]).

One intriguing feature of these results is that in almost every case, the family of languages in question can be characterized *algebraically,* in terms of the syntactic monoids or syntactic morphisms of the members the family. It is not particularly obvious why this should be, and the discovery of such an algebraic characterization always comes as something of a pleasant surprise.

The present paper gives a detailed explanation of this phenomenon. Section 2 presents a generalization of the notion of a pseudovariety of finite monoids that includes as special cases both the **M**-varieties and **S**-varieties of Eilenberg's theory [5]. We prove in this setting a generalization of Eilenberg's theorem giving the

S. Rajsbaum (Ed.): LATIN 2002, LNCS 2286, pp. 528–538, 2002.

correspondence between pseudovarieties of monoids and varieties of languages. In Section 3, we apply these ideas to the logical definitions of regular languages. We are able, in this way, to account for the algebraic characterizations of all of the families cited above, and to show that many other such families of languages admit characterizations in terms of the syntactic morphism, even though we do not, at present, possess an explicit description of the underlying class of homomorphisms.

In Section 4 we consider an assortment of related questions concerning locally testable languages, regular languages in circuit complexity classes, and possible extensions of our general theory.

Because of length limitations, we only give a broad outline of our argument. The complete proofs are, for the most part, fairly straightforward, but— especially in Section 3—rather long and repetitive. The full paper will contain all the details.

We assume that the reader is familiar with the fundamentals of the algebraic approach to finite automata, particularly with the notions of the syntactic monoid $M(L)$ and the syntactic morphism μ_L of a regular language L. See Pin [7] for the necessary background.

2 Pseudovarieties of Homomorphisms

2.1 Categories of Homomorphisms between Free Monoids

We consider classes $\mathcal{C}$ of homomorphisms between finitely-generated free monoids with the following properties:

(i) Let Σ, Γ, and Δ be finite alphabets. If $f : \Sigma^* \to \Gamma^*$ and $g : \Gamma^* \to \Delta^*$ are in $\mathcal{C}$, then $g \circ f : \Sigma^* \to \Delta^*$ is in $\mathcal{C}$.

(ii) For each finite alphabet Σ, the identity homomorphism $1_{\Sigma^*} : \Sigma^* \to \Sigma^*$ belongs to $\mathcal{C}$.

Such classes form the morphism classes of *categories* whose objects are all the finitely-generated free monoids. We will abuse terminology slightly by referring to the classes $\mathcal{C}$ themselves as categories. We are most interested in the following three categories:

(i) $\mathcal{C}_{all}$, which consists of all homomorphisms between finitely-generated free monoids.

(ii) $\mathcal{C}_{ne}$, which consists of all *nonerasing* homomorphisms; that is, homomorphisms $f : \Sigma^* \to \Gamma^*$ such that $f(\Sigma) \subseteq \Gamma^+$.

(iii) $\mathcal{C}_{lm}$, which consists of all the *length-multiplying* homomorphisms; that is, homomorphisms $f : \Sigma^* \to \Gamma^*$ for which there exists $k \geq 0$ with $f(\Sigma) \subseteq \Gamma^k$.

2.2 Pseudovarieties of Homomorphisms onto Finite Monoids

Let $\mathcal{C}$ be a category of homomorphisms between free monoids, as described above. A collection $\mathbf{V}$ of homomorphisms $\phi : \Sigma^* \to M$, where Σ is a finite alphabet, M is a finite monoid, and ϕ maps onto M, is a $\mathcal{C}$-*pseudovariety* if the following conditions hold:

(i) Let $\phi : \Sigma^* \to M$ be in $\mathbf{V}$, $f : \Gamma^* \to \Sigma^*$ in $\mathcal{C}$, and suppose there is a homomorphism α from $Im(\phi \circ f)$ onto a finite monoid N. Then $\alpha \circ \phi \circ f : \Gamma^* \to N$ is in $\mathbf{V}$.
(ii) If $\phi : \Sigma^* \to M$ and $\psi : \Sigma^* \to N$ belong to $\mathbf{V}$, then so does $\phi \times \psi : \Sigma^* \to Im(\phi \times \psi) \subseteq M \times N$.
Example. Let $\mathcal{C} = \mathcal{C}_{all}$, and let $\mathbf{V}$ be a $\mathcal{C}$-pseudovariety. Suppose $\phi : \Sigma^* \to M$ is in $\mathcal{C}$, and let $\psi : \Gamma^* \to M$ be a surjective homomorphism. Then there is a homomorphism $f : \Gamma^* \to \Sigma^*$ such that $\psi = \phi \circ f = 1_M \circ \phi \circ f$, and thus ψ is in $\mathbf{V}$. Consequently membership of ϕ in $\mathbf{V}$ depends only on $Im(\phi)$, so we can identify $\mathbf{V}$ with a collection of monoids. It is easy to see that this collection of monoids is closed under direct products and division, and so is a pseudovariety of finite monoids (an "**M**-variety" in Eilenberg's terminology). Conversely, if $\mathbf{V}$ is a pseudovariety of finite monoids, then the collection of all homomorphisms from finitely-generated free monoids onto members of $\mathbf{V}$ forms a $\mathcal{C}_{all}$-pseudovariety.
Example. In a like manner, if $\mathbf{V}$ is a $\mathcal{C}_{ne}$-pseudovariety, then the family of semigroups $\phi(\Sigma^+)$, where $\phi : \Sigma^+ \to S$ is in $\mathbf{V}$, forms a pseudovariety of finite semigroups (an "**S**-variety"). Conversely, if $\mathbf{V}$ is a pseudovariety of finite semigroups, then the family of homomorphisms $\phi : \Sigma^* \to M$, where $\phi(\Sigma^+) \in \mathbf{V}$, is a $\mathcal{C}_{ne}$-pseudovariety.
Example. We show how to generate $\mathcal{C}_{lm}$-pseudovarieties that are not $\mathcal{C}_{ne}$ pseudovarieties. These examples arise in the study of regular languages in circuit complexity classes (see Barrington, *et. al.* [2]); their logical theory is discussed at length in Straubing [10]. Let $\phi : \Sigma^* \to M$ be a homomorphism, with M a finite monoid. The sets $\phi(\Sigma^i) = \phi(\Sigma)^i$, for $i > 0$, form a finite subsemigroup $\mathcal{T}$ of the power semigroup of M, and consequently there is a unique idempotent $S = \phi(\Sigma^k)$ in $\mathcal{T}$. Since $S = S^2$, S is a subsemigroup of M, which we call the *stable subsemigroup* of ϕ. Let $\mathbf{V}$ be a pseudovariety of finite semigroups, and let $\mathbf{W}$ be the family of all surjective homomorphisms $\phi : \Sigma^* \to M$ whose stable subsemigroup belongs to $\mathbf{V}$. Then $\mathbf{W}$ is a $\mathcal{C}_{lm}$-pseudovariety. We omit the proof.

The following lemma, whose simple proof we omit, gives a description of the $\mathcal{C}$-pseudovariety generated by a collection of homomorphisms.

Lemma 1. *Let Φ be a collection of homomorphisms from finitely-generated free monoids onto finite monoids, and let $\mathcal{C}$ be a category of homomorphisms between free monoids. The smallest $\mathcal{C}$-pseudovariety containing Φ consists of all surjective homomorphisms $\alpha : \Sigma^* \to M$ for which there exist finite alphabets $\Sigma_1, \ldots, \Sigma_r$, homomorphisms $\phi_i : \Sigma_i^* \to M_i$ in Φ, and homomorphisms $f_i : \Sigma^* \to \Sigma_i^*$ in $\mathcal{C}$ such that whenever*

$$\phi_i \circ f_i(v) = \phi_i \circ f_i(w)$$

for all $i = 1, \ldots, r$, then $\alpha(v) = \alpha(w)$.

2.3 Varieties of Languages and the Eilenberg Correspondence

Let $\mathcal{C}$ be a category of homomorphisms between free monoids, and let $\mathbf{V}$ be a $\mathcal{C}$-pseudovariety. Let $\mathcal{V}$ be the mapping that associates to each finite alphabet Σ the family

$$\mathcal{V}(\Sigma) = \{L \subseteq \Sigma^* : \mu_L \in \mathbf{V}\},$$

where $\mu_L : \Sigma^* \to M(L)$ denotes the syntactic morphism of L. We call $\mathcal{V}$ the *$\mathcal{C}$-variety of languages* corresponding to $\mathbf{V}$.

The two theorems below are due to Eilenberg [5] for the case of **M**- and **S**-varieties, which, as we have seen, are identical to $\mathcal{C}_{all}$- and $\mathcal{C}_{ne}$-pseudovarieties. The generalization given here to arbitrary $\mathcal{C}$-pseudovarieties is entirely straightforward.

Theorem 1. *The mapping* $\mathbf{V} \mapsto \mathcal{V}$ *is one-to-one.*

Proof. Suppose $\mathbf{V}_1$ and $\mathbf{V}_2$ are $\mathcal{C}$-pseudovarieties of homomorphisms such that

$$\mathbf{V}_1 \mapsto \mathcal{V},$$

and

$$\mathbf{V}_2 \mapsto \mathcal{V}.$$

Let $\phi : \Sigma^* \to M$ be in $\mathbf{V}_1$. As is well known, for each $m \in M$, the syntactic morphism

$$\mu_{\phi^{-1}(m)} : \Sigma^* \to M(\phi^{-1}(m))$$

factors through ϕ, and ϕ in turn factors through

$$\prod_{m \in M} \mu_{\phi^{-1}(m)} : \Sigma^* \to \prod_{m \in M} M(\phi^{-1}(m)).$$

The first of these statements implies that each $\mu_{\phi^{-1}(m)}$ is in $\mathbf{V}_2$, and the second that ϕ is in $\mathbf{V}_2$. Thus $\mathbf{V}_1 \subseteq \mathbf{V}_2$. It follows by symmetry that $\mathbf{V}_1 = \mathbf{V}_2$.

Theorem 2. *Let $\mathcal{V}$ be a mapping that associates to each finite alphabet Σ a family $\mathcal{V}(\Sigma)$ of regular languages in Σ^*. $\mathcal{V}$ is a $\mathcal{C}$-variety of languages if and only if it satisfies the following properties:*
(i) $\mathcal{V}(\Sigma)$ is closed under boolean operations.
(ii) If $L \in \mathcal{V}(\Sigma)$ and $\sigma \in \Sigma$ then the quotients

$$\sigma^{-1}L = \{v \in \Sigma^* : \sigma v \in L\}$$

and

$$L\sigma^{-1} = \{v \in \Sigma^* : v\sigma \in L\}$$

are in $\mathcal{V}(\Sigma)$.
(iii) If $L \in \mathcal{V}(\Sigma)$ and $f : \Gamma^* \to \Sigma^*$ is in $\mathcal{C}$, then $f^{-1}(L) \in \mathcal{V}(\Gamma)$.

Proof. As is well known, the syntactic morphism of $L \subseteq \Sigma^*$ is identical to that of $\Sigma^* \backslash L$, $\mu_{L_1 \cup L_2}$ factors through $\mu_{L_1} \times \mu_{L_2}$, $\mu_{L\sigma^{-1}}$ and $\mu_{\sigma^{-1}L}$ both factor through μ_L, and $\mu_{f^{-1}(L)}$ factors through $\mu_L \circ f$. It follows that any $\mathcal{C}$-pseudovariety of languages satisfies the properties above.

For the converse, let $\mathcal{V}$ satisfy the properties listed in the lemma. Let $\mathbf{V}$ be the smallest $\mathcal{C}$-pseudovariety of homorphisms such that for each finite alphabet

Σ, $\mathbf{V}$ contains the syntactic morphisms of all languages in $\mathcal{V}(\Sigma)$. We claim $\mathcal{V}$ is the $\mathcal{C}$-variety of languages corresponding to $\mathbf{V}$. To prove this, it suffices to show that if $L \subseteq \Sigma^*$ and $\mu_L : \Sigma^* \to M(L)$ is in $\mathbf{V}$, then $L \in \mathcal{V}(\Sigma)$. By Lemma 1, if μ_L is in $\mathbf{V}$, then L is a union of sets of the form

$$\bigcap_{i=1}^{r} f_i^{-1} \circ \mu_{L_i}^{-1}(m_i),$$

where each $f_i : \Sigma^* \to \Sigma_i^*$ is in $\mathcal{C}$, each $L_i \subseteq \Sigma_i^*$ is in $\mathcal{V}(\Sigma_i)$, and where the union ranges over a finite set of r-tuples $(m_1, \ldots, m_r) \in M(L_1) \times \cdots \times M(L_r)$. From the definition of the syntactic monoid it follows that each of the sets $\mu_{L_i}^{-1}(m_i)$ is a finite boolean combination of sets of the form $u^{-1}L_i v^{-1}$, for various words u and v over Σ_i. It follows from the stated closure properties of $\mathcal{V}$ that $L \in \mathcal{V}(\Sigma)$.

3 Application to Logic

3.1 Defining Regular Languages in Formal Logic

We begin by giving an example of the kinds of logical formulas we are talking about. Consider the sentence

$$\neg\exists x \exists y (y = x + 1 \wedge Q_\sigma x \wedge Q_\sigma y).$$

We interpret this sentence in words over a fixed finite alphabet $\Sigma = \{\sigma, \tau\}$. The variables in the sentence represent positions in the word; *i.e.*, integers in the range from 1 to the length of the word. The formula $Q_\sigma x$ is interpreted to mean "the letter in position x is σ". Thus the sentence above says, "The word does not contain two consecutive occurrences of σ." The sentence thereby defines the language over Σ consisting of all such words.

We also consider sentences containing *modular quantifiers* $\exists^{r \bmod n}$, where $0 \leq r < n$. For example, the sentence

$$\exists x (Q_\sigma x \wedge \exists^{0 \bmod 2} y (y < x))$$

means that the word contains an occurrence of σ in an odd-numbered position (that is, a position x such that the number of positions strictly to the left of x is even).

In general we will consider families of regular languages in which we restrict (i) the kinds of quantifiers that appear in the sentence; (ii) the kinds of *numerical predicates* (*e.g.*, the formulas $y = x + 1$ and $y < x$ in the examples above) that appear in the sentence; (iii) the depth of nesting of the quantifiers; and (iv) the number of variables. A full account of this model-theoretic approach to defining regular languages can be found in [10].

For the class of quantifiers used, we will allow the following three possibilities: (i) ordinary first-order quantifiers; (ii) modular quantifiers with a fixed modulus n; (iii) both ordinary quantifiers and modular quantifiers of modulus n in combination. For the class of numerical predicates, we allow the following possibilities:

(i) $\{=\}$ (equality alone); (ii) $\{=,+1\}$ (successor and equality together; (iii) $\{<\}$ (ordering); (iv) $\{<,+1\}$ (ordering and successor); (v) $\{<,\equiv i \pmod{m}\}$ (ordering, together with the predicates $x \equiv i \pmod{m}$ for a fixed modulus m). We make the convention that this last class includes a 0-ary numerical predicate

$$length \equiv i \pmod{m}$$

which is satisfied by words whose length is congruent to i modulo m. This predicate can be defined by a depth 2 formula using the other predicates in the class, but as we shall see below, we get sharper results if we include it among the atomic formulas. Similarly, the successor relation can be defined in terms of $<$ and thus the class (iv) above might appear superfluous. But this definition requires us to introduce an extra level of quantification and new variables, and would thus affect the statement of our main theorem. On the other hand, $x = y$ can be defined in terms of $<$ (as $\neg((x < y) \vee (y < x))$ without introducing quantifiers or new variables, and thus it really would be superfluous to include equality in the class (iv).

Given one of our permitted choices $\mathcal{Q}$ for the class of quantifiers, $\mathcal{N}$ for the class of numerical predicates, $d \geq 0$ and $r \geq 0$ we define

$$\mathcal{V}_{\mathcal{Q},\mathcal{N},d,r}$$

to be the mapping that associates to each finite alphabet Σ the family

$$\mathcal{V}_{\mathcal{Q},\mathcal{N},d,r}(\Sigma)$$

of languages defined by sentences of quantifier depth no more than d, using quantifiers in $\mathcal{Q}$, numerical predicates in $\mathcal{N}$ and no more than r variables. Note that this notation suppresses mention of the fixed moduli n and m in the quantifier class and the class of numerical predicates.

Our main result is

Theorem 3. *$\mathcal{V}_{\mathcal{Q},\mathcal{N},d,r}$ is a $\mathcal{C}$-pseudovariety, where*
(i) $\mathcal{C} = \mathcal{C}_{all}$ if $\mathcal{N} = \{=\}$ or $\mathcal{N} = \{<\}$.
(ii) $\mathcal{C} = \mathcal{C}_{ne}$ if $\mathcal{N}$ is one of the classes that contains $+1$.
(ii) $\mathcal{C} = \mathcal{C}_{lm}$ if $\mathcal{N} = \{<, x \equiv i \bmod m\}$.

3.2 Sketch of the Proof of Theorem 3

Throughout this section we will suppose that the classes $\mathcal{Q}$ of quantifiers and $\mathcal{N}$ of numerical predicates, as well as the alphabet Σ, are fixed. The proof of the main theorem rests on two model-theoretic lemmas, which we shall state shortly; we omit their proofs. Readers familiar with Ehrenfeucht-Fraïssé games will recognize that if only ordinary quantifers were involved, then both lemmas could be proved fairly directly by the application of such games. But game-based arguments are difficult to adapt to modular quantifiers, and so we were obliged, in the full paper, to give a different argument.

In order to prove theorems about sentences (that is, formulas without free variables) safisfied by words, we need to establish results about formulas *with* free variables satisfied by *structures*. Here a structure is a pair (w, I), where $w \in \Sigma^*$ and I is a map from a finite set V of variable symbols into $\{1, \ldots, |w|\}$. A formula whose free variables are contained in $\mathrm{dom}(I)$ can be interpreted in such a structure. The formal semantics are defined in [10]; however, it should be clear from the context what we mean when we say that such a structure satisfies a formula.

Given $\mathcal{Q}$, $\mathcal{N}$, $r \geq 0$ and $d \geq 0$, we define an equivalence relation $\sim_{d,r}$ on structures as follows:

$$(w, I) \sim_{d,r} (w', I')$$

if $\mathrm{dom}(I) = \mathrm{dom}(I')$, and if the two formulas satisfy exactly the same formulas (over the given base of quantifiers and numerical predicates) of depth no more than d and with no more than r variables, whose sets of free variables are contained in $\mathrm{dom}(I)$. Because we fix the alphabet Σ and the moduli allowable in modular quantifiers and numerical predicates, there are only finitely many inequivalent formulas with a given depth and fixed set of free variables. Therefore $\sim_{d,r}$ is an equivalence relation of finite index on the set of structures (w, I) over a given alphabet Σ and finite set of variable symbols $\mathrm{dom}(I)$.

Our fist lemma says that, in a certain sense, $\sim_{d,r}$ is a congruence on structures.

Lemma 2. *Let* $d \geq 0$, $r \geq 0$. *Let* $(w, I) \sim_{d,r} (w', I')$, *and let* $\sigma \in \Sigma$. *Then*

$$(w\sigma, I) \sim_{d,r} (w'\sigma, I'),$$

and

$$(\sigma w, J) \sim_{d,r} (\sigma w', J'),$$

where $\mathrm{dom}(J) = \mathrm{dom}(J') = \mathrm{dom}(I)$, *and for all* $x \in \mathrm{dom}(I)$, $J(x) = I(x) + 1$, $J'(x) = I'(x) + 1$.

Now let (w, I) and (w', I') be $\sim_{d,r}$-equivalent structures, with $w, w' \in \Sigma^*$. Let $f : \Sigma^* \to \Gamma^*$ be a homomorphism. We will define a number of structures over Γ associated with these two structures. We begin with an example: Let $w = \sigma\tau$, $w' = \sigma\tau\tau$, $I(x) = I'(x) = 1$, $I(y) = 2$, and $I'(y) = 3$. The two structures are then $\sim_{0,2}$-equivalent over the base of ordinary quantifiers and $\mathcal{N} = \{<\}$. Let $f(\sigma) = \delta\gamma$ and $f(\tau) = \gamma\delta\gamma$. $f(w)$ then decomposes into the two factors $f(\sigma)f(\tau)$, and $f(w')$ likewise decomposes into three factors. A structure $(f(w), J)$ is *compatible* with (w, I) if the domains of I and J are the same, and if whenever $I(v) = i$, $J(v)$ is a position in the ith factor of $f(w)$. In the present example, $J(x)$ can be 1 or 2, and $J(y)$ can be 3,4 or 5; there are consequently six different such structures compatible with (w, I). Let us pick $J(x) = 2$ and $J(y) = 4$. There is then a unique structure $(f(w'), J')$ that is compatible with (w', I') and *consistent* with $(f(w), J)$ in that if J maps v to the jth position of one of the factors of $f(w)$, then J' maps v to the jth position of the corresponding factor. In our example we must have $J'(x) = 2$ and $J'(y) = 7$.

In general, set

$$w = \sigma_1 \cdots \sigma_k,$$

and

$$w' = \sigma'_1 \cdots \sigma'_{k'}.$$

So

$$f(w) = f(\sigma_1) \cdots f(\sigma_k),$$

and

$$f(w') = f(\sigma'_1) \cdots f(\sigma'_{k'}).$$

If i is a position in $f(w)$, then we set $p(i) = j$ if i belongs to the factor $f(\sigma_j)$, and $q(i) = t$, if i is the tth position in $f(\sigma_j)$. We define analogous mappings p' and q' on the positions of w'. We say $(f(w), J)$ is compatible with (w, I) if I and J have the same domain, and $p(J(v)) = I(v)$ for all v in this domain, and that $(f(w'), J')$ is consistent with $(f(w), J)$ if it is compatible with (w', I'), and if $q(J(v)) = q'(J'(v))$ for all variable symbols v.

Here is our second principal lemma.

Lemma 3. *Let $f : \Sigma^* \to \Gamma^*$ be a $\mathcal{C}$-homomorphism, where*
(i) if $\mathcal{N} = \{=\}$ or $\mathcal{N} = \{<\}$, then $\mathcal{C} = \mathcal{C}_{all}$;
(ii)if $\mathcal{N}$ contains $+1$, then $\mathcal{C} = \mathcal{C}_{ne}$;
(iii) if $\mathcal{N} = \{<, \equiv i \pmod m\}$, then $\mathcal{C} = \mathcal{C}_{lm}$.

Let $d, r \geq 0$ and let (w, I), (w', I') be $\sim_{d,r}$-equivalent structures, with $w, w' \in \Sigma^*$. Let $(f(w), J)$, $f(w'), J')$ be consistent structures compatible with (w, I) and (w', I'), respectively. Then

$$(f(w), J) \sim_{d,r} (f(w'), J').$$

We now complete the proof of Theorem 3. By Theorem 2, it suffices to show that $\mathcal{V}_{\mathcal{Q},\mathcal{N},d,r}$ is closed under boolean operations, quotients, and inverse images of homomorphisms in $\mathcal{C}$. Closure under boolean operations is trivial. To show closure under inverse images, let $L \in \mathcal{V}_{\mathcal{Q},\mathcal{N},d,r}(\Sigma)$, and let $f : \Gamma^* \to \Sigma^*$ be a homomorphism in $\mathcal{C}$. Let $w \in f^{-1}(L)$, and suppose $w \sim_{d,r} w'$, for some $d, r \geq 0$. Then by Lemma 3, $f(w) \sim_d f(w')$, and thus, since L is a union of $\sim_{d,r}$-classes, $f(w') \in L$, so $w' \in f^{-1}(L)$. It follows that $f^{-1}(L)$ is a union of $\sim_{d,r}$-classes, and thus in $\mathcal{V}_{\mathcal{Q},\mathcal{N},d,r}(\Gamma)$.

The identical argument, using Lemma 2, shows closure under quotients.

4 Related Results and Directions for Further Research

4.1 Explicit Characterization of Logically-Defined Classes

Which regular languages can be defined by 3-variable first-order sentences of quantifier depth no more than 7 over the base $\{=, +1\}$? We don't know, but we do know, thanks to Theorem 3, that a language belongs to this family if and only if its syntactic semigroup belongs to a particular pseudovariety of finite semigroups—that is, this family of languages forms a $\mathcal{C}_{ne}$-variety of languages. Our work leads to an infinite assortment of such questions, all of which, we now know, have algebraic answers, and some of these answers are likely to be compelling and interesting.

We do, in fact, possess an effective characterization of the pseudovariety of finite semigroups corresponding to the family of languages defined by first-order sentences over $\{=, +1\}$. (See [10].) We conjecture that there is an infinite hierarchy within this family based on the number of variables. (In contrast, if $<$ is included among the numerical predicates, three variables suffice to define all languages in the class.)

For technical reasons, our arguments do not apply to classes defined over the base $\{+1\}$, without the use of equality. In fact, $x = y$ is definable in terms of successor, by the 3-variable formula

$$\forall z(z = x + 1 \leftrightarrow z = y + 1).$$

This leaves open the characterization of the languages definable by 2-variable sentences over this base. In fact it is not difficult to show that this is precisely the class of locally testable languages, which is well known to form a $\mathcal{C}_{ne}$-pseudovariety of languages.(Brzozowski and Simon [3].)

4.2 Universal Algebra of Pseudovarieties of Homomorphisms

There is now a well-developed theory, rooted in universal algebra, concerning pseudovarieties of finite semigroups and monoids. This research, which centers around the use of a special kind of equational description—"pseudoidentities"—to define pseudovarieties, and computations in free profinite semigroups, has become an important tool in the study of finite semigroups. (See Almeida [1].) There is doubtless an extension of this theory to the pseudovarieties of homomorphisms that we study in this paper. As a first step in this direction we pose the problem of giving an equational description—whatever that might mean in this setting—of the $\mathcal{C}_{lm}$-pseudovariety of homomorphisms whose stable semigroups are aperiodic.

We also ask about an extension of our theory to include the *positive* varieties of Pin [8].

4.3 Applications outside of Logic

Circuit Complexity. The $\mathcal{C}_{lm}$-pseudovariety of homomorphisms whose stable semigroups are aperiodic first arose in the study of regular languages in circuit complexity classes [2]. It was this work on circuit complexity that to some degree motivated the present paper. Theorem 3 makes it easy to prove that a large assortment of families of regular languages defined by circuits are $\mathcal{C}_{lm}$-varieties of languages, and thus admit characterizations in terms of their syntactic morphisms. For example, let us fix $d > 0$, and consider the regular languages that are boolean combinations of depth d AC^0-languages—that is, the boolean closure of the family of languages recognized by polynomial-size, depth d families of circuits with unbounded fan-in AND and OR gates. It is easy to verify that the resulting family of regular languages satisfies the closure properties of Theorem 2 with $\mathcal{C} = \mathcal{C}_{lm}$, and is consequently a $\mathcal{C}_{lm}$-variety of languages. It would be quite interesting to precisely identify the corresponding pseudovariety of homomorphisms. (By way of comaparison, the union of these families over all d corresponds to the $\mathcal{C}_{lm}$-pseudovariety consisting of all homomorphisms whose stable subsemigroups are aperiodic.)

Generalized Star-Height. We can also consider the category $\mathcal{C}_{lp}$ of *length-preserving* homomorphisms between free monoids. It is a long-standing open problem whether the generalized star-height of regular languages is bounded, or indeed if there are any regular languages whose generalized star-height is strictly greater than 1. See Pin, *et. al.*, [9], where it is proved that the family of languages of generalized star-hight no greater than d, for each fixed d, satisfies the hypotheses of Theorem 2 for $\mathcal{C}_{lp}$. Thus membership of a language in this family is again determined by the syntactic morphism of the language.

Acknowledgments. I wish to thank Jorge Almeida, Jean-Eric Pin, and Denis Thérien for helpful conversations concerning this work.

References

1. J. Almeida, *Finite Semigroups and Universal Algebra*, World Scientific, Singapore, 1994.
2. D. Mix Barrington, K. Compton, H. Straubing, and D. Thérien, "Regular Languages in NC^1", *J. Comp. Syst. Sci.* **44** (1992) 478–499.
3. J. Brzozowski and I. Simon, "Characterizations of Locally Testable Events", *Discrete Math.* **4** (1973) 243–271.
4. J. R. Büchi, "Weak Second-order Arithmetic and Finite Automata",*Z. Math. Logik Grundl. Math.* **6** 66-92 (1960).
5. S. Eilenberg, *Automata, Languages and Machines,* vol. B,Academic Press, New York, 1976.
6. R. McNaughton and S. Papert, *Counter-Free Automata,* MIT Press, Cambridge, Massachusetts, 1971.
7. J. E. Pin, *Varieties of Formal Languages,* Plenum, London, 1986.
8. J.-E. Pin, A variety theorem without complementation, *Russian Mathematics (Izvestija vuzov.Matematika)* **39** (1995), 80–90.
9. J.-E. Pin, H. Straubing and D. Thrien, "New results on the generalized star-height problem", *Information and Computation* **101** 219-250 (1992).
10. H. Straubing, *Finite Automata, Formal Languages, and Circuit Complexity,* Birkhäuser, Boston, 1994.
11. H. Straubing and D. Thérien, "Regular Languages Defined by Generalized First-Order Sentences with a Bounded Number of Bound Variables", in *STACS 2001*, Springer, Berlin 551-562 (2001) (Lecture Notes in Computer Science 2010.)
12. H. Straubing, D. Thérien, and W. Thomas, "Regular Languages Defined by Generalized Quantifiers", *Information and Computation* **118** 289-301 (1995).

13. D. Thérien and T. Wilke, "Over Words, Two Variables are as Powerful as One Quantifier Alternation," *Proc. 30th ACM Symposium on the Theory of Computing* 256-263 (1998).
14. W. Thomas, "Classifying Regular Events in Symbolic Logic", *J. Computer and System Sciences* **25** (1982) 360–376.

Computing Boolean Functions from Multiple Faulty Copies of Input Bits

Mario Szegedy and Xiaomin Chen

Department of Computer Science, Rutgers, the State University of NJ
110 Frelinghuysen Road, Piscataway, NJ, USA 08854-8019
szegedy@cs.rutgers.edu, xiaomin@paul.rutgers.edu

Abstract. Suppose, we want to compute a Boolean function f, but instead of receiving the input, we only get l ϵ-faulty copies of each input bit. A typical solution in this case is to take the majority value of the faulty bits for each individual input bit and apply f on the majority values. We call this the *trivial construction.*
We show that if $f : \{0,1\}^n \rightarrow \{0,1\}$ and ϵ are known, the best construction function, F, is often not the trivial. In particular, in many cases the best F cannot be written as a composition of some functions with f, and in addition it is better to use a randomized F than a deterministic one. We also prove, that the trivial construction is optimal in some rough sense: if we denote by $l(f)$ the number of $\frac{1}{10}$-biased copies we need from each input to reliably compute f using the best (randomized) recovery function F, and we denote by $l_{triv}(f)$ the analogous number for the trivial construction, then $l_{triv}(f) = \Theta(l(f))$. Moreover, both quantities are in $\Theta(\log S(f))$, where $S(f)$ is the sensitivity of f.
A quantity related to $l(f)$ is $D^{rand}_{stat,\epsilon}(f) = \min \sum_{i=1}^{n} l_i$, where l_i is the number of 0.1-biased copies of x_i, such that the above number of readings is already sufficient to recover f with high certainty. This quantity was first introduced by Reischuk et al. [14] in order to provide lower bounds for the noisy circuit size of f. In this article we give a complete characterization of $D^{rand}_{stat,\epsilon}(f)$ through a combinatorial lemma, that can be interesting on its own right.

1 Introduction

Assume, we want to compute the majority funtion of three bits

$$MAJ(x_1, x_2, x_3) = \begin{cases} 1 \text{ if } x_1 + x_2 + x_3 \geq 2 \\ 0 \text{ if } x_1 + x_2 + x_3 \leq 1 \end{cases}$$

but we do not have access to x_1, x_2, x_3, only to three ϵ-biased copies of each x_i $(1 \leq i \leq 3)$. The original indices of the 9 faulty bits are known. This situation occurs when the nine velues come from inaccurate measurements that we cannot repeat. The goal is to minimize the maximal failure probability p of computing $MAJ(x_1, x_2, x_3)$ correctly, where p is maximized over all 8 evaluations of x_1, x_2, x_3. Different computations seem to make sense:

S. Rajsbaum (Ed.): LATIN 2002, LNCS 2286, pp. 539–553, 2002.

1. To compute the majority in each group, and take the majority of these;
2. To take the majority of all 9 bits;
3. To compute the majority function in three different ways, and then to vote on the most popular result.

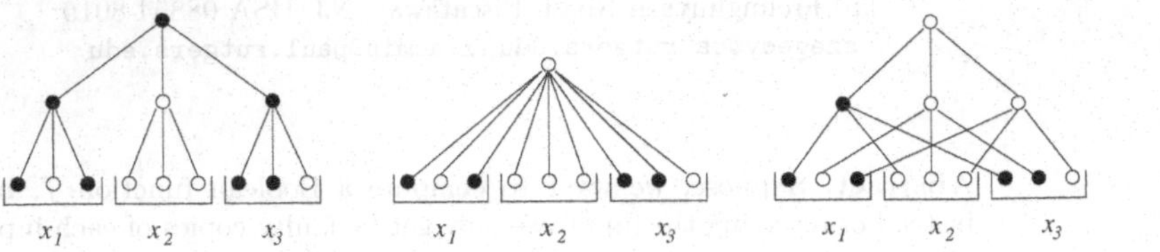

Fig. 1. Different ways of computing the $MAJ(x_1, x_2, x_3)$ function when we have a set of three faulty bits for each x_i

We have found to our surprise that if $\epsilon < 0.3673656...$, then the first construction is the best, while for greater than this value the second one.

In general, let f be a Boolean function on inputs $x_1, \ldots, x_n$. Let $l_1, l_2, \ldots, l_n$ be positive integers, and let $m = \sum_{i=1}^{n} l_i$. Let $0 \leq \epsilon \leq 0.5$ be an error probability, and let $\mathcal{B}_\epsilon^m$ be the space of m independent Bernoulli trials with $p = \epsilon$, $q = 1 - \epsilon$.

For $x = (x_1, \ldots, x_n)$ the vector $\overline{x}$ of length m is constructed by repeating x_1 l_1 times, x_2 l_2 times, etc. If $x = (x_1, \ldots, x_n)$ is an input-vector for f, and $r \in \mathcal{B}_\epsilon^m$, then

$$\overline{x} \oplus r = (y_{1,1}, \ldots, y_{1,l_1}, \ldots, y_{n,1}, \ldots, y_{n,l_n}),$$

is a random vector, where $y_{i,j}$ is interpreted as a (possibly) erroneous copy of x_i, such that the probability of the error is ϵ, and the errors are independent.

Our goal is to construct a Boolean function F on m inputs such that the expression:

$$\Delta(f, F, \epsilon) = \max_{x \in \{0,1\}^n} Prob_{r \in \mathcal{B}_\epsilon^m}(F(\overline{x} \oplus r) \neq f(x)) \tag{1}$$

is minimized. When F itself can be randomized, Equation (1) generalizes to:

$$\Delta(f; F, \epsilon) = \max_{x \in \{0,1\}^n} E_{r \in \mathcal{B}_\epsilon^m}(F(\overline{x} \oplus r)(1 - f(x)) + (1 - F(\overline{x} \oplus r))f(x) \tag{2}$$

The randomized version, besides often giving a better optimum, has the advantage that this optimum can be calculated by linear programming (see Section 2.1). Let us now assume that $l_1 = l_2 = \ldots = l_n = l$. We define the following quantities:

$\xi(f,l,\epsilon)=\min_F \Delta(f,F,\epsilon)$, where F runs through all randomized boolean functions on nl bits, and is applied on ϵ-faulty groups of the n input bits, each group having size l.

$\xi_{triv}(f,l,\epsilon)=\Delta(f,F,\epsilon)$, where F represents the trivial construction, i.e. $F=f(MAJ(y_{11},\ldots,y_{1,l}),\ldots,MAJ(y_{n1},\ldots,y_{n,l}))$.

$\xi_{comp}(f,l,\epsilon)=\min_F \Delta(f,F,\epsilon)$, where

$$F=f(h_1(y_{11},\ldots,y_{1,l}),\ldots,h_n(y_{n1},\ldots,y_{n,l}))$$

for some randomized boolean functions $h_1,\ldots,h_n$, each on l input bits. Note that the composition property is a serious structural restriction on F.

Clearly, $\xi(f,l,\epsilon)\le\xi_{comp}(f,l,\epsilon)\le\xi_{triv}(f,l,\epsilon)$. While it is easy to see that $\xi(f,l,\epsilon)$ and $\xi_{comp}(f,l,\epsilon)$ are monotone in ϵ, we cannot prove the same about $\xi_{triv}(f,l,\epsilon)$. We also define the following inverse quantities:

$$l(f)=\min\{l\in\mathbf{N}:\xi(f,l,0.1)\le 0.1\}$$
$$l_{triv}(f)=\min\{l\in\mathbf{N}:\xi_{triv}(f,l',0.1)\le 0.1,\forall l'\ge l\}$$
$$l_{comp}(f)=\min\{l\in\mathbf{N}:\xi_{comp}(f,l,0.1)\le 0.1\}$$

If in the definition of $l(f)$ or $l_{triv}(f)$ we replace 0.1 with some other positive constant less than 0.5, the formuli change only by a constant factor. One of our main results shows that $l(f)$, $l_{triv}(f)$, and $l_{comp}(f)$ are within a constant factor of each other. We cannot prove the same about $\xi(f,l,\epsilon)$, and its analogues, as these inverse quantities are far more sensitive to changes in the construction.

U. Feige, D. Peleg, P. Raghavan, E. Upfal in [4] study noisy boolean decision trees. These are like usual decision trees, except the answer to any query is incorrect with some fixed probability ϵ. All errors are independent. The model of Feige et. al. is very similar to ours, but in our case the algorithm corresponding to the tree is non-adaptive. The static model first appears in a paper of R. Reischuk and B. Schmeltz [14]. Although their model is equivalent to ours, the questions they raise are different, as they are more interested in estimating the l-quantities than the more sensitive ξ-quantities. The *static noisy decision tree complexity*, $D^{rand}_{stat,\epsilon}(f)$, of a boolean function f is defined as the minimum $l_1+\ldots+l_n$ such that there is an F that takes l_i ϵ-faulty bits of x_i and computes f with at least 90% certainty (in other words, $\Delta(F,f,\epsilon)\le 0.1$). The motivation of Reischuk et. al. was to give lower bounds on the noisy circuit size for the function f. This question goes back to von Neuman [11]. They prove that any function with sensitivity S has noisy circuit size $\Omega(S\log S)$, where the sensitivity of f is the number of bits, that if we flip individually for the most bit-sensitive input, the value of the function changes. The latter was also proven independently, and without noisy trees by P. Gács and A. Gál [5,6], and further related questions were studied by Pippenger [12,13].

In this paper we give a complete characterization of $D^{rand}_{stat,\epsilon}(f)$ in terms of the sensitivity sets of f. On the way to the characterization we present a lemma that gives a necessary and sufficient condition for the value of a Boolean function to remain constant under independent random changes of its input bits. This lemma is new to our knowledge, and may be interesting on its own right.

Other related results: I. Benjamini, G. Kalai, O. Schramm, [2] define the notion they call *sensitivity gauge*, which has some resemblance to our sensitivity measures. The biggest differences in their model compared to ours is that they have only one copy of each input bit, and they look for average error instead of the maximal one. The model of C. Kenyon and A. Yao in [8] is also somewhat similar to ours, but in their case the noise is non-deterministic rather than random, and their restriction is an upper limit on the number of incorrect bits.

2 Methodologies for Computing the Best Approximator

After we raised the question of computing $\xi(f,l,\epsilon)$ for a given f, (or, in general, $\xi(f,l_1,\ldots,l_n,\epsilon)$, if the number of samples differ for different variables) we sought to perform computer experiments. Linear programming was a straightforward tool, but the sheer size of the instance was an obstacle. Therefore we needed simplifications to decrease the instance size. Our investigations in the end also helped to resolve the problem for infinite families of functions.

2.1 The Linear Programming Approach

The extremal F that optimizes expression (2) can be computed via linear programming, where the unknowns are $F(y)$ $(y \in \{0,1\}^m)$ and p, the probability of error that F makes, and which is to be minimized. The instance consists of the obvious bounding inequalities,

$$0 \leq F(y) \leq 1 \quad \text{for } y \in \{0,1\}^m, \tag{3}$$

and:

$$\sum_{y\in\{0,1\}^m} P(x,y)F(y) \geq 1-p \qquad \text{for } x \in f^{-1}(1); \tag{4}$$

$$\sum_{y\in\{0,1\}^m} P(x,y)F(y) \leq p \qquad \text{for } x \in f^{-1}(0); \tag{5}$$

$$\min p \tag{6}$$

Here $P(x,y) = \epsilon^t(1-\epsilon)^{m-t}$, where t is the Hamming distance of $\overline{x}$ and y. It will often simplify our notations if we introduce:

$$\alpha = (1-\epsilon)^m; \qquad \beta = \frac{\epsilon}{1-\epsilon}.$$

With this notation $P(x,y) = \alpha\beta^t$, where t is as above.

2.2 Simplifying Symmetries

In the search of the optimal F satisfying (3)-(6) we have found some simplifying assumptions that helped both our CPLEX [16] experiments, and our calculations. In this section $[a, b]$ will denote the set $\{a, a+1, \ldots, b\}$

Our first observation is that we can replace the solution F of (3)-(6) with an F that depends only on the pattern $s = (s_1, \ldots, s_n)$, where

$$s_i = \text{weight}(y_{i,1}, \ldots, y_{i,l}) = |\{j \mid y_{i,j} = 1\}|.$$

Here $y = \overline{x} \oplus r$ as in Section 1. If we view F as a function over the patterns in $[0, l]^n$ then System (3) is replaced by

$$0 \leq F(s) \leq 1 \quad \text{for } s \in [0, l]^n, \qquad (3')$$

and (5) and (6) are replaced by (5') and (6'), where the coefficients of $F(s)$ are

$$P(x, s) = \binom{l}{s_1}\binom{l}{s_2}\cdots\binom{l}{s_n}\alpha\beta^{\sum_{i=1}^{n}|lx_i - s_i|}.$$

The fact that we can reach the optimal p with the symmetrized F comes from a standard argument taking the advantage that the feasible region of the system (3)-(5) is convex.

In order to find more symmetries, let G be a group acting on $\{0,1\}^n$ such that that for every $g \in G$, $x, z \in \{0,1\}^n$:

1. $dist(g(x), g(z)) = dist(x, z)$
2. If $f(x) = 1$ then $f(g(x)) = 1$. If $f(x) = 0$ then $f(g(x)) = 0$.

These definitions also work if f is a partial function (i.e. a promise problem).

Definition 1. *For a vector v (of arbitrary elements) of length n, and a permutation $\pi \in S_n$ we denote the vector we obtain from v by permuting its coordinates according to π by v^π.*

It is easy to see that the group of all distance preserving transformations of the hyper-cube is $Z_2^n \rtimes S_n$, i.e. the semi-direct product of Z_2^n with S_n. The action of an element (z, π) of this group can be described as $x \rightarrow (x \oplus z)^\pi$. Since G must obey Property 1, we have: $G \leqslant Z_2^n \rtimes S_n$. Given G, we can simplify our sets of equation further by letting G act on $[0, l]^n$ in the following way: Let $g = (z, \pi) \in G$. The action of g on a sequence $(s_1, \ldots, s_n)$ is described by:

$$(s_1, \ldots, s_n)^g \rightarrow (|z_1 l - s_1|, \ldots, |z_n l - s_n|)^\pi.$$

Lemma 1. *If G satisfies 1. and 2. then there is an optimum solution to the system (3')-(6') such that for every $g \in G$, and $s \in [0, l]^n$ we have $F(s) = F(s^g)$.*

The proof of the above lemma is again a symmetrization argument, that we omit. In the further simplified set of equations the coefficient, $P(x,S)$, of $F(S)$, where S is an orbit of G when acting on $[0,l]^n$ as above, is computed by $P(x,S) = \sum_{s\in S} P(x,s)$. Since for any $s,t \in S$ we have $\prod_{i=1}^n \binom{l}{s_i} = \prod_{i=1}^n \binom{l}{t_i}$, we can write:

$$P(x,S) = \left(\alpha \prod_{i=1}^n \binom{l}{t_i}\right) \sum_{s\in S} \beta^{\sum_{i=1}^n |lx_i - s_i|}, \qquad (7)$$

where $t = (t_1,\ldots,t_n) \in S$ is an arbitrary representative of the orbit. Our set of equations now simplifies in two ways: First, the number of variables decreases, since we make identifications among them. Second, the number of equations also decreases, because whenever x and z are in the same orbit of G, after the identifications of variables the inequality belonging to x becomes the same as the inequality that belonging to z.

We can simplify our equations even further if the 1 and 0 outputs for f play a symmetric role (like in the case of the parity or the majority functions). In this case we can include the set of those symmetries on the inputs of f under which the 0 and 1 outputs are switched, i.e. $f(g(x)) = 1 - f(x)$ for every $x \in \{0,1\}^n$. The group of those symmetries that obey rule 1, and either leave the values of f invariant or switch them as above we denote with G_1. For instance, if f is the parity function, G_1 can be taken the entire $Z_2^n \rtimes S_n$. We have that $G \leqslant G_1 \leqslant Z_2^n \rtimes S_n$. $|G_1/G| = 2$ iff switching symmetries exist.

Lemma 2. *Let $G \leqslant G_1$. For every $s \in [0,l]^n$ we can set*

$$F(s^g) = \begin{cases} F(s), & \text{if } g \in G \\ 1 - F(s) & \text{if } g \in G_1 \setminus G \end{cases}$$

Thus for every orbit of G_1, when acting on $[0,l]^n$ we have a single variable. Furthermore, all equations belonging to inputs corresponding to an orbit of G_1, when acting on $\{0,1\}^n$, collapse into a single equation.

The coefficient of $F(S)$ in the new system is:

$$P(x,S) = \alpha \prod_{i=1}^n \binom{l}{t_i} \left(\sum_{s\in S_0} \beta^{\sum_{i=1}^n |lx_i - s_i|} - \sum_{s\in S_1} \beta^{\sum_{i=1}^n |lx_i - s_i|} \right), \qquad (8)$$

where orbit S decomposes into $S = S_0 \cup S_1$, the two orbits according to G, (note that $G_1/G = Z_2$). We also get a constant term on each left hand side:

$$\sum_{S_1 : S \in orbits(G_1)} \left(\alpha \prod_{i=1}^n \binom{l}{t_i} \sum_{s\in S_1} \beta^{\sum_{i=1}^n |lx_i - s_i|} \right).$$

Note that for every orbit of G_1 on $[0,l]^n$ we need to choose S_0 and S_1. This freedom of choice corresponds to a linear change of variables, and does not effect the system essentially. The number of different equations we get is the number of orbits of G_1 acting on $\{0,1\}^n$. This reduces the case of XOR to a single non-trivial equation (see Section 3).

2.3 The Two-Equation Case

When the number of constraints is two, we can study the linear programming system from a geometric point of view. This leads us to several observations, especially lemma 4. A very similar result was proved in [14]. All the proofs of this section are elementary geometric considerations. In this section $[x, y]$ denotes the real closed interval in between x and y, and $[k]$ denotes the set $\{1, \ldots, k\}$.

Assume we have two probability distributions $a, b \in [0,1]^k$. First assume $a_i, b_i > 0$, and

$$\frac{a_1}{b_1} > \frac{a_2}{b_2} > \ldots > \frac{a_k}{b_k}.$$

For any $F \in [0,1]^k$, the map $\varphi(F) = (\langle a, F\rangle, \langle b, F\rangle)$ is a map from $[0,1]^k$ into $[0,1] \times [0,1]$. Let $\mathcal{D}$ be the image of φ. It is clear that, $\mathcal{D}$ is convex, and the extreme points of $\mathcal{D}$ is a subset of $\{\varphi(F) : F \in \{0,1\}^k\}$. We also know that $\varphi(\mathbf{0}) = (0,0)$ and $\varphi(\mathbf{1}) = (1,1)$.

As usual, the string 1^i0^{k-i} is understood as the vector where the first i entries are 1, and the rest are 0. For any $0 \le i \le k$, let F^i be the vector 1^i0^{k-i}. (So, $F^0 = \mathbf{0}$, and $F^k = \mathbf{1}$.)

Lemma 3. *$\varphi(F^i)$, $i = 0, 1, \ldots, k$, are the $k+1$ vertices of the lower boundary of $\mathcal{D}$, from left to right, in that order.*

The following theorem describes the region $\mathcal{D}$. It is a consequence of the lemma above.

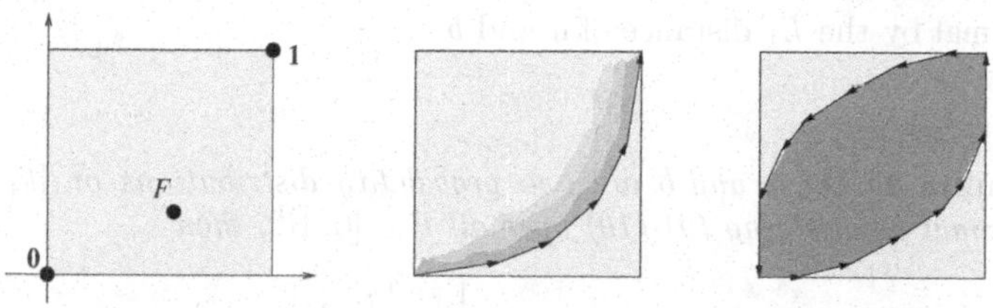

Fig. 2. The geometric place of all pairs $(\langle a, F\rangle, \langle b, F\rangle)$

Theorem 1. *Let a and b be two probability distributions on $[k]$, where a_i/b_i, $1 \le i \le k$, are in non-increasing order (we use $a/b = +\infty$ when $b = 0$), then the region*

$$\mathcal{D} = \{(\langle a, F\rangle, \langle b, F\rangle) : F \in [0,1]^k\}$$

is a convex polygon. The sequence 1^i0^{k-i}, $0 \le i \le k$ gives the vertices on the lower boundary of $\mathcal{D}$ from left to right; the sequence 0^i1^{k-i}, $0 \le i \le k$ gives the vertices on the upper boundary from right to left. On the lower boundary, from left to the right, the vector from the $(i-1)$-st vertex to the i-th vertex is (a_i, b_i); on the upper boundary, from right to the left, the vector from the $(i-1)$-st vertex to the i-th vertex is $(-a_i, -b_i)$.

It is clear from the Theorem above that $\mathcal{D}$ is central symmetric about the point $(1/2, 1/2)$.

Now, consider we have two inequalities in our system,

$$a_1F_1 + a_2F_2 + \ldots + a_kF_k \geq 1 - p \tag{9}$$

$$b_1F_1 + b_2F_2 + \ldots + b_kF_k \leq p \tag{10}$$

We may assume that the ratios a_i/b_i are non-increasing. Let $\mathcal{D}$ be the region defined above, let $(1 - p_0, p_0)$ be the intersection point of the line $x + y = 1$ and the lower boundary of the convex $\mathcal{D}$. (Figure 3.a) So, the best solution is always a sequence of 1's followed by a sequence of 0's, with the only possible fractional entry in the middle.

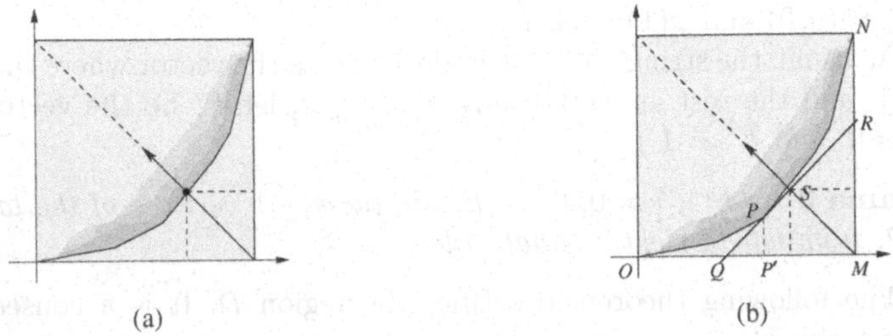

Fig. 3. (a) Find the optimal solution in two-constraints system (b) Estimate the optimal by the L_1 distance of a and b

Lemma 4. *Let a and b are two probability distributions on $[k]$, let p_0 be the minimal p satisfying (9)-(10) over all $F \in [0,1]^k$, then*

$$p_0 \geq \frac{1}{2} - \frac{1}{4}|a, b|_{L_1}$$

Proof. (See Figure 3.b) We may assume a_i/b_i are in non-increasing order. We create the region $\mathcal{D}$ as in Theorem 1. Let O, M, N be the points $(0,0)$, $(1,0)$, and $(1,1)$, respectively. Consider a line of slope 1 at the point M. We move it up until it touches the lower boundary of $\mathcal{D}$ at point P. (If it touches an edge of slope 1, let P be any end point of that edge.) Suppose this line intersects OM and MN at points Q and R, respectively. Let P' be the projection of P on OM.

3 Specific Functions

Perhaps the most interesting among the results we have on specific functions are the ones on the majority of 3, and on the general parity function.

Basically, we use the method introduced in the previous section to find the best solution.

For the majority function, after simplifying the linear programming system, we get a system of 10 variables and two constraints. The set of variables is $L = \{s \in \mathbf{N}^3 : 0 \le s_1 \le s_2 \le s_3 \le 3, s_2 \le 1\}$. We get two constraints $\sum_{s\in L} c_{i,s}F(s) + b_i$, $i = 1, 2$. Calculate the constants $c_{i,s}$'s and b_i's, we find all the coefficients are positive, except $c_{1,113} < 0$, and the sign of $c_{2,113}$ depends on the zero of $1-\beta-\beta^2-\beta^3+\beta^4$. We state the result, and omit the detailed proof. Notice that, in the second case, the best solution is not a composite function.

Theorem 2. *Let f be the function $MAJ(x_1, x_2, x_3)$ (the majority of three inputs), and $l = 3$. Then the construction that optimizes equations (3)-(6) is the trivial construction if*

$$\epsilon \le \frac{\frac{\sqrt{13}+1}{2} - \sqrt{\frac{\sqrt{13}-1}{2}}}{\frac{\sqrt{13}+5}{2} - \sqrt{\frac{\sqrt{13}-1}{2}}} \approx 0.3673656,$$

and the majority of all bits otherwise.

After simplify the system for parity function, we get only one constraint, so the value of each variable in the best solution depends only on the sign of its coefficient. By a combinatorial analysis, we get

Theorem 3. *For any n, l, and $0 \le \epsilon \le 0.5$,*

$$\xi(XOR(n), l, \epsilon) = \xi_{triv}(XOR(n), l, \epsilon) = \frac{1}{2} - \frac{1}{2}(1-2q)^n,$$

where

$$q = \begin{cases} \sum_{i>l/2} \binom{l}{i}\epsilon^i(1-\epsilon)^{l-i}, & \text{if } l \text{ is odd} \\ \sum_{i>l/2} \binom{l}{i}\epsilon^i(1-\epsilon)^{l-i} + \frac{1}{2}\binom{l}{l/2}\epsilon^{l/2}(1-\epsilon)^{l/2} & \text{if } l \text{ is even} \end{cases}$$

We have rewritten a theorem of Reischuk et. al. [14] about a promise problem that we call the FAN function (see the definition) into our language; and we reprove it here:

Theorem 4. *We define: $FAN(\mathbf{0}) = 0$; and $FAN(x) = 1$ for every x such that $\text{dist}(x, \mathbf{0}) = 1$. Every other value is undefined. We have: If $l \le 0.1 \log n$, then $\xi(FAN_n, l, 0.1) \ge \frac{5}{16}$.*

Proof. We show that with $\epsilon = 0.1$ and (for any other constant) $\xi(f, l, \epsilon)$ is very close to 0.5 unless l is at least logarithmic.

Let D_1 be the probability distribution on $[0, l]^n$:

$$P(\mathbf{0}, s) = \binom{l}{s_1}\binom{l}{s_2}\cdots\binom{l}{s_n}\alpha\beta^{\sum_{i=1}^n s_i},$$

and D_2 be the distribution

$$Q(s) = \frac{1}{n}\sum_{j=1}^{n} P(e_j, s) = \frac{1}{n}\sum_{j=1}^{n}\binom{l}{s_1}\binom{l}{s_2}\cdots\binom{l}{s_n}\alpha\beta^{l-2s_j+\sum_{i=1}^n s_i}.$$

Then for every $s \in [0, l]^n$:

$$\frac{Q(s)}{P(\mathbf{0}, s)} = \frac{1}{n} \sum_{j=1}^{n} \beta^{l-2s_j}$$

Let $n_i(s) = |\{j \mid s_j = i\}|$. Over distribution D_1 the expected value of $n_i(s)$ is $E_i = n\binom{l}{i}\epsilon^i(1-\epsilon)^{l-i}$. From the Chernoff inequality we have that

$$Prob_{s \in D_1}\left(n_i(s) \leq \frac{E_i}{2}\right) \leq e^{-\frac{1}{4}\frac{E_i}{3}} \leq \frac{1}{2l},$$

if $\epsilon = 0.1$, $l = 0.1 \log n$. Let s be in the complement of all the events $|n_i(s) - E_i| \geq \frac{E_i}{2}$. Then we can bound

$$\frac{1}{n}\sum_{j=1}^{n} \beta^{l-2s_j} = \frac{1}{n}\sum_{i=0}^{l} \beta^{l-2i} n_i(s) \geq \frac{1}{2n}\sum_{i=0}^{l} \beta^{l-2i} E_i$$

$$= \frac{1}{2}\sum_{i=0}^{l} \beta^{l-2i}\binom{l}{i}\epsilon^i(1-\epsilon)^{l-i} = 1/2.$$

Thus the L_1 distance of D_1 and D_2 is at most $1 - \left(\frac{1}{2}\right)^2 = \frac{3}{4}$. This implies That for every $F : \{0,1\}^{nl} \to [0,1]$ we have $\Delta(FAN_n, F, 0.1) \geq 5/16$ by the virtue of Lemma 4.

4 General Lower and Upper Bounds

Our general bounds are in one way or another related to a well known combinatorial parameter of Boolean functions called *sensitivity.* For each x the *sensitivity set* associated with S is $S_x = \{i \mid f(x) \neq f(x \oplus e_i)\}$. The following lemma, which is new in the literature to our best knowledge, gives a sufficient condition for the value of a boolean function to remain constant under a appropriately defined random change of input.

Lemma 5. *Let f be a Boolean function on n variables and $0 \leq \epsilon_i \leq \frac{1}{20}$ ($1 \leq i \leq n$) be a sequence of numbers that satisfy:*

$$\sum_{i \in S_x} \epsilon_i \leq \frac{1}{20} \text{ for every } x \in \{0,1\}^n. \tag{11}$$

Then for every $x \in \{0,1\}^n$ if we flip the ith bit of x with probability ϵ_i randomly and independently, the value that f attains on the new input equals to $f(x)$ with probability at least 90%.

Proof. Fix x. Let us denote the probability distribution on the inputs that we obtain by flipping the bits of x as described by the lemma by D. Let ϵ'_i be a probability related to ϵ_i to be determined later. For any $\omega \in \mathbf{N}$ we define a distribution D_ω by the following process:

We start from $x = x^{(0)}$. If $x^{(j)}$ is already created, we create $x^{(j+1)}$ by uniformly and randomly picking a (single) coordinate i, and flipping the bit $x_i^{(j)}$ with probability ϵ'_i/ω. The process ends after ωn steps, each new step being performed independently of the previous ones.

Lemma 6. *For an appropriate fixed sequence $\epsilon_i \leq \epsilon'_i \leq 2\epsilon_i$, $(1 \leq i \leq n)$ the distribution D_ω converges to the distribution D in the L_1 norm.*

Proof of the lemma: Let ω_i be the frequency with which the ith coordinate was selected during the process. We can interpret the above process so, that we first decide at a frequency list $(\omega_1, \ldots, \omega_n)$, and then for each i independently ω_i times we perform the experiment of flipping the ith bit with probability ϵ'_i/ω. For any $\delta > 0$, if ω is large enough ($\omega > \omega_\delta$), the probability of the event that the frequency list we select does not have the property, that $(1-\delta)\omega \leq \omega_i \leq (1+\delta)\omega$ for every $1 \leq i \leq n$, is less than δ. We call those lists *δ-typical* that have the property. Since for $\omega > \omega_\delta$ we have that D_ω is δ-close in L_1 norm to the convex combination of the distributions that come from the δ-typical lists, it is enough to show that for every fixed δ-typical frequency list the distribution that arises from that list, is very close to D. Observe, that the probability space arising from each fixed frequency list L is a product space where the ith bit of x is flipped with probability

$$p_{i,L} = \sum_{j=0}^{\omega_i/2} \binom{\omega_i}{2j+1} (1 - \epsilon'_i/\omega)^{\omega_i - 2j - 1} (\epsilon'_i/\omega)^{2j+1}$$
$$= \tfrac{1}{2} - \tfrac{1}{2}(1 - \tfrac{2\epsilon'_i}{\omega})^{\omega_i} \approx \tfrac{1}{2} - \tfrac{1}{2} e^{-2\epsilon' \frac{\omega_i}{\omega}}.$$

If $\epsilon'_i = -\frac{1}{2}\log(1 - 2\epsilon_i)$, and ω tends to infinity, and ω_i/ω tends to one, then the above probability tends to ϵ_i. □

Let us now continue the proof of the main lemma with estimating the probability that $f(x^{(j)}) \neq f(x)$. By induction on j we will show that

$$Prob(f(x^{(j)}) \neq f(x)) \leq \left(1 - \frac{1}{10\omega n}\right)^j.$$

Indeed, following the original definition of the random process, in the jth step ($j \geq 1$) we decide at an index $i \in [1, n]$ randomly. If $i \notin S_{x^{(j-1)}}$, then $f(x^{(j)}) = f(x^{(j-1)})$. The probability that our choice happens to fall on one of the of S_x, and this bit gets flipped is at most

$$\sum_{i \in S_x} \epsilon'_i/n\omega \leq \sum_{i \in S_x} 2\epsilon_i/n\omega \leq \frac{1}{10\omega n}$$

Since the jth step is independent from the previous ones, each branch of the process that results in $f(x^{(j-1)}) = f(x)$ yields an $f(x^{(j)})$ with $f(x^{(j)}) = f(x)$ with conditional probability at least $1 - \frac{1}{10\omega n}$. From here the induction is easily completed. The lemma now follows from $\left(1 - \frac{1}{10\omega n}\right)^{\omega n} \geq 0.9$.

Definition 2. $S(f) = \max_{x \in \{0,1\}^n} |S_x|$, *and is called the* sensitivity *of* f.

Theorem 5. *For some* $c_0, c_1 > 0$: $c_0 \log S(f) \leq l(f) \leq l_{triv}(f) \leq c_1 \log S(f)$.

For the lower bound let x be the most sensitive input of f. If we restrict f onto x, and its neighbors ($\{z \mid \mathrm{dist}(x,z) = 1\}$), we get a $FAN(S_n, n)$ problem. The bound now follows from Lemma 4

For the upper bound take $\epsilon_1 = \ldots = \epsilon_n = \frac{1}{20S(f)}$. By Lemma 5 if we start from any $x \in \{0,1\}^n$, flipping x_i with ϵ_i independently and randomly does not change the output with 90% probability. We can boost the accuracy of every bit to this level using the trivial construction as long as we elect l to be $\Theta(\log S(f))$. □

Assume now, that it cost unit price to obtain a single $\frac{1}{20}$-biased sample of an input bit. What is our total cost of computing a Boolean function f with 90% confidence if we charge only for buying the input, but not for the computation? Let $D^{rand}_{stat,0.05}(f)$ denote this minimal number. We determine $D^{rand}_{stat,0.05}(f)$ within a constant factor. For a set system $\mathcal{S}$ and a set H let us introduce:

$$\mathcal{S} \setminus \setminus H = \{S \setminus H \mid S \in \mathcal{S}\}$$

Define $\mathcal{S}_0 = \{S_x \mid x \in \{0,1\}^n\}$. We define a partition $S_1, S_2, \ldots$ of the set of bits on which f depends by

$$S_1 = S : \ S \in \mathcal{S}_0, \text{ and } |S| \text{ is maximal}; \ \mathcal{S}_1 = \mathcal{S}_0 \setminus \setminus S_1; \tag{12}$$

$$S_2 = S : \ S \in \mathcal{S}_1, \text{ and } |S| \text{ is maximal}; \ \mathcal{S}_2 = \mathcal{S}_1 \setminus \setminus S_2; \tag{13}$$

$$\vdots$$

Lemma 7. $D^{rand}_{stat,0.05}(f) = \Theta(\sum_i |S_i| \log |S_i|)$.

Proof. For the lower bound we use the simple conseqence of Theorem 4, that the total number of bits we need to buy for the variables belonging to each individual S_i is at least $\Omega(|S_i| \log |S_i|)$.

In order to prove the upper bound, we associate an $\epsilon_i = \frac{1}{100S_j^2}$ with every index i every index i in S_j. We show that for every $x \in \{0,1\}^n$ Equation 11 of lemma 5 holds. Let us enumerate the elements of S_x in the reverse order how they participate in the S_js, i.e. we take first those in the S_j with the highest index, etc. Elements in the same S_j can be taken in an arbitrary order. Let the enumeration of the elements of S_x we obtain this way be $k_1, k_2, \ldots$. We claim that $\epsilon_{k_i} \leq \frac{1}{100i^2}$. Indeed, if the opposite held, the S_j in which k_i participates were less than i. But then our greedy algorithm would select $\{k_1, k_2, \ldots, k_i\}$ in place of S_j in the jth step, since this set (or a super-set of it) is contained in $\mathcal{S}_{j-1}$.

The proof is now finished by two observations: First, by Lemma 5 if we flip bit i with probability ϵ_i independently and randomly, then, since $\sum \frac{1}{100i^2} \leq \frac{1}{20}$, we conclude that the output of f will not change with 90% certainty. Second, that by buying $O(\log |S_j|)$ copies of the input bits in S_j we can boost the certainty of our knowledge of the bits as to achieve failure probability at most $\frac{1}{100S_j^2}$, and therefore being able to implement our list of epsilons. This completes the proof. □

5 Counter-Examples and Computer Experiments

At the first sight to this topic, one might propose several reasonable conjectures. In this section, we provide several counter-examples, which helps us to build right intuitions. Some of the counter-examples are easily constructed and verified, some of them are verified by computer simulations. We wrote a C++ front end to CPLEX to generate the linear programming instances. CPLEX is a linear programming package sold by ILOG company.

A natual question is, whether the best construction is always a composition function. As we saw in MAJ function, the answer is negative. Another question could be: Does the best solution can always be achieved by integer solutions? If so, we lose nothing in the linear programming approach. As we can expect, the answer is again negative. Notice that a trivial fractional solution takes all F value to be 0.5 achieves $\Delta(f, F, \epsilon) = 0.5$. Now consider the AND function in two variables. It can be easily proved that the best integer approximation achieves 0.52, when it is the parity function or its negation. This example also shows that, if we have one copy of each input, then compute f directly from the input bits may be not the best solution to approximate f.

If the function f is monotone, one might natually think that best solution F is also monotone. The following example shows that, this is far away from the truth.

Table 1. A monotone function

0000 0	0001 1	0010 0	0011 1	0100 0	0101 1	0110 1	0111 1
1000 0	1001 1	1010 0	1011 1	1100 0	1101 1	1110 1	1111 1

The function shown in Table 1 is a monotone function. When $l = 3$, our simulation results show that the best solution is not monotone for ϵ equals to 0.1, 0.2, 0.3, and 0.4. When $\epsilon = 0.3$, there are 106 pairs of vectors in the best solution F violates the monotone property. Table 2 is a small part of the output.

Table 2. A small part of the best solution F. Each column shows F is not monotone

$F(y) = 1$	0002	0002	0102	3012	2102
$F(y) = 0$	0022	1032	2202	3022	2202

This example also serves another purpose. It is an evidence that, there are many fractional numbers in the best solution. This we believe is true for a general f. For $l = 3$, and $\epsilon = 0.3$, among all the 256 vectors, F does not take integer value on 10 of them.

By observing the truth table of the function above, it is quite reasonable why the best solution is not monotone. We can also show some similar yet smaller examples. Consider the monotone function $f(x_1, x_2, x_3) = x_3$. Table 3 is a set of values when $l = 3$ and $\epsilon = 0.3$.

Table 3. A small part of the best solution for $f(x_1, x_2, x_3) = x_3$. (The values are rounded)

y	002	012	102	202	212	302	312
$F(y)$	1.00	0.84	0.67	0.00	1.00	0.08	1.00

In all computer experiments, the best solution for FAN function and AND function are exactly the same.

Here are some results for the FAN function on 5 variables, with 3 copies for each bit. The results coincides with our theoretical result. The best solution is always monotone. There is at most one point y_0 where $F(y_0)$ is fractional. For small ϵ, the solution is in lexicographical order.

Table 4. The best solution for FAN function on 5 variables. There is a y_0 where $F(y_0)$ is fractional. For $\epsilon = 0.1$, 0.2, or 0.3, the best solution is in lexicographical order. For $\epsilon = 0.4$ or 0.45, there are y such that $y \leqslant_L y_0$ and $F(y) = 1$, as listed in last column

ϵ	y_0	$F(y_0)$	violation
0.1	12333	0.00406	
0.2	12233	0.52638	
0.3	11333	0.69614	
0.4	11222	0.52307	02333, 03333
0.45	11123	0.07840	01233, 01333, 02223, 02233, 02333, 03333

6 Conclusion

While we could determine $l(f)$ and related quantities within a constant factor, it is still open whether there is a fixed $c > 0$ such that $\xi(f, l, \epsilon) > c\xi_{triv}(f, l, \epsilon)$, for all f, l, and ϵ. We also succeeded in determining $D^{rand}_{stat,0.05}(f)$ (for the definition of $D^{rand}_{stat,0.05}$ see the previous section) within a constant factor. Our results have consequences on randomized decision trees, such as:

Lemma 8. *The randomized decision tree complexity of $f(g_1, \ldots, g_n)$ is at most $S(f) \cdot n$ times the maximal complexity of g_i.*

Since $l(f)$ gives an equivalent definition for the log-sensitivity of f, we may obtain a new approach to the sensitivity versus block-sensitivity question of Linial and Rubinstein [15]. Several open questions on constructions for specific functions remain unresolved, among which perhaps the most intriguing is: "Is the best construction for the FAN function and AND function coincide"?

Acknowledgements We would like to thank Noga Alon, Harry Buhrmann, Peter Gács, Mike Saks for conversations on the subject.

References

1. N. Alon, J. Spencer, *The Probabilistic Method*, Wisley, New York (2000).
2. I. Benjamini, G. Kalai, O. Schramm, Noise sensitivity of Boolean functions and applications to percolation, math.PR/9811157.
3. A. Bernasconi, Sensitivity vs. block sensitivity (an average-case study), *Information Processing Letters*, **59** (1996) 151-157.
4. U. Feige, D. Peleg, P. Raghavan, E. Upfal, Computing with unreliable information, *Proceedings of the 22nd Annual ACM Symposium on Theory of Computing* (1990), 128-137
5. P. Gács, A. Gál, Lower bounds for the complexity of reliable Boolean circuits with noisy gates, *IEEE Transactions on Information Theory*, Vol.40, (1994) pp.579-583.
6. A. Gál, Lower bounds for the complexity of reliable Boolean circuits with noisy gates, *Proceedings of the 32nd Annual IEEE Symposium on Foundations of Computer Science* (1991), 594–601.
7. J. Kahn, G. Kalai, N. Linial, The influence of variables on boolean functions, *Proceedings of the 29th Annual Symposium on Foundations of Computer Science* (1988), 68-80
8. C. Kenyon, A. C. Yao, On evaluating boolean functions with unreliable tests, *International Journal of Foundations of Computer Science* 1, 1 (1990), 1-10.
9. N. Nisan, M. Szegedy, On the degree of Boolean functions as real polynomials, *Proceedings of the Twenty Third Annual ACM Symposium on Theory of Computing* (1991), 419-429.
10. N. Nisan, CREW PRAMs and decision trees, *SIAM Journal on Computing*, **20** (1991), 999-1007.
11. J. Von Neumann, Probabilistic logics and the synthesis of reliable organisms from unreliable components, In *Automata Studies*, C. E. Shannon and J. McCarthy, eds. Princeton University Press (1956), 329-378.
12. N. Pippenger, On networks of noisy gates, *Proceedings of the 26th Annual Symposium on Foundations of Computer Science* (1985), 30-38.
13. N. Pippenger, Invariance of complexity measures for networks with unreliable gates, *Journal of the ACM* **36** (1989), 531-539.
14. Reliable computation with noisy circuits and decision trees, *Proceedings of the 32nd Annual Symposium on Foundations of Computer Science* (1991), 602-611.
15. D. Rubinstein, Sensitivity vs. block sensitivity of Boolean functions, *Combinatorica*, **15** (1995) 297–299.
16. CPLEX, http://www.ilog.com/products/cplex/

Inapproximability Results on Stable Marriage Problems

Magnús Halldórsson[1], Kazuo Iwama[2*], Shuichi Miyazaki[2], and Yasufumi Morita[2]

[1] Science Institute, University of Iceland
mmh@hi.is

[2] Graduate School of Informatics, Kyoto University
{iwama, shuichi, ymorita}@kuis.kyoto-u.ac.jp

Abstract. The stable marriage problem has received considerable attention both due to its practical applications as well as its mathematical structure. While the original problem has all participants rank all members of the opposite sex in a strict order of preference, two natural variations are to allow for incomplete preference lists and ties in the preferences. Both variations are polynomially solvable by a variation of the classical algorithm of Gale and Shapley. On the other hand, it has recently been shown to be NP-hard to find a maximum cardinality stable matching when both of the variations are allowed.
We show here that it is APX-hard to approximate the maximum cardinality stable matching with incomplete lists and ties. This holds for some very restricted instances both in terms of lengths of preference lists, and lengths and occurrences of ties in the lists. We also obtain an optimal $\Omega(N)$ hardness results for 'minimum egalitarian' and 'minimum regret' variants.

1 Introduction

An instance of the original *stable marriage problem* (*SM*) [5] consists of N men and N women, with each person having a preference list that totally orders all members of the opposite sex. A man and a woman form a *blocking pair* in a matching if both prefer each other to their current partners. A perfect matching is *stable* if it contains no blocking pair. For a matching M containing a pair (m, w), we write that $M(m) = w$ and $M(w) = m$. The stable marriage problem was first studied by Gale and Shapley [2], who showed that every instance contains a stable matching, and gave an $O(N^2)$-time algorithm, so-called the Gale-Shapley algorithm, to find one.

One natural relaxation is to allow for indifference [5,8], in which each person is allowed to include *ties* in his/her preference. This problem is denoted *SMT* (Stable Marriage with Ties). When ties are allowed, the definition of stability needs to be extended. A man and a woman form a blocking pair if each *strictly*

* Supported in part by Scientific Research Grant, Ministry of Japan, 13480081

S. Rajsbaum (Ed.): LATIN 2002, LNCS 2286, pp. 554–568, 2002.

prefers the other to his/her current partner. A matching without such a blocking pair is called *weakly stable* (or simply "stable"). Variations in which a blocking pair can involve non-strict preference suffer from the fact that a stable matching may not exist, whereas the Gale-Shapley algorithm can be modified to always find a weakly stable matching [5].

Another natural variation is to allow participants to declare one or more unacceptable partners. Thus each person's preference list may be incomplete. Again, the definition of a blocking pair is extended, so that each member of the pair prefers the other over the current partner *or* is currently single and acceptable. The Gale-Shapley algorithm can also be modified to find a stable matching of maximum size in this case [3].

The importance of stability in matchings has been clearly displayed by its success in assigning resident interns to hospitals. For instance, the National Resident Matching Program in the U.S. has used a modified Gale-Shapley algorithm to match more than 95% of its participants for over three decades [5]. Here, residents apply to a subset of hospitals (i.e. incomplete preference lists), with each hospital strictly ranking its applicants. A hospital-resident assignment is actually a many-one matching, but most algorithms and properties carry over from the one-one SM problem that we focus on here.

Strict ranking of all applicants is not reasonable for large hospitals; it is more that they would strictly rank the top candidates, leaving the remainder tied. Irving et. al. [9] report that in a planned Scottish matching scheme SPA, ties are allowed but then resolved using arbitrary tie-breaking. However, different tie-breakings can result in different sizes of stable matchings. Since the objective is to successfully assign as many of the candidates as possible (or to fill as many of the posts as possible, depending on viewpoint), it would be desirable to find an algorithm for maximum cardinality stable matching in the presence of ties and incomplete lists.

This problem *SMTI* (Stable Marriage with Ties and Incomplete lists) was recently considered by Iwama et al. [10] which resolved that it is NP-hard to find a maximum cardinality solution. Previously, such hardness results had been known only for non-bipartite stable matchings, known as the Stable Roommates problem [12]. This NP-hardness result for SMTI was further shown to hold for the restricted case when all ties occur only at the end of a list, occur only in one sex, and are of length only 2 [11]. 2-approximation is easy for this problem since any stable matching is maximal. However, no other approximability results have been shown up to the present.

We study in this paper the approximability of SMTI and several variants. First, we show that Max cardinality SMTI is APX-hard, i.e. hard to approximate within $1+\epsilon$, for some $\epsilon > 0$. The construction applies to a very restricted class of instances, where preference lists are of constant size and are either fully ordered or contain a single tied group. We can further modify it to make ties be of length 2. An important feature of the proof is to establish a "gap location at 1"; namely, that it is NP-hard to distinguish between instances that have a complete stable matching and those where any stable matching leaves a positive constant fraction of participants unmatched.

We then consider two variants: 'minimum egalitarian' SMT and 'minimum regret' SMT. Here preference lists are complete, but the quality of the stable matching depends on how much "less preferred" partner a participant receives, either in an average or a worst case sense. It is known that both problems are NP-hard and cannot be approximated within $N^{1-\epsilon}$ for any small ϵ unless P=NP [10,11]. We improve these results and show a worst possible $\Omega(N)$ lower bound on the approximability. Note that both problems are solvable in polynomial time if ties are not allowed [4,5,7].

Notation. Throughout this paper, instances contain equal number N of men and women. A goodness measure of an approximation algorithm T of an optimization problem is defined as usual: the *performance ratio* of T is the maximum over all instances x of size N of $\max\{T(x)/opt(x), opt(x)/T(x)\}$, where $opt(x)$ ($T(x)$) is the measure of the optimal (algorithm's) solution, respectively. A problem is *hard to approximate within* $f(N)$, if the existence of a polynomial-time algorithm with performance ratio $f(N)$ implies that P=NP.

2 Inapproximability of MAX SMTI

We focus on the maximum cardinality SMTI problem.

Problem: MAX SMTI.
Instance: N men, N women and each person's preference list which may be incomplete and may include ties.
Purpose: Find a stable matching of maximum cardinality.

Theorem 1. *MAX SMTI is hard to approximate within* $1+\epsilon$, *for some constant* $\epsilon > 0$.

Proof. Recall that MAX SAT is the problem of finding a truth assignment to the variables of a given propositional formula in CNF form that satisfies as many clauses as possible. MAX E3SAT(t) is a restriction of MAX SAT, where each clause has exactly three literals, and each variable appears at most t times. It is known that, if P≠NP, there exists a positive constant α and an integer t such that there is no approximation algorithm for MAX E3SAT(t) whose performance ratio is less than $1+\alpha$ [1,6]. More precisely, the problem has a useful 'gap location' property: Suppose that an instance f of SAT is translated into an instance g of MAX E3SAT(t). If f is satisfiable, then g is also satisfiable. Otherwise, i.e., if f is unsatisfiable, the number of unsatisfied clauses is more than $\delta(=\frac{\alpha}{1+\alpha})$ fraction of clauses of g in any assignment.

We translate an instance f of MAX E3SAT(t) having the above property into an instance $T(f)$ of MAX SMTI. Our reduction has the following property: If f is satisfiable, there is a stable matching for $T(f)$ in which all the people are matched (Lemma 2). If more than δ fraction of clauses of f are unsatisfied in any assignment, then more than $\frac{\delta}{9t}$ fraction of men are single in any stable matching for $T(f)$ (Lemma 4). Hence a polynomial-time $1/(1-\frac{\delta}{9t})$-approximation algorithm implies P=NP.

The reduction is similar to that of [10], with some important simplifications. Let n and l be the numbers of variables and clauses of f, respectively, and let $C_j (1 \le j \le l)$ be the jth clause of f. Let t_i be the number of appearances of the variable x_i. (Thus $t = \max\{t_1, t_2, \cdots, t_n\}$.) We construct an instance $T(f)$, namely, the set of men and women, each man's preference list, and each woman's preference list.

The Set of Men and Women $T(f)$ contains $2n + 6l$ men and women. We divide men and women into three groups, respectively, in the following way:

The Set of Men

Group (a): For each clause C_j, we introduce three men a_j, a'_j and a''_j.
Group (b): For each variable x_i, we introduce two men b_i and b'_i.
Group (c): For each literal x_i (or $\overline{x_i}$) in the clause C_j, we introduce a man $c_{i,j}$.

The Set of Women

Group (u): For each variable x_i, we introduce a woman u_i.
Group (v): For each variable x_i, we introduce a woman v_i.
Group (w): For each literal x_i (or $\overline{x_i}$) in the clause C_j, we introduce two women $w^1_{i,j}$ and $w^0_{i,j}$.

Since there are n variables, we have $2n$ group-(b) men, and n group-(u) and group-(v) women. Since there are l clauses and $3l$ literals, there are $3l$ group-(a) and group-(c) men and $6l$ group-(w) women.

Men's Preference Lists We then construct each man's preference list. For better exposition, we use an example of f,

$$f_0 = (x_1 + \overline{x_2} + x_3)(\overline{x_1} + x_2 + x_4)(x_2 + \overline{x_4} + \overline{x_5}).$$

For this instance, men's preference lists will turn out to be as illustrated in Table 1, in which t is equal to 3. Note that Table 1 contains several blanks defined for convenience for constructing the women's preference lists, as detailed later.

For each clause C_j, three men a_j, a'_j and a''_j in Group (a) are introduced. We show how to construct preference lists of men a_1, a'_1 and a''_1 who are associated with $C_1 = (x_1 + \overline{x_2} + x_3)$ of f_0. Since literals x_1, $\overline{x_2}$ and x_3 appear in C_1, six women $w^0_{1,1}$, $w^1_{1,1}$, $w^0_{2,1}$, $w^1_{2,1}$, $w^0_{3,1}$ and $w^1_{3,1}$ have been introduced. a_1 writes $w^1_{1,1}$, $w^0_{2,1}$ and $w^1_{3,1}$ at the first position. (These three women are tied in the list.) Generally speaking, the man a_j writes the woman $w^1_{i,j}$ if x_i appears positively in C_j, and writes $w^0_{i,j}$ if x_i appears negatively in C_j. Both a'_1 and a''_1 write all the above six women at the first position.

Then we construct preference lists of Group-(b) men. We show how to construct preference lists using men b_2 and b'_2 who are associated with the variable x_2 of f_0. The man b_2 writes the woman u_2 at the 2nd position. (The 2nd position is always determined without depending on f.) Then, b_2 writes the woman v_2 at the $t+4(=7)$th position. Since x_2 appears in clauses C_1, C_2 and C_3, three women $w^0_{2,1}$, $w^0_{2,2}$ and $w^0_{2,3}$ have been introduced. b_2 writes $w^0_{2,1}$, $w^0_{2,2}$ and $w^0_{2,3}$ at the 3rd, 4th and 5th positions, respectively. Generally speaking, there are t_i women of the form $w^0_{i,j}$ corresponding to the variable x_i. (Recall that t_i is the number of appearances of the variable x_i.) b_i writes these women at 3rd through (t_i+2)th positions. Since $t_i \leq t$, these women's positions never reach $(t+4)$th position which is already occupied by v_i.

b'_2's list is similarly constructed. b'_2 writes the woman u_2 at the 1st position and writes the woman v_2 at the $t+3(=6)$th position. There are three women $w^1_{2,1}$, $w^1_{2,2}$ and $w^1_{2,3}$ associated with the variable x_2 since x_2 appears in C_1, C_2 and C_3. b'_2 writes $w^1_{2,1}$, $w^1_{2,2}$ and $w^1_{2,3}$ at the 3rd, 4th and 5th positions, respectively.

Now we move to Group-(c) men. The man $c_{i,j}$ (associated with x_i in C_j) writes women $w^1_{i,j}$ and $w^0_{i,j}$ (associated with x_i in C_j) at the $t+3(=6)$th and $t+4(=7)$th positions, respectively. Now men's lists are completed. Table 1 shows the whole lists of men of $T(f_0)$. As we have mentioned before, men's lists currently contain blanks. Blanks will be removed after we construct women's lists.

Women's Preference Lists We construct women's preference lists automatically from men's preference lists. First, we determine the total order of all men; the position of each man in the order is called his *rank*. The rank of each man of our current example $T(f_0)$ is shown in Table 1, e.g., a_1 is the highest and $c_{5,3}$ is the lowest. Generally speaking, men are lexicographically ordered, where the significance of the indices of $\alpha^{\delta}_{\beta,\gamma}$ is in the order of α, β, γ and δ, e.g., α is the most significant index and δ is the least significant index. For α, the priority is given to a, b and c in this order. For β and γ, the smaller number precedes the larger number. For δ, fewer primes has more priority.

Women's lists are constructed based on this order. First of all, the preference list of a woman w does not include a man m if w does not appear on m's preference list. Then consider two men m_i and m_j who write w in the list. w strictly prefers m_i to m_j if and only if (1) the rank of m_i is higher than that of m_j, and (2) the position of w in m_i's list is higher than or equal to the position of w in m_j's list. One might think that women's lists can contain partial order in this construction. However, in our translation, each woman's list contains only ties.

It helps much to know that by our construction of women's preference lists, we can determine whether a matching includes a blocking pair only from men's lists. Consider men m_i and m_j matched with w_i and w_j, respectively. Then, (m_i, w_j) is a blocking pair if and only if (i) m_i strictly prefers w_j to w_i, (ii) m_i's rank is higher than m_j's rank, and (iii) the position of w_j in m_i's list is higher than or equal to the position of w_j in m_j's list. Observe that the combination of conditions (ii) and (iii) means that w_j strictly prefers m_i to m_j.

a_1	$w_{1,1}^1$	$w_{2,1}^0$	$w_{3,1}^1$				
a_1'	$w_{1,1}^0$	$w_{2,1}^0$	$w_{3,1}^0$	$w_{1,1}^1$	$w_{2,1}^1$	$w_{3,1}^1$	
a_1''	$w_{1,1}^0$	$w_{2,1}^0$	$w_{3,1}^0$	$w_{1,1}^1$	$w_{2,1}^1$	$w_{3,1}^1$	
a_2	$w_{1,2}^0$	$w_{2,2}^1$	$w_{4,2}^1$				
a_2'	$w_{1,2}^0$	$w_{2,2}^0$	$w_{4,2}^0$	$w_{1,2}^1$	$w_{2,2}^1$	$w_{4,2}^1$	
a_2''	$w_{1,2}^0$	$w_{2,2}^0$	$w_{4,2}^0$	$w_{1,2}^1$	$w_{2,2}^1$	$w_{4,2}^1$	
a_3	$w_{2,3}^1$	$w_{4,3}^0$	$w_{5,3}^0$				
a_3'	$w_{2,3}^0$	$w_{4,3}^0$	$w_{5,3}^0$	$w_{2,3}^1$	$w_{4,3}^1$	$w_{5,3}^1$	
a_3''	$w_{2,3}^0$	$w_{4,3}^0$	$w_{5,3}^0$	$w_{2,3}^1$	$w_{4,3}^1$	$w_{5,3}^1$	
b_1		u_1	$w_{1,1}^0$	$w_{1,2}^0$			v_1
b_1'	u_1		$w_{1,1}^1$	$w_{1,2}^1$		v_1	
b_2		u_2	$w_{2,1}^0$	$w_{2,2}^0$	$w_{2,3}^0$		v_2
b_2'	u_2		$w_{2,1}^1$	$w_{2,2}^1$	$w_{2,3}^1$	v_2	
b_3		u_3	$w_{3,1}^0$				v_3
b_3'	u_3		$w_{3,1}^1$			v_3	
b_4		u_4	$w_{4,2}^0$	$w_{4,3}^0$			v_4
b_4'	u_4		$w_{4,2}^1$	$w_{4,3}^1$		v_4	
b_5		u_5	$w_{5,3}^0$				v_5
b_5'	u_5		$w_{5,3}^1$			v_5	
$c_{1,1}$						$w_{1,1}^0$	$w_{1,1}^1$
$c_{1,2}$						$w_{1,2}^0$	$w_{1,2}^1$
$c_{2,1}$						$w_{2,1}^0$	$w_{2,1}^1$
$c_{2,2}$						$w_{2,2}^0$	$w_{2,2}^1$
$c_{2,3}$						$w_{2,3}^0$	$w_{2,3}^1$
$c_{3,1}$						$w_{3,1}^0$	$w_{3,1}^1$
$c_{4,2}$						$w_{4,2}^0$	$w_{4,2}^1$
$c_{4,3}$						$w_{4,3}^0$	$w_{4,3}^1$
$c_{5,3}$						$w_{5,3}^0$	$w_{5,3}^1$

Table 1. Preference lists of men of $T(f_0)$

$$f_0 = (x_1 + \overline{x_2} + x_3)(\overline{x_1} + x_2 + x_4)(x_2 + \overline{x_4} + \overline{x_5})$$

Finally, we remove blanks in men's lists. For each man's list, we simply slide women to the left until no blank remains.

Correctness of the reduction As mentioned before, the correctness follows from two Lemmas 2 and 4.

Lemma 2. *If f is satisfiable, then there is a perfect stable matching for $T(f)$.*

Proof. Suppose that f is satisfied by an assignment A and let $A(x_i) \in \{0, 1\}$ be the value assigned to x_i under A. We construct a stable matching M whose cardinality is $N(= 2n + 6l)$ as follows. Each man in Group (b) is matched according to the assignment A. If $A(x_i) = 0$, then let $M(b_i) = v_i$ and $M(b'_i) = u_i$, and if $A(x_i) = 1$, then let $M(b_i) = u_i$ and $M(b'_i) = v_i$. There is a Group-(c) man $c_{i,j}$ associated with a literal x_i (or $\overline{x_i}$) in the clause C_j. If $A(x_i) = 0$, then let $M(c_{i,j}) = w^1_{i,j}$, and if $A(x_i) = 1$, then let $M(c_{i,j}) = w^0_{i,j}$.

Now we go back to Group-(a) men. Recall that of the two women $w^0_{i,j}$ and $w^1_{i,j}$ associated with the literal x_i (or $\overline{x_i}$) in the clause C_j, one is matched with a man in Group (c) and the other one is still unmatched. Namely, if $A(x_i) = 0$ it is $w^0_{i,j}$ that is unmatched. These unmatched women will be matched with Group-(a) men. Consider a clause C_j with literals z_{i_1}, z_{i_2} and z_{i_3} (i.e., z_{i_k} is x_{i_k} or $\overline{x_{i_k}}$). Since C_j is satisfied by A, at least one of these three literals has the value 1. Without loss of generality, let the literal be z_{i_1}. If $z_{i_1} = x_{i_1}$ then $A(x_{i_1})$ must be 1 and hence, the woman $w^1_{i_1,j}$ is unmatched as mentioned above. By construction of preference lists of Group-(a) men, a_j writes $w^1_{i_1,j}$ in the list because x_{i_1} appears positively in C_j. Otherwise, i.e. if $z_{i_1} = \overline{x_{i_1}}$, then $w^0_{i_1,j}$ is unmatched and a_j writes $w^0_{i_1,j}$ in the list. In either case, a_j can be matched with the woman corresponding to the literal that makes the clause true. There are two other literals z_{i_2} and z_{i_3} in C_j. So there are two unmatched women; one is $w^0_{i_2,j}$ or $w^1_{i_2,j}$, and the other is $w^0_{i_3,j}$ or $w^1_{i_3,j}$, depending on which value x_{i_2} and x_{i_3} receive under A. a'_j and a''_j will be matched with those two women.

Now we have a perfect matching. Since we have shown how to detect blocking pairs, it is easy to check that this matching M is stable. □

Lemma 3. *Let M be an arbitrary stable matching for $T(f)$. If the number of unmatched men in M is k, then there is an assignment for f by which the number of unsatisfied clauses is at most tk.* (The proof is given in Sec. 2.1.)

Lemma 4. *If more than δ fraction of clauses of f are unsatisfied in any assignment, then more than $\frac{\delta}{9t}$ fraction of men are single in any stable matching for $T(f)$.*

Proof. Recall that the number of men in $T(f)$ is $2n+6l$, where n is the number of variables of f and l is the number of clauses of f. Since we can assume that each variable appears at least twice, we have $n \leq 3l/2$. Hence we have $2n + 6l \leq 9l$ men.

Suppose that there is a stable matching for $T(f)$ such that the number of single men is at most $\frac{\delta}{t}l$. Then, by Lemma 3, there must be an assignment for f such that the number of unsatisfied clauses is at most δl, a contradiction. Therefore, more than $\frac{\delta}{t}l$ men are single in any stable matching for $T(f)$. The fraction of men that are single exceeds $(\frac{\delta}{t}l)/9l = \frac{\delta}{9t}$. □

As we have mentioned in the beginning of this proof, there is a positive constant δ such that it is NP-hard to distinguish the following two cases for MAX E3SAT(t): (i) the formula is satisfiable, and (ii) the number of unsatisfied clauses is more than δ fraction in any assignment. By Lemmas 2 and 4, the theorem holds. □

The instances constructed in the proof above have quite restrictive properties. All preference lists are of constant size, or at most $t+2$, where t can be set as small as 5. Also, preference lists are either totally ordered, or totally unordered (i.e. a single tied list). In Sec. 2.2, we give a modification to ensure that ties are all of length 2.

2.1 Proof of Lemma 3

Proof. First of all, it should be noted that Group-(a) men are matched in any stable matching. The reason is as follows: Suppose there is a Group-(a) man m who is single in some stable matching M. Then women written on m's preference list cannot be single since otherwise, that single woman and m form a blocking pair. Thus every woman on m's list must be matched in M. Since we have assumed that m is single, at least one woman w on m's list cannot be matched with Group-(a) men, and hence she is matched with a man in Groups (b) or (c). This can be easily checked by construction. m and w form a blocking pair since Group-(w) women strictly prefers men in Group (a) to men in Groups (b) and (c).

Given an arbitrary stable matching M for $T(f)$, we first determine an *incomplete assignment* A_M which is an assignment to literals of f depending on how Group-(a) men are matched in M. A_M is not an usual truth assignment for variables but an assignment for literals which may contain several contradictions. We denote the literal x_i (resp. $\overline{x_i}$) in the clause C_j by x_i^j (resp. $\overline{x_i^j}$). Suppose x_i appears positively (resp. negatively) in C_j. Then we say that the value of the literal x_i^j (resp. $\overline{x_i^j}$) is *consistent* with the value of the variable x_i if $x_i^j = x_i$ (resp. $\overline{x_i^j} \neq x_i$). We say that two literals associated with x_i are consistent if we can assign the value to x_i so that both literals are consistent with x_i. Note that the consistency of two literals does not depend on the value of variable x_i.

Now we are ready to show how to construct A_M. As we have seen before, every man in Group (a) is matched in M. Suppose that the woman $w_{i,j}^d$ ($d \in \{0,1\}$) is matched with a man in Group (a). This woman exists in Group (w) because the variable x_i appears in the clause C_j. We assign the value to the literal x_i^j (or $\overline{x_i^j}$) so that the value of the literal is consistent with $x_i = d$. Note that, in an incomplete assignment, it can be the case that a literal receives both 0 and 1, or that a literal receives no value.

For example, recall the example in Table 1. Suppose that in a stable matching, say M_0, a_1, a_1' and a_1'' are matched with $w_{1,1}^1$, $w_{1,1}^0$ and $w_{2,1}^1$, respectively. Then, under the incomplete assignment A_{M_0}, x_1^1 receives both 0 and 1, $\overline{x_2^1}$ receives 0 (to be consistent with $x_2 = 1$), and x_3^1 receives no value. Observe that, by this incomplete assignment A_M, each clause C_j contains at least one literal whose value is 1, which corresponds to the woman who is matched with the man a_j.

For each variable x_i, define $CL_i = \{j | x_i \text{ appears in } C_j\}$. Partition CL_i into three subsets according to A_M: $CL_i^2(A_M) = \{j | j \in CL_i$ and x_i in C_j receives both 0 and 1 under $A_M\}$. $CL_i^1(A_M) = \{j | j \in CL_i$ and x_i in C_j receives exactly one value under $A_M\}$. $CL_i^0(A_M) = \{j | j \in CL_i$ and x_i in C_j receives no value under $A_M\}$.

We say that the variable x_i has *Type-I contradiction* if $CL_i^2(A_M) \neq \emptyset$. We say that x_i has *Type-II contradiction* if $CL_i^0(A_M) \neq \emptyset$. We say that x_i has *Type-III contradiction* if there are j_1 and j_2 such that $j_1, j_2 \in CL_i^1(A_M)$ and literals in C_{j_1} and C_{j_2}, associated with x_i, are not consistent. Before proving Lemma 3, we need to prove the following lemmas (Lemmas 5 through 8).

Lemma 5. *(1) If $i_1 \neq i_2$, then (i) $CL_{i_1}^2(A_M) \cap CL_{i_2}^2(A_M) = \emptyset$, and (ii) $CL_{i_1}^0(A_M) \cap CL_{i_2}^0(A_M) = \emptyset$. (2) (i) If $j \in CL_{i_1}^0(A_M)$ for some i_1, then there exists i_2 such that $j \in CL_{i_2}^2(A_M)$. (ii) If $j \in CL_{i_1}^2(A_M)$ for some i_1, then there exists i_2 such that $j \in CL_{i_2}^0(A_M)$.*

Proof. Note that every man in Group (a) is matched in M. Thus, for each clause C_j, the total number of values which three literals in C_j receive is exactly three. Then it is an easy calculation to see that the lemma holds. □

Lemma 6. *Let M be a stable matching for $T(f)$, and A_M be an incomplete assignment constructed from M. Suppose that x_i has Type-II contradiction under A_M and $j \in CL_i^0(A_M)$, namely, x_i in the clause C_j receives no value. Then the man $c_{p,j}$ is single in M. Here p is an integer such that $j \in CL_p^2(A_M)$, whose existence is guaranteed by Lemma 5 (2)-(i).*

Proof. Since $j \in CL_p^2(A_M)$, the literal z_p^j receives two values under A_M. This means that two women $w_{p,j}^1$ and $w_{p,j}^0$ associated with this literal are both matched with some men in Group (a). Then the man $c_{p,j}$ cannot be matched in M since this man writes only these two women in the list. □

Lemma 7. *Let M be a stable matching for $T(f)$, and A_M be an incomplete assignment constructed from M. Suppose that x_i has Type-III contradiction but no Type-II contradiction under A_M. Then at least one man among b_i, b_i' and $c_{i,j}$, where $j \in CL_i^1(A_M)$, is unmatched in M.*

Proof. Suppose that appearance of x_i in C_{j_1} and C_{j_2} causes Type-III contradiction. There are four cases according to the polarity of x_i in C_{j_1} and C_{j_2}. Assume that x_i appears positively in C_{j_1} and negatively in C_{j_2}, namely, each of these literals receive one value such that $x_i^{j_1} = \overline{x_i^{j_2}}$ under A_M. Other three cases are similar to this case and can be omitted. We still have two possibilities: (i) $x_i^{j_1} = \overline{x_i^{j_2}} = 1$, and (ii) $x_i^{j_1} = \overline{x_i^{j_2}} = 0$. We give the proof for (i). The other case is similar.

We assume that all b_i, b_i' and $c_{i,j}$ ($j \in CL_i^1(A_M)$) are matched in M and show a contradiction. Consider a man $c_{i,j}$ for each j such that $j \in CL_i^1(A_M)$. Since $j \in CL_i^1(A_M)$, one of $w_{i,j}^1$ and $w_{i,j}^0$ is matched with a man in Group (a) and the other is matched with the man $c_{i,j}$. Especially, $M(c_{i,j_1}) = w_{i,j_1}^0$, and $M(c_{i,j_2}) = w_{i,j_2}^1$ by our assumption that $x_i^{j_1} = \overline{x_i^{j_2}} = 1$.

Now we turn to two men b_i and b'_i. As discussed above, $w^1_{i,j}$ and $w^0_{i,j}$ ($j \in CL^1_i(A_M)$) are all matched with men in Groups (a) or (c). Also, for each j such that $j \in CL^2_i(A_M)$, $w^1_{i,j}$ and $w^0_{i,j}$ are both matched with men in Group (a). Note that there is no j such that $j \in CL^0_i(A_M)$ since x_i does not contain Type-II contradiction. As a result, all Group-(w) women written on the list of b_i or b'_i are matched with men in Group (a) or (c). Thus if both b_i and b'_i are matched in M, it must be one of the following two cases: (1) $M(b_i) = v_i$ and $M(b'_i) = u_i$, and (2) $M(b_i) = u_i$ and $M(b'_i) = v_i$. In case (1), b_i and w^0_{i,j_1} form a blocking pair. In case (2), b'_i and w^1_{i,j_2} form a blocking pair. In either case, it contradicts the fact that M is stable. □

Lemma 8. *Let M be a stable matching for $T(f)$, and A_M be an incomplete assignment constructed from M. If the number of variables that have Type-II and/or Type-III contradiction under A_M is k, then there are at least k single men in M.*

Proof. Suppose that variables x_{i_1} and x_{i_2} ($i_1 \neq i_2$) have Type-II or Type-III contradiction. By Lemmas 6 and 7, at least one man associated with each variable is single. Let them be m_{i_1} and m_{i_2}, respectively. All we have to show is that $m_{i_1} \neq m_{i_2}$. We consider the following four cases:

Case 1: Both x_{i_1} and x_{i_2} have Type-II contradiction. By Lemma 6, m_{i_1} is $c_{p,r}$ and m_{i_2} is $c_{q,s}$ for some p, q, r and s. Recall the proof of Lemma 6. This means that x_{i_1} in the clause C_r receives no value and so, x_p in C_r receives both 0 and 1. Also x_{i_2} in C_s receives no value and so, x_q in C_s receives both 0 and 1. Namely, $r \in CL^0_{i_1}(A_M)$ and $s \in CL^0_{i_2}(A_M)$. By Lemma 5 (1)-(ii), $r \neq s$ and hence $m_{i_1} \neq m_{i_2}$.

Case 2: Only x_{i_1} has Type-II contradiction. In this case, m_{i_1} is $c_{p,r}$ where $r \in CL^2_p(A_M)$, and m_{i_2} is one of b_{i_2}, b'_{i_2} and $c_{i_2,j}$, where $j \in CL^1_{i_2}(A_M)$. If m_{i_2} is b_{i_2} or b'_{i_2}, then clearly $m_{i_1} \neq m_{i_2}$. Suppose m_{i_2} is $c_{i_2,j}$ for some j such that $j \in CL^1_{i_2}(A_M)$. If $m_{i_1} = m_{i_2}$, p and r must be equal to i_2 and j, respectively. This is impossible because it results that $r \in CL^1_p(A_M)$ and $r \in CL^2_p(A_M)$.

Case 3: Only x_{i_2} has Type-II contradiction. Same to Case 2.

Case 4: Neither x_{i_1} nor x_{i_2} have Type-II contradiction. By Lemma 7, m_{i_1} must be one of b_{i_1}, b'_{i_1} and c_{i_1,j_1} for some j_1, and m_{i_2} must be one of b_{i_2}, b'_{i_2} and c_{i_2,j_2} for some j_2. Clearly $m_{i_1} \neq m_{i_2}$ because $i_1 \neq i_2$. □

Now we are ready to prove Lemma 3. Suppose there are k unmatched men in M. Then, by Lemma 8, the number of variables that have Type-II or III contradiction is at most k. We construct a truth assignment A'_M of f from A_M in the following way: If x_i contains Type-II or Type-III contradiction, determine $A'_M(x_i)$ arbitrarily. If x_i does not contain any type of contradiction, all literals associated with x_i are consistent under A_M. We determine $A'_M(x_i)$ so that all the literals become consistent with x_i. Otherwise, if x_i contains only Type-I contradiction, then $CL_i = CL^1_i(A_M) \cup CL^2_i(A_M)$, namely, each literal associated with x_i receives both 1 and 0, or exactly one value because x_i does not contain

Type-II contradiction. Furthermore, all literals which receive one value are consistent since x_i does not contain Type-III contradiction. We determine $A'_M(x_i)$ so that those literals become consistent with x_i.

Recall that, under the incomplete assignment A_M, every clause has at least one literal to which the value 1 is assigned. We count an upper bound on the number of clauses that become 0 by changing A_M into A'_M. Let L be the set of all clauses that contain a variable having Type-II or Type-III contradiction and let $\overline{L}$ be the set of all remaining clauses. Since there are at most k variables that have Type-II or III contradiction, and since each variable appears at most t times, it turns out that $|L| \leq tk$. We claim that clauses in $\overline{L}$ are all satisfied by A'_M. Let C_j be a clause in $\overline{L}$ and z_i^j (which is x_i^j or $\overline{x_i^j}$) be a literal in C_j. If x_i does not contain any contradiction under A_M, then the value of z_i^j is equivalent under A_M and A'_M. Suppose x_i contains only Type-I contradiction, i.e., $j \in CL_i^1(A_M)$ or $j \in CL_i^2(A_M)$. If $j \in CL_i^1(A_M)$, then again the value of z_i^j is equivalent under A_M and A'_M by definition of $A_{M'}$. We then claim that $j \notin CL_i^2(A_M)$: If $j \in CL_i^2(A_M)$ then there must be p such that $j \in CL_p^0(A_M)$ by Lemma 5 (2)-(ii). So C_j contains a variable containing Type-II contradiction and hence C_j must be in L.

Now, for any $C_j \in \overline{L}$, the value of every literal in C_j under A'_M is equivalent to the value of it under A_M. Again, recall that all the clauses have at least one literal having the value 1 under A_M. So every clause in $\overline{L}$ is satisfied by A'_M. □

2.2 Hardness for Restricted Instances

In this section, we show how to modify the construction in the proof of Theorem 1 to obtain the same hardness result for restricted instances where the length of ties is at most two.

Theorem 9. *MAX SMTI is hard to approximate within* $1+\epsilon$, *for some* $\epsilon > 0$, *even if each person writes at most one tie of length two.*

Proof. In the problem MAX ONE-IN-THREE E3SAT(t), we are given a CNF formula such that each clause contains exactly three literals and each variable appears at most t times. A clause is satisfied if and only if exactly one literal in the clause is true. The purpose of this problem is to find an assignment which satisfies a maximum number of clauses. By a simple polynomial-time reduction [13] from MAX E3SAT(t), we can show that there exists a constant $\delta > 0$ such that it is NP-hard to distinguish the following two cases: (i) There is an assignment that satisfies all the clauses of f. (ii) In any assignment, at least δ fraction of the whole clauses are unsatisfied.

We will slightly modify the reduction in the proof of Theorem 1 in the following way. In the reduction in the proof of Theorem 1, men a'_j and a''_j write six women corresponding to literals in the jth clause. In the new reduction, these two men write only three women, that is, three women among six, who do not appear in a_j's list. We can similarly show that the resulting MAX SMTI instance has a gap property. Observe that, in men's side, ties appear only in Group-(a) men's lists, each of length three.

Let I be an instance of SMTI constructed as above. We modify I and construct a new instance I' with preserving the gap property. Consider a Group-(a) man m who writes three women w_1, w_2 and w_3 in the list. We replace this man m with two men m_1 and m_2 and a woman y whose preference lists are as follows:

$$m_1\colon\ y\ \ (w_1\ \ w_2) \qquad\qquad y\colon\ (m_1\ \ m_2)$$
$$m_2\colon\ y\ \ (w_2\ \ w_3)$$

Here, persons within a parenthesis are tied. In w_1's list, m is replaced by m_1, and in w_3's list, m is replaced by m_2. In w_2's list, m is replaced by m_1 and m_2 in this order of perference. We call these three persons m_1, m_2 and y a *block* of m. The size of I' (i.e. the number of men in I') is bounded by a constant times of the size of I. The correctness follows from following two lemmas:

Lemma 10. *If there is a perfect stable matching for I, then there is a perfect stable matching for I'.*

Proof. Let M be a perfect stable matching for I. We construct a perfect stable matching M' for I'. Consider a Group-(a) man m of I who is replaced by m_1, m_2 and y as above. Recall that all Group-(a) men are matched in any stable matching. Hence m is matched in M. If m is matched with w_1 (w_2) in M, then m_1 is matched with w_1 (w_2) and m_2 is matched with y in M'. If m is matched with w_3 in M, then m_2 is matched with w_3 and m_1 is matched with y in M'. Men in Group (b) or (c) are matched in the same way as M. It is not hard to see that M' is stable in I'. □

Lemma 11. *If more than k men are unmatched in any stable matching for I, then more than k men are unmatched in any stable matching for I'.*

Proof. Suppose that there is a stable matching M' for I' in which k men are unmatched. We construct a stable matching M for I in which k men are unmatched.

Suppose that a man m of I is replaced by m_1, m_2 and y as above. Then, it is not hard to see that y is matched with m_1 or m_2 in any stable matching for I'. Hence exactly one man (m_1 or m_2) is unmatched with a woman in m's block. Although details are omitted, we can show that this man has a partner in any stable matching for I', namely, he is matched with a woman outside m's block.

Now we construct a matching M from M'. Each man except for Group-(a) men is matched with the same woman as M', or unmatched if he is unmatched in M'. Consider a Group-(a) man m. As discussed above, in M', there is one man, say m_i, who is matched with outside m's block. In M, m is matched with the woman with whom m_i is matched in M'. We can easily verify that M is stable. The number of unmatched men is same in M and M'. □

We can further show that the hardness result holds for instances in which ties appear only in one sex. Because of the space restriction, we give only a rough sketch of its proof.

Theorem 12. *MAX SMTI is hard to approximate within $1+\epsilon$, for some $\epsilon > 0$, even if ties appear in only men's lists, each man writes at most one tie of length at most three.*

Proof. As in the proof of Theorem 9, we modify an SMTI instance, say I, translated from MAX ONE-IN-THREE E3SAT(t). Recall that in I, ties of length three appear in Group-(a) men's lists and ties of length two appear in Group-(u) and Group-(v) women's lists. For each i ($1 \leq i \leq n$), consider two men b_i, b_i' and two women u_i, v_i:

$$\begin{array}{llll} b_i\colon & u_i \ \cdots \ v_i & \qquad u_i\colon & (b_i \ \ b_i') \\ b_i'\colon & u_i \ \cdots \ v_i & \qquad v_i\colon & (b_i \ \ b_i') \end{array}$$

In our reduction, these four people will be replaced by eight people whose preference lists are shown in the following figure:

$$\begin{array}{lll lll} s_u & : & (u_{i1} \ \ u_{i2}) & \qquad u_{i1} & : & s_u \ \ b_i \\ s_v & : & (v_{i1} \ \ v_{i2}) & \qquad u_{i2} & : & s_u \ \ b_i' \\ b_i & : & u_{i1} \ \cdots \ v_{i1} \ \ v_{i2} & \qquad v_{i1} & : & s_v \ \ b_i \ \ b_i' \\ b_i' & : & u_{i2} \ \cdots \ v_{i2} \ \ v_{i1} & \qquad v_{i2} & : & s_v \ \ b_i' \ \ b_i \end{array}$$

The correctness follows from a similar argument as in the proof of Theorem 9. □

3 Inapproximability of MIN Egalitarian SMT and MIN Regret SMT

The *regret* of a person p in a matching M is defined to be the rank (within p's preference list) of p's partner in M. Namely, $regret_M(p)$ is the number of persons that p *strictly prefers* to his/her current partner $M(p)$ plus one. Given that there can be many possible solutions to a stable marriage instance, it is natural to seek a solution that maximizes the overall satisfaction with the assignment, or alternatively minimizes the dissatisfaction.

Minimum egalitarian stable matching The cost of a stable matching M is defined to be $\sum_p regret_M(p)$, where the sum is taken over all persons. For the classical stable marriage instances without ties, a polynomial-time algorithm is known for finding an optimal stable matching [5,7]. However, when ties are allowed, the problem becomes intractable even with complete preference lists. Denote this problem, with ties but complete lists, as *MIN egalitarian SMT*. We give here a lower bound $\Omega(N)$ on the approximation ratio. Note that the cost of a matching is between $2N$ and $2N^2$. Hence an approximation ratio N is trivial. Our inapproximability result is optimal within a constant factor.

Theorem 13. *MIN egalitarian SMT is hard to approximate within* ϵN, *for some* $\epsilon > 0$.

Proof. Let I be an instance of SMTI constructed in the proof of Theorem 1. Let $X = \{m_1, m_2, \cdots m_N\}$ be the set of men and $Y = \{w_1, w_2, \cdots w_N\}$ be the set of women of I. Let P_i be the preference list of m_i and Q_i that of w_i. For each $1 \leq i \leq N$, we call women in P_i *proper women* for m_i and men in Q_i *proper men* for w_i.

We translate I into an instance I' of MIN egalitarian SMT. I' consists of X and Y, along with new men $X' = \{m'_1, m'_2, \cdots, m'_N\}$ and women $Y' = \{w'_1, w'_2, \cdots, w'_N\}$. Preference lists of I' are constructed as follows:

m'_i : w'_i [other $2N-1$ women arbitrarily] $(1 \leq i \leq N)$
m_i : P_i [women in Y' arbitrarily] [other women in Y arbitrarily] $(1 \leq i \leq N)$
w'_i : m'_i [other $2N-1$ men arbitrarily] $(1 \leq i \leq N)$
w_i : Q_i [men in X' arbitrarily] [other men in X arbitrarily] $(1 \leq i \leq N)$

Note that each m'_i is matched with w'_i in any stable matching for I' since they write each other strictly first in their lists. Thus, there is a one-to-one correspondence between stable matchings for I and I'. We know that either I has a stable matching of size N, or the size of any stable matching for I is at most $(1-\delta)N$ for a constant δ. If I has a stable matching of size N, then there is a stable matching, say M', for I', where each man in X is matched with a proper woman and each woman in Y is matched with a proper man. Since all preference lists in I are of constant length, say, at most d, the regret of each person with respect to M' is constant and the total cost of M' is at most $2N + 2dN$.

On the other hand, suppose that the size of any stable matching for I is at most $(1-\delta)N$. Then, any stable matching for I' has at least δN men and women that cannot be matched with proper persons. Since they cannot be matched with persons in $X' \cup Y'$, their regret must be larger than N, and hence the sum of their regrets is at least $2 \times \delta N^2$. Hence, a $\frac{\delta}{d+1}N$-approximation algorithm would solve an NP-complete problem. □

Minimum regret stable matching Another measure of general satisfaction with the assignment would be to measure the worst case *regret*, i.e. $\max_p regret_M(p)$. This problem is solvable in polynomial time for complete lists without ties [4]. Here we show an optimal inapproximability of problem when ties are allowed. Refer to this problem as *MIN regret SMT*.

Theorem 14. *MIN regret SMT is hard to approximate within* ϵN, *for some* $\epsilon > 0$.

Proof. We use the same reduction as described in the proof of Theorem 13. Let I and I' be as above. Hence I has N men and N women, and I' has $2N$ men and $2N$ women. If I has a perfect stable matching, then I' has a stable matching in which all persons are matched with proper persons. In this matching, every person's cost is constant and hence the optimal cost is constant, say, d. If I does not have a perfect stable matching, then there is at least one person who is not matched with a proper person and his/her cost is at least N. Therefore, a polynomial-time $\frac{N}{d}$-approximation algorithm implies P=NP. □

References

1. S. Arora and C. Lund, "Hardness of Approximations," Chapter in the book Approximation Algorithms for NP-hard problems, D. Hochbaum editor, PWS Publishing, 1996.
2. D. Gale and L. S. Shapley, "College admissions and the stability of marriage," *Amer. Math. Monthly*, Vol.69, pp.9-15, 1962.
3. D. Gale and M. Sotomayor, "Some remarks on the stable matching problem," *Discrete Applied Mathematics*, Vol.11, pp.223-232, 1985.
4. D. Gusfield, "Three fast algorithms for four problems in stable marriage," SIAM Journal on Computing, Vol. 16, pp. 111–128, 1987.
5. D. Gusfield and R. W. Irving, "The Stable Marriage Problem: Structure and Algorithms," MIT Press, Boston, MA, 1989.
6. J. Håstad, "Some optimal inapproximability results," *Proc. STOC 97*, pp. 1–11, 1997.
7. R. W. Irving, P. Leather and D. Gusfield, "An efficient algorithm for the "optimal" stable marriage," Journal of the A.C.M., Vol. 34, pp. 532–543, 1987.
8. R. W. Irving, "Stable marriage and indifference," *Discrete Applied Mathematics*, Vol.48, pp.261-272, 1994.
9. R. W. Irving, D. F. Manlove and S. Scott, "The hospital/residents problem with ties," *Proc. SWAT 2000*, LNCS 1851, pp. 259–271, 2000.
10. K. Iwama, D. Manlove, S. Miyazaki, and Y. Morita, "Stable marriage with incomplete lists and ties," *Proc. ICALP'99*, LNCS 1644, pp. 443-452, 1999.
11. D. Manlove, R. W. Irving, K. Iwama, S. Miyazaki, Y. Morita, "Hard variants of stable marriage," Technical Report TR-1999-43, Computing Science Department of Glasgow University, September 1999 (to appear in Theoretical Computer Science).
12. E. Ronn, "NP-complete stable matching problems," *J. Algorithms*, Vol.11, pp.285-304, 1990.
13. T. J. Schaefer, "The complexity of satisfiability problems," *Proc. STOC 78*, pp. 216–226, 1978.

Tight Bounds for Online Class-Constrained Packing

Hadas Shachnai[1*] and Tami Tamir[2]

[1] Bell Laboratories, Lucent Technologies, 600 Mountain Ave. Murray Hill, NJ 07974
hadas@research.bell-labs.com

[2] Department of Computer Science, Technion, Haifa 32000, Israel
tami@cs.technion.ac.il

Abstract. We consider *class constrained* packing problems, in which we are given a set of bins, each having a capacity v and c compartments, and n items of M different classes and the same (unit) size. We need to fill the bins with items, subject to capacity constraints, such that items of different classes are placed in separate compartments; thus, each bin can contain items of at most c distinct classes. We consider two optimization goals. In the *class-constrained bin-packing problem (CCBP)*, our goal is to pack all the items in a minimal number of bins; in the *class-constrained multiple knapsack problem (CCMK)*, we wish to maximize the total number of items packed in m bins, for $m > 1$. The CCBP and CCMK model fundamental resource allocation problems in computer and manufacturing systems. Both are known to be strongly NP-hard.

In this paper we derive tight bounds for the online variants of these problems. We first present a lower bound of $(1 + \alpha)$ on the competitive ratio of *any* deterministic algorithm for the online CCBP, where $\alpha \in (0, 1]$ depends on v, c, M and n. We show that this ratio is achieved by the algorithm *first-fit*.

We then consider the *temporary CCBP*, in which items may be packed for a *bounded* time interval (that is *unknown* in advance). We obtain a lower bound of v/c on the competitive ratio of *any* deterministic algorithm. We show that this ratio is achieved by all *any-fit* algorithms.

Finally, tight bounds are derived for the online CCMK and the *temporary* CCMK problems.

1 Introduction

In the well-known *bin packing (BP)* and *multiple knapsack (MK)* problems, a set of items of different sizes and values needs to be packed into bins of limited capacities; a packing is feasible if the total size of the items placed in a bin does not exceed its capacity. We consider the *class-constrained* variants of these problems, which model fundamental resource allocation problems in computer and manufacturing systems. Suppose that all items have the same (unit) size, and

* On leave from the Department of Computer Science, Technion, Haifa 32000, Israel.

S. Rajsbaum (Ed.): LATIN 2002, LNCS 2286, pp. 569–583, 2002.

the same value; however, the items may be of different *classes (colors)*. Each bin has a capacity and a limited number of compartments. Items of different colors cannot be placed in the same compartment. Thus, the number of compartments in each bin sets a bound on the number of *distinct* colors of items it can accommodate. A packing is feasible if it satisfies the traditional capacity constraint, as well as the *class constraint.*

Formally, the input to our packing problems is a set of items, I, of size $|I| = n$. Each item $a \in I$ has a unit size and a color. Thus, $I = I_1 \cup I_2 \cdots \cup I_M$, where all items in I_i are of color i, $1 \leq i \leq M$. The items need to be placed in identical bins, each having capacity v and c compartments. The output of our packing problems is a *placement*, which specifies the subset of items from each class to be placed in bin j, for any $j \geq 1$. In any feasible placement, at most v items of at most c distinct colors are placed in bin j, for all $j \geq 1$. We consider the following optimization problems:

The class-constrained bin-packing problem (CCBP), in which our goal is to find a feasible placement of all the items in a minimal number of bins.

The class-constrained multiple knapsack problem (CCMK), in which there are m bins (to which we refer as knapsacks). Our goal is to find a feasible placement, which maximizes the total number of packed items.

The CCMK problem is known to be NP-hard for $c = 2$, and strongly NP-hard for $c \geq 3$ [8]. These hardness results carry over to CCBP.

In this paper we study the *online* versions of these problems, in which the items arrive as a sequence, one at a time. In each step we get a unit size item of color i, $1 \leq i \leq M$. We need to pack this item before we know any of the subsequent items. Formally, I is given as a sequence $\sigma = a_1, a_2, \ldots$ of length n, such that $\forall k, a_k \in \{1, \ldots, M\}$. The algorithm can base its decision regarding the packing of a_k solely on the knowledge of $a_1, \ldots a_{k-1}$. The decisions of the algorithm are irrevocable, that is, packed items cannot be repacked at later times, and rejected items (in the knapsack problem) cannot be packed later.

Note that since all items have unit size, the non class-constrained versions of our problems can be solved optimally by a greedy algorithm that packs each arriving item in the 'currently filled' bin. For the BP problem, this algorithm uses $\lceil n/v \rceil$ bins, which is clearly optimal. For the MK problem this algorithm packs $\min(n, mv)$ items, which is also optimal. Since we are interested in instances in which the value of c imposes a restriction on the packing, we assume throughout this paper that $c < \min(M, v)$.

In our study of CCBP and CCMK, we first use the traditional assumption, that packed items remain permanently in the system. In the more general case, some items may be *temporary.* Thus, the input sequence may consist of (*i*) arrivals of items: each item colored with $i \in \{1, \ldots, M\}$; (*ii*) departures of items that were packed earlier. Generally, a departure is associated with specific *item* (of specific color, placed in specific bin). We also consider the special case where departure is associated with a *color*, and we may choose the item of that color to be removed. The resulting *temporary packing problems* are denoted by CCBP^t and CCMK^t. In the CCBP^t problem, our objective is to minimize the overall number of bins used throughout the execution of the algorithm. In CCMK^t our goal is to maximize the total number of items packed by the algorithm.

1.1 Applications

The CCMK and CCBP have several important applications, including storage management for continuous-media data (see, e.g.,[20,7]), scheduling jobs on parallel machines [17] and production planning [4].[1] These applications fall into a large set of resource allocation problems of the following form. Suppose that we have a set of devices, each possessing some amount of a *shared* resource. A user request for the resource can be satisfied by specific device if (*i*) there is sufficient amount of the resource available on that device; (*ii*) the device is in configuration to service this type of user. We represent a device as a bin, with the amount of shared resource given by its capacity; the number of compartments in each bin is the number of distinct configurations of the corresponding device. When our goal is to maximize the number of satisfied requests, we get an instance of CCMK. When we wish to minimize the number of (identical) devices needed for satisfying all the requests, we get an instance of CCBP.

The *temporary* CCBP and CCMK problems reflect the nature of user requests to hold the resource for a limited amount of time (which is often unknown in advance). In Section 3.2 we discuss a model in which a departure is associated with *a color*, and the algorithm may choose the item in this color to be omitted. This model captures the flexibility of many systems, in providing service from *any* idle device which is in appropriate configuration to handle users of a given type. Consider, for example, the transmission of video programs which reside on a centralized server, to a set of clients (on demand). Upon request for a video, any non-overloaded disk that holds the video file can provide the service. In particular, suppose that some video, f_1, is played from the disks d_1, d_2, to service the requests of the users u_1, u_2; the disk d_2 is overloaded while d_1 is underloaded. When the system completes the transmission from d_1, it may continue servicing u_2 from d_1, thus reducing the load on d_2. In terms of online class-constrained bin-packing, we may choose the bin from which we omit an item, as long as we take an item of the specified color (i.e., a copy of f_1).

1.2 Performance Measures

Let $\mathcal{A}$ be an algorithm for CCBP. We denote by $N_A(\sigma)$, $N_{opt}(\sigma)$, the number of bins used by $\mathcal{A}$, and an optimal (offline) algorithm, for packing the items in σ. By standard definition (see, e.g., [2]), we say that $\mathcal{A}$ is ρ-competitive, for some $\rho \geq 1$, if for any input sequence, σ, $r_A(\sigma) = N_A(\sigma)/N_{opt}(\sigma) \leq \rho$. Another measure of interest is the *asymptotic worst-case ratio*, given by

$$r_A = \lim_{N_0 \to \infty} \sup \; (\max_{\sigma} \frac{N_A(\sigma)}{N_0} \mid N_{opt}(\sigma) = N_0\}).$$

[1] A detailed survey is given in [16].

In some cases, we study the performance of a packing algorithm, $\mathcal{A}$, on *a set of inputs.* Formally, for a set S, $r_A(S)$ denotes the competitive ratio of $\mathcal{A}$ on inputs from S; that is, $\forall\sigma \in S$, $N_A(\sigma) \leq r_A(S)N_{opt}(\sigma)$. In particular, we will be interested in $r_A(S_{c,v}(k,h))$, where for given c, v, the set $S_{c,v}(k,h)$ consists of all the input sequences, σ, in which the number of colors, M_σ, satisfies $kc < M_\sigma \leq (k+1)c$, and $hv < |\sigma| \leq (h+1)v$. Our refined analysis for sets of inputs yields tighter performance bounds that depend on $v, c, |\sigma|$ and M_σ.

1.3 Our Results

In this paper we study the online (temporary) CCBP and CCMK problems. We obtain the best possible results for items of unit sizes and values, and arbitrary number of classes. Our first main contribution (in Section 2) is a tight lower and upper bound of 2 on the competitive ratio of *any* deterministic algorithm for the online CCBP. Tighter bounds are derived for subsets of instances, denoted by $S_{c,v}(k,h)$. Specifically, we show that for any deterministic algorithm, $\mathcal{A}_d$, $1 < c < v$, $k \geq 0$ and $h \geq k - 1 + \max\{\lceil k/(c-1)\rceil, \lceil (kc+1)/v\rceil\}$,

$$r_d(S_{c,v}(k,h)) \geq 1 + \frac{k+1-\lceil\frac{kc+1}{v}\rceil}{h+1}. \quad (1)$$

Our bound implies that an algorithm may be close to the optimal offline on long sequences that contain relatively *small* number of colors (e.g., $k = 2$ and $h \gg 1$); however, it may have a ratio of 2 when, on the average, any subsequence of length v contains items of c different colors (Take $h = k \gg 1$). We show that a variant of the *first-fit* algorithm, adapted to class-constrained packing, achieves the bound in (1). A greedy algorithm based on partitioning the items into *color-sets* is shown to be efficient as well.

Next, we examine the performance of the well-known set of *any-fit* algorithms in solving our problems. Recall that for classical BP, all *any-fit* algorithms, and some other greedy algorithms (such as *next-fit*) have constant competitive ratios [11,12]. We show that this is no longer true in the presence of class-constraints; that is, *next-fit*, and some any-fit algorithms, have competitive ratio v/c.

Our second main contribution (in Section 3) is a tight lower bound of v/c on the competitive ratio of any deterministic algorithm for the CCBP^t problem. We show that all any-fit algorithms achieve this bound. It may seem that online algorithms for CCBP^t perform better when departures are associated with a *color*, rather than *specific item.* Indeed, in this case each departure allows some re-packing of the items. However, we show a lower bound of $\min(v/c, c-1)$ on the competitive ratio of all any-fit algorithms. Thus, when $v < c^2 - c$, we get the ratio v/c obtained for departures of specific items. Also, the color-sets algorithm, which is optimal when no departures are allowed, is shown to achieve the worst possible ratio of v.

A natural question that we address here is whether a-priori knowledge of the *number of items* to be packed, or the number of *distinct colors* in the input sequence, can help. We answer this question in the negative and show that we gain no advantage from a-priori knowledge of these parameters. Specifically, our lower bounds hold for *any* deterministic algorithm, even one that knows n and M_σ in advance, and the algorithms whose competitive ratios are shown to match the lower bounds do not use the values of n, M_σ. This holds for all the problems considered in this paper.

Finally, in Section 4 we present the results for CCMK and CCMK^t. For CCMK we show a lower bound of c/v on the competitive ratio of any deterministic algorithm. Any greedy algorithm which never rejects items that can be packed achieves this bound, regardless of the way it packs the items. We also show that there is no competitive algorithm for CCMK^t.

1.4 Related Work

Online bin-packing has been studied since the early 1970's, starting with the classical works of Garey et al. [6] and Johnson [10]. The performance of *first-fit* (FF) and *best-fit* (BF) was analyzed in [12], where it was shown that $r_{FF}, r_{BF} \leq 1.7$. The first lower bound for any online bin packing algorithm was given by Yao in [21]. He showed that $r_A \geq 1.5$, for any algorithm $\mathcal{A}$. The current best lower bound, 1.540, is due to van Vliet [19]. Detailed surveys of online packing results can be found in [2,5].

There is a wide literature on the *offline* bin packing (BP) and the multiple knapsack (MK) problems (see comprehensive surveys in [9,15]). Shachnai and Tamir introduced in [18] the offline CCMK problem, and showed that it is NP-hard. The paper presents an approximation algorithm that achieves a factor of $c/(c+1)$ to the optimal. Golubchik et al. [8] derived a tighter bound of $1-1/(1+\sqrt{c})^2$ for this algorithm, with a matching lower bound for *any* algorithm for this problem. The paper [17] considers the offline CCMK and CCBP problems with items of arbitrary sizes.

When there is only one item (of arbitrary size) from each color, we get a *cardinality-constrained* packing problem. These problems were studied in [3,13] (the offline case) and in [1] (online). Finally, a recent paper by Krumke et al. [14] studies the online *bin coloring* problem, in which we need to pack unit sized items of different colors into a set of identical bins. However, this problem is essentially different than CCBP, both in the way the items are packed and the objective function (the goal is to minimize the maximal number of distinct colors in any of the bins, given that $c = v$).

1.5 Notation and Simple Facts

Given a (possibly partial) packing, we say that a bin is *full* if it contains v items; otherwise it is *non-full*. A bin is *occupied* if it contains items of c distinct colors; otherwise it is *non-occupied*. Denote by $C(B)$ the set of colors contained in the bin B. Initially, $\forall B, C(B) = \emptyset$. During the placement of the items, whenever B allocates a compartment to color i, i is added to $C(B)$.

Let $\sigma = a_1, a_2, \ldots$ be a sequence of items to be packed. Upon arrival, an item a_k of color i needs to be placed in some bin B such that (i) B is non-full, and (ii) $i \in C(B)$ or B is non-occupied. A bin satisfying these two requirement is *possible* for a_k. An online algorithm needs to determine in which possible bin a_k will be placed. In CCMK, an item can also be rejected. We first give a simple lower bound on $N_{opt}(\sigma)$.

Property 1. For any $v > c > 1$ and input sequence σ, $N_{opt}(\sigma) \geq \max\{\lceil \frac{n}{v} \rceil, \lceil \frac{M_\sigma}{c} \rceil\}$.

Note that when $c = 1$ we pack items of different colors in separate bins, and the greedy algorithm is optimal for CCBP. Thus, in our discussion of CCBP we assume that $c > 1$. An upper bound on $N_{opt}(\sigma)$ follows from the approximation algorithm presented in [18].

Property 2. For any $v > c > 1, \sigma$, $N_{opt}(\sigma) \leq \max\{\lceil \frac{n}{v} \rceil, \lceil \frac{M_\sigma - 1}{c-1} \rceil\}$.

An algorithm is called *any-fit* if it never opens a new bin for an item that can be packed in one of the open bins. Any-fit algorithms may differ in the way they choose the bin in which the item will be placed.

The *first-fit* algorithm always packs an arriving item into the first (lowest indexed) possible bin. That is, we place a new item, $a \in I_i$, in the first non-full bin, B, which is non-occupied, or $i \in C(B)$. If no such bin exists, we open for a a new bin.

2 Deterministic Algorithms for CCBP

In this section we present a general lower bound on the competitive ratio of *any* deterministic algorithm, and show that *first-fit* achieves this bound. Another greedy algorithm, based on partitioning the items into color-sets, is also shown to be efficient.

The algorithms that we analyze do not use the values of n, M_σ, while our lower bound holds for any deterministic algorithm, even one that knows n and M_σ in advance. We conclude that a-priori knowledge of these parameters cannot improve the performance of deterministic algorithms for CCBP.

2.1 A Lower Bound for Any Deterministic Algorithm

Recall that for $v, c > 1$, an input sequence σ is in $S_{c,v}(k,h)$ *iff* $kc < M_\sigma \leq (k+1)c$ and $hv < |\sigma| \leq (h+1)v$. In other words, $\lceil \frac{M_\sigma}{c} \rceil = k+1$ and $\lceil \frac{|\sigma|}{v} \rceil = h+1$.

Theorem 1. *For any deterministic algorithm, $\mathcal{A}_d$, and for any $v > c > 1$, $k \geq 0$, and $h \geq k - 1 + \max\{\lceil \frac{k}{c-1} \rceil, \lceil \frac{kc+1}{v} \rceil\}$,*

$$r_d(S_{c,v}(k,h)) \geq 1 + \frac{k + 1 - \lceil \frac{kc+1}{v} \rceil}{h+1}.$$

Proof: We show that there exists an input sequence, $\sigma \in S_{c,v}(k,h)$, such that $\mathcal{A}_d$ uses at least $k + h + 2 - \lceil \frac{kc+1}{v} \rceil$ bins to pack σ, while $N_{opt}(\sigma) = h + 1$. The sequence σ is constructed online by the adversary according to the way $\mathcal{A}_d$ packs the items. We denote by $B_1, B_2, \ldots$ the sequence of bins used by $\mathcal{A}_d$.

Let $x = h - k + 1 - \lceil \frac{kc+1}{v} \rceil \geq 0$. The length of σ is $n = (k+x)v + kc + 1$, thus $\lceil \frac{n}{v} \rceil = k + x + \lceil \frac{kc+1}{v} \rceil = h + 1$. The number of distinct colors in σ is $kc + 1$. Therefore, $\sigma \in S_{c,v}(k,h)$. We assume that this is known to the algorithm in advance; thus, $\mathcal{A}_d$ can use the values h, k in making its deterministic decisions. The sequence σ is constructed as follows.

Step 1. For $i = 1$ to kc: pack items of color i, until for the first time an item of color i is placed in one of the bins $B_{k+x+1}, B_{k+x+2}, \ldots$.
Step 2. Pack items of color $kc+1$, until $|\sigma| = (k+x)v + kc + 1$.

Claim 1. $\mathcal{A}_d$ uses at least $h + k + 2 - \lceil \frac{kc+1}{v} \rceil = 2k + x + 1$ bins to pack σ.

Proof: Clearly, if B_{2k+x+1} is used in Step 1, then at least $2k + x + 1$ bins are used for the whole sequence. Otherwise, the claim is proved by the following facts (See Figure 1):

- After Step 1, each of the bins $B_{k+x+1}, B_{k+x+2}, \ldots, B_{2k+x}$ contains exactly c items of different colors. This holds since the adversary moves to the next color whenever an item is placed in one of these bins, for each of the kc colors $1, 2, \ldots, kc$.
- The items packed in Step 2 cannot be placed in $B_{k+x+1}, B_{k+x+2}, \ldots, B_{2k+x}$ (which are occupied by items of colors different than $kc+1$); therefore, they must go to the bins $B_1, B_2, \ldots, B_{k+x}$ or $B_{2k+x+1}, B_{2k+x+2}, \ldots$.
- Since $|\sigma| = (k + x)v + kc + 1$ and exactly kc items are placed in the bins $B_{k+x+1}, B_{k+x+2}, \ldots, B_{2k+x}$, the remaining $(k + x)v + 1$ items will be placed in $B_1, B_2, \ldots, B_{k+x}$ or $B_{2k+x+1}, B_{2k+x+2}, \ldots$. Even if each of the bins $B_1, \ldots, B_{k+x}$ is full, at least one item is packed in B_{2k+x+1}. □

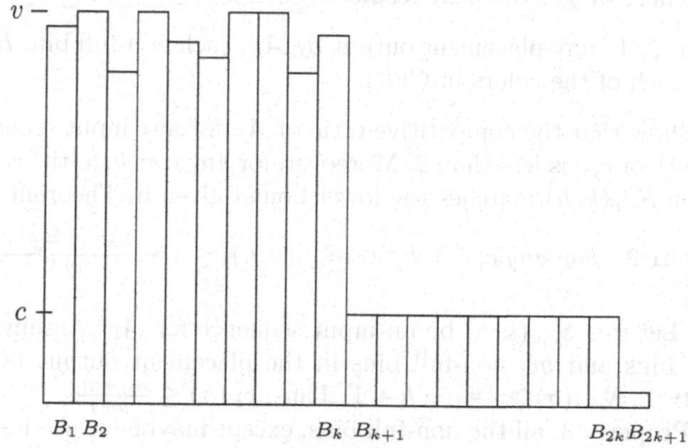

Fig. 1. The placement produced by $\mathcal{A}_d$

Claim 2. For any $h \geq k - 1 + \lceil \frac{k}{c-1} \rceil$, $N_{opt}(\sigma) = h + 1$.

Proof: The length of σ is $n = (k+x)v + kc + 1$. The number of distinct colors in σ is $M_\sigma = kc + 1$. By Property 2, $N_{opt}(\sigma) \leq \max\{\lceil \frac{n}{v} \rceil, \lceil \frac{M_\sigma - 1}{c-1} \rceil\}$. We have $\lceil \frac{n}{v} \rceil = h + 1$, and $\lceil \frac{M_\sigma - 1}{c-1} \rceil = \lceil \frac{kc}{c-1} \rceil$. Given that $h \geq k - 1 + \lceil \frac{k}{c-1} \rceil$, we get $h + 1 \geq \lceil \frac{kc}{c-1} \rceil$; that is, $N_{opt}(\sigma) \leq h + 1$. □

Combining Claims 1 and 2 we conclude that for any c, v, k and $h \geq k-1+\max\{\lceil\frac{k}{c-1}\rceil, \lceil\frac{kc+1}{v}\rceil\}$, $r_d(S_{c,v}(k,h)) \geq 1+\frac{k+1-\lceil\frac{kc+1}{v}\rceil}{h+1}$. □

Recall that $r_d = \limsup_{N_0\to\infty}(\max_\sigma\{N_d(\sigma)/N_0 \mid N_{opt}(\sigma)=N_0\})$. For any $N_0 > 1$, let $k = h = N_0 - 1$, and consider an instance with bins having $c \geq k+1$ and $v \geq kc+1$. Let σ be the input sequence constructed by the adversary for $S_{c,v}(k,h)$. Note that $h \geq k-1+\max\{\lceil\frac{k}{c-1}\rceil, \lceil\frac{kc+1}{v}\rceil\}$. By Claims 1 and 2, $\mathcal{A}_d$ uses at least $k+h+2-\lceil\frac{kc+1}{v}\rceil = 2N_0-1$ bins to pack σ, while $N_{opt}(\sigma) = h+1 = N_0$. That is, $r_d(\sigma) \geq \frac{2N_0-1}{N_0}$.

Corollary 1. *For any deterministic algorithm, $\mathcal{A}_d$, $r_d \geq 2$.*

2.2 First-fit, an Optimal Deterministic Algorithm

Consider the *first-fit* algorithm, $\mathcal{A}_F$. Clearly, $\mathcal{A}_F$ satisfies the following.

Property 3. At any time during the execution of $\mathcal{A}_F$, each bin, except maybe for the last one, is either full or occupied.

Note that, given a placement of the items by $\mathcal{A}_F$, for each bin, B, and color, $i \in C(B)$, B has a compartment of color i only if all the previous bins that contain items of I_i are full. Also, while B is non-full, any $a \in I_i$ can be placed in B. Hence, we get the next result.

Property 4. In any placement output by $\mathcal{A}_F$, each non-full bin, B, holds the last item of each of the colors in $C(B)$.

We show that the competitive ratio of $\mathcal{A}_F$ for any input sequence, σ, and for any values of v, c is less than 2. Moreover, for any c, v, k, h, the competitive-ratio of $\mathcal{A}_F$ on $S_{c,v}(k,h)$ matches the lower bound given in Theorem 1.

Theorem 2. *For any c, v, k, h, $r_F(S_{c,v}(k,h)) \leq 1+\frac{k+1-\lceil\frac{kc+1}{v}\rceil}{h+1}$.*

Proof: Let $\sigma \in S_{c,v}(k,h)$ be an input sequence for $\mathcal{A}_F$. Assume that there are m_1 full bins and m_2 non-full bins in the placement output by $\mathcal{A}_F$ for σ. By Property 1, $N_{opt}(\sigma) \geq \lceil\frac{n}{v}\rceil = h+1$. Thus, $r_F(\sigma) \leq \frac{m_1+m_2}{h+1}$.

By Property 3, all the non-full bins, except maybe for the last one, are occupied. Therefore, the length of σ is $n \geq m_1 v+(m_2-1)c+1$ (The last non-full bin contains at least one item). Since $\sigma \in S_{c,v}(k,h)$, $h+1 = \lceil\frac{n}{v}\rceil \geq m_1+\lceil\frac{(m_2-1)c+1}{v}\rceil$. The coefficients of m_1 and m_2 in this inequality are 1 and $\frac{c}{v} < 1$; thus, m_1+m_2 is maximized when m_2 gets a maximal value.

Claim 3. $m_2 \leq k+1$.

Proof: By Properties 3 and 4 each non-full bin, except maybe for the last one, contains items of c distinct colors. The last non-full bin contains items of at least one additional color. Thus, in order to have $k+2$ or more non-full bins we need items of at least $(k+1)c+1$ distinct colors; however, $M_\sigma \leq (k+1)c$. □

Setting $m_2 = k+1$, we get that $m_1 \le h+1-\lceil \frac{kc+1}{v} \rceil$. Thus, $m_1 + m_2 \le k+1+h+1-\lceil \frac{kc+1}{v} \rceil$, and $r_F(\sigma) = 1 + \frac{k+1-\lceil \frac{kc+1}{v} \rceil}{h+1}$. □

Next we show that for general input sequences, the competitive ratio of $\mathcal{A}_F$ may be arbitrarily close to 2, matching the lower bound in Corollary 1. For a given c, let $S_c(k)$ denote the set of input sequences σ such that $\sigma \in S_c(k)$ iff $kc < M_\sigma \le (k+1)c$ (that is, $\lceil M_\sigma / c \rceil = k+1$).

Theorem 3. *For any $k \ge 0$ and $c > 1$, $r_F(S_c(k)) \le 2 - \frac{1}{k+1} \lceil \frac{kc+1}{v} \rceil$.*

Proof: Let $\sigma \in S_{c,v}(k,h)$ be an input sequence for $\mathcal{A}_F$. If $h \ge k$, then by Theorem 2, $r_F(\sigma) \le 1 + \frac{k+1-\lceil \frac{kc+1}{v} \rceil}{h+1} \le 1 + \frac{k+1-\lceil \frac{kc+1}{v} \rceil}{k+1} = 2 - \frac{1}{k+1} \lceil \frac{kc+1}{v} \rceil$. If $h < k$, then by Property 1, $N_{opt}(\sigma) \ge \max\{h+1, k+1\} = k+1$. Assume that there are m_1 full bins and m_2 non-full bins in the placement produced by $\mathcal{A}_F$ for σ. Thus, $r_F(\sigma) \le \frac{m_1+m_2}{k+1}$. As in the proof of Theorem 2, we have that $m_1 + m_2$ gets its maximal value when $m_2 = k+1$, and $m_1 \le h+1-\lceil \frac{kc+1}{v} \rceil < k+1-\lceil \frac{kc+1}{v} \rceil$. Therefore, $m_1 + m_2 < 2(k+1) - \lceil \frac{kc+1}{v} \rceil$. □

2.3 Other Deterministic Algorithms

The Color-Sets Algorithm. Consider a simple algorithm, $\mathcal{A}_{CS}$, which partitions the M_σ colors in σ into $\lceil \frac{M_\sigma}{c} \rceil$ color-sets and packs the items of each color-set greedily. Each color-set consists of c colors (excluding the last color-set that may contain fewer colors).

The partition into color-sets is determined online by the input sequence. That is, the first set, C_1, consists of the first c colors in σ, the second set, C_2, of the next c colors in σ and so on. At any time there is one active bin for each color set. When an item a of color $i \in C_j$ arrives, it is placed in the active bin of C_j. If the active bin contains v items, we open a new bin for a and this is the new active bin of C_j. Since $|C_j| \le c$, the resulting placement is feasible.

Theorem 4. *For $\mathcal{A}_{CS}$, the color-sets algorithm, $r_{CS} < 2$.*

Proof: Assume that when $\mathcal{A}_{CS}$ terminates there are ℓ active bins, containing $x_1, \ldots, x_\ell$ items. Since we open a new active bin for some color-set only when the current active bin of that color set is full, we have

$$N_{CS}(\sigma) = \frac{n - (x_1 + x_2 +, \ldots, x_\ell)}{v} + \ell \le \frac{n}{v} + \ell(1 - \frac{1}{v}). \qquad (2)$$

Note that (2) is maximized when ℓ is. Since $\ell \le \lceil \frac{M_\sigma}{c} \rceil$, we have $N_{CS}(\sigma) \le \frac{n}{v} + \lceil \frac{M_\sigma}{c} \rceil (1 - \frac{1}{v}) \le 2N_{opt}(\sigma) - \frac{1}{v} \lceil \frac{M_\sigma}{c} \rceil$. Thus, $r_{CS} < 2$. □

Next-fit and Any-fit Algorithms. Property 3 asserts that, when using first-fit, we can partition the set of bins (except maybe for the last one), such that each bin is either full or occupied. We use this property to derive an upper bound on the competitive ratio of a large set of algorithms.

Theorem 5. *For any algorithm $\mathcal{A}$ that fulfills Property 3, and for any instance with bins having volume v and c compartments, $r_A \leq v/c$.*

Proof: We show that for any sequence σ, $N_A(\sigma) \leq \frac{v}{c}N_{opt}(\sigma) + 1$. By Property 3, when $\mathcal{A}$ terminates, each bin, except maybe for the last one, contains at least c items. Thus, $N_A(\sigma) \leq \frac{n}{c} + 1$. By Property 1, $N_{opt}(\sigma) \geq \frac{n}{v}$. Thus, $N_A(\sigma) \leq \frac{v}{c}N_{opt}(\sigma) + 1$, and $r_A \leq v/c$. □

Consider the well-known *next-fit* algorithm, denoted by $\mathcal{A}_N$, which always packs a newly arriving item into the currently *active* bin. If the item cannot be placed in the active bin, then the currently active bin is closed (and never used again) and an empty bin is opened and becomes the new active bin. Since $\mathcal{A}_N$ never opens a new bin if the active bin is non-full and non-occupied, it fulfills Property 3. Thus, $r_N \leq v/c$. This bound is tight, that is, $r_N = v/c$. Also, all any-fit algorithms fulfill Property 3 and their competitive ratio is at most v/c. For some any-fit algorithms, e.g. the algorithm that places a new item in the last (highest indexed) open bin into which it can fit, the ratio v/c is tight.

3 Online Class-Constrained Packing of Temporary Items

In this section we consider a generalization of online class-constrained bin-packing to the case where items may leave the system after a while. We first consider the case in which a departure is associated with specific item (of specific color, placed in specific bin). In Section 3.2 we discuss the case in which a departure is associated with a *color*; thus, we may choose which item of that color to remove.

3.1 Departures of Specific Items

For the model where only arrivals are allowed, we showed (in Section 2.2) that first-fit is superior to other any-fit algorithms, and that its competitive-ratio is less than 2. When some items are temporary, this is no longer true. We show a lower bound of v/c on the competitive ratio of any deterministic algorithm. Then, we show that all any-fit algorithms achieve this bound. Note that, as in Section 2, we conclude that a-priori knowledge of n, M_σ cannot help.

Theorem 6. *For any deterministic algorithm $\mathcal{A}_d$, $r_d \geq v/c$.*

Proof: Given v, c, for any N_0 such that $k = \frac{1}{c}(N_0 - v + 1)$ is an integer, we construct an input sequence, σ such that $N_d(\sigma) = kv$, while $N_{opt}(\sigma) = kc + v - 1 = N_0$. The construction of σ is done in kc iterations. In the jth iteration, $1 \leq j \leq kc$, we add and remove items of color j. Specifically, the jth iteration in σ consists of:

1. Arrivals of $v(v-1)+1$ items of color j. These items are placed in at least v distinct bins.
2. Departures of $v(v-2)+1$ items of color j, selected such that each of the v remaining items of color j is placed in a different bin.

Note that after the iteration kc, there are vkc items in the bins. Since the items of each color are placed in different bins, each bin holds at most c items. Thus, $\mathcal{A}_d$ uses at least kv bins.

Consider now an optimal algorithm, $\mathcal{A}_{opt}$, which knows the whole sequence, and in particular, the items that will leave during the second step in each iteration. For each color, $\mathcal{A}_{opt}$ will pack the v items that will not be removed in the first available bin, and the $v(v-2)+1$ items that are about to leave, in the next $v-1$ bins. Thus, after the first step of iteration j, $1 \le j \le ck$, $\mathcal{A}_{opt}$ uses $(j-1)+v$ bins, and during the second step of the jth iteration it retreats to j full bins. Hence, the maximal number of bins used by $\mathcal{A}_{opt}$ is $kc+v-1$.

We get that $r_d(\sigma) = (kv)/(kc+v-1)$. Recall that k was selected such that $N_0 = kc+v-1$. Thus, $r_d(\sigma) = v(N_0 - v + 1)/(cN_0) = (v/c)(1-(v+1)/N_0)$

$$\lim_{N_0 \to \infty} \sup \; (\max_{\sigma}\{\frac{N_d(\sigma)}{N_0} \mid N_{opt}(\sigma) = N_0\}) = \frac{v}{c}.$$

□

We now show that all any-fit algorithms achieve this ratio.

Theorem 7. *If $\mathcal{A}$ is an any-fit algorithm, then for any instance with bins having volume v and c compartments, $r_A \le v/c$.*

Proof: Let σ be an input sequence for $\mathcal{A}$. Let N_A be the number of bins used by $\mathcal{A}$ when packing σ. When analyzing the competitive ratio of $\mathcal{A}$, we can assume w.l.o.g. that σ ends when bin number N_A is opened. Indeed, after this point, the number of bins used by an optimal algorithm can only increase, and we need to bound the largest $N_A(\sigma)/N_{opt}(\sigma)$ ratio. Since $\mathcal{A}$ is an any-fit algorithm, when N_A is opened, all the other bins are either full or occupied, and therefore each open bin contains at least c items. Let n be the number of packed items at that time. Then $N_A(\sigma) \le \frac{n}{c}+1$. By Property 1, $N_{opt}(\sigma) \ge \frac{n}{v}$. Thus, $N_A(\sigma) \le \frac{v}{c}N_{opt}(\sigma)+1$, and $r_A \le v/c$. □

3.2 Departures of Items of Specific Colors

We now show that any-fit algorithms achieve a poor ratio, even if departures are associated only with a color, and the algorithm may select the item in this color to be removed. Our result holds for *any* removal policy.

Theorem 8. *If $\mathcal{A}$ is an any-fit algorithm, then $r_A \ge \min(v/c, c-1)$.*

Proof: Given v, c, let $x = v - c + 1$, and let $z = \max(\lceil \frac{xc}{v} \rceil, \lceil \frac{x}{c-1} \rceil)$. For any N_0 such that $k = \frac{1}{z}(N_0 + z - x)$ is an integer, we construct a sequence, σ such that $N_A(\sigma) = kx$, while $N_{opt}(\sigma) = kz - z + x = N_0$.

The construction of σ is done in k iterations. In the jth iteration $(1 \le j \le k)$ we handle items in the v colors $(j-1)v+1, \ldots, jv$. Specifically, the jth iteration in σ consists of:

1. Arrivals of xv items that fill the x bins $(j-1)x+1,\ldots,jx$ as follows. For $i=1$ to x: σ contains a sub-sequence of arrivals of $(c-1)$ items in colors $(j-1)v+1,\ldots,(j-1)v+c-1$, followed by $v-c+1$ arrivals of items of color $(j-1)v+c+i-1$. Since $\mathcal{A}$ is an any-fit algorithm, there is at most one possible open bin for each item as it arrives, thus, the bins $(j-1)x+1,\ldots,jx$ are filled sequentially one after the other.
2. For $i=1$ to x: remove $v-c$ items of color $(j-1)v+c+i-1$. Now, regardless of the removal policy of the algorithm, we end up with each of the x bins $(j-1)x+1,\ldots,jx$ containing exactly c items, one in each of the colors $(j-1)v+1,\ldots,(j-1)v+c-1$, and one item whose color is in $(j-1)v+c,\ldots,jv$.

Note that after each iteration, $1\le j\le k$, each bin contains c items of c different colors. Thus, the arriving items in each iteration, which are of new colors, must be packed in new bins. It follows that after the jth iteration $(1\le j\le k)$, $\mathcal{A}$ uses jx bins, each containing c items.

Claim 4. An optimal placement of σ uses $N_0=kz-z+x$ bins.

Proof: Consider an optimal algorithm, $\mathcal{A}_{opt}$, which knows the entire sequence, and in particular, the items that are will leave during the second step of each iteration. In each iteration $\mathcal{A}_{opt}$ can pack the items that are not removed in the first available bins. We show that these permanent items can be packed in z bins. Note that this set of items consists of xc items of $x+c-1$ colors. We distinguish between two types of colors: (*i*) *repeated*, from each of which there are x items (In the jth iteration, these are the $(c-1)$ colors $(j-1)v+1,\ldots,(j-1)v+c-1)$, and (*ii*) *single-items*, from each of which there is only one item (in the jth iteration, these are the x colors $(j-1)v+c,\ldots,jv)$.

Assume that $\lceil\frac{xc}{v}\rceil\ge\lceil\frac{x}{c-1}\rceil$, then $\mathcal{A}_{opt}$ can pack the permanent items in $\lceil\frac{xc}{v}\rceil$ bins as follows. At first, each bin contains $c-1$ items of single-item colors, and $x=v-c+1$ items of some repeated color. Once all the 'singles' are packed, we fill the remaining bins greedily with the remaining items of the repeated colors. Since all the bins are filled to capacity v, $\mathcal{A}_{opt}$ uses $\lceil\frac{xc}{v}\rceil$ bins.

Assume that $\lceil\frac{xc}{v}\rceil<\lceil\frac{x}{c-1}\rceil$, then we can pack the permanent items in $\lceil\frac{x}{c-1}\rceil$ bins as follows. First, we pack in each bin $c-1$ single items, and $x=v-c+1$ items of some repeated color. Once all the items of the repeated colors are packed, we use the remaining bins greedily to pack c single items in each. Since each bin contains at least $c-1$ single items, and there are x such items, all the single items are packed. Also, since $\lceil\frac{xc}{v}\rceil<\lceil\frac{x}{c-1}\rceil$, there is enough capacity for the other items.

Thus, after the first step of iteration j, $1\le j\le k$, $\mathcal{A}_{opt}$ uses $(j-1)z+x$ bins, and during the second step of the jth iteration it retreats to jz full bins. The maximal number of bins is $(k-1)z+x=N_0$, used during the last iteration. □

We get that $r_A(\sigma)\ge(kx)/(kz-z+x)$, and

$$r_A=\lim_{N_0\to\infty}\sup\ (\max_{\sigma}\{\frac{N_{CS}(\sigma)}{N_0}\mid N_{opt}(\sigma)=N_0\})\ge\frac{x}{z}\ge\min(\frac{v}{c},c-1)\ \square$$

The Color-Sets Algorithm. We now show that any algorithm based on packing by color sets achieves the worst possible ratio, which approaches v, even if departures are associated only with a color. Our result holds for any removal policy and an algorithm which packs items by color-sets.

Theorem 9. *Let $\mathcal{A}_{CS}$ be the color-sets algorithm, with any removal policy; then, $r_{CS} \geq v$.*

Proof: Recall that $\mathcal{A}_{CS}$ partitions the M_σ colors in σ into $\lceil \frac{M_\sigma}{c} \rceil$ color-sets, where the jth set, C_j, contains the jth set of c colors in σ. The items of each color-set are packed greedily.

Given v and c, for any N_0, let $k = N_0 - v + 1$. We construct a sequence, σ such that $N_A(\sigma) = kv$, while $N_{opt}(\sigma) = k + v - 1 = N_0$. W.l.o.g., the colors are numbered by the order of their first appearance in σ; thus, for each $1 \leq j \leq k$, $C_j = \{(j-1)c + 1, \ldots, jc\}$. The construction of σ is done in k iterations. In the jth iteration, $1 \leq j \leq k$, we add and remove items whose colors are in C_j. Specifically, the jth iteration consists of two steps:

1. Arrivals of v^2 items whose colors are in C_j. This sequence consists of repeating v times a sub-sequence of arrivals of a single item of color cj followed by $v-1$ arrivals of items of colors $(j-1)c+1, \ldots, jc-1$. This sequence contains at least one item in each of the colors in C_j (so $\mathcal{A}_{CS}$ can define C_j).
2. Departures of the $v(v-1)$ items in the colors $(j-1)c+1, \ldots, jc-1$.

Note that in the first step of iteration j, $\mathcal{A}_{CS}$ fills the bins $(j-1)v+1, \ldots, jv$ and after the second step of iteration j, regardless of the removal policy of $\mathcal{A}_{CS}$, we end up with each of these bins containing a single item (of color cj). However, these bins cannot be used by $\mathcal{A}_{CS}$ in the next iterations, since the arriving items belong to different color-sets. Thus, in each iteration, the arriving items are packed in new bins. It follows that after the kth iteration, $\mathcal{A}_{CS}$ has kv open bins, each contains a single item.

Consider now an optimal algorithm, $\mathcal{A}_{opt}$, which knows the entire sequence, and in particular, the items that will depart in the second step of each iteration. In each iteration, $\mathcal{A}_{opt}$ packs the v permanent items in the first available bin, and the temporary items in the next $v-1$ bins. Thus, after the first step of iteration j, $1 \leq j \leq k$, $\mathcal{A}_{opt}$ has $j - 1 + v$ full bins, and during the second step of the jth iteration it retreats to j full bins. The maximal number of bins is $k + v - 1$, used during the last iteration.

We get that $r_{CS}(\sigma) = \frac{kv}{k+v-1}$. Since $N_0 = k + v - 1$, $r_{CS}(\sigma) = \frac{N_0 v - v^2 + v}{N_0} = v(1 - \frac{v^2-v}{N_0})$, and $\limsup_{N_0 \to \infty}(\max_\sigma\{N_{CS}(\sigma)/N_0 \mid N_{opt}(\sigma) = N_0\}) = v$. □

4 Deterministic Algorithms for CCMK

Recall that in the CCMK problem we have m identical knapsacks, of volume v and c compartments, and our goal is to maximize the number of packed items from the input sequence σ. For a placement algorithm $\mathcal{A}$ and an input sequence σ, let $n_{\mathcal{A}}(\sigma, m)$ denote the number of items in σ packed by $\mathcal{A}$. Let $n_{opt}(\sigma, m)$ denote the maximal number of items in σ that can be packed in m knapsacks. The *competitive ratio* of $\mathcal{A}$ is $r_A = \min_{\sigma,m} n_A(\sigma, m)/n_{opt}(\sigma, m)$.

We derive a lower bound of c/v for deterministic algorithms. This bound is achieved by any greedy algorithm. For the temporary CCMK problem, we show that no deterministic algorithm is competitive.

Theorem 10. *For any $v \geq c$, and any deterministic algorithm $\mathcal{A}_d$ for CCMK, $r_d \leq c/v$.*

Proof: Consider the following sequence constructed online by the adversary, for an algorithm $\mathcal{A}_d$. (*i*) For all $i = 1, \ldots, mc$, repeat items of color i until one such item is packed, or until mv items of color i are rejected. In the latter case, the sequence ends. (*ii*) mv items of color $mc + 1$.

If no color is rejected mv times, then each of the knapsacks filled by $\mathcal{A}_d$ contains exactly c items of c distinct colors, thus, only mc items are packed. An optimal algorithm can pack the mv items of color $mc + 1$. If some color, i, is persistently rejected, then an optimal algorithm can pack the mv rejected items of color i, while $\mathcal{A}_d$ packs only $i - 1$ items of the colors $1, \ldots, i - 1$. In both cases $r_d \leq \frac{mc}{mv} = c/v$. □

A greedy algorithm never rejects an item that can be packed (at the time of its arrival). We show that the competitive ratio of any greedy algorithm matches our lower bound, regardless of the way we pack the items.

Theorem 11. *For any greedy algorithm $\mathcal{A}_G$, the competitive ratio of $\mathcal{A}_G$ is $r_G \geq c/v$.*

Proof: For any $m > 0$ and sequence σ, consider the knapsacks when σ terminates. If some knapsack contains less than c items, then, since $\mathcal{A}_G$ is greedy, no item was rejected and $r_G(\sigma, m) = 1$. Otherwise, there are at least c items in each knapsack, and $n_G(\sigma, m) \geq cm$. Since $n_{opt}(\sigma, m) \leq vm$, we get the desired ratio. □

Next we show that, when some of the items are temporary, no algorithm is competitive. Note that now the total number of packed items may be larger than mv.

Theorem 12. *For any $v \geq c$, any deterministic algorithm $\mathcal{A}_d$ for CCMK^t, and any $\rho > 0$, the competitive ratio of $\mathcal{A}_d$ is $r_d \geq \rho$.*

Proof: Consider a sequence, σ, which first blocks all the m knapsacks (as in the proof of Theorem 10). After all the knapsacks are full or occupied, σ continues with a sequence of arbitrary length, of pairs of arrival-departure of an item, (of a new color). Clearly, these repeated arrivals cannot be accepted by the algorithms, while an optimal algorithm can pack them all. Any ratio can be achieved by taking the sequence long enough. □

References

1. L. Babel, B. Chen, H. Kellerer, and V. Kotov. On-line algorithms for cardinality constraint bin packing problems. Technical report, Institut fuer Statistik und Operations Research, Universitaet Graz, 2001.
2. A. Borodin and R. El-Yaniv. *Online computation and competitive analysis*. Cambridge University Press, 1998.
3. A. Caprara, H. Kellerer, U. Pferschy, and D. Pisinger. Approximation algorithms for knapsack problems with cardinality constraints. *European Journal of Operations Research*, 123:333–345, 2000.
4. W. J. Davis, D. L. Setterdahl, J. G. Macro, V. Izokaitis, and B. Bauman. Recent advances in the modeling, scheduling and control of flexible automation. In *Proc. of the Winter Simulation Conference*, 143–155, 1993.
5. A. Fiat and G.J. Woeginger. *Online Algorithms: The State of the Art*. LNCS. 1442, Springer-Verlag, 1998.
6. M.R. Garey, R.L.Graham, and J.D.Ullman. Worst-case analysis of memory allocation algorithms. In *Proc. of STOC*, 143–150, 1972.
7. S. Ghandeharizadeh and R.R. Muntz. Design and implementation of scalable continuous media servers. *Parallel Computing Journal*, 24(1):91–122, 1998.
8. L. Golubchik, S. Khanna, S. Khuller, R. Thurimella, and A. Zhu. Approximation algorithms for data placement on parallel disks. In *Proc. of SODA*, 223–232, 2000.
9. D.S. Hochbaum. *Approximation Algorithms for NP-Hard Problems*. PUS Publishing Company, 1995.
10. D.S. Johnson. *Near-optimal bin packing algorithms*. PhD thesis, MIT, 1973.
11. D.S. Johnson. Fast algorithms for bin packing. *J. Comput. Sys. Sci.*, 8:272–314, 1974.
12. D.S. Johnson, A. Demers, J.D. Ullman, M.R. Garey, and R.L. Graham. Worst-case performance bounds for simple one-dimensional packing algorithm. *SIAM Journal of Computing*, 3:256–278, 1974.
13. H. Kellerer and U. Pferschy. Cardinality constrained bin-packing problems. *Annals of Operations Research*, 92:335–349, 1999.
14. S.O. Krumke, W. de Paepe, J. Rambau and L. Stougie. Online bin-coloring. In *Proc. of ESA*, 74–85, 2001.
15. S. Martello and P. Toth. Algorithms for knapsack problems. *Annals of Discrete Math.*, 31:213–258, 1987.
16. H. Shachnai and T. Tamir. Polynomial time approximation schemes for class-constrained packing problems. *Journal of Scheduling*, 4:313–338, 2001.
17. H. Shachnai and T. Tamir. Multiprocessor scheduling with machine allotment and parallelism constraints. 2001. Algorithmica, to appear.
18. H. Shachnai and T. Tamir. On two class-constrained versions of the multiple knapsack problem. *Algorithmica*, 29:442–467, 2001.
19. A. van Vliet. On the asymptotic worst case behavior of harmonic fit. *J. of Algorithms*, 20:113–136, 1996.
20. J.L. Wolf, P.S. Yu, and H. Shachnai. Disk load balancing for video-on-demand systems. *ACM Multimedia Systems Journal*, 5:358–370, 1997.
21. A. Yao. New algorithms for bin packing. *Journal of the ACM*, 27:207–227, 1980.

On-line Algorithms for Edge-Disjoint Paths in Trees of Rings

R. Sai Anand* and Thomas Erlebach

Computer Engineering and Networks Laboratory (TIK),
Eidgenössische Technische Hochschule Zürich,
CH-8092 Zürich, Switzerland,
{anand|erlebach}@tik.ee.ethz.ch

Abstract. A tree of rings is a graph that can be constructed by starting with a ring and then repeatedly adding a new disjoint ring to the graph and identifying one vertex of the new ring with a vertex of the existing graph. Trees of rings are a common topology for communication networks. We give randomized on-line algorithms for the problem of deciding for a sequence of requests (terminal pairs) in a tree of rings, which requests to accept and which to reject. Accepted requests must be routed along edge-disjoint paths. It is not allowed to reroute or preempt a request once it is accepted. The objective is to maximize the number of accepted requests. For the case that the paths for accepted requests can be chosen by the algorithm, we obtain competitive ratio $O(\log d)$, where d is the minimum possible diameter of a tree resulting from the tree of rings by deleting one edge from every ring. For the case where paths are pre-specified as part of the input, our algorithm achieves competitive ratio $O(\log \ell)$, where ℓ is the maximum length of a simple path in the given tree of rings.

1 Introduction

A fundamental problem in communication networks with bandwidth reservation is call admission control: If a request for a connection between two nodes arrives, the network must either accept the request and reserve the required bandwidth along a path between the endpoints of the request, or reject the request. This decision must be made without knowledge of requests that will arrive later on. The goal can be to maximize the profit gained from accepted requests or to maximize the throughput of the network, for example. This is an on-line problem, and one is interested in algorithms with good competitive ratios.

The optimization problem that lies at the heart of call control is the maximum edge-disjoint paths problem (MEDP): given an undirected graph $G = (V, E)$ and a (multi-)set R of requests (terminal pairs), select a subset $R' \subseteq R$ and connect all terminal pairs in R' along edge-disjoint paths, such that $|R'|$ is maximized.

* Supported by the joint Berlin/Zurich graduate program Combinatorics, Geometry, and Computation (CGC), financed by ETH Zurich and the German Science Foundation (DFG).

S. Rajsbaum (Ed.): LATIN 2002, LNCS 2286, pp. 584–597, 2002.

In the on-line setting, the terminal pairs arrive one by one, and the algorithm must accept (and route) or reject each pair without knowledge of the future. We assume that preemption is not allowed, i.e., if a request is accepted, it cannot be preempted (rejected) at a later time in favor of another request.

MEDP corresponds to the special case of call control where the bandwidth requirements of all connections are so high that no two connections can use the same link simultaneously. In fact, call control tends to be much easier to deal with algorithmically if the bandwidth requirements are small compared to the link capacities [2]. Thus, MEDP is among the harder subproblems of call control.

In this paper, we investigate randomized on-line algorithms for MEDP in undirected trees of rings. We deal with two versions of the problem. In the first version, each request is given by two nodes of the network and the algorithm must select the path to connect these nodes (in case the request is accepted). In the second version, the path for each request is specified as part of the input. We call the former version of the problem MEDP and the latter version MEDPwPP (MEDP with pre-specified paths).

1.1 Preliminaries

The *trees of rings* are the class of undirected graphs that can be constructed as follows.

1. A single ring (cycle) is a tree of rings.
2. If T is a tree of rings and R is a separate ring, the graph obtained by taking the union of T and R and then identifying one node of T with one node of R is a tree of rings.

Note that every ring of a tree of rings contains at least three nodes. For a given tree of rings T, a tree structure underlying T can be defined as follows. Pick an arbitrary ring R as the root and mark R as *ready*, while all other rings are marked *available*. Then repeat the following operation: Let v be a node in a ring C that is marked ready and that is contained in at least one ring that is marked available. Then make all available rings that contain v children of C and mark them ready as well. This procedure induces a unique parent ring to each ring (except the root ring), and the set of all rings forms a tree structure.

Let T be a tree of rings. We say that a path *touches* a ring if it contains at least one node of that ring. A path *passes* a ring if it uses at least one edge of that ring. If u and v are two nodes in a tree of rings, all simple paths from u to v pass the same rings, namely the rings on the path from the ring of u to the ring of v in the underlying tree structure. Two paths p and q *intersect* if they share an edge, i.e., if they are not edge-disjoint. The maximum length (number of edges) of a simple path in T is called the *maximum path length* of T, denoted by ℓ_T. The maximum path length of a tree is also called the *diameter* of the tree.

The on-line versions of the problems MEDP and MEDPwPP in trees of rings are defined as follows. Initially, a tree of rings T is given, and all its edges are available. Then a sequence of requests arrives. Each request is specified by two nodes u and v of T (in case of MEDP) or by a simple path p in T (in case of MEDPwPP). When the request arrives, the algorithm must either accept or reject the request without knowledge of future requests. In case of MEDP, the algorithm must route an accepted request along a simple path of available edges from u to v, and the edges on the path become unavailable. In case of MEDPwPP, a path p can be accepted only if all edges of p are still available, and these edges become unavailable after p is accepted. For both problems, the goal is to maximize the number of accepted requests.

For a given request sequence $\sigma_1, \sigma_2, \ldots, \sigma_s$, we denote by OPT the set of accepted requests in an optimal solution (computed by an all-powerful off-line algorithm with complete knowledge of the request sequence) and by A the set of accepted requests of an on-line algorithm. If $|OPT| \leq r|A|$ holds for all possible request sequences, we say that the on-line algorithm has competitive ratio r. For randomized algorithms, A is replaced by $\mathrm{E}[A]$ in this definition, where $\mathrm{E}[A]$ denotes the expected number of accepted requests of the on-line algorithm.

1.2 Related Work

Due to its great relevance to modern communication networks, call control has been studied intensively during the last decade. Pioneering results on on-line call control in chain networks can be found in [10], where a preemptive version of the problem was studied (i.e., a path that is accepted by the algorithm can still be rejected later on). Much of the later work has concentrated on the non-preemptive version of on-line MEDP. In [3], an $O(\log n)$-competitive randomized algorithm for trees has been proposed. In [4], an improved randomized algorithm is given that achieves competitive ratio $O(\log d)$ for trees with diameter d. This should be contrasted with the obvious lower bound of $\Omega(d)$ on the competitive ratio of any deterministic algorithm for trees (or even chains). An $O(\log \log n)$-competitive algorithm for $n \times n$ trees of meshes and an $O(\log n \log \log n)$- competitive algorithm for $n \times n$ meshes are given in [4] as well. For meshes, an improved algorithm with competitive ratio $O(\log n)$ is presented in [12]. In fact, their result holds for a class of *densely embedded, nearly-Eulerian graphs*, which contains the two-dimensional meshes. A survey of on-line algorithms for the disjoint paths problem and more general call admission control problems can be found in [7, Chapter 13]. Of immediate relevance to the problems we investigate here is a randomized $O(\log d)$-competitive algorithm for MEDP in trees, where d is the diameter of the tree, described in [7, Sect. 13.5.2].

The on-line edge-disjoint paths problem with pre-specified paths was studied in [6] for the case where the objective value is the number of *rejected* paths. It was shown that a simple on-line algorithm can achieve competitive ratio 2 for arbitrary topologies. For the case of larger edge capacities, this version of the problem turned out to be more difficult, and a 2-competitive algorithm was obtained only for chain networks. In this paper, we always consider the number of accepted requests as objective value.

For the preemptive version of on-line MEDP in chains, an $O(1)$-competitive randomized algorithm was given in [1]. It is an open question whether the combination of randomization and preemption can also lead to a constant competitive ratio for trees.

A number of researchers have considered problems in networks with tree of rings topology. All-to-all routing in symmetric directed trees of rings has been studied in [5], where it is shown that a routing that minimizes the maximum load (number of paths going through the same edge) can be computed in polynomial time and that the resulting paths can be colored with a number of colors that is equal to the maximum load. The off-line versions of the maximum edge-disjoint paths problem and the path coloring problem (assigning colors to paths such that paths with the same color are edge-disjoint) for trees of rings were considered in [9], where it was proved that these problems are *NP*-hard for trees of rings but can be approximated within a constant factor, no matter whether the paths are pre-specified or not. For the restricted class of undirected trees of rings in which each vertex is contained in at most two rings, a 2-approximation algorithm for path coloring with pre-specified paths was given in [8].

1.3 Our Results

For on-line MEDP in trees of rings, we give a simple algorithm that deletes one edge in every ring and then uses an $O(\log d)$-competitive call control algorithm for the resulting tree, where d is the diameter of that tree. This is shown to result in an $O(\log d_T)$-competitive algorithm for trees of rings, where d_T is the minimum possible diameter of a spanning tree of the given tree of rings T.

Our main result is a randomized $O(\log \ell_T)$-competitive on-line algorithm for MEDPwPP in trees of rings, where ℓ_T is the maximum length of a simple path in the given tree of rings T. In order to obtain this result, we adapt the idea of *roadblocks*, as presented in [4], to the tree of rings structure. We remark that competitive ratio $O(\log \ell_T)$ is optimal up to a constant factor for MEDPwPP in trees of rings due to the lower bound of $\Omega(\log n)$ on the competitive ratio of any randomized on-line algorithm for MEDP in a chain with n nodes [4,3].

2 The Algorithm for MEDP

In case of MEDP, we can reduce the problem for trees of rings to the problem for trees using a standard method: we simply remove an arbitrary edge from each ring of the given tree of rings and then handle all requests in the resulting tree using an algorithm for MEDP in trees. The following theorem, which has been proved in [9], implies that we lose only a constant factor by removing the edges.

Theorem 1. *Let S be a set of edge-disjoint paths assigned to requests in a tree of rings. If an arbitrary edge is removed from each ring in the tree of rings, at least one third of the requests with paths in S can still be connected along edge-disjoint paths in the resulting tree.*

It is easy to see that Theorem 1 holds: If all paths in S are rerouted in the resulting tree, the maximum number of paths going through the same edge is at most 2. Furthermore, any set of paths with maximum load 2 in a tree can be colored with 3 colors such that intersecting paths receive different colors. Therefore, one color class must contain edge-disjoint paths for at least one third of the requests in S.

If we use either of the $O(\log d)$-competitive algorithms in [7, Sect. 13.5.2] or [4] for MEDP in the tree obtained by deleting edges in the tree of rings, Theorem 1 implies that the resulting algorithm achieves competitive ratio $O(\log d)$ for MEDP in trees of rings as well. Here, d is the diameter of the tree resulting from the tree of rings after deleting one edge in every ring. In order to make d as small as possible, we choose the deleted edges in a way so that the diameter of the resulting spanning tree is minimized. Note that a spanning tree with minimum diameter can be computed efficiently, even in arbitrary graphs [11].

Theorem 2. *Given a tree of rings T, there is a randomized algorithm for MEDP that achieves competitive ratio $O(\log d_T)$, where d_T is the minimum possible diameter of a tree that can be obtained from T by deleting one edge from every ring.*

3 The Algorithm for MEDPwPP

In the case of MEDPwPP, we cannot use the cut-one-link heuristic as we did for MEDP, because in the worst case all requests would be for paths that use one of the deleted links, and so we cannot hope for a good competitive ratio. Therefore, we need a different approach.

In order to motivate our algorithm, we briefly explain why randomization is necessary and why placing "roadblocks" on the edges of a rejected path is helpful. It is easy to observe that no deterministic algorithm for MEDPwPP can achieve a good competitive ratio even in chain networks: Consider a chain with n nodes. Label the vertices $1, 2, \ldots, n$. Let the request sequence, specified as the end vertices of the requested paths, be $(1, n), (1, 2), (2, 3), \ldots (n-1, n)$. Any deterministic algorithm has to accept the first path presented to it. Otherwise, it will achieve a competitive ratio of ∞ on a request sequence that consists of this single path only (recall that in the on-line scenario, the algorithm has no knowledge of future requests). But accepting the first path in the above instance precludes the other $n-1$ paths being accepted, giving a competitive ratio of at least $n-1$ for any deterministic algorithm.

From the above, it is clear that only a randomized algorithm can hope to achieve a good competitive ratio for MEDPwPP. The basic idea of a randomized algorithm is to decide to accept or reject a path presented to it in a probabilistic manner. Since we consider non-preemptive algorithms, a path once accepted cannot be rejected later. However, a path that is rejected once by the algorithm might be presented to it again and again. If the probability of accepting that path is positive and depends only on the paths that have been accepted previously (but not on previously rejected paths), an adversary could force the algorithm to accept the path by presenting it sufficiently many times. Therefore, the randomized algorithm needs to remember (in some way) a path that it has rejected once. In the MEDP algorithm for trees given in [4], this is achieved by placing so-called *roadblocks* on some of the edges of a rejected path (cf. Subsection 3.3). The roadblock on an edge causes any future path touching it to be rejected immediately. While we use this idea for the tree of rings structure for the same reason, it has to be adapted in ways different from the one used for trees. Also, the topology of trees of rings presents various features in the analysis which are not encountered in trees. For example, it seems crucial to introduce the notion of *crossing* paths (see below) in order to carry out the analysis. The details of our algorithm and its analysis are presented in the following subsections.

3.1 Preliminaries & Terminology

We say that two paths *cross* if they share at least 2 vertices or if they pass through the same two consecutive rings. Note that any two paths that intersect also cross, but not vice versa (cf. Fig. 1). If a path p crosses a path q that was presented earlier, the *region of overlap* of p with q is defined as the subpath of q between the extreme shared vertices of p and q. If p and q cross and share only one vertex, then the region of overlap is just that shared vertex. Note that the region of overlap of p with q is not symmetric w.r.t. p and q.

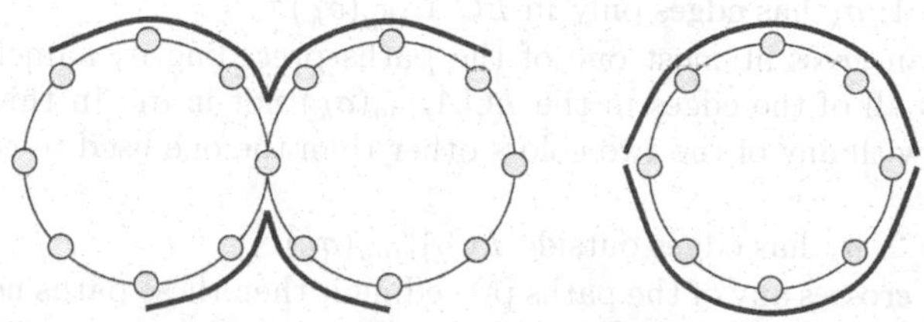

Fig. 1. Pairs of paths that cross, but do not intersect.

In the following, when we use terms like parent ring and ancestor ring, we refer to the underlying tree structure of the tree of rings T, rooted at an arbitrary ring. For a path σ_i, define $LCA_{ring}(\sigma_i)$ to be the ring R such that R contains at least one vertex of σ_i and no ancestor ring of R contains a vertex of σ_i. If a path σ_i passes more than one ring and the end points of σ_i are u_i and v_i, then call the subpath from u_i to $LCA_{ring}(\sigma_i)$ the *left side* of σ_i and the subpath from v_i to $LCA_{ring}(\sigma_i)$ the *right side* of σ_i. If u_i or v_i lie in the LCA_{ring} then the corresponding subpaths are treated as *empty.* (Terms like "left" and "right" can be taken to refer to some appropriate planar embedding of the tree of rings.)

3.2 Intersecting versus Crossing

Let OPT represent an optimal set of edge-disjoint paths for the given sequence of requests.

Lemma 1. *The paths in OPT can be colored with at most 3 colors such that no two paths with the same color cross.*

Proof. Consider the paths in OPT according to the descending LCA_{ring} of the paths. That is, paths that have the root ring as their LCA_{ring} are considered first, then the paths that have children of the root ring as their LCA_{ring} and so on. The order among paths that have the same LCA_{ring} is as follows: those that have at least one edge in the LCA_{ring} come before those that share only a vertex with the LCA_{ring}. For paths with sibling LCA_{ring}, we order from left to right. We prove the claim by induction on the number of paths in OPT when arranged in the above order.

Basis Step: The first path in the above order does not cross any other colored path. Thus, it can be colored with one color. Therefore, the claim holds for the basis step.

Induction Step: Consider a path σ_k in OPT. Assume that all paths preceding it in the above order have been colored with at most 3 colors and consider such a coloring. We will show how to extend this coloring such that all paths up to σ_k are colored with at most 3 colors such that no two paths with the same color cross. We consider the following cases:

Case 1: σ_k has edges only in $LCA_{ring}(\sigma_k)$.

σ_k can cross at most one of the paths preceding it, namely the one which contains all of the edges in the $LCA_{ring}(\sigma_k)$ not in σ_k. In this case, σ_k can be colored with any of the two colors other than the one used to color this crossing path.

Case 2: σ_k has edges outside $LCA_{ring}(\sigma_k)$.

If σ_k crosses any of the paths preceding it then these paths necessarily have to touch $LCA_{ring}(\sigma_k)$. The LCA_{ring} of such a preceding path is either $LCA_{ring}(\sigma_k)$ or its ancestor. Thus, σ_k crosses these paths in one of two possible ways: either, along the $LCA_{ring}(\sigma_k)$ and the "left" side of σ_k's path, or along the $LCA_{ring}(\sigma_k)$ and the "right" side of σ_k's path. Of course, it is possible that just one of the preceding paths meets both the "left" and "right" sides of σ_k's path. Since we consider an edge-disjoint optimal set, at most two of the preceding paths could have crossed σ_k. Thus, σ_k can be colored with the third color not used by these at most two crossing paths.

This covers all the cases for σ_k, and the induction step is proved. □

Our on-line algorithm for MEDPwPP, presented in the next subsection, will in fact accept a set of pairwise non-crossing paths. Lemma 1 implies that there exists a set of pairwise non-crossing paths that contains at least one third of the paths in OPT, so we lose at most a factor of 3 by considering non-crossing paths instead of edge-disjoint paths.

3.3 The Algorithm

When the algorithm rejects a path, it will sometimes place *roadblocks* on certain edges of the path, and two *extra roadblocks* above. If there is a roadblock on an edge, no path using that edge will be accepted later on. The subpath of a rejected path between two consecutive roadblocks is called a *segment*.

The algorithm is defined as follows. Let σ_i be the next path (request). If σ_i crosses a previously accepted path, reject it. If it crosses a previously rejected path such that the region of overlap of σ_i with the rejected path contains a roadblock, or if it passes through an *extra roadblock*, reject it as well. If σ_i is not rejected by the above conditions, it becomes a candidate. Accept a candidate with probability 1/2. If a candidate is not accepted, reject it and place roadblocks on it as follows: Number the edges from one end of the candidate to the other 1,2,3,... and so on. Choose an integer l (called level) uniformly at random from $[0, \lceil \log \ell_T \rceil + 1]$, where ℓ_T is the maximum path length in T. On each edge numbered $i \cdot 2^l$, $i = 1, 2, \ldots$, place a roadblock. Also, on the $LCA_{ring}(\sigma_i)$, place roadblocks on the edges that are incident on the extreme vertices of the candidate path in that ring but not contained in the candidate path. If the LCA_{ring} has only one vertex of the candidate, then the roadblocks are placed on the two edges incident on this vertex in the ring. Call these additional roadblocks *extra roadblocks.* The roadblocks placed by a rejected candidate path if l is chosen to be 1 are illustrated in Fig. 2.

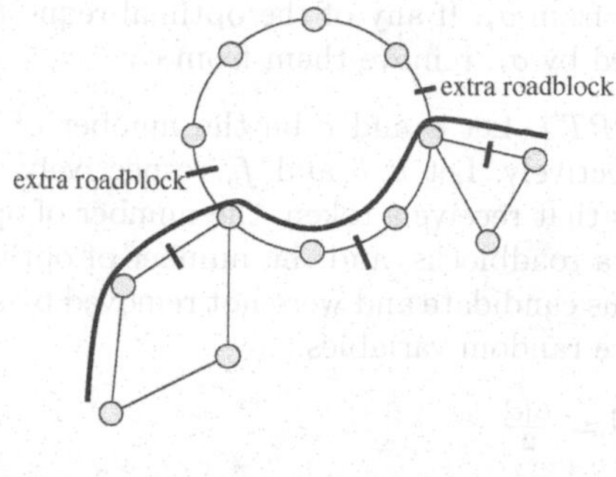

Fig. 2. Roadblocks placed by a rejected candidate for $l = 1$.

3.4 Analysis of the Algorithm

Consider any request sequence $\sigma_1, \sigma_2, \ldots, \sigma_s$. By Lemma 1, any optimal set of edge-disjoint paths can be colored with at most three colors such that no two paths with the same color cross. Thus, there exists a set of paths with the same color which has cardinality at least $\frac{|OPT|}{3}$. Call this set OPT'. We shall denote the paths in this set as "optimal requests" or "optimal paths" from now on. For the analysis, we distribute tokens to some of the optimal requests when a candidate is rejected. We maintain a subset of the optimal paths which is updated whenever a new request is processed. Initially, we set $C_0 = OPT'$. When request σ_i is processed, we compute C_i from C_{i-1}, $i = 1, 2, \ldots, s$, as follows:

Case 1: σ_i is not a candidate.
Set $C_i = C_{i-1}$.

Case 2: σ_i is an accepted candidate.
Remove from C_{i-1} all optimal requests that cross σ_i. Also, if there are optimal requests in C_{i-1} that cross a previously rejected candidate σ_j in the same segment as σ_i crosses σ_j and that received their last token from σ_j, remove them from C_{i-1}. Take the resulting set as C_i.

Case 3: σ_i is a rejected candidate.
The algorithm places roadblocks on some of the edges of σ_i. These roadblocks divide the path into segments. In each segment, we distribute a token to at most one optimal request that crosses σ_i in that segment. Define the level of an optimal request in C_{i-1} that crosses σ_i as j if the *region of overlap* of it with σ_i does not contain a roadblock when the roadblocks are placed 2^j apart, but contains one when placed 2^{j-1} apart. If the optimal request crosses σ_i and shares only one vertex with it, then its level is defined as 0. Let l be the level randomly chosen for σ_i. Then it is easy to see that there are at most two optimal requests in each segment of σ_i that have level l. If there is only one optimal request, give a token to it. If there are two, give a token to either of the optimal requests with probability 1/2. Define C_i as in Case 2. In addition, return to C_i all optimal requests that received a token from σ_i. If any of the optimal requests pass through the extra roadblocks placed by σ_i, remove them from C_i.

Let $a^* = |OPT'|$. Let a and c be the number of accepted candidates and candidates respectively. Let t, b and f, respectively, represent the number of optimal requests that receive a token, the number of optimal requests that were removed by extra roadblocks, and the number of optimal requests that did not cross any previous candidate and were not removed by an extra roadblock either. a, c, t, b and f are random variables.

Lemma 2. $\mathrm{E}[a] = \frac{E[c]}{2}$

Proof. A candidate is accepted with probability 1/2. Therefore,

$$\begin{aligned}\mathrm{E}[a] &= \textstyle\sum_{i=1}^{s} Pr(\sigma_i \text{ is a candidate})\cdot Pr(\sigma_i \text{ is accepted}|\sigma_i \text{ is a candidate}) \\ &= 1/2 \cdot \textstyle\sum_{i=1}^{s} Pr(\sigma_i \text{ is a candidate}) = \frac{\mathrm{E}[c]}{2}.\end{aligned}$$

□

Lemma 3. $\mathrm{E}[t + b + f] = \Omega(\frac{OPT}{\log \ell_T})$

Proof. Consider any optimal request ρ. Consider the event that it was not blocked by an extra roadblock, but crossed other candidates. Let σ_i be the first candidate that *crossed* ρ. ρ would receive a token from this candidate if (a) σ_i is rejected, (b) the level assigned to σ_i is its level, and (c) if there were another optimal request at the same level that crosses σ_i in the same segment, it is not given a token. The probability of event (a) is 1/2, that of event (b) is $1/(\log \ell_T + 2)$ and that of (c) is 1/2. Thus, the probability that ρ receives a token is at least $\frac{1}{4(\log \ell_T+2)}$. If ρ is blocked by an extra roadblock or if it does not cross any candidate, then it contributes 1 to the sum $b + f$. Now,

$$\mathrm{E}[t+b+f] \geq \sum_{\{\rho|\rho\in OPT'\}} \frac{1}{4(\log \ell_T+2)} = \frac{a^*}{4(\log \ell_T+2)} \geq \frac{OPT}{12(\log \ell_T+2)} = \Omega(\frac{OPT}{\log \ell_T})$$

Hence, the lemma follows. □

Lemma 4. *A rejected candidate that gives a token to an optimal path ρ must touch $LCA_{ring}(\rho)$. If a candidate σ_i crosses a previously rejected candidate then the previously rejected candidate must touch $LCA_{ring}(\sigma_i)$.*

Proof. Since a rejected candidate, call it σ_j, places two extra roadblocks in $LCA_{ring}(\sigma_j)$, neither an optimal path ρ that received a token from it nor a later candidate σ_i that crosses it can pass through any ancestor ring of $LCA_{ring}(\sigma_j)$. Now, consider σ_i. (The proof for ρ is analogous.) The crossing between σ_j and σ_i occurs somewhere in the sub-tree rooted at $LCA_{ring}(\sigma_j)$. Therefore, $LCA_{ring}(\sigma_i)$ is either $LCA_{ring}(\sigma_j)$ or its descendant. In the former case, we are done. In the latter case, recall that the graph T in which the paths lie is a tree of rings. Hence, any path that originates from the sub-tree rooted at $LCA_{ring}(\sigma_i)$ and passes through to $LCA_{ring}(\sigma_j)$ has to pass via $LCA_{ring}(\sigma_i)$. Thus, σ_j has to touch $LCA_{ring}(\sigma_i)$ after crossing σ_i, as claimed in the lemma. □

Now we want to bound the number of paths with tokens that are removed from C_{i-1} by a candidate σ_i.

Let σ_i be a candidate. If σ_i uses edges in $LCA_{ring}(\sigma_i)$, then let A and B be the two edges in $LCA_{ring}(\sigma_i)$ that are incident to vertices of σ_i but that are not contained in σ_i, and let C and D be the edges in $LCA_{ring}(\sigma_i)$ that are on σ_i and share an endpoint with A or B (see Fig. 3, left-hand side). If σ_i contains only one vertex v in $LCA_{ring}(\sigma_i)$, then let E and F be the edges incident to v that are used by σ_i, and let G and H be the other two edges incident to v that are contained in rings through which σ_i passes (see Fig. 3, right-hand side).

Lemma 5. *Let σ_i be a candidate. If σ_i uses edges in $LCA_{ring}(\sigma_i)$, then any previously rejected candidate σ_j that crosses σ_i contains A, or contains B, or contains C and D. If σ_i uses only one vertex in $LCA_{ring}(\sigma_i)$, then any previously rejected candidate σ_j that crosses σ_i contains E, or contains F, or contains G, or contains H. Furthermore, the edge in A, B, C, D or E, F, G, H contained in σ_j lies in the segment of σ_j where σ_i crosses σ_j.*

Proof. Case 1: $LCA_{ring}(\sigma_i)$ contains at least one edge of σ_i.

By Lemma 4, any previously rejected candidate σ_j that crosses σ_i touches $LCA_{ring}(\sigma_i)$. In fact, σ_j must use edges in $LCA_{ring}(\sigma_i)$, because otherwise its extra roadblocks would have led to the immediate rejection of σ_i. If σ_j does not use A or B, it must use all edges on the subpath of σ_i from C to D, because otherwise again its extra roadblocks would have led to the rejection of σ_i.

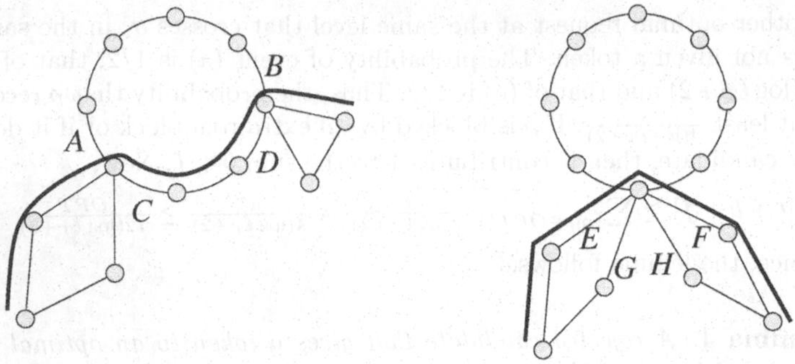

Fig. 3. Crossing previously rejected candidates must use one of the edges A, B, C, D or E, F, G, H.

Case 2: $LCA_{ring}(\sigma_i)$ has no edges of σ_i in it.

Let v be the vertex in $LCA_{ring}(\sigma_i)$ that is contained in σ_i. Any previously rejected candidate σ_j that crosses σ_i must touch v (by Lemma 4) and must use edges in one of the two rings containing v through which σ_i passes. But then σ_j must reach v through one of the edges E, F, G, or H.

In both cases, it is easy to see that the respective edge must lie in the segment of σ_j where σ_i crosses σ_j. Otherwise, the region of overlap of σ_i with σ_j would contain a roadblock. □

Lemma 5 shows that the set of all previously rejected candidates that cross the current candidate σ_i can be grouped into at most four different classes such that the rejected candidates in the same class cross each other in the same segment in which they cross σ_i. For each class, there can be at most one optimal path in C_{i-1} that has received its last token from that segment of a previously rejected candidate in that class. Therefore, we have the following lemma.

Lemma 6. *At most* 4 *optimal paths must be removed from* C_{i-1} *because they received their last token from a previously rejected candidate that crossed them in the same segment as it crosses* σ_i.

It remains to bound the number of optimal paths that have tokens and cross σ_i.

Lemma 7. *There can be at most* 10 *optimal paths with tokens that cross* σ_i.

Proof. If σ_i itself has tokens, no other path with tokens can cross σ_i. So assume that σ_i does not have a token itself, and let ρ be a call with token that crosses σ_i.

Case 1: ρ uses edges in $LCA_{ring}(\sigma_i)$.

If σ_i uses edges in $LCA_{ring}(\sigma_i)$ as well, then ρ must either contain edge A or edge B (there can be at most two such paths), or it uses only edges on the subpath of σ_i from C to D. In the latter case, the previously rejected candidate that gave the last token to ρ must use edge A, or edge B, or edges C and D. Therefore, ρ is already accounted for in Lemma 6 in this case. In total, there can be at most two calls with tokens that cross σ_i, use edges in $LCA_{ring}(\sigma_i)$, and are not accounted for already in Lemma 6. The same conclusion is reached if σ_i does not use edges in $LCA_{ring}(\sigma_i)$, because then ρ must use one of the two edges of $LCA_{ring}(\sigma_i)$ incident to the vertex on σ_i in that ring.

Case 2: $LCA_{ring}(\rho)$ is a descendant of $LCA_{ring}(\sigma_i)$ or ρ contains only one vertex in $LCA_{ring}(\sigma_i)$.

Consider the rejected candidate σ_j that gave ρ its last token. By Lemma 4 we know that σ_j touches $LCA_{ring}(\rho)$. But then σ_j must also touch $LCA_{ring}(\sigma_i)$, because otherwise the extra roadblocks placed by σ_j would not have allowed σ_i to become a candidate in the first place. Therefore, σ_j must in fact use one of the edges A, B, or C and D (if σ_i uses edges in its LCA_{ring}), or of the edges E, F, G, H (if σ_i does not use edges in its LCA_{ring}), so σ_j belongs to one of the at most four classes of previously rejected candidates resulting from Lemma 5. Now, if the segment in which ρ crosses σ_j is the same as that in which the respective edge A, B, ... lies, the path ρ is already accounted for by Lemma 6.

The only problematic case is when ρ crosses σ_j in a different segment. Call any path ρ for which this happens a *bad path.* Among all bad paths crossing the left side of σ_i, consider one whose LCA_{ring} is furthest from the root. We will show that in addition to this bad path ρ there can be at most one other bad path crossing the left side of σ_i. (The same argument can then be applied to the right side of σ_i.) Let σ_j be the previously rejected candidate that gave the last token to ρ. Since the segement of σ_j in which σ_j crosses ρ is different from the segment of σ_j in which σ_j reaches $LCA_{ring}(\sigma_i)$, there must be at least one roadblock on σ_j separating these two segments. Furthermore, this roadblock must in fact be on an edge in $LCA_{ring}(\rho)$, so the situation must be exactly as shown in Fig. 4: σ_i begins at w or enters $LCA_{ring}(\rho)$ at w and leaves $LCA_{ring}(\rho)$ at u; σ_j meets σ_i for the first time at vertex u; ρ crosses σ_j, meets σ_i for the first time at vertex w, and either uses the edge incident to w in $LCA_{ring}(\rho)$ that is contained in σ_i (Fig. 4, left-hand side), or uses edges in the ring below $LCA_{ring}(\rho)$ where σ_i comes from (Fig. 4, right-hand side); and there is at least one roadblock on σ_j somewhere between the crossing with ρ and the vertex u.

We note that there cannot be any other bad path whose LCA_{ring} is equal to $LCA_{ring}(\rho)$, because any such bad path would also have to contain the edge in $LCA_{ring}(\rho)$ that is not in σ_i and that is incident on the vertex labeled w in Fig. 4. So assume that there is a bad path ρ' crossing the left side of σ_i and whose LCA_{ring} is an ancestor of $LCA_{ring}(\rho)$. Let u' be the vertex of $LCA_{ring}(\rho')$ where σ_i enters that ring from below, and let v' be the vertex where σ_i leaves that ring going up. Then σ_j must also touch u' and v'. Furthermore, ρ' must either contain both edges incident to u' in $LCA_{ring}(\rho')$ or it must use edges in $LCA_{ring}(\rho')$ and the child ring of $LCA_{ring}(\rho')$ through which σ_i passes. In both cases, ρ' crosses σ_j in the same segment in which σ_j contains one of the edges named A, B, etc. At the time when σ_j was processed by the algorithm, at most one optimal path with tokens crossing that segment of σ_j was not removed from

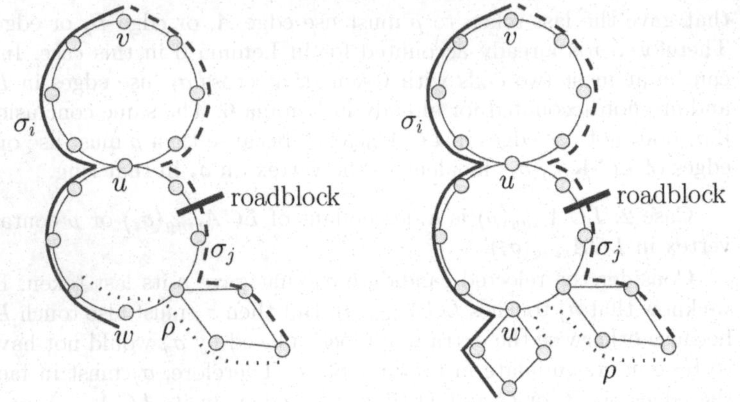

Fig. 4. The two possibilities for a "bad path" ρ, which receives a token from σ_j in a different segment than where σ_j reaches $LCA_{ring}(\sigma_i)$.

C_{j-1}. Therefore, at most one bad path that crosses σ_i (in addition to the bad path ρ) can still be in C_{i-1} when σ_i is processed.

In summary, the number of paths with tokens that cross σ_i can be bounded as follows. There are at most 4 paths with tokens that are already accounted for in Lemma 6. Then there are at most 2 additional paths with tokens that use edges in $LCA_{ring}(\sigma_i)$ and at most 4 paths with tokens that are bad paths. In total there are at most 10 paths with tokens that cross σ_i. □

Lemma 8. $\mathrm{E}[c] \geq \frac{\mathrm{E}[t]}{16}$.

Proof. We show that any candidate σ_i removes at most 16 calls with tokens. By Lemma 6, at most 4 paths with tokens are removed because they received their last token from a previously rejected candidate that crossed them in the same segment as it crosses σ_i. By Lemma 7, there are at most 10 paths with tokens that cross σ_i. Finally, if σ_i is rejected, there are at most 2 paths with tokens that are removed because of the extra roadblocks. In total, at most 16 paths with tokens are removed. □

Theorem 3. *There is a randomized algorithm for MEDPwPP in trees of rings that achieves competitive ratio* $O(\log \ell_T)$.

Proof. Every rejected candidate places two extra roadblocks. Hence, $c \geq \frac{b}{2}$. Also, every optimal request that was not removed by an extra roadblock and did not cross any candidate must have been a candidate itself. Therefore, $c \geq f$. From this and the above lemmas we get:

$$\mathrm{E}[a] = \frac{\mathrm{E}[c]}{2} \geq \frac{1}{2}\max\{\frac{\mathrm{E}[t]}{16}, \frac{\mathrm{E}[b]}{2}, \mathrm{E}[f]\} = \Omega(\mathrm{E}[t+b+f]) = \Omega(\frac{OPT}{\log \ell_T})$$

□

References

1. Adler, R., Azar, Y.: Beating the logarithmic lower bound: randomized preemptive disjoint paths and call control algorithms. In: Proceedings of the 10th Annual ACM–SIAM Symposium on Discrete Algorithms SODA. (1999) 1–10
2. Awerbuch, B., Azar, Y., Plotkin, S.: Throughput-competitive on-line routing. In: Proceedings of the 34th Annual Symposium on Foundations of Computer Science FOCS. (1993) 32–40
3. Awerbuch, B., Bartal, Y., Fiat, A., Rosén, A.: Competitive non-preemptive call control. In: Proceedings of the 5th Annual ACM–SIAM Symposium on Discrete Algorithms SODA. (1994) 312–320
4. Awerbuch, B., Gawlick, R., Leighton, T., Rabani, Y.: On-line admission control and circuit routing for high performance computing and communication. In: Proceedings of the 35th Annual Symposium on Foundations of Computer Science FOCS. (1994) 412–423
5. Beauquier, B., Pérennes, S., Tóth, D.: All-to-all routing and coloring in weighted trees of rings. In: Proceedings of the 11th Annual ACM Symposium on Parallel Algorithms and Architectures SPAA. (1999) 185–190
6. Blum, A., Kalai, A., Kleinberg, J.: Admission control to minimize rejections. In: Proceedings of the 7th International Workshop on Algorithms and Data Structures WADS. LNCS 2125 (2001) 155–164
7. Borodin, A., El-Yaniv, R.: Online Computation and Competitive Analysis. Cambridge University Press (1998)
8. Deng, X., Li, G., Zang, W., Zhou, Y.: A 2-approximation algorithm for path coloring on trees of rings. In: Proceedings of the 11th Annual International Symposium on Algorithms and Computation ISAAC. LNCS 1969 (2000) 144–155
9. Erlebach, T.: Approximation algorithms and complexity results for path problems in trees of rings. In: Proceedings of the 26th International Symposium on Mathematical Foundations of Computer Science MFCS. LNCS 2136 (2001) 351–362
10. Garay, J.A., Gopal, I.S., Kutten, S., Mansour, Y., Yung, M.: Efficient on-line call control algorithms. Journal of Algorithms **23** (1997) 180–194
11. Hassin, R., Tamir, A.: On the minimum diameter spanning tree problem. Information Processing Letters **53** (1995) 109–111
12. Kleinberg, J., Tardos, É.: Disjoint paths in densely embedded graphs. In: Proceedings of the 36th Annual Symposium on Foundations of Computer Science FOCS. (1995) 52–61

Massive Quasi-Clique Detection

James Abello[1], Mauricio G.C. Resende[1], and Sandra Sudarsky[2*]

[1] AT&T Labs Research, Florham Park, NJ 07032 USA
{abello,mgcr}@research.att.com

[2] Siemens Corporate Research, Inc, Princeton, NJ 08540 USA
sudarsky@scr.siemens.com

Abstract. We describe techniques that are useful for the detection of dense subgraphs (quasi-cliques) in massive sparse graphs whose vertex set, but not the edge set, fits in RAM. The algorithms rely on efficient semi-external memory algorithms used to preprocess the input and on greedy randomized adaptive search procedures (GRASP) to extract the dense subgraphs. A software platform was put together allowing graphs with hundreds of millions of nodes to be processed. Computational results illustrate the effectiveness of the proposed methods.

1 Introduction

A variety of massive data sets can be modeled as very large multi-digraphs M with a special set of edge attributes that represent special characteristics of the application at hand [1]. Understanding the structure of the underlying digraph $D(M)$ is useful for storage organization and information retrieval. The availability of computers with gigabytes of RAM allows us to make the realistic assumption that the vertex set (but not the edge set) of $D(M)$ fits in main memory. Moreover, it has been observed experimentally, in data gathered in the telecommunications industry, the internet, and geographically based information systems, that $D(M)$ is a sparse graph with very skewed distribution and low undirected diameter [1]. These observations have made the processing of $D(M)$ feasible. We present here an approach for discovering large dense subgraphs (quasi-cliques) in such large sparse multi-digraphs with millions of vertices and edges. We report a sample of our current experimental results.

Before proceeding any further let us agree on the following notational conventions.

Let $G = (V, E)$ be a graph where V is the set of vertices and E is the set of edges in G. A multi-graph M is just a graph with an integer multiplicity associated with every edge. Whenever it becomes necessary to emphasize that the underlying graph is directed we use the term multi-digraph.

For a subset $S \subseteq V$, we let G_S denote the subgraph induced by S.

A graph $G = (V, E)$ is γ-dense if $|E(G)| \geq \gamma\binom{|V(G)|}{2}$. A γ-clique S, also called a quasi-clique, is a subset of V such that the induced graph G_S is connected

* Work completed as an AT&T consultant and DIMACS visitor.

S. Rajsbaum (Ed.): LATIN 2002, LNCS 2286, pp. 598–612, 2002.

and γ-dense. The maximum γ-clique problem is to find a γ-clique of maximum cardinality in a graph G.

For a graph $G = (V, E)$, $S \subset V$, $E' \subset E$, and $x \in V$, let

- $\deg(x) = |\mathcal{N}(x)|$ where $\mathcal{N}(x) = \{y \in V(G) \big| (x, y) \in E(G)\}$;
- $\deg(x)\big|_S = |\mathcal{N}(x)\big|_S|$ where $\mathcal{N}(x)\big|_S = \mathcal{N}(x) \cap S$;
- $\mathcal{N}(S) = \cup_{x \in S} \mathcal{N}(x)$;
- $E(S) = \{(x, y) \in E(G) : x \in S, y \in S\}$;
- $G_S = (S, E(S))$;
- For $S, R \subset V(G)$, $E(S, R) = \{(x, y) \in E(G) : x \in S, y \in R\}$;

Finding a maximum 1-clique is a classical *NP*-hard problem and therefore one can expect exact solution methods to have limited performance on large instances. In terms of approximation, a negative breakthrough result by Arora et al. [4,5] together with results of Feige et al. [6], and more recently Håstad [9], prove that no polynomial time algorithm can approximate the maximum clique size within a factor of n^ϵ ($\epsilon > 0$), unless $P = NP$. Given this current state of knowledge, it is very unlikely (with our current models of computation) that general heuristics exist which can provide answers with certifiable measures of optimality. A suitable approach is then to devise heuristics equipped with mechanisms that allow them to escape from poor local optimal solutions. These include heuristics such as simulated annealing [11], tabu search [8], and genetic algorithms [10], that move from one solution to another, as well as the multistart heuristic GRASP [7,12], which samples different regions of the solution space, finding a local minimum each time.

In this paper, we use the concept of quasi-cliques as a unifying notion that drives heuristics towards the detection of dense subgraphs in very large but sparse multi-digraphs. (It is assumed here that the vertex set of the graph, but not the edge set, fits in RAM. These graphs are termed semi-external in [2]).

Our main contributions include the introduction of a very intuitive notion of potential of a vertex set (edge set) with respect to a given quasi-clique (bi-clique), the use of edge pruning and an external memory breadth first search (BFS) traversal to decompose a disk resident input digraph into smaller subgraphs to make the search feasible. We remark that the techniques described here are being applied to very large daily telecommunications traffic graphs containing hundreds of millions of vertices.

The paper is organized as follows. In Section 2, we consider several graph decomposition schemes and pruning approaches used in our computations to reduce the search space. Section 3 and Section 4 present basic quasi-clique notions and define the potential function that is central to our approach for discovering maximal quasi-cliques. Section 5 contains a brief discussion of the main ingredients necessary to design greedy randomized adaptive search procedures (GRASP). Section 6 uses the machinery presented in the preceeding sections to describe a GRASP tailored for finding maximal quasi-cliques in both bipartite and non-bipartite graphs. Section 7 contains the description of our semi-external approach to handle very large sparse multi-digraphs. Sample computational results and concluding remarks appear in Section 8 and Section 9.

2 Graph Decomposition and Pruning

We introduce in this section two decomposition schemes that make the processing of very large graphs feasible.

First, we identify sources, sinks and transmitter vertices in the input graph. Namely, consider the underlying directed graph

$$D(M) = \{(x, y) \mid (x, y) \text{ is an edge in } M\}$$

and the corresponding underlying undirected graph

$$U(M) = \{\{x, y\} \mid (x, y) \text{ is an edge in } D(M)\}.$$

It is worthwhile recalling here that for data gathered in certain telecommunications applications (such as phone calls), internet data (such as URL links), and geographical information systems, $U(M)$ is a sparse graph with very low diameter [1].

For a vertex $u \in M$, let

$$\text{out}(u) = \{x \mid (u, x) \in D(M)\} \text{ and } \text{in}(u) = \{y \mid (y, u) \in D(M)\}.$$

Furthermore, let $\text{outdeg}(u) = |\text{out}(u)|$, and $\text{indeg}(u) = |\text{in}(u)|$.

In a preprocessing phase, we use efficient external memory algorithms for computing the undirected connected components of $U(M)$ [2]. For each connected component, consider the sub-digraph of $D(M)$ induced by its vertex set and classify its vertices as sources (indeg $= 0$), sinks (outdeg $= 0$) and transmitters (indeg and outdeg > 0). We then partition the edge set of each connected component by traversing the corresponding subgraph in a breadth first search manner. The assumption that the vertex set fits in main memory together with the fact that $U(M)$ has low diameter allow us to perform the Breadth First Search in few passes over the data. We store each connected component as a collection of subgraphs according to the BFS order. Namely, the subgraphs induced by vertices at the same undirected distance, from the root of the corresponding BFS tree, are stored contiguously. The edges between vertices at adjacent levels are also stored contigously. Clearly, 1-cliques can appear only in these subgraphs. We explot this fact to localize (and parallelize) the quasi-clique search by using maximal 1-cliques as seeds.

A complementary but very effective processing scheme that also helps to localize quasi-clique detection, consists in pruning those edges that do not contribute to a better solution. These edges are called *peelable*. In the case of 1-cliques, if a lower bound k on the cardinality of maximum clique is known, we can delete all those edges incident with a vertex of degree less than k. This process affects the degrees of other vertices, hence, further simplification is possible by reapplying the same reduction scheme. To control the pruning we recursively delete edges incident to vertices of degree i, from $i = 1$ to $k - 1$, in that order, updating the degree of both endpoints.

When γ is less than one, we use the notion of *γk-peelable* vertices. Namely, a vertex v is *γk-peelable* if v and all its neighbors have degree smaller than γk. *γk-peelable* vertices cannot be added to a quasi-clique of density γ and cardinality at least k to obtain a larger quasiclique with density greater than or equal to γ. Because of this, if we know the existence of a quasi-clique of density γ and of cardinality at least k, then we can prune all those edges incident to *γk peelable* vertices, in increasing degree order, updating every time the degrees of both endpoints.

We want to remark here that designing a good and efficient peeling schedule is by itself a problem that could be useful in the exploration of massive data sets. Our experimental results indicate that the proposed approach works well for sparse and low diameter graphs that are reducible to trees by a sequence of dense subgraph contractions. In Section 8 we report results obtained by a combination of these techniques when applied to multi-graphs extracted from telecommunications data sets. From now on, we assume that we have an index structure to the subgraphs induced by vertices on the same level of the BFS and to the subgraphs induced by the union of the vertices in two consecutive BFS levels.

3 Quasicliques

Quasicliques are subgraphs with specified edge density. Two optimization problems related to quasicliques arise naturally. In the first, we fix an edge density and seek a quasiclique of maximum cardinality with at least the specified edge density. In the other, we specify a fixed cardinality and seek a quasiclique of maximum edge density. In this section, we describe general properties about quasicliques in an undirected graph $G = (V, E)$. We denote by S the set of vertices of the subgraph G_S we wish to find, i.e. the subgraph induced by S on G. Let γ be a real parameter such that $0 < \gamma \leq 1$. Recall that a set of vertices $S \subseteq V(G)$ is a γ-clique if G_S is connected and $|E(G_S)| \geq \gamma\binom{|S|}{2}$. We also refer to γ-cliques as quasicliques. A vertex $x \in \bar{S}$ is called a γ-vertex, with respect to a γ-clique S, if $G_{S\cup\{x\}}$ is a γ-clique. Similarly, a set of vertices $R \subseteq \bar{S}$ is called a γ-set with respect to S if $S \cup R$ is a γ-clique. The set of γ-vertices with respect to S is denoted by $\mathcal{N}_\gamma(S)$. Notice that $\mathcal{N}_\gamma(S)$ is not necessarily a γ-clique.

One basic property of γ-cliques is the following.

Lemma 1. *Let S and R be disjoint γ-cliques with*

$$\frac{|E(G_S)|}{\binom{|S|}{2}} = \gamma_S \text{ and } \frac{|E(G_R)|}{\binom{|R|}{2}} = \gamma_R.$$

$S \cup R$ is a γ-clique, if and only if

$$|E(S,R)| \geq \gamma(|R||S|) - (\gamma_R - \gamma)\binom{|R|}{2} - (\gamma_S - \gamma)\binom{|S|}{2}. \tag{1}$$

The proof of this simple but fundamental lemma follows easily from the definitions. It provides a general framework to find quasicliques. Namely, given a γ-clique S, find another γ-clique R with $S \cap R = \emptyset$, such that (1) holds. In order to guarantee that the joint condition (1) is satisfied one can restrict R to be a γ-clique in $\mathcal{N}_\gamma(S)$. More generally, the objective is to find large γ-sets with respect to S. One approach to achieve this is to use the notion of set potential. Define the potential of a set R to be

$$\phi(R) = |E(R)| - \gamma\binom{|R|}{2}$$

and the potential of a set R with respect to a disjoint set S to be

$$\phi_S(R) = \phi(S \cup R).$$

Sets with nonnegative potential are precisely γ-sets. We seek γ-sets R with large potential $\phi_S(R)$. Ideally, in a construction algorithm, one would prefer sets R of maximum cardinality. Finding such sets is computationally intractable. Our approach is to build a maximal γ-clique incrementally. In the next section, we describe such an algorithm.

4 Finding Maximal Quasicliques

Assume S is a γ-clique. We seek a vertex $x \in \mathcal{N}_\gamma(S)$ to be added to S. One strategy for selecting x is to measure the effect of its selection on the potential of the other vertices in $\mathcal{N}_\gamma(S)$. To accomplish this, define the potential difference of a vertex $y \in \mathcal{N}_\gamma(S) \setminus \{x\}$ to be

$$\delta_{S,x}(y) = \phi_{S\cup\{x\}}(\{y\}) - \phi_S(\{y\}).$$

It follows from the definitions that

$$\delta_{S,x}(y) = \deg(x)\big|_S + \deg(y)\big|_{\{x\}} - \gamma(|S| + 1).$$

The above equation shows that the potential of γ-neighbors of x does not decrease with the inclusion of x. On the other hand, the potential of the non-γ-neighbors of x may decrease when the potential of the γ-neighbors of x increases by less than a unit. If the increase in the potential of a γ-neighbor is exactly one unit, there is no change in the potential of the non-γ-neighbors. It also follows that if x and y are adjacent γ-vertices and x is added to S, then y remains a γ-vertex with respect to $S \cup \{x\}$.

The total effect on the potentials, caused by the selection of x, on the remaining vertices of $\mathcal{N}_\gamma(S)$ is

$$\Delta_{S,x} = \sum_{y \in \mathcal{N}_\gamma(S)\setminus\{x\}} \delta_{S,x}(y) = \left|\mathcal{N}_\gamma(\{x\})\right| + \left|\mathcal{N}_\gamma(S)\right| \left(\deg(x)\big|_S - \gamma(|S| + 1)\right). \tag{2}$$

Note that $|\mathcal{N}_\gamma(\{x\})| = \deg(x)|_{\mathcal{N}_\gamma(S)}$. The vertex x that maximizes $\Delta_{S,x}$ is one with a high number of γ neighbors and with high degree with respect to S. A greedy algorithm that recursively selects such a vertex will eventually terminate with a maximal γ-set.

A construction procedure builds a quasiclique, one vertex at a time. Ideally, the cardinality of the set of γ-neighbors of a given partial solution S should not increase. This could occur in the case that a vertex with nonpositive potential gains enough potential with the inclusion of a vertex to make it a γ-vertex. To insure that the set of γ-neighbors does not increase, we use a control variable $\gamma^* \geq \gamma$ that corresponds to the density of the current partial solution. The construction procedure, whose pseudo-code is shown in Figure 1, incorporates this idea. Its time complexity is $O(|S|\ |V|^2)$, where S is the set of vertices of the constructed maximal quasiclique.

```
procedure construct_dsubg(I:γ,V,E,O:S)
  γ* = 1;
  Select x ∈ V;
  S* = {x};
  while γ* ≥ γ do
      S = S*;
      if (N_γ*(S) ≠ ∅) then
          Select x ∈ N_γ*(S);
      else
          if (N(S) \ S = ∅) return(S);
          Select x ∈ N(S) \ S;
      end;
      S* = S ∪ {x};
      γ* = |E(S*)|/(|S*| choose 2);
  end while;
  return(S);
end construct_dsubg;
```

Fig. 1. Pseudo-code of construction procedure for maximal dense subgraphs

Notice that by modifying the stopping criterion of the while loop (to $|S| < K$), the procedure can be used to find a K cardinality quasiclique of large density.

Given a maximal γ-clique S and an objective function $f(S)$, define

$$H(S; f) = \{X \subseteq S;\ Y \subseteq \bar{S} \mid f((S \setminus X) \cup Y) > f(S)\}$$

to be the set of subset exchanges that improve the objective function. A local improvement procedure that makes use of this neighborhood structure can further improve the quasiclique. A restricted version of this local search with $|X| = 1$ and $|Y| = 2$ called a $(1, 2)$ local exchange procedure, is used to potentially improve the γ-clique in the case that $f(S) = |S|$. When $f(S) = |E(S)|/\binom{|S|}{2}$, we use local search with $|X| = |Y| = 1$ to try to find a denser K cardinality quasiclique.

This local search, coupled with a greedy randomized version of the construction procedure `construct_dsubg` is described in Section 6. That type of algorithm is called a greedy randomized adaptive search procedure (GRASP), and we review the overall idea in the next section.

5 GRASP

A GRASP, or greedy randomized adaptive search procedure [7], is a multi-start or iterative process, in which each GRASP iteration consists of two phases, a construction phase, in which a feasible solution is produced, and a local search phase, in which a local optimum in the neighborhood of the constructed solution is sought. The best overall solution is kept as the result. The pseudo-code in Figure 2 illustrates a GRASP procedure for maximization in which `maxitr` GRASP iterations are done. For a recent survey of GRASP, see [12].

```
procedure grasp(f(·),g(·),maxitr,x*)
  f* = -∞;
  for k = 1,2,... ,maxitr do
      grasp_construct(g(·),α,x);
      local(f(·),x);
      if f(x) > f* do
          x* = x;
          f* = f(x*);
      end if;
  end for;
end grasp;
```

Fig. 2. GRASP pseudo-code

In the construction phase, a feasible solution is iteratively constructed, one element at a time. At each construction iteration, the choice of the next element to be added is determined by ordering all candidate elements (i.e. those that can be added to the solution) in a restricted candidate list (RCL) C with respect to a greedy function $g : C \to \mathbb{R}$, and randomly choosing one of the candidates in the list. Let $\alpha \in [0, 1]$ be a given real parameter. The pseudo code in Figure 3 describes a basic GRASP construction phase. The pseudo-code shows that the parameter α controls the amounts of greediness and randomness in the algorithm. A value $\alpha = 1$ corresponds to a purely greedy construction procedure, while $\alpha = 0$ produces a purely random construction.

A solution generated by a GRASP construction is not guaranteed to be locally optimal with respect to simple neighborhood definitions. It is almost always

```
procedure grasp_construct(g(·),α,x)
  x = ∅;
  Initialize candidate set C;
  while C ≠ ∅ do
      s̲ = min{g(t) | t ∈ C};
      s̄ = max{g(t) | t ∈ C};
      RCL = {s ∈ C | g(s) ≥ s̲ + α(s̄ − s̲)};
      Select s, at random, from the RCL;
      x = x ∪ {s};
      Update candidate set C;
  end while;
end grasp_construct;
```

Fig. 3. GRASP construction pseudo-code

beneficial to apply a local improvement procedure to each constructed solution. A local search algorithm works in an iterative fashion by successively replacing the current solution by a better solution in the neighborhood of the current solution. It terminates when no better solution is found in the neighborhood. The *neighborhood structure* N for a problem P relates a solution s of the problem to a subset of solutions $N(s)$. A solution s is said to be *locally optimal* if there is no better solution in $N(s)$. The pseudo-code in Figure 4 describes a basic local search procedure.

```
procedure local(f(·),N(·),x)
  H = {y ∈ N(x) | f(y) > f(x)};
  while |H| > 0 do
      Select x ∈ H;
      H = {y ∈ N(x) | f(y) > f(x)};
  end while;
end local;
```

Fig. 4. GRASP local search pseudo-code

6 GRASP for Finding Quasicliques

In this section, we describe a GRASP for finding γ-cliques. First, we describe a procedure for the nonbipartite case. Then, we describe a procedure for the bipartite case.

6.1 Nonbipartite Case

This GRASP uses a greedy randomized version of the construction procedure `construct_dsubg` described in Section 4 and a (2, 1)-exchange local search.

First, we consider the problem of finding a high cardinality quasiclique of specified density γ. We need to specify how the selections are made in lines 2, 7, and 10 of the procedure `construct_dsubg`. In line 2, the greedy function used is the vertex degree $g(x) = \deg(x)\big|_V$. In line 7, we use the total potential difference $g(x) = \Delta_{S,x}$, as given in equation (2). In line 10, the greedy function is the vertex degree with respect to the current solution S, i.e. $g(x) = \deg(x)\big|_S$.

For the problem of finding a high density quasiclique of specified cardinality K, the only difference is that in line 7 the greedy function becomes the vertex degree with respect to the current solution S, i.e. $g(x) = \deg(x)\big|_S$.

6.2 Bipartite Case

For a bipartite graph $G = (V, E)$, where $V(G) = L \cup R$ and $L \cap R = \emptyset$, the procedure builds a quasiclique one edge at a time. A balanced bipartite quasiclique (γ-biclique) is a subgraph G_S such that $V(G_S) = S_1 \cup S_2$ and $S_1 \cap S_2 = \emptyset$ such that $|S_1| = |S_2|$ and $|E(S_1, S_2)| \geq \gamma|L||R|$. We seek a balanced bipartite quasiclique with large cardinality.

In this case, we consider the following potential function that takes as arguments the pair of disjoint sets of vertices L and R. The potential $\phi((L, R)) = |E(L, R)| - \gamma|L||R|$. The potential of a γ-biclique is nonnegative. For a γ-biclique S with $V(S) = L \cup R$ and $L \cap R = \emptyset$, an edge $(x, y) \in E(\mathcal{N}_\gamma(S) \setminus S$ is a γ-neighbor of S if $|E(S \cup \{x, y\})| \geq \gamma((|L| + 1)(|R| + 1))$.

The notions of potential and potential difference of an edge, and the total effect on the potentials of the remaining edges by the inclusion of an edge can be extended to this case using the modified potential given above. Likewise, the construction procedure `construct_dsubg` can be adapted to find large balanced bipartite quasicliques. The edge selection greedy function is now

$$g(x, y) = \deg(x)\big|_{\mathcal{N}_\gamma(S)} \deg(y)\big|_{\mathcal{N}_\gamma(S)} / (\deg(x)\big|_{\mathcal{N}_\gamma(S)} + \deg(y)\big|_{\mathcal{N}_\gamma(S)}).$$

The complexity in this case becomes $O(|Ma||V|^2)$ where $|Ma|$ is the size of a maximum matching in G.

7 A Semi External Memory Approach for Finding Quasicliques

The procedure described in the previous sections requires access to the edges and vertices of the input graph. This limits its use to graphs small enough to fit in memory. We now propose a semi-external procedure [2,3] that works only with vertex degrees and a subset of the edges in main memory, while most of the edges can be kept in secondary disk storage. Besides enabling its use on smaller memory machines, the procedure we describe below also speeds up the computation. We assume that very large graphs appearing in massive datasets are sparse. The approach described next takes advantage of this situation in two respects. First, the largest cardinality clique is bounded from above by approximately the square root of the number of edges. Second, exploring the vertices of the graph in a breadth first search manner will eventually get to a maximal γ-clique.

In order to proceed, we first define a `peel` operation on a graph. Given a parameter q, `peel`(G, q) recursively deletes from G all vertices having degree less than q, along with their incident edges. This peeling operation is used in the case of cliques, i.e. quasicliques with $\gamma = 1$.

In the case of γ cliques when γ is strictly less than 1, we use a more constrained peeling operation `localpeel`.

`localpeel`(G, q) recursively deletes from G only those vertices with degree less than q, all of whose neighbors have also degree less than q, along with their incident edges.

We proceed now with the details for cliques and later on we discuss how to extend them to general quasicliques.

First, we sample a subset $\mathcal{E}$ of edges of E such that $\mid \mathcal{E} \mid < \mathcal{T}_G$, where $\mathcal{T}_G$ is a threshhold function of the graph G. The subgraph corresponding to $\mathcal{E}$ is denoted by $\mathcal{G} = (\mathcal{V}, \mathcal{E})$, where $\mathcal{V}$ is the vertex set of $\mathcal{G}$. The procedure `grasp` is applied to $\mathcal{G}$ to produce a clique Q. Let the size of the clique found be $q = |Q|$. Since q is a lower bound on the largest clique in G, any vertex in G with degree less than q cannot possibly be in a maximum clique and can be therefore discarded from further consideration. This is done by applying `peel` to G with parameter q. Procedures `grasp` and `peel` are reapplied until no further reduction is possible. The aim is to delete irrelevant vertices and edges, allowing `grasp` to focus on the subgraph of interest. Reducing the size of the graph allows the GRASP to explore portions of the solution space at greater depth, since GRASP iterations are faster on smaller graphs. If the reduction results in a subgraph smaller than the specified threshhold, the GRASP can be made to explore the solution space in more detail by increasing the number of iterations to `maxitr`$_l$. This is what usually occurs, in practice, when the graph is very sparse. However, it may be possible that the repeated applications of `grasp` and `peel` do not reduce the graph to the desired size. In this case, we partition the edges that remain into sets that are smaller than the threshhold and apply `grasp` to each resulting subgraph. The size of the largest clique found is used as parameter q in a `peel` operation and if a reduction is achieved the procedure `clique` is recursively called. Figure 5 shows pseudo-code for this semi-external approach.

In procedure `clique`, edges of the graph are sampled. As discussed earlier, we seek to find a clique in a connected component, examining one component at a time. Within each component, we wish to maintain edges that share vertices close together so that when they are sampled in `clique` those edges are likely to be selected together. To do this, we perform a semi-external breadth first search on the subgraph in the component and store the edges for sampling in the order determined by the search.

```
procedure clique(V,E,maxitr_s, maxitr_l,T_G,Q)
1     Let 𝒢 = (𝒱,ℰ) be a subgraph of G = (V,E) such that | ℰ | ≤ T_G;
2     while 𝒢 ≠ G do
3           𝒢⁺ = 𝒢;
4           grasp(𝒱,ℰ,maxitr_s,Q);
5           q = |Q|;
6           peel(V,E,q);
7           Let 𝒢 = (𝒱,ℰ) be a subgraph of G = (V,E) such that | ℰ | ≤ T_G;
8           if 𝒢⁺ == 𝒢 break;
9     end while;
10    Partition E into E_1,... ,E_k such that | E_j | ≤ T_G, for j = 1,... ,k;
11    for j = 1,... ,k do
12          Let V_j be the set of vertices in E_j;
13          grasp(V_j,E_j,maxitr_l,Q_j);
14    end for;
15    𝒢⁺ = 𝒢;
16    q = max {|Q_1|,|Q_2|,... ,|Q_k|};
17    peel(V,E,q);
18    if 𝒢⁺ ≠ G then
19          clique(V,E,maxitr_s,maxitr_l, T_G,Q);
20    end if;
end clique;
```

Fig. 5. Semi-external approach for maximum clique

A semi-external procedure similar to the one proposed for finding large cliques can be derived for quasi-cliques. As before, the procedure samples edges from the original graph such that the subgraph induced by the vertices of those edges is of a specified size. The GRASP for quasi-cliques is applied to the subgraph producing a quasi-clique $\mathcal{Q}$ of size k and density γ. In the original subgraph, edges adjacent to at least one γk-peelable vertex are removed in increasing degree order as justified in Section 2. To prevent the peeling process from bypassing a large portion of the search space, we generate disjoint maximal γ-cliques, using the `construct-dsubg` procedure of Figure 1. In this case, we set the peeling parameter

$$q = \min\{q_1\gamma_1, q_2\gamma_2, \dots, q_i\gamma_i, \dots, q^*\gamma\},$$

where q_i and γ_i denote the cardinalities of the obtained maximal quasi-cliques and their densities, respectively, and q^* is the minimum cardinality of the maximal quasi-cliques under consideration. Therefore, lines 16 and 17 of Figure 5, are modified to adapt the clique semi-external approach to general quasi-cliques. Namely, the peeling parameter q is chosen more conservatively (as indicated above) and the procedure `peel` is substituted by the procedure `localpeel`. Recall that in this case a vertex is peeled only if its degree and that of all its neighbors is smaller than q. It is important to point out that we use different threads on the different subgraphs obtained by the external memory BFS. In general, different quasi-cliques are obtained by starting the construction from different vertices in an independent manner. It is with regard to maximality within the local vicinity of the current solution that peeling becomes very effective. Global peeling as proposed above, i.e. with respect to the current set of obtained quasi-cliques, becomes a necessity, specially when we are dealing with graphs with hundreds of millions of vertices. The main idea is to use the obtained quasicliques as a guide to the exploration of the remaining portions of the graph. In the case of 1-cliques, information among the threads is used to peel off vertices that do not satisfy the lower bound degree. In this case, the peeling can be done in a more aggressive manner specially if the aim is to maximize cardinality.

8 Experiments with a Large Graph

In this section, we report typical results obtained with telecommunications data sets. The experiments were done on a Silicon Graphics Challenge computer (20 MIPS 196MHz R10000 processors with 6.144 Gbytes of main memory). A substantial amount of disk space was also used. Our current data comes from telecommunications traffic. The corresponding multi-graph has 53,767,087 vertices and over 170 million edges. We found 3,667,448 undirected connected components out of which only 302,468 were of size greater than 3 (there were 255 self-loops, 2,766,206 pairs and 598,519 triplets). A giant component with 44,989,297 vertices was detected. The giant component has 13,799,430 Directed Depth First Search Trees (DFSTs) and one of them is a giant DFST (it has 10,355,749 vertices and 19,072,448 edges). Most of the DFSTs have no more than 5 vertices. The interesting trees have sizes between 5 and 100. Their corresponding induced subgraphs are most of the time very sparse ($|E| < |V| \log |V|$), except for some occasional dense subgraphs ($|E| > |V| \sqrt{|V|}$) with 11 to 32 vertices. By counting the edges in the trees, one observes that there are very few edges that go between trees and consequently it is more likely that cliques are within the subgraphs induced by the nodes of a tree. To begin our experimentation, we considered 10% of the edges in the large component from which we recursively removed all vertices of degree one by applying $\mathtt{peel}(V, E, 1)$. This resulted in a graph with 2,438,911 vertices and 5,856,224 edges, which fits in memory. In this graph we searched for large cliques. The GRASP was repeated 1000 times, with each iteration producing a locally optimal clique. Because of the independent nature of the GRASP iterations and since our computer is configured with 20 processors, we created 10 threads, each independently running GRASP starting from a different random number generator seed. It is interesting to observe that the cliques found, even though distinct, share a large number of vertices. Next, we considered 25% of the edges in the large component from which we recursively removed all vertices of degree 10 or less. The resulting graph had 291,944 vertices and 2,184,751 edges. 12,188 iterations of GRASP produced cliques of size 26.

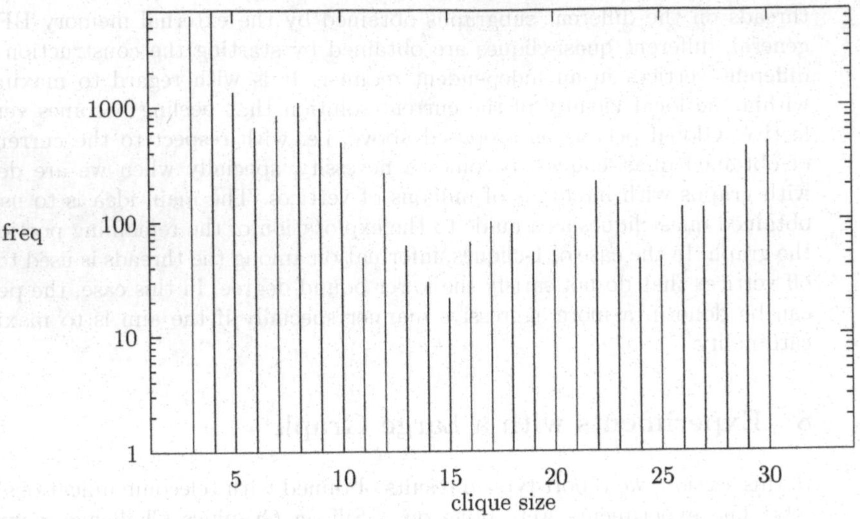

Fig. 6. Frequency of clique sizes found on entire dataset

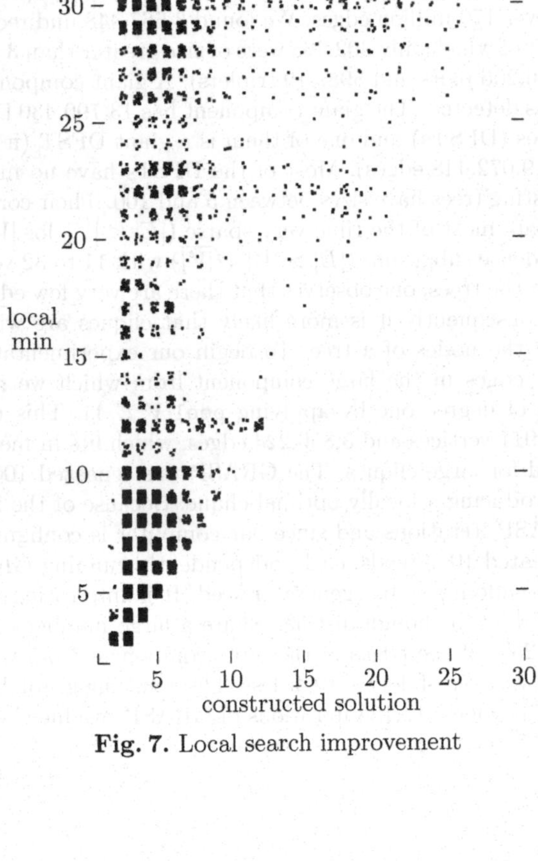

Fig. 7. Local search improvement

Having found cliques of size 26 in a quarter of the graph, we next intensified our search on the entire giant connected component. In this component, we recursively removed all vertices of degree 20 or less. The resulting graph has 27,019 vertices and 757,876 edges. Figure 6 shows the frequencies of cliques of different sizes found by the algorithm. Figure 7 shows the statistics of the improvement attained by local search. Over 20,000 GRASP iterations were carried out on the 27,019 vertex – 757,876 edge graph. Cliques of 30 vertices were found. These cliques are very likely to be close to optimal. The local search can be seen to improve the constructed solution not only for large constructed cliques, but also for small cliques. In fact, in 26 iterations, constructed cliques of size 3 were improved by the local search to size 30. To increase our confidence that the cliques of size 30 were maximum, we applied $\texttt{peel}(V, E, 30)$, resulting in a graph with 8724 vertices and about 320 thousand edges. We ran 100,000 GRASP iterations on the graph taking 10 parallel processors about one and a half days to finish. The largest clique found had 30 vertices. Of the 100,000 cliques generated, 14,141 were distinct, although many of them share one or more vertices. Finally, to compute quasi-cliques on this test data, we looked for large quasi-cliques with density parameters $\gamma = .9, .8, .7$, and $.5$. Quasi-cliques of sizes 44, 57, 65, and 98, respectively, were found.

9 Concluding Remarks

We introduced a very intuitive notion of potential of a vertex set with respect to a given quasi-clique. This potential is used to devise a local search procedure that finds either a larger quasi-clique with the same density or a denser quasi-clique with the same cardinality. Iterating this procedure eventually produces maximal quasi-cliques. In a similar vein, a potential function of a set of edges with respect to the set of edges in a given bi-clique was used to find balanced and maximally dense subgraphs of a given bipartite graph.

In order to make these procedures applicable to very large but sparse multi-digraphs, we used a specialized graph edge pruning and an external memory breadth first search traversal to decompose the input graph into a smaller collection of subgraphs on which the detection of quasi-cliques becomes possible.

We presented a sample of our experimental results when the algorithms were applied to very large graphs collected in the telecommunications industry. Our main intention was to show the feasibility of massive multi-digraph processing. In fact, our platform is currently being used to experimentally process phone data. We expect to be able to report, in the full journal version, our quasi-clique analysis of the largest AT&T telephone traffic day in history which occurred on September 11, 2001. Given the similarities exhibited by telephone traffic and internet data (i.e. skew distribution, sparsity and low diameter) we are currently applying the techniques described here to internet routing data.

There are many natural questions being raised by the processing of massive multi-digraphs. One is how to devise a greedy randomized adaptive search procedure when the input is a semi-external but weighted multi-digraph. Similarly, it is tantalizing to study the case when the input graph is fully external, i.e. not even the vertex set fits in RAM.

10 Acknowledgements

We thank the members of the Network Services Research Center at AT&T Labs Research for maintaining a reservoir of challenging research problems. We also acknowledge the suggestions made by three anonymous referees that helped to improve the paper presentation.

References

1. J. Abello, P. Pardalos and M. Resende, editors. *Handbook of Massive Data Sets*, Kluwer Academic Publishers, 2002.
2. J. Abello, A. Bushbaum, and J. Westbrook. A functional approach to external memory algorithms. In *European Symposium on Algorithms*, volume 1461 of *Lecture Notes in Computer Science*, pages 332–343. Springer-Verlag, 1998.
3. J. Abello and J. Vitter, editors. *External Memory Algorithms*, volume 50 of *DIMACS Series on Discrete Mathematics and Theoretical Computer Science*. American Mathematical Society, 1999.
4. S. Arora, C. Lund, R. Motwani, M. Sudan, and M. Szegedy. Proof verification and hardness of approximation problems. *Proc. 33rd IEEE Symp. on Foundations of Computer Science*, pages 14–23, 1992.
5. S. Arora and S. Safra. Probabilistic checking of proofs: A new characterization of NP. *J. of the ACM*, volume 45, pages 70–122, 1998.
6. U. Feige, S. Goldwasser, L. Lovász, S. Safra, and M. Szegedy. Approximating the maximum clique is almost *NP*-complete. In *Proc. 32nd IEEE Symp. on Foundations of Computer Science*, pages 2–12, 1991.
7. T.A. Feo and M.G.C. Resende. A probabilistic heuristic for a computationally difficult set covering problem. *Operations Research Letters*, volume 8, pages 67–71, 1989.
8. F. Glover. Tabu search. Part I. *ORSA J. Comput.*, volume 1, pages 190–206, 1989.
9. J. Håstad, Clique is hard to approximate within $n^{1-\epsilon}$. *Acta Mathematica*, volume 182, pages 105–142, 1999.
10. J. H. Holland. *Adaptation in Natural and Artificial Systems*. University of Michigan Press, Ann Arbor, MI, 1975.
11. S. Kirkpatrick, C. D. Gellat Jr., and M. P. Vecchi. Optimization by simulated annealing. *Science*, volume 220, 671–680, 1983.
12. L.S. Pitsoulis and M.G.C. Resende. Greedy randomized adaptive search procedures. In P. M. Pardalos and M. G. C. Resende, editors, *Handbook of Applied Optimization*. Oxford University Press, pages 168–182, 2002.

Improved Tree Decomposition Based Algorithms for Domination-like Problems

Jochen Alber* and Rolf Niedermeier

Wilhelm-Schickard-Institut für Informatik,
Universität Tübingen,
Sand 13, D-72076 Tübingen, Fed. Rep. of Germany,
{alber,niedermr}@informatik.uni-tuebingen.de

Abstract. We present an improved dynamic programming strategy for DOMINATING SET and related problems on graphs that are given together with a tree decomposition of width k. We obtain an $O(4^k n)$ algorithm for DOMINATING SET, where n is the number of nodes of the tree decomposition. This result improves the previously best known algorithm of Telle and Proskurowski running in time $O(9^k n)$. The key to our result is an argument on a certain "monotonicity" in the table updating process during dynamic programming.
Moreover, various other domination-like problems as discussed by Telle and Proskurowski are treated with our technique. We gain improvements on the base of the exponential term in the running time ranging between 55% and 68% in most of these cases. These results mean significant breakthroughs concerning practical implementations.

Classification: algorithms and data structures, combinatorics and graph theory, computational complexity.

1 Introduction

Solving domination-like problems on graphs developed into a research field of its own [12,13]. According to a year 1998 survey [12, Chapter 12], more than 200 research papers have been published on the algorithmic complexity of domination and related graph problems. Since these problems turn out to be very hard and are even NP-complete for most special graph classes (cf. [8,16]), a main road of attack against their intractability has been through studying their complexity for graphs of bounded treewidth (or, equivalently, partial k-trees), see, e.g., [1,2,10,20,21]. It is well-known that for graphs of bounded treewidth k, DOMINATING SET can be solved in time exponential in k, but linear in the size of the tree decomposition [7]. Hence, a central question is to get the combinatorial explosion in k as small as possible—this is what we investigate here, significantly improving previous work [10,20,21].

* Supported by the Deutsche Forschungsgemeinschaft (DFG), research project PEAL (Parameterized complexity and Exact ALgorithms), NI 369/1-1.

S. Rajsbaum (Ed.): LATIN 2002, LNCS 2286, pp. 613–627, 2002.

Before describing previous work and our results in more detail, let us briefly define the core problem DOMINATING SET. An *h-dominating set* D of an undirected graph G is a set of h vertices of G such that each of the rest of the vertices has at least one neighbor in D. The minimum h such that the graph G has an h-dominating set is called the *domination number* of G, denoted by $\gamma(G)$. The DOMINATING SET problem is to decide, given a graph $G = (V, E)$ and a positive integer h, whether or not there exists an h-dominating set. In the course of this paper, firstly, we will concentrate on DOMINATING SET and, secondly, we show how to extend our findings to several other domination-like problems. To better understand our results, a few words about tree decomposition (formally defined in Section 2) based algorithms seem appropriate. Typically, treewidth based algorithms proceed according to the following scheme in two stages:

1. Find a tree decomposition of bounded width for the input graph, and then
2. solve the problem using dynamic programming approaches on the tree decomposition (see [7]).

As to the first stage, when given a graph G and an integer k, the problem to determine whether the treewidth of G is at most k, is NP-complete. When the parameter k is a fixed constant, however, a lot of work was done on polynomial time solutions, culminating in Bodlaender's celebrated linear time algorithm [6]—with a hidden constant factor exponential in k that still seems too large for practical purposes. That is why also heuristic approaches for constructing tree decompositions are in use, see [15] for an up-to-date account. In our own recent research on fixed parameter algorithms for problems on planar graphs [1,2] together with corresponding implementation work [3] (based on LEDA [17]) we are currently doing, it turned out that, the first stage above usually can be done very quickly, the computational bottleneck being the second stage. More precisely, in [2] we showed that planar graph input instances $G = (V, E)$ of a parameterized problem that satisfies the so-called Layerwise Separation Property (these problems include, e.g., DOMINATING SET, VERTEX COVER, INDEPENDENT SET) allow for tree decompositions of width $k = O(\sqrt{h})$, where h is the problem parameter (such as the size of the dominating set, vertex cover, independent set, etc.). This tree decomposition can be constructed in time $O(\sqrt{h}|V|)$. Our work in progress concerning experimental studies on random planar graph instances shows that, in this context, for parameter values $h \approx 1000$ (independent of the graph size), we are confronted with tree decompositions of width approximately 15 to 20 [3]. This stirred our interest in research on "dynamic programming on tree decompositions," the results of which we report upon in the following.

To our knowledge, the best previous results for (dynamic programming) algorithms on tree decompositions applied to domination-like problems were obtained by Telle and Proskurowski [20,21]. For the time being, let us concentrate on DOMINATING SET itself. For a graph with a tree decomposition of n tree nodes and width k, Telle and Proskurowski solve DOMINATING SET in time $O(9^k n)$. Ten years earlier, Corneil and Keil [10] presented an $O(4^k n^{k+2})$ algorithm for k-trees. Observe, however, that the latter algorithm is *not* "fixed parameter tractable" in the sense of parameterized complexity theory [11], since

Algorithm	$k = 5$	$k = 10$	$k = 15$	$k = 20$
$9^k n$	0.05 sec	1 hour	6.5 years	$3.9 \cdot 10^5$ years
$4^k n$	0.001 sec	1 sec	18 minutes	13 days

Table 1. Comparing our $O(4^k n)$ algorithm for DOMINATING SET with the $O(9^k n)$ algorithm of Telle and Proskurowski in the case $n = 1000$ (number of nodes of the tree decomposition), we assume a machine executing 10^9 instructions per second and we neglect the constants hidden in the O-terms (which are comparable in both cases).

its running time is not of the form "$f(k)n^{O(1)}$," where f may be an arbitrary function depending only on k. Our new result is to improve the running time from $O(9^k n)$ to $O(4^k n)$, a significant reduction of the combinatorial explosion.[1] We achieve this by introducing a new, in a sense general concept of "monotonicity" for dynamic programming for domination-like problems. Using this concept, we can improve basically all running times for domination-like problems given by Telle and Proskurowski (see Table 2 in the concluding section for a complete overview). For example, we can solve the so-called TOTAL DOMINATING SET problem in time $O(5^k n)$, where Telle and Proskurowski had $O(16^k n)$.

To illustrate the significance of our results, in Table 1, we compare (hypothetical) running times of the $O(9^k n)$ algorithm of Telle and Proskurowski [20,21] to our $O(4^k n)$ algorithm for some realistic values of k and $n = 1000$; we assume a fixed underlying machine with 10^9 instructions per second. It is worth noting that improving exponential terms is a "big issue" for fixed parameter algorithms. For example, much attention was paid to lowering the exponential term for VERTEX COVER (for general graphs) from a trivial 2^h to below 1.3^h [9,18], where h is the size of a minimum vertex cover. By way of contrast, here we obtain much more drastic improvements, which, additionally, even apply to a whole class of problems.

2 Preliminaries

The main tool we use in our algorithm is the concept of tree decompositions as, e.g., described in [7,14].

Definition 2.1. *Let $G = (V, E)$ be a graph. A* tree decomposition *of G is a pair $\langle\{X_i \mid i \in I\}, T\rangle$, where each X_i is a subset of V, called a* bag, *and T is a tree with the elements of I as nodes. The following three properties must hold:*

1. *$\bigcup_{i \in I} X_i = V$;*
2. *for every edge $\{u, v\} \in E$, there is an $i \in I$ such that $\{u, v\} \subseteq X_i$;*
3. *for all $i, j, k \in I$, if j lies on the path between i and k in T, then $X_i \cap X_k \subseteq X_j$.*

[1] Observe that in our work [1] we gave wrong bounds for the dynamic programming algorithm for DOMINATING SET, claiming a running time $O(3^k n)$. This was due to a misinterpretation of [21, Theorem 5.7] where the correct base of the exponential term in the running time is $3^2 = 9$ instead of 3.

The width *of* $\langle\{X_i \mid i \in I\}, T\rangle$ *equals* $\max\{|X_i| \mid i \in I\} - 1$. *The* treewidth *of* G *is the minimum* k *such that* G *has a tree decomposition of width* k.

A tree decomposition with a particularly simple structure is given by the following.

Definition 2.2. *A tree decomposition* $\langle\{X_i \mid i \in I\}, T\rangle$ *is called a* nice tree decomposition *if the following conditions are satisfied:*

1. *Every node of the tree* T *has at most 2 children.*
2. *If a node* i *has two children* j *and* k, *then* $X_i = X_j = X_k$ *(in this case* i *is called a* JOIN NODE*).*
3. *If a node* i *has one child* j, *then one of the following situations must hold*
   (a) $|X_i| = |X_j| + 1$ *and* $X_j \subset X_i$ *(in this case* i *is called an* INTRODUCE NODE*), or*
   (b) $|X_i| = |X_j| - 1$ *and* $X_i \subset X_j$ *(in this case* i *is called a* FORGET NODE*).*

It is not hard to transform a given tree decomposition into a nice tree decomposition. More precisely, the following result holds (see [14, Lemma 13.1.3]).

Lemma 2.3. *Given a width* k *and* n *nodes tree decomposition of a graph* G, *one can find a width* k *and* $O(n)$ *nodes nice tree decomposition of* G *in linear time.* □

3 Dynamic Programming Based on "Monotonicity"

In this section, we present our main algorithm which is based on a fresh view on dynamic programming. Compared to previous work, we perform the updating process of the tables in a more careful, less time consuming way by making use of the "monotonous structure of the tables." Our main result is as follows.

Theorem 3.1. *If a width* k *tree decomposition of a graph is known, then a minimum dominating set can be determined in time* $O(4^k n)$, *where* n *is the number of nodes of the tree decomposition.*

The outline of the corresponding algorithm and its proof of correctness fill the rest of this section. From now on suppose that the given tree decomposition of our input graph $G = (V, E)$ is $\mathcal{X} = \langle\{X_i \mid i \in I\}, T\rangle$. By Lemma 2.3, we can assume that $\mathcal{X}$ is a nice tree decomposition.

3.1 Colorings and Monotonicity

Suppose that $V = \{x_1, \ldots, x_n\}$. We assume that the vertices in the bags are ordered by their indices, i.e., $X_i = (x_{i_1}, \ldots, x_{i_{n_i}})$ with $i_1 \leq \ldots \leq i_{n_i}$ for all $i \in I$.

Colorings. In the following, we use three different "colors" that will be assigned to the vertices in a bag:

- "black" (represented by 1, meaning that the vertex belongs to the dominating set),
- "white" (represented by 0, meaning that the vertex is already dominated at the current stage of the algorithm), and
- "grey" (represented by $\hat{0}$, meaning that, at the current stage of the algorithm, we still ask for a domination of this vertex).

A vector $c = (c_1, \ldots, c_{n_i}) \in \{0, \hat{0}, 1\}^{n_i}$ will be called a *coloring* for the bag $X_i = (x_{i_1}, \ldots, x_{i_{n_i}})$, and the *color* assigned to vertex x_{i_t} by the coloring c is given by the coordinate c_t.

For each bag X_i (with $|X_i| = n_i$), we will use a mapping

$$A_i : \{0, \hat{0}, 1\}^{n_i} \longrightarrow \mathbb{N} \cup \{+\infty\}.$$

For a coloring $c = (c_1, \ldots, c_{n_i}) \in \{0, \hat{0}, 1\}^{n_i}$, the value $A_i(c)$ stores how many vertices are needed for a minimum dominating set (of the graph visited up to the current stage of the algorithm) under the restriction that the color assigned to vertex x_{i_t} is c_t $(t = 1, \ldots, n_i)$.

A coloring $c \in \{0, \hat{0}, 1\}^{n_i}$ is *locally invalid* for a bag X_i if

$$(\exists s \in \{1, \ldots, n_i\} : c_s = 0) \wedge (\nexists t \in \{1, \ldots, n_i\} : (x_{i_t} \in N(x_{i_s}) \wedge c_t = 1)).$$

In other words, a coloring is locally invalid if there is some vertex x_{i_s} in the bag that is colored white, but this color is not "justified" within the bag, i.e., x_{i_s} is not dominated by a vertex within the bag using this coloring.[2] Also, for a coloring $c = (c_1, \ldots, c_m) \in \{0, \hat{0}, 1\}^m$ and a color $d \in \{0, \hat{0}, 1\}$, we use the notation

$$\#_d(c) := |\{t \in \{1, \ldots, m\} : c_t = d\}|.$$

Monotonicity. On the color set $\{0, \hat{0}, 1\}$, let $\prec$ be the partial ordering given by $\hat{0} \prec 0$ and $d \prec d$ for all $d \in \{0, \hat{0}, 1\}$. This ordering naturally extends to colorings: For $c = (c_1, \ldots, c_m), c' = (c'_1, \ldots, c'_m) \in \{0, \hat{0}, 1\}^m$, we let $c \prec c'$ iff $c_t \prec c'_t$ for all $t = 1, \ldots, m$.

We call a mapping

$$A_i : \{0, \hat{0}, 1\}^{n_i} \longrightarrow \mathbb{N} \cup \{+\infty\}$$

monotonous from the partially ordered set $(\{0, \hat{0}, 1\}^{n_i}, \prec)$ to $(\mathbb{N} \cup \{+\infty\}, \leq)$ if for $c, c' \in \{0, \hat{0}, 1\}^{n_i}$, $c \prec c'$ implies $A(c) \leq A(c')$. It is very essential for the correctness of our algorithm as well as for the claimed running time that all the mappings A_i will be monotonous.

[2] A locally invalid coloring still may be a correct coloring if the white vertex whose color is not "justified" *within* the bag already is dominated by a vertex from other bags.

3.2 The Algorithm

We use the mappings introduced above to perform a dynamic programming approach. Note that at each stage of the algorithm the mappings visited up to that stage are monotonous. This is guaranteed by Lemmas 3.2, 3.3, 3.4, and 3.5.

Step 1 (initialization). In the first step of the algorithm, for each leaf node i of the tree decomposition, we initialize the mapping A_i:

$$\texttt{for all } c \in \{0, \hat{0}, 1\}^{n_i} \texttt{ do}$$

$$A_i(c) \leftarrow \begin{cases} +\infty & \text{if } c \text{ is locally invalid for } X_i \\ \#_1(c) & \text{otherwise} \end{cases} \tag{1}$$

By this initialization step, we make sure that only colorings are taken into consideration where an assignment of color 0 is justified.

It is trivial to observe the following.

Lemma 3.2. *1. The evaluations in (1) can be carried out in time $O(3^{n_i} n_i)$. 2. The mapping A_i is monotonous.* □

Step 2 (updating process). After the initialization, we visit the bags of our tree decomposition bottom-up from the leaves to the root, evaluating the corresponding mappings in each step according to the following rules.

FORGET NODES: Suppose i is a FORGET NODE with child j and suppose that $X_i = (x_{i_1}, \ldots, x_{i_{n_i}})$. W.l.o.g.[3], we may assume that $X_j = (x_{i_1}, \ldots, x_{i_{n_i}}, x)$. Evaluate the mapping A_i of X_i as follows:

$$\texttt{for all } c \in \{0, \hat{0}, 1\}^{n_i} \texttt{ do}$$

$$A_i(c) \leftarrow \min_{d \in \{0,1\}} A_j(c \times \{d\}) \tag{2}$$

Note that a coloring $c \times \{\hat{0}\}$ for X_j means that the vertex x is assigned color $\hat{0}$, i.e., not yet dominated by a vertex. Since, by condition 3. of Definition 2.1, the vertex x will never appear in a bag for the rest of the algorithm, a coloring $c \times \{\hat{0}\}$ will remain unresolved and it will not lead to a dominating set. That is why the minimum in the assignment (2) is taken over colors 1 and 0 only.

The following lemma is easy to see.

Lemma 3.3. *1. The evaluations in (2) can be carried out in time $O(3^{n_i})$. 2. If the mapping A_j is monotonous, then mapping A_i also is monotonous.* □

INTRODUCE NODES: Suppose that i is an INTRODUCE NODE with child j and suppose, furthermore, that $X_j = (x_{j_1}, \ldots, x_{j_{n_j}})$. W.l.o.g.[4], we may assume

[3] Possibly after rearranging the vertices in X_j and the entries of A_j accordingly.
[4] Possibly after rearranging the vertices in X_i and the entries of A_i accordingly.

that $X_i = (x_{j_1}, \ldots, x_{j_{n_j}}, x)$. Let $N(x) \cap X_i = \{x_{j_{p_1}}, \ldots, x_{j_{p_s}}\}$ be the neighbors of the "introduced" vertex x that appear in the bag X_i. We now define a function $\phi : \{0, \hat{0}, 1\}^{n_j} \to \{0, \hat{0}, 1\}^{n_j}$ on the set of colorings of X_j. For $c = (c_1, \ldots, c_{n_j}) \in \{0, \hat{0}, 1\}^{n_j}$, let $\phi(c) := (c'_1, \ldots, c'_{n_j})$ such that

$$c'_t = \begin{cases} \hat{0} & \text{if } t \in \{p_1, \ldots, p_s\} \text{ and } c_t = 0, \\ c_t & \text{otherwise.} \end{cases}$$

Then, evaluate the mapping A_i of X_i as follows:

for all $c = (c_1, \ldots, c_{n_j}) \in \{0, \hat{0}, 1\}^{n_j}$ **do**

$$A_i(c \times \{0\}) \leftarrow \begin{cases} A_j(c) & \text{if } x \text{ has a neighbor } x_{j_q} \text{ in } X_i \text{ with } c_q = 1, \\ +\infty & \text{otherwise} \end{cases} \quad (3)$$

$$A_i(c \times \{1\}) \leftarrow A_j(\phi(c)) + 1 \quad (4)$$

$$A_i(c \times \{\hat{0}\}) \leftarrow A_j(c) \quad (5)$$

For the correctness of the assignments (3) and (4), we remark the following: It is clear that, if we assign color 0 to vertex x (see assignment (3)), we again (as already done in the initializing step of assignment (1)) have to check whether this color can be justified at the current stage of the algorithm. Such a justification is given if and only if the coloring under examination already assigns a 1 to some neighbor of x in X_i. This is true, since condition 3. of Definition 2.1 implies that no neighbor of x has been considered in previous bags, and, hence, up to the current stage of the algorithm, x can only be dominated by a vertex in X_i (as checked in assignment (3)).
If we assign color 1 to vertex x (see assignment (4)), we already dominate all vertices $\{x_{j_{p_1}}, \ldots, x_{j_{p_s}}\}$. Suppose now we want to evaluate $A_i(c \times \{1\})$ and suppose some of these vertices are assigned color 0 by c, say $c_{p'_1} = \ldots = c_{p'_q} = 0$ (where $(p'_1, \ldots, p'_q)$ is a subsequence of $(p_1, \ldots, p_s)$). Since the "1-assignment" of x already justifies the "0-values" of $c_{p'_1}, \ldots, c_{p'_q}$, and since our mapping A_j is monotonous, we obtain $A_i(c \times \{1\})$ by taking entry $A_j(c')$, where $c'_{p'_1} = \ldots = c'_{p'_q} = \hat{0}$, i.e., where $c' = \phi(c)$.
Again, it is not hard to verify the following statements.

Lemma 3.4. *1. The evaluations of (3),(4), and (5) can be carried out in time $O(3^{n_i} n_i)$.*

2. If the mapping A_j is monotonous, then mapping A_i also is monotonous. □

JOIN NODES: Suppose i is a JOIN NODE with children j and k and suppose that $X_i = X_j = X_k = (x_{i_1}, \ldots, x_{i_{n_i}})$. Let $c = (c_1, \ldots, c_{n_i}) \in \{0, \hat{0}, 1\}^{n_i}$ be a coloring for X_i. We say that $c' = (c'_1, \ldots, c'_{n_i})$, $c'' = (c''_1, \ldots, c''_{n_i}) \in \{0, \hat{0}, 1\}^{n_i}$ *divide* c if

1. $(c_t \in \{1, \hat{0}\} \Rightarrow c'_t = c''_t = c_t)$, and
2. $(c_t = 0 \Rightarrow [(c'_t, c''_t \in \{0, \hat{0}\}) \wedge (c'_t = 0 \vee c''_t = 0)])$.

Then, evaluate the mapping A_i of X_i as follows:

$$\texttt{for all } c \in \{0,\hat{0},1\}^{n_i} \texttt{ do}$$
$$A_i(c) \leftarrow \min\{A_j(c') + A_k(c'') - \#_1(c) \mid c' \text{ and } c'' \text{ divide } c\} \qquad (6)$$

In other words, in order to determine the value $A_i(c)$, we look up the corresponding values for coloring c in A_j (which gives us the minimum dominating set for c needed for the bags considered up to this stage in the left subtree) and in A_k (the minimum dominating set for c needed according to the right subtree), add the corresponding values, and subtract the number of "1-assignments" in c, since they would be counted twice, otherwise. Clearly, if coloring c of node i assigns the colors 1 or $\hat{0}$ to a vertex x in X_i, we have to make sure that we use colorings c' and c'' of the children j and k that also assign the same color to x. However, if c assigns color 0 to x, it is sufficient to justify this color by *at least one* of the colorings c' or c''. Observe that, by the monotonicity of A_j and A_k we obtain the same "min" in assignment (6), by replacing condition 2. in the definition of "divide" by:

$$2'.\ (c_t = 0 \Rightarrow (c'_t, c''_t \in \{0,\hat{0}\} \wedge c'_t \neq c''_t)).$$

This observation will be the key to prove the following statement on the running time for this step.

Lemma 3.5. *1. The evaluations in (6) can be carried out in time $O(4^{n_i})$.*
2. If the mappings A_j and A_k are monotonous, then mapping A_i also is monotonous.

Proof. The second statement basically follows from the definition of "divide." As to the time complexity, note that the running time of this step is given by

$$\sum_{c\in\{0,\hat{0},1\}^{n_i}} \left|\{(c',c'') \mid c' \text{ and } c'' \text{ divide } c\}\right|. \qquad (7)$$

For given $c \in \{0,\hat{0},1\}^{n_i}$, with $z := \#_0(c)$, we have 2^z many pairs (c',c'') that divide c (if we use condition 2'. instead of 2. (sic!)). Since there are $2^{n_i-z}\binom{n_i}{z}$ many colorings c with $\#_0(c) = z$, again using condition 2'. instead of 2., the expression in (7) equates to

$$\sum_{z=0}^{n_i} 2^{n_i-z}\binom{n_i}{z}\cdot 2^z \quad = \quad 4^{n_i}.$$

□

Step 3 (final evaluation). Let r denote the root of T. For the domination number $\gamma(G)$, we finally get

$$\gamma(G) = \min\{A_r(c) \mid c \in \{0,1\}^{n_r}\}. \qquad (8)$$

The minimum in Equation (8) is taken only over colorings containing colors 1 and 0, since a valid dominating set does not contain "unresolved" vertices of color $\hat{0}$. Also, note that, when bookkeeping how the minima in the assignments (2), (6), and (8) of Step 2 and Step 3 were obtained, this algorithm constructs a dominating set D corresponding to $\gamma(G)$.

3.3 Correctness and Time Complexity

For the correctness of the algorithm, we observe the following. Firstly, property 1. of a tree decomposition (see Definition 2.1) guarantees that each vertex is assigned a color. Secondly, in our initialization Step 1, as well as in the updating process for INTRODUCE NODES and JOIN NODES of Step 2, we made sure that the assignment of color 0 to a vertex x always guarantees that, at the current stage of the algorithm, x is already dominated by a vertex from previous bags. Since, by property 2. of a tree decomposition (see Definition 2.1), any pair of neighbors appears in at least one bag, the validity of the colorings was checked for each such pair of neighbors. And, thirdly, property 3. of a tree decomposition (see Definition 2.1), together with the comments given in Step 2 of the algorithm, implies that the updating of each mapping is done consistently with all mappings that have been visited earlier in the algorithm.

Lemma 3.6. *The total running time of the algorithm is $O(4^k n)$.*

Proof. This follows directly from Lemmas 3.2, 3.3, 3.4, and 3.5. □

This finishes the proof for Theorem 3.1.

We remark that Aspvall *et al.* [4] addressed the memory requirement problem arising in the type of algorithms described above.

4 Further Applications and Extensions

In this section, we describe how our new dynamic programming strategy can be applied to further "domination-like" problems as, e.g., treated in [19,20,21].

4.1 DOMINATING SET WITH PROPERTY P

In the following, a *property* P of a vertex set $V' \subseteq V$ of an undirected graph $G = (V, E)$ will be a Boolean predicate which yields true or false values when given as input V, E, and V'. Since V and E will always be clear from the context, we will simply write $P(V')$ instead of $P(V, E, V')$.

The DOMINATING SET WITH PROPERTY P problem is the task to find a minimum size dominating set D with property P, i.e., such that $P(D)$ is true.

Examples for such problems (all also appearing in [20,21]) are:

- the INDEPENDENT DOMINATING SET problem, where the property $P(D)$ of the dominating set D is that D is independent,

- the TOTAL DOMINATING SET problem, where the property $P(D)$ of the dominating set D is that each vertex of D has a neighbor in D,
- the PERFECT DOMINATING SET problem, where the property $P(D)$ is that each vertex which is not in D has *exactly* one neighbor in D,
- the PERFECT INDEPENDENT DOMINATING SET problem, also known as the PERFECT CODE problem, where the dominating set has to be perfect and independent, and
- the TOTAL PERFECT DOMINATING SET problem, where the dominating set has to be total and perfect.

For all those instances, we obtain algorithms where the base q_i of the exponential term and the number λ_i of colors needed for the mappings in the dynamic programming are as follows:

Theorem 4.1. *If a width k and n nodes tree decomposition of a graph is known, then we can solve the subsequent problems $\mathcal{P}_i$ in time $O(q_i^k n)$, using λ_i colors in the dynamic programming step:*

- $\mathcal{P}_1$ = INDEPENDENT DOMINATING SET: $q_1 = 4$, $\lambda_1 = 3$;
- $\mathcal{P}_2$ = TOTAL DOMINATING SET: $q_2 = 5$, $\lambda_2 = 4$;
- $\mathcal{P}_3$ = PERFECT DOMINATING SET: $q_3 = 4$, $\lambda_3 = 3$;
- $\mathcal{P}_4$ = PERFECT CODE: $q_4 = 4$, $\lambda_4 = 3$;
- $\mathcal{P}_5$ = TOTAL PERFECT DOMINATING SET: $q_5 = 5$, $\lambda_5 = 4$.

Proof. (Sketch) For problem $\mathcal{P}_1$, in contrast to the algorithm given in the proof of Theorem 3.1 (see Subsection 3.2), after each update of a mapping A_i for bag X_i, we check, for each coloring $c \in \{0, \hat{0}, 1\}^{n_i}$, if there exist two vertices $x, y \in X_i$ that both are assigned color 1 by c, and, if so, set $A_i(c) \leftarrow +\infty$.

For problem $\mathcal{P}_2$, one must also distinguish for the vertices in the domination set whether or not they have been dominated by other vertices from the dominating set. We may use 4 colors:

- 1, meaning that the vertex is in the dominating set and it is already dominated;
- $\hat{1}$, meaning that the vertex is in the dominating set and it still needs to be dominated;
- 0, meaning that the vertex is not in the dominating set and it is already dominated;
- $\hat{0}$, meaning that the vertex is not in the dominating set and it still needs to be dominated.

The partial ordering $\prec$ on $C := \{0, \hat{0}, 1, \hat{1}\}$, according to which our mappings will be monotonous, is given by $\hat{1} \prec 1$, $\hat{0} \prec 0$, and $d \prec d$ for all $d \in C$.

The various steps of the algorithm for updating the mappings are similar the ones given in the algorithm of Theorem 3.1 (see Subsection 3.2). The most cost-expensive part again is a JOIN NODE. Here, in the assignment (6), we have to adapt the definition of "divide" as follows: For a coloring $c = (c_1, \ldots, c_{n_i}) \in \{0, \hat{0}, 1, \hat{1}\}^{n_i}$ for X_i, we say that the two colorings $c' = (c'_1, \ldots, c'_{n_i})$, $c'' = (c''_1, \ldots, c''_{n_i}) \in \{0, \hat{0}, 1, \hat{1}\}^{n_i}$ *divide* c if

1. $(c_t = 0 \Rightarrow (c'_t, c''_t \in \{0, \hat{0}\} \wedge c'_t \neq c''_t))$, and
2. $(c_t = 1 \Rightarrow (c'_t, c''_t \in \{1, \hat{1}\} \wedge c'_t \neq c''_t))$.

Similar to the proof of Lemma 3.5 the running time for updating a JOIN NODE is given by

$$\sum_{c \in \{0,\hat{0},1,\hat{1}\}^{n_i}} \left|\{(c', c'') \mid c' \text{ and } c'' \text{ divide } c\}\right|. \tag{9}$$

We use a combinatorial argument to compute this expression. For a fixed coloring $c \in \{0, \hat{0}, 1, \hat{1}\}^{n_i}$, we have $\#_0(c) \in \{0, \ldots, n_i\}$, and $\#_1(c) \in \{0, \ldots, n_i - \#_0(c)\}$. The number of colorings $c \in \{0, \hat{0}, 1, \hat{1}\}^{n_i}$ with $\#_0(c) = z_0$ and $\#_1(c) = z_1$ is given by $\binom{n_i}{z_0}\binom{n_i - z_0}{z_1}$. Since, by definition of "divide," for each position in c with $c_t = 0$ or $c_t = 1$, we have to consider two different divide pairs, we get

$$\begin{aligned} &\sum_{c \in \{0,\hat{0},1,\hat{1}\}^{n_i}} \left|\{(c', c'') \mid c' \text{ and } c'' \text{ divide } c\}\right| \\ &= \sum_{\#_0(c)=0}^{n_i} \sum_{\#_1(c)=0}^{n_i - \#_0(c)} \binom{n_i}{\#_0(c)} \binom{n_i - \#_0(c)}{\#_1(c)} 2^{\#_0(c)} 2^{\#_1(c)} \\ &= \sum_{\#_0(c)=0}^{n_i} \binom{n_i}{\#_0(c)} 2^{\#_0(c)} 3^{n_i - \#_0(c)} \quad = \quad 5^{n_i} \end{aligned}$$

This determines the running time of the algorithm.

For problem $\mathcal{P}_3$, in contrast to the algorithm given in the proof of Theorem 3.1 (see Subsection 3.2), a vertex is colored "grey" if it is dominated by exactly one "black" vertex which either lies in the "current" bag of the tree decomposition algorithm or in one of its child bags.

For problem $\mathcal{P}_4$ we can use appropriate combinations of the arguments given for problems $\mathcal{P}_1$, $\mathcal{P}_3$.

For problem $\mathcal{P}_5$ we can use appropriate combinations of the arguments given for problems $\mathcal{P}_2$, $\mathcal{P}_3$. □

Note that our updating technique which makes strong use of the monotonicity of the mappings involved yields a basis in the exponential term of the running time which outperforms the results of Telle and Proskurowski. The corresponding constants q'_i for the above listed problems that were derived in [20, Theorem 4, Table 1], and [21, Theorem 5.7] are $q'_1 = 9$, $q'_2 = 16$, $q'_3 = 9$, $q'_4 = 9$, and $q'_5 = 16$. (See Table 2 for an overview.)

4.2 Weighted Versions of DOMINATING SET

Our algorithm can be adapted to the weighted version of DOMINATING SET (and its variants): Take a graph $G = (V, E)$ together with a positive integer weight function $w : V \to \mathbb{N}$. The weight of a vertex set $D \subseteq V$ is defined as

$w(D) = \sum_{v \in D} w(v)$. The WEIGHTED DOMINATING SET problem is the task to determine, given a graph $G = (V, E)$ and a weight function $w : V \to \mathbb{N}$, a dominating set with minimum weight.

Only small modifications in the bookkeeping technique used in Theorem 3.1 (or Theorem 4.1) are necessary in order to solve the weighted version of DOMINATING SET (and its variations). More precisely, we have to adapt the initialization (1) of the mappings A_i for the bag $X_i = (x_{i_1}, \ldots, x_{i_{n_i}})$ according to:

$$\texttt{for all } c = (c_1, \ldots, c_{n_i}) \in \{0, \hat{0}, 1\}^{n_i} \texttt{ do}$$

$$A_i(c) \leftarrow \begin{cases} +\infty & \text{if } c \text{ is locally invalid for } X_i \\ w(c) & \text{otherwise,} \end{cases} \tag{10}$$

where $w(c) := \sum_{t=0, c_t=1}^{n_i} w(x_{i_t})$. The updating of the mappings A_i in Step 2 in the algorithm of Theorem 3.1 (or Theorem 4.1) is adapted similarly.

4.3 RED-BLUE DOMINATING SET

We finally turn our attention to the following version of DOMINATING SET, called RED-BLUE DOMINATING SET[5] ([11, Exercise 3.1.5]):

An instance of RED-BLUE DOMINATING SET is given by a (planar) bipartite graph $G = (V, E)$, where the bipartition is given by $V = V_{red} \cup V_{blue}$. The question is to determine a set $V' \subseteq V_{red}$ of minimum size such that every vertex of V_{blue} is adjacent to at least one vertex of V'.

This problem is directly related to the FACE COVER problem (see [5,11]). A *face cover* C of an undirected plane graph $G = (V, E)$ (i.e., a planar graph with a fixed embedding) is a set of faces that cover all vertices of G, i.e., for every vertex $v \in V$, there exists a face $f \in C$ so that v lies on the boundary of f. The FACE COVER problem is the task to find a minimum size face cover for a given plane graph.

The relation between FACE COVER and RED-BLUE DOMINATING SET is as follows. For an instance $G = (V, E)$ of the FACE COVER problem, consider the following graph: Add a vertex to each face of G, and make each such "face vertex" adjacent to all vertices that are on the boundary of that face. These are the only edges of the bipartite graph $G' = (V', E')$. Write $V' = V \cup V_F$, where V_F is the set of face vertices, i.e., each $v \in V_F$ represents a face f_v in G. In other words, V and V_F form the bipartition of G'. Observe that G' can be viewed as an instance of RED-BLUE DOMINATING SET.

Theorem 4.2. *Let a bipartite graph $G = (V, E)$ with bipartition $V = V_{red} \cup V_{blue}$ be given together with a tree decomposition of width k. Then,* RED-BLUE DOMINATING SET *can be solved in time $O(3^k n)$, where n is the number of nodes of the tree decomposition.*

---

[5] Observe that RED-BLUE DOMINATING SET is *not* a variant of DOMINATING SET in the sense of the first subsection, because a solution V' is not a dominating set, since red vertices cannot and hence need not be dominated by red vertices.

Proof. (Sketch) Basically, the technique exhibited in Theorem 3.1 (see Subsection 3.2) can be applied. Due to the bipartite nature of the graph, only two "states" have to be stored for each vertex: red vertices are either within the dominating set or not (represented by colors 1_{red} and 0_{red}, respectively), and blue vertices are either already dominated or not yet dominated (represented by colors 0_{blue} and $\hat{0}_{\text{blue}}$, respectively).

We consider our bags as bipartite sets, i.e.,

$$X_i := X_{i,\text{red}} \cup X_{i,\text{blue}},$$

where $X_{i,\text{red}} := X_i \cap V_{\text{red}}$ and $X_{i,\text{blue}} := X_i \cap V_{\text{blue}}$. Let $n_{i,\text{red}} := |X_{i,\text{red}}|$ and $n_{i,\text{blue}} := |X_{i,\text{blue}}|$, i.e., $|X_i| =: n_i = n_{i,\text{red}} + n_{i,\text{blue}}$.

The partial ordering $\prec$ on the color set $C = C_{\text{red}} \cup C_{\text{blue}}$, where $C_{\text{red}} := \{1_{\text{red}}, 0_{\text{red}}\}$ and $C_{\text{blue}} := \{0_{\text{blue}}, \hat{0}_{\text{blue}}\}$, is given by $\hat{0}_{\text{blue}} \prec 0_{\text{blue}}$ and $d \prec d$ for all $d \in C$.

A *valid* coloring for X_i is a coloring where we assign colors from C_{red} to vertices in $X_{i,\text{red}}$ and colors from C_{blue} to vertices in $X_{i,\text{blue}}$. The various steps of the algorithm for updating the mappings are similar to the ones given in the algorithm of Theorem 3.1 (see Subsection 3.2).

Again, the most cost-expensive part is the updating of a JOIN NODE. For a correct updating of JOIN NODES, we adapt the definition of "divide" that appears in the assignment (6) according to: For a valid coloring $c = (c_1, \dots, c_{n_i}) \in C^{n_i}$ of X_i, we say that the valid colorings $c' = (c'_1, \dots, c'_{n_i}), c'' = (c''_1, \dots, c''_{n_i}) \in C^{n_i}$ *divide* c if

1. $(c_t \neq 0_{\text{blue}} \Rightarrow (c'_t, c''_t = c_t))$, and
2. $(c_t = 0_{\text{blue}} \Rightarrow (c'_t, c''_t \in \{0_{\text{blue}}, \hat{0}_{\text{blue}}\} \wedge c'_t \neq c''_t))$.

For a fixed valid coloring c that contains $z := \#_{0_{\text{blue}}}(c)$ many colors 0_{blue}, the number of pairs that divide c is 2^z. Since there are $2^{n_{i,\text{red}}}\binom{n_{i,\text{blue}}}{z}$ many colorings with $\#_{0_{\text{blue}}}(c) = z$, the total number of pairs that divide a fixed coloring c is upper-bounded by

$$\sum_{z=0}^{n_{i,\text{blue}}} 2^{n_{i,\text{red}}} \binom{n_{i,\text{blue}}}{z} \cdot 2^z \;=\; 2^{n_{i,\text{red}}} \cdot 3^{n_{i,\text{blue}}} \;\leq\; 3^{n_i}.$$

This determines the running time of the algorithm. Note that in the worst case, for a bag X_i, we may have $n_{i,\text{red}} = 0$ and $n_{i,\text{blue}} = n_i$. □

5 Conclusion

In this paper, we focused on solving domination-like problems for graphs that are given together with a tree decomposition of width k. We presented a new "monotonicity" argument for the usual dynamic programming procedure. This

Problem	λ	q	q'
(WEIGHTED) DOMINATING SET	3	4	9
INDEPENDENT DOMINATING SET	3	4	9
PERFECT DOMINATING SET	3	4	9
PERFECT CODE	3	4	9
TOTAL DOMINATING SET	4	5	16
TOTAL PERFECT DOMINATING SET	4	5	16
RED-BLUE DOMINATING SET	4	4	-
VERTEX COVER	2	2	4
INDEPENDENT SET	2	2	4

Table 2. Summary of our results (Theorems 3.1, 4.1, 4.2) **and comparison with previous work.** The entries in the second column give the number λ of colors used in our dynamic programming step. The third column gives the base q for our $O(q^k n)$ time algorithm (k being the width of the given tree decomposition). The entries of the fourth column give the corresponding base values q' of the so far best known algorithms by Telle and Proskurowski [20, Theorem 4, Table 1], and [21, Theorem 5.7]. The results for VERTEX COVER and INDEPENDENT SET can be obtained by a strategy similar to the one for DOMINATING SET. However, in contrast to the other problems, the up-dating process is straight forward and much less involved.

new strategy yields significant improvements over the so far best known algorithms. The results, and the corresponding improvements over previous work are summarized in Table 2.

Observe that we obtained "base values" q which are at most $\lambda + 1$, where λ denotes the number of colors needed in the dynamic programming. The corresponding base values q' obtained by Telle and Proskurowski all are of the form $q' = \lambda^2$ (see Table 2).

It remains a challenge for future research whether the "base values" q in Table 2 all can be lowered to *match* the corresponding value λ of colors needed in the dynamic programming. For instance, in the case of DOMINATING SET we only need three colors in the dynamic programming tables, but the updating process so far invokes the exponential term 4^k. Such an improvement would imply a significant speed-up of our software package described in [3].

Acknowledgments. We profited from constructive and inspiring collaboration with Hans L. Bodlaender and Henning Fernau. We are grateful to Frederic Dorn for his careful (and successful) work on implementing the presented algorithms. We thank Ute Schmid for her helpful hints concerning some combinatorial arguments.

References

1. J. Alber, H. L. Bodlaender, H. Fernau, and R. Niedermeier. Fixed parameter algorithms for planar dominating set and related problems. In *Proceedings 7th SWAT 2000*, Springer-Verlag LNCS 1851, pp. 97–110, 2000.
2. J. Alber, H. Fernau, and R. Niedermeier. Parameterized complexity: exponential speed-up for planar graph problems. In *Proceedings 28th ICALP 2001*, Springer-Verlag LNCS 2076, pp. 261–272, 2001.
3. J. Alber, F. Dorn, and R. Niedermeier. Experiments on optimally solving parameterized problems on planar graphs. Manuscript, December 2001.
4. B. Aspvall, A. Proskurowski, and J. A. Telle. Memory requirements for table computations in partial k-tree algorithms. *Algorithmica* **27**: 382–394, 2000.
5. D. Bienstock and C. L. Monma. On the complexity of covering vertices by faces in a planar graph. *SIAM J. Comput.* **17**:53–76, 1988.
6. H. L. Bodlaender. A linear time algorithm for finding tree-decompositions of small treewidth. *SIAM J. Comput.* **25**:1305–1317, 1996.
7. H. L. Bodlaender. Treewidth: Algorithmic techniques and results. In *Proceedings 22nd MFCS'97*, Springer-Verlag LNCS 1295, pp. 19–36, 1997.
8. A. Brandstädt, V. B. Le, and J. P. Spinrad. *Graph Classes: a Survey.* SIAM Monographs on Discrete Mathematics and Applications. Society for Industrial and Applied Mathematics, 1999.
9. J. Chen, I. Kanj, and W. Jia. Vertex cover: further observations and further improvements. In *Proceedings 25th WG*, Springer-Verlag LNCS 1665, pp. 313–324, 1999.
10. D. G. Corneil and J. M. Keil. A dynamic programming approach to the dominating set problem on k-trees. *SIAM J. Alg. Disc. Meth.*, **8**: 535–543, 1987.
11. R. G. Downey and M. R. Fellows. *Parameterized Complexity.* Monographs in Computer Science. Springer-Verlag, 1999.
12. T. W. Haynes, S. T. Hedetniemi, and P. J. Slater. *Fundamentals of Domination in Graphs.* Monographs and textbooks in Pure and Applied Mathematics Vol. 208, Marcel Dekker, 1998.
13. T. W. Haynes, S. T. Hedetniemi, and P. J. Slater (eds.). *Domination in Graphs; Advanced Topics.* Monographs and textbooks in Pure and Applied Mathematics Vol. 209, Marcel Dekker, 1998.
14. T. Kloks. *Treewidth. Computations and Approximations.* Springer-Verlag LNCS 842, 1994.
15. A. M. C. A. Koster, H. L. Bodlaender, and S. P. M. Hoesel. Treewidth: Computational Experiments. *Electronic Notes in Discrete Mathematics 8*, Elsevier Science Publishers, 2001.
16. D. Kratsch. Algorithms. Chapter 8 in [13].
17. K. Mehlhorn and S. Näher. *LEDA: A Platform of Combinatorial and Geometric Computing.* Cambridge University Press, Cambridge, England, 1999.
18. R. Niedermeier and P. Rossmanith. Upper bounds for Vertex Cover further improved. In *Proc. 16th STACS'99*, Springer-Verlag LNCS 1563, pp. 561–570, 1999.
19. J. A. Telle. Complexity of domination-type problems in graphs. *Nordic J. Comput.* **1**:157–171, 1994.
20. J. A. Telle and A. Proskurowski. Practical algorithms on partial k-trees with an application to domination-like problems. In *Proceedings 3rd WADS'93*, Springer-Verlag LNCS 709, pp. 610–621, 1993.
21. J. A. Telle and A. Proskurowski. Algorithms for vertex partitioning problems on partial k-trees. *SIAM J. Discr. Math.* **10(4)**:529–550, 1997.

Author Index

Lecture Notes in Computer Science

For information about Vols. 1–2216
please contact your bookseller or Springer-Verlag

Vol. 2217: T. Gomi (Ed.), Evolutionary Robotics. Proceedings, 2001. XI, 139 pages. 2001.

Vol. 2218: R. Guerraoui (Ed.), Middleware 2001. Proceedings, 2001. XIII, 395 pages. 2001.

Vol. 2219: S.T. Taft, R.A. Duff, R.L. Brukardt, E. Ploedereder (Eds.), Consolidated Ada Reference Manual. XXV, 560 pages. 2001.

Vol. 2220: C. Johnson (Ed.), Interactive Systems. Proceedings, 2001. XII, 219 pages. 2001.

Vol. 2221: D.G. Feitelson, L. Rudolph (Eds.), Job Scheduling Strategies for Parallel Processing. Proceedings, 2001. VII, 207 pages. 2001.

Vol. 2222: M.J. Wooldridge, G. Weiß, P. Ciancarini (Eds.) Agent-Oriented Software Engineering II. Proceedings, 2001. X, 319 pages. 2002.

Vol. 2223: P. Eades, T. Takaoka (Eds.), Algorithms and Computation. Proceedings, 2001. XIV, 780 pages. 2001.

Vol. 2224: H.S. Kunii, S. Jajodia, A. Sølvberg (Eds.), Conceptual Modeling – ER 2001. Proceedings, 2001. XIX, 614 pages. 2001.

Vol. 2225: N. Abe, R. Khardon, T. Zeugmann (Eds.), Algorithmic Learning Theory. Proceedings, 2001. XI, 379 pages. 2001. (Subseries LNAI).

Vol. 2226: K.P. Jantke, A. Shinohara (Eds.), Discovery Science. Proceedings, 2001. XII, 494 pages. 2001. (Subseries LNAI).

Vol. 2227: S. Boztaş, I.E. Shparlinski (Eds.), Applied Algebra, Algebraic Algorithms and Error-Correcting Codes. Proceedings, 2001. XII, 398 pages. 2001.

Vol. 2228: B. Monien, V.K. Prasanna, S. Vajapeyam (Eds.), High Performance Computing – HiPC 2001. Proceedings, 2001. XVIII, 438 pages. 2001.

Vol. 2229: S. Qing, T. Okamoto, J. Zhou (Eds.), Information and Communications Security. Proceedings, 2001. XIV, 504 pages. 2001.

Vol. 2230: T. Katila, I.E. Magnin, P. Clarysse, J. Montagnat, J. Nenonen (Eds.), Functional Imaging and Modeling of the Heart. Proceedings, 2001. XI, 158 pages. 2001.

Vol. 2231: A. Pasetti, Software Frameworks and Embedded Control Systems. XIV, 293 pages. 2002.

Vol. 2232: L. Fiege, G. Mühl, U. Wilhelm (Eds.), Electronic Commerce. Proceedings, 2001. X, 233 pages. 2001.

Vol. 2233: J. Crowcroft, M. Hofmann (Eds.), Networked Group Communication. Proceedings, 2001. X, 205 pages. 2001.

Vol. 2234: L. Pacholski, P. Ružička (Eds.), SOFSEM 2001: Theory and Practice of Informatics. Proceedings, 2001. XI, 347 pages. 2001.

Vol. 2235: C.S. Calude, G. Păun, G. Rozenberg, A. Salomaa (Eds.), Multiset Processing. VIII, 359 pages. 2001.

Vol. 2236: K. Drira, A. Martelli, T. Villemur (Eds.), Cooperative Environments for Distributed Systems Engineering. IX, 281 pages. 2001.

Vol. 2237: P. Codognet (Ed.), Logic Programming. Proceedings, 2001. XI, 365 pages. 2001.

Vol. 2238: G.D. Hager, H.I. Christensen, H. Bunke, R. Klein (Eds.), Sensor-Based Intelligent Robots. Proceedings, 2000. VIII, 375 pages. 2002.

Vol. 2239: T. Walsh (Ed.), Principles and Practice of Constraint Programming – CP 2001. Proceedings, 2001. XIV, 788 pages. 2001.

Vol. 2240: G.P. Picco (Ed.), Mobile Agents. Proceedings, 2001. XIII, 277 pages. 2001.

Vol. 2241: M. Jünger, D. Naddef (Eds.), Computational Combinatorial Optimization. IX, 305 pages. 2001.

Vol. 2242: C.A. Lee (Ed.), Grid Computing – GRID 2001. Proceedings, 2001. XII, 185 pages. 2001.

Vol. 2243: G. Bertrand, A. Imiya, R. Klette (Eds.), Digital and Image Geometry. VII, 455 pages. 2001.

Vol. 2244: D. Bjørner, M. Broy, A.V. Zamulin (Eds.), Perspectives of System Informatics. Proceedings, 2001. XIII, 548 pages. 2001.

Vol. 2245: R. Hariharan, M. Mukund, V. Vinay (Eds.), FST TCS 2001: Foundations of Software Technology and Theoretical Computer Science. Proceedings, 2001. XI, 347 pages. 2001.

Vol. 2246: R. Falcone, M. Singh, Y.-H. Tan (Eds.), Trust in Cyber-societies. VIII, 195 pages. 2001. (Subseries LNAI).

Vol. 2247: C. P. Rangan, C. Ding (Eds.), Progress in Cryptology – INDOCRYPT 2001. Proceedings, 2001. XIII, 351 pages. 2001.

Vol. 2248: C. Boyd (Ed.), Advances in Cryptology – ASIACRYPT 2001. Proceedings, 2001. XI, 603 pages. 2001.

Vol. 2249: K. Nagi, Transactional Agents. XVI, 205 pages. 2001.

Vol. 2250: R. Nieuwenhuis, A. Voronkov (Eds.), Logic for Programming, Artificial Intelligence, and Reasoning. Proceedings, 2001. XV, 738 pages. 2001. (Subseries LNAI).

Vol. 2251: Y.Y. Tang, V. Wickerhauser, P.C. Yuen, C.Li (Eds.), Wavelet Analysis and Its Applications. Proceedings, 2001. XIII, 450 pages. 2001.

Vol. 2252: J. Liu, P.C. Yuen, C. Li, J. Ng, T. Ishida (Eds.), Active Media Technology. Proceedings, 2001. XII, 402 pages. 2001.

Vol. 2253: T. Terano, T. Nishida, A. Namatame, S. Tsumoto, Y. Ohsawa, T. Washio (Eds.), New Frontiers in Artificial Intelligence. Proceedings, 2001. XXVII, 553 pages. 2001. (Subseries LNAI).

Vol. 2254: M.R. Little, L. Nigay (Eds.), Engineering for Human-Computer Interaction. Proceedings, 2001. XI, 359 pages. 2001.

Vol. 2255: J. Dean, A. Gravel (Eds.), COTS-Based Software Systems. Proceedings, 2002. XIV, 257 pages. 2002.

Vol. 2256: M. Stumptner, D. Corbett, M. Brooks (Eds.), AI 2001: Advances in Artificial Intelligence. Proceedings, 2001. XII, 666 pages. 2001. (Subseries LNAI).

Vol. 2257: S. Krishnamurthi, C.R. Ramakrishnan (Eds.), Practical Aspects of Declarative Languages. Proceedings, 2002. VIII, 351 pages. 2002.

Vol. 2258: P. Brazdil, A. Jorge (Eds.), Progress in Artificial Intelligence. Proceedings, 2001. XII, 418 pages. 2001. (Subseries LNAI).

Vol. 2259: S. Vaudenay, A.M. Youssef (Eds.), Selected Areas in Cryptography. Proceedings, 2001. XI, 359 pages. 2001.

Vol. 2260: B. Honary (Ed.), Cryptography and Coding. Proceedings, 2001. IX, 416 pages. 2001.

Vol. 2261: F. Naumann, Quality-Driven Query Answering for Integrated Information Systems. X, 166 pages. 2002.

Vol. 2262: P. Müller, Modular Specification and Verification of Object-Oriented Programs. XIV, 292 pages. 2002.

Vol. 2263: T. Clark, J. Warmer (Eds.), Object Modeling with the OCL. VIII, 281 pages. 2002.

Vol. 2264: K. Steinhöfel (Ed.), Stochastic Algorithms: Foundations and Applications. Proceedings, 2001. VIII, 203 pages. 2001.

Vol. 2265: P. Mutzel, M. Jünger, S. Leipert (Eds.), Graph Drawing. Proceedings, 2001. XV, 524 pages. 2002.

Vol. 2266: S. Reich, M.T. Tzagarakis, P.M.E. De Bra (Eds.), Hypermedia: Openness, Structural Awareness, and Adaptivity. Proceedings, 2001. X, 335 pages. 2002.

Vol. 2267: M. Cerioli, G. Reggio (Eds.), Recent Trends in Algebraic Development Techniques. Proceedings, 2001. X, 345 pages. 2001.

Vol. 2268: E.F. Deprettere, J. Teich, S. Vassiliadis (Eds.), Embedded Processor Design Challenges. VIII, 327 pages. 2002.

Vol. 2270: M. Pflanz, On-line Error Detection and Fast Recover Techniques for Dependable Embedded Processors. XII, 126 pages. 2002.

Vol. 2271: B. Preneel (Ed.), Topics in Cryptology – CT-RSA 2002. Proceedings, 2002. X, 311 pages. 2002.

Vol. 2272: D. Bert, J.P. Bowen, M.C. Henson, K. Robinson (Eds.), ZB 2002: Formal Specification and Development in Z and B. Proceedings, 2002. XII, 535 pages. 2002.

Vol. 2273: A.R. Coden, E.W. Brown, S. Srinivasan (Eds.), Information Retrieval Techniques for Speech Applications. XI, 109 pages. 2002.

Vol. 2274: D. Naccache, P. Paillier (Eds.), Public Key Cryptography. Proceedings, 2002. XI, 385 pages. 2002.

Vol. 2275: N.R. Pal, M. Sugeno (Eds.), Advances in Soft Computing – AFSS 2002. Proceedings, 2002. XVI, 536 pages. 2002. (Subseries LNAI).

Vol. 2276: A. Gelbukh (Ed.), Computational Linguistics and Intelligent Text Processing. Proceedings, 2002. XIII, 444 pages. 2002.

Vol. 2277: P. Callaghan, Z. Luo, J. McKinna, R. Pollack (Eds.), Types for Proofs and Programs. Proceedings, 2000. VIII, 243 pages. 2002.

Vol. 2281: S. Arikawa, A. Shinohara (Eds.), Progress in Discovery Science. XIV, 684 pages. 2002. (Subseries LNAI).

Vol. 2282: D. Ursino, Extraction and Exploitation of Intensional Knowledge from Heterogeneous Information Sources. XXVI, 289 pages. 2002.

Vol. 2284: T. Eiter, K.-D. Schewe (Eds.), Foundations of Information and Knowledge Systems. Proceedings, 2002. X, 289 pages. 2002.

Vol. 2285: H. Alt, A. Ferreira (Eds.), STACS 2002. Proceedings, 2002. XIV, 660 pages. 2002.

Vol. 2286: S. Rajsbaum (Ed.), LATIN 2002: Theoretical Informatics. Proceedings, 2002. XIII, 630 pages. 2002.

Vol. 2287: C.S. Jensen, K.G. Jeffery, J. Pokorny, Saltenis, E. Bertino, K. Böhm, M. Jarke (Eds.), Advances in Database Technology – EDBT 2002. Proceedings, 2002. XVI, 776 pages. 2002.

Vol. 2288: K. Kim (Ed.), Information Security and Cryptology – ICISC 2001. Proceedings, 2001. XIII, 457 pages. 2002.

Vol. 2289: C.J. Tomlin, M.R. Greenstreet (Eds.), Hybrid Systems: Computation and Control. Proceedings, 2002. XIII, 480 pages. 2002.

Vol. 2291: F. Crestani, M. Girolami, C.J. van Rijsbergen (Eds.), Advances in Information Retrieval. Proceedings, 2002. XIII, 363 pages. 2002.

Vol. 2292: G.B. Khosrovshahi, A. Shokoufandeh, A. Shokrollahi (Eds.), Theoretical Aspects of Computer Science. IX, 221 pages. 2002.

Vol. 2293: J. Renz, Qualitative Spatial Reasoning with Topological Information. XVI, 207 pages. 2002. (Subseries LNAI).

Vol. 2296: B. Dunin-Kęplicz, E. Nawarecki (Eds.), From Theory to Practice in Multi-Agent Systems. Proceedings, 2001. IX, 341 pages. 2002. (Subseries LNAI).

Vol. 2300: W. Brauer, H. Ehrig, J. Karhumäki, A. Salomaa (Eds.), Formal and Natural Computing. XXXVI, 431 pages. 2002.

Vol. 2301: A. Braquelaire, J.-O. Lachaud, A. Vialard (Eds.), Discrete Geometry for Computer Imagery. Proceedings, 2002. XI, 439 pages. 2002.

Vol. 2302: C. Schulte, Programming Constraint Services. XII, 176 pages. 2002. (Subseries LNAI).

Vol. 2305: D. Le Métayer (Ed.), Programming Languages and Systems. Proceedings, 2002. XII, 331 pages. 2002.

Vol. 2309: A. Armando (Ed.), Frontiers of Combining Systems. Proceedings, 2002. VIII, 255 pages. 2002. (Subseries LNAI).

Vol. 2314: S.-K. Chang, Z. Chen, S.-Y. Lee (Eds.), Recent Advances in Visual Information Systems. Proceedings, 2002. XI, 323 pages. 2002.